Thermodynamics

and

Chemical Equilibrium

Paul C. Ellgen

Thermodynamics and Chemical Equilibrium

ISBN-13:
978-1492114277

ISBN-10:
1492114278

Preface

This book began as lecture notes for a one-semester course, which the author presented for several years at the Oklahoma School of Science and Mathematics, OSSM. OSSM is located in Oklahoma City. It is a publicly-supported two-year residential school, offering a college level curriculum in mathematics and science to able and motivated high school juniors and seniors. Since students taking the course were expecting to study chemistry in college, the course objective was to present an introduction to physical chemistry that would give them a strong foundation for further study.

As an organizing principle, the course introduced physical chemistry by developing the principles of chemical equilibrium from the perspectives of chemical kinetics, classical thermodynamics, and statistical thermodynamics. The central objective was to develop the ideas that are necessary to produce the equilibrium constant expression for a chemical reaction from each of these perspectives. Since these ideas comprise the core of each of these subjects, this device necessarily selects subject matter that goes to the heart of these subjects.

This book emphasizes the basic concepts that underlie these subjects. Although it assumes familiarity with the calculus and the basics of chemistry and physics, development of the scientific concepts proceeds from a reasonably rudimentary foundation.

Students taking the course had studied mathematics through multivariate calculus. They had taken courses at the college-freshman level in chemistry and physics. Most had taken organic chemistry and advanced physics courses.

In the usual manner, the book attempts to present basic ideas and then to develop the logical consequences of these ideas into a consistent scientific theory. Two particular efforts are made in this direction.

The first is to present ideas in a quasi-historical sequence whenever possible. Such presentations have been called "physicist's history," recognizing that they tend to omit or gloss over many significant facts. Such history is, of course, the recapitulation of the experimental observations that have led to our core concepts—and the elaboration of the way in which our current theories explain these observations. Often, appreciation of our current theories is enhanced by characterizing reasons that predecessor theories have fallen from favor. These presentations clarify important ideas, and they give the student an appreciation for the reasons that the ideas arose and became accepted.

The second is to present—to the maximum extent possible—a logical rational for the ideas that underlie our theories. Often enough the rational is a simple restatement of a commonplace observation, but one whose significance may otherwise go unappreciated. Sometimes more detailed rationalization is valuable. For example, discussion of how the basic ideas of quantum mechanics "might have" arisen has the great value of demystifying ideas that otherwise appear to have sprung from nowhere. Mysteries have entertainment value, but the objective of science is to banish them from our understanding of the natural world.

This book presents several important subjects in probability and statistics. For most of the students in the OSSM course, these were familiar topics. Nevertheless, because some aspects of this presentation may be a little unusual, they are given considerable attention. Two central objectives are to provide a heuristic appreciation of distribution functions and to fully appreciate the central limit theorem. The "total probability sum"—the sum of the probabilities of the possible outcomes from multiple trials—has a prominent role. The approach that is taken to these subjects segues into a presentation of the basic ideas of statistical thermodynamics that emphasizes basic assumptions on the one hand and affords several alternative derivations of the essential results on the other.

The bulk of the book is a self-contained development of the basic ideas of chemical thermodynamics. The presentation stresses that every change in a thermodynamic function models a change in a physical system; to understand the model, we must be explicit about the definition of the system and the change that it undergoes.

The writing presents core ideas from multiple perspectives. A reader can find ambiguity in an overly concise account. When more than one perspective is presented, he is less likely to fall into an unwarranted inference or to miss the core concept. When the ideas are complex, a modest increase in the number of words used to explain them can produce a dramatic decrease in the time required to understand them. In many cases, the same idea is addressed several times, at increasing levels of sophistication.

Equations are not numbered. When a second, or third, or n^{th} instance of an equation is needed, it is repeated in the text. The goal is to eliminate the need to refer to previous statements of the equation. Similarly, a effort is made to exhibit all but the most obvious steps in most derivations. Doubtless this is often unnecessary; however, the time the reader spends reading through "obvious" steps may well be saved many times over by the one instance in which the obvious step is, for some reason, just not obvious.

An effort has been made to address subsidiary questions that, left unexplored, can undermine the reader's understanding. The intent is to achieve an intermediate-level exposition. However, the presentation is neither comprehensive nor particularly rigorous.

The book attempts to clarify two kinds of potentially confusing statements that are in common use because they are useful invocations of more complex ideas. One is the use of phrases in which words are understood to carry meaning that isn't obvious from the statement. (If we say that the first law of thermodynamics is that "Energy is conserved," we expect the reader to understand that our definition of energy includes that it is a state function.) The second is the use of figures of speech whose meaning is clear once understood, but which may well be misunderstood by a student encountering them for the first time. (We often speak of a reaction "occurring at constant temperature" when we don't really mean literally "constant" but only that the initial and final temperatures of the reacting system are the same.)

The evolution of this book has benefitted from comments by Terry Andersen, Walter Deal, Clifford Ellgen, John Gerlach, and the late F. G. Helfferich.

The text makes extensive use of the first person plural. This is a common textbook device. Nevertheless, it raises a question: Just who are the "we" who are expounding? The ideas presented in this book have been developed over a long time by many individuals whose contributions, large and small, are woven into our present understanding. The author invites the reader to join him in adopting the view that "we" encompasses everyone who has ever labored to understand these ideas.

Table of Contents

1

Introduction

Physical chemistry encompasses a wide variety of ideas that are intimately linked. For the most part, we cannot understand one without having some understanding of many others. We overcome this problem by looking at the same idea from a series of slightly different and increasingly sophisticated perspectives. This book focuses on the theories of physical chemistry that describe and make predictions about chemical equilibrium. We omit many topics that are usually understood to be included in the subject of physical chemistry. In particular, we treat quantum mechanics only briefly and spectroscopy not at all.

The goals of chemistry are to predict molecular structures and chemical reactivity. For a given empirical formula, we want to be able to predict all of the stable, three-dimensional atomic arrangements that can exist. We also want to be able to predict all of the reactions in which each such molecule can participate. And, while we are drawing up our wish list, we want to be able to predict how fast each reaction goes at any set of reaction conditions that may happen to be of interest.

Many different kinds of theories enable us to make useful predictions about chemical reactivity. Sometimes we are able to make predictions based on detailed quantum mechanical calculations. When thermodynamic data are available, we can make precise predictions about the extent to which a particular reaction can occur. Lacking such data about a compound, but given its structure, we can usually make some worthwhile predictions based on generalizations (models) that reflect our accumulated experience with particular classes of compounds and their known reactions. Usually these predictions are qualitative, and they fail to include many noteworthy features that emerge when experimental studies are made.

Physical chemistry is the general theory of the properties of chemical substances, with chemical reactivity being a pivotally important property. This book focuses on the core ideas in the subjects of ***chemical kinetics***, ***chemical thermodynamics***, and ***statistical thermodynamics***. These ideas apply to the characterization, correlation, and prediction of the extent to which any chemical reaction can occur. Because predicting how a system will react is substantially equivalent to predicting its equilibrium position, we direct our efforts to understanding how each of these subjects contributes to our understanding of the equilibrium processes that are important in chemistry. In this chapter, we review the general characteristics of these subjects.

The study of chemical kinetics gives us one way to think about chemical equilibrium that is simple and direct. Classical chemical thermodynamics gives us a way to predict chemical equilibria for a wide range of reactions from experimental observations made under a much smaller range of conditions. That is, once we have measured the thermodynamic functions that characterize a compound, we can use these values to predict the behavior of the compound in a wide variety of reactions. Statistical mechanics gives us a conceptual basis for understanding why the laws of chemical thermodynamics take the form that they do. It also provides a way to obtain accurate values for the thermodynamic properties of many compounds.

§1 The role of the ideal gas

The concept of ideal gas behavior plays a pivotal role in the development of science and particularly in the development of thermodynamics. As we shall emphasize, intermolecular forces do not influence the behavior of an ideal gas. Ideal gas molecules are neither attracted to one another nor repelled by one another. For this reason, the properties of an ideal gas are particularly simple. Because ideal gas behavior is so important, we begin by studying ideal gases from both an experimental and a theoretical perspective.

In Chapter 2, we review the experimental observations that we can make on gases and the idealizations that we introduce to extrapolate the behavior of ideal gases from the observations we make on real gases. We also develop Boyle's law from a very simple model for the interactions between point-mass gas molecules and the walls of their container.

In Chapter 4, we develop a detailed model for the behavior of an ideal gas. The physical model is the one we use in Chapter 2, but the mathematical treatment is much more sophisticated. For this treatment we need to develop a number of ideas about probability, distribution functions, and statistics. Chapter 3 introduces these topics, all of which again play important roles when we turn to the development of statistical thermodynamics in Chapter 19.

§2 Chemical kinetics

Chemical kinetics is the study of how fast chemical reactions occur. In Chapter 5, we see that there is a unique way to specify what we mean by "how fast." We call this specification the ***reaction rate***. Chemical kinetics is the study of the factors that determine the rate of a particular reaction. There are many such factors, among them:

- temperature
- pressure
- concentrations of the reactants and products
- nature and concentrations of "spectator species" like a solvent or dissolved salts
- isotopic substitution
- presence or absence of a catalyst.

We will look briefly at all of these, but the thrust of our development will be to understand how the rate of a reaction depends on the concentrations of the reaction's reactants and products.

Many reactions that we observe actually occur as a sequence of more simple reactions. Such a sequence of simple reaction steps is called a ***reaction mechanism***. Our principal goal is to understand the relationships among concentrations, reaction rates, reaction mechanisms, and the conditions that must be satisfied when a particular reaction reaches equilibrium. We will find that two related ideas characterize equilibrium from a reaction-rate perspective. One is that concentrations no longer change with time. The other is a fundamental postulate, called ***the principle of microscopic reversibility***, about the relative rates of individual steps in an overall chemical reaction mechanism when the reacting system is at equilibrium.

§3 Classical thermodynamics

One goal of chemical thermodynamics is to predict whether a particular chemical reaction can occur. We say can, not will, because chemical thermodynamics is unable to make predictions about reaction rates. If we learn from our study of chemical thermodynamics that a particular reaction can occur, we still do not know whether it will occur in a millisecond—or so slowly that no change is detectable. The science of thermodynamics builds on the idea that a particular chemical system can be characterized by the values of certain ***thermodynamic functions***. These ***state functions*** include such familiar quantities as ***pressure***, ***temperature***, ***volume***, ***concentrations***, and ***energy***, as well as some that are not so well known, notably ***enthalpy***, ***entropy***, ***Gibbs free energy***, ***Helmholtz free energy***, ***chemical potential***, ***fugacity***, and ***chemical activity***. We can think of a state function as a quasi-mathematical function whose argument is a physical system. That is, a state function maps a real system onto a real number. When we insert a thermometer into a mixture, the measurement that we make maps the state of the mixture onto the real number that we call temperature.

The word "thermodynamics" joins roots that convey the ideas of heat and motion. In general, motion involves kinetic energy and mechanical work. The interconversion of heat and mechanical work is the core concern of the science of thermodynamics. We are familiar with the idea that kinetic energy can be converted into work; given a suitable arrangement of ropes and pulleys, a falling object can be used to lift another object. Kinetic energy can also be converted—or, as we often say, degraded—into heat by the effects of friction. We view such processes as the conversion of the kinetic energy of a large object into increased kinetic energy of the atoms and molecules that comprise the warmed objects. We can say that easily visible mechanical motions are converted into invisible mechanical motions. The idea that heating an object increases the kinetic energy of its component atoms is called ***the kinetic theory of heat***.

It is often convenient to use the term ***microscopic process*** to refer to an event that occurs at the atomic or molecular level. We call a process that occurs on a larger scale a ***macroscopic process***, although the usual connotation is that a macroscopic process is observable in a quantity of bulk matter. When friction causes the degradation of macroscopic motion to heat, we can say that macroscopic motion is converted to microscopic motion.

While this terminology is convenient, it is not very precise. Changes visible under an optical microscope are macroscopic processes. Of course, all macroscopic changes are ultimately attributable to an accumulation of molecular-level processes. The Brownian motion of a colloidal particle suspended in a liquid medium is noteworthy because this relationship is visible. Viewed with an optical microscope, a suspended, macroscopic, colloidal particle is seen to undergo a rapid and random jiggling motion. Each jiggle is the accumulated effect of otherwise invisible collisions between the particle and the molecules of the liquid. Each collision imparts momentum to the particle. Over long times, the effects average out; momentum transfer is approximately equal in every direction. During the short time of a given jiggle, there is an imbalance of collisions such that more momentum is transferred to the particle in the direction of the jiggle than in any other.

We are also familiar with the idea that heat can be converted into mechanical motion. In an earlier era, steam engines were the dominant means by which heat was converted to work. Steam turbines remain important in large stationary facilities like power plants. For applications we encounter in daily life, the steam engine has been replaced by the internal-combustion engine. When we want to create mechanical motion (do work) with a heat engine, it is important to know how much heat we need in order to produce a given quantity of work. Sadi Carnot was the first to analyze this problem theoretically. In doing so, he discovered the idea that we call the second law of thermodynamics.

The interconversion of heat and work involves an important asymmetry. We readily appreciate that the conversion of kinetic energy to heat can be complete, because we have seen countless examples of objects coming to a complete standstill as the result of frictional forces. However, ordinary experience leaves us less prepared to deal with the question of whether heat can be completely converted into work. Possibly, we remember hearing that it cannot be done and that the reason has something to do with the second law of thermodynamics. If we have heard more of the story, we may remember that it is slightly more complicated. Under idealized circumstances, heat can be converted into work completely. If we confine an ideal gas in a frictionless piston and arrange to add heat to the gas while increasing the volume of the piston in a coordinated way, such that the temperature of the gas remains constant, the expanding piston will do work on some external entity, and the amount of this work will be just equal to the thermal energy added to the gas. We call this process a reversible isothermal expansion. This process does not involve a cycle; the volume of the gas at the end of the process is greater than its volume at the start.

What Carnot realized is that an engine must operate in a cyclic fashion, and that no device—not even an idealized frictionless device—operating around a cycle can convert heat to work with 100% efficiency. Carnot analyzed the process of converting heat into work in terms of an ideal engine that accepts thermal energy (heat) at a high temperature, uses some of this thermal energy to do work on its surroundings, and rejects the rest of its thermal-energy intake to the surroundings in the form of thermal energy at a lower temperature. Carnot's analysis preceded the development of our current ideas about the nature of thermal energy. He expressed his ideas using a now-abandoned theory of heat. In this theory, heat is considered to be a fluid-like quantity—called ***caloric***. Transfers of heat comprise the flow of caloric from one object to another. Carnot's ideas originated as an analogy between the flow of caloric through a steam engine and the flow of water through a water wheel. In this view, the temperature of the steam, entering and leaving the engine, is analogous to the altitude of the water entering and leaving the wheel[1].

Such considerations are obviously relevant if we are interested in building engines, but we are interested in chemical reactivity. How does chemical change relate to engines and the conversion of heat into work? Well, rather directly, actually; after all, a chemical reaction usually liberates or absorbs heat. If we can relate mechanical work to heat, and we can relate the amount of heat liberated to the extent of a chemical reaction, then we can imagine allowing the reaction to go to equilibrium in a machine that converts heat to work. We can expect that the amount of work produced will have some relationship to the extent of the reaction. The nature of this relationship is obscure at this point, but we can reasonably expect that one exists.

§4 Statistical thermodynamics

Statistical thermodynamics is a theory that uses molecular properties to predict the behavior of macroscopic quantities of compounds. While the origins of statistical thermodynamics predate the development of quantum mechanics, the modern development of statistical thermodynamics assumes that the quantized energy levels associated with a particular system are known. From these energy-level data, a temperature-dependent quantity called the ***partition function*** can be calculated. From the partition function, all of the thermodynamic properties of the system can be calculated. We begin our development of statistical thermodynamics by using the energy levels of an individual molecule to find its molecular partition function and the thermodynamic properties of a system that contains N non-interacting molecules of that substance. Later, we see that the partition function of a system containing molecules that do interact with one another can be found by very similar arguments.

Statistical thermodynamics has also been applied to the general problem of predicting reaction rates. This application is called ***transition state theory*** or ***the theory of absolute reaction rates***. In principle, we should be able to predict the rate of any reaction. To do so, we need only to solve the quantum mechanical equations that give the energy levels associated with the reactants and the energy levels associated with a transitory chemical structure called the ***transition state*** for the reaction. From the energy levels we calculate partition functions; from partition functions we calculate thermodynamic functions; and from these thermodynamic functions we obtain the reaction rate. There is a big difference between "in principle" and "in practice." While increases in computer speed make it increasingly feasible to do quantum mechanical calculations to useful degrees of accuracy, the results of such calculations remain too inaccurate to give generally reliable reaction rate predictions. The theory of absolute reaction rates is an important application of statistical thermodynamics. However, it is not included in this book.

Quantum mechanical calculations are not the only way to obtain the energy-level information that is needed to evaluate partition functions. Particularly for small molecules, these energy levels can be deduced from spectroscopic data. In these cases, the theory of statistical thermodynamics enables us to calculate thermodynamic properties from spectroscopic measurements. Excellent agreement is obtained between the values of thermodynamic functions obtained from classical thermodynamic (thermochemical) measurements and those obtained from statistical-thermodynamic calculations based on energy levels derived from spectroscopic measurements. In Chapter 24, we consider a particular example to illustrate this point.

§5 Heat transfer in practical devices

The amount of heat transferred to or from a system undergoing change is an important thermodynamic variable. In practical devices, the rate at which heat can be transferred to or from a system plays a very important role also. Consider again the work produced by heating a gas that is confined in a cylinder that is closed by a piston. Clearly, the rate at which heat can be transferred from the outside to the gas determines the rate at which the piston moves outward and thus the rate at which work is done on the environment.

Does it matter whether the heat-transfer process is fast or slow? If the heat cost nothing, would we care if our engine produced work only very slowly? After all, if we want more work and the heat is free, we need only build more engines; eventually we will have enough of them to produce any required amount of work. Of course, heat is not free; more significantly for our present considerations, the engines are not free either. Engineers and accountants call the cost of heat an operating cost. There are many other operating costs, like labor, supplies, insurance, and taxes. The cost of the engine is called a capital cost. To find the total cost of a unit of work, we need to add up the various operating costs and a part of the cost of the engine.

cost of a unit of work
= fuel cost + other operating costs + capital cost

The difference between an operating cost and a capital cost is that an operating cost is incurred at (about) the same time that the product, in this case a unit of work, is created. In contrast, a capital cost is incurred well before the product is created. The purchase of a machine is a typical capital expense. The cost of the machine is incurred long before the machine makes its last product. This occurs because the machine must be paid for when it is acquired, but it continues to function over a useful lifetime that is typically many years. For example, if an engine that costs $1,000,000 can produce a maximum of 1,000,000 units of work before it wears out, the minimum contribution that the cost of the engine makes to the cost of the work it produces is $1 per unit. The life of the engine also enters into the estimation of capital cost. If some of the work done by the engine will be produced ten years in the future, we will be foregoing the interest that we could otherwise have earned on the money that we invested in the engine while we wait around to get the future work. Operating costs are well defined because they are incurred here and now. Capital costs are more problematic, because they depend upon assumptions about things like the life of the machine and the variation in interest rates during that life.

Suppose that we are developing a new engine. All else being equal, we can decrease the capital-cost component of the work our engine produces by decreasing the time it needs to produce a unit of work. The savings occurs because we can get the same amount of work from a smaller and hence less-costly engine. Since each unit of work requires that the same amount of heat be moved, we can make the engine smaller only if we can move heat around more quickly. In internal combustion engines, we get heat into the engine with a combustion reaction (an explosion) and take most of it out again by venting the combustion products (the exhaust gas). So internal combustion engines have the great advantage that both of these steps can be fast. Steam engines are successful because we can get heat into the engine quickly by allowing steam to flow from a boiler into the engine. We can remove heat from the steam engine quickly by venting the spent steam, which is feasible because the working fluid is water. The Stirling engine is a type of external combustion engine that works by alternately heating (expanding) and cooling (compressing) an enclosed working fluid. Stirling engines have theoretical advantages, but they are not economically competitive, essentially because heat transfer to and from the working fluid cannot be made fast enough.

Why does anyone care about capital cost? Well, we can be sure that the owner of an engine will be keenly interested in minimizing the dollars that come out of his pocket. But capital cost is also a measure of the consumption of resources—resources that may have more valuable alternative uses. So if any segment of an economy uses resources inefficiently, other segments of that economy must give up other goals that could have been achieved using the wasted resources. Economic activity benefits many people besides the owners of capital. If capital is used inefficiently, society as a whole is poorer as a result.

Heat transfer has a profound effect also on the design of the machines that manufacture chemicals. This occurs most conspicuously in processes that involve very exothermic reactions. If heat cannot be removed from the reacting material fast enough, the temperature of the material rises. The higher temperature may cause side reactions that decrease the yield of the product. If the temperature rises enough, there may be an explosion. For such reactions, the equipment needed to achieve rapid heat transfer, and to manage the rate of heat production and dissipation, may account for a large fraction of the cost of the whole plant. In some cases, chemical reactions used for the production of chemicals produce enough heat that it is practical to use this "waste heat" for the production of electricity.

§6 The concept of equilibrium

We are familiar with the idea that a system undergoing change eventually reaches equilibrium. We say that a system is at equilibrium when no further change is possible. When we talk about change, we always have in mind some particular property. We measure the change in the system by the amount of change in this property. When the property stops changing, we infer that the system has stopped changing, and we say that the system has come to equilibrium. Of course, we may interest ourselves in a system in which many properties undergo change. In such cases, we recognize that the system as a whole cannot be at equilibrium until all of these properties stop changing.

On the other hand, we also recognize that the absence of observable change is not enough to establish that a system is at equilibrium with respect to all of the possible changes that it could undergo. We know that hydrogen burns readily in oxygen to form water, but a mixture of hydrogen and oxygen undergoes no change under ordinary conditions. This unchanging mixture is plainly not at equilibrium with respect to the combustion reaction. Only when a catalyst or an ignition source is introduced does reaction begin.

It is also possible, indeed probable, that a system can be at equilibrium with respect to one process and not be at equilibrium with respect to other processes, which, while possible, simply do not occur under the conditions at hand. For example, if an aqueous solution of the oxygen-carrying protein hemoglobin is added to the hydrogen—air system, the protein will add or lose coordinated oxygen molecules until the equilibrium composition is reached. If our investigation is focused on the protein–oxygenation reaction, we do not hesitate to say that the system is at equilibrium. The non-occurrence of the oxygen—hydrogen reaction is not relevant to the phenomenon we are studying.

It is even possible to reach a non-equilibrium state in which the concentrations of the reactants and products are constant. Such a system is said to have reached a ***steady state***. In order for this to occur, the reaction must occur in an ***open system***; that is, one in which materials are being added or removed; there must be continuous addition of reactants, and continuous removal of products. In Chapter 5, we discuss a simple system in which this can be achieved. A ***closed system*** is one that can neither gain nor lose material. An ***isolated system*** is a closed system that can neither gain nor lose energy; in consequence, its volume is fixed. In an isolated system, change ceases when equilibrium is reached, and conversely.

We will consider several commonly encountered kinds of change, including mechanical motions, heat transfers, phase changes, partitioning of a solute between two phases, and chemical reactions. Here we review briefly what occurs in each of these kinds of change. In Chapter 6, we review the characteristics that each of these kinds of change exhibits at equilibrium.

A system in ***mechanical equilibrium*** is stationary because the net force acting on any macroscopic portion of the system is zero. Another way of describing such a situation is to say that the system does not move because of the presence of constraints that prevent movement.

Two macroscopic objects are in ***thermal equilibrium*** if they are at the same temperature. We take this to be equivalent to saying that, if the two objects are in contact with one another, no heat flows between them. Moreover, if object A is in thermal equilibrium with each of two other objects, B and C, then we invariably find that objects B and C are in thermal equilibrium with one another. This observation is sometimes called the ***zeroth law of thermodynamics***. It justifies the concept of temperature and the use of a standard system—a thermometer—to measure temperature.

For an isolated system to be in ***phase equilibrium***, it must contain macroscopic quantities of two or more phases, and the amount of each phase present must be unchanging. For example, at 273.15 K and 1 bar, and in the presence of one atmosphere of air, liquid water and ice are in equilibrium; the amounts of water and ice remain unchanged so long as the system remains isolated. Similarly, a saturated aqueous solution of copper sulfate is in equilibrium with solid copper sulfate; if the system is isolated, the amounts of solid and dissolved copper sulfate remain constant.

If a system is in phase equilibrium, we can remove a portion of any phase without causing any change in the other phases. At equilibrium, the concentrations of species present in the various phases are independent of the absolute amount of each phase present. It is only necessary that some amount of each phase be present. To describe this property, we say that the ***condition for equilibrium*** is the same irrespective of the amounts of the phases present in the particular system. For example, if one of the species is present in both a gas phase and a condensed phase, we can specify the equilibrium state by specifying the pressure and temperature of the system. However, we can change the relative amounts of the phases present in this equilibrium state by changing the volume of the system. (If its volume can change, the system is not isolated.)

Partitioning of a solute between two immiscible condensed phases is important in many chemical systems. If we add water and chloroform to the same vessel, two immiscible liquid phases are formed. Elemental iodine is very sparingly soluble in water and substantially more soluble in chloroform. If we add a small amount of iodine to the water–chloroform system, some of the iodine dissolves in the water and the remainder dissolves in the chloroform layer. We say that the iodine is distributed between the two phases. When the iodine concentrations become constant, we say that the system has reached ***distribution equilibrium***.

In a ***chemical reaction***, one or more chemical substances (reactants) undergo a change to produce one or more new chemical substances (products). We are accustomed to representing chemical substances by symbols and representing their reactions by chemical equations. Thus, for the hydrolysis of ethyl acetate, we write

$$CH_3CO_2CH_2CH_3 + H_2O \rightleftharpoons CH_3CO_2H + CH_3CH_2OH$$

A chemical equation like this expresses a stoichiometric relationship between reactants and products. Often we invoke it as a symbol for various distinctly different physical situations. For example:

- We may view the equation as a symbolic representation of a *single solution* that contains the four compounds ethyl acetate, water, acetic acid, and ethanol—and possibly other substances.
- We may view the equation as a symbolic representation of a *relationship between two systems* whose proportions are arbitrary. The first system comprises ethyl acetate and water. The second system comprises acetic acid and ethanol. The equation represents the idea that the first system can be converted into the second.
- We may view the symbols on each side of the equation as representing mixtures of the indicated chemical substances in the specified stoichiometric proportions.

- We may view the equation as representing the specified stoichiometric proportions of pure, unmixed chemical substances. When we are discussing changes in "standard" thermodynamic properties that accompany a chemical reaction, this is the interpretation that we have in mind.

When we discuss a chemical equation, the intended interpretation is normally evident from the context. Indeed, we often skip back and forth among these interpretations in the course of a single discussion. Nevertheless, it is important to avoid confusing them.

By doing experiments, we can discover that there is an equation that uniquely defines the position of a chemical reaction at equilibrium, an equation that we usually think of as the definition of the ***equilibrium constant***. If our measurements are not too accurate, or we confine our study to a limited range of concentrations, or the system is particularly well behaved, we can express the equilibrium constant as a function of concentrations[2]. For the hydrolysis of ethyl acetate, we find

$$K = \frac{[CH_3CO_2H][HOCH_2CH_3]}{[CH_3CO_2CH_2CH_3][H_2O]}$$

In general, for the reaction

$$a\,A + b\,B \rightleftharpoons c\,C + d\,D$$

we find

$$K = \frac{[C]^c[D]^d}{[A]^a[B]^b}$$

That is, at equilibrium the indicated function of reactant concentrations always computes to approximately the same numerical value.

When our concentration measurements are more accurate, we find that we must introduce new quantities that we call ***chemical activities***. We can think of an activity as a corrected concentration. The correction compensates for the effects of intermolecular attraction and repulsion. Denoting the activity of substance X as $\tilde{a}_x$, we find that

$$K_a = \frac{\tilde{a}_C^c \tilde{a}_D^d}{\tilde{a}_A^a \tilde{a}_B^b}$$

gives a fully satisfactory characterization of the equilibrium states that are possible for systems in which this reaction occurs. K_a, the equilibrium constant computed as a function of reactant activities, always has exactly the same numerical value.

We can develop the equilibrium constant expression from three distinctly different theoretical treatments. We develop it first from some basic ideas about the rates of chemical reactions. Then we obtain same result from both the macroscopic-behavior considerations of classical thermodynamics and the molecular-property considerations of statistical thermodynamics.

Our most basic concept of equilibrium is based on the observation that change in an isolated system eventually ceases; once change ceases, it never resumes. In this book, we call the idea of a static state of an isolated system the ***primitive equilibrium*** concept. We also observe that change eventually ceases in a closed system that is not isolated but whose temperature, pressure, and volume are kept constant. Conversely, if a system is at equilibrium, its temperature, pressure, and volume are necessarily constant; all interactions between such a system and its surroundings can be severed without changing any of the properties of the system. We can view any particular equilibrium state as a primitive equilibrium state.

A system whose temperature, pressure, or volume is established by interactions between the system and its surroundings is inherently more variable than an isolated system. For a given isolated system, only one equilibrium state exists; for a system that interacts with its surroundings, many different equilibrium states may be possible. In chemical thermodynamics, our goal is to develop mathematical models that specify the equilibrium states available to a system; we seek models in which the independent variables include pressure, temperature, volume, and other conditions that can be imposed on the system by its surroundings. In this conception, an equilibrium system is characterized by a set of points in a variable space. We can think of this set of points as a surface or a manifold in the variable space; every point in the set is a different primitive-equilibrium state of the system. By imposing particular changes on some variables, a particular equilibrium system can be made to pass continuously through a series of primitive-equilibrium states.

For reasons that become apparent as we proceed, we use the name ***Gibbsian equilibrium*** to denote this more general conception. When we talk about equilibrium in thermodynamics, we usually mean Gibbsian equilibrium. In Chapter 6, we see that the idea of (Gibbsian) equilibrium is closely related to the idea of a ***reversible process***. We also introduce Gibbs' phase rule, which amounts to a more precise definition, from the perspective of classical thermodynamics, of what we mean by (Gibbsian) equilibrium in chemical systems.

§7 Chemical equilibrium and predicting chemical change

When we talk about predicting chemical reactions, we imagine taking quantities of various pure compounds and mixing them under some set of conditions. We suppose that they react until they reach a position of equilibrium in which one or more new compounds are present. We want to predict what these new compounds are and how much of each will be produced.

For any given set of reactants, we can accomplish this predictive program in two steps. First we find all of the sets of products that can be obtained from the given reactants. Each such set represents a possible reaction. We suppose that, for each set of possible products, we are able to predict the equilibrium composition. Predicting which reaction will occur is equivalent to finding the reaction whose position of equilibrium lies farthest in the direction of its products. From this perspective, being able to predict the position of equilibrium for the reactants and any stoichiometrically consistent set of products is the same thing as being able to predict what reaction will occur. (If there is no single reaction whose position of equilibrium is much further in the direction of its products than that of any other reaction, multiple reactions can occur simultaneously.)

This two-step procedure corresponds to the sense in which chemical thermodynamics enables us to predict reaction products. We measure values for certain characteristic thermodynamic functions for all relevant compounds. Given the values of these functions for all of the compounds involved in a hypothesized reaction, we calculate the position of equilibrium. That we must begin by guessing the products makes this approach cumbersome and uncertain. We can never be positive that the true products are among the possibilities that we consider. Nevertheless, as a practical matter for most combinations of reactants, the number of plausible product sets is reasonably small.

§8 Equilibrium and classical thermodynamics

We develop classical thermodynamics by reasoning about ***reversible processes***—processes in which a system passes through a series of equilibrium states. Any such process corresponds to a path on one or more of the Gibbsian manifolds that are available to the system. The resulting theory consists of equations that relate the changes in the values of the system's state functions as the system undergoes a reversible change. For this reason, the body of theory that we are calling classical thermodynamics is often called ***equilibrium thermodynamics*** or ***reversible thermodynamics***.

As we discuss further in Chapter 6, any change that we can actually observe in a real system must be the result of a ***spontaneous process***. In a reversible process, both the initial and the final states are equilibrium states. In a spontaneous process, the initial state of the system is not an equilibrium state. A spontaneous process begins with the system in a non-equilibrium state and proceeds until an equilibrium state is reached.

The domain of classical thermodynamics—reversible processes—is distinct from the domain of real observations, because real observations can be made only for spontaneous processes. We bridge this gap by careful selection of real-world systems to serve as models for the reversible systems that inhabit our theory. That is, we find that we can make measurements on non-equilibrium systems and irreversible processes from which we can estimate the properties of equilibrium systems and reversible processes. Saying almost the same thing from another perspective, we find that the classical thermodynamic equations that apply to equilibrium states can also be approximately valid for non-equilibrium states. For many non-equilibrium states, notably those whose individual phases are homogenous, the approximations can be very good. For other non-equilibrium states, notably those whose individual phases are markedly inhomogeneous, these approximations may be very poor.

§9 A few ideas from the philosophy of science

The goal of science is to create theories that accurately describe physical reality. In this book, we explore some of the most useful scientific theories that exist. They have been tested extensively. We know that there are limits to their applicability. We expect that further thought and experimentation will expand their scope. We expect that some elements of these theories will need to be modified in ways that we cannot anticipate, but we do not expect that the core concepts will be invalidated.

We love theories because they rationalize our environment. This is true not only of scientific theories but also of the panoply of conceptual frameworks that we use to organize our views about—and responses to— all of life's issues. We are addicted to theories. We find nothing more disconcerting than information that we cannot put into a coherent context. Indeed, the term ***cognitive dissonance*** has entered the language to describe the feeling of disorientation that we experience when "things just don't add up."

Logically, confronted with a fact that contradicts one of our theories, we are compelled to give up the theory. We are not always logical. A fact that contradicts a pet theory is unlikely to be accepted at face value. We challenge it, as indeed we should. We scrutinize the offending fact and try to convince ourselves that it is no fact at all, merely a spurious artifact. Often, of course, this proves to be the case. Sometimes we conclude that the offending fact is spurious when it is not. We get stuffy about our theories. When we find one that suits us, we resist giving it up. It has been observed that a revolutionary scientific theory often achieves universal acceptance only after all those who grew up with the predecessor theory have died.

Science is ultimately a social enterprise. To develop and test theories about physical reality, participants in this enterprise must be in general agreement about the criteria that are to be applied. These criteria are frequently called "the philosophy of science." To summarize the philosophy of science, we begin by observing that the goal of science is to explain the world that we experience through our sensory perceptions. It is easy to generate putative explanations that have little or no real value. Unfortunate experiences with past explanations have led to a broad consensus that scientific theories must have the following properties:

- operational definitions,
- logical structure,
- predictive capability and testability,
- internal consistency, and
- consistency with any and all experimental observations.

The theory must be about the properties of some set of things. By ***operational definitions***, we mean that the subjects of the theory must measurable, and the theory must specify a set of operations for making each of these measurements. By ***logical structure***, we mean that a satisfactory theory must include well-defined rules to specify how the subjects of the theory relate to one another. By ***predictive capability*** we mean that a satisfactory theory must be capable of predicting the results of experiments that have not been performed. By ***internal consistency*** we mean that a satisfactory theory must not allow us to logically derive contradictory conclusions, which also means that it must not predict more than one outcome for any particular experiment. Because a satisfactory theory makes predictions, it is also ***testable***. It is possible to check whether the predictions correspond to reality. We require that the theory's ***predictions be consistent with the results*** that we observe when we do the experiment.

The first four of these requirements really detail the characteristics that a theory must have in order to be considered a proper subject for scientific investigation. Only the last requirement speaks to the all-important issue of whether the theory accurately mirrors physical reality. We can never prove that any theory is true. What we can prove is that a theory fails to meet one of our criteria. Science progresses when we discover a fatal flaw in a currently accepted theory.

Let us think further about what we mean when we say that a theory must be a logical structure. Consider a simple classical syllogism.

Major premise: All dogs are cats.
Minor premise: All cats are white.
Conclusion: All dogs are white.

As a logical structure, this seems to be satisfactory. We can represent the whole of its content in a simple diagram (See Figure 1), so if we want to view this syllogism as a theory about nature, its internal consistency is more or less self-evident. Moreover, viewed as a theory, it makes a prediction: All dogs are white. If we have operational definitions for "dog", "cat", and "white" that conform to customary usage, we can say that this syllogism meets our criteria for a proper subject for scientific investigation. Of course, as a mirror of reality, it fails.

To see the issue of logical structure from another perspective, let us consider the theory of evolution. Some people summarize the theory of evolution as teaching that the fittest individuals survive and defining survivors as those individuals who are most fit. They then point out that these are circular statements and proceed from this observation to the conclusion that the theory of evolution is devoid of real content. So it can be dismissed. Now, if we are not closed-minded about evolution, this analysis looks like a case of throwing out the baby with the bath water. Even so, we are likely to be troubled, because the circularity is undeniable. Does this circularity mean that the theory of evolution is bad science?

A ***tautology*** is a statement that must be true. Our analysis attempts to recast the entire content of the theory of evolution as one tautologous statement. If the whole

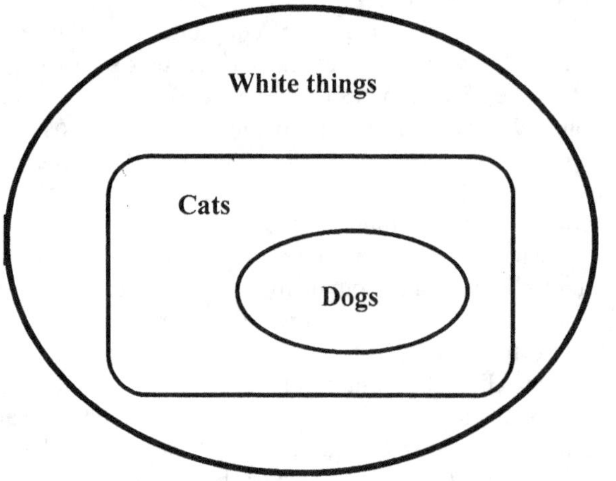

Figure 1. Venn diagram of a classical syllogism

of a theory is a single statement, and that statement must be true, then the theory cannot be tested. Our rules require that we reject it. However, our tautologous summary fails to capture the whole of the theory of evolution. If a theory that contains tautologous statements also makes predictions that are not tautologous, then it can be tested. In the present example, we can predict from the theory of evolution that selection of a particular trait through any process will cause increased expression of that trait in succeeding generations. Evolution is based on natural selection, but it postulates a mechanism that must be valid for any consistent selection process. It predicts that a farmer who selects for cows that produce more milk will eventually get cows that produce more milk. Thus, attempts to apply selective breeding are tests of the central element of the theory of evolution, and the success of selective breeding in every aspect of agriculture verifies a prediction of the theory.

There is no reason to object to a theory that has tautological elements so long as the content of the theory has real substance. What we require is that a theory's predictions be non-trivial. We object when substantially all of a theory's purported predictions are merely restatements of its premises, so that the whole of the theory is an exercise in verbiage.

We require scientific theories to be internally consistent. (Normally, we do not expect to be able to prove internal consistency. What we really mean is that we will discard any theory that we can show to be internally inconsistent.) The presence of tautologous statements cannot make a theory internally inconsistent. Indeed, we can expect any internally consistent theory to have tautological elements. After all, if we try to define all of the subjects of a theory, at least some of our definitions will inevitably be circular.

Another way to describe the logical structure of a physical theory is to say that a theory is a model for some observable part of the world. We want the model to include things and rules. The rules should specify how the things of the model change. When we talk about comparing predictions of the theory to the results of experiment, we mean that the changes that occur in the model

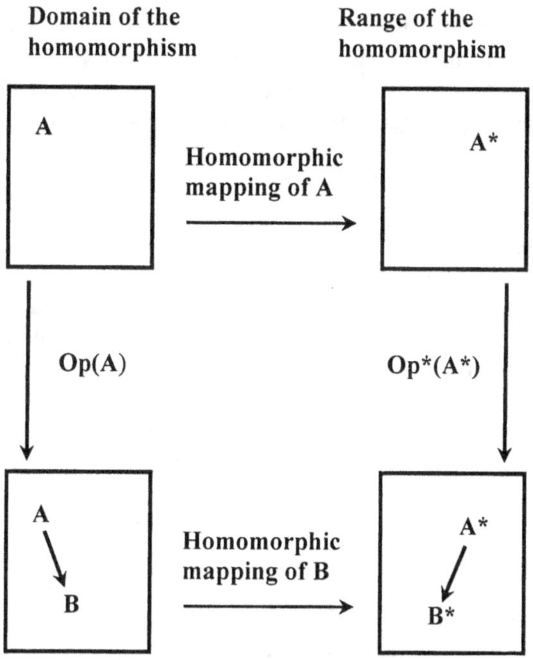

Figure 2. A homomorphism.

when we apply the rules of the theory should parallel the changes that occur in the real world when we do the corresponding experiment. The idea is analogous to the mathematical concept of a ***homomorphism***. A dictionary definition of a homomorphism[3] is "a mapping of a mathematical group, ring, or vector space onto another in such a way that the result obtained by applying an operation to elements of the domain is mapped onto the result of applying the same or a corresponding operation to their images in the range." Stated more picturesquely, the idea of a homomorphism is that two mathematical structures have the same form: If one is "laid on top of the other", there is a perfect correspondence between them. (See Figure 2.)

The idea of a one-to-one correspondence between two structures can be extended to the case where one structure is a logical structure and the other is a physical structure. Consider the logical structure, usually called a truth table, associated with the truth of the proposition "sentence A is true and sentence B is true." The proposition is false unless both A and B are true. Now consider the performance of an electrical circuit that consists of a battery, switches A and B, and a light bulb, all connected in series. The light is off unless both A and B are on. There is a perfect correspondence between the elements of the logical structure and the performance of the circuit. (See Figure 3.) In fact, this is a special case of a much more extensive parallelism. For any sentence in propositional logic, there is a corresponding circuit—involving batteries, switches, and a light bulb—such that the bulb is on if the corresponding sentence is true and off if the sentence is false. Moreover, there is a mathematical structure, called Boolean algebra, which is homomorphic to propositional logic and exhibits the same parallelism to circuits. Digital computers carry this parallelism to structures of great complexity.

If we stretch our definition of a homomorphism to include comparisons of logical constructs with physical things, we can view a scientific theory as a logical construct that we are trying to make homomorphic with physical reality. We want our theory to map onto reality in such a way that changes in the logical construct map onto changes in physical reality. Of course, what we mean by this is the same thing we meant previously when we said that the predictions of our theory should agree with the results of experiment.

Truth of "A is true and B is true

	A true	A false
B true	**True**	**False**
B false	**False**	**False**

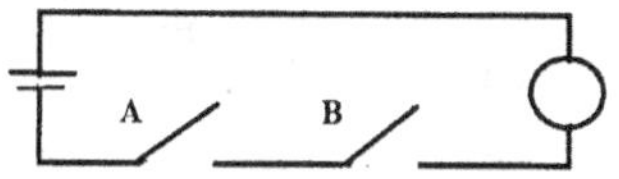

Status of light on series switches

	A closed	A open
B closed	**On**	**Off**
B open	**Off**	**Off**

Figure 3. Mapping a physical system onto a logical construct.

It is implicit in this view that any scientific theory describes an abstraction from reality. If, for example, we say that a physical system is at a particular temperature, this statement represents an approximation in at least two respects. First, we believe that we can measure temperature in continually more refined ways. If we know the temperature to six significant figures, this represents a very good approximation, but we believe that the "true" temperature can only be expressed using indeterminably many significant figures. To say that the temperature of the system has some value expressed to six significant figures is an approximation, albeit a very good one. Second, it is unlikely that all parts of any real system are actually at the same temperature. When we say that a physical system has a particular temperature, we mean that any differences in temperature between different parts of the system are too small to affect the behavior in which we are currently interested. In our discussions and deliberations about the physical system, we replace its actual properties by our best measurement. In doing so, we abstract something that we believe to be an essential feature from the reality we actually observe.

Implicit also is the idea that our theories about reality are subject to inevitable limitations. In the example above, any theory that, however accurately, predicts the single temperature that we use to describe the real system is inadequate to describe whatever small variations there may be from point to point within it. If we expand our theory to encompass variations over, say, distances of millimeters, then the expanded theory will be inadequate to describe variations over some smaller distance. Any "perfect" theory must exactly describe the motions of all of the system's constituent particles. This is impossible not only because it conflicts with the basis premises of quantum mechanics, but also because it requires the theory to contain information at a level of detail equal to that in the physical system itself.

One way to express all of the same information at exactly the same level of detail is to have an exact replica of the system. This idea has been expressed by saying that an absolutely accurate map of Oklahoma City would have to be as large as Oklahoma City itself. For use as a map, such a thing would be useless. Similarly, a theory that predicts temperature is useful only if it predicts the temperature we measure (measurement being a part of the process by which we effect abstraction) experimentally. We can make use of multiple theories of the same phenomenon, if each of them has advantages and limitations that we recognize and respect. We see however, that in the end, there can be no single, all-encompassing theory. Any theory must model an approximation to reality. In the final analysis, reality is ... reality.

§10 A few ideas from formal logic

Formal logic deals with relationships among propositions, where a ***proposition*** is any statement of (alleged) fact. Any proposition can be expressed as an ordinary English sentence, although it may be more convenient to use mathematical symbols or some other notation. The following are all propositions:

Albert Einstein is deceased.
Tulsa is in Oklahoma.
Two plus two equals four.
$2 + 2 = 4$
$\int x^2 dx = x + 1$

A proposition need not be true. The last of these examples is a false proposition. We represent an arbitrary proposition by any convenient symbol, usually a letter of the alphabet. Thus, we could stipulate that "p" represents any of the propositions above. Once we have associated a symbol with a particular proposition, the symbol itself is taken to represent an assertion that the proposition is true. It is an axiom of ordinary logic that any proposition must be either true or false. If we associate the symbol "p" with a particular proposition, we write "$\sim p$" to represent the statement: "The proposition represented by the symbol 'p' is false." $\sim p$ is called the ***negation of p***. We can use the negation of p, $\sim p$, to state the axiom that a proposition must be either true or false. To do so, we write: Either p or $\sim p$ is true. We can write this as the proposition "p or $\sim p$". The negation of the negation of p is an assertion that p is true; that is, $\sim\sim p = p$.

Logic is concerned with relationships among propositions. One important relationship is that of ***implication***. If a proposition, q, follows logically from another proposition, p, we say that q is implied by p. Equivalently, we say that proposition p implies proposition q. The double-shafted arrow, $\Rightarrow$, is used to symbolize this relationship. We write "$p \Rightarrow q$" to mean, "That proposition p is true implies that proposition q is true." We usually read this more tersely, saying, "p implies q." Of course, "$p \Rightarrow q$" is itself a proposition; it asserts the truth of a particular logical relationship between propositions p and q.

For example, let p be the proposition, "Figure A is a square." Let q be the proposition, "Figure A is a rectangle." Then, writing out the proposition, $p \Rightarrow q$, we have: Figure A is a square implies figure A is a rectangle. This is, of course, a valid implication; for this example, the proposition $p \Rightarrow q$ is true. For reasons that will become clear shortly, $p \Rightarrow q$ is called the ***conditional*** of p and q. Proposition p is often called a ***sufficient condition***, while proposition q is called a ***necessary condition***. That is, the truth of p is sufficient to establish the truth of q.

sufficient condition $\Rightarrow$ necessary condition

Now, if proposition $p \Rightarrow q$ is true, and proposition q is also true, can we infer that proposition p is true? We most certainly cannot! In the example we just considered, the fact that figure A is a rectangle does not prove that figure A is a square. We call $q \Rightarrow p$ the converse of $p \Rightarrow q$. The conditional of p and q can be true while the converse is false. Of course, it can happen that both $p \Rightarrow q$ and $q \Rightarrow p$ are true. We often write "$p \Leftrightarrow q$" to express this relationship of mutual implication. We say that, "p implies q and conversely."

What if $p \Rightarrow q$, and q is false? That is, $\sim q$ is true. In this case, p must be false! If $\sim q$ is true, it must also be that $\sim p$ is true. Using our notation, we can express this fact as

$$(p \Rightarrow q \text{ and } \sim q) \implies \sim p$$

Equivalently, we can write

$$(p \Rightarrow q) \Leftrightarrow (\sim q \Rightarrow \sim p)$$

That is, $p \Rightarrow q$ and $\sim q \Rightarrow \sim p$ are equivalent propositions; if one is true, the other must be true. $\sim q \Rightarrow \sim p$ is called the ***contrapositive*** of $p \Rightarrow q$. The equivalence of the conditional and its contrapositive is a theorem that can be proved rigorously in an axiomatic formulation of logic. In our later reasoning about thermodynamic principles, we use the equivalence of the conditional and the contrapositive of p and q.

The equivalence of the conditional, $p \Rightarrow q$, and the contrapositive, $\sim q \Rightarrow \sim p$, is the reason that q is called a necessary condition. If $p \Rightarrow q$, it is necessary that q be true for p to be true. (If figure A is to be a square, it must be a rectangle.)

It is also intimately related to proof by contradiction. Suppose that we know p to be true. If, by assuming that q is false ($\sim q$ is true), we can validly demonstrate that p must also be false ($\sim q \Rightarrow \sim p$, so that $\sim p$ is true), we have the contradiction that p is both true and false (p and $\sim p$). Since p cannot be both true and false, it must be false that q is false ($\sim\sim q = q$). Otherwise stated, the equivalence of the conditional and the contrapositive leads not only to (p and $\sim p$) but also to (q and $\sim q$). [($\sim q \Rightarrow \sim p$) implies ($p \Rightarrow q$)].

In summary, since we know p to be true, our assumption that q is false, together with the valid implication $\sim q \Rightarrow \sim p$, leads to the conclusion that q is true, which contradicts our original assumption, so that the assumption is false, and q is true.

Problems

1. Philosophers argue about the feasibility of a private language, a language that is known by only one individual. The case against the possibility of a private language is based on the assumption that language comes into existence as a tool for communication in a society comprising two or more individuals. Solipsism is a philosophical conception in which your sensory perceptions are internally generated; they are not the result of your interactions with the world. Solipsism assumes that the world you think you perceive does not in fact exist. Since a solipsistic individual is the only being that exists, any language he uses is necessarily a private language. Evidently, the existence of a solipsistic individual who uses some language to think and the impossibility of a private language are mutually exclusive: If a private language is impossible, a solipsistic individual cannot use any language. Since you are reading this, you are using a language, and therefore you cannot be a solipsistic individual. Does this argument convince you that you are not a solipsistic individual; that is, does this argument convince you of the existence of a physical reality that is external to yourself? Why or why not? Can a solipsistic individual engage in scientific inquiry?

2. Many people find the theory of evolution deeply repugnant. Some argue that the theory of evolution has not been proved and that ***creationism*** is an alternative scientific theory, where creationism is the Biblical description of God's creation of the world in six days.

(a) Is it valid to say, "The theory of evolution is unproved."?

(b) Comment very briefly on whether or not the theory of evolution meets each of our criteria for a scientific theory.

(c) Comment very briefly on whether or not creationism meets each of our criteria for a scientific theory.

3. Use an ordinary English sentence to state the meaning of propositions (a) and (b):

(a) $\sim(p \text{ and } q) \Rightarrow (\sim p \text{ or } \sim q)$

(b) $\sim(p \text{ or } q) \Rightarrow (\sim p \text{ and } \sim q)$

Are propositions (a) and (b) true or false?

Using propositions (a) and (b), prove propositions (c) and (d):

(c) $[(p \text{ and } q) \Rightarrow r] \Rightarrow [\sim r \Rightarrow (\sim p \text{ or } \sim q)]$

(d) $[(p \text{ or } q) \Rightarrow r] \Rightarrow [\sim r \Rightarrow (\sim p \text{ and } \sim q)]$

Notes

[1] R. Clausius, ***The Mechanical Theory of Heat***, translated by Walter R. Browne, Macmillan and Co., London, 1879, p. 76.

[2] We use square brackets around the symbol for a chemical substance to denote the concentration of that substance in molarity (moles per liter of solution) units.

[3] ***Webster's Ninth New Collegiate Dictionary***, Merriam-Webster, Inc., Springfield, Massachusetts, 1988.

2

Gas Laws

Early experimenters discovered that the pressure, volume, and temperature of a gas are related by simple equations. The classical gas laws include ***Boyle's law***, ***Charles' law***, ***Avogadro's hypothesis***, ***Dalton's law of partial pressures***, and ***Amagat's law of partial volumes***. These laws were inferred from experiments done at relatively low pressures and at temperatures well above those at which the gases could be liquefied. We begin our discussion of gas laws by reviewing the experimental results that are obtained under such conditions. As we extend our experiments to conditions in which gas densities are greater, we find that the accuracy of the classical gas laws decreases.

§1 Boyle's law

Robert Boyle discovered Boyle's law in 1662. Boyle's discovery was that the pressure, P, and volume, V, of a gas are inversely proportional to one another if the temperature, T, is held constant. We can imagine rediscovering Boyle's law by trapping a sample of gas in a tube and then measuring its volume as we change the pressure. We would observe behavior like that in Figure 1. We can represent this behavior mathematically as

$$PV = \alpha^*(n, T)$$

where we recognize that the "constant", α^*, is actually a function of the temperature and of the number of moles, n, of gas in the sample. That is, the product of pressure and volume is constant for a fixed quantity of gas at a fixed temperature.

A little thought convinces us that we can be more specific about the dependence on the quantity of gas. Suppose that we have a volume of gas at a fixed pressure and temperature, and imagine that we introduce a very thin barrier that divides the volume into exactly equal halves, without changing anything else. In this case, the pressure and temperature of the gas in each of the new containers will be the same as they were originally. But the volume is half as great, and the number of moles in each of the half-size containers must also be half of the original number. That is, the pressure–volume product must be directly proportional to the number of moles of gas in the sample:

$$PV = n\alpha(T)$$

where $\alpha(T)$ is now a function only of temperature. When we repeat this experiment using different gaseous substances, we discover a further remarkable fact: Not only do they all obey Boyle's law, but also the value of $\alpha(T)$ is the same for any gas.

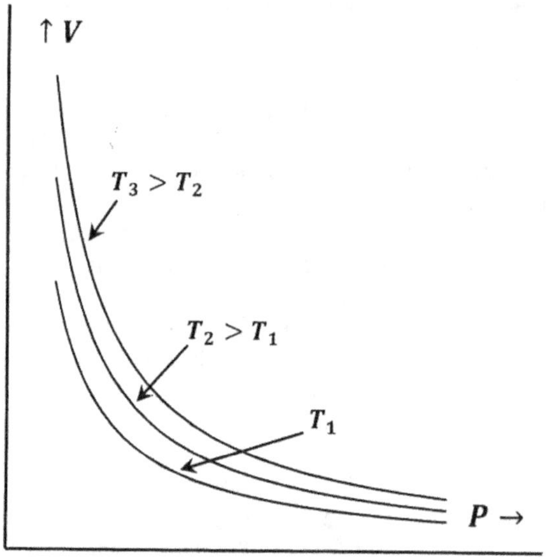

Figure 1. Gas volume *versus* pressure.

§2 Charles' law

Quantitative experiments establishing the law were first published in 1802 by Gay-Lussac, who credited Jacques Charles with having discovered the law earlier. Charles' law relates the volume and temperature of a gas when measurements are made at constant pressure. We can imagine rediscovering Charles' law by trapping a sample of gas in a tube and measuring its volume as we change the temperature, while keeping the pressure constant. This presumes that we have a way to measure temperature, perhaps by defining it in terms of the volume of a fixed quantity of some other fluid—like liquid mercury. At a fixed pressure, P_1, we observe a linear relationship between the volume of a sample of gas and its temperature, like that in Figure 2. If we repeat this experiment with the same gas sample at a higher pressure, P_2, we observe a second linear relationship between the volume and the temperature of the gas. If we extend these lines to their intersection with the temperature axis at zero volume, we make a further important discovery: Both lines intersect the temperature axis at the same point.

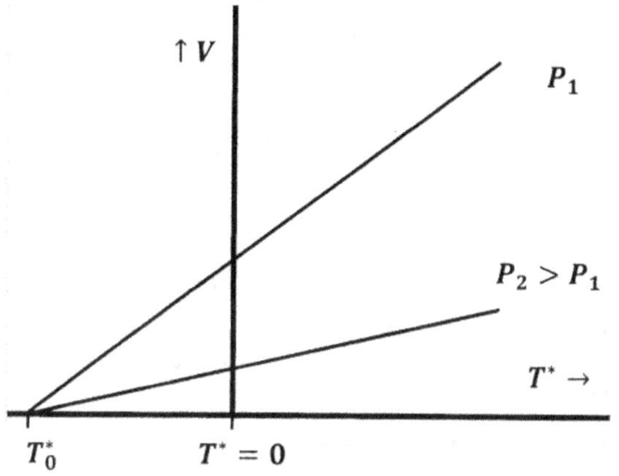

Figure 2. Gas volume *versus* temperature.

We can represent this behavior mathematically as

$$V = \beta^*(n,P)T^* + \gamma^*(n,P)$$

where we recognize that both the slope and the V-axis intercept of the graph depend on the pressure of the gas and on the number of moles of gas in the sample. A little reflection shows that here too the slope and intercept must be directly proportional to the number of moles of gas, so that we can rewrite our equation as

$$V = n\beta(P)T^* + n\gamma(P)$$

When we repeat these experiments with different gaseous substances, we discover an additional important fact: $\beta(P)$ and $\gamma(P)$ are the same for any gas. This means that the temperature at which the volume extrapolates to zero is the same for any gas and is independent of the constant pressure we maintain as we vary the temperature.

§3 Avogadro's hypothesis

Avogadro's hypothesis is another classical gas law. It can be stated: At the same temperature and pressure, equal volumes of different gases contain the same number of molecules.

When the mass, in grams, of an ideal gas sample is equal to the ***gram molar mass*** (traditionally called the ***molecular weight***) of the gas, the number of molecules in the sample is equal to ***Avogadro's number***, $\bar{N}$[1]. Avogadro's number is the number of molecules in a ***mole***. In the modern definition, one mole is the number of atoms of C^{12} in exactly 12 g of C^{12}. That is, the number of atoms of C^{12} in exactly 12 g of C^{12} is Avogadro's number. The currently accepted value is 6.02214199×10^{23} molecules per mole. We can find the gram atomic mass of any other element by finding the mass of that element that combines with exactly 12 g of C^{12} in a compound whose molecular formula is known.

The validity of Avogadro's hypothesis follows immediately either from the fact that the Boyle's law constant, $\alpha(T)$, is the same for any gas or from the fact that the Charles' law constants, $\beta(P)$ and $\gamma(P)$, are the same for any gas. However, this entails a significant circularity; these experiments can show that $\alpha(T)$, $\beta(P)$, and $\gamma(P)$ are the same for any gas only if we know how to find the number of moles of each gas that we use. To do so, we must know the molar mass of each gas. Avogadro's hypothesis is crucially important in the history of chemistry: Avogadro's hypothesis made it possible to determine relative molar masses. This made it possible to determine molecular formulas for gaseous substances and to create the atomic mass scale.

§4 Finding Avogadro's number

This use of Avogadro's number raises the question of how we know its value. There are numerous ways to measure Avogadro's number. One such method is to divide the charge of one mole of electrons by the charge of a single electron. We can obtain the charge of a mole of electrons from electrolysis experiments. The charge of one electron can be determined in a famous experiment devised by Robert Millikan, the "Millikan oil-drop experiment". The charge on a mole of electrons is called the ***faraday***. Experimentally, it has the value $96{,}485\ \text{C mol}^{-1}$ (coulombs per mole). As determined by Millikan's experiment, the charge on one electron is 1.6022×10^{-19} C. Then

$$\left(\frac{96{,}485\ C}{\text{mole electrons}}\right)\left(\frac{1\ \text{electron}}{1.6022\times10^{-19}\ C}\right)$$

$$= 6.022\times10^{23}\ \frac{\text{electrons}}{\text{mole electrons}}$$

§5 The Kelvin temperature scale

Thus far, we have assumed nothing about the value of the temperature corresponding to any particular volume of our standard fluid. We could define one unit of temperature to be any particular change in the volume of our standard fluid. Historically, Fahrenheit defined one unit (degree) of temperature to be one one-hundredth of the increase in volume of a fixed quantity of standard fluid as he warmed it from the lowest temperature he could achieve, which he elected to call 0 degrees, to the temperature of his body, which he elected to call 100 degrees. Fahrenheit's zero of temperature was achieved by mixing salt with ice and water. This is not a very reproducible condition, so the temperature of melting ice (with no salt present), soon became the calibration standard. Fahrenheit's experiments put the melting point of ice at 32 F. The normal temperature for a healthy person is now taken to be 98.6 F; possibly Fahrenheit had a slight fever when he was doing his calibration experiments. In any case, human temperatures vary enough so that Fahrenheit's 100-degree point was not very practical either. The boiling point of water, which Fahrenheit's experiments put at 212 F, became the calibration standard. Later, the centigrade scale was developed with fixed points at 0 degrees and 100 degrees at the melting point of ice and the boiling point of water, respectively. The centigrade scale is now called the Celsius scale after Anders Celsius, a

Swedish astronomer. In 1742, Celsius proposed a scale on which the temperature interval between the boiling point and the freezing point of water was divided into 100 degrees; however, a more positive number corresponded to a colder condition.

Further reflection convinces us that the Charles' law equation can be simplified by defining a new temperature scale. When we extend the straight line in any of our volume-*versus*-temperature plots, it always intersects the zero-volume horizontal line at the same temperature. Since we cannot associate any meaning with a negative volume, we infer that the temperature at zero volume represents a natural minimum point for our temperature scale. Let the value of T^* at this intersection be T_0^*. Substituting into our volume-temperature relationship, we have

$$0 = n\beta(P)T_0^* + n\gamma(P) \quad \text{or} \quad \gamma(P) = -\beta(P)T_0^*$$

So that

$$\begin{aligned} V &= n\beta(P)T^* - n\beta(P)T_0^* \\ &= n\beta(P)[T^* - T_0^*] \\ &= n\beta(P)T \end{aligned}$$

where we have created a new temperature scale. Temperature values on our new temperature scale, T, are related to temperature values on the old temperature scale, T^*, by the equation

$$T = T^* - T_0^*$$

When the size of one unit of temperature is defined using the Celsius scale (*i.e.*, T^* is the temperature in degrees Celsius), this is the origin of the ***Kelvin temperature scale***[2]. Then, on the Kelvin temperature scale, T_0^* is -273.15 degrees. (That is, $T = 0$ when $T_0^* = 273.15$; 0 K is -273.15 degrees Celsius.) The temperature at which the volume extrapolates to zero is called the ***absolute zero*** of temperature. When the size of one unit of temperature is defined using the Fahrenheit scale and the zero of temperature is set at absolute zero, the resulting temperature scale is called the ***Rankine scale***, after Wllliam Rankine, a Scottish engineer who proposed it in 1859.

§6 Deriving the ideal gas law from Boyle's law and Charles' law

We can solve Boyle's law and Charles' law for the volume. Equating the two, we have

$$\frac{n\alpha(T)}{P} = n\beta(P)T$$

The number of moles, n, cancels. Rearranging gives

$$\frac{\alpha(T)}{T} = P\beta(P)$$

In this equation, the left side is a function only of temperature, the right side only of pressure. Since pressure and temperature are independent of one another, this can be true only if each side is in fact constant. If we let this constant be R, we have

$$\alpha(T) = RT \qquad \text{and} \quad \beta(P) = R/P$$

Since the values of $\alpha(T)$ and $\beta(P)$ are independent of the gas being studied, the value of R is also the same for any gas. R is called the ***gas constant***, the ***ideal gas constant***, or the ***universal gas constant***. Substituting the appropriate relationship into either Boyle's law or Charles' law gives the ***ideal gas equation***

$$PV = nRT$$

The product of pressure and volume has the units of work or energy, so the gas constant has units of energy per mole per degree. (Remember that we simplified the form of Charles's law by defining the Kelvin temperature scale; temperature in the ideal gas equation is in degrees Kelvin.)

§7 The ideal gas constant and Boltzmann's constant

Having developed the ideal gas equation and analyzed experimental results for a variety of gases, we will have found the value of R. It is useful to have R expressed using a number of different energy units. Frequently useful values are

$$\begin{aligned} R &= 8.314 \text{ Pa m}^3 \text{ K}^{-1} \text{ mol}^{-1} \\ &= 8.314 \text{ J K}^{-1} \text{ mol}^{-1} \\ &= 0.08314 \text{ L bar K}^{-1} \text{ mol}^{-1} \\ &= 1.987 \text{ cal K}^{-1} \text{ mol}^{-1} \\ &= 0.08205 \text{ L atm K}^{-1} \text{ mol}^{-1} \end{aligned}$$

We also need the gas constant expressed per molecule rather than per mole. Since there is Avogadro's number of molecules per mole, we can divide any of the values above by $\overline{N}$ to get R on a per-molecule basis. Traditionally, however, this constant is given a different name; it is ***Boltzmann's constant***, usually given the symbol k.

$$k = R/\overline{N} = 1.381 \times 10^{-23} \text{ J K}^{-1} \text{ molecule}^{-1}$$

This means that we can also write the ideal gas equation as $PV = nRT = n\overline{N}kT$. Because the number of molecules in the sample, N, is $N = n\overline{N}$, we have

$$PV = NkT.$$

§8 Real gases *versus* ideal gases

Now, we need to expand on the qualifications with which we begin this chapter. We imagine that the results of a large number of experiments are available for our analysis. Our characterization of these results has been that all gases obey the same equations—Boyle's law,

Charles' law, and the ideal gas equation—and do so exactly. This is an oversimplification. In fact they are always approximations. They are approximately true for all gases under all "reasonable" conditions, but they are not exactly true for any real gas under any condition. It is useful to introduce the idea of hypothetical gases that obey the classical gas equations exactly. In the previous section, we call the combination of Boyle's law and Charles' law the ideal gas equation. We call the hypothetical substances that obey this equation ***ideal gases***. Sometimes we refer to the classical gas laws collectively as the ***ideal gas laws***.

At very high gas densities, the classical gas laws can be very poor approximations. As we have noted, they are better approximations the lower the density of the gas. In fact, experiments show that the pressure—volume—temperature behavior of any real gas becomes arbitrarily close to that predicted by the ideal gas equation in the limit as the pressure goes to zero. This is an important observation that we use extensively.

At any given pressure and temperature, the ideal gas laws are better approximations for a compound that has a lower boiling point than they are for a compound with a higher boiling point. Another way of saying this is that they are better approximations for molecules that are weakly attracted to one another than they are for molecules that are strongly attracted to one another.

Forces between molecules cause them to both attract and repel one another. The net effect depends on the distance between them. If we assume that there are no intermolecular forces acting between gas molecules, we can develop exact theories for the behavior of macroscopic amounts of the gas. In particular, we can show that such substances obey the ideal gas equation. (We shall see that a complete absence of repulsive forces implies that the molecules behave as point masses.) Evidently, the difference between the behavior of a real gas and the behavior it would exhibit if it were an ideal gas is just a measure of the effects of intermolecular forces.

The ideal gas equation is not the only equation that gives a useful representation for the interrelation of gas pressure–volume–temperature data. There are many such ***equations of state***. They are all approximations, but each can be a particularly useful approximation in particular circumstances . We discuss ***van der Waal's equation*** and the ***virial equations*** later in this chapter. Nevertheless, we use the ideal gas equation extensively.

We will see that much of chemical thermodynamics is based on the behavior of ideal gases. Since there are no ideal gases, this may seem odd, at best. If there are no ideal gases, why do we waste time talking about them? After all, we don't want to slog through tedious, long-winded, pointless digressions. We want to understand how real stuff behaves! Unfortunately, this is more difficult. The charm of ideal gases is that we can understand their behavior; the ideal gas equation expresses this understanding in a mathematical model. Real gases are another story. We can reasonably say that we can best understand the behavior of a real gas by understanding how and why it is different from the behavior of a (hypothetical) ideal gas that has the same molecular structure.

§9 Temperature and the ideal gas thermometer

In §2 we suppose that we have a thermometer that we can use to measure the temperature of a gas. We suppose that this thermometer uses a liquid, and we define an increase in temperature by the increase in the volume of this liquid. Our statement of Charles' law asserts that the volume of a gas is a linear function of the volume of the liquid in our thermometer, and that the same linear function is observed for any gas. As we note in §8, there is a problem with this statement. Careful experiments with such thermometers produce results that deviate from Charles' law. With sufficiently accurate volume measurements, this occurs to some extent for any choice of the liquid in the thermometer. If we make sufficiently accurate measurements, the volume of a gas is not exactly proportional to the volume of any liquid (or solid) that we might choose as the working substance in our thermometer. That is, if we base our temperature scale on a liquid or solid substance, we observe deviations from Charles' law. There is a further difficulty with using a liquid as the standard fluid on which to base our temperature measurements: temperatures outside the liquid range of the chosen substance have to be measured in some other way.

Evidently, we can choose to use a gas as the working fluid in our thermometer. That is, our gas-volume measuring device is itself a thermometer. This fact proves to be very useful because of a further experimental observation. To a very good approximation, we find: If we keep the pressures in the thermometer and in some other gaseous system constant at low enough values, both gases behave as ideal gases, and we find that the volumes of the two gases are proportional to each other over any range of temperature. Moreover, this proportionality is observed for any choice of either gas. This means that we can define temperature in terms of the expansion of any constant-pressure gas that behaves ideally. In principle, we can measure the same temperature using any gas, so long as the constant operating pressure is low enough. When we do so, our device is called the ***ideal gas thermometer***. In so far as any gas behaves as an ideal gas at a sufficiently low pressure, any real gas can be used in an ideal gas thermometer and to measure any temperature accurately. Of course, practical problems emerge when we attempt to make such measurements at very high and very low temperatures.

The (very nearly) direct proportionality of two low-pressure real gas volumes contrasts with what we observe for liquids and solids. In general, the volume of a given liquid (or solid) substance is not exactly proportional to the volume of a second liquid (or solid) substance over a wide range of temperatures.

In practice, the ideal-gas thermometer is not as convenient to use as other thermometers—like the mercury-in-glass thermometer. However, the ideal-gas thermometer is used to calibrate other thermometers. Of course, we have to calibrate the ideal-gas thermometer itself before we can use it.

We do this by assigning a temperature of 273.16 K to the triple point of water. (It turns out that the melting point of ice isn't sufficiently reproducible for the most precise work. Recall that the triple point is the temperature and pressure at which all three phases of water are at equilibrium with one another, with no air or other substances present. The triple-point pressure is 611 Pa or 6.03×10^{-3} atm. See §6-3.) From both theoretical considerations and experimental observations, we are confident that no system can attain a temperature below absolute zero. Thus, the size[3] of the kelvin (one degree on the Kelvin scale) is fixed by the difference in temperature between a system at the triple point of water and one at absolute zero. If our ideal gas thermometer has volume V at thermal equilibrium with some other constant-temperature system, the proportionality of V and T means that

$$\frac{T}{V} = \frac{273.16}{V_{273.16}}$$

With the triple point fixed at 273.16 K, experiments find the freezing point of air-saturated water to be 273.15 K when the system pressure is 1 atmosphere. (So the melting point of ice is 273.15 K, and the triple-point is 0.10 C. We will find two reasons for the fact that the melting point is lower than the triple point: In §6-3 we find that the melting point of ice decreases as the pressure increases. In §16-10 we find that solutes usually decrease the temperature at which the liquid and solid states of a substance are in phase equilibrium.)

If we could use an ideal gas in our ideal-gas thermometer, we could be confident that we had a rigorous operational definition of temperature. However, we note in §8 that any real gas will exhibit departures from ideal gas behavior if we make sufficiently accurate measurements. For extremely accurate work, we need a way to correct the temperature value that we associate with a given real-gas volume. The issue here is the value of the partial derivative

$$\left(\frac{\partial V}{\partial T}\right)_P$$

For one mole of an ideal gas,

$$\left(\frac{\partial V}{\partial T}\right)_P = \frac{R}{P} = \frac{V}{T}$$

is a constant. For a real gas, it is a function of temperature. Let us assume that we know this function. Let the molar volume of the real gas at the triple point of water be $V_{273.16}$ and its volume at thermal equilibrium with a system whose true temperature is T be V_T. We have

$$\int_{273.16}^{T} \left(\frac{\partial V}{\partial T}\right)_P dT = \int_{V_{273,16}}^{V_T} dV = V_T - V_{273.16}$$

When we know the integrand on the left as a function of temperature, we can do the integration and find the temperature corresponding to any measured volume, V_T.

When the working fluid in our thermometer is a real gas we make measurements to find $(\partial V/\partial T)_P$ as a function of temperature. Here we encounter a circularity: To find $(\partial V/\partial T)_P$ from pressure-volume-temperature data we must have a way to measure temperature; however, this is the very thing that we are trying to find.

In principle, we can surmount this difficulty by iteratively correcting the temperature that we associate with a given real-gas volume. As a first approximation, we use the temperatures that we measure with an uncorrected real-gas thermometer. These temperatures are a first approximation to the ideal-gas temperature scale. Using this scale, we make non-pressure-volume-temperature measurements that establish $(\partial V/\partial T)_P$ as a function of temperature for the real gas. [This function is

$$\left(\frac{\partial V}{\partial T}\right)_P = \frac{V + \mu_{JT} C_P}{T}$$

where C_P is the ***constant-pressure heat capacity*** and μ_{JT} is the ***Joule-Thomson coefficient***. Both are functions of temperature. We introduce C_P in §7-9. We discuss the Joule-Thomson coefficient further in §10 below, and in detail in §10-14. Typically $V \gg C_P$, and the value of $(\partial V/\partial T)_P$ is well approximated by $V/T = R/P$. With $(\partial V/\partial T)_P$ established using this scale, integration yields a second-approximation to the ideal-gas temperatures. We could repeat this process until successive temperature scales converge at the number of significant figures that our experimental accuracy can support.

In practice, there are several kinds of ideal-gas thermometers, and numerous corrections are required for very accurate measurements. There are also numerous other ways to measure temperature, each of which has its own complications. Our development has considered some of the ideas that have given rise to the concept[4] that temperature is fundamental property of nature that can be measured using a thermodynamic-temperature scale on which values begin at zero and increase to arbitrarily high values. This thermodynamic temperature scale is a creature of theory, whose real-world counterpart would be the scale established by an ideal-gas thermometer whose gas actually obeyed $PV = nRT$ at all conditions. We have seen that such an ideal-gas thermometer is itself a creature of theory.

The current real-world standard temperature scale is the ***International Temperature Scale of 1990*** **(ITS-90)**. This defines temperature over a wide range in terms of the pressure-volume relationships of helium isotopes and the triple points of several selected elements. The triple points fix the temperature at each of several conditions up to 1357.77 K (the freezing point of copper). Needless to say, the temperatures assigned at the fixed points are the results of painstaking experiments designed to give the closest possible match to the thermodynamic scale. A variety of measuring devices—thermometers—can be used to interpolate temperature values between different pairs of fixed points.

§10 Deriving Boyle's law from Newtonian mechanics

We can derive Boyle's law from Newtonian mechanics. This derivation assumes that gas molecules behave like point masses that do not interact with one another. The pressure of the gas results from collisions of the gas molecules with the walls of the container. The contribution of one collision to the force on the wall is equal to the change in the molecule's momentum divided by the time between collisions. The magnitude of this force depends on the molecule's speed and the angle at which it strikes the wall. Each such collision makes a contribution to the pressure that is equal to the force divided by the area of the wall. To find the pressure from this model, it is necessary to average over all possible molecular speeds and all possible collision angles. In Chapter 4, we derive Boyle's law in this way.

We can do a simplified derivation by making a number of assumptions. We assume that all of the molecules in a sample of gas have the same speed. Let us call it u. As sketched in Figure 3, we assume that the container is a cubic box whose edge length is d. If we consider all of the collisions between molecules and walls, it is clear that each wall will experience $1/6$ of the collisions; or, each pair of opposing walls will experience $1/3$ of the collisions. Instead of averaging over all of the possible angles at which a molecule could strike a wall and all of the possible times between collisions, we assume that the molecules travel at constant speed back and forth between opposite faces of the box. Since they are point masses, they never collide with one another. If we suppose that $1/3$ of the molecules go back and forth between each pair of opposite walls, we can expect to accomplish the same kind of averaging in setting up our artificial model that we achieve by averaging over the real distribution of angles and speeds. In fact, this turns out to be the case; the derivation below gets the same result as the rigorous treatment we develop in Chapter 4.

Since each molecule goes back and forth between opposite walls, it collides with each wall once during each round trip. At each collision, the molecule's speed remains constant, but its direction changes by 180°; that is, the molecule's velocity changes from $\vec{u}$ to $-\vec{u}$. Letting Δt be the time required for a round trip, the distance traversed in a round trip is

$$\begin{aligned} 2d &= |\vec{u}|\Delta t \\ &= u\Delta t \end{aligned}$$

The magnitude of the momentum change for a molecule in one collision is

$$\begin{aligned} |\Delta(m\vec{u})| &= |m\vec{u}_{final} - m\vec{u}_{initial}| \\ &= |m\vec{u}_{final} - (-m\vec{u}_{final})| \\ &= 2mu \end{aligned}$$

The magnitude of the force on the wall from one collision is

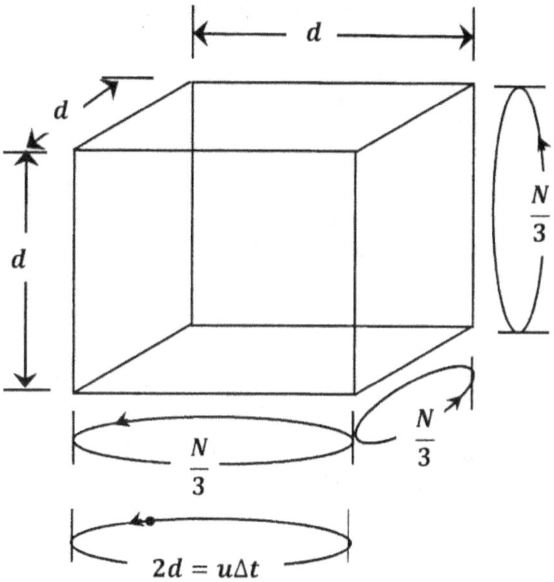

Figure 3. Simplified model for velocities of gas molecules in a cubic box.

$$\begin{aligned} F &= \frac{|\Delta(m\vec{u})|}{\Delta t} \\ &= \frac{2mu}{(2d/u)} \\ &= \frac{mu^2}{d} \end{aligned}$$

and the pressure contribution from one collision on the wall, of area d^2, is

$$\begin{aligned} P &= \frac{F}{A} \\ &= \frac{mu^2}{d \cdot d^2} \\ &= \frac{mu^2}{d^3} \\ &= \frac{mu^2}{V} \end{aligned}$$

so that we have

$$PV = mu^2$$

from the collision of one molecule with one wall.

If the number of molecules in the box is N, $N/3$ of them make collisions with this wall, so that the total pressure on one wall attributable to all N molecules in the box is

$$P = \frac{mu^2}{V}\frac{N}{3}$$

or

$$PV = \frac{Nmu^2}{3}$$

Since the ideal gas equation can be written as $PV = NkT$

we see that $Nmu^2/3 = NkT$ so that $mu^2 = 3kT$ and

$$u = \sqrt{\frac{3kT}{m}}$$

Thus we have found a relationship between the molecular speed and the temperature of the gas. (The actual speed of a molecule, v, can have any value between zero and—for present purposes—infinity. When we average the values of v^2 for many molecules, we find the average value of the squared speeds, $\overline{v^2}$. In Chapter 4, we find that $u^2 = \overline{v^2}$. That is, the average speed we use in our derivation turns out to be a quantity called the ***root-mean-square speed***, $v_{rms} = u = \sqrt{\overline{v^2}}$.) This result also gives us the (average) kinetic energy of a single gas molecule:

$$KE = \frac{mu^2}{2} = \frac{3kT}{2}$$

From this derivation, we have a simple mechanical model that explains Boyle's law as the logical consequence of point-mass molecules colliding with the walls of their container. By combining this result with the ideal gas equation, we find that the average speed of ideal gas molecules depends only on the temperature. From this we have the very important result that ***the translational kinetic energy of an ideal gas depends only on temperature.***

Since our non-interacting point-mass molecules have no potential energy arising from their interactions with one another, their translational kinetic energy is the whole of their energy. (Because two such molecules neither attract nor repel one another, no work is required to change the distance between them. The work associated with changing the volume of a confined sample of an ideal gas arises because of the pressure the molecules exert on the walls of the container; the pressure arises because of the molecules' kinetic energy.) The energy of one mole of monatomic ideal gas molecules is

$$KE = (3/2)RT$$

When we expand our concept of ideal gases to include molecules that have rotational or vibrational energy, but which neither attract nor repel one another, it remains true that the energy of a macroscopic sample depends only on temperature. However, the molar energy of such a gas is greater than $(3/2)RT$, because of the energy associated with these additional motions.

We make extensive use of the conclusion that the energy of an ideal gas depends only on temperature. As it turns out, this conclusion follows rigorously from the second law of thermodynamics. In Chapter 10, we show that

$$\left(\frac{\partial E}{\partial V}\right)_T = \left(\frac{\partial E}{\partial P}\right)_T = 0$$

for a substance that obeys the ideal gas equation; at constant temperature, the energy of an ideal gas is independent of the volume and independent of the pressure. So long as pressure, volume, and temperature are the only variables needed to specify its state, the laws of thermodynamics imply that the energy of an ideal gas depends only on temperature.

While the energy of an ideal gas is independent of pressure, the energy of a real gas is a function of pressure at a given temperature. At ordinary pressures and temperatures, this dependence is weak and can often be neglected. The first experimental investigation of this issue was made by James Prescott Joule, for whom the SI unit of energy is named. Beginning in 1838, Joule did a long series of careful measurements of the mechanical equivalent of heat. These measurements formed the original experimental basis for the kinetic theory of heat. Among Joule's early experiments was an attempt to measure the heat absorbed by a gas as it expanded into an evacuated container, a process known as a ***free expansion***. No absorption of heat was observed, which implied that the energy of the gas was unaffected by the volume change. However, it is difficult to do this experiment with meaningful accuracy.

Subsequently, Joule collaborated with William Thomson (Lord Kelvin) on a somewhat different experimental approach to essentially the same question. The ***Joule-Thomson experiment*** provides a much more sensitive measure of the effects of intermolecular forces of attraction and repulsion on the energy of a gas during its expansion. Since our definition of an ideal gas includes the stipulation that there are no intermolecular forces, the Joule-Thomson experiment is consistent with the conclusion that the energy of an ideal gas depends only on temperature. However, since intermolecular forces are not zero for any real gas, our analysis reaches this conclusion in a somewhat indirect way. The complication arises because the Joule-Thomson results are not entirely consistent with the idea that all properties of a real gas approach those of an ideal gas at a sufficiently low pressure. (The best of models can have limitations.) We discuss the Joule-Thomson experiment in §10-14.

§11 The barometric formula

We can measure the pressure of the atmosphere at any location by using a ***barometer***. A mercury barometer is a sealed tube that contains a vertical column of liquid mercury. The space in the tube above the liquid mercury is occupied by mercury vapor. Since the vapor pressure of liquid mercury at ordinary temperatures is very low, the pressure at the top of the mercury column is very low and can usually be ignored. The pressure at the bottom of the column of mercury is equal to the pressure of a column of air extending from the elevation of the barometer all the way to the top of the earth's atmosphere. As we take the barometer to higher altitudes, we find that the height of the mercury column decreases, because less and less of the atmosphere is above the barometer.

If we assume that the atmosphere is composed of an

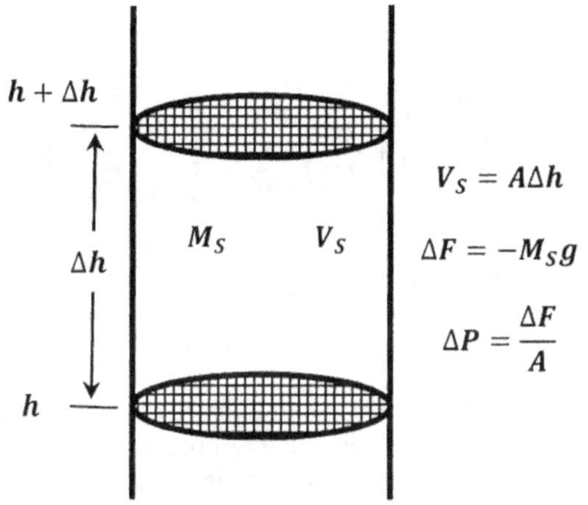

Figure 4. Atmospheric pressure *versus* altitude.

ideal gas and that its temperature is constant, we can derive an equation for atmospheric pressure as a function of altitude. Imagine a cylindrical column of air extending from the earth's surface to the top of the atmosphere. (See Figure 4.) The force exerted by this column at its base is the weight of the air in the column; the pressure is this weight divided by the cross-sectional area of the column. Let the cross-sectional area of the column be A.

Consider a short section of this column. Let the bottom of this section be a distance h from the earth's surface, while its top is a distance $h + \Delta h$ from the earth's surface. The volume of this cylindrical section is then $V_S = A\Delta h$. Let the mass of the gas in this section be M_S. The pressure at $h + \Delta h$ is less than the pressure at h by the weight of this gas divided by the cross-sectional area. The weight of the gas is $M_S g$. The pressure difference is $\Delta P = -M_S g/A$. We have

$$\frac{P(h+\Delta h) - P(h)}{\Delta h} = \frac{\Delta P}{\Delta h} = \frac{-M_S g}{A\Delta h} = \frac{-M_S g}{V_S}$$

Since we are assuming that the sample of gas in the cylindrical section behaves ideally, we have $V_S = n_S RT/P$. Substituting for V_S and taking the limit as $\Delta h \to 0$, we find

$$\frac{dP}{dh} = \left(\frac{-M_S g}{n_S RT}\right)P = \left(\frac{-n_S \overline{M} g}{n_S RT}\right)P = \left(\frac{-mg}{kT}\right)P$$

where we introduce n_S as the number of moles of gas in the sample, $\overline{M}$ as the molar mass of this gas, and m as the mass of an individual atmosphere molecule. The last equality on the right makes use of the identities $\overline{M} = m\overline{N}$ and $R = \overline{N}k$. Separating variables and integrating between limits $P(0) = P_0$ and $P(h) = P$, we find

$$\int_{P_0}^{P} \frac{dP}{P} = \left(\frac{-mg}{kT}\right)\int_0^h dh$$

so that

$$\ln\left(\frac{P}{P_0}\right) = \frac{-mgh}{kT}$$

and

$$P = P_0 \exp\left(\frac{-mgh}{kT}\right)$$

Either of the latter relationships is frequently called the ***barometric formula***.

If we let η be the number of molecules per unit volume, $\eta = N/V$, we can write $P = NkT/V = \eta kT$ and $P_0 = \eta_0 kT$ so that the barometric formula can be expressed in terms of these number densities as

$$\eta = \eta_0 \exp\left(\frac{-mgh}{kT}\right)$$

§12 van der Waals' equation

An equation due to van der Waals extends the ideal gas equation in a straightforward way. ***Van der Waals' equation*** is

$$\left(P + \frac{an^2}{V^2}\right)(V - nb) = nRT$$

It fits pressure-volume-temperature data for a real gas better than the ideal gas equation does. The improved fit is obtained by introducing two parameters (designated "a" and "b") that must be determined experimentally for each gas. Van der Waals' equation is particularly useful in our effort to understand the behavior of real gases, because it embodies a simple physical picture for the difference between a real gas and an ideal gas.

In deriving Boyle's law from Newton's laws, we assume that the gas molecules do not interact with one another. Simple arguments show that this can be only approximately true. Real gas molecules must interact with one another. At short distances they repel one another. At somewhat longer distances, they attract one another. The ideal gas equation can also be derived from the basic assumptions that we make in §10 by an application of the theory of statistical thermodynamics. By making different assumptions about molecular properties, we can apply statistical thermodynamics to derive[5] van der Waals' equation. The required assumptions are that the molecules occupy a finite volume and that they attract one another with a force that varies as the inverse of a power of the distance between them. (The attractive force is usually assumed to be proportional to r^{-6}.)

To recognize that real gas molecules both attract and repel one another, we need only remember that any gas can be liquefied by reducing its temperature and increasing the pressure applied to it. If we cool the liquid further, it freezes to a solid. Now, two distinguishing features of a solid are that it retains its shape and that it is almost incompressible. We attribute the incompressibility of a solid to repulsive forces between its constituent molecules; they have come so close to one another that repulsive forces between them have become important. To compress the solid, the molecules must be pushed still closer together, which requires inordinate force. On the other hand, if we throw an ice cube across the room, all of its constituent water molecules fly across the room together. Evidently, the water molecules in the solid are attracted to one another, otherwise they would all go their separate ways—throwing the ice cube would be like throwing a handful of dry sand. But water molecules are the same molecules whatever the temperature or pressure, so if there are forces of attraction and repulsion between them in the solid, these forces must be present in the liquid and gas phases also.

In the gas phase, molecules are far apart; in the liquid or the solid phase, they are packed together. At its boiling point, the volume of a liquid is much less than the volume of the gas from which it is condensed. At the freezing point, the volume of a solid is only slightly different from the volume of the liquid from which it is frozen, and it is certainly greater than zero. These commonplace observations are readily explained by supposing that any molecule has a characteristic volume. We can understand this, in turn, to be a consequence of the nature of the intermolecular forces; evidently, these forces become stronger as the distance between a pair of molecules decreases. Since a liquid or a solid occupies a definite volume, the repulsive force must increase more rapidly than the attractive force when the intermolecular distance is small. Often it is useful to talk about the molar volume of a condensed phase. By molar volume, we mean the volume of one mole of a pure substance. The molar volume of a condensed phase is determined by the intermolecular distance at which there is a balance between intermolecular forces of attraction and repulsion.

Evidently molecules are very close to one another in condensed phases. If we suppose that the empty spaces between molecules are negligible, the volume of a condensed phase is approximately equal to the number of molecules in the sample multiplied by the volume of a single molecule. Then the molar volume is Avogadro's number times the volume occupied by one molecule. If we know the density, D, and the molar mass, $\overline{M}$, we can find the molar volume, $\overline{V}$, as

$$\overline{V} = \frac{\overline{M}}{D}$$

The volume occupied by a molecule, $V_{molecule}$, becomes

$$V_{molecule} = \overline{V}/\overline{N}$$

The pressure and volume appearing in van der Waals' equation are the pressure and volume of the real gas. We can relate the terms in van der Waals' equation to the ideal gas equation: It is useful to think of the terms $(P + an^2/V^2)$ and $(V - nb)$ as the pressure and volume of a *hypothetical ideal gas*. That is

$$P_{ideal\ gas}V_{ideal\ gas} = \\ = \left(P_{real\ gas} + \frac{an^2}{V_{real\ gas}^2}\right)(V_{real\ gas} - nb) \\ = nRT$$

Then we have

$$V_{real\ gas} = V_{ideal\ gas} + nb$$

We derive the ideal gas equation from a model in which the molecules are non-interacting point masses. So the volume of an ideal gas is the volume occupied by a gas whose individual molecules have zero volume. If the individual molecules of a real gas effectively occupy a volume $b/\overline{N}$, then n moles of them effectively occupy a volume $(b/\overline{N})(n\overline{N}) = nb$. Van der Waals' equation says that the volume of a real gas is the volume that would be occupied by non-interacting point masses, $V_{ideal\ gas}$, plus the effective volume of the gas molecules themselves. (When data for real gas molecules are fit to the van der Waals' equation, the value of b is usually somewhat greater than the volume estimated from the liquid density and molecular weight. See problem 24.)

Similarly, we have

$$P_{real\ gas} = P_{ideal\ gas} - \frac{an^2}{V_{real\ gas}^2}$$

We can understand this as a logical consequence of attractive interactions between the molecules of the real gas. With $a > 0$, it says that the pressure of the real gas is less than the pressure of the hypothetical ideal gas, by an amount that is proportional to $(n/V)^2$. The proportionality constant is a. Since n/V is the molar density (moles per unit volume) of the gas molecules, it is a measure of concentration. The number of collisions between molecules of the same kind is proportional to the square of their concentration. (We consider this point in more detail in Chapters 4 and 5.) So $(n/V)^2$ is a measure of the frequency with which the real gas molecules come into close contact with one another. If they attract one another when they come close to one another, the effect of this attraction should be proportional to $(n/V)^2$. So van der Waals' equation is consistent with the idea that the pressure of a real gas is different from the pressure of the hypothetical ideal gas by an amount that is proportional to the frequency and strength of attractive interactions.

But why should attractive interactions have this effect; why should the pressure of the real gas be less than that of the hypothetical ideal gas? Perhaps the best way to develop a qualitative picture is to recognize that attractive intermolecular forces tend to cause the gas molecules to clump up. After all, it is these attractive forces that cause the molecules to aggregate to a liquid at low temperatures. Above the boiling point, the ability of gas molecules to go their separate ways limits the effects of this tendency; however, even in the gas, the attractive forces must act in a way that tends to reduce the volume occupied by the molecules. Since the volume occupied by the gas is dictated by the size of the container—not by the properties of the gas itself—this clumping-up tendency finds expression as a decrease in pressure.

It is frequently useful to describe the interaction between particles or chemical moieties in terms of a potential energy versus distance diagram. The van der Waals' equation corresponds to the case that the repulsive interaction between molecules is non-existent until the molecules come into contact. Once they come into contact, the energy required to move them still closer together becomes arbitrarily large. Often this is described by saying that they behave like "hard spheres". The attractive force between two molecules decreases as the distance between them increases. When they are very far apart the attractive interaction is very small. We say that the energy of interaction is zero when the molecules are infinitely far apart. If we initially have two widely separated, stationary, mutually attracting molecules, they will spontaneously move toward one another, gaining kinetic energy as they go. Their potential energy decreases as they approach one another, reaching its smallest value when the molecules come into contact. Thus, van der Waals' equation implies the potential energy *versus* distance diagram sketched in Figure 5.

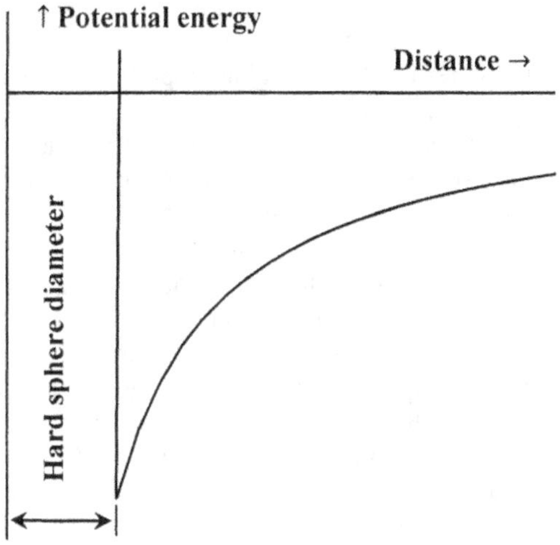

Figure 5. Potential energy *versus* distance for "hard sphere molecules."

§13 Virial equations

It is often useful to fit accurate pressure-volume-temperature data to polynomial equations. The experimental data can be used to compute a quantity called the ***compressibility factor***, Z. Z is defined as the pressure–volume product for the real gas divided by the pressure–volume product for an ideal gas at the same temperature. We have

$$(PV)_{ideal\ gas} = nRT$$

Letting P and V represent the pressure and volume of the real gas, and introducing the molar volume, $\overline{V} = V/n$, we have

$$Z = \frac{(PV)_{real\ gas}}{(PV)_{ideal\ gas}} = \frac{PV}{nRT} = \frac{P\overline{V}}{RT}$$

Since $Z = 1$ if the real gas behaves exactly like an ideal gas, experimental values of Z will tend toward unity under conditions in which the density of the real gas becomes low and its behavior approaches that of an ideal gas. At a given temperature, we can conveniently ensure that this condition is met by fitting the Z values to a polynomial in P or a polynomial in $\overline{V}^{-1}$. The coefficients are functions of temperature. If the data are fit to a polynomial in the pressure, the equation is

$$Z = 1 + B^*(T)P + C^*(T)P^2 + D^*(T)P^3 + \cdots$$

For a polynomial in $\overline{V}^{-1}$, the equation is

$$Z = 1 + \frac{B(T)}{\overline{V}} + \frac{C(T)}{\overline{V}^2} + \frac{D(T)}{\overline{V}^3} + \cdots$$

These empirical equations are called ***virial equations***. As indicated, the parameters are functions of temperature. The values of $B^*(T)$, $C^*(T)$, $D^*(T)$, …, and $B(T)$, $C(T)$, $D(T)$,…, must be determined for each real gas at every temperature. (Note also that $B^*(T) \neq B(T)$, $C^*(T) \neq C(T)$, $D^*(T) \neq D(T)$, *etc.* However, it is true that $B^* = B/RT$.) Values for these parameters are tabulated in various compilations of physical data. In these tabulations, $B(T)$ and $C(T)$ are called the second and third virial coefficients, respectively.

§14 Gas mixtures: Dalton's law of partial pressures

Thus far, our discussion of the properties of a gas has implicitly assumed that the gas is pure. We turn our attention now to mixtures of gases—gas samples that contain molecules of more than one compound. Mixtures of gases are common, and it is important to understand their behavior in terms of the properties of the individual gases that make it up. The ideal-gas laws we have for mixtures are approximations. Fortunately, these approximations are often very good. When we think about it, this is not surprising. After all, the distinguishing feature of a gas is that its molecules do not interact with one another very much. Even if the gas is composed of molecules of different kinds, the unimportance of molecule—molecule interactions means that the properties of one kind of molecules should be nearly independent of the properties of the other kinds.

Consider a sample of gas that contains a fixed number of moles of each of two or more compounds. This sample has a pressure, a volume, a temperature, and a specified composition. Evidently, the challenge here is to describe the pressure, volume, and temperature of the mixture in terms of measurable properties of the component compounds.

There is no ambiguity about what we mean by the pressure, volume, and temperature of the mixture; we can measure these properties without difficulty. Given the nature of temperature, it is both reasonable and unambiguous to say that the temperature of the sample and the temperature of its components are the same. However, we cannot measure the pressure or volume of an individual component in the mixture. If we hope to describe the properties of the mixture in terms of properties of the components, we must first define some related quantities that we can measure. The concepts of a component partial pressure and a component partial volume meet this need.

We define the ***partial pressure*** of a component of a gas mixture as the pressure exerted by the same number of moles of the pure component when present in the volume occupied by the mixture, $V_{mixture}$, at the temperature of the mixture. In a mixture of n_A moles of component A, n_B moles of component B, etc., it is customary to designate the partial pressure of component A as P_A. It is important to appreciate that the partial pressure of a real gas can only be determined by experiment.

We define the ***partial volume*** of a component of a gas mixture as the volume occupied by the same number of moles of the pure component when the pressure is the same as the pressure of the mixture, $P_{mixture}$, at the temperature of the mixture. In a mixture of components A, B, etc., it is customary to designate the partial volume of component A as V_A. The partial volume of a real gas can only be determined by experiment.

Dalton's law of partial pressures asserts that the pressure of a mixture is equal to the sum of the partial pressures of its components. That is, for a mixture of components A, B, C, etc., the pressure of the mixture is

$$P_{mixture} = P_A + P_B + P_C + \cdots$$

Under conditions in which the ideal gas law is a good approximation to the behavior of the individual components, Dalton's law is usually a good approximation to the behavior of real gas mixtures. For mixtures of ideal gases, it is exact. To see this, we recognize that, for an ideal gas, the definition of partial pressure becomes

$$P_A = \frac{n_A RT}{V_{mixture}}$$

The ideal-gas mixture contains
$n_{mixture} = n_A + n_B + n_C + \cdots$ moles, so that

$$\begin{aligned} P_{mixture} &= \frac{n_{mixture} RT}{V_{mixture}} \\ &= \frac{(n_A + n_B + n_C + \cdots)RT}{V_{mixture}} \\ &= \frac{n_A RT}{V_{mixture}} + \frac{n_B RT}{V_{mixture}} + \frac{n_C RT}{V_{mixture}} + \cdots \\ &= P_A + P_B + P_C + \cdots \end{aligned}$$

Applied to the mixture, the ideal-gas equation yields Dalton's law. When x_A is the mole fraction of A in a mixture of ideal gases, $P_A = x_A P_{mixture}$.

§15 Gas mixtures: Amagat's law of partial volumes

Amagat's law of partial volumes asserts that the volume of a mixture is equal to the sum of the partial volumes of its components. For a mixture of components A, B, C, etc., Amagat's law gives the volume as

$$V_{mixture} = V_A + V_B + V_C + \cdots$$

For real gases, Amagat's law is usually an even better approximation than Dalton's law[6]. Again, for mixtures of ideal gases, it is exact. For an ideal gas, the partial volume is

$$V_A = \frac{n_A RT}{P_{mixture}}$$

Since $n_{mixture} = n_A + n_B + n_C + \cdots$, we have, for a mixture of ideal gases,

$$\begin{aligned} V_{mixture} &= \frac{n_{mixture} RT}{P_{mixture}} \\ &= \frac{(n_A + n_B + n_C + \cdots)RT}{P_{mixture}} \\ &= V_A + V_B + V_C + \cdots \end{aligned}$$

Applied to the mixture, the ideal-gas equation yields Amagat's law. Also, we have $V_A = x_A V_{mixture}$.

Problems

1. If A is an ideal gas in a mixture of ideal gases, prove that its partial pressure, P_A, is given by $P_A = x_A P_{mixture}$.

2. If A is an ideal gas in a mixture of ideal gases, prove that its partial volume, V_A, is given by $V_A = x_A V_{mixture}$.

3. A sample of hydrogen chloride gas, HCl, occupies 0.932 L at a pressure of 1.44 bar and a temperature of 50 C. The sample is dissolved in 1 L of water. What is the resulting hydronium ion, H_3O^+, concentration?

4. Ammonia gas, NH_3, also dissolves quantitatively in water. If it is measured at 0.720 bar and 50 C, what volume of NH_3 gas is required to neutralize the solution prepared in problem 3? For present purposes, assume that the neutralization reaction occurs quantitatively.

5. Two pressure vessels are separated by a closed valve. One contains 10.0 moles of helium, He, at 5.00 bar. The other contains 5.00 moles of neon, Ne, at 20.0 bar. Both vessels are at the same temperature. The valve is opened and the gases are allowed to mix. The temperature remains constant. What is the final pressure?

6. What is the average velocity of a molecule of nitrogen, N_2, at 300 K? Of a molecule of hydrogen, H_2, at the same temperature?

7. The Homestake gold mine near Lead, South Dakota, is excavated to 8000 feet below the surface. Lead is nearly a mile high; the bottom of the Homestake is about 900 m below sea level. Nearby Custer Peak is about 2100 m above sea level. What is the ratio of the barometric pressure on top of Custer Peak to the barometric pressure at the bottom of the Homestake? Assume that the entire atmosphere is at 300 K and that it behaves as a single ideal gas whose molar mass is 29.

8. On the sidewalk in front of a tall building, the barometric pressure is 740 torr and the temperature is 25 C. On the roof of this building, the barometric pressure is 732 torr. Assuming that the entire atmosphere behaves as an ideal gas of molecular weight 29 at 25 C, estimate the height of the building. Comment on the likely accuracy of this estimate.

9. At 1 bar, the boiling point of water is 372.78 K. At this temperature and pressure, the density of liquid water is 958.66 kg m^{-3} and that of gaseous water is 0.59021 kg m^{-3}. What are the molar volumes, in m^3 mol^{-1}, of liquid and gaseous water at this temperature and pressure? In L mol^{-1}?

10. Refer to your results in Problem 9. Assuming that a water molecule excludes other water molecules from a cubic region centered on itself, estimate the average distance between nearest-neighbor water molecules in the liquid and in the gas.

11. Calculate the molar volume of gaseous water at 1 bar and 372.78 K from the ideal gas equation. What is the error, expressed as a percentage of the value you calculated in Problem 9?

12. At 372.78 K, the virial coefficient B* for water is -1.487×10^{-7} Pa^{-1}. Calculate the molar volume of gaseous water at 1 bar and 372.78 K from the virial equation: $Z = P\overline{V}/RT = 1 + B^{*}P$. What is the error, expressed as a percentage of the value you calculated in Problem 9?

13. Calculate the molar volume of gaseous water at 1 bar and 372.78 K from van der Waals' equation. The van der Waals' parameters for water are $a =$ 5.537 bar L^2 mol^{-1} and $b =$ 0.0305 L mol^{-1}. What is the error, expressed as a percentage of the value you calculated in Problem 9?

14. Comment on the results in Problems 11 – 13. At this temperature, would you expect the accuracy to increase or decrease at lower pressures?

15. The critical temperature for water is 647.1 K. At 10^3 bar and 700 K, the density of supercritical water is 651.37 kg m^3. Note that this is about 68% of the value for liquid water at the boiling point at 1 bar. What is the molar volume, in m^3 mol^{-1}, of water at this temperature and pressure? In L mol^{-1}?

16. Refer to your results in Problem 15. Assuming that a water molecule excludes other water molecules from a cubic region centered on itself, estimate the average distance between nearest-neighbor water molecules in supercritical water at 10^3 bar and 700 K.

17. Calculate the molar volume of supercritical water at 10^3 bar and 700 K from the ideal gas equation. What is the error, expressed as a percentage of the value you calculated in Problem 15?

18. At 700 K, the virial coefficient B* for water is -1.1512×10^{-8} Pa^{-1} Calculate the molar volume of supercritical water at 10^3 bar and 700 K from the virial equation. (See Problem 12.) What is the error, expressed as a percentage of the value you calculated in Problem 15?

19. Calculate the molar volume of supercritical water at 10^3 bar and 700 K from van der Waals' equation. (See Problem 13.) What is the error, expressed as a percentage of the value you calculated in Problem 15?

20. Comment on the results in Problems 16 – 19.

21. Comment on the results in Problems 10 – 13 *versus* the results in Problems 16 – 19.

22. A 1.000 L combustion bomb is filled with natural gas at 2.00 bar and 300 K. Pure oxygen is then pressured into the bomb until the pressure reaches 7.00 bar, at 300 K. Combustion is initiated. When reaction is complete, the bomb is thermostatted at 500 K, and the pressure is measured to be 12.08 bar. Thereafter, the bomb is cooled to 260 K, so that all of the water freezes. The pressure is then found to be 2.812 bar. The natural gas is a mixture of helium, methane, and ethane. How many moles of each gas are in the original sample?

23. An unknown liquid compound boils at 124 C. A classical method is used to find the approximate molecular weight of this compound. This method uses a glass bulb whose only opening is a long thin capillary tube, so that a gas sample inside the bulb can mix with the air outside only slowly. Filled with water, the bulb weighs 102.7535 grams. Empty, it weighs 50.0230 grams. A quantity of the unknown liquid is put into the bulb, and the body of the bulb is immersed in an oil bath at 150 C. The end of the capillary tube extends out of the oil bath. The liquid vaporizes filling the bulb with its gas. The total amount of vapor generated is large compared to the volume of the bulb, so the escaping vapor effectively sweeps all of the air out of the bulb, leaving the bulb filled with just the vapor of the unknown compound. When the last drop of liquid has just vaporized, the bulb is filled with the vapor of the unknown substance at the ambient atmosphere pressure, which is 0.980 bar, and a temperature of 150 C. The bulb is then removed from the oil bath and allowed to cool quickly so that the vapor condenses to a liquid film on the inside of the bulb. The oil is cleaned from the outside of the bulb, and the bulb is reweighed. The bulb and the liquid inside weigh 50.1879 grams. What is the approximate molecular weight of the liquid?

24. From the data below, calculate the molar volume, in liters, of each substance. For each substance, divide van der Waals' *b* by the molar volume you calculate. Comment.

Compound	**Mol Mass, g mol^{-1}**	**Density, g mL^{-1}**	**Van der Waals *b*, L mol^{-1}**
Acetic acid	60.05	1.0491	0.10680
Acetone	58.08	0.7908	0.09940
Acetonitrile	41.05	0.7856	0.11680
Ammonia	17.03	0.7710	0.03707
Aniline	93.13	1.0216	0.13690
Benzene	78.11	0.8787	0.11540
Benzonitrile	103.12	1.0102	0.17240
iso-Butylbenzene	134.21	0.8621	0.21440
Chlorine	70.91	3.2140	0.05622
Durene	134.21	0.8380	0.24240
Ethane	30.07	0.5720	0.06380
Hydrogen chloride	36.46	1.1870	0.04081
Mercury	200.59	13.5939	0.01696
Methane	16.04	0.4150	0.04278
Nitrogen dioxide	46.01	1.4494	0.04424
Silicon tetrafluoride	104.08	1.6600	0.05571
Water	18.02	1.0000	0.03049

Notes

[1] We use the over-bar to indicate that the quantity is per mole of substance. Thus, we write $\overline{N}$ to indicate the number of particles per mole. We write $\overline{M}$ to represent the gram molar mass. In Chapter 14, we introduce the use of the over-bar to denote a partial molar quantity; this is consistent with the usage introduced here, but carries the further qualification that temperature and pressure are constant at specified values. We also use the over-bar to indicate the arithmetic average; such instances will be clear from the context.

[2] The unit of temperature is named the kelvin, which is abbreviated as K.

[3] A redefinition of the size of the unit of temperature, the kelvin, is under consideration. The practical effect will be inconsequential for any but the most exacting of measurements.

[4] For a thorough discussion of the development of the concept of temperature, the evolution of our means to measure it, and the philosophical considerations involved, see Hasok Chang, ***Inventing Temperature***, Oxford University Press, 2004.

[5] See T. L. Hill, ***An Introduction to Statistical Thermodynamics***, Addison-Wesley Publishing Company, 1960, p 286.

[6] See S. M. Blinder, ***Advanced Physical Chemistry***, The Macmillan Company, Collier-Macmillan Canada, Ltd., Toronto, 1969, pp 185-189

3

Distributions, Probability, and Expected Values

§1 The distribution function as a summary of experimental results

In §2-10, we derive Boyle's law from Newton's laws using the assumption that all gas molecules move at the same speed at a given temperature. This is a poor assumption. Individual gas molecules actually have a wide range of velocities. In Chapter 4, we derive the Maxwell–Boltzmann distribution law for the distribution of molecular velocities. This law gives the fraction of gas molecules having velocities in any range of velocities. Before developing the Maxwell–Boltzmann distribution law, we need to develop some ideas about distribution functions. Most of these ideas are mathematical. We discuss them in a non-rigorous way, focusing on understanding what they mean rather than on proving them.

The overriding idea is that we have a real-world source of data. We call this source of data the ***distribution***. We can collect data from this source to whatever extent we please. The datum that we collect is called the distribution's ***random variable***. We call each possible value of the random variable an ***outcome***. The process of gathering a set of particular values of the random variable from a distribution is often called ***sampling*** or ***drawing a sample***. The set of values that is collected is called ***the sample***. The set of values that comprise the sample is often called "the data." In scientific applications, the random variable is usually a number that results from making a measurement on a physical system. Calling this process "drawing a sample" can be inappropriate. Often we call the process of getting a value for the random variable "doing an experiment", "doing a test", or "making a trial".

As we collect increasing amounts of data, the accumulation quickly becomes unwieldy unless we can reduce it to a mathematical model. We call the mathematical model we develop a ***distribution function***, because it is a function that expresses what we are able to learn about the data source—the distribution. A distribution function is an equation that summarizes the results of many measurements; it is a mathematical model for a real-world source of data. Specifically, it models the ***frequency*** with which we obtain a particular outcome. We usually believe that we can make our mathematical model behave as much like the real-world data source as we want if we use enough experimental data in developing it.

Often we talk about ***statistics***. By a statistic, we mean any mathematical entity that we can calculate from data. Broadly speaking a distribution function is a statistic, because it is obtained by fitting a mathematical function to data that we collect. Two other statistics are often used to characterize experimental data: the ***mean*** and the ***variance***. The mean and variance are defined for any distribution. We want to see how to estimate the mean and variance from a set of experimental data collected from a particular distribution.

We distinguish between discrete and continuous distributions. A ***discrete distribution*** is a real-world source of data that can produce only particular data values. A coin toss is a good example. It can produce only two outcomes—heads or tails. A ***continuous distribution*** is a real-world source of data that can produce data values in a continuous range. The speed of an automobile is a good example. An automobile can have any speed within a rather wide range of speeds. For this distribution, the random variable is automobile speed. Of course we can generate a discrete distribution by aggregating the results of sampling a continuous distribution; if we lump all automobile speeds between 20 mph and 30 mph together, we lose the detailed information about the speed of each automobile and retain only the total number of automobiles with speeds in this interval.

§2 Outcomes, events, and probability

We also need to introduce the idea that a function that successfully models the results of past experiments can be used to predict some of the characteristics of future results.

We reason as follows: We have results from drawing many samples of a random variable from some distribution. We suppose that a mathematical representation has been found that adequately summarizes the results of these experiences. If the underlying distribution—the physical system in scientific applications—remains the same, we expect that a long series of future results would give rise to essentially the same mathematical representation. If 25% of many previous results have had a particular characteristic, we expect that 25% of a large number of future trials will have the same characteristic. We also say that there is one chance in four that the next individual result will have this characteristic; when we say this, we mean that 25% of a large number of future trials will have this characteristic, and the next trial has as good a chance as any other to be among those that do. ***The***

probability that an outcome will occur in the future is equal to the frequency with which that outcome has occurred in the past.

Given a distribution, the possible outcomes must be mutually exclusive; in any given trial, the random variable can have only one of its possible values. Consequently, a discrete distribution is completely described when the probability of each of its outcomes is specified. Many distributions are comprised of a finite set of *N* mutually exclusive possible outcomes. If each of these outcomes is equally likely, the probability that we will observe any particular outcome in the next trial is $1/N$.

We often find it convenient to group the set of possible outcomes into subsets in such a way that each outcome is in one and only one of the subsets. We say that such assignments of outcomes to subsets are ***exhaustive***, because every possible outcome is assigned to some subset; we say that such assignments are ***mutually exclusive***, because no outcome belongs to more than one subset. We call each such subset an ***event***. When we partition the possible outcomes into exhaustive and mutually exclusive events, we can say the same things about the probabilities of events that we can say about the probabilities of outcomes. In our discussions, the term "events" will always refer to an exhaustive and mutually exclusive partitioning of the possible outcomes. Distinguishing between outcomes and events just gives us some language conventions that enable us to create alternative groupings of the same set of real world observations.

Suppose that we define a particular event to be a subset of outcomes that we denote as *U*. If in a large number of trials, the fraction of outcomes that belong to this subset is *F*, we say that the probability is *F* that the outcome of the next trial will belong to this event. To express this in more mathematical notation, we write $P(U) = F$. When we do so, we mean that the fraction of a large number of future trials that belong to this subset will be *F*, and the next trial has as good a chance as any other to be among those that do. In a sample comprising *M* observations, the best forecast we can make of the number of occurrences of U is $M \times P(U)$, and we call this the ***expected number of occurrences*** of *U* in a sample of size *M*.

The idea of grouping real world observations into either outcomes or events is easy to remember if we keep in mind the example of tossing a die. The die has six faces, which are labeled with 1, 2, 3, 4, 5, or 6 dots. The dots distinguish one face from another. On any given toss, one face of the die must land on top. Therefore, there are six possible outcomes. Since each face has as good a chance as any other of landing on top, the six possible outcomes are equally probable. The probability of any given outcome is $1/6$. If we ask about the probability that the next toss will result in one of the even-numbered faces landing on top, we are asking about the probability of an event—the event that the next toss will have the characteristic that an even-numbered face lands on top. Let us call this event *X*. That is, event *X* occurs if the outcome is a 2, a 4, or a 6. These are three of the six equally likely outcomes. Evidently, the probability of this event is $3/6 = 1/2$.

Having defined event *X* as the probability of an even-number outcome, we still have several alternative ways to assign the odd-number outcomes to events. One assignment would be to say that all of the odd-number outcomes belong to a second event—the event that the outcome is odd. The events "even outcome" and "odd outcome" are exhaustive and mutually exclusive. We could create another set of events by assigning the outcomes 1 and 3 to event *Y*, and the outcome 5 to event *Z*. Events *X*, *Y*, and Z are also exhaustive and mutually exclusive.

We have a great deal of latitude in the way we assign the possible outcomes to events. If it suits our purposes, we can create many different exhaustive and mutually exclusive partitionings of the outcomes of a given distribution. We require that each partitioning of outcomes into events be exhaustive and mutually exclusive, because we want to apply the laws of probability to events.

§3 Some important properties of events

If we know the probabilities of the possible outcomes of a trial, we can calculate the probabilities for combinations of outcomes. These calculations are based on two rules, which we call ***the laws of probability***. If we partition the outcomes into exhaustive and mutually exclusive events, the laws of probability also apply to events. Since, as we define them, "events" is a more general term than "outcomes," we call them the law of the ***probability of alternative events*** and the law of the ***probability of compound events***. These laws are valid so long as three conditions are satisfied. We have already discussed the first two of these conditions, which are that the outcomes possible in any individual trial must be exhaustive and mutually exclusive. The third condition is that, if we make more than one trial, the outcomes must be ***independent***; that is, the outcome of one trial must not be influenced by the outcomes of the others.

We can view the laws of probability as rules for inferring information about combinations of events. The law of the probability of alternative events applies to events that belong to the same distribution. The law of the probability of compound events applies to events that can come from one or more distributions. An important special case occurs when the compound events are *N* successive samplings of a given distribution that we identify as the ***parent distribution***. If the random variable is a number, and we average the numbers that we obtain from *N* successive samplings of the parent distribution, these "averages-of-*N*" themselves constitute a distribution. If we know certain properties of the parent distribution, we can calculate corresponding properties of the "distribution of averages-of-*N* values obtained by sampling the parent distribution." These calculations are specified by the ***central limit theorem***, which we discuss in §11.

In general, when we combine events from two distributions, we can view the result as an event that belongs to a third distribution. At first encounter, the idea of combining events and distributions may seem esoteric. A few examples serve to show that what we have in mind is very simple.

Since an event is a set of outcomes, an event occurs whenever any of the outcomes in the set occurs. Partitioning the outcomes of tossing a die into "even outcomes" and "odd outcomes" illustrates this idea. The event "even outcome" occurs whenever the outcome of a trial is 2, 4, or 6. The probability of an event can be calculated from the probabilities of the underlying outcomes. We call the rule for this calculation the law of the probabilities of alternative events. (We create the opportunity for confusion here because we are illustrating the idea of alternative events by using an example in which we call the alternatives "alternative outcomes" rather than "alternative events." We need to remember that "event" is a more general term than "outcome." One possible partitioning is that which assigns every outcome to its own event.) We discuss the probabilities of alternative events further below.

To illustrate the idea of compound events, let us consider a first distribution that comprises "tossing a coin" and a second distribution that comprises "drawing a card from a poker deck." The first distribution has two possible outcomes; the second distribution has 52 possible outcomes. If we combine these distributions, we create a third distribution that comprises "tossing a coin and drawing a card from a poker deck." The third distribution has 104 possible outcomes. If we know the probabilities of the outcomes of the first distribution and the probabilities of the outcomes of the second distribution, and these probabilities are independent of one another, we can calculate the probability of any outcome that belongs to the third distribution. We call the rule for this calculation the law of the probability of compound events. We discuss it further below.

A similar situation occurs when we consider the outcomes of tossing two coins. We assume that we can tell the two coins apart. Call them coin 1 and coin 2. We designate heads and tails for coins 1 and 2 as H_1, T_1, H_2, and T_2, respectively. There are four possible outcomes in the distribution we call "tossing two coins:" H_1H_2, H_1T_2, T_1H_2, and T_1T_2. (If we could not tell the coins apart, H_1T_2 would be the same thing as T_1H_2; there would be only three possible outcomes.) We can view the distribution "tossing two coins" as being a combination of the two distributions that we can call "tossing coin 1" and "tossing coin 2." We can also view the distribution "tossing two coins" as a combination of two distributions that we call "tossing a coin a first time" and "tossing a coin a second time." We view the distribution "tossing two coins" as being equivalent to the distribution "tossing one coin twice." This is an example of ***repeated trials***, which is a frequently encountered type of distribution. In general, ***we call such a distribution a "distribution of events from a trial repeated N times," and we view this distribution as being completely equivalent to N simultaneous trials of the same kind***. Chapter 19 considers the distribution of outcomes when a trial is repeated many times. Understanding the properties of such distributions is the single most essential element in understanding the theory of statistical thermodynamics. The central limit theorem relates properties of the repeated-trials distribution to properties of the parent distribution.

The probability of alternative events. If we know the probability of each of two mutually exclusive events that belong to an exhaustive set, the probability that one or the other of them will occur in a single trial is equal to the sum of the individual probabilities. Let us call the independent events A and B, and represent their probabilities as $P(A)$ and $P(B)$, respectively. The probability that one of these events occurs is the same thing as the probability that either A occurs or B occurs. We can represent this probability as $P(A\ or\ B)$. The probability of this combination of events is the sum: $P(A) + P(B)$. That is,

$$P(A\ or\ B) = P(A) + P(B)$$

Above we define Y as the event that a single toss of a die comes up either 1 or 3. Because each of these outcomes is one of six, mutually-exclusive, equally-likely outcomes, the probability of either of them is $1/6$: $P(tossing\ a\ 1) = P(tossing\ a\ 3) = 1/6$. From the law of the probability of alternative events, we have

$$\begin{aligned} P(event\ Y) &= (tossing\ a\ 1\ or\ tossing\ a\ 3) \\ &= P(tossing\ a\ 1)\ or P(tossing\ a\ 3) \\ &= 1/6 + 1/6 \\ &= 2/6 \end{aligned}$$

We define X as the event that a single toss of a die comes up even. From the law of the probability of alternative events, we have

$$\begin{aligned} P(event\ X) &= P(tossing\ 2\ or\ 4\ or\ 6) \\ &= P(tossing\ a\ 2) + P(tossing\ a\ 4) \\ &\quad + P(tossing\ a\ 6) \\ &= 3/6 \end{aligned}$$

We define Z as the event that a single toss comes up 5.

$$P(event\ Z) = P(tossing\ a\ 5) = 1/6$$

If there are ω independent events (denoted $E_1, E_2, \dots, E_i, \dots, E_\omega$), the law of the probability of alternative events asserts that the probability that one of these events will occur in a single trial is

$$\begin{aligned} &P(E_1\ or\ E_2\ or\ \dots E_i \dots or\ E_\omega) \\ &\quad = P(E_1) + P(E_2) + \cdots + P(E_i) + \cdots + P(E_\omega) \\ &\quad = \sum_{i=1}^{\omega} P(E_i) \end{aligned}$$

If these ω independent events encompass all of the possible outcomes, the sum of their individual probabilities must be unity.

The probability of compound events. Let us now suppose that we make two trials in circumstances where event A is possible in the first trial and event B is possible in the second trial. We represent the probabilities of these events by $P(A)$ and $P(B)$ and stipulate that they are independent of one another; that is, the probability that B occurs in the second trial is independent of the outcome of the first trial. Then, the probability that A occurs in the first trial and B occurs in the second trial, $P(A\ and\ B)$, is equal to the product of the individual probabilities.

$$P(A\ and\ B) = P(A) \times P(B)$$

To illustrate this using outcomes from die-tossing, let us suppose that event A is tossing a 1 and event B is tossing a 3. Then, $P(A) = 1/6$ and $P(B) = 1/6$. The probability of tossing a 1 in a first trial and tossing a 3 in a second trial is then

$$\begin{aligned} &P(tossing\ a\ 1\ first\ and\ tossing\ a\ 3\ second) \\ &\quad = P(tossing\ a\ 1) \times P(tossing\ a\ 3) \\ &\quad = 1/6 \times 1/6 \\ &\quad = 1/36 \end{aligned}$$

If we want the probability of getting one 1 and one 3 in two tosses, we must add to this the probability of tossing a 3 first and a 1 second.

If there are ω independent events (denoted $E_1, E_2, \ldots, E_i, \ldots, E_\omega$), the law of the probability of compound events asserts that the probability that E_1 will occur in a first trial, and E_2 will occur in a second trial, *etc.*, is

$$\begin{aligned} &P(E_1\ and\ E_2\ and \ldots E_i \ldots and\ E_\omega) \\ &\quad = P(E_1) \times P(E_2) \times \ldots \times P(E_i) \times \ldots \times P(E_\omega) \\ &\quad = \prod_{i=1}^{\omega} P(E_i) \end{aligned}$$

§4 Applying the laws of probability

The laws of probability apply to events that are independent. If the result of one trial depends on the result of another trial, we may still be able to use the laws of probability. However, to do so, we must know the nature of the interdependence.

If the activity associated with event C precedes the activity associated with event D, the probability of D may depend on whether C occurs. Suppose that the first activity is tossing a coin and that the second activity is drawing a card from a deck; however, the deck we use depends on whether the coin comes up heads or tails. If the coin is heads, we draw a card from an ordinary deck; if the coin is tails, we draw a coin from a deck with the face cards removed. Now we ask about the probability of drawing an ace. If the coin is heads, the probability of drawing an ace is $4/52 = 1/13$. If the coin is tails, the probability of drawing an ace is $4/40 = 1/10$. The combination coin is heads and card is ace has probability: $(1/2)(1/13) = 1/26$. The combination coin is tails and card is ace has probability $(1/2)(1/10) = 1/20$. In this case, the probability of drawing an ace depends on the modification we make to the deck based on the outcome of the coin toss.

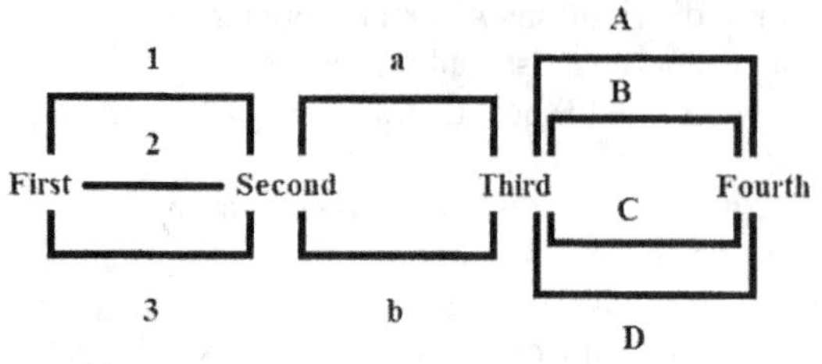

Figure 1. A simple case that illustrates the laws of probability.

Applying the laws of probability is straightforward. An example that illustrates the application of these laws in a transparent way is provided by villages First, Second, Third, and Fourth, which are separated by rivers. (See Figure 1.) Bridges 1, 2, and 3 span the river between First and Second. Bridges a and b span the river between Second and Third. Bridges A, B, C, and D span the river between Third and Fourth. A traveler from First to Fourth who is free to take any route he pleases has a choice from among $3 \times 2 \times 4 = 24$ possible combinations. Let us consider the probabilities associated with various events:

- There are 24 possible routes. If a traveler chooses his route at random, the probability that he will take any particular route is $1/24$. This illustrates our assumption that each event in a set of N exhaustive and mutually exclusive events occurs with probability $1/N$.

- If he chooses a route at random, the probability that he goes from First to Second by either bridge 1 or bridge 2 is $P(1) + P(2) = 1/3 + 1/3 = 2/3$. This illustrates the calculation of the probability of alternative events.

- The probability of the particular route $2 \to a \to C$ is $P(2) \times P(a) \times P(C) = (1/3)(1/2)(1/4) = 1/24$, and we calculate the same probability for any other route from First to Fourth. This illustrates the calculation of the probability of a compound event.

- If he crosses bridge 1, the probability that his route will be $2 \to a \to C$ is zero, of course. The probability of an event that has already occurred is 1, and the probability of any alternative is zero. If he crosses bridge 1, $P(1) = 1$, and $P(2) = P(3) = 0$.

- Given that a traveler has used bridge 1, the probability of the route $1 \to a \to C$ becomes the probability of path $a \to C$, which is $P(a) \times P(C) = (1/2)(1/4) = 1/8$. Since $P(1) = 1$, the probability of the compound event $1 \to a \to C$ is the probability of the compound event $a \to C$.

The outcomes of rolling dice provide more

Table 1. Outcomes from tossing two dice.

Outcome for Second die		Outcome for first die					
		1	2	3	4	5	6
	1	2	3	4	5	6	7
	2	3	4	5	6	7	8
	3	4	5	6	7	8	9
	4	5	6	7	8	9	10
	5	6	7	8	9	10	11
	6	7	8	9	10	11	12

illustrations. If we roll two dice, we can classify the possible outcomes according to the sums of the outcomes for the individual dice. There are thirty-six possible outcomes. They are displayed in Table 1. Let us consider the probabilities associated with various dice-throwing events:

- The probability of any given outcome, say the first die shows 2 and the second die shows 3, is $1/36$.

- Since the probability that the first die shows 3 while the second die shows 2 is also $1/36$, the probability that one die shows 2 and the other shows 3 is
$$\begin{aligned} &P(3) \times P(2) + P(2) \times P(3) \\ &\quad = (1/36) + (1/36) \\ &\quad = 1/18. \end{aligned}$$

- Four different outcomes correspond to the event that the score is 5. Therefore, the probability of rolling 5 is
$$\begin{aligned} &P(1) \times P(4) + P(2) \times P(3) \\ &\quad + P(3) \times P(2) + P(4) \times P(1) = 1/9 \end{aligned}$$

- The probability of rolling a score of three or less is the probability of rolling 2, plus the probability of rolling 3 which is $(1/36) + (2/36) = 3/36 = 1/12$

- Suppose we roll the dice one at a time and that the first die shows 2. The probability of rolling 7 when the second die is thrown is now $1/6$, because only rolling a 5 can make the score 7, and there is a probability of $1/6$ that a 5 will come up when the second die is thrown.

- Suppose the first die is red and the second die is green. The probability that the red die comes up 2 and the green die comes up 3 is $(1/6)(1/6) = 1/36$.

Above we looked at the number of outcomes associated with a score of 3 to find that the probability of this event is $1/18$. We can use another argument to get this result. The probability that two dice roll a score of three is equal to the probability that the first die shows 1 or 2 times the probability that the second die shows whatever score is necessary to make the total equal to three. This is:

$$\begin{aligned} &P(first\ die\ shows\ 1\ or\ 2) \times (1/6) \\ &\quad = [(1/6) + (1/6)] \times 1/6 \\ &\quad = 2/36 \\ &\quad = 1/18 \end{aligned}$$

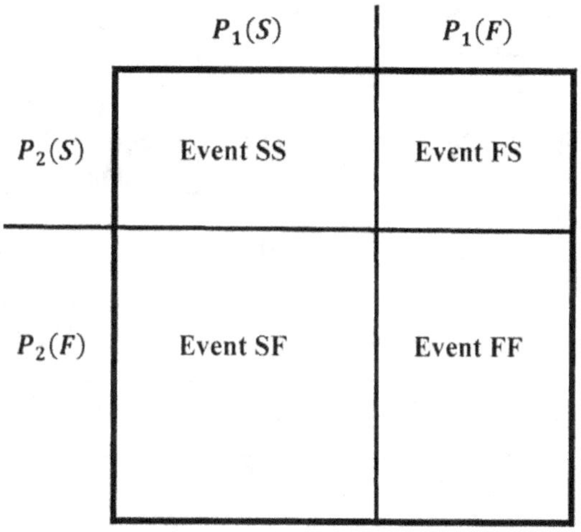

Figure 2. Success and failure in successive trials.

Application of the laws of probability is frequently made easier by recognizing a simple restatement of the requirement that events be mutually exclusive. In a given trial, either an event occurs or it does not. Let the probability that an event A occurs be $P(A)$. Let the probability that event A does not occur be $P(\sim A)$. Since in any given trial, the outcome must belong either to event A or to event $\sim A$, we have

$$P(A) + P(\sim A) = 1$$

For example, if the probability of success in a single trial is $2/3$, the probability of failure is $1/3$. If we consider the outcomes of two successive trials, we can group them into four events.

Event SS: First trial is a success; second trial is a success.
Event SF: First trial is a success; second trial is a failure.
Event FS: First trial is a failure; second trial is a success.
Event FF: First trial is a failure; second trial is a failure.

Using the laws of probability, we have

$$\begin{aligned} 1 &= P(Event\ SS) + P(Event\ SF) + P(Event\ FS) \\ &\quad + P(Event\ FF) \\ &= P_1(S) \times P_2(S) + P_1(S) \times P_2(F) \\ &\quad + P_1(F) \times P_2(S) + P_1(F) \times P_2(F) \end{aligned}$$

where $P_1(X)$ and $P_2(X)$ are the probability of event X in the first and second trials, respectively.

This situation can be mapped onto a simple diagram. We represent the possible outcomes of the first trial by line segments on one side of a unit square $P_1(S) + P_1(F) = 1$. We represent the outcomes of the second trial by line segments along an adjoining side of the unit square. The four possible events are now represented by the areas of four mutually exclusive and exhaustive portions of the unit square as shown in Figure 2.

§5 Bar graphs and histograms

Since a discrete distribution is completely specified

Table 2. Vehicle speed data.			
Speed (mph)	**Number of cars**	**Fraction of cars**	**Height for bar area to equal fraction**
−10			
	200	0.020	0.20/20 = 0.0010
10			
	800	0.08	0.08/20 = 0.0040
30			
	2500	0.25	0.25/20 = 0.0125
50			
	5500	0.55	0.55/20 = 0.0275
70			
	1000	0.10	0.10/20 = 0.0050
90			

by the probabilities of each of its events, we can represent it by a bar graph. The probability of each event is represented by the height of one bar. We can generalize this graphical representation to represent continuous distributions. To see what we have in mind, let us consider a particular example.

Let us suppose that we have a radar gun and that we decide to interest ourselves in the typical speeds of cars on a highway just outside of town. As we think about this project, we recognize that speeds might vary with the time of day and the day of the week. Random variations in many other factors might also be important; these include weather conditions and accidents in the vicinity. To eliminate as many atypical factors as possible, we might decide that typical speeds are those of cars going north between 1:00 pm and 4:00 pm on weekdays when the road surface is dry and there are no disabled vehicles in view. If we have a lot of time and the road is busy, we could collect a lot of data. Let us suppose that we record the speeds of 10,000 cars. Each datum would be the speed of a car on the road at a time when the selected conditions are satisfied.

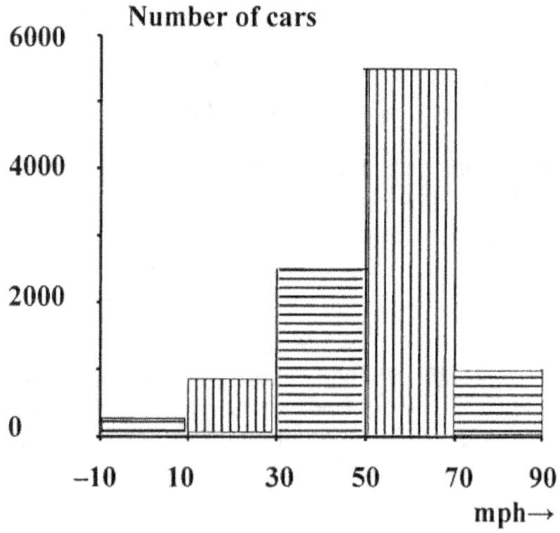

Figure 3. Number of cars *versus* speed.

To use this data, we want to summarize it in a form that is easy to visualize. One way to do this is to aggregate the data to give the number of cars in each 20 mph range; the results might look something like the data in Table 2. Figure 3 is a five-channel bar graph that displays the number of cars in each 20 mph range. A great deal of information is lost in the aggregating process. In particular, nothing on the graph represents the number of automobiles in narrower speed intervals.

Now, suppose that we repeat this task, but that we do not have enough time to collect data on as many as 10,000 more cars. We will be curious about the extent to which our two samples agree with one another. Since the total number of vehicles will be different, the appropriate way to go about this is obviously to compare the fraction of cars in each speed range. In fact, using fractions enables us to compare any number of such studies. To the extent that these studies measure the same thing—typical speeds under the specified conditions—the fraction of automobiles in any particular speed interval should be approximately constant. Dividing the number of automobiles in each speed interval by the total number of automobiles gives a representation that focuses attention on the proportion of automobiles with various speeds. The shape of the bar graph remains the same; all that changes is the scale we use to label the ordinate. (See Figure 4.)

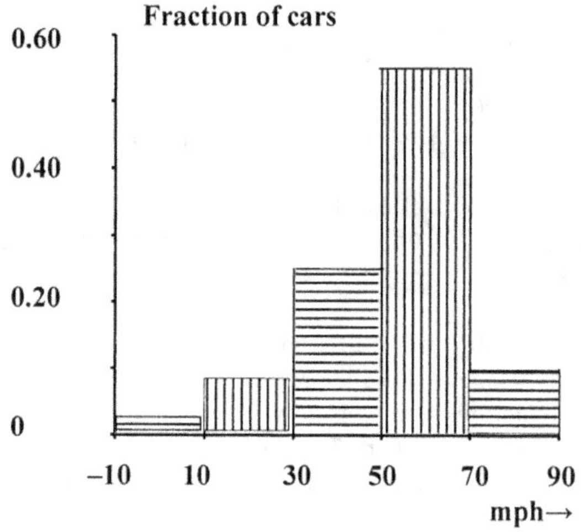

Figure 4. Fraction of cars *versus* speed.

Insofar as any repetition of this experiment gives nearly the same results, this is a useful change. However, the fundamental limitations of the graph remain. For example, if we want to use the graph to estimate how speeds are distributed in any other set of intervals, we have to read values off the ordinate and manipulate them in ways that may not be very satisfactory. To estimate the fraction with speeds between 20 mph and 40 mph, we might assign half of the automobiles in the 10 – 30 mph interval and half of those in the 30 – 50 mph interval to the new interval. This enables us to estimate that the fraction in the 20 – 40 mph interval is 0.165. This estimate is much less reliable than one that could be made by going back to the raw data for all 10,000 automobiles.

The data can also be represented as a ***histogram***. In a histogram, the information is represented by the area rather than the height of the bar. In the present case, the only visible change to the graph is another change in the numerical values on the ordinate. In Figure 5, the area of a bar represents the fraction of automobiles with speeds in the given interval. As the speed interval is made smaller, any of these bar graphs looks increasingly like a continuous curve. (See Figure 6.) The histogram has the advantage that, as the curve becomes continuous, the interpretation remains constant: the area under the curve between any two speeds always represents the fraction of automobiles with speeds in this interval. It turns out that we are adept at visually estimating the relative areas of different parts of a histogram. That is, from a quick glance at a histogram, we are able to obtain a good semi-quantitative appreciation of the significance of the underlying data.

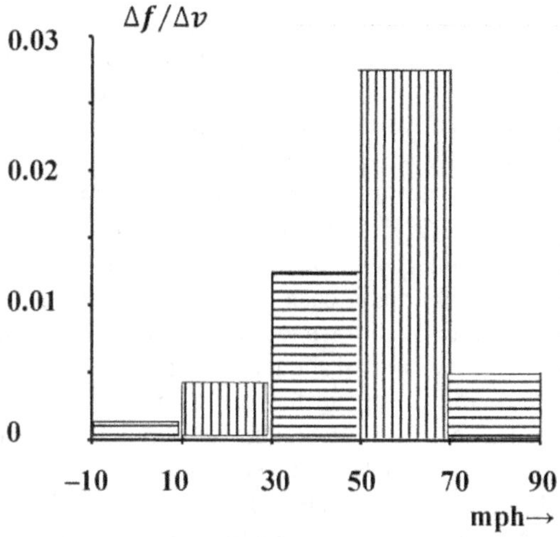

Figure 5. Speed data presented as a histogram.

If the histogram captures our experience, and we expect future events to have the same characteristics, the histogram becomes an expression of probability. All that is necessary is that we construct the histogram so that the total area under the graph is unity. If we let $f(u)$ be the area under the graph from $u = -\infty$ to $u = u$, then $f(u)$

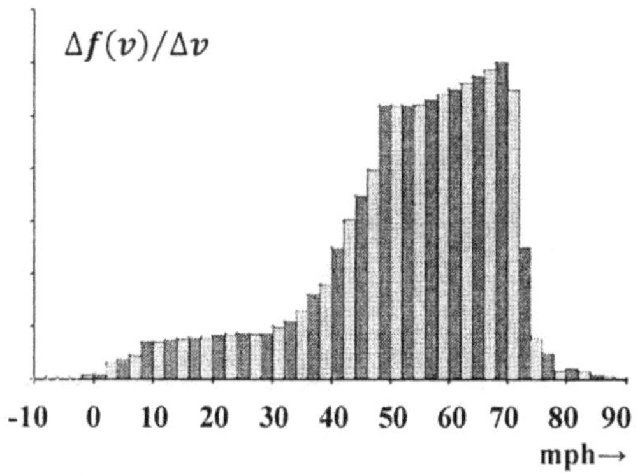

Figure 6. Histogram with narrower speed intervals.

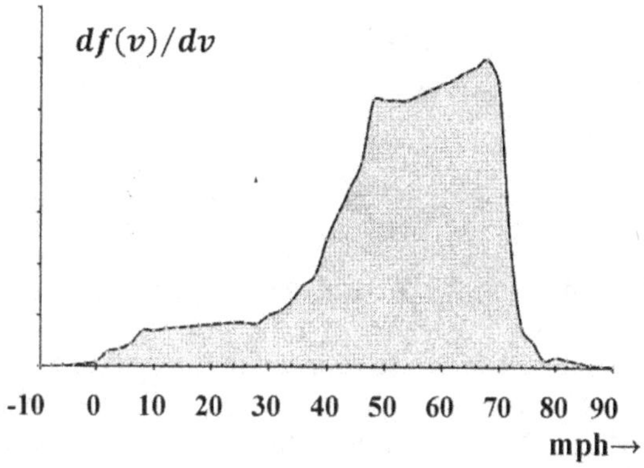

Figure 7. The histogram can be a continuous function.

represents the probability that the speed of a randomly selected automobile will lie between $-\infty$ and u. For any a and b, the probability that u lies in the interval $a < u < b$ is $f(b) - f(a)$. The function $f(u)$ is called **the *cumulative probability distribution function***, because its value for any u is the fraction of automobiles that have a speed less than u. $f(a)$ is the frequency with which we observe values of the random variable, u, that are less than a. Equivalently, we can say that $f(u)$ is the probability that any randomly selected automobile will have a speed less than u. If we let the width of every interval go to zero, the bar graph representation of the histogram becomes a curve, and the histogram becomes a continuous function of the random variable, u. (See Figure 7.) ***Note that the curve—the enclosing envelope—is not $f(u)$. $f(u)$ is the area under the enclosing envelope curve.***

§6 Continuous distribution functions; the envelope function is the derivative of the area

When we can represent the envelope curve as a continuous function, the envelope curve is the derivative of the cumulative probability distribution function: The cumulative distribution function is $f(u)$; the envelope function is $df(u)/du$. The envelope function is a ***probability density***, and we will refer to the envelope function, $df(u)/du$, as the ***probability density function***. The probability density function is the derivative, with respect to the random variable, of the cumulative distribution function. This is an immediate consequence of the fundamental theorem of calculus.

If $H(u)$ is the anti-derivative of a function $h(u)$, we have $dH(u)/du = h(u)$, and the fundamental theorem of calculus asserts that the area under $h(u)$, from $u = a$ to $u = b$ is

$$\int_a^b h(u)du = \int_a^b \left(\frac{dH(u)}{du}\right) du$$
$$= H(b) - H(a)$$

In the present instance, $H(u) = f(u)$, so that

$$\int_a^b \left(\frac{df(u)}{du}\right) du = f(b) - f(a)$$

and

$$h(u) = \frac{df(u)}{du}$$

The envelope function, $h(u)$, and $df(u)/du$ are the same function.

This point is also apparent if we consider the incremental change in the area, dA, under a histogram as the variable u increases from u to $u + du$. If we let the envelope function be $h(u)$, we have

$$dA = h(u)du \qquad \text{or} \qquad h(u) = dA/du$$

That is, the envelope function is the derivative of the area with respect to the random variable, u. The area is $f(u)$, so the envelope function is $h(u) = df(u)/du$.

Calling the envelope curve the probability density function emphasizes that it is analogous to a function that expresses the density of matter. That is, for an incremental change in u, the incremental change in probability is

$$\Delta(probability) = \frac{df}{du}\Delta u$$

analogous to the incremental change in mass accompanying an incremental change in volume

$$\Delta(mass) = density \times \Delta(volume)$$

where $density = d(mass)/d(volume)$. In this analogy, we suppose that mass is distributed in space with a density that varies from point to point in the space. The mass enclosed in any particular volume is given by the integral of the density function over the volume enclosed; that is, $mass = \int_V (density)dV$. Conversely, the density at any given point is the limit, as the enclosing volume shrinks to zero, of the enclosed mass divided by the magnitude of the enclosing volume.

Similarly, for any value of the random variable, ***the probability density is the limit, as an interval spanning the value of the random variable shrinks to zero, of the probability that the random variable is in the interval, divided by the magnitude of the interval.***

§7 A heuristic view of the probability density function

Suppose that we have a probability density function like that sketched in Figure 8 and that the area under the curve in the interval $a < u < b$ is 0.25. If we draw a large number of samples from the distribution, our definitions of probability and the probability density function mean that about 25% of the values we draw will lie in the interval $a < u < b$. We expect the percentage to become closer and closer to 25% as the total number of samples drawn becomes very large. The same would be true of any other interval, $c < u < d$, where the area under the curve in the interval $c < u < d$ is 0.25.

If we draw exactly four samples from this distribution, the values can be anywhere in the domain of u. However, if we ask what arrangement of four values best approximates the result of drawing a large number of samples, it is clear that this arrangement must have a value in each of the four, mutually-exclusive, 25% probability zones. We can extend this conclusion to any number of representative points. If we ask what arrangement of *N* points would best represent the arrangement of a large number of points drawn from the distribution, the answer is clearly that one of the *N* representative points should lie within each of *N*, mutually-exclusive, equal-area segments that span the domain of u.)

We can turn this idea around. In the absence of information to the contrary, the best assumption we can make about a set of *N* values of a random variable is that each represents an equally probable outcome. If our entire store of information about a distribution consists of four data points drawn from the distribution, the best description that we can give of the probability density function is that one-fourth of the area under the curve lies above a segment of the domain that is associated with each point. If we have *N* points, the best estimate we can make of the distribution from which the *N* points are drawn is that $(1/N)^{th}$ of the area lies above each of them.

This view tells us to associate a probability of $1/N$ with an interval around each data point, but it does not tell us where to begin or end the interval. If we could decide where the interval about each data point began and ended, we could estimate the shape of the probability density function. For a small number of points, we could not expect this estimate to be very accurate, but it would be the best possible estimate based on the given data.

Now, instead of trying to find the best interval to associate with each data point, let us think about the intervals into which the data points divide the domain. This small change of perspective leads us to a logical way to

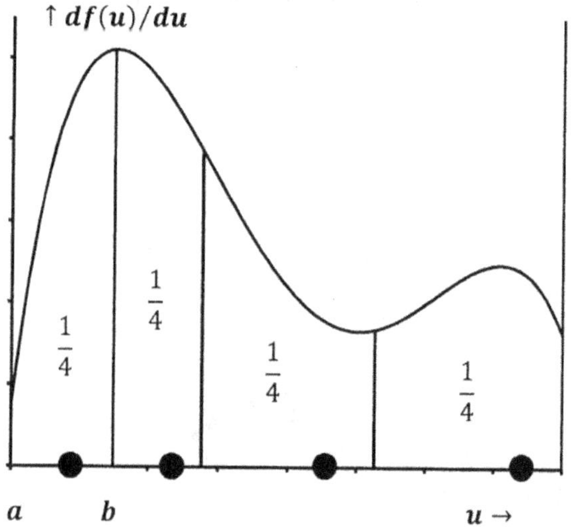

Figure 8. A sample of four that approximates its distribution.

divide the domain of u into specific intervals of equal probability. If we put N points on any line, these points divide the line into $N+1$ segments. There is a segment to the left of every point; there are N such segments. There is one final segment to the right of the right-most point, and so there are $N+1$ segments in all.

In the absence of information to the contrary, the best assumption we can make is that N data points divide their domain into $N+1$ segments, each of which is associated with equal probability. The fraction of the area above each of these segments is $1/(N+1)$; also, the probability associated with each segment is $1/(N+1)$. If, as in the example above, there are four data points, the best assumption we can make about the probability density function is that 20% of its area lies between the left boundary and the left-most data point, and 20% lies between the right-most data point and the right boundary. The three intervals between the four data points each represent an additional 20% of the area. Figure 9 indicates the N data points that best approximate the distribution sketched in Figure 8.

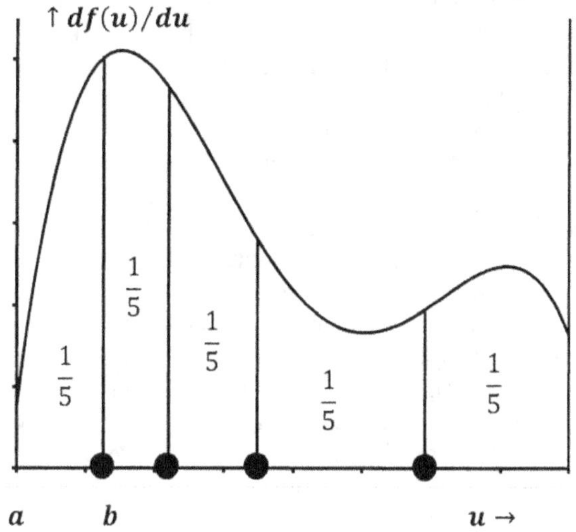

Figure 9. The sample of four that best approximates its distribution.

The sketches in Figure 10 describe the probability density functions implied by the indicated sets of data points.

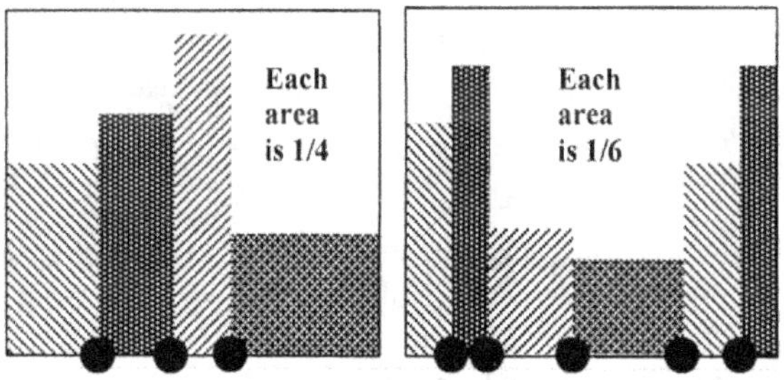

Figure 10. Approximate probability density functions.

§8 A heuristic view of the cumulative probability distribution function

We can use these ideas to create a plot that approximates the cumulative probability distribution function given any set of N measurements of a random variable u. To do so, we put the u_i values found in our N measurements in order from smallest to largest. We label the ordered values $u_1, u_2, \ldots, u_i, \ldots, u_N$, where u_1 is the smallest. By the argument that we develop in the previous section, the probability of observing a value less than u_1 is about $1/(N+1)$. If we were to make a large number of additional measurements, a fraction of about $1/(N+1)$ of this large number of additional measurements would be less than u_1. This fraction is just $f(u_1)$, so we reason that $f(u_1) \approx 1/(N+1)$. The probability of observing a value between u_1 and u_2 is also about $1/(N+1)$; so the probability of observing a value less than u_2 is about $2/(N+1)$, and we expect $f(u_2) \approx 2/(N+1)$. In general, the probability of observing a value between u_{i-1} and u_i is also about $1/(N+1)$, and the probability of observing a value less than u_i is about $i/(N+1)$. In other words, we expect the cumulative probability distribution function for u_i to be such that the ith smallest observation corresponds to $f(u_i) \approx i/(N+1)$. The quantity $i/(N+1)$ is often called the ***rank probability*** of the ith data point.

Figure 11 is a sketch of the sigmoid shape that we usually expect to find when we plot $i/(N+1)$ versus the ith value of u. This plot approximates the cumulative probability distribution function, $f(u)$. We expect the sigmoid shape because we expect the observed values of u to bunch up around their average value. (If, within some domain of u values, all possible values of u were equally likely, we would expect the difference between successive observed values of u to be roughly constant, which would make the plot look approximately linear.) At any value of u, the slope of the curve is just the

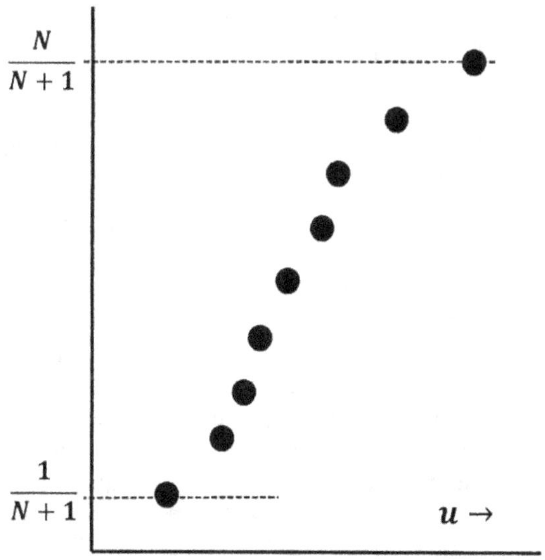

Figure 11. An approximate cumulative probability distribution function.

probability-density function, $df(u)/du$.

These ideas mean that we can test whether the experimental data are described by any particular mathematical model, say $F(u)$. To do so, we use the mathematical model to predict each of the N rank probability values: $1/(N+1)$, $2/(N+1)$,..., $i/(N+1)$,..., $N/(N+1)$. That is to say, we calculate $F(u_1)$, $F(u_2)$,..., $F(u_i)$, ..., $F(u_N)$; if $F(u)$ describes the data well, we will find, for all i, $F(u_i) \approx i/(N+1)$. Graphically, we can test the validity of the relationship by plotting $i/(N+1)$ *versus* $F(u_i)$. If $F(u)$ describes the data well, this plot will be approximately linear, with a slope of one.

In §12, we introduce the ***normal distribution***, which is a mathematical model that describes a great many sources of experimental observations. The normal distribution is a distribution function that involves two parameters, the mean, μ, and the standard deviation, σ. The ideas we have discussed can be used to develop a particular graph paper—usually called normal probability paper. If the data are normally distributed, plotting them on this paper produces an approximately straight line.

We can do essentially the same test without benefit of special graph paper, by calculating the average, $\bar{u} \approx \mu$, and the estimated standard deviation, $s \approx \sigma$, from the experimental data. (Calculating $\bar{u}$ and s is discussed below.) Using $\bar{u}$ and s as estimates of μ and σ, we can find the model-predicted probability of observing a value of the random variable that is less than u_i. This value is $f(u_i)$ for a normal distribution whose mean is $\bar{u}$ and whose standard deviation is s. We can find $f(u_i)$ by using standard tables (usually called the ***normal curve of error*** in mathematical compilations), by numerically integrating the normal distribution's probability density function, or by using a function embedded in a spreadsheet program, like Excel®. If the data are described by the normal distribution function, this value must be approximately equal to the rank probability; that is, we expect $f(u_i) \approx i/(N+1)$. A plot of $i/(N+1)$ *versus* $f(u_i)$ will be approximately linear with a slope of about one.

§9 Random variables, expected values, and population sets

When we sample a particular distribution, the value that we obtain depends on chance and on the nature of the distribution described by the function $f(u)$. The probability that any given trial will produce u in the interval $a < u < b$ is equal to $f(b) - f(a)$. We often find situations in which a second function of u, call it $g(u)$, is also of interest. If we sample the distribution and obtain a value of the random variable, u_k, then the value of g associated with that trial is $g(u_k)$. The question arises: Given $g(u)$ and the distribution function $f(u)$, what should we expect the value of $g(u_k)$ to be? That is, if we get a value of u from the distribution and then find $g(u)$, what value should we expect to find for $g(u)$? While this seems like a reasonable question, it is obvious that we can give a meaningful answer only when we can define more precisely just what we mean by "expect."

To understand our definition of the ***expected value*** (sometimes called the ***expectation value***) of $g(u)$, let us consider a game of chance. Suppose that we have a needle that rotates freely on a central axis. When spun, the needle describes a circular path, and its point eventually comes to rest at some point on this path. The location at which the needle stops is completely random. Imagine that we divide the circular path into six equal segments, which we number from one to six. When we spin the needle, it is equally likely to stop over any of these segments. Now, let us suppose that we conduct a lottery by selling six tickets, also numbered from one to six. We decide the winner of the lottery by spinning the needle. The holder of the ticket whose number matches the number on which the needle stops receives a payoff of \$6000. After the spin, one ticket is worth \$6000, and the other five are valueless. We ask: Before the spin, what is any one of the lottery tickets worth?

In this context, it is reasonable to define the expected value of a ticket as the amount that we should be willing to pay to buy a ticket. If we buy them all, we receive \$6000 when the winning ticket is selected. If we pay \$1000 per ticket to buy them all, we get our money back. If we buy all the tickets, the expected value of each ticket is \$1000. What if we buy only one ticket? Is it reasonable to continue to say that its expected value is \$1000? We argue that it is. One argument is that the expected value of a ticket should not depend on who owns the ticket; so, it should not depend on whether we buy one, two, or all of them. A more general argument supposes that repeated lotteries are held under the same rules. If we spend \$1000 to buy one ticket in each of a very large number of such lotteries, we expect that we will eventually "break even." Since the needle comes to rest at each number with equal probability, we reason that

$$\begin{aligned}&\textit{Expected value of a ticket}\\&= \$6000(\textit{fraction of times our ticket would be selected})\\&= \$6000(1/6)\\&= \$1000\end{aligned}$$

Since we assume that the fraction of times our ticket would be selected in a long series of identical lotteries is the same thing as the probability that our ticket will be selected in any given drawing, we can also express the expected value as

$$\begin{aligned}&\textit{Expected value of a ticket}\\&= \$6000(\textit{probability that our ticket will be be selected})\\&= \$6000(1/6)\\&= \$1000\end{aligned}$$

Clearly, the ticket is superfluous. The game depends on obtaining a value of a random variable from a distribution. The distribution is a spin of the needle. The random variable is the location at which the needle comes to rest. We can conduct essentially the same game by allowing any number of participants to bet that the needle will come to rest on any of the six equally probable segments of the circle. If an individual repeatedly bets on the same segment in many repetitions of this game, the total of his winnings eventually matches the total amount that he has wagered. (More precisely, the total of his winnings divided by the total amount he has wagered becomes arbitrarily close to one.)

Suppose now that we change the rules. Under the new rules, we designate segment 1 of the circle as the payoff segment. Participants pay a fixed sum to be eligible for the payoff for a particular game. Each game is decided by a spin of the needle. If the needle lands in segment 1, everyone who paid to participate in that game receives $\$6000$. Evidently, the new rules have no effect on the value of participation. Over the long haul, a participant in a large number of games wins $\$6000$ in one-sixth of these games. We take this to be equivalent to saying that he has a probability of one-sixth of winning $\$6000$ in a given game in which he participates. His expected payoff is

$$\begin{aligned}&\textit{Expected value of game}\\ &= \$6000(\textit{probability of winning } \$6000)\\ &= \$6000(1/6)\\ &= \$1000\end{aligned}$$

Let us change the game again. We sub-divide segment 2 into equal-size segments $2A$ and $2B$. The probability that the needle lands in $2A$ or $2B$ is $1/12$. In this new game, the payoff is $\$6000$ when the needle lands in either segment 1 or segment $2A$. We can use any of the arguments that we have made previously to see that the expected payoff game is now $\$6000(1/4) = \$1500$. However, the analysis that is most readily generalized recognizes that the payoff from this game is just the sum of the payout from the previous game plus the payout from a game in which the sole payout is $\$6000$ whenever the needle lands in segment $2A$. For the new game, we have

$$\begin{aligned}&\textit{Expected value of a game}\\ &= \$6000 \times P(\textit{segment } 1) + \$6000 \times P(\textit{segment } 2A)\\ &= \$6000(1/6) + \$6000(1/12)\\ &= \$1500\end{aligned}$$

We can devise any number of new games by dividing the needle's circular path into Ω non-overlapping segments. Each segment is a possible outcome. We number the possible outcomes 1, 2, ..., i, ..., Ω, label these outcomes $u_1, u_2, \dots, u_i, \dots, u_\Omega$, and denote their probabilities as $P(u_1), P(u_2), \dots, P(u_i), \dots, P(u_\Omega)$. We say that the probability of outcome u_i, $P(u_i)$, is the ***expected frequency*** of outcome u_i. We denote the respective payoffs as $g(u_1), g(u_2), \dots, g(u_i), \dots, g(u_\Omega)$. Straightforward generalization of our last analysis shows that the expected value for participation in any game of this type is

$$\sum_{i=1}^{\Omega} g(u_i) \times P(u_i)$$

Moreover, the spinner is representative of any distribution, so it is reasonable to generalize further. We can say that the expected value of the outcome of a single trial is always the probability-weighted sum, over all possible outcomes, of the value of each outcome. A common notation uses angular brackets to denote the expected value for a function of the random variable; the expected value of $g(u)$ is $\langle g(u) \rangle$. For a discrete distribution with Ω exhaustive mutually-exclusive outcomes u_i, probabilities $P(u_i)$, and outcome values (payoffs) $g(u_i)$, we define the ***expected value*** of $g(u)$ to be

$$\langle g(u) \rangle = \sum_{i=1}^{\Omega} g(u_i) \times P(u_i)$$

Now, let us examine the expected value of $g(u)$ from a slightly different perspective. Let the number of times that each of the various outcomes is observed in a particular sample of N observations be $N_1, N_2, \dots, N_3, \dots, N_\Omega$. We have $N = N_1 + N_2 + \cdots + N_i + \cdots + N_\Omega$. The set $\{N_1, N_2, \dots, N_i, \dots, N_\Omega\}$ specifies the way that the possible outcomes are populated in this particular series of N observations. We call $\{N_1, N_2, \dots, N_i, \dots, N_\Omega\}$ a ***population set***. If we make a second series of N observations, we obtain a second population set. We infer that the best forecast we can make for the number of occurrences of outcome u_i in any future series of N observations is $N \times P(u_i)$. We call $N \times P(u_i)$ the ***expected number*** of observations of outcome u_i in a sample of size N.

In a particular series of N trials, the number of occurrences of outcome u_i, and hence of $g(u_i)$, is N_i. For the set of outcomes $\{N_1, N_2, \dots, N_3, \dots, N_\Omega\}$, the average value of $g(u)$ is

$$\overline{g(u)} = \frac{1}{N} \sum_{i=1}^{\Omega} g(u_i) \times N_i$$

Collecting a second sample of N observations produces a second estimate of $\overline{g(u)}$. If N is small, successive estimates of $\overline{g(u)}$ may differ significantly from one another. If we make a series of N observations multiple times, we obtain multiple population sets. In general, the population set from one series of N observations is different from the population set for a second series of N observations. If $N \gg \Omega$, collecting such samples of N a sufficiently large number of times must produce some population sets more than once, and among those that are observed more than once, one must occur more often than any other. We call it the ***most probable*** population set. Let the elements of the most probable population set be $\{N_1^\bullet, N_2^\bullet, \dots, N_i^\bullet, \dots, N_\Omega^\bullet\}$. We infer that the most probable population set is the best forecast we can make about the outcomes of any future sample of N from this distribution. Moreover, we infer that the best estimate we can make of $N_i^\bullet$ is that it equals the expected number of observations of outcome u_i; that is,

$$N_i^\bullet \approx N \times P(u_i)$$

Now, N_i and $N_i^\bullet$ must be natural numbers, while $N \times P(u_i)$ need only be real. In particular, we can have $0 < N \times P(u_i) < 1$, but $N_i^\bullet$ must be 0 or 1 (or some higher integer). This is a situation of practical importance, because circumstances may limit the sample size to a number, N, that is much less than the number of possible outcomes, Ω. (We encounter this situation in our discussion of statistical thermodynamics in Chapter 21. We find that the number of molecules in a system can be much smaller than the number of outcomes—observable energy levels—available to any given molecule.)

If many more than N outcomes have about the same probability, repeated collection of samples of N observations can produce a series of population sets (each population set different from all of the others) in each of which every element is either zero or one. When this occurs, it may be that no single population set is significantly more probable than any of many others. Nevertheless, every outcome occurs with a well-defined probability. We infer that the set $\{N \times P(u_1), N \times P(u_2), \ldots, N \times P(u_i), \ldots, N \times P(u_\Omega)\}$ is always an adequate proxy for calculating the expected value for the most probable population set.

To illustrate this kind of distribution, suppose that there are 3000 possible outcomes, of which the first and last thousand have probabilities that are so low that they can be taken as zero, while the middle 1000 outcomes have approximately equal probabilities. Then $P(u_i) \approx 0$ for $1 < i < 1000$ and $2001 < i < 3000$, while $P(u_i) \approx 10^{-3}$ for $1001 < i < 2000$. We are illustrating the situation in which the number of outcomes we can observe, N, is much less than the number of outcomes that have appreciable probability, which is 1000. So let us take the number of trials to be $N = 4$. If the value of $g(u)$ for each of the 1000 middle outcomes is the same, say $g(u_i) = 100$ for $1001 < i < 2000$, then our calculation of the expected value of $g(u)$ will be

$$\begin{aligned}\langle g(u)\rangle &= \frac{1}{4}\sum_{i=1}^{3000} g(u_i) \times N \times P(u_i) \\ &= \frac{1}{4}\sum_{i=1001}^{2000} 100 \times N_i \\ &= \frac{400}{4} \\ &= 100\end{aligned}$$

regardless of which population set results from the four trials. That is, because all of the populations sets that have a significant chance to be observed have $N_i = 1$ and $g(u_i) = 100$ for exactly four values of i in the range $1001 < i < 2011$, all of the population sets that have a significant chance to be observed give rise to the same expected value.

Let us compute the arithmetic average, $\overline{g(u)}$, using the most probable population set for a sample of N trials. In this case, the number of observations of the outcome u_i is $N_i^\bullet = N \times P(u_i)$.

$$\begin{aligned}\overline{g(u)} &= \frac{1}{N}\sum_{i=1}^{\Omega} g(u_i) \times N_i^\bullet \\ &= \frac{1}{N}\sum_{i=1}^{\Omega} g(u_i) \times N \times P(u_i) \\ &= \sum_{i=1}^{\Omega} g(u_i) \times P(u_i) \\ &= \langle g(u)\rangle\end{aligned}$$

For a discrete distribution, $\langle g(u)\rangle$ is the value of $\overline{g(u)}$ that we calculate from the most probable population set, $\{N_1^\bullet, N_2^\bullet, \ldots, N_i^\bullet, \ldots, N_\Omega^\bullet\}$, or its proxy $\{N \times P(u_1), N \times P(u_2), \ldots, N \times P(u_i), \ldots, N \times P(u_\Omega)\}$.

We can extend the definition of the expected value, $\langle g(u)\rangle$, to cases in which the cumulative probability distribution function, $f(u)$, and the outcome-value function, $g(u)$, are continuous in the domain of the random variable, $u_{min} < u < u_{max}$. To do so, we divide this domain into a finite number, Ω, of intervals, Δu_i. We let u_i be the lower limit of u in the interval Δu_i. Then the probability that a given trial yields a value of the random variable in the interval Δu_i is $P(\Delta u_i) = f(u_i + \Delta u_i) - f(u_i)$, and we can approximate the expected value of $g(u)$ for the continuous distribution by the finite sum

$$\begin{aligned}\langle g(u)\rangle &= \sum_{i=1}^{\Omega} g(u_i) \times P(\Delta u_i) \\ &= \sum_{i=1}^{\Omega} g(u_i) \times [f(u_i + \Delta u_i) - f(u_i)] \\ &= \sum_{i=1}^{\Omega} g(u_i) \times \left[\frac{f(u_i + \Delta u_i) - f(u_i)}{\Delta u_i}\right] \times \Delta u_i\end{aligned}$$

In the limit as Ω becomes arbitrarily large and all of the intervals Δu_i become arbitrarily small, the expected value of $g(u)$ for a continuous distribution becomes

$$\langle g(u)\rangle = \int_{-\infty}^{\infty} g(u)\left[\frac{df(u)}{du}\right] du$$

This integral is the value of $\langle g(u)\rangle$, where $df(u)/du$ is the probability density function for the distribution. If c is a constant, we have

$$\langle g(cu)\rangle = c\langle g(u)\rangle$$

If $h(u)$ is a second function of the random variable, we have

$$\langle g(u) + h(u)\rangle = \langle g(u)\rangle + \langle h(u)\rangle$$

§10 Statistics: The mean and the variance of a distribution

There are two important statistics associated with any probability distribution, the ***mean*** and the ***variance***. The mean is defined as the expected value of the random variable itself. The Greek letter μ is usually used to represent the mean. If $f(u)$ is the cumulative probability distribution, the mean is the expected value for $g(u) = u$. From our definition of expected value, the mean is

$$\mu = \int_{-\infty}^{\infty} u\left(\frac{df}{du}\right)du$$

The variance is defined as the expected value of $(u-\mu)^2$. The variance measures how dispersed the data are. If the variance is large, the data are—on average—farther from the mean than they are if the variance is small. The ***standard deviation*** is the square root of the variance. The Greek letter σ is usually used to denote the standard deviation. Then, σ^2 denotes the variance, and

$$\sigma^2 = \int_{-\infty}^{\infty} (u-\mu)^2\left(\frac{df}{du}\right)du$$

If we have a small number of points from a distribution, we can estimate μ and σ by approximating these integrals as sums over the domain of the random variable. To do this, we need to estimate the probability associated with each interval for which we have a sample point. By the argument we make in §7, the best estimate of this probability is simply $1/N$, where N is the number of sample points. We have therefore

$$\mu = \int_{-\infty}^{\infty} u\left(\frac{df}{du}\right)du \approx \sum_{1}^{N} u_i\left(\frac{1}{N}\right) = \bar{u}$$

That is, the best estimate we can make of the mean from N data points is $\bar{u}$, where $\bar{u}$ is the ordinary arithmetic average. Similarly, the best estimate we can make of the variance is

$$\sigma^2 = \int_{-\infty}^{\infty} (u-\mu)^2\left(\frac{df}{du}\right)du \approx \sum_{i=1}^{N} (u_i-\mu)^2\left(\frac{1}{N}\right)$$

Now a complication arises in that we usually do not know the value of μ. The best we can do is to estimate its value as $\mu \approx \bar{u}$. It turns out that using this approximation in the equation we deduce for the variance gives an estimate of the variance that is too small. A more detailed argument (see §14) shows that, if we use $\bar{u}$ to approximate the mean, the best estimate of σ^2, usually denoted s^2, is

$$estimated\ \sigma^2 = s^2 = \sum_{i=1}^{N} (u_i-\bar{u})^2\left(\frac{1}{N-1}\right)$$

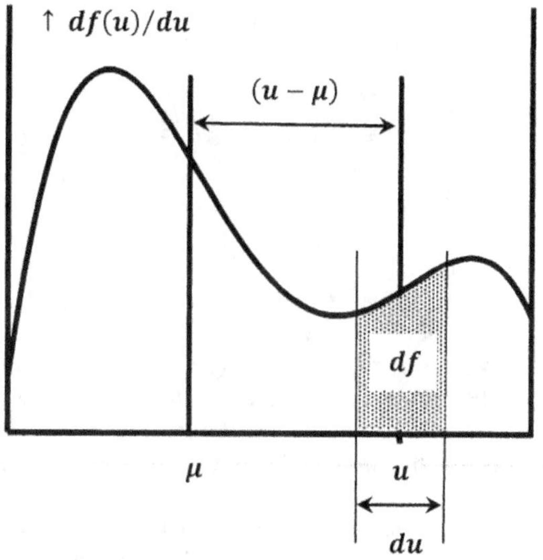

Figure 12. Variance is analogous to a moment of inertia.

Dividing by $N-1$, rather than N, compensates exactly for the error introduced by using $\bar{u}$ rather than μ.

The mean is analogous to a center of mass. The variance is analogous to a moment of inertia. For this reason, the variance is also called ***the second moment about the mean***. To show these analogies, let us imagine that we draw the probability density function on a uniformly thick steel plate and then cut along the curve and the u-axis. (See Figure 12.) Let M be the mass of the cutout piece of plate; M is the mass below the probability density curve. Let dA and dm be the increments of area and mass in the thin slice of the cutout that lies above a small increment, du, of u. Let ρ be the density of the plate, expressed as mass per unit area. Since the plate is uniform, ρ is constant. We have $dA = (df/du)du$ and $dm = \rho dA$ so that

$$dm = \rho\left(\frac{df}{du}\right)du$$

The mean of the distribution corresponds to a vertical line on this cutout at $u = \mu$. If the cutout is supported on a knife-edge along the line $u = \mu$, gravity induces no torque; the cutout is balanced. Since the torque is zero, we have

$$0 = \int_{m=0}^{M} (u-\mu)dm = \int_{-\infty}^{\infty} (u-\mu)\rho\left(\frac{df}{du}\right)du$$

Since μ is a constant property of the cut-out, it follows that

$$\mu = \int_{-\infty}^{\infty} u\left(\frac{df}{du}\right)du$$

The cutout's moment of inertia about the line $u = \mu$ is

$$I = \int_{m=0}^{M} (u-\mu)^2 dm$$
$$= \int_{-\infty}^{\infty} (u-\mu)^2 \rho \left(\frac{df}{du}\right) du$$
$$= \rho\sigma^2$$

The moment of inertia about the line $u - \mu$ is simply the mass per unit area, ρ, times the variance of the distribution. If we let $\rho = 1$, we have $I = \sigma^2$.

We define the mean of $f(u)$ as the expected value of u. It is the value of u we should "expect" to get the next time we sample the distribution. Alternatively, we can say that the mean is the best prediction we can make about the value of a future sample from the distribution. If we know μ, the best prediction we can make is $u_{predicted} = \mu$. If we have only the estimated mean, $\bar{u}$, then $\bar{u}$ is the best prediction we can make. Choosing $u_{predicted} = \bar{u}$ makes the difference, $|u - u_{predicted}|$, as small as possible.

These ideas relate to another interpretation of the mean. We saw that the variance is the second moment about the mean. The first moment about the mean is

$$1^{st}\ moment = \int_{-\infty}^{\infty} (u-\mu)\left(\frac{df}{du}\right) du$$
$$= \int_{-\infty}^{\infty} u\left(\frac{df}{du}\right) du - \mu \int_{-\infty}^{\infty} \left(\frac{df}{du}\right) du$$
$$= \mu - \mu$$
$$= 0$$

Since the last two integrals are μ and 1, respectively, ***the first moment about the mean is zero***. We could have defined the mean as the value, μ, for which the first moment of u about μ is zero.

The first moment about the mean is zero. The second moment about the mean is the variance. We can define third, fourth, and higher moments about the mean. Some of these higher moments have useful applications.

§11 The variance of the average: the central limit theorem

The ***central limit theorem*** establishes very important relationships between the statistics for two distributions that are related in a particular way. It enables us to understand some important features of physical systems.

The central limit theorem concerns the distribution of averages. If we have some original distribution and sample it three times, we can calculate the average of these three data points. Call this average $A_{3,1}$. We could repeat this activity and obtain a second average of three values, $A_{3,2}$. We can do this repeatedly, generating averages $A_{3,3},\ldots, A_{3,n}$. Several things will be true about these averages:

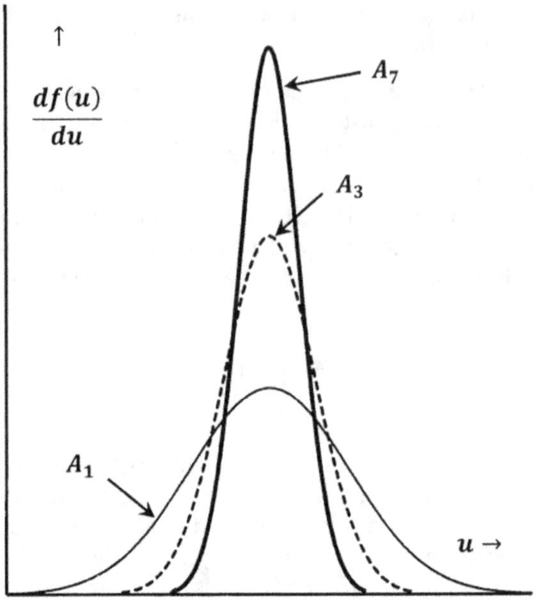

Figure 13. The variance of an average of *N* is proportional to $1/N$.

- ***The set of all of the possible averages-of-three, $\{A_{3,i}\}$, is itself a distribution. This averages-of-three distribution is different from the original distribution. Each average-of-three is a value of the random variable associated with the averages-of-three distribution.***

- ***Each of the $A_{3,i}$ is an estimate of the mean of the original distribution.***

- ***The distribution of the $A_{3,i}$ will be less spread out than the original distribution.***

There is nothing unique about averaging three values. We could sample the original distribution seven times and compute the average of these seven values, calling the result $A_{7,1}$. Repeating, we could generate averages $A_{7,2},\ldots, A_{7,m}$. All of the things we say about the averages-of-three are also true of these averages-of-seven. However, we can now say something more: The distribution of the $A_{7,i}$ will be less spread out than the distribution of the $A_{3,i}$. The corresponding probability density functions are sketched in Figure 13.

The central limit theorem relates the mean and variance of the distribution of averages to the mean and variance of the original distribution:

If random samples of N values are taken from a distribution, whose mean is μ and whose variance is σ^2, averages of these N values, A_N, are approximately normally distributed with a mean of μ and a variance of σ^2/N. The approximation to the normal distribution becomes better as N becomes larger.

It turns out that the number, *N*, of trials that is needed to get a good estimate of the variance is substantially larger than the number required to get a good estimate of the mean.

§12 The normal distribution

The ***normal distribution*** is very important. The central limit theorem says that if we average enough values from any distribution, the distribution of the averages we calculate will be the normal distribution. The probability density function for the normal distribution is

$$\frac{df}{du} = \frac{1}{\sigma\sqrt{2\pi}} \exp\left[\frac{-(u-\mu)^2}{2\sigma^2}\right]$$

The integral of the normal distribution from $u = -\infty$ to $u = \infty$ is unity. However, the definite integral between arbitrary limits cannot be obtained as an analytical function. This turns out to be true for some other important distributions also; this is one reason for working with probability density functions rather than the corresponding cumulative probability functions. Of course, the definite integral can be calculated to any desired accuracy by numerical methods, and readily available tables give values for definite integrals from $u = -\infty$ to $u = u$. (We mention normal curve of error tables in §8, where we introduce a method for testing whether a given set of data conforms to the normal distribution equation.)

§13 The expected value of a function of several variables and the central limit theorem

We can extend the idea of an expected value to a function of multiple random variables. Let U and V be distributions whose random variables are u and v, respectively. Let the probability density functions for these distributions be $df_u(u)/du$ and $df_v(v)/dv$. In general, these probability density functions are different functions; that is, U and V are different distributions. Let $g(u,v)$ be some function of these random variables. The probability that an observation made on U produces a value of u in the range $u^* < u < u^* + du$ is

$$P(u^* < u < u^* + du) = \frac{df_u(u^*)}{du} du$$

and the probability that an observation made on V produces a value of v in the range $v^* < v < v^* + dv$ is

$$P(v^* < v < v^* + dv) = \frac{df_v(v^*)}{dv} dv$$

The probability that making one observation on each of these distributions produces a value of u that lies in the range $u^* < u < u^* + du$ and a value of v that lies in the range $v^* < v < v^* + dv$ is

$$\frac{df_u(u^*)}{du}\frac{df_v(v^*)}{dv} dudv$$

In a straightforward generalization, we define the expected value of $g(u,v)$, $\langle g(u,v) \rangle$, as

$$\langle g(u,v) \rangle = \int_{v=-\infty}^{\infty} \int_{u=-\infty}^{\infty} g(u,v) \frac{df_u(u)}{du}\frac{df_v(v)}{dv} dudv$$

If $g(u,v)$ is a sum of functions of independent variables, $g(u,v) = h(u) + k(v)$, we have

$$\begin{aligned}\langle g(u,v)\rangle &= \int_{-\infty}^{\infty}\int_{-\infty}^{\infty} [h(u)+k(v)] \frac{df_u(u)}{du}\frac{df_v(v)}{dv} dudv \\ &= \int_{-\infty}^{\infty} h(u)\frac{df_u(u)}{du} du + \int_{-\infty}^{\infty} k(v)\frac{df_v(v)}{dv} dv \\ &= \langle h(u) \rangle + \langle k(v) \rangle\end{aligned}$$

If $g(u,v)$ is a product of independent functions, $g(u,v) = h(u)k(v)$, we have

$$\begin{aligned}\langle g(u,v)\rangle &= \int_{-\infty}^{\infty}\int_{-\infty}^{\infty} h(u)k(v) \frac{df_u(u)}{du}\frac{df_v(v)}{dv} dudv \\ &= \int_{-\infty}^{\infty} h(u)\frac{df_u(u)}{du} du \times \int_{-\infty}^{\infty} k(v)\frac{df_v(v)}{dv} d \\ &= \langle h(u)\rangle \langle k(v)\rangle\end{aligned}$$

We can extend these conclusions to functions of the random variables of any number of distributions. If u_i is the random variable of distribution U_i whose probability density function is $df_i(u_i)/du_i$, the expected value of

$$\begin{aligned}g(u_1, \dots, u_i, \dots, u_N) &= \\ &= h_1(u_1) + \cdots + h_i(u_i) + \cdots + h_N(u_N)\end{aligned}$$

becomes

$$\langle g(u_1, \dots, u_i, \dots, u_N)\rangle = \sum_{i=1}^{N} \langle h_i(u_i)\rangle$$

and the expected value of

$$g(u_1, \dots, u_i, \dots, u_N) = h_1(u_1) \dots h_i(u_i) \dots h_N(u_N)$$

becomes

$$\langle g(u_1, \dots, u_i, \dots, u_N)\rangle = \prod_{i=1}^{N} \langle h_i(u_i)\rangle$$

We are particularly interested in expected values for repeated trials made on the same distribution. We consider distributions for which the outcome of one trial is independent of the outcome of any other trial. The probability density function is the same for every trial, so we have $f(u) = f_1(u_1) = \cdots = f_i(u_i) = \cdots = f_N(u_N)$. Let the values obtained for the random variable in a series of trials on the same distribution be $u_1, \dots, u_i, \dots, u_N$. For each trial, we have

$$\langle h_i(u_i)\rangle = \int_{-\infty}^{\infty} h_i(u_i) \frac{df_i(u_i)}{du_i} du_i$$

If we consider the special case of repeated trials in which the functions $h_i(u_i)$ are all the same function, so that $h(u) = h_1(u_1) = \cdots = h_i(u_i) = \cdots = h_N(u_N)$, the expected value of

$$g(u_1, \dots, u_i, \dots, u_N) = h_1(u_1) + \cdots + h_i(u_i) + \cdots + h_N(u_N)$$

becomes

$$\langle g(u_1, \dots, u_i, \dots, u_N) \rangle = \sum_{i=1}^{N} \langle h_i(u_i) \rangle = N\langle h(u) \rangle$$

and the expected value of

$$g(u_1, \dots, u_i, \dots, u_N) = h_1(u_1) \dots h_i(u_i) \dots h_N(u_N)$$

becomes

$$\langle g(u_1, \dots, u_i, \dots, u_N) \rangle = \prod_{i=1}^{N} \langle h_i(u_i) \rangle = \langle h(u) \rangle^N$$

Now let us consider N independent trials on the same distribution and let $h_i(u_i) = h(u_i) = u_i$. Then, the expected value of

$$g(u_1, \dots, u_i, \dots, u_N) = \\ = h_1(u_1) + \cdots + h_i(u_i) + \cdots + h_N(u_N)$$

becomes

$$\langle u_1 + \cdots + u_i + \cdots + u_N \rangle = \sum_{i=1}^{N} \langle u_i \rangle = N\langle u \rangle = N\mu$$

By definition, the average of N repeated trials is $\bar{u}_N = (u_1 + \cdots + u_i + \cdots + u_N)/N$, so that the expected value of the mean of a distribution of an average-of-N repeated trials is

$$\langle \bar{u}_N \rangle = \frac{\langle u_1 + \cdots + u_i + \cdots + u_N \rangle}{N} = \mu$$

This proves one element of the central limit theorem: The mean of a distribution of averages-of-N values of a random variable drawn from a parent distribution is equal to the mean of the parent distribution.

The variance of these averages-of-N is

$$\begin{aligned}
\sigma_N^2 &= \langle (\bar{u}_N - \mu)^2 \rangle \\
&= \left\langle \left[\left(\frac{1}{N}\sum_{i=1}^{N} u_i\right) - \mu\right]^2 \right\rangle \\
&= \left\langle \left[\left(\frac{1}{N}\sum_{i=1}^{N} u_i\right) - \frac{N\mu}{N}\right]^2 \right\rangle \\
&= \frac{1}{N^2} \left\langle \left[\left(\sum_{i=1}^{N} u_i\right) - N\mu\right]^2 \right\rangle \\
&= \frac{1}{N^2} \left\langle \left[\left(\sum_{i=1}^{N} (u_i - \mu)\right)\right]^2 \right\rangle \\
&= \frac{1}{N^2} \left\langle \sum_{i=1}^{N} (u_i - \mu)^2 \right\rangle + \frac{2}{N^2} \left\langle \sum_{i=1}^{N-1} (u_i - \mu) \sum_{j=i+1}^{N} (u_j - \mu) \right\rangle \\
&= \frac{1}{N^2} \sum_{i=1}^{N} \langle (u_i - \mu)^2 \rangle + \frac{2}{N^2} \left\langle \sum_{i=1}^{N-1} (u_i - \mu) \right\rangle \left\langle \sum_{j=i+1}^{N} (u_j - \mu) \right\rangle
\end{aligned}$$

Where the last term is zero, because

$$\left\langle \sum_{i=1}^{N-1} (u_i - \mu) \right\rangle = \sum_{i=1}^{N-1} \langle (u_i - \mu) \rangle$$

and

$$\langle (u_i - \mu) \rangle = 0$$

By definition, $\sigma^2 = \langle (u_i - \mu)^2 \rangle$, so that we have

$$\sigma_N^2 = \frac{N\sigma^2}{N^2} = \frac{\sigma^2}{N}$$

This proves a second element of the central limit theorem: The variance of an average of N values of a random variable drawn from a parent distribution is equal to the variance of the parent distribution divided by N.

§14 Where does the $N - 1$ come from?

If we know μ and we have a set of N data points, the best estimate we can make of the variance is

$$\sigma^2 = \int_{u_{min}}^{u_{max}} (u - \mu)^2 \left(\frac{df}{du}\right) du \approx \sum_{i=1}^{N} (u_i - \mu)^2 \left(\frac{1}{N}\right)$$

We have said that if we must use $\bar{u}$ to approximate the mean, the best estimate of σ^2, usually denoted s^2, is

$$\text{estimated } \sigma^2 = s^2 = \sum_{i=1}^{N} (u_i - \bar{u})^2 \left(\frac{1}{N-1}\right)$$

The use of $N - 1$, rather than N, in the denominator is distinctly non-intuitive; so much so that this equation often causes great irritation. Let us see how this equation comes about.

Suppose that we have a distribution whose mean is μ and variance is σ^2. Suppose that we draw N values of the random variable, u, from the distribution. We want to think about the expected value of $(u-\mu)^2$. Let us write $(u-\mu)$ as $(u-\mu)=(u-\bar{u})+(\bar{u}-\mu)$. Squaring this gives

$$(u-\mu)^2=(u-\bar{u})^2+(\bar{u}-\mu)^2+2(u-\bar{u})(\bar{u}-\mu).$$

From our definition of expected value, we can write:

$$\begin{aligned}&Expected\ value\ of\ (u-\mu)^2=\\ &\quad = expected\ value\ of\ (u-\bar{u})^2\\ &\quad\quad +expected\ value\ of\ (\bar{u}-\mu)^2\\ &\quad\quad +expected\ value\ of\ 2(u-\bar{u})(\bar{u}-\mu)\end{aligned}$$

From our discussion above, we can recognize each of these expected values:

- The expected value of $(u-\mu)^2$ is the variance of the original distribution, which is σ^2. Since this is a definition, it is exact.

- The best possible *estimate* of the expected value of $(u-\bar{u})^2$ is

$$\sum_{i=1}^{N}(u_i-\bar{u})^2\left(\frac{1}{N}\right)$$

- The expected value of $(\bar{u}-\mu)^2$ is the expected value of the variance of averages of N random variables drawn from the original distribution. That is, the expected value of $(\bar{u}-\mu)^2$ is what we would get if we repeatedly drew N values from the original distribution, computed the average of each set of N values, and then found the variance of this new distribution of average values. By the central limit theorem, this variance is σ^2/N. Thus, the expected value of $(\bar{u}-\mu)^2$ is exactly σ^2/N.

- Since $(\bar{u}-\mu)$ is constant, the expected value of $2(u-\bar{u})(\bar{u}-\mu)$ is

$$2(\bar{u}-\mu)\left[\frac{1}{N}\sum_{i-1}^{N}(u_i-\bar{u})\right]$$

which is equal to zero, because

$$\sum_{i-1}^{N}(u_i-\bar{u})=\left(\sum_{i=1}^{N}u_i\right)-N\bar{u}=0$$

by the definition of $\bar{u}$.

Substituting, our expression for the expected value of $(u-\mu)^2$ becomes:

$$\sigma^2\approx\sum_{i=1}^{N}(u_i-\bar{u})^2\left(\frac{1}{N}\right)+\frac{\sigma^2}{N}$$

so that

$$\sigma^2\left(1-\frac{1}{N}\right)=\sigma^2\left(\frac{N-1}{N}\right)\approx\sum_{i=1}^{N}\frac{(u_i-\bar{u})^2}{N}$$

and

$$\sigma^2\approx\sum_{i=1}^{N}\frac{(u_i-\bar{u})^2}{N-1}$$

That is, as originally stated, when we must use $\bar{u}$ rather than the true mean, μ, in the sum of squared differences, the best possible *estimate* of σ^2, usually denoted s^2, is obtained by dividing by $N-1$, rather than by N.

Problems

1. At each toss of a die, the die lands with one face on top. This face is distinguished from the other five faces by the number of dots that appear on it. Tossing a die produces data. What is the distribution? What is the random variable of this distribution? What outcomes are possible for this distribution? How would we collect a sample of ten values of the random variable of this distribution?

2. Suppose that we toss a die three times and average the results observed. How would you describe the distribution from which this average is derived? What is the random variable of this distribution? What outcomes are possible for this distribution? What would we do to collect a sample of ten values of the random variable of this distribution?

3. Suppose that we toss three dice simultaneously and average the results observed. How would you describe the distribution from which this average is derived? What is the random variable of this distribution? What outcomes are possible for this distribution? What would we do to collect a sample of ten values of the random variable of this distribution? Suppose that some third party collects a set, call it A, of ten values from this distribution and a second set, call it B, of values from the distribution in problem 2. If we are given the data in each set but are not told which label goes with which set of data, can we analyze the data to determine which set is A and which is B?

4. The manufacturing process for an electronic component produces 3 bad components in every 1000 components produced. The bad components appear randomly. What is the probability that
(a) a randomly selected component is bad?
(b) a randomly selected component is good?
(c) 2 bad components are produced in succession?
(d) 100 good components are produced in succession?

5. A product incorporates two of the components in the previous problem. What is the probability that
(a) both components are good?
(b) both components are bad?
(c) one component is good and one component is bad?
(d) at least one component is good?

6. A card is selected at random from a well-shuffled deck. A second card is then selected at random from among the remaining 51 cards. What is the probability that
(a) the first card is a heart?
(b) the second card is a heart?
(c) neither card is a heart?
(d) both cards are hearts?
(e) at least one card is a heart?

7. A graduating class has 70 men and 77 women. How many combinations of homecoming king and queen are possible?

8. After the queen is selected from the graduating class of problem 7, one woman is selected to "first attendant" to the homecoming queen. Thereafter, another woman is selected to be "second attendant." After the queen is selected, how many ways can two attendants be selected?

9. A red die and a green die are rolled. What is the probability that
(a) both come up 3?
(b) both come up the same?
(c) they come up different?
(d) the red die comes up less than the green die?
(e) the red die comes up exactly two less than the green die?
(f) together they show 5?

10. A television game show offers a contestant a new car as the prize for correctly guessing which of three doors the car is behind. After the contestant selects a door, the game-show host opens an incorrect door. The host then gives the contestant the option of switching from the door he originally chose to the other door that remains unopened. Should the contestant change his selection?

[Hint: Consider the final set of outcomes to result from a sequence of three choices. First, the game-show producer selects a door and places the car behind this door. Diagram the possibilities. What is the probability of each? Second, the contestant selects a door. There are now nine possible outcomes. Diagram them. What is the probability of each? Third, the host opens a door. There are now twelve possible outcomes. Diagram them. What is the probability of each? Note that these twelve possibilities are not all equally probable.]

11. For a particular distribution, possible values of the random variable, x, range from zero to one. The probability density function for this distribution is $df/dx = 1$.
(a) Show that the probability of finding x in the range $0 \leq x \leq 1$ is one.
(b) What is the mean of this distribution?
(c) What is the variance of this distribution? The standard deviation?
(d) A quantity, g, is a function of x: $g(x) = x^2$. What is the expected value of g?

12. For a particular distribution, possible values of the random variable, x, range from one to three. The probability density function for this distribution is $df/dx = cx$, where c is a constant.
(a) What is the value of the constant, c?
(b) What is the mean of this distribution?
(c) What is the variance of this distribution? The standard deviation?
(d) If $g(x) = x^2$, what is the expected value of g?

13. For a particular distribution, possible values of the random variable, x, range from two to four. The probability density function for this distribution is $df/dx = cx^3$, where c is a constant.
(a) What is the value of the constant, c?
(b) What is the mean of this distribution?
(c) What is the variance of this distribution? The standard deviation?
(d) If $g(x) = x^2$, what is the expected value of g?

14. For a particular distribution, possible values of the random variable, x, range from zero to four. For $0 \leq x \leq$ ≤ 1, the probability density function is $df/dx = x/2$. For $1 < x \leq 4$, the probability density function is $df/dx = (4 - x)/6$.
(a) Show that the area under this probability distribution function is one.
(b) What is the mean of this distribution?
(c) What is the variance of this distribution? The standard deviation?
(d) If $g(x) = x^2$, what is the expected value of g?

15. The following values, x_i, of the random variable, x, are drawn from a distribution: 9.63, 9.00, 11.87, 10.13, 10.83, 9.50, 10.40, 9.83, and 10.09.
(a) Arrange these values in increasing order and calculate the "rank probability," $i/(N + 1)$, associated with each of the x_i values.
(b) Plot the rank probability (on the ordinate) versus the random-variable value (on the abscissa). Sketch a smooth curve through the points on this plot.
(c) What function is approximated by the curve sketched in part b?

(d) Plot the data points along a horizontal axis. Then create a bar graph (histogram) by erecting bars of equal area between each pair of data points.
(e) What function is approximated by the tops of the bars erected in part d?

16. For a particular distribution, possible values of the random variable range from zero to four. The following values of the random variable are drawn from this distribution: 0.1, 1.0, 1.1, 1.5, 2.1. Sketch an approximate probability density function for this distribution.

17. The possible values for the random variable of a particular distribution lie in the range $0 \leq x \leq 10$. In six trials, the following values are obtained: 1.0, 1.9, 2.3, 2.7, 3.0, 3.8.
(a) Sketch an approximate probability density function for this distribution.
(b) What is the best estimate we can make of the mean of this distribution?
(c) What is the best estimate we can make of the variance of this distribution?
(d) What is the best estimate we can make of the variance of averages-of-six drawn from this distribution?
(e) What is the best estimate we can make of the variance of averages-of-sixteen drawn from this distribution?

18. A computer program generates numbers from a normal distribution with a mean of zero and a standard deviation of 10. Also, for any integer N, the program will generate and average N values from this distribution. It will repeat this operation until it has produced 100 such averages. It will then compute the estimated standard deviation of these 100 average values. The table below gives various values of N and the estimated standard deviation, s, that was found for 100 averages of that N. Plot these data in a way that tests the validity of the central limit theorem.

N	s
4	5.182
9	2.794
16	2.206
25	2.152
36	1.689
49	1.092
64	1.001
81	1.004
100	1.074
144	0.601
196	0.546
256	0.690
324	0.545

19. If $f(u)$ is the cumulative probability distribution function for a distribution, what is the expected value of $f(u)$? What interpretation can you place on this result?

20. Five replications of a volumetric analysis yield concentration estimates of 0.3000, 0.3008, 0.3012, 0.3014, and 0.3020 mol L^{-1}. Calculate the rank probability of each of these results. Sketch, over the concentration range $0.3000 < x_i < 0.3020$ mol L^{-1}, an approximation of the cumulative probability distribution function for the distribution that yielded these data.

21. The Louisville Mudhens play on a square baseball field that measures 100 meters on a side. Casey's hits always fall on the field. (He never hits a foul ball or hits one out of the park.) The probability density function for the distance that a Casey hit goes parallel to the first-base line is $df_x(x)/dx = (2 \times 10^{-4})x$. (That is, we take the first-base line as our x-axis; the third-base line as our y-axis; and home plate is at the origin. $df_x(x)/dx$ is independent of the distance that the hit goes parallel to the third-base line, our y-axis.) The probability density function for the distance that a Casey hit goes parallel to the third-base line is $df_y(y)/dy = (3 \times 10^{-6})y^2$. ($df_y(y)/dy$ is independent of the distance that the hit goes parallel to the first-base line, our x-axis.)
(a) What is the probability that a Casey hit lands at a point (x, y) such that $x^* < x < x^* + dx$ and $y^* < y < y^* + dy$?
(b) What is the two-dimensionally probability density function that describes Casey's hits, expressed in this Cartesian coordinate system?
(c) Recall that polar coordinates transform to Cartesian coordinates according to $x = r\cos\theta$ and $y = r\sin\theta$. What is the probability density function for Casey's hits expressed using polar coordinates?
(d) Recall that the differential element of area in polar coordinates is $rdrd\theta$. Find the probability that a Casey hit lands within the pie-shaped area bounded by $0 < r < 50$ m and $0 < \theta < \pi/4$.

22. In Chapter 2, we derived the Barometric Formula, $\eta(h) = \eta(0)\exp(-mgh/kT)$ for molecules of mass m in an isothermal atmosphere at a height h above the surface of the earth. $\eta(h)$ is the number of molecules per unit volume at height h ; $\eta(0)$ is the number of molecules per unit volume at the earth's surface, where $h = 0$. Consider a vertical cylinder of unit cross-sectional area, extending from the earth's surface to an infinite height. Let $f(h)$ be the fraction of the molecules in this cylinder that is at a height less than h. Prove that the probability density function is $df/dh = (mg/kT)\exp(-mgh/kT)$. [Hint: $\eta(h)dh$ is the number of molecules, dn, in a cylindrical segment of unit cross-sectional area, of thickness dh, located at height h. How many molecules are found in the cylinder below h? In the entire cylinder?]

23. A particular distribution has six outcomes. These outcomes and their probabilities are a (0.1); b (0.2); c (0.3); d (0.2); e (0.1); and f (0.1).
(a) Partitioning I assigns these outcomes to a set of three events: Event $A =$ a or b or c; Event $B = d$;

and Event $C = e$ or f. What are the probabilities of Events A, B, and C?

(b) Partitioning II assigns the outcomes to two events: Event $D = a$ or b or c; and Event $E = d$ or e or f. What are the probabilities of Events D and E? Express the probabilities of Events D and E in terms of the probabilities of Events A, B, and C.

(c) Partitioning III assigns the outcomes to three events: Event $F = a$ or b; Event $G = c$ or d; and Event $H = e$ or f. What are the probabilities of Events F, G, and H? Can the probabilities of Events F, G, and H be expressed in terms of the probabilities of Events A, B, and C?

24. Consider a partitioning of outcomes into events that is not exhaustive; that is, not every outcome is assigned to an event. What problem arises when we want to describe the probabilities of these events?

25. Consider a partitioning of outcomes into events that is not mutually exclusive; that is, one (or more) outcome is assigned to two (or more) events. What problem arises when we want to describe the probabilities of these events?

26. For integer values of p ($p \neq 1$), we find

$$\int x^p \ln(x)\, dx = \left(\frac{x^{p+1}}{p+1}\right)\ln(x) - \frac{x^{p+1}}{(p+1)^2}$$

(a) Sketch the function, $h(x) = df(x)/dx = -4x\ln(x)$, over the interval $0 \le x \le 1$.

(b) Show that we can consider $h(x) = df(x)/dx = -4x\ln(x)$ to be a probability density function over this interval; that is, show $f(1) - f(0) = 1$. Let us name the corresponding distribution "Sam."

(c) What is the mean, μ, of Sam?

(d) What is the variance, σ^2, of Sam?

(e) What is the standard deviation, σ, of Sam?

(f) What is the variance of averages-of-four samples taken from Sam?

(g) The following four values are obtained in random sampling of an unknown distribution: 0.050; 0.010; 0.020; and 0.040. Estimate the mean, μ, variance (σ^2or s^2), and the standard deviation (σ or s) for this unknown distribution.

(h) What is the probability that a single sample drawn from Sam will lie in the interval $0 \le x \le 0.10$? Note: The upper limit of this interval is 0.10, not 1.0 as in part (a).

(i) Is it likely that the unknown distribution sampled in part g is in fact the distribution we named Sam? Why, or why not?

27. We define the mean, μ, as the expected value of the random variable: $\mu = \int_{-\infty}^{\infty} u(df/du)\, du$. Define $\bar{u} = \sum_{i=1}^{N}(u_i/N)$, where the u_i are N independent values of the random variable. Show that the expected value of $\bar{u}$ is μ.

28. A box contains a large number of plastic balls. An integer, W, in the range $1 \le W \le 20$ is printed on each ball. There are many balls printed with each integer. The integer specifies the mass of the ball in grams. Six random samples of three balls each are drawn from the box. The balls are replaced and the box is shaken between drawings. The numbers on the balls in drawings I through VI are:

I: 3, 4, 9
II: 1, 6, 17
III: 2, 5, 8
IV: 2, 6, 7
V: 3, 5, 6
VI: 2, 3, 10

(a) What are the population sets represented by the samples I through VI?

(b) Sketch the probability density function as estimated from sample I.

(c) Sketch the probability density function as estimated from sample II

(d) Using the data from samples I through VI, estimate the probability of drawing a ball of each mass in a single trial.

(e) Sketch the probability density function as estimated from the probability values in part (d).

(f) From the data in sample I, estimate the average mass of a ball in the box.

(g) From the data in sample II, estimate the average mass of a ball in the box.

(h) From the probability values calculated in part (d), estimate the average mass of a ball in the box.

4

The Distribution of Gas Velocities

§1 Distribution functions for gas-velocity components

In Chapter 2, we assume that all of the molecules in a gas move with the same speed and use a simplified argument to conclude that this speed depends only on temperature. We now recognize that the individual molecules in a gas sample have a wide range of speeds; the velocities of gas molecules must be described by a distribution function. It is true, however, that the average speed depends only on temperature.

James Clerk Maxwell was the first to derive the distribution function for gas velocities. He did it about 1860. We follow Maxwell's argument. For a molecule moving in three dimensions, there are three velocity components. Maxwell's argument uses only one assumption: the speed of a gas molecule is independent of the direction in which it is moving. Equivalently, we can say that the components of the velocity of a gas molecule are independent of one another; knowing the value of one component of a molecule's velocity does not enable us to infer anything about the values of the other two components. When we use Cartesian coordinates, Maxwell's assumption means also that the same mathematical model must describe the distribution of each of the velocity components.

Since the velocity of a gas molecule has three components, we must treat the velocity distribution as a function of three random variables. To understand how this can be done, let us consider how we might find probability distribution functions for velocity components. We need to consider both spherical and Cartesian coordinate systems.

Let us suppose that we are able to measure the Cartesian-coordinate components v_x, v_y, and v_z of the velocities of a large number of randomly selected gas molecules in a particular constant-temperature sample. Then we can transform each set of Cartesian components to spherical-coordinate velocity components v, θ, and φ. We imagine accumulating the results of these measurements in a table like Table 1. As a practical matter, of course, we cannot make the measurements to complete such a table. However, there is no doubt that, at every instant, every gas molecule can be characterized by a set of such velocity components; the values exist, even if we cannot measure them. We imagine that we have such data only as a way to clarify the properties of the distribution functions that we need.

Table 1. Molecular velocity components

Molecule Number	v_x	v_x	v_x	v	θ	φ
1	$v_x(1)$	$v_y(1)$	$v_z(1)$	$v(1)$	$\theta(1)$	$\varphi(1)$
2	$v_x(2)$	$v_y(2)$	$v_z(2)$	$v(2)$	$\theta(2)$	$\varphi(2)$
3	$v_x(3)$	$v_y(3)$	$v_z(3)$	$v(3)$	$\theta(3)$	$\varphi(3)$
4	$v_x(4)$	$v_y(4)$	$v_z(4)$	$v(4)$	$\theta(4)$	$\varphi(4)$
…	…	…	…	…	…	…
N	$v_x(N)$	$v_x(N)$	$v_z(N)$	$v(N)$	$\theta(N)$	$\varphi(N)$

These data have several important features. The scalar velocity, v, ranges from 0 to $+\infty$; v_x, v_y, and v_z range from $-\infty$ to $+\infty$. In §2, we see that θ varies from 0 to π; and φ ranges from 0 to 2π. Each column represents data sampled from the distribution of the corresponding random variable. In Chapter 3, we find that we can use such data to find mathematical models for such distributions. Here, we can find mathematical models for the cumulative distribution functions $f_x(v_x)$, $f_y(v_y)$, and $f_z(v_z)$. We can approximate the graph of $f_x(v_x)$ by plotting the rank probability of v_x versus v_x. We expect this plot to be sigmoid; at any v_x, the slope of this plot is the probability-density function, $df_x(v_x)/dv_x$. The probability density function for v_x depends only on v_x, because the value measured for v_x is independent of the values measured for v_y and v_z. However, by Maxwell's assumption, the functions describing the distribution of v_y and v_z are the same as those describing the distribution of v_x. While redundant, it is convenient to introduce additional symbols to represent these probability density functions. We define $\rho_x(v_x) = df_x(v_x)/dv_x$, $\rho_y(v_y) = df_y(v_y)/dv_y$, and $\rho_z(v_z) = df_z(v_z)/dv_z$.

When we find these one-dimensional distribution functions by modeling the experimental data in this way, each v_x datum that we use in our analysis comes from an observation on a molecule and is associated with particular v_y and v_z values. These values of v_y and v_z can be anything from $-\infty$ to $+\infty$. This is a significant point. The functions $f_x(v_x)$ and $df_x(v_x)/dv_x$ are independent of v_y and v_z. We can also say that $df_x(v_x)/dv_x$ describes the distribution of v_x when v_y and v_z are averaged over all the values it is possible for them to have.

To clarify this, let us consider another cumulative probability distribution function, $f_{xyz}(v_x, v_y, v_z)$, which is just the fraction of all molecules whose respective Cartesian velocity components are less than v_x, v_y, v_z. Since $f_x(v_x)$, $f_y(v_y)$, and $f_z(v_z)$ are the fractions whose components are less than v_x, v_y, and v_z, respectively, their product is equal to $f_{xyz}(v_x, v_y, v_z)$ We have $f_{xyz}(v_x, v_y, v_z) = f_x(v_x)f_y(v_y)f_z(v_z)$. For the velocity of a randomly selected molecule, (v_x^*, v_y^*, v_z^*), to be included in the fraction represented by $f_{xyz}(v_x, v_y, v_z)$, the velocity must be in the particular range $-\infty < v_x^* < v_x$, $-\infty < v_y^* < v_y$, and $-\infty < v_z^* < v_z$.

However, for a velocity v_x^* to be included in $f_x(v_x)$, we must have $v_x^* < v_x$, $v_y^* < \infty$, and $v_z^* < \infty$; that is, the components v_y^* and v_z^* can have any values. Since the probability that v_x, v_y, and v_z satisfy $v_x^* < v_x$, $v_y^* < v_y$, and $v_z^* < v_z$ is

$$\begin{aligned} P(v_x^* < v_x, v_y^* < v_y, v_z^* < v_z) &= f_{xyz}(v_x, v_y, v_z) \\ &= f_x(v_x)f_y(v_y)f_z(v_z) \end{aligned}$$

the probability that v_x^* is included in $f_x(v_x)$ becomes

$$\begin{aligned} P(v_x^* < v_x, v_y^* < \infty, v_z^* < \infty) &= f_{xyz}(v_x, \infty, \infty) \\ &= f_x(v_x)f_y(\infty)f_z(\infty) \\ &= f_x(v_x) \end{aligned}$$

For our purposes, we need to be able to express the probability that the velocity lies within any range of velocities. Let us use υ to designate a particular "volume" region in velocity space and use $P(\upsilon)$ to designate the probability that the velocity of a randomly selected molecule is in this region. When we let υ be the region in velocity space in which x-components lie between v_x and $v_x + dv_x$, y-components lie between v_y, and $v_y + dv_y$, and z-components lie between v_z and $v_z + dv_z$, $dP(\upsilon)$ denotes the probability that the velocity of a randomly chosen molecule, (v_x^*, v_y^*, v_z^*), satisfies the conditions $v_x < v_x^* < v_x + dv_x$, $v_y < v_y^* < v_y + dv_y$, and $v_z < v_z^* < v_z + dv_z$.

$dP(\upsilon)$ is an increment of probability. The dependence of $dP(\upsilon)$ on v_x, v_y, v_z, dv_x, dv_y, and dv_zcan be made explicit by introducing a new function, $\rho(v_x, v_y, v_z)$, defined by

$$dP(\upsilon) = \rho(v_x, v_y, v_z)dv_x dv_y dv_z$$

Since $dv_x dv_y dv_z$ is the volume available in velocity space for velocities whose x-components are between v_x and $v_x + dv_x$, whose y-components are between v_y, and $v_y + dv_y$, and whose z-components are between v_z and $v_z + dz$, we see that $\rho(v_x, v_y, v_z)$ is a probability density function in three dimensions. The value of $\rho(v_x, v_y, v_z)$ is the probability, per unit volume in velocity space, that a molecule has the velocity (v_x, v_y, v_z). For any velocity, (v_x, v_y, v_z), there is a value of $\rho(v_x, v_y, v_z)$; this value is just a number. If we want the probability of finding a velocity within some small volume of velocity space around (v_x, v_y, v_z), we can find it by multiplying $\rho(v_x, v_y, v_z)$ by this volume.

From the one-dimensional probability-density functions, the probability that the x-component of a molecular velocity lies between v_x and $v_x + dv_x$, is just $(df_x(v_x)/dv_x)dv_x$, whatever the values of v_y and v_z. The probability that the y-component lies between v_y and $v_y + dv_y$, is just $(df_y(v_y)/dv_y)dv_y$, whatever the values of v_x and v_z. The probability that the z-component lies between v_z and $v_z + dv_z$, is just $(df_z(v_z)/dv_z)dv_z$, whatever the values of v_x and v_y. When we interpret Maxwell's assumption to mean that these are independent probabilities, the probability that all three conditions are realized simultaneously is

$$\begin{aligned} dP(\upsilon) &= \left(\frac{df_x(v_x)}{dv_x}\right)\left(\frac{df_y(v_y)}{dv_y}\right)\left(\frac{df_z(v_z)}{dv_z}\right)dv_x dv_y dv_z \\ &= \rho(v_x, v_y, v_z)dv_x dv_y dv_z \end{aligned}$$

Evidently, the product of these three one-dimensional probability densities is the three-dimensional probability density. We have

$$\begin{aligned} \rho(v_x, v_y, v_z) &= \left(\frac{df_x(v_x)}{dv_x}\right)\left(\frac{df_y(v_y)}{dv_y}\right)\left(\frac{df_z(v_z)}{dv_z}\right) \\ &= \rho_x(v_x)\rho_y(v_y)\rho_z(v_z) \end{aligned}$$

From Maxwell's assumption, we have derived the conclusion that $\rho(v_x, v_y, v_z)$ can be expressed as a product of the one-dimensional probability densities $(df(v_x)/dv_x)dv_x$, $(df(v_y)/dv_y)dv_y$, and $(df(v_z)/dv_z)dv_z$. Since these are probability densities, we have

$$\begin{aligned} \int_{-\infty}^{\infty}\left(\frac{df_x(v_x)}{dv_x}\right)dv_x &= \int_{-\infty}^{\infty}\left(\frac{df_y(v_y)}{dv_y}\right)dv_y \\ &= \int_{-\infty}^{\infty}\left(\frac{df_z(v_z)}{dv_z}\right)dv_z \\ &= 1 \end{aligned}$$

and

$$\iiint_{-\infty}^{\infty} \rho(v_x, v_y, v_z)dv_x dv_y\, dv_z = 1$$

Moreover, because the Cartesian coordinates differ from one another only in orientation, $(df(v_x)/dv_x)dv_x$, $(df(v_y)/dv_y)dv_y$, and $(df(v_z)/dv_z)dv_z$ must all be the same function.

To summarize the development above, we define $\rho(v_x, v_y, v_z)$ independently of $df_x(v_x)/dv_x$, $df_y(v_y)/dv_y$, and $df_z(v_z)/dv_z$. Then, from Maxwell's

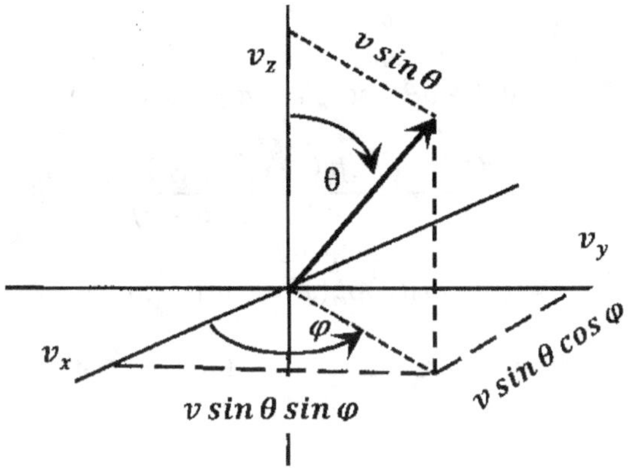

Figure 1. Transformation from Cartesian to spherical coordinates.

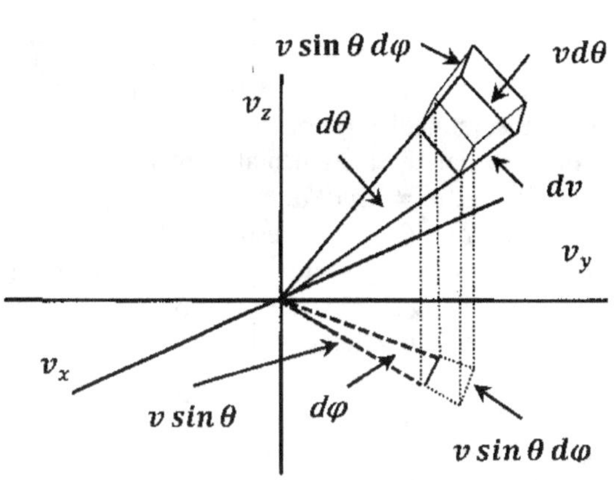

Figure 2. The differential volume element in spherical coordinates.

assumption that the three one-dimensional probabilities are independent, we find

$$\rho(v_x, v_y, v_z) = \left(\frac{df_x(v_x)}{dv_x}\right)\left(\frac{df_y(v_y)}{dv_y}\right)\left(\frac{df_z(v_z)}{dv_z}\right)$$
$$= \rho_x(v_x)\rho_y(v_y)\rho_z(v_z)$$

Alternatively, we could take Maxwell's assumption to be that the three-dimensional probability density function is expressible as a product of three one-dimensional probability densities:

$$\rho(v_x, v_y, v_z) = \rho_x(v_x)\rho_y(v_y)\rho_z(v_z)$$

In this case, the relationships of $\rho_x(v_x)$, $\rho_y(v_y)$, and $\rho_z(v_z)$, to the one-dimensional cumulative probabilities ($f_x(v_x)$, *etc.*) must be deduced from the properties of $\rho(v_x, v_y, v_z)$. As emphasized above, our deduction of $f_x(v_x)$ from experimental data uses v_x values that are associated with all possible values of v_y and v_z. That is, what we determine in our (hypothetical) experiment is

$$f_x(v_x) = \int_{v_x=-\infty}^{v_x} \iint_{v_{y,z}=-\infty}^{\infty} \rho(v_x, v_y, v_z)dv_x dv_y dv_z$$
$$= \int_{-\infty}^{v_x} \rho_x(v_x)dv_x \int_{-\infty}^{\infty} \rho_y(v_y)\,dv_y \int_{-\infty}^{\infty} \rho_z(v_z)\,dv_z$$
$$= \int_{-\infty}^{v_x} \rho_x(v_x)dv_x$$

from which it follows that

$$\frac{df_x(v_x)}{dv_x} = \rho_x(v_x)$$

§2 Probability density functions for velocity components in spherical coordinates

We introduce the idea of a three-dimensional probability-density function by showing how to find it from data referred to a Cartesian coordinate system. The probability density associated with a particular molecular velocity is just a number—a number that depends only on the velocity. Given a velocity, the probability density associated with that velocity must be independent of our choice of coordinate system. We can express the three-dimensional probability density using any coordinate system. We turn now to expressing velocities and probability density functions using spherical coordinates.

Just as we did for the Cartesian velocity components, we deduce the cumulative probability functions $f_v(v)$, $f_\theta(\theta)$, and $f_\varphi(\varphi)$ for the spherical-coordinate components. Our deduction of $f_v(v)$ from the experimental data uses v-values that are associated with all possible values of θ and φ. Corresponding statements apply to our deductions of $f_\theta(\theta)$, and $f_\varphi(\varphi)$. We also obtain their derivatives, the probability-density functions $df_v(v)/dv$, $df_\theta(\theta)/d\theta$, and $df_\varphi(\varphi)/d\varphi$. From the properties of probability-density functions, we have

$$\int_0^\infty \left(\frac{df_v(v)}{dv}\right)dv = \int_0^\pi \left(\frac{df_\theta(\theta)}{d\theta}\right)d\theta$$
$$= \int_0^{2\pi} \left(\frac{df_\varphi(\varphi)}{d\varphi}\right)d\varphi$$
$$= 1$$

Let υ' be the arbitrarily small increment of volume in velocity space in which the v-, θ-, and φ-components of velocity lie between v and $v + dv$, θ and $\theta + d\theta$, and φ and $\varphi + d\varphi$. Then the probability that the velocity of a randomly selected molecule lies within υ' is

$$dP(\upsilon') = \left(\frac{df_v(v)}{dv}\right)\left(\frac{df_\theta(\theta)}{d\theta}\right)\left(\frac{df_\varphi(\varphi)}{d\varphi}\right)dv d\theta d\varphi$$

Note that the product $\left(\frac{df_v(v)}{dv}\right)\left(\frac{df_\theta(\theta)}{d\theta}\right)\left(\frac{df_\varphi(\varphi)}{d\varphi}\right)$ is not a three-dimensional probability density function. This is most immediately appreciated by recognizing that $dvd\theta d\varphi$ is not an incremental "volume" in velocity space. That is, $\upsilon' \neq dvd\theta d\varphi$

We let $\rho(v,\theta,\varphi)$ be the probability-density function for the velocity vector in spherical coordinates. When v, θ, and φ specify the velocity, $\rho(v,\theta,\varphi)$ is the probability per unit volume at that velocity. We want to use $\rho(v,\theta,\varphi)$ to express the probability that an arbitrarily selected molecule has a velocity vector whose magnitude lies between v and $v + dv$, while its θ-component lies between θ and $\theta + d\theta$, and its φ-component lies between φ and $\varphi + d\varphi$. This is just $\rho(v,\theta,\varphi)$ times the velocity-space "volume" included by these ranges of v, θ, and φ.

When we change from Cartesian coordinates, $\vec{v} = (v_x, v_y, v_z)$, to spherical coordinates, $\vec{v} = (v,\theta,\varphi)$, the transformation is $v_x = v \sin\theta \cos\varphi$, $v_y = v \sin\theta \sin\varphi$, $v_z = v\cos\theta$. (See Figure 1.) As sketched in Figure 2, an incremental increase in each of the coordinates of the point specified by the vector (v,θ,φ) advances the vector to the point $(v + dv, \theta + d\theta, \varphi + d\varphi)$. When dv, $d\theta$, and $d\varphi$ are arbitrarily small, these two points specify the diagonally opposite corners of a rectangular parallelepiped, whose edges have the lengths dv, $vd\theta$, and $v \sin\theta\, d\varphi$. The volume of this parallelepiped is $v^2 \sin\theta\, dvd\theta d\varphi$. Hence, the differential volume element in Cartesian coordinates, $dv_x dv_y dv_z$, becomes $v^2 \sin\theta\, dvd\theta d\varphi$ in spherical coordinates.

Mathematically, this conversion is obtained using the absolute value of the ***Jacobian***, $J\left(\frac{v_x, v_y, v_z}{v,\theta,\varphi}\right)$, of the transformation. That is,

$$dv_x dv_y dv_z = \left|J\left(\frac{v_x, v_y, v_z}{v,\theta,\varphi}\right)\right| dvd\theta d\varphi$$

where the Jacobian is a determinate of partial derivatives

$$J\left(\frac{v_x, v_y, v_z}{v,\theta,\varphi}\right) = \begin{vmatrix} \partial v_x/\partial v & \partial v_x/\partial\theta & \partial v_x/\partial\varphi \\ \partial v_y/\partial v & \partial v_y/\partial\theta & \partial v_y/\partial\varphi \\ \partial v_z/\partial v & \partial v_z/\partial\theta & \partial v_z/\partial\varphi \end{vmatrix}$$
$$= v^2 \sin\theta$$

Since the differential unit of volume in spherical coordinates is $v^2 \sin\theta\, dvd\theta d\varphi$, ***the probability that the velocity components lie within the indicated ranges*** is

$$dP(\upsilon') = \rho(v,\theta,\varphi)v^2 \sin\theta\, dvd\theta d\varphi$$

We can develop the next step in Maxwell's argument by taking his assumption to mean that the three-dimensional probability density function is expressible as a product of three one-dimensional functions. That is, we take Maxwell's assumption to assert the existence of independent functions $\rho_v(v)$, $\rho_\theta(\theta)$, and $\rho_\varphi(\varphi)$ such that $\rho(v,\theta,\varphi) = \rho_v(v)\rho_\theta(\theta)\rho_\varphi(\varphi)$. The probability that the v-, θ-, and φ-components of velocity lie between v and $v + dv$, θ and $\theta + d\theta$, and φ and $\varphi + d\varphi$ becomes

$$dP(\upsilon') = \left(\frac{df_v(v)}{dv}\right)\left(\frac{df_\theta(\theta)}{d\theta}\right)\left(\frac{df_\varphi(\varphi)}{d\varphi}\right) dvd\theta d\varphi$$
$$= \rho(v,\theta,\varphi)v^2 \sin\theta dvd\theta d\varphi$$
$$= \rho_v(v)\rho_\theta(\theta)\rho_\varphi(\varphi)v^2 \sin\theta dvd\theta d\varphi$$

Since v, θ, and φ are independent, it follows that

$$\frac{df_v(v)}{dv} = v^2\rho_v(v)$$

$$\frac{df_\theta(\theta)}{d\theta} = \rho_\theta(\theta)\sin\theta$$

$$\frac{df_\varphi(\varphi)}{d\varphi} = \rho_\varphi(\varphi)$$

Moreover, the assumption that velocity is independent of direction means that $\rho_\theta(\theta)$ must actually be independent of θ; that is, $\rho_\theta(\theta)$ must be a constant. We let this constant be α_θ; so $\rho_\theta(\theta) = \alpha_\theta$. By the same argument, we set $\rho_\varphi(\varphi) = \alpha_\varphi$. Each of these probability-density functions must be normalized. This means that

$$1 = \int_0^\infty v^2\rho_v(v)\, dv$$

$$1 = \int_0^\pi \alpha_\theta \sin\theta\, d\theta = 2\alpha_\theta$$

$$1 = \int_0^{2\pi} \alpha_\varphi d\varphi = 2\pi\alpha_\varphi$$

from which we see that $\rho_\theta(\theta) = \alpha_\theta = 1/2$ and $\rho_\varphi(\varphi) = \alpha_\varphi = 1/2\pi$. It is important to recognize that, while $\rho_x(v_x)$, $\rho_y(v_y)$, and $\rho_z(v_z)$ are probability density functions, $\rho_\theta(\theta)$ and $\rho_v(v)$ are not. (However, $\rho_\varphi(\varphi)$ is a probability density function.) We can see this by noting that, if $\rho_\theta(\theta)$ were a probability density, its integral over all possible values of θ $(0 < \theta < \pi)$would be one. Instead, we find

$$\int_0^\pi \rho_\theta(\theta)d\theta = \int_0^\pi d\theta/2 = \pi/2$$

Similarly, when we find $\rho_v(v)$, we can show explicitly that

$$\int_0^\infty \rho_v(v)dv \neq 1$$

Our notation now allows us to express the probability that an arbitrarily selected molecule has a velocity vector whose magnitude lies between v and $v + dv$, while its θ-component lies between θ and $\theta + d\theta$, and its φ-component lies between φ and $\varphi + d\varphi$ using three equivalent representations of the probability density function:

$$\begin{aligned} dP(\mathrm{v}') &= \rho(v,\theta,\varphi)v^2 \sin\theta dv d\theta d\varphi \\ &= \rho_v(v)\rho_\theta(\theta)\rho_\varphi(\varphi)v^2 \sin\theta dv d\theta d\varphi \\ &= \left(\frac{1}{4\pi}\right)\rho_v(v)v^2 \sin\theta\, dv d\theta d\varphi \end{aligned}$$

The three-dimensional probability-density function in spherical coordinates is

$$\begin{aligned} \rho(v,\theta,\varphi) &= \rho_v(v)\rho_\theta(\theta)\rho_\varphi(\varphi) \\ &= \frac{\rho_v(v)}{4\pi} \end{aligned}$$

This shows explicitly that $\rho(v,\theta,\varphi)$ is independent of θ and φ; if the speed is independent of direction, the probability density function that describes velocity must be independent of the coordinates, θ and φ, that specify its direction.

§3 Maxwell's derivation of the gas-velocity probability-density function

To this point, we have been developing our ability to characterize the gas-velocity distribution functions. We now want to use Maxwell's argument to find them. We have already introduced the first step, which is the recognition that three-dimensional probability-density functions can be expressed as products of independent one-dimensional functions, and that $\rho_\theta(\theta)$, and $\rho_\varphi(\varphi)$ are the constants $1/2$ and $1/2\pi$. Now, because the probability density associated with any given velocity is just a number that is independent of the coordinate system, we can equate the three-dimensional probability-density functions for Cartesian and spherical coordinates: $\rho(v_x,v_y,v_z) = \rho(v,\theta,\varphi)$ so that

$$\rho_x(v_x)\rho_y(v_y)\rho_z(v_z) = \frac{\rho_v(v)}{4\pi}$$

We take the partial derivative of this last equation with respect to v_x. The probability densities $\rho_y(v_y)$ and $\rho_z(v_z)$ are independent of v_x. However, v is a function of v_x, because $v^2 = v_x^2 + v_y^2 + v_z^2$. We find

$$\begin{aligned} \frac{d\rho_x(v_x)}{dv_x}\rho_y(v_y)\rho_z(v_z) &= \frac{1}{4\pi}\left(\frac{\partial\rho_v(v)}{\partial v_x}\right)_{v_y v_v} \\ &= \frac{1}{4\pi}\left(\frac{d\rho_v(v)}{dv}\right)\left(\frac{\partial v}{\partial v_x}\right)_{v_y v_z} \end{aligned}$$

Since $v^2 = v_x^2 + v_y^2 + v_z^2$, $2v(\partial v/\partial v_x)_{v_y v_z} = 2v_x$ and

$$\left(\frac{\partial v}{\partial v_x}\right)_{v_y v_z} = \frac{v_x}{v}$$

Making this substitution and dividing by the original equation gives

$$\frac{d\rho_x(v_x)}{dv_x}\frac{\rho_y(v_y)\rho_z(v_z)}{\rho_x(v_x)\rho_y(v_y)\rho_z(v_z)} = \frac{v_x}{v}\frac{1}{\rho_v(v)}\frac{d\rho_v(v)}{dv}$$

Cancellation and rearrangement of the result leads to an equation in which the independent variables v_x and v are separated. This means that each term must be equal to a constant, which we take to be $-\lambda$. We find

$$\begin{aligned} \left(\frac{1}{v_x\rho_x(v_x)}\right)\frac{d\rho_x(v_x)}{dv_x} &= \left(\frac{1}{v\rho_v(v)}\right)\frac{d\rho_v(v)}{dv} \\ &= -\lambda \end{aligned}$$

so that

$$\frac{d\rho_x(v_x)}{\rho_x(v_x)} = -\lambda v_x dv_x$$

and

$$\frac{d\rho_v(v)}{\rho_v(v)} = -\lambda v dv$$

From the first of these equations, we obtain the probability density function for the distributions of one-dimensional velocities. (See §4.) The three-dimensional probability density function can be deduced from the one-dimensional function. (See §5.)

From the second equation, we obtain the three-dimensional probability-density function directly. Integrating from $v = 0$, where $\rho_v(0)$ has a fixed value, to an arbitrary scalar velocity, v, where the scalar-velocity function is $\rho_v(v)$, we have

$$\int_{\rho_v(0)}^{\rho_v(v)} \frac{d\rho_v(v)}{\rho_v(v)} = -\lambda\int_0^v v dv$$

or

$$\rho_v(v) = \rho_v(0)exp\left(\frac{-\lambda v^2}{2}\right)$$

The probability-density function for the scalar velocity becomes

$$\begin{aligned} \frac{df_v(v)}{dv} &= v^2\rho_v(v) \\ &= \rho_v(0)v^2 exp\left(\frac{-\lambda v^2}{2}\right) \end{aligned}$$

This is the result we want, except that it contains the unknown parameters $\rho_v(0)$ and λ. The value of $\rho_v(0)$

must be such as to make the integral over all velocities equal to unity. We require

$$1 = \int_0^\infty \left(\frac{df_v(v)}{dv}\right) dv$$
$$= \rho_v(0) \int_0^\infty v^2 \exp\left(\frac{-\lambda v^2}{2}\right) dv$$
$$= \frac{\rho_v(0)}{4\pi}\left(\frac{2\pi}{\lambda}\right)^{3/2}$$

so that

$$\rho_v(0) = 4\pi\left(\frac{\lambda}{2\pi}\right)^{3/2}$$

where we use the definite integral $\int_0^\infty x^2 \exp(-ax^2)dx = (1/4)\sqrt{\pi/a^3}$. (See Appendix D.) The scalar-velocity function in the three-dimensional probability-density function becomes

$$\rho_v(v) = 4\pi\left(\frac{\lambda}{2\pi}\right)^{3/2} \exp\left(\frac{-\lambda v^2}{2}\right)$$

The probability-density function for the scalar velocity becomes

$$\frac{df_v(v)}{dv} = v^2 \rho_v(v)$$
$$= 4\pi\left(\frac{\lambda}{2\pi}\right)^{3/2} v^2 \exp\left(\frac{-\lambda v^2}{2}\right)$$

The three-dimensional probability density in spherical coordinates becomes

$$\rho(v,\theta,\varphi) = \rho_v(v)\rho_\theta(\theta)\rho_\varphi(\varphi)$$
$$= \left(\frac{\lambda}{2\pi}\right)^{3/2} \exp\left(\frac{-\lambda v^2}{2}\right)$$

The probability that an arbitrarily selected molecule has a velocity vector whose magnitude lies between v and $v + dv$, while its θ-component lies between θ and $\theta + d\theta$, and its φ-component lies between φ and $\varphi + d\varphi$ becomes

$$dP(\mathrm{v}') = \left(\frac{df_v(v)}{dv}\right)\left(\frac{df_\theta(\theta)}{d\theta}\right)\left(\frac{df_\varphi(\varphi)}{d\varphi}\right) dv d\theta d\varphi$$
$$= \rho(v,\theta,\varphi) v^2 \sin\theta dv d\theta d\varphi$$
$$= \left(\frac{1}{4\pi}\right)\rho_v(v) v^2 \sin\theta dv d\theta d\varphi$$
$$= \left(\frac{\lambda}{2\pi}\right)^{3/2} v^2 exp\left(\frac{-\lambda v^2}{2}\right) \sin\theta dv d\theta d\varphi$$

In §6, we again derive Boyle's law and use the ideal gas equation to show that $\lambda = m/kT$.

§4 The probability density function for gas velocities in one dimension

In §3, we find a differential equation in the function $\rho_x(v_x)$. Unlike the velocity, which takes values from zero to infinity, the x-component, v_x, takes values from minus infinity to plus infinity. The probability density at an infinite velocity, in either direction, is necessarily zero. Therefore, we cannot evaluate the integral of $d\rho_x(v_x)/\rho_x(v_x)$ from $v_x = -\infty$ to an arbitrary velocity, v_x. However, we know from Maxwell's assumption that the probability density for v_x must be independent of whether the molecule is traveling in the direction of the positive x-axis or the negative x-axis. That is, $\rho_x(v_x)$ must be an even function; the probability density function must be symmetric around $v_x = 0$; $\rho_x(v_x) = \rho_x(-v_x)$. Hence, we can express $\rho_x(v_x)$ relative to its fixed value, $\rho_x(0)$, at $v_x = 0$. We integrate $d\rho_x(v_x)/\rho_x(v_x)$ from $\rho_x(0)$ to $\rho_x(v_x)$ as v_x goes from zero to an arbitrary velocity, v_x, to find

$$\int_{\rho_x(0)}^{\rho_x(v_x)} \frac{d\rho_x(v_x)}{\rho_x(v_x)} = -\lambda \int_0^{v_x} v_x dv_x$$

or

$$\rho_x(v_x) = \frac{df_x(v_x)}{dv_x}$$
$$= \rho_x(0)\exp\left(\frac{-\lambda v_x^2}{2}\right)$$

The value of $\rho_x(0)$ must be such as to make the integral of $\rho_x(v_x)$ over all possible values of v_x, $-\infty < v_x < \infty$, equal to unity. That is, we must have

$$1 = \int_{-\infty}^{\infty} \rho_x(v_x)\, dv_x$$
$$= \int_{-\infty}^{\infty} \frac{df_x(v_x)}{dv_x} dv_x$$
$$= \rho_x(0) \int_{-\infty}^{\infty} \exp\left(\frac{-\lambda v_x^2}{2}\right) dv_x$$
$$= \rho_x(0)\sqrt{\frac{2\pi}{\lambda}}$$

where we use the definite integral $\int_{-\infty}^{\infty} \exp(-ax^2)\, dx = \sqrt{\pi/a}$. (See Appendix D.) It follows that $\rho_x(0) = (\lambda/2\pi)^{1/2}$. The one-dimensional probability-density function becomes

$$\rho_x(v_x) = \frac{df_x(v_x)}{dv_x}$$
$$= \left(\frac{\lambda}{2\pi}\right)^{1/2} \exp\left(\frac{-\lambda v_x^2}{2}\right)$$

Note that this is the normal distribution with $\mu = 0$ and $\sigma^2 = \lambda^{-1}$. So λ^{-1} is the variance of the normal one-dimensional probability-density function. As noted above, in §6 we find that $\lambda = m/kT$.

§5 Combining the one-dimensional probability density functions

In §4, we derive the probability density function for one Cartesian component of the velocity of a gas molecule. The probability density functions for the other two Cartesian components are the same function. For $\vec{v} = (v_x, v_y, v_z)$, we have $v^2 = v_x^2 + v_y^2 + v_z^2$, and

$$\frac{df_x(v_x)}{dv_x} = \left(\frac{\lambda}{2\pi}\right)^{1/2} \exp\left(\frac{-\lambda v_x^2}{2}\right)$$
$$\frac{df_y(v_y)}{dv_y} = \left(\frac{\lambda}{2\pi}\right)^{1/2} \exp\left(\frac{-\lambda v_y^2}{2}\right)$$
$$\frac{df_z(v_z)}{dv_z} = \left(\frac{\lambda}{2\pi}\right)^{1/2} \exp\left(\frac{-\lambda v_z^2}{2}\right)$$

We now want to derive the three-dimensional probability density function from these relationships. Given these probability density functions for the Cartesian components of $\vec{v}$, we can find the probability density function in spherical coordinates

$$\left(\frac{df_x(v_x)}{dv_x}\right)\left(\frac{df_y(v_y)}{dv_y}\right)\left(\frac{df_z(v_z)}{dv_z}\right)$$
$$= \left(\frac{\lambda}{2\pi}\right)^{3/2} \exp\left(\frac{-\lambda v_x^2}{2}\right) exp\left(\frac{-\lambda v_y^2}{2}\right) exp\left(\frac{-\lambda v_z^2}{2}\right)$$
$$= \left(\frac{\lambda}{2\pi}\right)^{3/2} \exp\left(\frac{-\lambda v^2}{2}\right)$$
$$= \rho(v, \theta, \varphi)$$

Since the differential volume element in spherical coordinates is $v^2 \sin\theta\, dv d\theta d\varphi$, the probability that a molecule has a a velocity vector whose magnitude lies between v and $v + dv$, while its θ-component lies between θ and $\theta + d\theta$, and its φ-component lies between φ and $\varphi + d\varphi$ becomes

$$\left(\frac{df_v(v)}{dv}\right)\left(\frac{df_\theta(\theta)}{d\theta}\right)\left(\frac{df_\varphi(\varphi)}{d\varphi}\right) dv d\theta d\varphi$$
$$= \rho(v, \theta, \varphi) v^2 \sin\theta\, dv d\theta d\varphi$$
$$= \left(\frac{\lambda}{2\pi}\right)^{3/2} v^2 \exp\left(\frac{-\lambda v^2}{2}\right) \sin\theta\, dv d\theta d\varphi$$

(We found the same result in §3, of course.) We can find the probability-density function for the scalar velocity by eliminating the dependence on the angular components. To do this, we need only sum up, at a given value of v, the contributions from all possible values of θ and φ, recalling that $0 \le \theta < \pi$ and $0 \le \varphi < 2\pi$. This sum is just

$$\frac{df_v(v)}{dv} \int_{\theta=0}^{\pi} \left(\frac{df_\theta(\theta)}{d\theta}\right) d\theta \int_{\varphi=0}^{2\pi} \left(\frac{df_\varphi(\varphi)}{d\varphi}\right) d\varphi =$$
$$= \left(\frac{\lambda}{2\pi}\right)^{3/2} v^2 exp\left(\frac{-\lambda v^2}{2}\right) \int_{\theta=0}^{\pi} \sin\theta\, d\theta \int_{\varphi=0}^{2\pi} d\varphi$$

Since $\int_{\theta=0}^{\pi} \left(\frac{df_\theta(\theta)}{d\theta}\right) d\theta = \int_{\varphi=0}^{2\pi} \left(\frac{df_\varphi(\varphi)}{d\varphi}\right) d\varphi = 1$, $\int_0^\pi \sin\theta\, d\theta = 2$, and $\int_0^{2\pi} d\varphi = 2\pi$, we again obtain the Maxwell-Boltzmann probability-density function for the scalar velocity:

$$\frac{df_v(v)}{dv} = 4\pi \left(\frac{\lambda}{2\pi}\right)^{3/2} v^2 exp\left(\frac{-\lambda v^2}{2}\right)$$

Unlike the distribution function for the Cartesian components of velocity, the Maxwell-Boltzmann distribution for scalar velocities is not a normal distribution. Possible speeds lie in the interval $0 \le v < \infty$. Because of the v^2 term, the Maxwell-Boltzmann equation is asymmetric; it has a pronounced tail at high velocities.

§6 Boyle's law from the Maxwell-Boltzmann probability density

In Chapter 2, we derive Boyle's law using simplifying assumptions. We are now able to do this derivation much more rigorously. We consider the collisions of gas molecules with a small portion of the wall of their container. We suppose that the wall is smooth, so that we can select a small and compact segment of it that is arbitrarily close to being planar. We denote both the segment of the wall and its area as A. A can have any shape so long as it is a smooth, flat surface enclosed by a smooth curve.

Let the volume of the container be V and the number of gas molecules in the container be N. We imagine that we follow the trajectory of one particular molecule as it moves to hit the wall somewhere within A. We begin our observations at time $t = 0$ and suppose that the collision occurs at time t.

As sketched in Figure 3, we erect a Cartesian coordinate system with its origin at the location in space of the molecule at time $t = 0$. We orient the axes of this coordinate system so that the xy-plane is parallel to the plane of A, and the z-axis is pointed toward the wall. Then the unit vector along the z-axis and a vector perpendicular to A are parallel to one another. It is convenient to express the velocity of the selected molecule in spherical coordinates. We suppose that, referred to the Cartesian coordinate system we have erected, the velocity vector of the selected molecule is $(v^*, \theta^*, \varphi^*)$. The vector $\overrightarrow{v^*}t$, drawn from the origin of our Cartesian system to the point of impact on the wall, follows the trajectory of the molecule from time zero to time t. The z-component of the molecular velocity vector is normal to the plane of A at the point of impact; the magnitude of the z-component

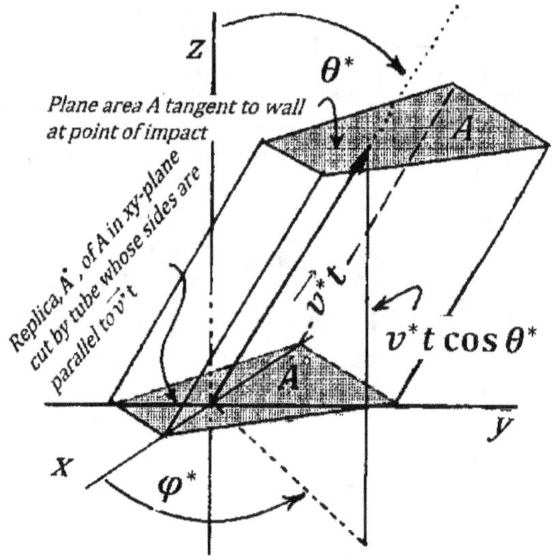

Figure 3. Trajectory of a molecule colliding with a wall of its container.

is $v^* \cos\theta^*$. The perpendicular distance from the plane of A to the xy-plane of the Cartesian system is $v^* t \cos\theta^*$.

We assume that the collision is perfectly elastic. Before collision, the velocity component perpendicular to the wall is $v_z = v^* \cos\theta^*$. Afterward, it is $v_z = -v^* \cos\theta^*$. Only this change in the v_z component contributes to the force on the wall within A. (The v_x and v_y components are not changed by the collision.) During the collision, the molecule's momentum change is $-2mv^* \cos\theta^*$. During our period of observation, the average force on the molecule is thus $(-2mv^* \cos\theta^*)/t$. The force that the molecule exerts on the wall is $(2mv^* \cos\theta^*)/t$, and hence the contribution that this particular collision—by one molecule traveling at velocity v^*— makes to the pressure on the wall is

$$P_1(v^*) = \frac{2mv^* \cos\theta^*}{At}$$

We want to find the pressure on segment A of the wall that results from all possible impacts. To do so, we recognize that any other molecule whose velocity components are v^*, θ^*, and φ^*, and whose location at time $t = 0$ enables it to reach A within time t, makes the same contribution to the pressure as the selected molecule does. Let us begin by assuming that the velocities of all N of the molecules in the volume, V, are the same as that of the selected molecule. In this case, we can find the number of the molecules in the container that can reach A within time t by considering a tubular segment of the interior of the container. The long axis of this tube is parallel to the velocity vector of the selected molecule. The sides of this tube cut the container wall along the perimeter of A. This tube also cuts the xy-plane (the $z = 0$ plane) of our coordinate system in such a way as to make an exact replica of A in this plane. Call this replica A^o.

The area of A^o is A; the plane of A^o is parallel to the plane of A; and the perpendicular distance between the plane of A and the plane of A^o is $v^* t \cos\theta^*$. The volume of this tube is therefore $Av^* t \cos\theta^*$. Since there are N/V molecules per unit volume, the total number of molecules in the tube is $(ANv^* t \cos\theta^*)/V$. When we assume that every molecule has velocity components v^*, θ^*, and φ^*, all of the molecules in the tube reach A within time t, because each of them travels parallel to the selected molecule, and each of them is initially at least as close to A as is the selected molecule. Therefore, each molecule in the tube contributes $P_1(v^*) = 2mv^* \cos\theta^*/At$ to the pressure at A. The total pressure is the pressure per molecule multiplied by the number of molecules:

$$\left(\frac{2mv^* \cos\theta^*}{At}\right)\left(\frac{ANv^* t \cos\theta^*}{V}\right) = \frac{2mN(v^* \cos\theta^*)^2}{V}$$

However, the molecular velocities are not all the same, and the pressure contribution $2mN(v^* \cos\theta^*)^2/V$ is made only by that fraction of the molecules whose velocity components lie in the intervals $\theta^* < \theta < \theta^* + d\theta$ and $\varphi^* < \varphi < \varphi^* + d\varphi$. This fraction is

$$\rho(v^*, \theta^*, \varphi^*)(v^*)^2 \sin\theta^* dv d\theta d\varphi =$$
$$= \left(\frac{\lambda}{2\pi}\right)^{3/2} (v^*)^2 \exp\left(\frac{-\lambda(v^*)^2}{2}\right) \sin\theta^* dv d\theta d\varphi$$

so that the pressure contribution from molecules whose velocity components lie in these ranges is

$$dP = \frac{2mN(v^* \cos\theta^*)^2}{V} \times$$
$$\left(\frac{\lambda}{2\pi}\right)^{3/2} (v^*)^2 \exp\left(\frac{-\lambda(v^*)^2}{2}\right) \sin\theta^* dv d\theta d\varphi$$

The total pressure at A is just the sum of the contributions from molecules with all possible combinations of velocities v^*, θ^*, and φ^*. To find this sum, we integrate over all possible velocity vectors. The allowed values of v are $0 \le v < \infty$. There are no constraints on the values of φ; we have $0 \le \varphi < 2\pi$. However, since all of the impacting molecules must have a velocity component in the positive z-direction, the possible values of θ lie in the interval $0 \le \theta < \pi/2$. We designate the velocity of the original molecule as $(v^*, \theta^*, \varphi^*)$ and retain this notation to be as specific as possible in describing the tube bounded by A and A^o. However, the velocity components of an arbitrary molecule can have any of the allowed values. To integrate (See Appendix D) over the allowed values, we drop the superscripts. The pressure at A becomes

$$P = \frac{2mN}{V}\left(\frac{\lambda}{2\pi}\right)^{3/2} \times$$
$$\int_0^\infty v^4 exp\left(\frac{-\lambda v^2}{2}\right) dv \int_0^{\pi/2} \cos^2\theta \sin\theta \, d\theta \int_0^{2\pi} d\varphi$$
$$= \frac{2mN}{V}\left(\frac{\lambda}{2\pi}\right)^{3/2} \left[\frac{3}{8}\left(\frac{2}{\lambda}\right)^2 \left(\frac{2\pi}{\lambda}\right)^{1/2}\right] \left[\frac{1}{3}\right] [2\pi]$$
$$= mN/V\lambda$$

and the pressure–volume product becomes

$$PV = \frac{mN}{\lambda}$$

Since m, N, and λ are constants, this is Boyle's law. Equating this pressure–volume product to that given by the ideal gas equation, we have $mN/\lambda = NkT$ so that

$$\lambda = \frac{m}{kT}$$

Finally, the ***Maxwell-Boltzmann equation*** becomes

$$\frac{df_v(v)}{dv} = 4\pi\left(\frac{m}{2\pi kT}\right)^{3/2} v^2 \exp\left(\frac{-mv^2}{2kT}\right)$$

and the probability density becomes

$$\rho(v,\theta,\varphi) = \left(\frac{m}{2\pi kT}\right)^{3/2} v^2 \exp\left(\frac{-mv^2}{2kT}\right)$$

This derivation can be recast as a computation of the expected value of the pressure. To do so, we rephrase our description of the system: A molecule whose velocity components are (v^*,θ^*,φ^*) creates a pressure $2mv^*\cos\theta^*/At$ on the area A with a probability of $Av^*t\cos\theta^*/V$. (The latter term is the probability that a molecule, whose velocity is (v^*,θ^*,φ^*), is, at time $t = 0$, in a location from which it can reach A within time t. If the molecule is to hit the wall within time t, at time $t = 0$ the molecule must be within the tubular segment of volume is $Av^*t\cos\theta^*$. The probability that the molecule is within this tubular segment is equal to the fraction of the total volume that this segment occupies.) Therefore, the product

$$\left(\frac{2mv^*\cos\theta^*}{At}\right)\left(\frac{Av^*t\cos\theta^*}{V}\right) = \frac{2m}{V}(v^*\cos\theta^*)^2$$

is the pressure contribution of a molecule with velocity (v^*,θ^*,φ^*), when θ^* is in the interval $0 \le \theta^* < \pi/2$. The total pressure per molecule is the expected value of this pressure contribution; the expected value is the integral, over the entire volume of velocity space, of the pressure contribution times the probability density function for velocities.

It is useful to view the Maxwell-Boltzmann equation as the product of a term

$$\exp(-mv^2/2kT)$$

—called the ***Boltzmann factor***—and a pre-exponential term that is proportional to the ***number of ways*** that a molecule can have a given velocity, v. If there were no constraints on a molecule's speed, we would expect that the number of molecules with speeds between v and $v + dv$ would increase as v increases, because the probability that a molecule has a speed between v and $v + dv$ is proportional to the volume in velocity space of a spherical shell of thickness dv. The volume of a spherical shell of thickness dv is $4\pi v^2 dv$, which increases as the square of v. However, the number of molecules with large values of v is constrained by the conservation of energy. Since the total energy of a collection of molecules is limited, only a small proportion of the molecules can have very large velocities. The Boltzmann factor introduces this constraint. A molecule whose mass is m and whose scalar velocity is v has kinetic energy $\epsilon = mv^2/2$. The Boltzmann factor is often written as $\exp(-\epsilon/kT)$.

§7 Experimental tests of the Maxwell-Boltzmann probability distribution

There are numerous applications of the Maxwell-Boltzmann equation. These include predictions of collision frequencies, mean-free paths, effusion and diffusion rates, the thermal conductivity of gases, and gas viscosities. These applications are important, but none of them is a direct test of the validity of the Maxwell-Boltzmann equation.

The validity of the equation has been demonstrated directly in experiments in which a gas of metal atoms is produced in an oven at a very high temperature. As sketched in Figure 4, the gas is allowed to escape into a vacuum chamber through a very small hole in the side of the oven. The escaping atoms impinge on one or more metal plates. Narrow slits cut in these plates stop any metal atoms whose flight paths do not pass though the slits. This produces a beam of metal atoms whose velocity distribution is the same as that of the metal-atom gas inside the oven. The rate at which metal atoms arrive at a detector is measured. Various methods are used to translate the atom-arrival rate into a measurement of their speed.

One device uses a solid cylindrical drum, which rotates on its cylindrical axis. As sketched in Figure 5, a spiral groove is cut into the cylindrical face of this drum. This groove is cut with a constant pitch. When the drum rotates at a constant rate, an atom traveling at a constant velocity parallel to the cylindrical axis can traverse the length of the drum while remaining within the groove. That is, for a given rotation rate, there is one critical velocity at which an atom can travel in a straight line while remaining in the middle of the groove all the way from one end of the drum to the other. If the atom moves significantly faster or slower than this critical velocity, it

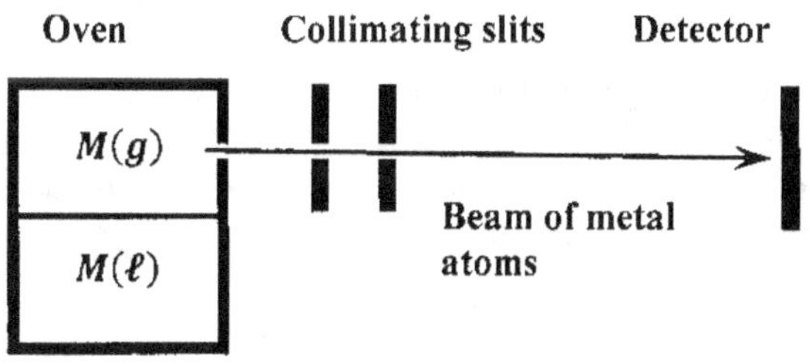

Figure 4. Producing a beam of metal atoms.

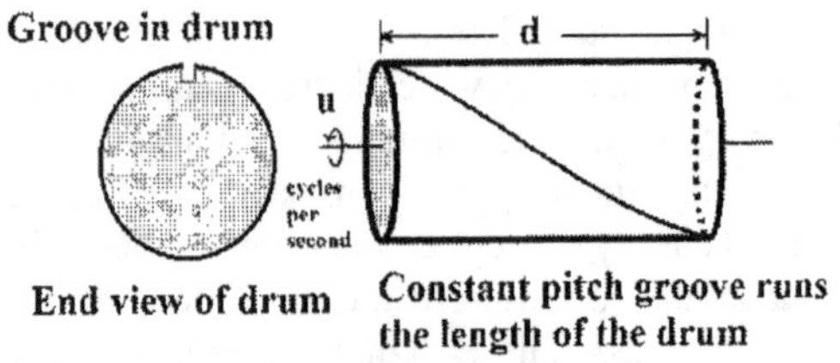

Figure 5. Device to select metal atoms having a specified velocity.

collides with—and sticks to—one side or the other of the groove.

Since the groove has a finite width, atoms whose velocities lie in a narrow range about the critical velocity can traverse the groove without hitting one of the sides. Let us assume that the groove is cut so that the spiral travels half way around the cylinder. That is, if we project the spiral onto one of the circular faces of the drum, the projection traverses an angle of 180° on the face. In order to remain in the middle of this groove all the way from one end of the drum to the other, the atom must travel the length of the cylindrical drum in exactly the same time that it takes the drum to make a half-rotation. Let the critical velocity be $v_{critical}$. Then the time required for the atom to traverse the length, *d*, of the drum is $d/v_{critial}$. If the drum rotates at u cycles/sec, the time required for the drum to make one-half rotation is $1/2u$. Thus, the atom will remain in the middle of the groove all the way through the drum if

$$v_{critical} = 2ud$$

By varying the rotation rate, we can vary the critical velocity.

Because the groove has a finite width, atoms whose velocities are in a range $v_{min} < v < v_{max}$ can successfully traverse the groove. Whether or not a particular atom can do so depends on its velocity, where it enters the groove, and the width of the groove. Let the width of the groove be w and the radius of the drum be r, where the drum is constructed with $r \gg w$. A slower atom that enters the groove at the earliest possible time—when the leading edge of the groove first encounters the beam of atoms—can traverse the length of the groove in a longer time, t_{max}. A point on the circumference of the drum travels with speed $2\pi ru$. The slowest atom traverses the length of the drum while a point on the circumference of the drum travels a distance $\pi r + w$. (To intercept the slowest atom, the trailing edge of the groove must travel a distance equal to half the circumference of the drum, πr, plus the width of the groove, w .) The time required for this rotation is the maximum time a particle can take to traverse the length, so

$$t_{max} = (\pi r + w)/(2\pi ru)$$

and

$$v_{min} = d/t_{max} = 2\pi rud/(\pi r + w)$$

A fast atom that enters the groove at the last possible moment—when the trailing edge of the grove just leaves the beam of atoms—can still traverse the groove if it does so in the time, t_{min} that it takes the trailing edge of the groove to travel a distance $\pi r - w$. So,

$$t_{min} = (\pi r - w)/(2\pi ru)$$

and

$$v_{max} = d/t_{min} = 2\pi rud/(\pi r - w)$$

At a given rotation rate, the drum will pass atoms whose speeds are in the range

$$\begin{aligned}\Delta v &= v_{max} - v_{min} \\ &= 2ud\left(\frac{\pi r}{\pi r - w} - \frac{\pi r}{\pi r + w}\right) \\ &= 2ud\left(\frac{2\pi r w}{(\pi r)^2 - w^2}\right) \\ &\approx v_{critical}\left(\frac{2w}{\pi r}\right)\end{aligned}$$

So that

$$\frac{\Delta v}{v_{critical}} \approx \frac{2w}{\pi r}$$

The fraction of the incident atoms that successfully travers the groove is equal to the fraction that have velocities in the interval Δv centered on the critical velocity, $v_{critical} = 2ud$.

§8 Statistics for molecular speeds

Expected values for several quantities can be calculated from the Maxwell-Boltzmann probability density function. The required definite integrals are tabulated in Appendix D.

The ***most probable speed***, v_{mp}, is the speed at which the Maxwell-Boltzmann equation takes on its maximum value. At this speed, we have

$$\begin{aligned}0 &= \frac{d}{dv}\left(\frac{df(v)}{dv}\right) \\ &= \frac{d}{dv}\left[4\pi\left(\frac{m}{2\pi kT}\right)^{3/2} v^2 exp\left(\frac{-mv^2}{2kT}\right)\right] \\ &= \left[4\pi\left(\frac{m}{2\pi kT}\right)^{3/2} exp\left(\frac{-mv^2}{2kT}\right)\right]\left[2v - \frac{mv^3}{kT}\right]\end{aligned}$$

from which

$$v_{mp} = \sqrt{\frac{2kT}{m}} \approx 1.414\sqrt{\frac{kT}{m}}$$

The ***average speed***, $\overline{v}$ or $\langle v\rangle$, is the expected value of the scalar velocity ($g(v) = v$). We find

$$\overline{v} = \langle v\rangle = \int_0^\infty 4\pi\left(\frac{m}{2\pi kT}\right)^{3/2} v^3 exp\left(\frac{-mv^2}{2kT}\right) dv$$

$$= \sqrt{\frac{8kT}{\pi m}} \approx 1.596\sqrt{\frac{kT}{m}}$$

The ***mean-square speed***, $\overline{v^2}$ or $\langle v^2 \rangle$, is the expected value of the velocity squared $(g(v) = v^2)$:

$$\overline{v^2} = \langle v^2 \rangle = \int_0^\infty 4\pi \left(\frac{m}{2\pi kT}\right)^{3/2} v^4 exp\left(\frac{-mv^2}{2kT}\right) dv = \frac{3kT}{m}$$

and the ***root mean-square speed***, v_{rms}, is

$$v_{rms} = \sqrt{\langle v^2 \rangle} = \sqrt{\frac{3kT}{m}} \approx 1.732\sqrt{\frac{kT}{m}}$$

Figure 6 shows the velocity distribution for nitrogen molecules at 300 K.

Finally, let us find the variance of the velocity; that is, the expected value of $(v - \langle v \rangle)^2$:

$$\text{variance}(v) = \sigma_v^2 = \int_0^\infty (v - \langle v \rangle)^2 \left(\frac{df(v)}{dv}\right) dv = \int_0^\infty v^2 \left(\frac{df}{dv}\right) dv - 2\langle v \rangle \int_0^\infty v \left(\frac{df}{dv}\right) dv + \langle v \rangle^2 \int_0^\infty \left(\frac{df}{dv}\right) = \langle v^2 \rangle - 2\langle v \rangle\langle v \rangle + \langle v \rangle^2 = \langle v^2 \rangle - \langle v \rangle^2$$

For N_2 at 300 K, we calculate:

$v_{mp} = 422 \text{ m s}^{-1}$
$\langle v \rangle = \bar{v} = 476 \text{ m s}^{-1}$
$v_{rms} = 517 \text{ m s}^{-1}$
variance $(v) = \sigma_v^2 = 40.23 \times 10^{-3} \text{ m s}^{-1}$
$\sigma_v = 201 \text{ m s}^{-1}$

§9 Pressure variations for macroscopic samples

At 300 K, the standard deviation of N_2 speeds is about 40% of the average speed. Clearly the relative variation among molecular speeds in a sample of ordinary gas is very large. Why do we not observe macroscopic effects from this variation? In particular, if we measure the pressure at a small area of the container wall, why do we not observe pressure variations that reflect the wide variety of speeds with which molecules strike the wall?

Qualitatively, the answer is obvious. A single molecule whose scalar velocity is v contributes $P_1(v) = mv^2/3V$ to the pressure on the walls of its container.

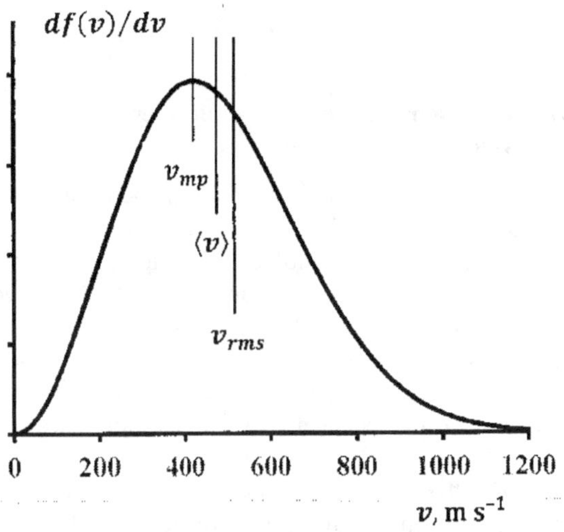

Figure 6. The Maxwell-Boltzmann distribution function for N_2 at 300 K.

(See problem 20.) When we measure pressure, we measure an average squared velocity. Even if we measure the pressure over a very small area and a very short time, the number of molecules striking the wall during the time of the measurement is very large. Consequently, the average speed of the molecules hitting the wall during any one such measurement is very close to the average speed in any other such measurement.

We are now able to treat this question quantitatively. For N_2 gas at 300 K and 1 bar, roughly 3×10^{15} molecules collide with a square millimeter of wall every microsecond. (See problem 12.) The standard deviation of the velocity of an N_2 molecule is 201 m s^{-1}. Using the central limit theorem, the standard deviation of the average of 3×10^{15} molecular speeds is

$$\frac{201 \text{ m s}^{-1}}{\sqrt{3 \times 10^{15}}} \approx 4 \times 10^{-6} \text{ m s}^{-1}$$

The distribution of the average of 3×10^{15} molecular speeds is very narrow indeed.

Similarly, when molecular velocities follow the Maxwell-Boltzmann distribution function, we can show that the expected value of the pressure for a single-molecule collision is $\langle P_1(v) \rangle = kT/V$. (See problem 21.) The variance of the distribution of these individual pressure measurements is $\sigma_{P_1(v)}^2 = 2k^2T^2/3V^2$, so that the magnitude of the standard deviation is comparable to that of the average:

$$\sigma_{P_1(v)}/\langle P_1(v) \rangle = \sqrt{2/3}$$

For the distribution of averages of 3×10^{15} pressure contributions, we find

$$P_{avg} = \langle P_1(v) \rangle = \sqrt{3/2}\,\sigma_{P_1(v)}$$

and

$$\sigma_{avg} = \sigma_{P_1(v)}/\sqrt{3 \times 10^{15}}$$

$$\frac{\sigma_{avg}}{P_{avg}} \approx 1.5 \times 10^{-8}$$

§10 Collisions between gas molecules: relative velocity coordinates

The pressure of a gas depends on the frequency with which molecules collide with the wall of their container. The rate at which gas molecules escape through a very small opening in their container is called the ***effusion rate***. The effusion rate also depends on the frequency of collisions with the wall. (See problem 12.) Other gas properties depend not on the rate of collision with the wall, but on the rate with which gas molecules collide with one another. We turn now to some of these properties. For these considerations, we need to describe the motion of one molecule relative to another. We need the probability density function for the ***relative velocity*** of two particles.

To describe the relative velocity of two particles, we introduce ***relative velocity coordinates***. Let us begin by considering a Cartesian coordinate frame, with x-, y-, and z-axes, whose origin is at a point O; we will use $Oxyz$ to designate this set of axes. We specify the location of particle 1 by the vector $\vec{r}_1 = (x_1, y_1, z_1) = x_1\vec{\imath} + y_1\vec{\jmath} + z_1\vec{k}$ and that of particle 2 by $\vec{r}_2 = (x_2, y_2, z_2)$. We let the location of the center of mass of this two-particle system be specified by $\vec{r}_0 = (x_0, y_0, z_0)$. The vector from particle 1 to particle 2, $\vec{r}_{12} = (x_{12}, y_{12}, z_{12})$, is the vector difference

$$\vec{r}_{12} = \vec{r}_2 - \vec{r}_1 = (x_2 - x_1, y_2 - y_1, z_2 - z_1)$$

When the particles are moving, these vectors and their components are functions of time. Using the notation $\dot{x}_1 = dx_1/dt$, we can specify the velocity of particle 1, for example, as $\vec{v}_1 = d\vec{r}_1/dt = (\dot{x}_1, \dot{y}_1, \dot{z}_1)$. Our goal is to find the relative velocity vector, $\vec{v}_{12} = d\vec{r}_{12}/dt$. We call the components of $\vec{v}_{12}$ the relative velocity coordinates.

The essential idea underlying relative velocity coordinates is that the vectors $\vec{r}_0$ and $\vec{r}_{12}$ contain the same information as the vectors $\vec{r}_1$ and $\vec{r}_2$. This is equivalent to saying that we can transform the locations as specified by (x_1, y_1, z_1) and (x_2, y_2, z_2) to the same locations as specified by (x_0, y_0, z_0) and (x_{12}, y_{12}, z_{12}), and *vice versa*. To accomplish this, we write the equation defining the x-component of the center of mass, x_0:

$$m_1(x_1 - x_0) + m_2(x_2 - x_0) = 0$$

which we rearrange to

$$\frac{x_1}{m_2} + \frac{x_2}{m_1} = \left(\frac{1}{m_1} + \frac{1}{m_2}\right) x_0$$

Corresponding relationships can be written for the y- and z-components. It proves to be useful to introduce the ***reduced mass***, μ, defined by

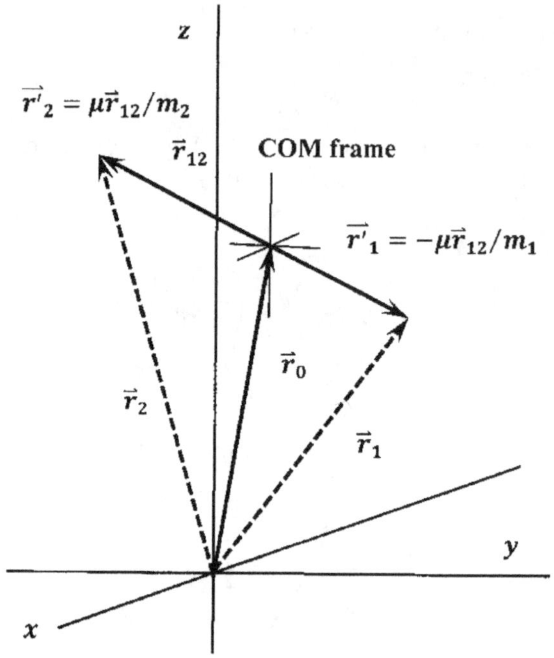

Figure 7. The center of mass frame.

$$\frac{1}{\mu} = \frac{1}{m_1} + \frac{1}{m_2}$$

Using the reduced mass, we can express the coordinates of the center of mass in terms of the coordinates of the individual particles. That is,

$$x_0 = \left(\frac{\mu}{m_2}\right) x_1 + \left(\frac{\mu}{m_1}\right) x_2$$
$$y_0 = \left(\frac{\mu}{m_2}\right) y_1 + \left(\frac{\mu}{m_1}\right) y_2$$
$$z_0 = \left(\frac{\mu}{m_2}\right) z_1 + \left(\frac{\mu}{m_1}\right) z_2$$

Since, by definition, we also have

$$x_{12} = x_2 - x_1$$
$$y_{12} = y_2 - y_1$$
$$z_{12} = z_2 - z_1$$

we have developed the transformation from (x_0, y_0, z_0) and (x_{12}, y_{12}, z_{12}) to (x_1, y_1, z_1) and (x_2, y_2, z_2). The inverse transformation is readily found to be

$$x_1 = x_0 - (\mu/m_1)x_{12}$$
$$y_1 = y_0 - (\mu/m_1)y_{12}$$
$$z_1 = z_0 - (\mu/m_1)z_{12}$$

$$x_2 = x_0 + (\mu/m_2)x_{12}$$
$$y_2 = y_0 + (\mu/m_2)y_{12}$$
$$z_2 = z_0 + (\mu/m_2)z_{12}$$

Now we can create two new Cartesian coordinate frames. Which of these is more useful depends on the

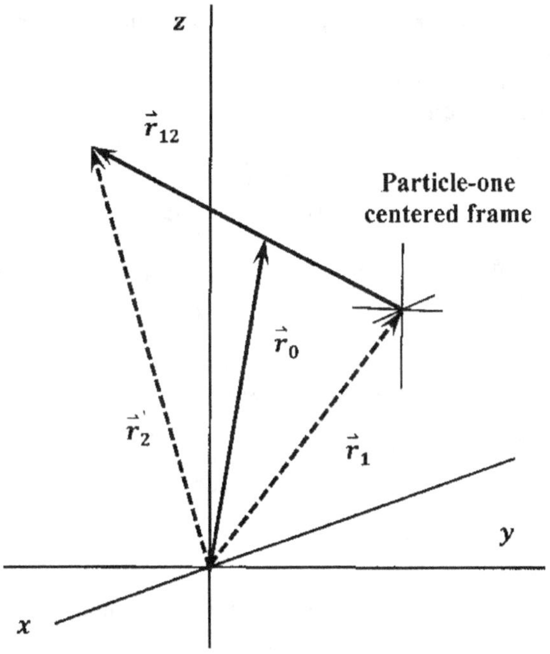

Figure 8. The particle-one centered frame.

objective of the particular analysis we have at hand. We call the first one the ***center of mass frame***, $O_0x'y'z'$. It is sketched in Figure 7. The x'-, y'-, and z'-axes of $O_0x'y'z'$ are parallel to the corresponding axes of $Oxyz$, but their origin, O_0, is always at the point occupied by the center of mass of the two-particle system. In this reference frame, the coordinates of particles 1 and 2 are their displacements from the center of mass:

$$x_1' = x_1 - x_0 = -(\mu/m_1)x_{12}$$
$$y_1' = y_1 - y_0 = -(\mu/m_1)y_{12}$$
$$z_1' = z_1 - z_0 = -(\mu/m_1)z_{12}$$

$$x_2' = x_2 - x_0 = (\mu/m_2)x_{12}$$
$$y_2' = y_2 - y_0 = (\mu/m_2)y_{12}$$
$$z_2' = z_2 - z_0 = (\mu/m_2)z_{12}$$

The center of mass frame is particularly useful for analyzing interactions between colliding particles.

For our purposes, a third Cartesian coordinate frame, which we will denote the $O_1x''y''z''$ frame, is more useful. It is sketched in Figure 8. The x''-, y''-, and z''-axes of $O_1x''y''z''$ are parallel to the corresponding axes of $Oxyz$, but their origin, O_1, is always at the point occupied by particle 1. In this reference frame, the coordinates of particles 1 and 2 are

$$x_1'' = 0$$
$$y_1'' = 0$$
$$z_1'' = 0$$

$$x_2'' = x_2 - x_1 = x_{12}$$
$$y_2'' = y_2 - y_1 = y_{12}$$
$$z_2'' = z_2 - z_1 = z_{12}$$

and the coordinates of the center of mass are

$$x_0'' = x_0 - x_1 = \mu x_{12}/m_1$$
$$y_0'' = y_0 - y_1 = \mu y_{12}/m_1$$
$$z_0'' = z_0 - z_1 = \mu z_{12}/m_1$$

The $O_1x''y''z''$ frame is sometimes called the center of mass frame also. To avoid confusion, we call $O_1x''y''z''$ the ***particle-one centered frame***. In the particle-one centered frame, particle 1 is stationary at the origin. With its tail at the origin, the vector $\vec{r}_{12} = (x_{12}, y_{12}, z_{12})$ specifies the position of particle 2.

We are interested in the relative velocity of particles 1 and 2. The velocity components for particles 1 and 2, and for their relative velocity, are obtained by finding the time-derivatives of the corresponding displacement components. Since the transformations of the displacement coordinates are linear, the velocity components transform from one reference frame to another in exactly the same way that the displacement components do. We have

$$\vec{v}_0 = d\vec{r}_0/dt = (\dot{x}_0, \dot{y}_0, \dot{z}_0)$$

and

$$\vec{v}_{12} = d\vec{r}_{12}/dt = (\dot{x}_{12}, \dot{y}_{12}, \dot{z}_{12})$$

The vector $\vec{v}_{12}$ specifies the velocity of particle 2, relative to a stationary particle 1. Just as $\vec{r}_0$ and $\vec{r}_{12}$ contain the same information as the vectors $\vec{r}_1$ and $\vec{r}_2$, the vectors $\vec{v}_0$ and $\vec{v}_{12}$ contain the same information as $\vec{v}_1$ and $\vec{v}_2$. Since a parallel displacement leaves a vector unchanged, each of these vectors is the same in any of the three reference frames. In §11, we find the probability density function for the magnitude of the scalar relative velocity, $v_{12} = |\vec{v}_{12}|$. Since the probability is independent of direction, the probability that two molecules have relative velocity $\vec{v}_{12}$ is the same as that they have relative velocity $-\vec{v}_{12}$. (In spherical coordinates, if $\vec{v}_{12} = (v_{12}, \theta, \varphi)$, then $-\vec{v}_{12} = (v_{12}, \theta + \pi, \varphi + \pi)$.) The probability and magnitude of the relative velocity are independent of which particle—if either—we choose to view as being stationary; they are independent of whether the particles are approaching or receding from one another.

§11 The probability density function for the relative velocity

From our development of the Maxwell-Boltzmann probability density functions, we can express the probability that the velocity components of particle 1 lie in the intervals v_{1x} to $v_{1x} + dv_{1x}$; v_{1y} to $v_{1y} + dv_{1y}$; v_{1z} to $v_{1z} + dv_{1z}$; while those of particle 2 simultaneously lie in the intervals v_{2x} to $v_{2x} + dv_{2x}$; v_{2y} to $v_{2y} + dv_{2y}$; v_{2z} to $v_{2z} + dv_{2z}$ as

$$\left(\frac{df(v_{1x})}{dv_{1x}}\right)\left(\frac{df(v_{1y})}{dv_{1y}}\right)\left(\frac{df(v_{1z})}{dv_{1z}}\right)\left(\frac{df(v_{2x})}{dv_{2x}}\right)\left(\frac{df(v_{2y})}{dv_{2y}}\right)\left(\frac{df(v_{2z})}{dv_{2z}}\right) \times dv_{1x}dv_{1y}dv_{1z}dv_{2x}dv_{2y}dv_{2z}$$

$$= \left(\frac{df(v_1)}{dv_1}\right)\left(\frac{df(v_2)}{dv_2}\right)dv_1dv_2$$

We want to express this probability using the relative velocity coordinates. Since the velocity of the center of mass and the relative velocity are independent, we might expect that the Jacobian of this transformation is just the product of the two individual Jacobians. This turns out to be the case. The Jacobian of the transformation

$$(\dot{x}_1, \dot{y}_1, \dot{z}_1, \dot{x}_2, \dot{y}_2, \dot{z}_2) \rightarrow (\dot{x}_0, \dot{y}_0, \dot{z}_0, \dot{x}_{12}, \dot{y}_{12}, \dot{z}_{12})$$

is a six-by-six determinate. It is messy, but straightforward, to show that it is equal to the product of two three-by-three determinants and that the absolute value of this product is one. Therefore, we have

$$dv_{1x}dv_{1y}dv_{1z}dv_{2x}dv_{2y}dv_{2z} = d\dot{x}_1 d\dot{y}_1 d\dot{z}_1 d\dot{x}_2 d\dot{y}_2 d\dot{z}_2 = d\dot{x}_0 d\dot{y}_0 d\dot{z}_0 d\dot{x}_{12} d\dot{y}_{12} d\dot{z}_{12}$$

We transform the probability density by substituting into the one-dimensional probability density functions. That is,

$$\left(\frac{df(v_1)}{dv_1}\right)\left(\frac{df(v_2)}{dv_2}\right) =$$

$$= \left(\frac{m_1}{2\pi kT}\right)^{3/2} \exp\left(\frac{-m_1\left(v_{1x}^2 + v_{1y}^2 + v_{1z}^2\right)}{2kT}\right) \times \left(\frac{m_2}{2\pi kT}\right)^{3/2} \exp\left(\frac{-m_2\left(v_{2x}^2 + v_{2y}^2 + v_{2z}^2\right)}{2kT}\right)$$

$$= \left(\frac{m_1 m_2}{4\pi^2 k^2 T^2}\right)^{3/2} \times \exp\left(\frac{-m_1\left(v_{1x}^2 + v_{1y}^2 + v_{1z}^2\right) - m_2\left(v_{2x}^2 + v_{2y}^2 + v_{2z}^2\right)}{2kT}\right)$$

$$= \left(\frac{m_1 m_2}{4\pi^2 k^2 T^2}\right)^{3/2} \times \exp\left(\frac{-\frac{m_1 m_2}{\mu}(\dot{x}_0^2 + \dot{y}_0^2 + z_0^2) - \mu(\dot{x}_{12}^2 + \dot{y}_{12}^2 + \dot{z}_{12}^2)}{2kT}\right)$$

where the last expression specifies the probability density as a function of the relative velocity coordinates.

Next, we make a further transformation of variables. We convert the velocity of the center of mass, $(\dot{x}_0, \dot{y}_0, \dot{z}_0)$, and the relative velocity, $(\dot{x}_{12}, \dot{y}_{12}, \dot{z}_{12})$, from Cartesian coordinates to spherical coordinates, referred to the $Oxyz$ axis system. (The motion of the center of mass is most readily visualized in the original frame $Oxyz$. The relative motion, $\vec{v}_{12}$, is most readily visualized in the Particle-One Centered Frame, $O_1x''y''z''$. In $O_1x''y''z''$, the motion of particle 2 is specified by $\dot{x}_2'' = \dot{x}_{12}$, $\dot{y}_2'' = \dot{y}_{12}$, and $\dot{z}_2'' = \dot{z}_{12}$. The motion of the center of mass is specified by $\dot{x}_0'' = \mu\dot{x}_{12}/m_1$, $\dot{y}_0'' = \mu\dot{y}_{12}/m_1$, and $\dot{z}_0'' = \mu\dot{z}_{12}/m_1$. Since it is the relative motion that is actually of interest, it might seem that we should refer the spherical coordinates to the $O_1x''y''z''$ frame. This is an unnecessary distinction because all three coordinate frames are parallel to one another, and $\vec{r}_0$ and $\vec{r}_{12}$ are the same vectors in all three frames.) Letting

$$v_0^2 = \dot{x}_0^2 + \dot{y}_0^2 + z_0^2$$
$$v_{12}^2 = \dot{x}_{12}^2 + \dot{y}_{12}^2 + z_{12}^2$$

the Cartesian velocity components are expressed in spherical coordinates by

$$\dot{x}_0 = v_0 \sin\theta_0 \cos\varphi_0$$
$$\dot{y}_0 = v_0 \sin\theta_0 \sin\varphi_0$$
$$\dot{z}_0 = v_0 \cos\theta_0$$

$$\dot{x}_{12} = v_{12} \sin\theta_{12} \cos\varphi_{12}$$
$$\dot{y}_{12} = v_{12} \sin\theta_{12} \sin\varphi_{12}$$
$$\dot{z}_{12} = v_{12} \cos\theta_{12}$$

The angles θ_0, θ_{12}, φ_0, and φ_{12} are defined in the usual manner relative to the $Oxyz$ axis system. The Jacobian of this transformation is a six-by-six determinate; which can again be converted to the product of two three-by-three determinates. We find

$$d\dot{x}_0 d\dot{y}_0 d\dot{z}_0 d\dot{x}_{12} d\dot{y}_{12} d\dot{z}_{12} = v_0^2 \sin\theta_0 \, dv_0 d\theta_0 d\varphi_0 v_{12}^2 \sin\theta_{12} \, dv_{12} d\theta_{12} d\varphi_{12}$$

The probability that the components of the velocity of the center of mass lie in the intervals v_0 to $v_0 + dv_0$; θ_0 to $\theta_0 + d\theta_0$; φ_0 to $\varphi_0 + d\varphi_0$; while the components of the relative velocity lie in the intervals v_{12} to $v_{12} + dv_{12}$; θ_{12} to $\theta_{12} + d\theta_{12}$; φ_{12} to $\varphi_{12} + d\varphi_{12}$; becomes

$$\left(\frac{m_1 m_2}{4\pi^2 k^2 T^2}\right)^{3/2} exp\left(\frac{-m_1 m_2 v_0^2}{2\mu kT}\right) exp\left(\frac{-\mu v_{12}^2}{2kT}\right) \times v_0^2 \sin\theta_0 \, dv_0 d\theta_0 d\varphi_0 v_{12}^2 \sin\theta_{12} \, dv_{12} d\theta_{12} d\varphi_{12}$$

We are interested in the probability increment for the relative velocity irrespective of the velocity of the center of mass. To sum the contributions for all possible motions of the center of mass, we integrate this expression over the possible ranges of v_0, θ_0, and φ_0. We have

$$\left(\frac{df(v_{12})}{dv_{12}}\right)\left(\frac{df(\theta_{12})}{d\theta_{12}}\right)\left(\frac{df(\varphi_{12})}{d\varphi_{12}}\right) dv_{12} d\theta_{12} d\varphi_{12} =$$

$$= \left(\frac{m_1 m_2}{4\pi^2 k^2 T^2}\right)^{3/2} \int_0^\infty v_0^2 exp\left(\frac{-m_1 m_2 v_0^2}{2\mu kT}\right) dv_0 \times \int_0^\pi \sin\theta_0 \, d\theta_0 \int_0^{2\pi} d\varphi_0 \times \left[v_{12}^2 exp\left(\frac{-\mu v_{12}^2}{2kT}\right) \sin\theta_{12} \, dv_{12} d\theta_{12} d\varphi_{12}\right]$$

$$= \left(\frac{\mu}{2\pi kT}\right)^{3/2} v_{12}^2 exp\left(\frac{-\mu v_{12}^2}{2kT}\right) \sin\theta_{12} \, dv_{12} d\theta_{12} d\varphi_{12}$$

This is the same as the probability increment for a single-particle velocity—albeit with μ replacing m; v_{12} replacing v; θ_{12} replacing θ; and φ_{12} replacing φ. As in the single-particle case, we can obtain the probability increment for the scalar component of the relative velocity by integrating over all possible values of θ_{12} and φ_{12}. We find

$$\frac{df(v_{12})}{dv_{12}} = 4\pi\left(\frac{\mu}{2\pi kT}\right)^{3/2} v_{12}^2 exp\left(\frac{-\mu v_{12}^2}{2kT}\right) dv_{12}$$

In §8, we find the most probable velocity, the mean velocity, and the root-mean-square velocity for a gas whose particles have mass m. By identical arguments, we obtain the most probable relative velocity, the mean relative velocity, and the root-mean-square relative velocity. To do so, we can simply substitute μ for m in the earlier results. In particular, the mean relative velocity is

$$\begin{aligned}\overline{v}_{12} &= \langle v_{12}\rangle \\ &= \left(\frac{8kT}{\pi\mu}\right)^{1/2} \\ &\approx 1.596\left(\frac{kT}{\pi\mu}\right)^{1/21}\end{aligned}$$

If particles 1 and 2 have the same mass, m, the reduced mass becomes $\mu = m/2$. In this case, we have

$$\begin{aligned}\langle v_{12}\rangle &= \left(\frac{2(8kT)}{\pi m}\right)^{1/2} \\ &= \sqrt{2}\langle v\rangle\end{aligned}$$

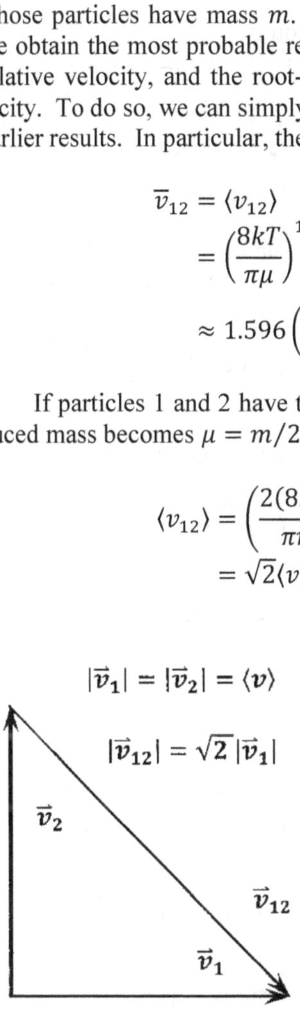

Figure 9. The average relative velocity.

We can arrive at this same conclusion by considering the relative motion of two particles that represents the average case. As illustrated in Figure 9, this occurs when the two particles have the same speed, $\langle v\rangle$, but are moving at 90-degree angles to one another. In this situation, the length of the resultant vector—the relative speed— is just $|\overline{v}_{12}| = \langle v_{12}\rangle = \sqrt{2}\langle v\rangle$.

§12 The frequency of collisions between unlike gas molecules

Thus far in our theoretical development of the properties of gases, we have assumed that ideal gas molecules are point masses. While they can collide with the walls of their container, point masses cannot collide with one another. As we saw in our discussion of van der Waals equation, the deviation of real gases from ideal gas behavior is one indication that an individual gas molecule occupies a finite volume.

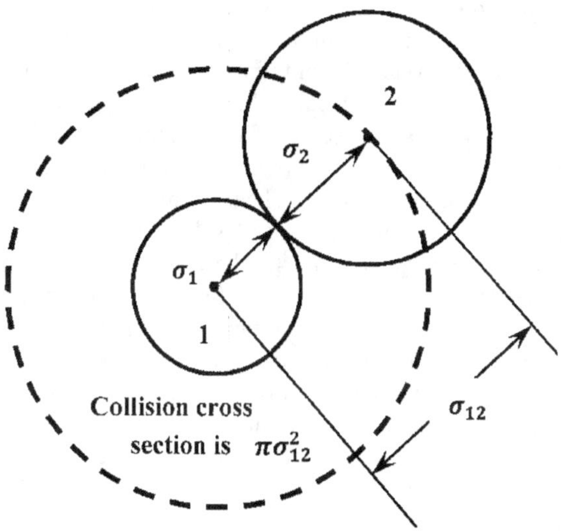

Figure 10. The molecular collision radius.

To develop a model for molecular collisions, we need to know the size and shape of the colliding molecules. For a general model, we want to use the simplest possible size and shape. Accordingly, we consider a model in which gas molecules are spheres with well-defined radii. We let the radii of molecules 1 and 2 be σ_1 and σ_2, respectively. See Figure 10. When such molecules collide, their surfaces must come into contact, and the distance between their centers must be $\sigma_{12} = \sigma_1 + \sigma_2$. We call σ_{12} the ***collision radius***.

Let us consider a molecule of type 1 in a container with a large number of molecules of type 2. We suppose that there are N_2 molecules of type 2 per unit volume. Every molecule of type 2 has some velocity, v_{12}, relative to the molecule of type 1. From our development above, we know both the probability density function for v_{12} and the expected value $\langle v_{12}\rangle$. Both molecule 1 and all of the molecules of type 2 are moving with continuously varying speeds. However, it is reasonable to suppose that—on average—the encounters between molecule 1 and molecules of type 2 are the same as they would be if all of the type 2 molecules were fixed at random locations in the volume, and molecule 1 moved among them with a

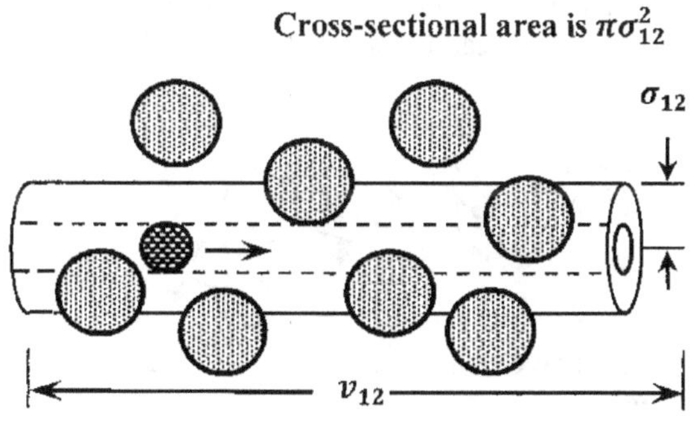

Figure 11. The collision volume of a gas molecule in unit time.

speed equal to the average relative velocity, $\langle v_{12} \rangle$.

Under this assumption, a molecule 1 travels a distance equal to $\langle v_{12} \rangle$ in unit time. As it does so, it collides with any type 2 molecule whose center is within a distance σ_{12} of its own center. For the moment, let us suppose that the trajectory of molecule 1 is unaffected by the collisions it experiences. Then, in unit time, molecule 1 sweeps out a cylinder whose length is $\langle v_{12} \rangle$ and whose cross-sectional area is $\pi\sigma_{12}^2$. The volume of this cylinder is $\pi\sigma_{12}^2\langle v_{12} \rangle$. (See Figure 11.) Since there are N_2 molecules of type 2 per unit volume, the number of type 2 molecules in the cylinder is $N_2\pi\sigma_{12}^2\langle v_{12} \rangle$. Each of these molecules is a molecule of type 2 that experiences a collision with molecule 1 in unit time. Letting $\tilde{\nu}_{12}$ be the frequency (number of collisions per unit time) with which molecule 1 collides with molecules of type 2, we have

$$\tilde{\nu}_{12} = N_2\pi\sigma_{12}^2\langle v_{12} \rangle = N_2\pi\sigma_{12}^2\left(\frac{8kT}{\pi\mu}\right)^{1/2}$$
$$= N_2\sigma_{12}^2\left(\frac{8\pi kT}{\mu}\right)^{1/2}$$

Two additional parameters that are useful for characterizing molecular collisions are τ_{12}, the ***mean time between collisions***, and λ_{12}, the ***mean distance*** that molecule 1 travels between collisions with successive molecules of type 2. λ_{12} is called the ***mean free path***. The mean time between collisions is simply the reciprocal of the collision frequency,

$$\tau_{12} = 1/\tilde{\nu}_{12}$$

and the mean free path for molecule 1 is the distance that molecule 1 actually travels in this time, which is $\langle v_1 \rangle$, not $\langle v_{12} \rangle$, so that

$$\lambda_{12} = \langle v_1 \rangle\tau_{12} = \frac{(\mu/m)^{1/2}}{N_2\pi\sigma_{12}^2}$$

Now, we need to reevaluate the assumption that the trajectory of a molecule 1 is unaffected by its collisions with molecules of type 2. Clearly, this is not the case. The path of molecule 1 changes abruptly at each collision. The actual cylinder that molecule 1 sweeps out will have numerous kinks, as indicated in Figure 12. The kinked cylinder can be produced from a straight one by making a series oblique cuts (one for each kink) across the straight cylinder and then rotating the ends of each cut into convergence. If we think of the cylinder as a solid rod, its volume is unchanged by these cuttings and rotations. The volume of the kinked cylinder is the same as that of the straight cylinder. Thus, our conclusions about the collision frequency, the mean time between collisions, and the mean free path are not affected by the fact that the trajectory of molecule 1 changes at each collision.

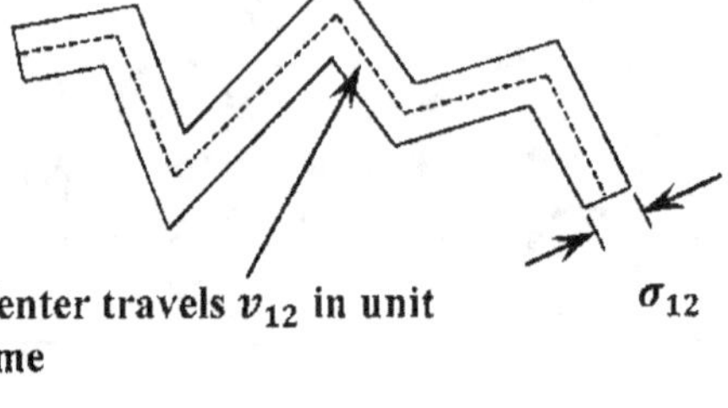

Figure 12. The collision volume is unaffected by collisions.

§13 The rate of collisions between unlike gas molecules

We define the collision frequency, $\tilde{\nu}_{12}$, as the number of collisions per unit time between a single molecule of type 1 and any of the molecules of type 2 present in the same container. We find $\tilde{\nu}_{12} = N_2\pi\sigma_{12}^2\langle v_{12} \rangle$. If there are N_1 molecules of type 1 present in a unit volume of the gas, the total number of collisions between type 1 molecules and type 2 molecules is N_1 times greater. For clarity, let us refer to the total number of such collisions, per unit volume and per unit time, as the ***collision rate***, ρ_{12}. We have

$$\rho_{12} = N_1\tilde{\nu}_{12} = N_1N_2\pi\sigma_{12}^2\langle v_{12} \rangle = N_1N_2\sigma_{12}^2\left(\frac{8\pi kT}{\mu}\right)^{1/2}$$

§14 Collisions between like gas molecules

When we consider collisions between different gas molecules of the same substance, we can denote the relative velocity and the expected value of the relative velocity as v_{11} and $\langle v_{11} \rangle$, respectively. By the argument we make above, we can find the number of collisions between any one of these molecules and all of the others. Letting this collision frequency be $\tilde{\nu}_{11}$, we find $\tilde{\nu}_{11} = N_1\pi\sigma_{11}^2\langle v_{11} \rangle$, where $\sigma_{11} = 2\sigma_1$. Since we have $\langle v_{11} \rangle = \sqrt{2}\langle v_1 \rangle$, while $\langle v_1 \rangle = \sqrt{8kT/\pi m_1}$, we have $\langle v_{11} \rangle = 4\sqrt{kT/\pi m_1}$. The frequency of collisions between molecules of the same substance becomes

$$\tilde{\nu}_{11} = N_1\pi\sigma_{11}^2\langle v_{11} \rangle = 4N_1\sigma_{11}^2\left(\frac{\pi kT}{m_1}\right)^{1/2}$$

The mean time between collisions, τ_{11}, is

$$\tau_{11} = 1/\tilde{\nu}_{11}$$

and the mean free path, λ_{11},

$$\lambda_{11} = \langle v_1 \rangle\tau_{11} = \frac{1}{\sqrt{2}N_1\pi\sigma_{11}^2}$$

When we consider the rate of collisions between all of the molecules of type 1 in a container, ρ_{11}, there is a minor complication. If we multiply the collision frequency per molecule, $\tilde{\nu}_{11}$, by the number of molecules available to undergo such collisions, N_1, we count each collision twice, because each such collision involves two type 1 molecules. To find the collision rate among like molecules, we must divide this product by 2. That is,

$$\begin{aligned}\rho_{11} &= \frac{N_1\tilde{v}_{11}}{2} \\ &= 2N_1^2\sigma_{11}^2\left(\frac{\pi kT}{m_1}\right)^{1/2}\end{aligned}$$

§15 The geometry of a collision between spherical molecules

Thus far we have not concerned ourselves with the relative orientation of a pair of colliding molecules. We want to develop a more detailed model for the collision process[1] itself, and the first step is to specify what we mean by relative orientation.

As before, we consider a molecule of type 1 moving with the relative velocity v_{12} through a gas of stationary type 2 molecules. In unit time, molecule 1 travels a distance v_{12} and collides with many molecules of type 2. We can characterize each such collision by the angle, θ, between the velocity vector and the line of centers of the colliding pair. For glancing collisions, we have $\theta = \pi/2$. For head-on collisions, we have $\theta = 0$. All else being equal, the collision will be more violent the smaller the angle θ. Evidently, we can describe the average effect of collisions more completely if we can specify the frequency of collisions as a function of θ. More precisely, we want to find the frequency of collisions in which this angle lies between θ and $\theta + d\theta$.

When a collision occurs, the distance between the molecular centers is σ_{12}. We can say that the center of molecule 2 is at a particular point on the surface of a sphere, of radius σ_{12}, circumscribed about molecule 1. As sketched in Figure 13, we can rotate the line of centers around the velocity vector, while keeping the angle between them constant at θ. As we do so, the line of centers traces out a circle on the surface of the sphere; collisions that put the center of molecule 2 at any two points on this circle are completely equivalent. Letting the radius of this circle be r, we see that $r = \sigma_{12}\sin\theta$. Evidently, for spherical molecules, specifying θ specifies the relative orientation at the time of collision.

If we now allow θ to vary by $d\theta$, the locus of equivalent points on the circumscribed sphere expands to a band. Measured along the surface of the sphere, the width of this band is $\sigma_{12}d\theta$. As molecule 1 moves through the gas of stationary type 2 molecules, this band sweeps out a cylindrical shell. Molecule 1 collides, at an angle between θ and $\theta + d\theta$, with every type 2 molecule in this cylindrical shell. Conversely, every type 2 molecule in this cylindrical shell collides with molecule 1 at an angle between θ and $\theta + d\theta$. (Molecule 1 also collides with many other type 2 molecules, but those collisions are at other angles; they have different orientations.) In unit time, the length of the cylindrical shell is v_{12}. The volume of the cylindrical shell is its length times its cross-sectional area.

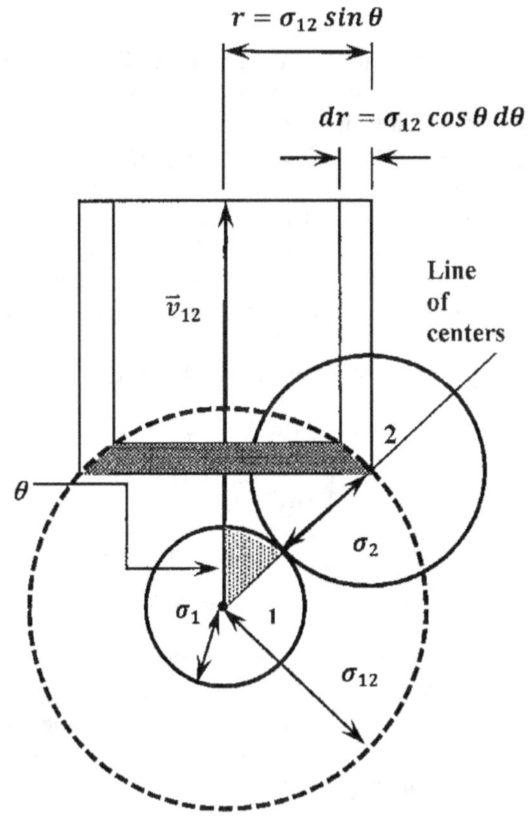

Figure 13. The geometry of collisions between spheres.

The cross-section of the cylindrical shell is a circular annulus. Viewing the annulus as a rectangular strip whose length is the circumference of the shell and whose width is the radial thickness of the annulus, the area of the annulus is the circumference times the radial thickness. Since the radius of the shell is $r = \sigma_{12}\sin\theta$, its circumference is $2\pi\sigma_{12}\sin\theta$. The radial thickness of the annulus is just the change in the distance, $r = \sigma_{12}\sin\theta$, between the velocity vector and the wall of the cylinder when θ changes by a small amount $d\theta$. This is

$$\begin{aligned}dr &= \left(\frac{dr}{d\theta}\right)d\theta \\ &= \sigma_{12}\cos\theta\, d\theta\end{aligned}$$

Therefore, the area of the annulus is

$$2\pi\sigma_{12}^2\sin\theta\cos\theta\, d\theta$$

and the volume of the cylindrical shell swept out by a type 1 molecule (traveling at exactly the speed v_{12}) in unit time is

$$2\pi\sigma_{12}^2 v_{12}\sin\theta\cos\theta\, d\theta$$

We again let N_2 be the number of molecules of type 2 per unit volume. The number of collisions, per unit time, between a molecule of type 1, traveling at exactly v_{12}, and molecules of type 2, in which the collision angle lies between θ and $\theta + d\theta$ is

$$2\pi N_2\sigma_{12}^2 v_{12} \sin\theta \cos\theta \, d\theta$$

We need to find the number of such collisions in which the relative velocity lies between v_{12} and $v_{12} + dv_{12}$. The probability of finding v_{12} in this interval is $(df(v_{12})/dv_{12})dv_{12}$. Let $d\tilde{v}_{12}(v_{12}, \theta)$ be the number of collisions made in unit time, by a type 1 molecule, with molecules of type 2, in which the collision angle is between θ and $\theta + d\theta$, and the scalar relative velocity is between v_{12} and $v_{12} + dv_{12}$. This is just the number of collisions when the relative velocity is v_{12} multiplied by the probability that the relative velocity is between v_{12} and $v_{12} + dv_{12}$. We have the result we need:

$$\begin{aligned} d\tilde{v}_{12}(v_{12}, \theta) &= \\ &= 2\pi N_2\sigma_{12}^2 v_{12} \left(\frac{df(v_{12})}{dv_{12}}\right) \sin\theta \cos\theta \, d\theta dv_{12} \\ &= 8\pi^2 N_2\sigma_{12}^2 \left(\frac{\mu}{2\pi kT}\right)^{3/2} v_{12}^3 exp\left(\frac{-\mu v_{12}^2}{2kT}\right) \\ &\qquad \times \sin\theta \cos\theta \, d\theta dv_{12} \end{aligned}$$

Recognizing that possible values of θ lie in the range $0 \le \theta < \pi/2$ and that possible values of v_{12} lie in the range $0 \le v_{12} < \infty$, we can find the frequency of all possible collisions, $\tilde{v}_{12}$, by summing over all possible values of θ and v_{12}. That is,

$$\begin{aligned} \tilde{v}_{12} &= 8\pi^2 N_2\sigma_{12}^2 \left(\frac{\mu}{2\pi kT}\right)^{3/2} \int_0^\infty v_{12}^3 exp\left(\frac{-\mu v_{12}^2}{2kT}\right) dv_{12} \\ &\qquad \times \int_0^{\pi/2} \sin\theta \cos\theta \, d\theta \\ &= 8\pi^2 N_2\sigma_{12}^2 \left(\frac{\mu}{2\pi kT}\right)^{3/2} \left[2\left(\frac{kT}{\mu}\right)^2\right]\left[\frac{1}{2}\right] \\ &= N_2\sigma_{12}^2 \left(\frac{8\pi kT}{\mu}\right)^{1/2} \end{aligned}$$

In §12, we obtained this result by a slightly different argument, in which we did not explicitly consider the collision angle, θ.

§16 The energy of a collision between gas molecules

It is useful to extend our model of molecular collisions to suppose that one or both of the molecules can undergo chemical change as a result of the collision. In doing so, we are introducing some ideas that we develop further in Chapter 5.

When we ask about the factors that determine whether such a reaction can occur, there can be several possibilities. We want to focus on one such factor—the violence of the collision. We expect that a collision is more likely to result in a reaction the harder the two molecules hit one another. When we try to formulate our basis for this expectation, we see that the underlying idea is that a collision deforms the colliding molecules. The more violent the collision, the greater the deformation, and the greater the likelihood of reaction becomes.

To proceed, we need to be more precise about what we mean by the violence of the collision. Evidently, what we have in mind has two components: the relative velocity and the collision angle. If the collision is a glancing one, $\theta = \pi/2$, we expect the effect on the molecules to be minimal, even if the relative velocity is high. On the other hand, a direct collision, $\theta \approx 0$, might lead to reaction even if the relative velocity is comparatively low. With these ideas in mind, we see that a reasonable model is to suppose that forces acting along the line of centers can lead to reaction, whereas forces acting perpendicular to the line of centers cannot. If the colliding molecules have complex shapes, this may be a poor assumption.

We also need a way to specify how much deformation occurs in a collision. If we want to specify the deformation by describing specific changes in the molecular structures, this is a complex problem. For a general model, however, we can avoid this level of detail. To do so, we recognize that any deformation can proceed only until the work done in deforming the molecules equals the energy that can be expended to do this work. As the molecules are deformed, their potential energies change. The maximum change in this potential energy is just the amount of kinetic energy that the colliding molecules can use to effect this deformation. We can identify this amount of kinetic energy with the component of the molecules' kinetic energy that is associated with their relative motion along the line of centers.

If we now associate a threshold level of deformation with the occurrence of a chemical change, the kinetic energy required to effect this deformation determines whether the change can occur. If the available kinetic energy is less than that required to achieve the threshold level of deformation, reaction cannot occur. If the available kinetic energy exceeds this minimum, reaction takes place. We call the minimum kinetic energy the ***activation energy*** and usually represent it by the symbol ϵ_a. (In discussing reaction rates, we usually express the activation energy per mole and represent it as E_a, where $E_a = \overline{N}\epsilon_a$.)

We can apply these ideas to our model for collisions between spherical molecules. In §10, we develop relative velocity coordinates. It follows that we can partition kinetic energy of the two-particle system into a component that depends on the velocity of the center of mass and a component that depends on the relative velocity. That is, we have

$$\begin{aligned} KE &= \frac{m_1 v_1^2}{2} + \frac{m_2 v_2^2}{2} \\ &= \frac{m_1 m_2}{2\mu}(\dot{x}_0^2 + \dot{y}_0^2 + z_0^2) + \mu(\dot{x}_{12}^2 + \dot{y}_{12}^2 + \dot{z}_{12}^2) \\ &= \frac{m_1 m_2 v_0^2}{2\mu} + \frac{\mu v_{12}^2}{2} \end{aligned}$$

Only the component that depends on the relative velocity can contribute to the deformation of the colliding molecules. The relative velocity can be resolved into components parallel and perpendicular to the line of centers. The parallel component is the projection of the velocity vector onto the line of centers. This is $v_{12}\cos\theta$, and the perpendicular component is $v_{12}\sin\theta$. We see that the kinetic energy associated with the relative motion of particles 1 and 2 has a component

$$\frac{\mu v_{12}^2 \cos^2\theta}{2}$$

parallel to the line of centers and a component

$$\frac{\mu v_{12}^2 \sin^2\theta}{2}$$

perpendicular to it.

The idea that the kinetic energy parallel to the line of centers must exceed ε_a for reaction to occur can now be expressed as the requirement that

$$\epsilon_a < \frac{\mu v_{12}^2 \cos^2\theta}{2}$$

When we consider all possible collisions between molecules 1 and 2, the collision angle varies from 0 to $\pi/2$. However, only those collisions for which v_{12} satisfies the inequality above will have sufficient kinetic energy along the line of centers for reaction to occur. The smallest value of v_{12} that can satisfy this inequality occurs when $\theta = 0$. This minimum relative velocity is

$$v_{12}^{mininum} = (2\epsilon_a/\mu)^{1/2}$$

For relative velocities in excess of this minimum, collisions are effective only when

$$\cos\theta > (2\epsilon_a/\mu v_{12}^2)^{1/2}$$

so that

$$\theta < \cos^{-1}(2\epsilon_a/\mu v_{12}^2)^{1/2}$$

Let us designate the frequency of collisions satisfying these constraints as $\tilde{v}_{12}(\epsilon_a)$. Recalling that

$$d\tilde{v}_{12}(v_{12},\theta) = = 8\pi^2 N_2 \sigma_{12}^2 \left(\frac{\mu}{2\pi kT}\right)^{3/2} v_{12}^3 \exp\left(\frac{-\mu v_{12}^2}{2kT}\right) \times \sin\theta\cos\theta\, d\theta dv_{12}$$

we see that

$$\tilde{v}_{12}(\epsilon_a) = 8\pi^2 N_2 \sigma_{12}^2 \left(\frac{\mu}{2\pi kT}\right)^{3/2} \times \int_{v_{12}=(2\epsilon_a/\mu)^{1/2}}^{\infty} \int_{\theta=0}^{\cos^{-1}(2\epsilon_a/\mu v_{12}^2)^{1/2}} v_{12}^3 \exp\left(\frac{-\mu v_{12}^2}{2kT}\right) \times \sin\theta\cos\theta\, d\theta dv_{12}$$

The integral involving θ is

$$\int_{\theta=0}^{\cos^{-1}(2\epsilon_a/\mu v_{12}^2)^{1/2}} \sin\theta\cos\theta\, d\theta = \left[\frac{\sin^2\theta}{2}\right]_0^{\cos^{-1}(2\epsilon_a/\mu v_{12}^2)^{1/2}} = \frac{1}{2}\left[1 - \frac{2\epsilon_a}{\mu v_{12}^2}\right]$$

where, to evaluate the integral at its upper limit, we note that the angle $\theta = \cos^{-1}(2\epsilon_a/\mu v_{12}^2)^{1/2}$ lies in a triangle whose sides have lengths as indicated in Figure 14.

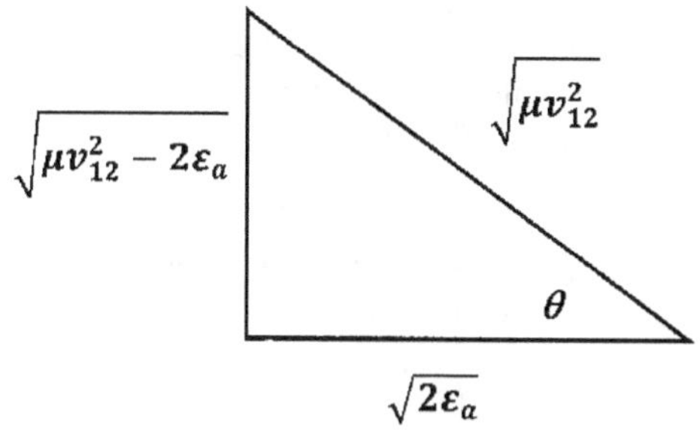

Figure 14. Maximum angle for an effective collision.

The collision frequency becomes

$$\tilde{v}_{12}(\epsilon_a) = 4\pi^2 N_2 \sigma_{12}^2 \left(\frac{\mu}{2\pi kT}\right)^{3/2} \times \int_{v_{12}=(2\epsilon_a/\mu)^{1/2}}^{\infty} \left[1 - \frac{2\epsilon_a}{\mu v_{12}^2}\right] v_{12}^3 \exp\left(\frac{-\mu v_{12}^2}{2kT}\right) dv_{12}$$

This integral can be evaluated by making the substitution $v_{12} = (2\epsilon/\mu)^{1/2}$. The lower limit of integration becomes ϵ_a; we have

$$\int_{v_{12}=(2\epsilon_a/\mu)^{1/2}}^{\infty} \left[1 - \frac{2\epsilon_a}{\mu v_{12}^2}\right] v_{12}^3 \exp\left(\frac{-\mu v_{12}^2}{2kT}\right) dv_{12} = = \frac{2}{\mu^2}\int_{\epsilon_a}^{\infty} (\epsilon - \epsilon_a)\exp\left(\frac{-\epsilon}{kT}\right) d\epsilon = 2\left(\frac{kT}{\mu}\right)^2 \exp\left(\frac{-\epsilon_a}{kT}\right)$$

Then

$$\tilde{v}_{12}(\epsilon_a) = 4\pi^2 N_2 \sigma_{12}^2 \left(\frac{\mu}{2\pi kT}\right)^{3/2} \times 2\left(\frac{kT}{\mu}\right)^2 \exp\left(\frac{-\epsilon_a}{kT}\right) = N_2 \sigma_{12}^2 \left(\frac{8\pi kT}{\mu}\right)^{1/2} \exp\left(\frac{-\epsilon_a}{kT}\right)$$

Note that when $\epsilon_a = 0$, this reduces to the same expression for $\tilde{\nu}_{12}$ that we have obtained twice previously. The frequency of collisions having kinetic energy along the line of centers in excess of ϵ_a depends exponentially on $-\epsilon_a/kT$. All else being equal, this frequency increases as the temperature increases; it decreases as the activation energy increases.

Problems

1. For an oxygen molecule at 25 C, calculate (a) the most probable velocity, (b) the average velocity, (c) the root-mean-square velocity.

2. For a gas of oxygen molecules at 25 C and 1.00 bar, calculate (a) the collision frequency, (b) the mean time between collisions, (c) the mean free path. The diameter of an oxygen molecule, as estimated from gas-viscosity measurements, is 3.55×10^{-10} m.

3. For oxygen molecules at 25 C, calculate (a) the fraction with speeds between 150 and 151 m s^{-1}, (b) the fraction with speeds between 400 and 401 m s^{-1}, (c) the fraction with speeds between 550 and 551 m s^{-1}.

4. For a hydrogen molecule at 100 C, calculate (a) the most probable velocity, (b) the average velocity, (c) the root-mean-square velocity.

5. For a gas of hydrogen molecules at 100 C and 1.00 bar, calculate (a) the collision frequency, (b) the mean time between collisions, (c) the mean free path. The diameter of a hydrogen molecule, as estimated from gas-viscosity measurements, is 2.71×10^{-10} m.

6. For a uranium hexafluoride (UF_6) molecule at 100 C, calculate (a) the most probable velocity, (b) the average velocity, (c) the root-mean-square velocity.

7. For a gas of uranium hexafluoride molecules at 100 C and 1.00 bar, calculate (a) the collision frequency, (b) the mean time between collisions, (c) the mean free path. Assume that the diameter of a uranium hexafluoride molecule is 7.0×10^{-10} m.

8. What is the average kinetic energy of hydrogen molecules at 100 C? What is the average kinetic energy of uranium hexafluoride (UF_6) molecules at 100 C?

9. Assuming the temperature in interstellar space is 2.73 K, calculate, for a hydrogen atom, (a) the most probable velocity, (b) the average velocity, (c) the root-mean-square velocity.

10. Assuming that interstellar space is occupied entirely by hydrogen atoms at a particle density of 10^2 molecules m^{-3}, calculate (a) the collision frequency, (b) the mean number of years between collisions, (c) the mean free path. Assume that the diameter of a hydrogen atom is 2.40×10^{-10} m.

11. Ignoring any effects attributable to its charge and assuming that the temperature is 2.73 K, calculate, for an electron in interstellar space, (a) the most probable velocity, (b) the average velocity, (c) the root-mean-square velocity.

12. If a wall of a gas-filled container contains a hole, gas molecules escape through the hole. If all of the molecules that hit the hole escape, but the hole is so small that the number escaping has no effect on the velocity distribution of the remaining gas molecules, we call the escaping process **effusion**. That is, we call the process effusion only if it satisfies three rather stringent criteria. First, the hole must be large enough (and the wall must be thin enough) so that most molecules passing through the hole do not hit the sides of the hole. Second, a molecule that passes through the hole must not collide with anything on the other side that can send it back through the hole into the original container. Third, the hole must be small enough so that the escaping molecules do not create a pressure gradient; the rate at which gas molecules hit the hole and escape must be determined entirely by the equilibrium distribution of gas velocities and, of course, the area of the hole. Show that the number of molecules effusing through a hole of area A in time t is

$$At\left(\frac{N}{V}\right)\left(\frac{kT}{2\pi m}\right)^{1/2}$$

where (N/V) is the number density of molecules in the container, and m is their molecular mass.

13. A vessel contains hydrogen and oxygen at 350 K and partial pressures of 0.50 bar and 1.50 bar, respectively. These gases effuse into a vacuum. What is the ratio of hydrogen to oxygen in the escaping gas?

14. How could we use effusion to estimate the molecular weight of an unknown substance?

15. An equimolar mixture of $^{235}UF_6$ and $^{238}UF_6$ is subjected to effusion. What is the ratio of ^{235}U to ^{238}U in the escaping gas?

16. Calculate the number of nitrogen molecules that collide with 10^{-6} m^2 of wall in 10^{-6} s, if the pressure is 1.00 bar and the temperature is 300 K.

17. Air is approximately 20% oxygen and 80% nitrogen by volume. Assume that oxygen and nitrogen molecules both have a radius of 1.8×10^{-8} m. For air at 1.0 bar and 298 K, calculate:
(a) The number of collisions that one oxygen molecule makes with nitrogen molecules every second.
(b) The number of collisions that occur between oxygen and nitrogen molecules in one cubic meter of air every second.
(c) The number of collisions that one oxygen molecule makes with other oxygen molecules every second.

(d) The number of collisions that occur between oxygen molecules in one cubic meter of air every second.

(e) The number of collisions that occur between oxygen and nitrogen molecules in one cubic meter each second in which the kinetic energy along the line of centers exceeds 100 kJ mol^{-1} or 1.66×10^{-19} J per collision.

(f) The number of oxygen-nitrogen collisions that occur in which the kinetic energy along the line of centers exceeds 50 kJ mol^{-1}.

18. Show that $\int_0^\infty \rho_v(v)\, dv \neq 1$.

19. For what volume element, υ, is
$$P(\upsilon) = f_{xyz}(v_x, v_y, v_z)?$$

20. Using the model we develop in §2-10:

(a) Show that the pressure, $P_1(v)$, attributable to a single molecule of mass m and velocity v in a container of volume V is

$$P_1(v) = \frac{mv^2}{3V}$$

(b) In §6, we find that this pressure is

$$\delta P_1(v) = \frac{2mv^2 \cos^2\theta}{V}$$

for a molecule whose velocity vector lies between θ and $\theta + d\theta$ and between φ and $\varphi + d\varphi$. This angular region comprises a solid angle whose magnitude is $d\Omega = \sin\theta d\theta d\varphi$. Since the solid angle surrounding a given point is 4π, the probability that a randomly oriented velocity vector lies between θ and $\theta + d\theta$ and between φ and $\varphi + d\varphi$ is

$$\frac{d\Omega}{4\pi} = \frac{\sin\theta d\theta d\varphi}{4\pi}$$

Therefore, given that the scalar component of a molecule's velocity is v, its contribution to the pressure at A is

$$dP_1(v) = \left(\frac{mv^2}{2\pi V}\right)\cos^2\theta \sin\theta\, d\theta d\varphi$$

To find the pressure contribution made by this molecule irrespective of the values of θ and φ, we must integrate $dP_1(v)$ over all values of θ and φ that allow the molecule to impact the wall at A. Recalling that these ranges are $0 \le \theta < \pi/2$ and $0 \le \varphi < 2\pi$, show that

$$P_1(v) = \frac{mv^2}{3V}$$

21. Taking $P_1(v) = mv^2/3V$ as the contribution made to the pressure by one molecule whose velocity is v:

(a) Show that the expected value for the contribution made to the pressure by one molecule when the Maxwell–Boltzmann distribution function describes the distribution of molecular velocities is

$$\langle P_1(v)\rangle = \frac{kT}{V}$$

(b) Show that the variance of the contribution made to the pressure by one molecule is

$$\sigma^2_{P_1(v)} = \frac{2k^2T^2}{3V^2}$$

What is the standard deviation, $\sigma_{P_1(v)}$?

(c) What is the value of the ratio

$$\frac{\sigma_{P_1(v)}}{\langle P_1(v)\rangle}$$

(d) Taking 3×10^{15} as the number of collisions of N_2 molecules at 1 bar and 300 K with one one square millimeter per microsecond, what pressure, P_{avg}, would we find if we could measure the individual contribution made by each collision and compute their average? What would be the variance, σ^2_{avg}, of this average? The standard deviation, σ_{avg}? The ratio σ_{avg}/P_{avg}?

22. Let $\epsilon = mv^2/2$ be the translational kinetic energy of a gas molecule whose mass is m. Show that the probability density function for ϵ is

$$\frac{df}{d\epsilon} = 2\pi\left(\frac{1}{\pi kT}\right)^{3/2} \epsilon^{1/2} exp\left(\frac{-\epsilon}{kT}\right)$$

Letting the translational kinetic energy per mole be $E = \overline{N}\epsilon$, show that

$$\frac{df}{dE} = 2\pi\left(\frac{1}{\pi RT}\right)^{3/2} E^{1/2} exp\left(\frac{-E}{RT}\right)$$

Notes

[1] Our collision model and quantitative treatment of the role of activation energy in chemical reaction rates follow those given by Arthur A. Frost and Ralph G. Pearson, ***Kinetics and Mechanism***, 2nd Ed., John Wiley and Sons, New York, 1961, pp 65-68. See also R. H. Fowler, ***Statistical Mechanics***, Cambridge University Press, New York, 1936.

5

Chemical Kinetics, Reaction Mechanisms, and Chemical Equilibrium

§1 Chemical kinetics

Chemical kinetics is the study of how fast chemical reactions occur and of the factors that affect these rates. The study of reaction rates is closely related to the study of ***reaction mechanisms***, where a reaction mechanism is a theory that explains how a reaction occurs.

A reaction mechanism describes the sequence of bond-making, bond-breaking, and intramolecular-rearrangement steps that results in the overall chemical change. These individual steps are called ***elementary reactions*** or ***elementary processes***. In an elementary reaction, no intermediate species is formed; that is, none of the arrangements of atoms that occur during the elementary reaction has a lifetime greater than the duration of a molecular vibration, which is typically from 10^{-12} to 10^{-14} seconds. We expand on this point in §6.

We can distinguish two levels of detail in a chemical reaction mechanism: The first is the series of elementary processes that occurs for a given net reaction. This is called the ***stoichiometric mechanism***. Frequently it is also possible to infer the relative positions of all of the atoms during the course of a reaction. This sort of model is called an ***intimate*** or ***detailed mechanism***.

A ***rate law*** is an equation that describes how the observed reaction rate depends on the concentrations of the species involved in the reaction. This concentration dependence can be determined experimentally. We will see that any series of elementary reactions predicts the dependence of reaction rates on concentrations, so one of the first tests of a proposed mechanism is that it be consistent with the rate law that is observed experimentally. (If the overall reaction proceeds in more than one step and the concentration of an intermediate species becomes significant, we may need more than one equation to adequately describe the rates of all of the reactions that occur.)

The rate law plays a central role in our study of reaction rates and mechanisms. We infer the rate law from experimental measurements. We must be able to prove that the experimental rate law is consistent with any mechanism that we propose. The rate law that we deduce from experimental rate data constitutes an experimental fact. Our hypothesized mechanism is a theory. We can entertain the idea that the theory may be valid only so long as its predictions about the rate law are consistent with the experimental result. We can predict rate laws for elementary processes by rather simple arguments. For a mechanism involving a series of elementary processes, we can often predict rate laws by making simplifying assumptions. When simplifying assumptions are inadequate, we can use numerical integration to test agreement between the proposed mechanism and experimental observations of the dependence of the reaction rate on the concentrations of the species involved in the reaction. We will see that a given experimental rate law may be consistent with any of several mechanisms. In such cases, we must develop additional information in order to discriminate among the several mechanisms.

§2 Reaction rates and rate laws

Chemical reactions occur under a wide variety of circumstances: Many chemicals are manufactured by passing a homogeneous mixture of gaseous reactants through a bed of solid catalyst pellets. Corrosion of metals is a reaction between the solid metal and oxygen from the air, often catalyzed by other common chemical species like water or chloride ion. An enormous number of biological reactions occur within the cells of living organisms. In the laboratory, we typically initiate a reaction by mixing the reactants, often with a solvent, in a temperature-controlled vessel.

Chemical reactions can be carried out in ***batch reactors*** and in a wide variety of ***flow reactors***. A batch reactor is simply a container, in which we initiate the reaction by mixing the reactants with one another (and any additional ingredients), and in which the reaction occurs and the products remain—until we get around to removing them. A reaction carried out under these conditions is called a ***batch reaction***. If any of the reactants are gases, a batch reactor must be sealed to prevent their escape. Otherwise, we may leave the reactor open to the atmosphere as the reaction occurs.

Flow reactors have been designed to achieve a variety of objectives. Nevertheless, they have a number of characteristics in common. A flow reactor is a container, into which reactants and other ingredients are injected. The products are recovered by withdrawing portions of the reaction mixture from one or more locations within the reactor. The rates at which materials are injected or withdrawn are usually constant. In the simplest case, the reactants are mixed at one end of a long tube. The reacting mixture flows through the tube. If the tube is long enough, the mixture emerging from the other end contains the equilibrium concentrations of reactants and products. In such a ***tubular reactor***, it is usually a good approximation to assume that the material injected during one short time interval does not mix with the material injected during the next time interval as they pass through the tube. We view the contents of the reactor as a series of fluid "plugs" that traverse the reactor independently of one another and call this behavior ***plug flow***.

In §11, we discuss another simple flow reactor, called a ***continuous stirred-tank reactor*** (CSTR) or a ***capacity-flow reactor***. A CSTR consists of a single constant-volume vessel into which reactants are continuously injected and from which reaction mixture is continuously withdrawn. The contents of this container are constantly stirred. In our discussion, we assume that the reactor is completely filled with a homogeneous liquid solution. We express the rate of reaction within the CSTR in moles per liter of reactor volume per second, $\text{mol L}^{-1}\text{s}^{-1}$.

When we talk about the rate of a particular reaction, we intend to specify the amount of chemical change that occurs in unit time because of that reaction. It is usually advantageous to specify the amount of chemical change in units of moles. We can specify the amount of chemical change by specifying the number of moles of a reactant that are consumed, or the number of moles of a product that are produced, per second, by that reaction. If we do so, the amount of chemical change depends on the stoichiometric coefficient of the reactant or product that we choose. Moreover, the rate is proportional to the size of the system. Since the properties of reaction rates that are of interest to us are usually independent of the size of the system, we find it convenient to express reaction rates as moles per second per unit system size, so that the most convenient units are usually concentration per second.

For reactors containing heterogeneous catalysts, we typically express the reaction rate in moles per unit volume of catalyst bed per second. For corrosion of a metal surface, we often express the rate in moles per unit area per second. For biological reactions, we might express the reaction rate in moles per gram of biological tissue per second. For reactions in liquid solutions, we typically express the rate in moles per liter of reaction mixture per second, $\text{mol L}^{-1}\text{s}^{-1}$ or $\underline{M}\,\text{s}^{-1}$.

Evidently, we need to express the rate of a reaction in a way that accounts for the stoichiometry of the reaction and is independent of the size of the system. Moreover, we must distinguish the effect of the reaction on the number of moles of a reagent present from the effects of other processes, because competing reactions and mechanical processes can affect the amount of a substance that is present.

To develop the basic idea underlying our definition of reaction rate, let us consider a chemical substance, A, that undergoes a single reaction in a closed system whose volume is V. For a gas-phase reaction, this volume can vary with time, so that $V = V(t)$. (The volume of any open system can vary with time.) Since the system is closed, the reaction is the is the only process that can change the amount of A that is present. Let Δn_A be the increase in the number of moles of A in a short interval, Δt, that includes time t. Let the average rate at which the number of moles of A increases in this interval be $\overline{r}(A) = \Delta n_A/\Delta t$. The corresponding instantaneous rate is

$$\begin{aligned} r(A) &= \lim_{\Delta t \to 0} \left(\frac{\Delta n_A}{\Delta t}\right) \\ &= \frac{dn_A}{dt} \end{aligned}$$

To express this information per unit volume, we can define the instantaneous rate of reaction of A, at time t, as

$$\begin{aligned} R(A) &= \frac{1}{V(t)} \lim_{\Delta t \to 0} \left(\frac{\Delta n_A}{\Delta t}\right) \\ &= \frac{1}{V(t)} \frac{dn_A}{dt} \end{aligned}$$

Experimental studies of reaction rates are typically done in constant-volume closed systems under conditions in which only one reaction occurs. Most of the discussion in this chapter is directed toward reactions in which these conditions are satisfied, and the system comprises a single homogeneous phase. In the typical case, we mix reactants with a liquid solvent in a reactor and immerse the reactor in a constant-temperature bath. Under these conditions, the rate at which a particular substance reacts is equal to the rate at which its concentration changes. Writing $[A]$ to designate the molarity, mol L^{-1}, of A, we have

$$\begin{aligned} R(A) &= \frac{1}{V} \frac{dn_A}{dt} \\ &= \frac{d(n_A/V)}{dt} \\ &= \frac{d[A]}{dt} \end{aligned}$$

If we express a reaction rate as a rate of concentration change, it is essential that both conditions be satisfied. If both n_A and V vary with time, we have

$$\begin{aligned} \frac{d[A]}{dt} &= \frac{d(n_A/V)}{dt} \\ &= \frac{1}{V}\frac{dn_A}{dt} - \frac{n_A}{V^2}\frac{dV}{dt} \end{aligned}$$

The instantaneous rate at which substance A undergoes a particular reaction is equal to $V^{-1}(dn_A/dt)$ only if the reaction is the sole process that changes n_A; the contribution to $d[A]/dt$ made by $(n_A V^{-2})\, dV/dt$ vanishes only if the volume is constant.

If a single reaction is the only process that occurs in a particular system, the rate at which the number of moles of any reactant or product changes is a measure of the rate of the reaction. However, these rates depend on the stoichiometric coefficients and the size of the system. For a reaction of specified stoichiometry, we can use the ***extent of reaction***, ξ, to define a unique reaction rate, R. The amounts of reactants and products present at any time are fixed by the initial conditions and the stoichiometry of the reaction. Let us write n_A to denote the number of moles of reagent A present at an arbitrary time and n_A^o to denote the number of moles of A present at the time $(t = 0)$ that the reaction is initiated. We define the extent of reaction as the change in the number of moles of a product divided by the product's stoichiometric coefficient or as the change in the number of moles of a reactant divided by the negative of the reactant's stoichiometric coefficient. For the stoichiometry

$$aA + bB + \cdots \to cC + dD + \cdots$$

we have

$$\xi = \frac{n_A - n_A^o}{-a} = \frac{n_B - n_B^o}{-b} = \cdots$$

$$= \frac{n_C - n_C^o}{c} = \frac{n_D - n_D^o}{d} = \cdots$$

If A is the limiting reagent, ξ varies from zero, when the reaction is initiated with $n_A = n_A^o$, to n_A^o/a, when $n_A = 0$. At any time, t, we have

$$\frac{d\xi}{dt} = -\frac{1}{a}\frac{dn_A}{dt} = -\frac{1}{b}\frac{dn_B}{dt} = \cdots$$

$$= \frac{1}{c}\frac{dn_C}{dt} = \frac{1}{d}\frac{dn_D}{dt} = \cdots$$

and we can define a unique ***reaction rate*** as

$$R = \frac{1}{V(t)}\frac{d\xi}{dt}$$

$$= -\frac{1}{aV(t)}\frac{dn_A}{dt} = -\frac{1}{bV(t)}\frac{dn_B}{dt} = \cdots$$

$$= \frac{1}{cV(t)}\frac{dn_C}{dt} = \frac{1}{dV(t)}\frac{dn_D}{dt} = \cdots$$

The relationship between the instantaneous rate at which reactant A undergoes this reaction, $R(A)$, and the reaction rate, R, is

$$R(A) = \frac{1}{V(t)}\frac{dn_A}{dt} = -\frac{a}{V(t)}\frac{d\xi}{dt} = -aR$$

If the volume is constant, we have

$$\frac{\xi}{V} = \frac{[A] - [A]_0}{-a} = \frac{[B] - [B]_0}{-b} = \cdots$$

$$= \frac{[C] - [C]_0}{c} = \frac{[D] - [D]_0}{d} = \cdots$$

and the reaction rate is

$$R = \frac{1}{V}\frac{d\xi}{dt} = -\frac{1}{a}\frac{d[A]}{dt} = -\frac{1}{b}\frac{d[B]}{dt} = \cdots$$

$$= \frac{1}{c}\frac{d[C]}{dt} = \frac{1}{d}\frac{d[D]}{dt} = \cdots$$

The name "extent of reaction" is sometimes given to the fraction of the stoichiometrically possible reaction that has occurred. To distinguish this meaning, we call it the ***fractional conversion***, χ. When A is the stoichiometrically limiting reactant, the fractional conversion is

$$\chi = -\frac{n_A - n_A^o}{n_A^o}$$

The extent of reaction, ξ, and the fractional conversion, χ, are related as

$$\chi = \frac{a\xi}{n_A^o}$$

We have $0 \le \xi \le n_A^o/a$ and $0 \le \chi \le 1$.

The rate of a reaction usually depends on the concentrations of some or all of the substances involved. The dependence of reaction rate on concentrations is the rate law. It must be determined by experiment. For reaction

$$aA + bB + \cdots \to cC + dD + \cdots$$

the observed rate law is often of the form

$$R = \frac{1}{V}\frac{d\xi}{dt} = k[A]^m[B]^n \ldots [C]^p[D]^q$$

where $m, n, \ldots, p, q, \ldots$ are small positive or negative integers or (less often) simple fractions.

We use a conventional terminology to characterize rate laws like this one. We talk about the ***order in a chemical species*** and the ***order of the reaction***. To characterize the rate law above, we say that the reaction is "m^{th} order in compound A," "n^{th} order in compound B," "p^{th} order in compound C," and "q^{th} order in compound D." We also say that the reaction is "$(m + n + \cdots + p + q + \cdots)^{\text{th}}$ order overall." Here k is an experimentally determined parameter that we call the ***rate constant*** or ***rate coefficient***.

It frequently happens that we are interested in an overall chemical change whose stoichiometric mechanism involves two or more elementary reactions. In this case, an exact rate model includes a differential equation for each elementary reaction. Nevertheless, it is often possible to approximate the rate of the overall chemical change by a single differential equation, which may be a relatively complex function of the concentrations of the species present in the reaction mixture. For the reaction above, the experimental observation might be

$$R = \frac{1}{V}\frac{d\xi}{dt} = \frac{k_1[A][B]}{k_2[C] + k_3[D]}$$

In such cases, we continue to call the differential equation the rate law. The concept of an overall order is no longer defined. The constants (k_1, k_2, and k_3) may or may not be rate constants for elementary reactions that are part of the overall process. Nevertheless, it is common to call any empirical constant that appears in a rate law a rate constant. In a complex rate law, the constants can often be presented in more than one way. In the example above, we can divide numerator and denominator by, say, k_3, to obtain a representation in which the constant coefficients have different values, one of which is unity.

Most of the rest of this chapter is devoted to understanding the relationship between an observed overall reaction rate and the rates of the elementary processes that contribute to it. Our principal objective is to understand chemical equilibrium in terms of competing forward and reverse reactions. At equilibrium, chemical reactions may be occurring rapidly; however, no concentration changes can be observed because each reagent is produced by one set of reactions at the same rate as it is consumed by another set. For the most part, we focus on reactions that occur in closed constant-volume systems.

§3 Simultaneous processes

The number of moles of a substance in a system can change with time because several processes occur simultaneously. Not only can a given substance participate in more than one reaction, but also the amount of it that is present can be affected by processes that are not chemical reactions. A variety of transport processes can operate to increase or decrease the amount of the substance that is present in the reaction mixture: A pure solid reactant could dissolve in a reacting solution, or a product could precipitate from it, as the reaction proceeds. A reacting species could diffuse into a reactor across a semi-permeable membrane. Controlled amounts of a reacting species could be added, either continuously or at specified intervals.

Each of the simultaneous processes contributes to the change in the number of moles of A present. At every instant, each of these contributions can be characterized by a rate. Over a short time interval, Δt, let $\Delta_i n_A$ be the contribution that the ith process makes to the change in the amount of A in volume V. If, even though the ith process may not be a reaction, we use $R_i(A)$ to represent its rate, its contribution to the rate at which the amount of A changes is

$$R_i(A) = \lim_{\Delta t \to 0} \left(\frac{\Delta_i n_A}{V(t)\Delta t} \right) = \frac{1}{V(t)} \frac{d_i n_A}{dt}$$

If there are numerous such processes, whose rates are $R_1(A)$, $R_2(A)$, ..., $R_i(A)$,..., $R_\omega(A)$, the observed overall rate is

$$R(A) = \sum_{i=1}^{\omega} R_i(A) = \frac{1}{V(t)} \sum_{i=1}^{\omega} \frac{d_i n_A}{dt}$$

If the volume is constant,

$$R_i(A) = \frac{1}{V} \frac{d_i n_A}{dt} = \frac{d_i[A]}{dt}$$

and

$$R(A) = \sum_{i=1}^{\omega} R_i(A) = \sum_{i=1}^{\omega} \frac{d_i[A]}{dt} = \frac{d[A]}{dt}$$

To illustrate these ideas, let us consider the base hydrolyses of methyl and ethyl iodide. No intermediates are observed in these reactions. If we carry out the base hydrolysis of methyl iodide,

$$CH_3I + OH^- \rightarrow CH_3OH + I^-$$

in a closed constant-volume system, we can express the reaction rate in several equivalent ways:

$$\begin{aligned} R(CH_3I) &= \frac{1}{V} \frac{d\xi(CH_3I)}{dt} = \frac{d[CH_3OH]}{dt} = \frac{d[I^-]}{dt} \\ &= -\frac{d[CH_3I]}{dt} = -\frac{d[OH^-]}{dt} \end{aligned}$$

If a mixture of methyl and ethyl iodide is reacted with aqueous base, both hydrolysis reactions consume hydroxide ion and produce iodide ion. The rates of these individual processes can be expressed as

$$R(CH_3I) = \frac{1}{V} \frac{d\xi(CH_3I)}{dt} = \frac{d[CH_3OH]}{dt} = -\frac{d[CH_3I]}{dt}$$

and

$$\begin{aligned} R(CH_3CH_2I) &= \frac{1}{V} \frac{d\xi(CH_3CH_2I)}{dt} = \frac{d[CH_3CH_2OH]}{dt} \\ &= -\frac{d[CH_3CH_2I]}{dt} \end{aligned}$$

but the rates at which the concentrations of hydroxide ion and iodide ion change depend on the rates of both reactions. We have

$$\begin{aligned} \frac{d[I^-]}{dt} &= -\frac{d[OH^-]}{dt} \\ &= R(CH_3I) + R(CH_3CH_2I) \\ &= -\frac{1}{V} \frac{d\xi(CH_3I)}{dt} - \frac{1}{V} \frac{d\xi(CH_3CH_2I)}{dt} \end{aligned}$$

In principle, either of the reaction rates can be measured by finding the change, over a short time interval, in the number of moles of a particular substance present.

Simultaneous processes occur when a reaction does not go to completion. The hydrolysis of ethyl acetate,

$$CH_3CO_2CH_2CH_3 + H_2O \rightleftharpoons CH_3CO_2H + CH_3CH_2OH$$

can reach equilibrium before the limiting reactant is completely consumed. The reaction rate, defined as

$R = V^{-1}(d\xi/dt)$, falls to zero. However, ethyl acetate molecules continue to undergo hydrolysis; the extent of reaction becomes constant because ethyl acetate molecules are produced from acetic acid and ethanol at the same rate as they are consumed by hydrolysis. Evidently, the rate of the forward reaction does not fall to zero even though the net reaction rate does.

Let R_f represent the number of moles of ethyl acetate undergoing hydrolysis per unit time per unit volume. Let R_r represent the number of moles of ethyl acetate being produced per unit time per unit volume. The net rate of consumption of ethyl acetate is $R = R_f - R_r$. At equilibrium, $R = 0$, and $R_f = R_r > 0$. In such cases, it can be ambiguous to refer to "the reaction" or "the rate of reaction." The rates of the forward and of the net reaction are distinctly different things.

So long as no intermediate species accumulate to significant concentrations in the reaction mixture, we can find the forward and reverse rates for a reaction like this, at any particular equilibrium composition, in a straightforward way. When we initiate reaction with no acetic acid or ethanol present, the rate of the reverse reaction must be zero. We can find the rate law for the forward reaction by studying the rate of the hydrolysis reaction when the product concentrations are low. Under these conditions, $R_r = 0$ and $R \approx R_f$. From the rate law that we find and the equilibrium concentrations, we can calculate the rate of the forward reaction at equilibrium. Likewise, when the ethyl acetate concentration is low, the rate of the hydrolysis reaction is negligible in comparison to that of the esterification reaction. We have $R_f \approx 0$ and $R = -R_r$, and we can find the rate law for the esterification reaction by studying the rate of the esterification reaction when the concentration of ethyl acetate is negligible. From this rate law, we can calculate the rate of the reverse reaction at the equilibrium concentrations.

§4 The effect of temperature on reaction rates

In practice, rate constants vary in response to changes in several factors. Indeed, they are usually the same in two experiments only if we keep everything but the reagent concentrations the same. Another way of saying this is that the rate law captures the dependence of reaction rate on concentrations, while the dependence of reaction rate on any other variable appears as a dependence of rate constants on that variable.

Temperature usually has a big effect. The experimentally observed dependence of rate constants on temperature can be expressed in a compact fashion. Over small temperature ranges it can usually be expressed adequately by the ***Arrhenius equation***:

$$k = A \exp\left(\frac{-E_a}{RT}\right)$$

E_a and A are called the ***Arrhenius activation energy*** and the ***frequency factor*** (or ***pre-exponential factor***), respectively.

The Arrhenius equation is an empirical relationship. As we see below for our collision-theory model, theoretical treatments predict that the pre-exponential term, A, is weakly temperature dependent. When we investigate reaction rates experimentally, the temperature dependence of A is usually obscured by the uncertainties in the measured rate constants. It is often said, as a rough rule of thumb, that the rate of a chemical reaction doubles if the temperature increases by 10 K. However, this rule can fail spectacularly. A reaction can even proceed more slowly at a higher temperature, and there are multi-step reactions for which this is observed.

§5 Other factors that affect reaction rates

A reaction that occurs in one solvent usually occurs also in a number of similar solvents. For example, a reaction that occurs in water will often occur with a low molecular weight alcohol—or an alcohol-water mixture—as the solvent. Typically, the same rate law is observed in a series of solvents, but the rate constants are solvent-dependent.

Other chemical species that are present in the reaction medium (but which are neither products nor reactants) can also affect observed reaction rates. Any such species meets the usual definition of a ***catalyst***. However, common practice restricts use of the word "catalyst" to a chemical species that substantially increases the rate of the reaction. A chemical species that decreases the rate of the reaction is usually called an ***inhibitor***. If we think that the rate effect of the non-reacting species results from a non-specific or a greater-than-bonding-distance interaction with one or more reacting species, we call the phenomenon a ***medium effect***. A solvent effect is a common kind of medium effect; altering the solvent affects the reaction rate even though the solvent does not form a chemical bond to any of the reactants or products. Dissolved salts can affect reaction rates in a similar way. Such effects often occur when the degree of charge separation along the path of an elementary reaction is significantly different from that in the reactants.

Isotopic substitution in a reactant can affect the reaction rate. (Replacement of a hydrogen atom with a deuterium atom is the most common case.) The effect of an isotopic substitution on a reaction rate is called a ***kinetic isotope effect***. Kinetic isotope effects can provide valuable information about the reaction mechanism. A kinetic isotope effect is expected if the energy needed make or break a chemical bond to the isotopically substituted atom is a significant component of the activation energy for the reaction. Kinetic isotope effects are usually small in comparison to other factors that affect reaction rates. A ten-fold change in the reaction rate is a big kinetic isotope effect. Effects much smaller than this are often useful; indeed, the absence of a kinetic isotope effect can help distinguish among alternative mechanisms.

In studies of reaction rates that are focused on finding the reaction mechanism, many characteristics of the reaction that are not strictly rate-related can be important. These include the stereochemistry of the product; the Walden inversion that accompanies S_N2 reactions at tetrahedral carbon centers is a notable example. Isotopic substitution that occurs incidental to a reaction can help establish that an intermediate is formed. The effects of competing reactions are often significant. The study of competing reactions is frequently helpful when the reaction involves a short-lived and otherwise undetectable intermediate. The use of isotopic substitution and competing reactions is illustrated in §16, in which we review the base hydrolysis of cobalt (III) pentaammine complexes, $Co(NH_3)_5X^{n+}$.

§6 Mechanisms and elementary processes

To see what we mean by an elementary process, let us consider some possible mechanisms for the base hydrolysis of methyl iodide: $CH_3I + OH^- \rightarrow CH_3OH + I^-$. In this reaction, a carbon–iodide bond is broken and a carbon–oxygen bond is formed. While any number of reaction sequences sum to this overall equation, we can write down three that are reasonably simple and plausible. The $C{-}I$ could be broken first and the $C{-}OH$ bond formed thereafter. Alternatively, the $C{-}OH$ bond could be formed first and the $C{-}I$ bond broken thereafter. In the first case, we have an intermediate species, CH_3^+, of reduced coordination number, and in the second we have an intermediate, $HO{-}CH_3{-}I^-$, of increased coordination number. Finally, we can suppose that the bond-forming and bond-breaking steps occur simultaneously, so that no intermediate species is formed at all.

(a) Heterolytic bond-breaking precedes bond-making

$$\begin{aligned} CH_3I &\rightarrow CH_3^+ + I^- \\ CH_3^+ + OH^- &\rightarrow CH_3OH \end{aligned}$$

(b) Bond-making precedes bond-breaking

$$\begin{aligned} CH_3I + OH^- &\rightarrow HO{-}CH_3{-}I^- \\ HO{-}CH_3{-}I^- &\rightarrow CH_3OH + I^- \end{aligned}$$

(c) Bond-breaking and bond-making are simultaneous

$$CH_3I + OH^- \rightarrow [\,HO{-}CH_3{-}I^-\,]^\ddagger \rightarrow CH_3OH + I^-$$

The distinction between mechanism (b) and mechanism (c) is that an intermediate is formed in the former but not in the latter. Nevertheless, mechanism (c) clearly involves an intermediate structure in which both the incoming and the leaving group are bonded to the central carbon atom. The distinction between mechanisms (b) and (c) depends on the nature of the intermediate structure. In mechanism (b), we suppose that the intermediate is a *bona fide* chemical entity; once a molecule of it is formed, that molecule has a finite lifetime. In (c), we suppose that the intermediate structure is transitory; it does not correspond to a molecule with an independent existence.

For this distinction to be meaningful, we must have a criterion that establishes the shortest lifetime we are willing to associate with "real molecules." It might seem that any minimum lifetime we pick must be wholly arbitrary. Fortunately this is not the case; there is a natural definition for a minimum molecular lifetime. The definition arises from the fact that molecules undergo vibrational motions. If a collection of atoms retains a particular relative orientation for such a short time that it never undergoes a motion that we would recognize as a vibration, it lacks an essential characteristic of a normal molecule. This means that the period of a high-frequency molecular vibration (roughly 10^{-14} s) is the shortest time that a collection of atoms can remain together and still have all of the characteristics of a molecule. If a structure persists for more than a few vibrations, it is reasonable to call it a molecule, albeit a possibly very unstable one.

In mechanism (c) the structure designated $[\,HO{-}CH_3{-}I^-\,]^\ddagger$ depicts a transitory arrangement of the constituent atoms. The atomic arrangement does not persist long enough for the $HO{-}CH_3$ bond or the $CH_3{-}I$ bond to undergo vibrational motion. A structure with these characteristics is called an ***activated complex*** or a ***transition state*** for the reaction, and a superscript double dagger, ‡, is conventionally used to signal that a structure has this character. The distinction between a *bona fide* intermediate and a transition state is clear enough in principle, but it can be very difficult to establish experimentally.

These considerations justify our earlier definition: An elementary reaction is one in which there are no intermediates. Any atomic arrangement that occurs during an elementary reaction does not persist long enough to vibrate before the arrangement goes on to become products or reverts to reactants.

We can distinguish a small number of possible kinds of elementary reactions: A single molecule can spontaneously rearrange to a new structure or break into smaller pieces. Two molecules can react to form one or more products. Three molecules can react to produce products. Or we can imagine that some larger number of molecules reacts. We refer to these possibilities as ***unimolecular***, ***bimolecular***, ***termolecular***, and ***higher-molecularity*** processes.

The stoichiometry of many reactions is so complicated as to preclude the possibility that they could occur as a single elementary process. For example, the reaction

$$3Fe^{2+} + HCrO_4{}^- + 7H^+ \rightarrow 3Fe^{3+} + Cr^{3+} + 4H_2O$$

can not plausibly occur in a single collision of three ferrous ions, one chromate ion, and seven hydronium ions. It is just too unlikely that all of these species could find themselves in the same place, at the same time, in the proper orientation, and with sufficient energy to react. In such cases, the stoichiometric mechanism must be a series of elementary steps. For this reaction, a skeletal representation of one plausible series is

$$\begin{aligned} Fe(II) + Cr(VI) &\rightarrow Fe(III) + Cr(V) \\ Fe(II) + Cr(V) &\rightarrow Fe(III) + Cr(IV) \\ Fe(II) + Cr(IV) &\rightarrow Fe(III) + Cr(III) \end{aligned}$$

§7 Rate Laws for Elementary Processes

If we think about an elementary bimolecular reaction between molecules A and B, we recognize that the reaction can occur only when the molecules come into contact. They must collide before they can react. So the probability that they react must be proportional to the probability that they collide, and the number of molecules of product formed per unit time must be proportional to the number of A— B collisions that occur in unit time. In our development of the collision theory for bimolecular reactions in the gas phase, (§4-12 to §4-16), we find that the number of such collisions is proportional to the concentration of each reactant. It is clear that this conclusion must apply to any bimolecular reaction.

If we have a vessel containing some concentration of A molecules and some concentration of B molecules, the collection experiences some number of A— B collisions per unit time. If we double the concentration of B molecules, each A molecule is twice as likely as before to encounter a B molecule. Indeed, for any increase in the concentration of B molecules, the number of collisions of an A molecule with B molecules increases in the same proportion. The number of A— B collisions must be proportional to the concentration of B molecules. Likewise, increasing the concentration of A molecules must increase the number of A— B collisions proportionately; the number of A— B collisions must also be proportional to the concentration of A molecules. We conclude that the rate for any bimolecular reaction between molecular substances A and B is described by the equations

$$\begin{aligned} R &= \frac{1}{V}\frac{dn_A}{dt} \\ &= \frac{1}{V}\frac{dn_B}{dt} \\ &= -k_2[A][B] \end{aligned}$$

This is a second-order rate law, and the proportionality constant, k_2, is called a second-order rate constant.

In §4-16, we derive an equation for the frequency with which a type 1 molecule collides with type 2 molecules in the gas phase when the concentration of type 2 molecules is N_2 and the kinetic energy along the line of centers exceeds a threshold value, ϵ_a per molecule, or E_a per mole. The rate at which such collisions occur is

$$\rho_{12}(\epsilon_a) = N_1 N_2 \sigma_{12}^2 \left(\frac{8\pi kT}{\mu}\right)^{1/2} \exp\left(\frac{-\epsilon_a}{kT}\right)$$

which is just $\exp(-\epsilon_a/kT)$ times the rate at which collisions of any energy occur between molecules of type 1 and molecules of type 2. If reaction occurs at every collision between a molecule 1 and a molecule 2 in which the kinetic energy along the line of centers exceeds ϵ_a, the collision rate, $\rho_{12}(\epsilon_a)$, equals the reaction rate. We have $R = \rho_{12}(\epsilon_a)$.

If the temperature-dependence of the rate constant is given by the Arrhenius equation, the rate of the bimolecular reaction between species 1 and species 2 is

$$\begin{aligned} R &= k_2 N_1 N_2 \\ &= \left[A \exp\left(\frac{-E_a}{RT}\right)\right] N_1 N_2 \end{aligned}$$

where A is independent of temperature and $E_a = \overline{N}\epsilon_a$. The collision-theory model for the bimolecular reaction is almost the same; the difference being a factor of $T^{1/2}$ in the pre-exponential factor, $\sigma_{12}^2(8\pi kT/\mu)^{1/2}$. The effect of the $T^{1/2}$ term is usually small in comparison to the effect of temperature in the exponential term. Thus, the temperature dependence predicted by collision theory, which is a highly simplified theoretical model, and that predicted by the Arrhenius equation, which is an empirical generalization usually used to describe data taken over a limited temperature range, are in substantial agreement.

Experimentally determined values of the pre-exponential factor for gas-phase bimolecular reactions can approach the value calculated from collision theory. However, particularly for reactions between polyatomic molecules, the experimental value is often much smaller than the calculated collision frequency. We rationalize this observation by recognizing that our colliding-spheres model provides no role for the effect of molecular structures. When the colliding molecules are not spherical, the collision angle is an incomplete description of their relative orientation. If the relative orientation of two colliding molecules is unfavorable to reaction, it is entirely plausible that they can fail to react no matter how energetic their collision. To recognize this effect, we suppose that the reaction rate is proportional to a ***steric factor***, γ, where γ represents the probability that a colliding pair of molecules have the relative orientation that is necessary for reaction to occur. Of course, γ must be less than one. Taking this amplification of the collision model into account, the relationship between the reaction rate and the collision frequency becomes

$$\frac{d\xi}{dt} = \gamma\rho_{12}(\varepsilon_a)$$

When we consider reactions in solution, we recognize that there are usually many more solvent molecules than reactant molecules. As a result, collisions of a reactant molecule with solvent molecules are much more frequent than collisions of a molecule of one reactant with a molecule of another reactant. The high frequency of collisions with solvent molecules means that the net distance moved by a reactant molecule in unit time is much less in solution than in a gas. This decreases the probability that two reactant molecules will meet. On the other hand, once two reactant molecules near one another, the solvent molecules tend to keep them together, and they are likely to collide with one another many times before they finally drift apart. (This is known as the ***solvent-cage effect.***) We can expect these effects to roughly offset one another.

A termolecular elementary process is a reaction in which three reactant molecules collide. For this to happen, an A molecule and a B molecule must be very close to one another at exactly the time that a C molecule encounters the pair of them. If the reactants are not very concentrated, the probability that a given A molecule is very close to a B molecule during any short time interval is small. The probability that this A molecule will be hit by a C molecule during the same time interval is also very small. The probability that all three species will collide at the same time is the product of two small probabilities; under any given set of conditions, the number of collisions involving three molecules is smaller than the number of collisions between two molecules. The probability of a termolecular collision and hence the rate of a termolecular elementary process is proportional to the concentrations of all three reacting species

$$R = \frac{1}{V}\frac{dn_A}{dt} = \frac{1}{V}\frac{dn_B}{dt} = \frac{1}{V}\frac{dn_C}{dt} = -k_2[A][B][C]$$

However, the low probability of a termolecular collision means that we can expect the termolecular rate constant, k_3, to be very small. If termolecular mechanisms are rare, higher-molecularity mechanisms must be exceedingly rare, if, indeed, any occur at all. For most chemical reactions, the mechanism is a series of unimolecular and bimolecular elementary reactions.

A unimolecular elementary process is one in which a molecule spontaneously undergoes a chemical change. If we suppose that there is a constant probability that any given A molecule undergoes reaction in unit time, then the total number reacting in unit time is proportional to the number of A molecules present. Let the average number of moles reacting in unit time be $\overline{\Delta n_A}$, the number of molecules in the system be n_A, and the proportionality constant be k. (We choose a unit of time that is small enough to insure that $\overline{\Delta n_A} \ll n_A$.) If the probability of reaction is constant, we have $\overline{\Delta n_A} = -kn_A$. Since $\overline{\Delta n_A}$ is the number of moles that react in unit time, the number of moles that react in time Δt is $\Delta n_A = \overline{\Delta n_A}\Delta t$, so that

$$\overline{\Delta n_A} = \frac{\Delta n_A}{\Delta t} = -kn_A$$

Dividing by the volume of the system, we have

$$\frac{1}{V}\frac{\Delta n_A}{\Delta t} = -k\frac{n_A}{V}$$

In the limit that $\Delta t \to 0$, the term on the left becomes the reaction rate, $R = V^{-1}(dn_A/dt)$, and since $n_A/V = [A]$, we have

$$R = \frac{1}{V}\frac{dn_A}{dt} = -k[A]$$

Thus, a constant reaction probability implies that a unimolecular reaction has a first-order rate law. If the volume is constant, we have

$$\frac{d[A]}{dt} = -k[A]$$

The idea that a unimolecular reaction corresponds to a constant reaction probability can be rationalized by introducing a simple model of the reaction process. This model assumes that reactant molecules have a distribution of energies, that only molecules whose energies exceed some minimum can react, and that this excess energy must be in some specific internal motion before the reaction can occur. Molecules exchange energy by colliding with one another. When a molecule acquires excess energy as the result of a collision, redistribution of this energy among the motions available to the molecule is not instantaneous. A characteristic length of time is required for excess energy to reach the specific internal motion that leads to reaction. Any given molecule can retain excess energy only for the short time between two collisions. The molecule gains excess energy in one collision and loses it in a subsequent one. Reaction can occur only if the excess energy reaches the specific internal motion before the molecule undergoes a deactivating collision. (We return to these ideas in §14 and §15.)

In summary, only two kinds of elementary processes are needed to develop a mechanism for nearly any chemical change. These elementary processes and their rate laws are:

Unimolecular	**A →**	$\dfrac{d[A]}{dt} = -k[A]$
Bimolecular	**A + B →**	$\dfrac{d[A]}{dt} = -k[A][B]$

Finally, we should note that we develop these rate laws for elementary processes under the assumption that the rate at which molecules collide is proportional to the concentrations of the colliding species. In doing so, we implicitly assume that intermolecular forces of attraction or repulsion have no effect on this rate. When our goal is to predict rate laws from reaction mechanisms, this assumption is almost always an adequate approximation. However, when we study chemical equilibria, we often find that we must allow for the effects of intermolecular forces in order to obtain an adequate description. In chemical thermodynamics, we provide for the effects of such forces by introducing the idea of a ***chemical activity***. The underlying idea is that the chemical activity of a compound is the effective concentration of the compound—we can view it as the concentration "corrected" for the effects of intermolecular forces.

§8 Experimental determination of rate laws

The determination of a rate law is a matter of finding an empirical equation that adequately describes reaction-rate data. We can distinguish two general approaches to this task. One approach is to measure reaction rate directly. That is, for $A + B \to C$, we measure the reaction rate in experiments where the concentrations, $[A]$, $[B]$, and $[C]$, of reactants and products are known. The other is to measure a concentration at frequent time intervals as a batch reaction goes nearly to completion. We then seek a differential equation that is consistent with this concentration-*versus*-time data.

If the reaction is the only process that affects Δn_C, direct measurement of the reaction rate can be effected by measuring Δn_C over a short time interval, Δt, in which the concentrations, $[A]$, $[B]$, and $[C]$, do not change appreciably. This is often difficult to implement experimentally, primarily because it is difficult to measure small values of Δn_C with the necessary accuracy at known values of $[A]$, $[B]$, and $[C]$. The ***method of initial rates*** is an experimentally simple method in which the reaction rate is measured directly. Initial-rate measurements are extensively used in the study of enzyme-catalyzed reactions. Direct measurement of reaction rate can also be accomplished using a flow reactor. We discuss the method of initial rates, a particular kind of flow reactor known as a CSTR, and enzyme catalysis in §10, §11, and §13, respectively.

The most common reaction-rate experiment is a batch reaction in which we mix the reactants as rapidly as possible and then monitor the concentration of one (or more) of the reactants or products as the reaction proceeds. We do the mixing so that the initially mixed reactants are at a known temperature, which can be maintained constant for the remainder of the experiment. The data from such an experiment are a set of concentrations and the times at which they are measured. To find the rate law corresponding to these concentration-versus-time data, we employ a trial-and-error procedure. We guess what the rate law is likely to be. We then obtain a general solution for this differential equation. This solution predicts the dependence of concentrations *versus* time as a function of one or more rate constants. If we can obtain a satisfactory fit of experimental concentration-*versus*-time data to the concentration-*versus*-time equation predicted by the rate law, we conclude that the rate law is a satisfactory representation of the experimental data.

For a reaction $A \rightarrow C$ in a closed constant-volume system, we would want to test a first-order rate law, which we can express in several alternative ways:

$$\frac{1}{V}\frac{d\xi}{dt} = -\frac{d[A]}{dt} = \frac{d[C]}{dt} = k[A]$$

Using the changing concentration of A to express the rate, separating variables, and integrating between the initial concentration $[A] = [A]_0$ at $t = 0$ and concentration $[A]$ at time t gives

$$\int_{[A]_0}^{[A]} \frac{d[A]}{[A]} = -k \int_0^t dt$$

so that

$$\ln\frac{[A]}{[A]_0} = -kt$$

or

$$[A] = [A]_0 \exp(-kT)$$

Frequently it is convenient to introduce the extent of reaction or the concentration of a product as a parameter. In the present instance, if the initial concentration of C is zero, $[C] = \xi/V = x$. Then at any time, t, we have $[A] = [A]_0 - x$, and the first-order rate equation can be written as

$$\frac{dx}{dt} = k\left([A]_0 - x\right)$$

which we rearrange and integrate between the limits $x(0) = 0$ and $x(t) = x$ as

$$\int_0^x \frac{-dx}{[A]_0 - x} = -k \int_0^t dt$$

To give

$$\ln\left(\frac{[A]_0 - x}{[A]_0}\right) = -kt$$

It is easy to test whether concentration versus time data conform to the first-order decay model. If they do, a plot of $\ln\left([A]_0 - x\right)$ or $\ln[A]$, versus time, t, is a straight line.

For a reaction $2A \rightarrow C$, we would want to test a rate law of the form

$$\frac{1}{V}\frac{d\xi}{dt} = -\frac{1}{2}\frac{d[A]}{dt} = \frac{d[C]}{dt} = k[A]^2$$

If the initial concentration of C is zero, $[C] = \xi/V = x$, and $[A] = [A]_0 - 2x$ at any time t. The rate law can be written as

$$\frac{dx}{dt} = k\left([A]_0 - 2x\right)^2$$

and rearranged and integrated as

$$\int_0^x \frac{-dx}{\left([A]_0 - 2x\right)^2} = -k \int_0^t dt$$

to give

$$\frac{1}{[A]_0 - 2x} - \frac{1}{[A]_0} = 2kt$$

or

$$\frac{1}{[A]} - \frac{1}{[A]_0} = 2kt$$

If concentration-*versus*-time data conform to this second-order rate law, a plot of $[A]^{-1}$ *versus* time is a straight line.

For a reaction $A + B \rightarrow C$, we would want to test a rate law of the form

$$\frac{1}{V}\frac{d\xi}{dt} = -\frac{d[A]}{dt} = -\frac{d[B]}{dt} = \frac{d[C]}{dt} = k[A][B]$$

If the initial concentration of C is again zero,

$[C] = \xi/V = x,\ \ [A] = [A]_0 - x.$ and $[B] = [B]_0 - x.$ at any time t. The rate law can be written as

$$\frac{dx}{dt} = k\,([A]_0 - x)\,([B]_0 - x)$$

If $[A]_0 \neq [B]_0$, this can be integrated (by partial fractions) to give

$$\frac{1}{[B]_0 - [A]_0} \ln \frac{[A]_0([B]_0 - x)}{[B]_0([A]_0 - x)} = kt$$

If experimental data conform to this equation, a plot of

$$\ln \frac{([B]_0 - x)}{([A]_0 - x)}$$

versus time is linear. In practice, this often has disadvantages, and experiments to study reactions like this typically exploit the technique of ***flooding***.

Flooding is a widely used experimental technique that enables us to simplify a complex rate law in a way that makes it more convenient to test experimentally. In the case we are considering, we can often arrange to carry out the reaction with the initial concentration of B much greater than the initial concentration of A. Then the change that occurs in the concentration of B during the reaction has much less effect on the reaction rate than the change that occurs in the concentration of A; in the rate equation, it becomes a good approximation to let $[B] = [B]_0$ at all times. (For a fuller consideration of this point, see problem 23.) The second-order rate equation simplifies to

$$\frac{d[A]}{dt} = -\frac{d[C]}{dt} = -k_{obs}[A]$$

where

$$k_{obs} = k[B]_0$$

Since the simplified rate equation is approximately first order, the *observed* rate constant, k_{obs}, is the slope of a plot of $\ln[A]$ *versus* t. k_{obs} is called a ***pseudo-first-order rate constant***.

Of course, one such experiment tests only whether the true rate law is first order in $[A]$. It tells nothing about the dependence on $[B]$. If we do several such experiments at different initial concentrations of B, the resulting set of k_{obs} values must be directly proportional to the corresponding $[B]_0$ values. This can be tested graphically by plotting k_{obs} *versus* $[B]_0$. If the rate law is first order in $[B]$, the resulting plot is linear with an intercept of zero. The slope of this plot is the second-order rate constant, k.

Flooding works by simplifying the rate law that is observed in a given experiment. Similar simplification can be achieved by designing the experiment so that the initial concentrations of two or more reactants are proportional to their stoichiometric coefficients. For the reaction $A + B \rightarrow C$ and the expected rate law

$$\frac{d[C]}{dt} = k[A][B]$$

we would initiate the experiment with equal concentration of reactants A and B. Letting $[A]_0 = [B]_0 = \alpha$ and $[C] = \xi/V = x$, the concentrations of A and B at longer times become $[A] = [B] = \alpha - x$. The rate law becomes effectively second order.

$$\frac{dx}{dt} = k(\alpha - x)^2$$

§9 First-order rate processes

First-order rate processes are ubiquitous in nature—and commerce. In chemistry we are usually interested in first-order decay processes; in other subjects, first-order growth is common. We can develop our appreciation for the dynamics—and mathematics—of first-order processes by considering the closely related subject of compound interest.

When a bank says that it pays 5% annual interest, ***compounded annually***, on a deposit, it means that for every \$1.00 we deposit at the beginning of a year, the bank will add 5% or \$0.05 to our account at the end of the year, making our deposit worth \$1.05. If we let the value of our deposit at the end of year n be $P(n)$, and the interest rate (expressed as a fraction) be r, with $r > 0$, we can write

$$P(1) = P(0) + \Delta P = P(0) + rP(0) = (1 + r)P(0)$$

where we represent the first year's interest by $\Delta P = rP(0)$. If we leave all of the money in the account for an additional year, we will have

$$P(2) = (1 + r)P(1) = (1 + r)^2 P(0)$$

and after t years we will have

$$P(t) = (1 + r)^t P(0)$$

Sometimes a bank will say that it pays 5% annual interest, ***compounded monthly***. Then the bank means that it will compute a new balance every month, based on $r = 0.05\ \text{year}^{-1} = (0.05/12)\ \text{month}^{-1}$. After one month

$$P(1\ \text{month}) = \left(1 + \frac{0.05}{12}\right) P(0)$$

and after n months

$$P(n\ \text{months}) = \left(1 + \frac{0.05}{12}\right)^n P(0)$$

If we want the value of the account after t years, we have, since $n = 12t$,

$$P(t) = \left(1 + \frac{0.05}{12}\right)^{12t} P(0)$$

If the bank were to say that it pays interest at the rate r, ***compounded daily***, the balance at the end of t years would be

$$P(t) = \left(1 + \frac{r}{365}\right)^{365t} P(0)$$

For any number of compoundings, m, at rate r, during a year, the balance at the end of t years would be

$$P(t) = \left(1 + \frac{r}{m}\right)^{mt} P(0)$$

Sometimes banks speak of ***continuous compounding***, which means that they compute the value of the account at time t as the limit of this equation as m becomes arbitrarily large. That is, for continuous compounding, we have

$$P(t) = \lim_{m\to\infty} \left[\left(1 + \frac{r}{m}\right)^{mt}\right] P(0)$$

Fortunately, we can think about the continuous compounding of interest in another way. What we mean is that the change in the value of the account, ΔP, over a short time interval, Δt, is given by

$$\Delta P = rP\Delta t$$

where P is the (initial) value of the account for the interval Δt, and r is the fractional change in P during one unit of time. So we can write

$$\lim_{\Delta t\to 0} \left(\frac{\Delta P}{\Delta t}\right) = \frac{dP}{dt} = rP$$

Separating variables to obtain $dP/P = rdt$ and integrating between the limits $P = P(0)$ at $t = 0$ and $P = P(t)$ at $t = t$, we obtain

$$\ln \frac{P(t)}{P(0)} = rt$$

or

$$P(t) = P(0)\exp(rt)$$

Comparing the two equations we have derived for continuous compounding, we see that

$$\exp(rt) = \lim_{m\to\infty} \left(1 + \frac{r}{m}\right)^{mt}$$

Continuous compounding of interest is an example of ***first-order*** or ***exponential growth***. Other examples are found in nature; the growth of bacteria normally follows such an equation. Reflection suggests that such behavior should not be considered remarkable. It requires only that the increase per unit time in some quantity, P, be proportional to the amount of P that is already present: $\Delta P = rP\Delta t$. Since P measures the number of items (dollars, molecules, bacteria) present, this is equivalent to our observation in §7 that a first-order process corresponds to a constant probability that a given individual item will disappear (***first-order decay***) or reproduce (***first-order growth***) in unit time. For a first-order decay we have, keeping $r > 0$,

$$\Delta P = -rP\Delta t$$

In the limit as $\Delta t \to 0$,

$$\frac{dP}{dt} = -rP$$

which has solution

$$P(t) = P(0)\exp(-rt)$$

First-order growth and first-order decay both depend exponentially on rt. The difference is in the sign of the exponential term. For exponential growth, $P(t)$ becomes arbitrarily large as $t \to \infty$; for exponential decay, $P(t)$ goes to zero. If the concentration of a chemical species A decreases according to a first-order rate law, we have

$$\ln \frac{[A]}{[A]_0} = -kt$$

The units of the rate constant, k, are s^{-1}. The ***half-life*** of a chemical reaction is the time required for one-half of the stoichiometrically possible change to occur. For a first-order decay, the half-life, $t_{1/2}$, is the time required for the concentration of the reacting species to decrease to one-half of its value at time zero; that is, when the time is $t_{1/2}$, the concentration is $[A] = [A]_0/2$. Substituting into the integrated rate law, we find that the half-life of a first-order decay is independent of concentration; the half-life is

$$t_{1/2} = \frac{\ln 2}{k}$$

§10 Rate laws by the study of initial rates

In concept, the most straightforward way to measure reaction rate directly is to measure the change in the concentration of one reagent in a short time interval immediately following initiation of the reaction. The initial concentrations are known from the way the reaction mixture is prepared. If necessary, the initial mixture can be prepared so that known concentrations of products or reaction intermediates are present. The initial reaction rate is approximated as the measured concentration change divided by the elapsed time. The accuracy of initial-rate measurements is often poor. This can result from concentration variations associated with initiation of the reaction; the actual mixing process is not instantaneous and significant reaction can occur before the mixture becomes truly homogeneous. Measuring small changes in concentration with sufficient accuracy can also be difficult.

Enzymes are naturally occurring catalysts for biochemical reactions. In the study of enzyme-catalyzed reactions, it is usually possible to select the enzyme concentration and other reaction conditions so that the initial rate can be measured with adequate accuracy. For such studies, initial-rate measurements are used extensively. For other types of reactions, the method of initial rates is usually less effective than alternative methods.

To illustrate the application of the method, suppose we have a reaction

$$A + B + C \rightarrow Products$$

and that we are able to measure small changes in $[A]$ with good accuracy. We seek a rate law of the form

$$-\frac{d[A]}{dt} = f([A],[B],[C])$$

For any given experiment we approximate $d[A]/dt$ by $\Delta[A]/\Delta t$, and approximate the average concentrations of the reagents over the interval Δt by their initial values: $[A] = [A]_0$, $[B] = [B]_0$, and $[C] = [C]_0$. By carrying out an number of such experiments with suitably chosen initial concentrations, we can determine the functional form of the rate law and evaluate the rate constants that appear in it.

Table 1 Hypothetical reaction rate data

$\Delta[A]/\Delta t$, $\underline{M}$	Δt,s^{-1}	$[A]_0$	$[B]_0$	$[C]_0$
-2×10^{-7}	1000	0.010	0.010	0.010
-4×10^{-7}	500	0.010	0.010	0.020
-2×10^{-7}	1000	0.010	0.020	0.010
-8×10^{-7}	2500	0.020	0.010	0.010

Table 1 presents data for a hypothetical reaction that serve to illustrate the basic concept. We suppose that initial rates have been determined for four different combinations of initial concentrations. Comparison of the first and second experiments indicates that doubling $[C]$ doubles the reaction rate, indicating that the rate depends on $[C]$ to the first power. Comparison of the first and third experiments indicates that doubling $[B]$ leaves the reaction rate unchanged, implying that the rate is independent of $[B]$. Comparison of the first and fourth experiments indicates that doubling $[A]$ increases the reaction rate by a factor of four, implying that the rate is proportional to the second power of $[A]$. We infer that the rate law is

$$-\frac{d[A]}{dt} = k[A]^2[B]^0[C]^1$$

Given the form of the rate law, an estimate of the value of the rate constant, k, can be obtained from the data for each experiment. For this illustration, we calculate $k = 0.20\ \underline{M}\ s^{-1}$ from each of the experiments.

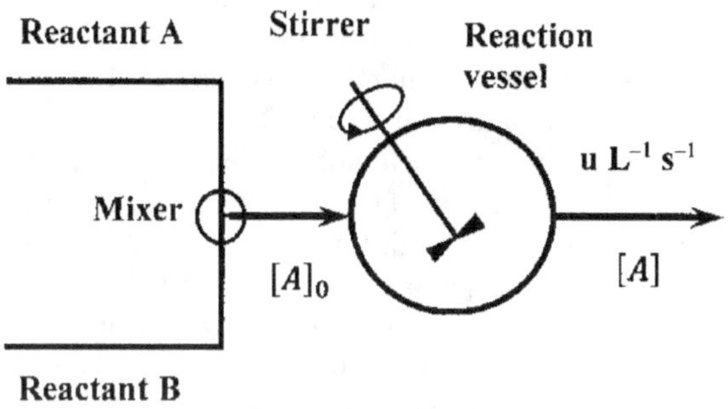

Figure 1. A continuous stirred-tank reactor.

§11 Rate laws from experiments in a CSTR

A continuous stirred tank reactor (CSTR)—or capacity-flow reactor—is a superior method of collecting kinetic data when the rate law is complex. Unfortunately, a CSTR tends to be expensive to construct and complex to operate. Figure 1 gives a schematic representation of the essential features of a CSTR. Fresh reagents are fed to a reactor vessel of volume V at a constant rate. A portion of the reactor contents is continuously removed at the same volumetric flow rate. Because the addition and removal of material occur at the same rate, the reactor is always filled with a fixed volume of reaction mixture. The reaction vessel and its contents are maintained at a constant temperature. The vessel contains a stirrer, which operates continuously and at a high enough speed to keep the contents of the vessel homogeneous (free of concentration and temperature gradients) at all times.

The essential idea involved in the operation of a CSTR is that, after the passage of sufficient time, the concentrations of the various species present in the reactor become constant. We say that the reactor contains steady-state concentrations of the reactants and products. When the reactor reaches this steady state, processes that increase reagent concentrations are occurring at the same rate as processes that decrease them.

Let the reaction be of the form: $A + B \rightarrow$ Products. The concentrations of the reagents in the feed solution are known. Let the concentration of A in the fresh feed solution be $[A]_0$. Let the rate at which fresh reagent-containing solution is fed to the reactor be $u\ L^{-1}s^{-1}$. Homogeneous reaction mixture is withdrawn from the vessel at the same flow rate. The amount of A in the reactor is increased by the flow of fresh reactant solution into the reactor. It is decreased both by reaction and by the flow of solution out of the reactor. The steady-state reaction rate, $R = R(A)$, is the number of moles of reactant A consumed by the reaction per unit time per unit volume of reaction vessel after all of the reagent concentrations have become constant. Since A is a reactant, this rate is

$$R = -\frac{1}{V}\frac{d_r n_A}{dt}$$

where $d_r n_A/dt$ is the contribution that the reaction makes to the rate at which the number of moles of A in the reactor changes. Since all of the reaction occurs within the vessel, and the vessel is entirely filled with the solution, R is also the number of moles of A consumed by reaction per unit time per unit volume of solution.

At steady state, the number of moles of A in the reactor is determined by (1) the number of moles of A entering the reactor per unit time, (2) the number of moles of A being consumed by reaction per unit time, and (3) the number of moles of A leaving the reactor in the effluent stream per unit time. In unit time, the number of moles entering with the feed is given by $u[A]_0$; the number leaving with the effluent is given by $u[A]$. In unit time, the contribution that the reaction makes to the change in the number of moles of A present is $-RV$. When the steady state is reached, the number of moles entering, plus the change due to reaction, must equal the number of moles leaving:

moles flowing in + change in moles due to reaction - moles flowing out = 0

or, in unit time,

$$u[A]_0 - RV - u[A] = 0$$

Solving for R, we have

$$R = -\frac{u}{V}([A] - [A]_0)$$

(We define reaction rate so that $R > 0$. If A is produced by the reaction, the mass-balance equation is $u[A]_0 + RV - u[A] = 0$.)

As with the method of initial rates, the rate law is determined by measuring reaction rates in a series of experiments in which the steady-state concentrations of the various reactants and products vary. For each experiment it is necessary to determine both the reaction rate and the steady-state concentration of each reagent that might be involved in the rate law. Using the equation above, the rate is calculated from the difference between a reagent concentration in the feed solution and its steady-state concentration in the reactor. The concentration of each reagent in the effluent is the same as its concentration in the reactor, so the necessary concentration information can be obtained by chemical analysis of the effluent solution. The chemical analysis must be done in such a way that no significant reaction occurs between the time the material leaves the reaction vessel and the time the analysis is completed.

§12 Predicting rate laws from proposed mechanisms

Because a proposed mechanism can only be valid if it is consistent with the rate law found experimentally, the rate law plays a central role in the investigation of chemical reaction mechanisms. The discussion above introduces the problems and methods associated with collecting rate data and with finding an empirical rate law that fits experimental concentration-*versus*-time data. We turn now to finding the rate laws that are consistent with a particular proposed mechanism. For simplicity, we consider reactions in closed constant-volume systems.

In principle, numerical integration can be used to predict the concentration at any time of each of the species in any proposed reaction mechanism. This prediction can be compared to experimental observations to see whether they are consistent with the proposed mechanism. To do the numerical integration, it is necessary to know the initial concentrations of all of the chemical species and to know, or assume, values of all of the rate constants. The initial concentrations are known from the procedure used to initiate the reaction. However, the rate constants must be determined by some iterative procedure in which initial estimates of the rate constants are used to predict concentration-*versus*-time data that can be compared to the experimental results to produce refined estimates.

In practice, we tailor our choice of reaction conditions so that we can use various approximations to test whether a proposed mechanism can explain the data. We now consider the most generally useful of these approximations.

In this discussion, we assume that the overall reaction goes to completion; that is, at equilibrium the concentration of the reactant whose concentration is limiting has become essentially zero. If the overall reaction involves more than one elementary step, then an intermediate compound is involved. A valid mechanism must include this intermediate, and more than one differential equation may be needed to characterize the time rate of change of all of the species involved in the reaction. We focus on conditions and approximations under which the rate of appearance of the final products in a multi-step reaction mechanism can be described by a single differential equation, the rate law.

We examine the application of these approximations to a particular reaction mechanism. When we understand the application of these approximations to this mechanism, the ways in which they can be used in other situations are clear.

Consider the following sequence of elementary steps

$$A + B \underset{k_2}{\overset{k_1}{\rightleftharpoons}} C \overset{k_3}{\rightarrow} D$$

whose kinetics are described by the following simultaneous differential equations:

$$\frac{d[A]}{dt} = \frac{d[B]}{dt} = -k_1[A][B] + k_2[C]$$

$$\frac{d[C]}{dt} = k_1[A][B] - k_2[C] - k_3[C]$$

$$\frac{d[D]}{dt} = k_3[C]$$

The general analytical solution for this system of coupled differential equations can be obtained, but it is rather complex, because $[C]$ increases early in the reaction, passes through a maximum, and then decreases at long times. In principle, experimental data could be fit to these equations. The numerical approach requires that we select values for k_1, k_2, k_3, $[A]_0$, $[B]_0$, $[C]_0$, and $[D]_0$, and then numerically integrate to get $[A]$, $[B]$, $[C]$, and $[D]$ as functions of time. In principle, we could refine our estimates of k_1, k_2, and k_3 by comparing the calculated values of one or more concentrations to the experimental ones. In practice, the approximate treatments we consider next are more expedient.

When we begin a kinetic study, we normally have a working hypothesis about the reaction mechanism, and we design our experiments to simplify the differential equations that apply to it. For the present example, we will assume that we always arrange the experiment so that $[C]_0 = 0$ and $[D]_0 = 0$. In consequence, $[A]_0 = [A] + [C] + [D]$ at all times.

Also, we restrict our considerations to experiments in which $[B]_0 \gg [A]_0$. This exemplifies the use of flooding. The practical effect is that the concentration of B remains effectively constant at its initial value throughout the entire reaction, which simplifies the differential equations significantly. In the present instance, setting $[B]_0 \gg [A]_0$ means that the rate-law term $k_1[A][B]$ can be replaced, to a good approximation, by $k_{obs}[A]$, where $k_{obs} = k_1[B]_0$.

Once we have decided upon the reaction conditions we are going to use, whether the resulting concentration-*versus*-time data can be described by a single differential equation depends on the relative magnitudes of the rate constants in the several steps of the overall reaction. Particular combinations of relationships that lead to simplifications are often referred to by particular names; we talk about a combination that has a ***rate-determining step***, or one that involves a ***prior equilibrium***, or one in which a ***steady-state approximation*** is applicable. To see what we mean by these terms, let us consider some particular relationships that can exist among the rate constants in the mechanism above.

Case I. Suppose that $k_1[A][B] \gg k_2[C]$ and $k_3 \gg k_2$. We often describe this situation by saying, rather imprecisely, that the reaction to convert C to D is very fast and that the reaction to convert C back to A and B is very slow—compared to the reaction that forms C from A and B. When C is produced in these circumstances, it is converted to D so rapidly that we never observe a significant concentration of C in the reaction mixture. The formation of a molecule of C is tantamount to the formation of a molecule of D, and the reaction produces D at essentially the same rate that it consumes A or B. We say that the first step, $A + B \rightarrow C$, is the rate-determining step in the reaction. We have

$$-\frac{d[A]}{dt} = -\frac{d[B]}{dt} \approx \frac{d[D]}{dt}$$

The assumption that $k_1[A][B] \gg k_2[C]$ means that we can neglect the smaller term in the equation for $d[A]/dt$, giving the approximation

$$\frac{d[A]}{dt} = \frac{d[B]}{dt} = -\frac{d[D]}{dt} = -k_1[A][B]$$

Letting $[D] = x$ and recognizing that our assumptions make $[C] \approx 0$, the mass-balance condition, $[A]_0 = [A] + [C] + [D]$, becomes $[A] = [A]_0 - x$. Choosing $[B]_0 \gg [A]_0$ means that $k_1[B] \approx k_1[B]_0 = k_{I,obs}$. The rate equation becomes first-order:

$$\frac{dx}{dt} = k_{I,obs}(\,[A]_0 - x)$$

Since $k_{I,obs}$ is not strictly constant, it is a pseudo-first-order rate constant. The disappearance of A is said to follow a pseudo-first-order rate equation.

The concept of a rate-determining step is an approximation. In general, the consequence we have in mind when we invoke this approximation is that no intermediate species can accumulate to a significant concentration if it is produced by the rate-determining step or by a step that occurs after the rate-determining step. We do not intend to exclude the accumulation of a species that is at equilibrium with another product. Thus, in the mechanism

$$A \xrightarrow{k\,(rds)} B \rightleftharpoons C$$

we suppose that the conversion of A to B is rate-determining and that the interconversion of B and C is so rapid that their concentrations always satisfy the equilibrium relationship $K = [C]/[B]$. For the purpose at hand, we do not consider B to be an intermediate; B is a product that happens to be at equilibrium with the co-product, C.

Case II. Suppose that $k_1[A][B] \gg k_3[C]$. In this case $A + B \rightarrow C$ is fast compared to the rate at which C is converted to D, and we say that $C \rightarrow D$ is the rate-determining step. We can now distinguish three sub-cases depending upon the way $[C]$ behaves during the course of the reaction.

Case IIa. Suppose that $k_1[A][B] \gg k_3[C]$ and $k_3 \gg k_2$. Then $A + B \rightarrow C$ is rapid and essentially quantitative. That is, within a short time of initiating the reaction, all of the stoichiometrically limiting reactant is converted to C. Letting $[D] = x$ and recognizing that our assumptions make $[A] \approx 0$, the mass-balance condition, $[A]_0 = [A] + [C] + [D]$, becomes $[C] = [A]_0 - x$. After a short time, the rate at which D is formed becomes

$$\frac{d[D]}{dt} = k_3[C]$$

or

$$\frac{dx}{dt} = k_3(\,[A]_0 - x)$$

The disappearance of C and the formation of D follow a

first-order rate law.

Case IIb. If the forward and reverse reactions in the first elementary process are rapid, then this process may be effectively at equilibrium during the entire time that D is being formed. (This is the case that $k_1[A][B] \gg k_3[C]$ and $k_2 \gg k_3$.) Then, throughout the course of the reaction, we have

$$K_{eq} = [C]/[A][B]$$

Letting $[D] = x$ and making the further assumption that $[A] \gg [C] \approx 0$ throughout the reaction, the mass-balance condition, $[A]_0 = [A] + [C] + [D]$, becomes $[A] = [A]_0 - x$. Substituting into the equilibrium-constant expression, we find

$$[C] = K_{eq}[B]_0\,([A]_0 - x)$$

Substituting into $d[D]/dt = k_3[C]$ we have

$$\begin{aligned}\frac{dx}{dt} &= k_3 K_{eq}[B]_0\,([A]_0 - x)\\ &= k_{IIa,obs}([A]_0 - x)\end{aligned}$$

where $k_{IIa,obs} = k_3 K_{eq}[B]_0$. The disappearance of A and the formation of D follow a pseudo-first-order rate equation. The pseudo-first-order rate constant is a composite quantity that is directly proportional to $[B]_0$.

Case IIc. If we suppose that the first step is effectively at equilibrium during the entire time that D is being produced (as in case IIb) but that $[C]$ is not negligibly small compared to $[A]$, we again have $K_{eq} = [C]/[A][B]$. With $[D] = x$, the mass-balance condition becomes $[A] = [A]_0 - [C] - x$. Eliminating $[A]$ between the mass-balance and equilibrium-constant equations gives

$$[C] = \frac{K_{eq}[B]_0([A]_0 - x)}{1 + K_{eq}[B]_0}$$

so that $d[D]/dt = k_3[C]$ becomes

$$\begin{aligned}\frac{dx}{dt} &= \left(\frac{k_3 K_{eq}[B]_0}{1 + K_{eq}[B]_0}\right)([A]_0 - x)\\ &= k_{IIc,obs}([A]_0 - x)\end{aligned}$$

The formation of D follows a pseudo-first-order rate equation. (The disappearance of A is also pseudo-first-order, but the pseudo-first-order rate constant is different.) As in Case IIb, the pseudo-first-order rate constant, $k_{IIc,obs}$, is a composite quantity, but now its dependence on $[B]_0$ is more complex. The result for Case IIc reduces to that for Case IIb if $K_{eq}[B]_0 \ll 1$.

Case III. In the cases above, we have assumed that one or more reactions are intrinsically much slower than others are. The differential equations for this mechanism can also become much simpler if all three reactions proceed at similar rates, but do so in such a way that the concentration of the intermediate is always very small, $[C] \approx 0$. If the concentration of C is always very small, then we expect the graph of $[C]$ versus time to have a slope, $d[C]/dt$, that is approximately zero. In this case, we have

$$\frac{d[C]}{dt} = k_1[A][B] - k_2[C] - k_3[C] \approx 0$$

so that

$$[C] = \frac{k_1[A][B]}{k_2 + k_3}$$

With $[D] = x$, $d[D]/dt = k_3[C]$ becomes

$$\begin{aligned}\frac{dx}{dt} &= \left(\frac{k_1 k_3[B]_0}{k_2 + k_3}\right)([A]_0 - x)\\ &= k_{III,obs}([A]_0 - x)\end{aligned}$$

As in the previous cases, the disappearance of A and the formation of D follow a pseudo-first-order rate equation. The pseudo-first-order rate constant is again a composite quantity, which depends on $[B]_0$ and the values of all of the rate constants.

Case III illustrates the ***steady-state approximation***, in which we assume that the concentration of an intermediate species is much smaller than the concentrations of other species that affect the reaction rate. Under these circumstances, we can be confident that the time-derivative of the intermediate's concentration is negligible compared to the reaction rate, so that it is a good approximation to set it equal to zero. The idea is simply that, if the concentration is always small, its time-derivative must also be small. If the graph of the intermediate's concentration versus time is always much lower than that of other participating species, then its slope will be much less.

Equating the time derivative of the steady-state intermediate's concentration to zero produces an algebraic expression that involves the intermediate's concentration. Solving this expression for the concentration of the steady-state intermediate makes it possible to greatly simplify the set of simultaneous differential equations that is predicted by the mechanism. When there are multiple intermediates to which the approximation is applicable, remarkable simplifications can result. This often happens when the mechanism involves free-radical intermediates.

The name "steady-state approximation" is traditional. When we use it, we do so on the understanding that the "state" which is approximately "steady" is the concentration of the intermediate, not the state of the system. Since a net reaction is occurring, the state of the system is distinctly not constant.

§13 The Michaelis-Menten mechanism for enzyme-catalyzed reactions

An enzyme is a molecule produced by a living organism. Its enzymes are essential to the life processes of the organism. Often enzymes are large molecules that are catalytically active only when folded into a particular conformation. Molecules whose reactions are catalyzed by an enzyme are customarily referred to as substrates for that enzyme. Two aspects of enzymatic catalysis are remarkable. First, it is often found that an enzyme can discriminate between two very similar substrates, greatly enhancing the reaction rate of one while having a much smaller effect on the rate of the other. Second, the rate enhancements achieved by enzyme catalysis are often very large; rate increases by a factor of 10^6 are observed.

Simple mechanistic and kinetic models are sufficient to explain these essential features of enzyme catalysis. We consider the simplest case. Both the mechanisms and the rate laws for enzymatic reactions can become much more complex than the model we develop. (The literature of enzyme catalysis uses a specialized vocabulary. Problem 32 introduces some of this terminology and some of the mechanistic complications that can be observed.)

The catalytic specificity of enzymes is explained by the idea that the enzyme and the substrate have complex three-dimensional structures. These structures complement one another in the sense that enzyme and substrate can fit together tightly, bringing the catalytically active parts of the enzyme's structure into close proximity with those substrate chemical bonds that are changed in the reaction. This is often called the ***lock and key model*** for enzyme specificity, invoking the idea that the detailed features of the enzyme's structure are shaped to fit into the structure of its substrate, just as a key is machined to match the arrangement of tumblers in the lock that it opens. Figure 2 illustrates this idea.

The rates of enzyme-catalyzed reactions can exhibit complex dependence on the relative concentrations of enzyme and substrate. Most of these features are explained by the Michaelis-Menten mechanism, which postulates a rapid equilibration of enzyme and substrate with their enzyme-substrate complex. Transformation of the substrate occurs within this complex. The reaction products do not complex strongly with the enzyme. After the substrate has been transformed, the products diffuse away. The enzyme can then complex with another substrate molecule and catalyze its reaction. Representing the enzyme, substrate, enzyme–substrate complex, and products as E, S, ES, and P, respectively, the simplest-case Michaelis-Menten mechanism is

Complexation equilibrium: $$ES \overset{K}{\rightleftharpoons} E + S$$

Substrate transformation: $$ES \overset{k}{\rightarrow} E + P$$

where K is the equilibrium constant for the ***dissociation*** of the enzyme–substrate complex, and k is the rate constant for the rate-determining transformation of the enzyme–substrate complex into products.

Since the first-order decay of the enzyme–substrate complex is the product-forming step, the reaction rate is expected to be

$$\frac{d[P]}{dt} = k[ES]$$

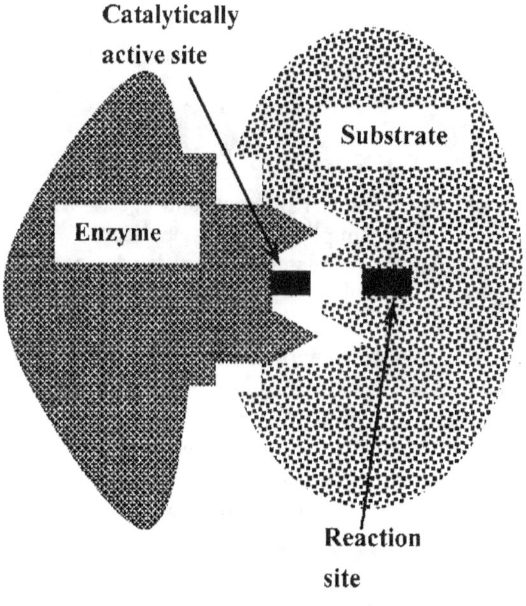

Figure 2. The lock and key model.

Letting the product concentration be $[P] = x$ and the initial concentrations of enzyme and substrate be E_0 and S_0, respectively, material balance requires

$$E_0 = [E] + [ES] \quad \text{and} \quad S_0 = [S] + [ES] + x$$

Most experiments are done with the substrate present at a much greater concentration than the enzyme, $S_0 \gg E_0$. In this case, $[ES]$ is negligible in the S_0 equation, and the concentration of free substrate becomes $[S] = S_0 - x$. The enzyme-substrate equilibrium imposes the relationship

$$K = \frac{[E][S]}{[ES]} = \frac{[E](S_0 - x)}{[ES]}$$

Using this relationship to eliminate $[E]$ from the expression for E_0 gives

$$E_0 = \frac{K[ES]}{S_0 - x} + [ES] = \left(\frac{K}{S_0 - x} + 1\right)[ES]$$

The rate law for product formation becomes

$$\frac{dx}{dt} = k[ES] = \frac{kE_0(S_0 - x)}{K + (S_0 - x)}$$

If $K \ll S_0 - x$, the dependence on substrate cancels, and the rate law becomes pseudo-zero-order

$$\frac{dx}{dt} = kE_0 = k_{obs}$$

where the observed zero-order rate constant depends on

E_0. If $K \gg S_0 - x$, the rate law becomes pseudo-first-order

$$\frac{dx}{dt} = \frac{kE_0}{K}(S_0 - x)$$
$$= k_{obs}(S_0 - x)$$

with pseudo-first-order rate constant $k_{obs} = kE_0/K$. If neither of these simplifications is applicable, fitting experimental data to the integrated rate law becomes inconvenient.

In practice, the integrated form of the rate law is seldom used. As noted earlier, enzyme catalysis is usually studied using the method of initial rates. That is, the rate is approximated as $dx/dt \approx \Delta x/\Delta t$ by measuring the amount of product formed over a short time interval, Δt, in which $x \ll S_0$. The rate and equilibrium constants can then be found by fitting the rate measured at various initial concentrations, E_0 and S_0, to the equation

$$\frac{\Delta x}{\Delta t} \approx \frac{kE_0S_0}{K + S_0}$$

Since the concentration of the enzyme–substrate complex is normally small compared to the concentration of the substrate, the Michaelis-Menten mechanism can also be analyzed by applying the steady-state approximation to $[ES]$. If the product-forming step occurs so rapidly that the concentration of the enzyme–substrate complex is not maintained at the equilibrium value, the steady-state treatment becomes necessary.

§14 The Lindemann-Hinshelwood mechanism for first-order decay

First-order kinetics for a unimolecular reaction corresponds to a constant probability that a given molecule reacts in unit time. In §7, we outline a simple mechanism that rationalizes this fact. This mechanism assumes that the probability of reaction is zero unless the molecule has some minimum energy. For molecules whose energy exceeds the threshold value, we assume that the probability of reaction is constant. However, when collisions are frequent, a molecule can have excess energy only for brief intervals.

The ***Lindemann-Hinshelwood mechanism*** for gas-phase unimolecular reactions provides a mathematical model of these ideas. Since molecules exchange energy via collisions, any given molecule acquires excess energy by collisions with other molecules, and loses it within a short time through other collisions. If it retains its excess energy long enough, it will react. If collisions are very infrequent, every molecule that acquires excess energy reacts before it undergoes a deactivating collision. In this case the reaction rate is proportional to the rate at which molecules acquire excess energy, which is proportional to the number of collisions. In a collection of A molecules, the total number of A—A collisions is proportional to $[A]^2$ not $[A]$ and so the reaction rate depends on $[A]^2$ not $[A]$.

We represent molecules with excess energy as A^*, and assume that all A^* molecules undergo reaction with a constant probability. A^* molecules are formed in collisions between A molecules, and they are deactivated by subsequent collisions with A molecules.

$$A + A \underset{k_2}{\overset{k_1}{\rightleftharpoons}} A^* + A$$

$$A^* \xrightarrow{k_3} Products\ (P)$$

The rate at which the number of moles of A^* molecules changes is

$$\frac{1}{V}\frac{dn_{A^*}}{dt} = k_1[A]^2 - k_2[A^*][A] - k_3[A^*]$$

and since we suppose that $[A^*]$ is always very small, the steady-state approximation applies, so that $dn_{A^*}/dt = 0$, and

$$[A^*] = \frac{k_1[A]^2}{k_2[A] + k_3}$$

The reaction rate is given by

$$\frac{1}{V}\frac{d\xi}{dt} = k_3[A^*]$$
$$= \frac{k_1k_3[A]^2}{k_2[A] + k_3}$$

When $k_2[A] \gg k_3$, the rate of deactivating collisions between A^* and A is greater than the rate at which A^* molecules go on to become products. The rate law for consumption of $[A]$ becomes first order. This is termed the rate law's ***high-pressure limit***. The ***low-pressure limit*** occurs when $k_2[A] \ll k_3$. The rate law becomes second order in $[A]$. The rate of product formation becomes equal to the rate at which A^* molecules are formed.

§15 Why unimolecular reactions are first order

In the discussion above, we assume that a molecule with energy in excess of the minimum activation energy undergoes reaction with some fixed probability, represented by the rate constant k_3. A complete answer to the question of why unimolecular processes are characteristically first-order (in the high pressure limit) requires that we rationalize this assumption. Another way of phrasing this question is to ask why the activated molecule does not react immediately: Why isn't $k_3 \approx \infty$?

The total energy of a molecule is distributed among numerous degrees of freedom. The molecule has translational kinetic energy, rotational kinetic energy, and vibrational energy. When it acquires excess energy through a collision with another molecule, the additional energy could go directly into any of these modes. However, before the molecule can react, enough energy must find its way into some rather particular mode. If, for example, the reaction involves the breaking of a chemical bond, and the collision puts excess kinetic energy into the molecule's translational modes, the reaction can occur only after some part of the excess translational energy has been converted to excess vibrational energy in the breaking bond. This intramolecular transfer of energy among the molecule's various internal modes is time-dependent.

From this perspective, the probability that an excited molecule will react in unit time is the probability that the necessary energy will reach the critical locus in unit time. The reshuffling of energy among the molecule's internal modes is a stochastic process, and the probability that the reshuffling will put the necessary energy where it is needed is a constant characteristic of the molecule.

§16 The mechanism of the base hydrolysis[1] of $Co(NH_3)_5X^{n+}$

This chapter focuses on the relationship between rate laws and reaction mechanisms. We have noted that the rate law is rarely sufficient to establish the mechanism of a particular reaction. The base hydrolysis of cobalt pentammine complexes is a reaction for which numerous lines of evidence converge to establish the mechanism. To illustrate the range of data that can be useful in the determination of reaction mechanisms, we summarize this evidence here.

Cobalt(III) complexes usually undergo substitution reactions at readily measurable rates. Cobalt pentammine complexes, $Co(NH_3)_5X^{n+}$, have been studied extensively. In acidic aqueous solutions, the reaction

$$Co(NH_3)_5X^{n+} + Y^{p-} \rightarrow Co(NH_3)_5Y^{m+} + X^{q-}$$

(X^{q-}, $Y^{p-} = Cl^-, Br^-, NO_2^-, SCN^-, CH_3CO_2^-$, *etc.*) usually proceeds exclusively through the aquo complex, $Co(NH_3)_5OH_2{}^{3+}$. The first step in the reaction is the breaking of a $Co—X$ bond and the formation of a $Co—OH_2$ bond. Subsequently, a Y^{p-} moiety can replace the aquo group. In aqueous solution, water is always present at a much higher concentration than the various possible entering groups Y^{p-}, so it is reasonable that it should be favored in the competition to form the new bond to Co(III). Nevertheless, we expect the strength of the $Co—Y$ bond to be an indicator of the nucleophilicity of Y^{p-} in these substitution reactions. The fact that the aquo complex is the predominant reaction product strongly suggests that the energetics of the reaction are dominated by the breaking of the $Co—X$ bond; formation of the new bond to the incoming ligand apparently has little effect. Whether the old $Co—X$ bond has been completely broken (so that $Co(NH_3)_5{}^{3+}$ is a true intermediate) before the new $Co—OH_2$ bond has begun to form remains an issue on which it is possible to disagree.

There is a conspicuous exception to the description given above. When the entering group, Y^{p-}, is the hydroxide ion, the reaction is

$$Co(NH_3)_5X^{n+} + OH^- \rightarrow Co(NH_3)_5OH^{2+} + X^{q-}$$

This is called the ***base-hydrolysis reaction***. It is faster than the formation of the aquo complex in acidic solutions, and the rate is law found to be

$$\frac{d[Co(NH_3)_5OH^{2+}]}{dt} = k\,[Co(NH_3)_5X^{n+}][OH^-]$$

This rate law is consistent with S_N2 nucleophilic attack by the hydroxide ion at the cobalt center, so that $Co—OH$ bond formation occurs simultaneously with breaking of the $Co—X$ bond. However, this interpretation means that the hydroxide ion is a uniquely effective nucleophile toward cobalt(III). Nucleophilic displacements have been investigated on many other electrophiles. In general, hydroxide is not a particularly effective nucleophile toward other electrophilic centers. So, assignment of an S_N2 mechanism to this reaction is reasonable only if we can explain why hydroxide is uniquely reactive in this case and not in others.

An alternative mechanism, usually labeled the S_N1CB (**S**ubstitution, **N**ucleophilic, first-order in the **C**onjugate **B**ase) mechanism, is also consistent with the second-order rate law. In this mechanism, hydroxide removes a proton from one of the ammine ligands, to give a six-coordinate intermediate, containing an amido (NH_2^-) ligand. This intermediate loses the leaving group X^{q-} in the rate determining step to form a five-coordinate intermediate, $Co(NH_3)_4NH_2^{2+}$. This intermediate picks up a water molecule to give the aquo complex. In a series of proton transfers to and from the aqueous solvent, the aquo complex rearranges to the final product. With Cl^- as the leaving group, the S_N1CB mechanism is

$$Co(NH_3)_5Cl^{2+} + OH^- \rightleftharpoons Co(NH_3)_4(NH_2)Cl^+ + H_2O$$

$$Co(NH_3)_4(NH_2)Cl^+ \rightleftharpoons Co(NH_3)_4(NH_2)^{2+} + Cl^-$$

$$Co(NH_3)_4(NH_2)^{2+} + H_2O \rightleftharpoons Co(NH_3)_4(NH_2)(OH_2)^{2+}$$

$$Co(NH_3)_4(NH_2)(OH_2)^{2+} + OH^- \rightleftharpoons Co(NH_3)_4(NH_2)OH^+ + H_2O$$

$$Co(NH_3)_4(NH_2)OH^+ + H_2O \rightleftharpoons Co(NH_3)_5OH^{2+} + OH^-$$

The evidence in favor of the S_N1CB mechanism is persuasive. It requires that the ammine protons be acidic, so that they can undergo the acid–base reaction in the first step. That this reaction occurs is demonstrated by proton-exchange experiments. In basic D_2O, the ammine protons undergo $H—D$ exchange according to

$$Co(NH_3)_5Cl^{2+} + D_2O \rightleftharpoons Co(ND_3)_5Cl^{2+} + HDO$$

The ammine protons are also necessary; base hydrolysis does not occur for similar compounds, like $Co(2,2'\text{-}bipyridine)_2(O_2CCH_3)_2^+$, in which there are no protons on the nitrogen atoms that are bound to cobalt (*i.e.*, there are no $H-N-Co$ moieties).

The evidence that $Co(NH_3)_4(NH_2)^{2+}$ is an intermediate is also persuasive. When the base hydrolysis reaction is carried out in the presence of other possible entering groups, Y^{p-}, the *rate* at which $Co(NH_3)_5X^{n+}$ is consumed is unchanged, but the product is a mixture of $Co(NH_3)_5OH^{2+}$ and $Co(NH_3)_5Y^{n+}$. If this experiment is done with a variety of leaving groups, X^{q-}, the proportions of $Co(NH_3)_5OH^{2+}$ and $Co(NH_3)_5Y^{n+}$ are constant—independent of which leaving group the reactant molecule contains. These observations are consistent with the hypothesis that all reactants, $Co(NH_3)_5X^{n+}$, give the same intermediate, $Co(NH_3)_4(NH_2)^{2+}$. The product distribution is always the same, because it is always the same species undergoing the product-forming reaction.

§17 Chemical equilibrium as the equality of rates for opposing reactions

Suppose that the reaction $A + B \underset{k_r}{\overset{k_f}{\rightleftharpoons}} C + D$ occurs as an elementary process. From our conclusions about the concentration dependencies of elementary reactions, the rate of the net reaction is

$$R = \frac{1}{V}\frac{d\xi}{dt} = -\frac{1}{V}\frac{dn_A}{dt} = k_f[A][B] - k_r[C][D]$$

at any time. In particular, this rate equation must remain true at arbitrarily long times—times at which the reaction has reached equilibrium and at which $dn_A/dt = 0$. Therefore, at equilibrium we have

$$K = \frac{k_f}{k_r} = \frac{[C]_{eq}[D]_{eq}}{[A]_{eq}[B]_{eq}}$$

where the concentration-term subscripts serve to emphasize that the concentration values correspond to the reaction being in a state of equilibrium. We see that the ratio of rate constants, k_f/k_r, characterizes the equilibrium state. This constant is so useful we give it a separate name and symbol, the ***equilibrium constant***, K.

Now, let us consider the possibility that the reaction is not an elementary process, but instead proceeds by a two-step mechanism involving an intermediate, E:

$$A + B \underset{k_2}{\overset{k_1}{\rightleftharpoons}} E$$

$$E \underset{k_4}{\overset{k_3}{\rightleftharpoons}} C + D$$

The sum of these elementary processes yields the same overall reaction as before. This mechanism implies the following differential equations:

$$\frac{1}{V}\frac{dn_A}{dt} = -k_1[A][b] + k_2[E]$$

$$\frac{1}{V}\frac{dn_D}{dt} = k_3[E] - k_4[C][D]$$

At equilibrium, both n_A and n_D must be constant, so both differential equations must be equal to zero. Hence, at equilibrium,

$$\frac{k_1}{k_2} = \frac{[E]_{eq}}{[A]_{eq}[B]_{eq}}$$

$$\frac{k_3}{k_4} = \frac{[C]_{eq}[D]_{eq}}{[E]_{eq}}$$

Multiplying these, we have

$$K = \frac{k_1 k_3}{k_2 k_4} = \frac{[C]_{eq}[D]_{eq}}{[A]_{eq}[B]_{eq}}$$

The concentration dependence of the equilibrium constant for the two-step mechanism is the same as for case that the reaction is an elementary process. As far as the description of the equilibrium system is concerned, the only difference is that the equilibrium constant is interpreted as a function of different rate constants.

For the general reaction

$$aA + bB + \cdots \rightleftharpoons fF + gG + \cdots$$

we see that any sequence of elementary reactions will give rise to the same concentration expression for the equilibrium system. Whatever the mechanism, reactant A must appear a times more often on the left side of elementary reactions than it does on the right. Product F must appear f times more often on the right side of elementary reactions than it does on the left. Any intermediates must appear an equal number of times on the left and on the right in the various elementary reactions. As a result, when we form the ratio of forward to reverse rate constants for each of the elementary reactions and multiply them, the concentration of reactant A must appear in the product to the $-a$ power, the concentration of product F must appear to the $+f$ power, and the concentrations of the intermediates must all cancel out. We conclude that the condition for equilibrium in the general case is

$$K = \frac{[F]^f[G]^g \ldots}{[A]^a[B]^b \ldots}$$

where we drop the "eq" subscripts, trusting ourselves to remember that the equation is valid only when the concentration terms apply to the equilibrated system.

When the reaction involves a pure phase as a reactant or product, we observe experimentally that the amount of the pure phase present in the reaction mixture does not affect the position of equilibrium. The composition of the reaction solution is the same so long as the solution is in contact with a finite amount of the pure phase. This means that we can omit the concentration of the substance that makes up the pure phase when we write the equilibrium-constant expression. In writing the equilibrium constant expression, we can take the concentration the substance to be an arbitrary constant. Unity is usually the most convenient choice for this constant.

To rationalize this experimental observation within our kinetic model for equilibrium, we postulate that the rate at which molecules leave the pure phase is proportional to the area, S, of the phase that is in contact with the reaction solution; that is, $R_{leaving} = k_1 S$. We postulate that the rate at which molecules return to the pure phase from the reaction solution is proportional to both the area and the concentration of the substance in the reaction solution. If the pure phase consists of substance A, we have $R_{returning} = k_2 S[A]$. At equilibrium, we have $R_{leaving} = R_{returning}$, so that $k_1 S = k_2 S[A]$, and $[A] = k_1/k_2 =$ constant

§18 The principle of microscopic reversibility[2,3,4,5]

The equilibrium constant expression is an important and fundamental relationship that relates the concentrations of reactants and products at equilibrium. We deduce it above from a simple model for the concentration dependence of elementary-reaction rates. In doing so, we use the criterion that the time rate of change of any concentration must be zero at equilibrium. Clearly, this is a necessary condition; if any concentration is changing with time, the reaction is not at equilibrium. However, our deduction uses another assumption that we have not yet emphasized. We assume that ***the forward and reverse rates of each elementary step are equal when the overall reaction is at equilibrium***. This is a special case of the ***principle of microscopic reversibility:***

> **Any molecular process and its reverse occur with equal rates at equilibrium**

The principle of microscopic reversibility applies to any molecular process; it is inferred from the fact that such processes can be described by their equations of motion if the initial state of the constituent particles can be specified. The equations of motion can be either classical mechanical or quantum mechanical. We consider the implications of the principle for molecular processes that constitute elementary reactions. However, the principle also applies to equilibria in other molecular processes, notably the absorption and emission of radiation.

When we apply it to elementary reactions, we see that the principle of microscopic reversibility provides a necessary and sufficient condition for equilibrium from a reaction-mechanism perspective. The principle also imposes several significant conditions on the sequences of elementary processes that constitute a mechanism and on their relative rates.

In the previous section, we see that microscopic reversibility provides a ***sufficient*** basis for deducing the relationship relating reactant and product concentrations at

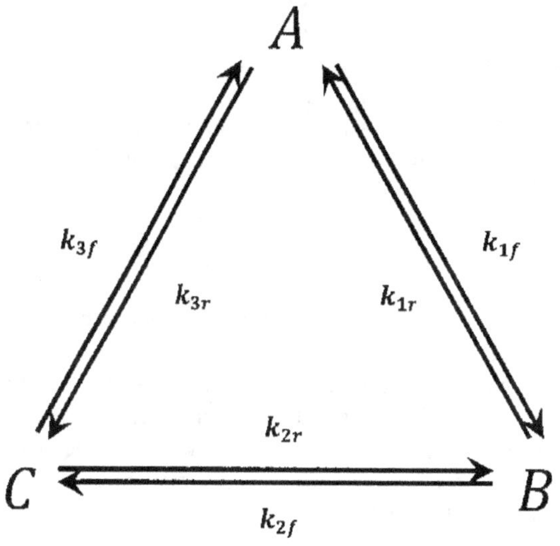

Figure 3. Cyclic equilibrium.

equilibrium—the equilibrium constant expression—from our rate equations for elementary reactions. We now want to see that the principle of microscopic reversibility is indeed ***necessary***. That is, **setting $dn_x/dt = 0$ for all species, X, involved in the reaction is not in itself *sufficient* to assure that the system is at equilibrium.**

We consider the triangular network of elementary reactions[3,4] shown in Figure 3. This network gives rise to the following reaction-rate equations:

$$V^{-1}\, dn_A/dt = -k_{1f}[A] + k_{1r}[B] + k_{3f}[C] - k_{3r}[A]$$

$$V^{-1}\, dn_B/dt = +k_{1f}[A] - k_{1r}[B] + k_{2r}[C] - k_{2f}[B]$$

$$V^{-1}\, dn_C/dt = +k_{2f}[B] - k_{2r}[C] + k_{3r}[A] - k_{3f}[C]$$

At equilibrium each of these equations must equal zero. Since we have three equations in three unknowns, it might at first appear that we can solve for the three unknowns $[A]$, $[B]$, and $[C]$. We can see, however, either from the equations themselves or by considering the physical situation that they represent, that only two of these equations are independent. That is, we have

$$\frac{dn_A}{dt} + \frac{dn_B}{dt} + \frac{dn_C}{dt} = 0$$

While we cannot solve for $[A]$, $[B]$, and $[C]$ independently, we can solve for their ratios, which are

$$\frac{[B]}{[A]} = \frac{k_{1f}k_{2r} + k_{2r}k_{3r} + k_{1f}k_{3f}}{k_{2f}k_{3f} + k_{1f}k_{3f} + k_{1r}k_{2r}}$$

$$\frac{[C]}{[B]} = \frac{k_{1f}k_{2f} + k_{1r}k_{3r} + k_{2f}k_{3r}}{k_{1f}k_{2r} + k_{2r}k_{3r} + k_{1f}k_{3f}}$$

$$\frac{[A]}{[C]} = \frac{k_{2f}k_{3f} + k_{1r}k_{3f} + k_{1r}k_{2r}}{k_{1f}k_{2f} + k_{1r}k_{3r} + k_{2f}k_{3r}}$$

Since we deduce these equations from the condition that

all the time derivatives are zero, it might seem that they should represent the criteria for the system of reactions to be at equilibrium. Purely as a name for easy reference, let us call these equations the cyclic set.

When we consider the reactions one at a time, we deduce the following equilibrium relationships:

$$\frac{[B]}{[A]} = \frac{k_{1f}}{k_{1r}} \qquad \frac{[C]}{[B]} = \frac{k_{2f}}{k_{2r}} \qquad \frac{[A]}{[C]} = \frac{k_{3f}}{k_{3r}}$$

For easy reference, let us call these equations the one-at-a-time set.

Now, it cannot be true that both sets of relationships specify a sufficient condition for the system to be at equilibrium. To see this, let us first suppose that the principle of microscopic reversibility is a sufficient condition for equilibrium. Then the one-at-a-time set of equations must be sufficient to uniquely specify the position of equilibrium. It is easy to show that a set of rate constants that satisfies the one-at-a-time set also satisfies the cyclic set. Therefore, if microscopic reversibility is a sufficient condition for equilibrium, the cyclic network rate equations are necessarily equal to zero at equilibrium. In short, if we assume that microscopic reversibility is a sufficient condition for equilibrium, we encounter no inconsistencies, because the cyclic set of equations is satisfied by the same equilibrium-concentration ratios.

On the other hand, if we suppose that setting $dn_X/dt = 0$ for all species, X, is a sufficient condition for equilibrium, then the cyclic set of equations must be sufficient to uniquely specify the position of equilibrium. Let us consider a particular set of rate constants: $k_{1f} = k_{2f} = k_{3f} = 1$ and $k_{1r} = k_{2r} = k_{3r} = 2$. This set of rate constants satisfies the cyclic set of equations and requires that each of the equilibrium-concentration ratios be equal to 1. In this case, the one-at-a-time set of equations implied by microscopic reversibility cannot be satisfied. (We have $[B]/[A] = 1$ and $k_{1f}/k_{1r} = 1/2$. Therefore, $[B]/[A] \neq k_{1f}/k_{1r}$.) That is, if we assume that setting $dn_X/dt = 0$, for all species, X, is a sufficient condition for equilibrium, we must conclude that the principle of microscopic reversibility is false. Using the contrapositive: If the principle of microscopic reversibility is true, it is false that setting $dn_X/dt = 0$ for all species, X, is a sufficient condition for equilibrium.

Setting the derivatives for the reaction network equal to zero is not sufficient to assure that the system is at equilibrium. It is merely necessary. To assure that the network is at equilibrium, we must apply the principle of microscopic reversibility and require that each elementary process in the network be at equilibrium.

The principle of microscopic reversibility requires that any elementary process occur via the same sequence of transitory molecular structures in both the forward and reverse directions. Consequently, if a sequence of elementary steps is a mechanism for a forward reaction, the same sequence of steps—traversed backwards—must be a mechanism for the reverse reaction. The principle does not exclude the possibility that a given reaction can occur simultaneously by two different mechanisms. However, it does mean that a given reaction cannot have one mechanism in the forward direction and a second, different mechanism in the reverse direction.

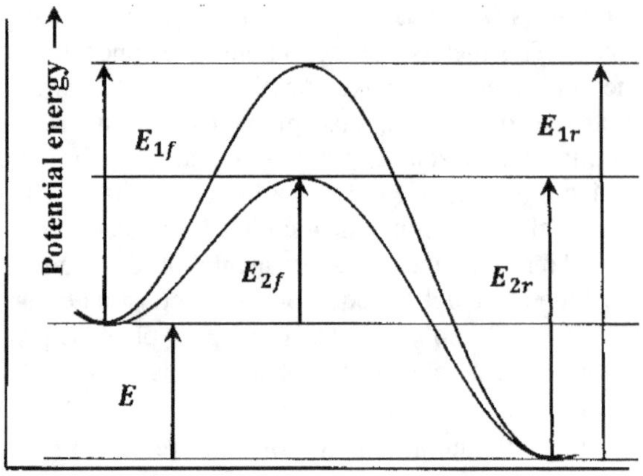

Figure 4. Potential energy versus reaction coordinate.

In describing reaction mechanisms, we assume that the energy of the reacting molecules depends on their progress along the path that they follow during the course of the reaction. We call this path the reaction coordinate. We suppose that we can plot the energy of the system as a function of the system's position on the path, or displacement along the reaction coordinate. In the context of such a graph, the principle of microscopic reversibility is essentially the observation that the path is the same irrespective of the direction in which it is traversed. Two such paths are sketched in Figure 4. In this sketch, E_{1f} and E_{2f} are the activation energies for the two forward reactions; E_{1r} and E_{2r} are the activation energies for the reverse reactions.

§19 Microscopic reversibility and the second law

The principle of microscopic reversibility requires that parallel mechanisms give rise to the same expression for the concentration-dependence of the equilibrium constant. That is, the ***function*** that characterizes the equilibrium composition must be the same for each mechanism. If, for the reaction $aA + bB \rightleftharpoons cC + dD$, the equilibrium composition for mechanism 1 is $[A]_1, [B]_1, [C]_1, [D]_1$, and that for mechanism 2 is $[A]_2, [B]_2, [C]_2, [D]_2$, microscopic reversibility asserts that

$$K_1 = \frac{[C]_1^c[D]_1^d}{[A]_1^a[B]_1^b}$$

and

$$K_2 = \frac{[C]_2^c[D]_2^d}{[A]_2^a[B]_2^b}$$

In and of itself, microscopic reversibility makes no assertion about the value of $[A]_1$ compared to that of $[A]_2$. While microscopic reversibility asserts that the same function characterizes the concentration relationships for parallel mechanisms, it does not assert that the numerical value of this function is necessarily the same for each of the mechanisms.

However, that these numerical values must be equal follows directly when we introduce another of our most basic observations. No matter how many mechanisms may be available to a reaction in a particular system, the concentration of any reagent can have only one value in an equilibrium state. At equilibrium, $[A]_1 = [A]_2$, *etc.*; therefore, the numerical values of the equilibrium constants must be the same: $K_1 = K_2$.

The uniqueness of the equilibrium composition is a fundamental feature of our ideas about what chemical equilibrium means. Nevertheless, it is of interest to show that we can arrive at this conclusion from a different perspective: We can use an idealized machine to show that the second law of thermodynamics requires that parallel mechanisms must produce the same the equilibrium composition. Our argument is a proof by contradiction.

Let us suppose that A, B, and C are gases. Suppose that the reaction $A \rightarrow B + C$ occurs in the absence of a catalyst, but that reaction occurs in the opposite direction, $C + B \rightarrow A$, when a catalyst is present. More precisely, we assume that the position of equilibrium $A \rightleftharpoons B + C$ lies to the right in the absence of the catalyst and to the left in its presence, while all other reaction conditions are maintained constant. These assumptions mean that the equilibrium composition for the catalyzed mechanism is different from that of the mechanism that does not involve the catalyst.

We can show that these assumptions imply that the second law of thermodynamics is false. If we accept the validity of the second law, this violation means that the assumptions cannot in fact describe any real system. (We are getting a bit ahead of ourselves here, inasmuch as our detailed consideration of the laws of thermodynamics begins in Chapter 6.)

Given our assumptions, we can build a machine consisting of a large cylinder, closed by a frictionless piston. The cylinder contains a mixture of A, B, and C, and a quantity of the catalyst. We provide a container for the catalyst, and construct the device so that the catalyst container can be opened and closed from outside the cylinder. Finally, we immerse the entire cylinder in a fluid, which we maintain at a constant temperature.

When the catalyst container is sealed, so that the gaseous contents of the cylinder are not in contact with the catalyst, reaction occurs according to $A \rightarrow B + C$, and the piston moves outward, doing work on the surroundings. When the catalyst container is open, reaction occurs according to $C + B \rightarrow A$, and the piston moves inward. Figure 5 shows these changes schematically. At the end of a cycle, the machine is in exactly the same state as it was in the beginning, and the temperature of the reaction mixture is the same at the end of a cycle as it was at the beginning. By connecting the piston to a load, we can do net work on the load as the machine goes through a cycle. For example, if we connect the piston to a mechanical device that converts the reciprocating motion of the piston into rotary motion, we can wind a rope around an axle and thereby lift an attached weight.

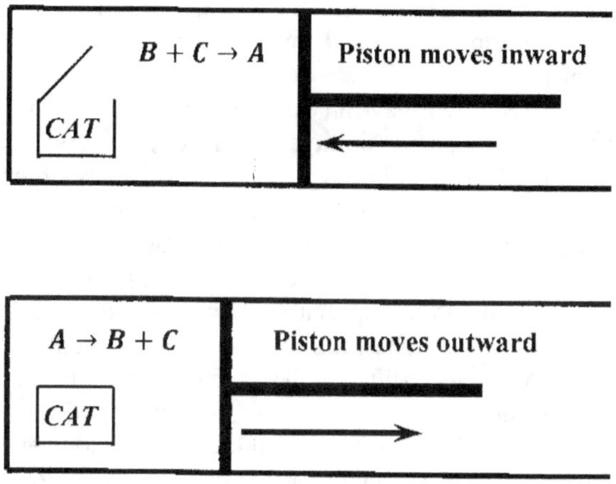

Figure 5. A machine that violates the second law.

We can operate this machine as an engine by alternately opening and closing the catalyst container. We can make the cylinder as large as we want, so the energy we expend in opening and closing the catalyst container can be made arbitrarily small compared to the amount of work we get out of the machine in a given cycle. All of this occurs with the machine maintained at a constant temperature. If energy is conserved, the machine must absorb heat from the bath during the cycle; otherwise, the machine would be doing work with no offsetting consumption of energy. This would be a violation of the first law of thermodynamics. (See §7-10,11.)

From experience, we know that this machine cannot function in the manner we have described. This experience is embodied in the ***second law of thermodynamics***, from which we know that **it is impossible to construct a machine that operates in a cycle, exchanges heat with its surroundings at only one temperature, and produces work in the surroundings**. (See §9-1.) Our argument assumes that two reaction mechanisms are available in a particular physical system, that they consume the same reactants, that they produce the same products, and that the equilibrium compositions are different. These assumptions imply that the second law is false. Since we are confident that it is possible for some system to satisfy the first three of these assumptions, the second law requires that the last one be false: the equilibrium compositions must be the same.

We see that there is a complementary relationship between microscopic reversibility and this statement of the second law. Microscopic reversibility asserts that a unique function of concentrations characterizes the equilibrium state for any reaction mechanism, but does not require that every mechanism reach the same state at equilibrium. This statement of the second law implies that a reaction's equilibrium composition unique, but it does not specify a law relating the equilibrium concentrations of the reacting species. (In Chapters 9 and 13, we see that, by augmenting this statement of the second law with some additional ideas, we are led to a more rigorous statement, from which we are eventually able to infer the same functional form for the equilibrium constant.)

G. N. Lewis gave an early statement of the principle of microscopic reversibility. He called it "the law of entire equilibrium," and observed[5] that it is "a law which in its general form is not deducible from thermodynamics, but proves to be compatible with the laws of thermodynamics in all cases where a comparison is possible."

It is worth noting that we have not shown that the existence of a unique equilibrium state implies either microscopic reversibility or the second law. Also, even though the principle of microscopic reversibility is inferred from the laws of mechanics, our development of the equilibrium constant relationship—which we do view as a law of thermodynamics—depends on our equations for the rates of elementary reactions. Our rate equations are not logical consequences of the laws of motion. Rather, they follow from assumptions we make about the average behavior of systems that contain many molecules. Consequently, we should not suppose that we have deduced a thermodynamic result (the condition for chemical equilibrium) solely from the laws of mechanics. In §12-2, we give brief additional consideration to the relationship between the theories of mechanics and thermodynamics. Beginning in Chapter 20, we develop thermodynamic equations by applying statistical models to the distribution of molecular energy levels.

Problems

1. A dimeric molecule, A_2, dissociates in aqueous solution according to $A_2 \rightarrow 2A$. One millimole of A_2 is dissolved rapidly in one liter of pure water. After 100 seconds, the concentration of A_2 is 8.0×10^{-4} $\underline{M}$.
(a) What is the concentration of A?
(b) What is the average rate at which A has been formed, in moles per liter per second?
(c) What is the average rate at which A_2 has reacted?
(d) What is the average reaction rate?

2. The initial concentration of A in a solution is 10^{-2} $\underline{M}$. The initial concentrations of B and C are both 10^{-3} $\underline{M}$. The volume of the solution is 2 L. Reaction occurs according to the stoichiometry: $A + 2B \rightarrow 3C$. After 50 seconds, the concentration of C is 1.3×10^{-3} $\underline{M}$.
(a) What is the concentration of A?
(b) What is the concentration of B?
(c) What is the change in the extent of this reaction during this 50 s period?
(d) What is the average reaction rate?

3. When A_2 dissociates according to $A_2 \rightarrow 2A$, the observed rate law is

$$\frac{1}{V}\frac{dn_A}{dt} = \left(\frac{k}{2}\right)[A_2]^{3/2}$$

(a) What is the order of the reaction in $[A_2]$?
(b) What is the order of the reaction overall?

4. For the reaction $A + 2B \rightarrow 3C$, the observed rate law is

$$\frac{1}{V}\frac{d\xi}{dt} = \frac{k[A][B]^2}{[C]}$$

(a) What is the order of the reaction in $[A]$?
(b) In$[B]$?
(c) In $[C]$?
(d) What is the order of the reaction overall?

5. You deposit \$1000 in a bank that pays interest at a 5% annual rate. How much will your account be worth after one year if the bank compounds interest annually? How much if it compounds interest monthly? Daily? Continuously?

6. We deduced that $\exp(rt) = \lim_{m\to\infty}\left(1 + \frac{r}{m}\right)^{mt}$. Take $r = 0.2$ and $t = 10$. Calculate $\exp(rt)$. Calculate $\left(1 + \frac{r}{m}\right)^{mt}$ for $m = 1, 10, 100, 10^3, 10^4$. Do the same for $r = -0.2$ and $t = 10$.

7. Suppose that you invest \$10,000 in the stock market and that your nest egg grows at the rate of 11% per year. (This number, or something close to it, is often cited as the historical long-term average performance of equities traded on the New York Stock Exchange.) Assuming continuous compounding, what will be the value of your nest egg at the end of 30 years?

8. Suppose instead that you "invest" your \$10,000 in an automobile. The value of the automobile will most likely decrease with time, by, say, roughly 20% per year. Assuming continuous decay at this rate, what will be the value of the automobile at the end of 5 years? 30 years?

9. At particular reaction conditions, a compound C decays in a first-order reaction with rate constant 10^{-3} s^{-1}. If the initial concentration of C, is $[C]_0 = 10^{-2}$ mol L^{-1}, how much does $[C]$ change in the first second? What is $[C]$ after 100 seconds? 1000 seconds? 2000 seconds? 4000 seconds?

10. In problem 9, how long does it take for one-half of the original concentration of C to disappear? How does the half-life depend on the initial concentration of C?

11. C^{14} is produced continuously in the upper atmosphere. It decays with a half-life of 5715 years. Since this has been going on for a long time, the concentration of C^{14} in the atmosphere has reached a steady-state value. Living things continuously exchange carbon with the atmosphere, so the concentration of C^{14} (i.e., the fraction of the C that is C^{14}) in the biosphere is the same as it is in the atmosphere. When an organism dies, it ceases to exchange carbon with the biosphere, and the concentration of C^{14} in its remains begins to decrease. Charcoal found at an ancient campsite during an archeological dig has a C^{14} content that is 22% of the atmospheric value.

(a) What is the rate constant, in y^{-1}, for C^{14} decay?

(b) How old is the charcoal?

(c) Reverend Smith tells his parishioners that God created the universe about 4000 B.C. If Smith accepts that the atmospheric concentration and the decay rate of C^{14} have been constant since the time of creation, what would he conclude about the C^{14} content of the charcoal when God created it?

12. Mordred has introduced an exotic fungus that is growing on the surface of King Arthur's favorite pond at Camelot. Merlin has calculated that the area covered by the fungus increases by 10% per day. That is,

$$d(area)/dt = (0.10/day) \times area$$

Fortunately, a trained Knight of the Round Table can clear 100 m^2 of fungus per day. Arthur has six trained knights who would cheerfully perform this remediation work, but all six are committed to out-of-town dragon-slaying activities for the next 10 days. Merlin says that the fungus covers 2874 m^2 at 8:00 a.m. this morning. (The total area of the pond is about 11,200 m^2.) Can Arthur wait for the dragon slayers to return, or does he need to develop an alternative effective management action plan?

13. What happens to the balance in a bank account, $P(t)$, if $P(0) < 0$? What do bankers call this sort of account?

14. Suppose that you have an account whose initial balance is $-\$1000$. The bank will continuously compound interest on this account at the annual rate of 11%. You make continuous payments to this account at a rate of q dollars/year. What must q be if you want to increase the value of the account to exactly zero at the end of 10 years?

For problems 15 – 18, prove that your conclusion is correct by making an appropriate plot. In each case, the reaction occurs at constant volume.

15. The reaction $A + B \rightarrow C$ is studied with a large excess of B. ($[A]_0 = 10^{-2}$ M. $[B]_0 = 10^{-1}$ M. $[C]_0 = 0.0$ M) Concentration *versus* time data are given in the table below. What is the order of the reaction in the concentration of A, and what is the rate constant?

Time, s	$[A]$, M
100	9.1×10^{-3}
300	7.4×10^{-3}
500	6.1×10^{-3}
800	4.6×10^{-3}
1000	3.6×10^{-3}
1500	2.3×10^{-3}
2000	1.3×10^{-3}
2500	8.3×10^{-4}

16. The following data are collected for a reaction in which A dimerizes: $2A \rightarrow A_2$. What is the order of the reaction in $[A]$, and what is the rate constant?

Time, hr	$[A]$, M
0.0	1.0×10^{-2}
0.28	9.1×10^{-3}
0.56	8.3×10^{-3}
1.39	6.7×10^{-3}
2.78	5.0×10^{-3}
5.56	3.3×10^{-3}
11.10	2.0×10^{-3}
16.70	1.4×10^{-3}

17. In the reaction $A + B \rightarrow C$, the rate at which B is consumed is first-order in $[B]$. In a series of experiments whose results are tabulated below, the observed first-order rate constant, k_{obs} is measured for the disappearance of B in the presence of large excesses of A. What is the order of the reaction in $[A]$? The rate law? The rate constant? .($[B]_0 = 10^{-4}$ M in all experiments.)

$[A]_0$, M	k_{obs}, s^{-1}
2.0×10^{-1}	2.6×10^{-5}
1.1×10^{-1}	1.4×10^{-5}
6.3×10^{-2}	8.2×10^{-6}
2.5×10^{-2}	3.3×10^{-6}
9.1×10^{-3}	1.2×10^{-6}

18. In the reaction $A + B \rightarrow C$, the rate at which B is consumed is first-order in $[B]$. The table below presents first-order rate constants for the disappearance of B in the presence of large excesses of A. Plot these data to test the hypothesis that

$$k_{obs} = \frac{k_1[A]_0}{1 + k_2[A]_0}$$

What are the values of k_1 and k_2?

$[A]_0$, M	k_{obs}, s^{-1}
5.0×10^{-1}	8.3×10^{-5}
2.0×10^{-1}	6.7×10^{-5}
1.0×10^{-1}	5.0×10^{-5}
5.0×10^{-2}	3.3×10^{-5}
1.4×10^{-2}	1.2×10^{-5}
7.6×10^{-3}	7.1×10^{-6}
3.0×10^{-3}	3.0×10^{-6}

19. For the reaction $A + 2B \rightarrow C + D$, the rate law is

$$\frac{d[C]}{dt} = k[A][B]^2$$

The volume is constant. Suggest a mechanism for this reaction that ***does not*** include a termolecular elementary process. Show that this mechanism is consistent with the rate law.

20. For the reaction $X + 2Y \rightarrow W + Z$, the rate law is

$$\frac{d[W]}{dt} = k[X]$$

The volume is constant. Suggest a mechanism for this reaction. Show that this mechanism is consistent with the rate law.

21. For the reaction $A_2 + 2B \rightarrow 2C$, the rate law is

$$\frac{d[C]}{dt} = k[A_2]^{1/2}[B]$$

The volume is constant. Suggest a mechanism for this reaction. Show that this mechanism is consistent with the rate law.

22. For the reaction $AB + C \rightarrow A + D$, the rate law is

$$\frac{d[D]}{dt} = \frac{k_u[AB][C]}{k_v[A] + k_w[C]}$$

The volume is constant. Suggest a mechanism for this reaction. Show that this mechanism is consistent with the rate law.

23. When we use the flooding technique to study a reaction rate, we often say that the concentrations of species present in great stoichiometric excess are essentially constant. This is a convenient but rather imprecise way to describe a useful approximation. Consider the reaction $A + B \rightarrow C$. Over any time interval, Δt, we have $\Delta[B] = \Delta[A]$. In absolute terms, $[B]$ is no more constant than $[A]$. Suppose that the reaction rate is described by $R = k[A][B]$ and that $[B]_0 = 100[A]_0$. Define the extent of reaction by $\xi = [A]_0 - [A] = [B]_0 - [B]$. Find

$$\frac{\partial R/\partial[B]}{\partial R/\partial[A]}$$

and evaluate this relative concentration dependence at 0% conversion, $\xi = 0$ (where $[A] = [A]_0$), and at 90% conversion, $\xi = 0.9[A]_0$ (where $[A] = 0.9[A]_0$). Give a more precise statement of what we mean when we say that "the concentration of B is essentially constant" in such circumstances.

24. For the reaction $aA + bB \rightleftharpoons cC + dD$, we define the extent of reaction $\xi = -(n_A - n_A^o)/a$. When the reaction reaches equilibrium (at $t = \infty$), the extent of reaction becomes $\xi = -(n_A^\infty - n_A^o)/a$. If A is the limiting reagent and the reaction goes to completion, the theoretical extent of reaction is $\xi_{theoretical} = n_A^o/a$. Why? It is often useful to describe the amount of reaction that has occurred as a dimensionless fraction. If the reaction does not go to completion, $n_A^\infty > 0$. Use ξ and ξ_∞ to express the "extent of equilibration," $f_{equilibrium}$, as a dimensionless fraction. Use ξ and $\xi_{theoretical}$ to express the "conversion," $f_{conversion}$, as a dimensionless fraction. How would you define the "equilibrium conversion"?

25. We often exercise a degree of poetic license in talking about "fast" and "slow" steps in reaction mechanisms. In §12, Case I, for example, we say that the step that consumes A to produce intermediate C is "slow" but the step that consumes C to produce D is "fast." We then write $-d[A]/dt \approx d[D]/dt$. Discuss.

26. Find the rate law for the simplest-case Michaelis-Menten mchanism by applying the steady-state approximation to the concentration of the enzyme–substrate complex. Under what conditions do this treatment and the result developed in the text converge to the same rate law?

27. What is the half-life of a constant-volume second-order reaction, $2A \rightarrow C$, for which

$$\frac{d[A]}{dt} = -\frac{2}{V}\frac{d\xi}{dt} = -2k[A]^2$$

28. For the reaction between oxygen and nitric oxide, $2NO + O_2 \rightarrow 2NO_2$, the observed rate law, at constant volume, is

$$\frac{d[NO_2]}{dt} = k[NO]^2[O_2]$$

Show that this rate law is consistent with either of the following mechanisms:

(i) $2NO \rightleftharpoons N_2O_2$ (fast equilibrium)
$N_2O_2 + O_2 \rightarrow 2NO_2$ (rate-determining step)

(ii) $NO + O_2 \rightleftharpoons NO_3$ (fast equilibrium)
$NO_3 + NO \rightarrow 2NO_2$ (rate-determining step)

29. For the reaction between gaseous chlorine and nitric oxide, $2NO + Cl_2 \rightarrow 2NOCl$, doubling the nitric oxide concentration quadruples the rate, and doubling the chlorine concentration doubles the rate.

(a) Deduce the rate law for this reaction.
(b) Keeping mind the mechanisms in problem 28, write down two possible mechanisms that are consistent with the rate law you deduced in part (a). Show that each of these mechanisms is consistent with the rate law in part (a).

30. Nitric oxide reacts with hydrogen according to the equation, $2NO + 2H_2 \rightarrow N_2 + 2H_2O$. At constant volume, the following kinetic data have been obtained for this reaction at 1099 K. [1 mm = 1 torr = (1/760) atm.] C. N. Hinshelwood and T. Green, ***J. Chem. Soc.***, 730 (1926)]

$P^0(H_2)$, mm	$P^o(NO)$, mm	Initial reaction rate, mm s^{-1}
289	400	0.162
205	400	0.110
147	400	0.079
400	359	0.150
400	300	0.103
400	152	0.025

(a) What is the rate law for this reaction?
(b) Suggest two mechanisms for this reaction that are consistent with the rate law you deduce in part (a).

31. Review the reactions, rate laws, and mechanisms that you considered in problems 28, 29, and 30.
(a) Does comparing these three reactions and their rate laws provide any basis for preferring one set of mechanisms to the other?
(b) Which set of mechanisms do you prefer; that is, which mechanism in each of problems 29, 30, and 31 seems more likely to you? Why?

32. The rate of an enzyme-catalyzed reaction, commonly called the velocity, v, is measured directly as $v = d[P]/dt \approx \Delta[P]/\Delta t = -\Delta[S]/\Delta t$. For small S_0, a plot of v *versus.* the initial substrate concentration, S_0, increases with increasing S_0. For large values of S_0, v reaches a constant value, v_{max}. The substrate concentration at which the reaction rate is equal to $v_{max}/2$ is defined to be the Michaelis constant, K_M. It is customary to express the equilibrium constant as the dissociation constant for the enzyme–substrate complex. Let the total enzyme concentration be E_0. For the mechanism

$$ES \overset{K_S}{\rightleftharpoons} E + S \qquad K_S = [E][S]/[ES]$$

$$ES \overset{k}{\rightarrow} E + P \qquad \text{(rate-determining step)}$$

(a) Show that the velocity is given by

$$v = kE_0/\left(1 + \frac{K_S}{S_0}\right)$$

(b) What is v_{max}?
(c) What is the Michaelis constant, K_M?
(d) Does a larger value of K_M correspond to stronger or weaker complexation of the substrate by the enzyme?
(e) Sketch the curve of v *versus.* S_0 for the reaction rate described in (a). On this sketch, identify v_{max}, $v_{max}/2$, and K_M.
(f) If a second substrate, I, can form a complex with the enzyme, the reaction rate for substrate S decreases in the presence of I. Such substrates, I, are called inhibitors. Many kinds of inhibition are observed. One common distinction is between inhibitors that are competitive and inhibitors that are not competitive. Competitive inhibition can be explained in terms of a mechanism in which the enzyme equilibrates with both substrates.

$$ES \overset{K_S}{\rightleftharpoons} E + S \qquad K_S = [E][S]/[ES]$$

$$EI \overset{K_I}{\rightleftharpoons} E + I \qquad K_I = [E][I]/[EI]$$

$$ES \overset{k}{\rightarrow} E + P \qquad \text{(rate-determining step)}$$

Show that the velocity is given by

$$v = kE_0/\left[1 + \frac{K_S}{S_0}\left(1 + \frac{I_0}{K_I}\right)\right]$$

(g) A series of experiments is done in which S_0 is varied, while I_0 is maintained constant. The results are described by the equation in (f). What is v_{max} in this series of experiments?
(h) For the series of experiments done in (g), what is the Michaelis constant, K_M?

33. Consider a bimolecular reaction between molecules of substances A and B. If there are no forces of attraction or repulsion between A molecules and B molecules, we expect their collision rate to be $k[A][B]$, where k is a constant whose value is independent of the values of $[A]$ and $[B]$. Now suppose that molecules of A and B experience a strong attractive force whenever their intermolecular separation becomes comparable to, say, twice the diameter of an A molecule. Will the value of k be different when there is a strong force of attraction than when there is no such force?

Notes

[1] See Fred Basolo and Ralph G. Pearson, ***Mechanisms of Inorganic Reactions***, 2nd Ed., John Wiley & Sons, Inc., New York, 1967, pp 177-193.
[2] R.C. Tolman, ***The Principles of Statistical Thermodynamics***, Dover Publications, 1979, (published originally in 1938 by Oxford University Press), p 163.
[3] R. L. Burwell and R. G. Pearson, ***J. Phys. Chem.***, **79**, 300, (1966).
[4] George M. Fleck, ***Chemical Reaction Mechanisms***, Holt, Rinehard, and Winston, Inc., New York, NY, 1971, pp 104-112.
[5] G. N. Lewis, ***Proc. Nat. Acad. Sci. U. S.***, **11**, 179 (1925).

6

Equilibrium States and Reversible Processes

§1 The thermodynamic perspective

Classical thermodynamics does not consider the atomic and molecular characteristics of matter. In developing it, we focus exclusively on the measurable properties of macroscopic quantities of matter. In particular, we study the relationship between the thermodynamic functions that characterize a system and the increments of heat and work that the system receives as it undergoes some change of state. In doing so, we adopt some particular perspectives. The first is to imagine that we can segregate the macroscopic sample that we want to study from the rest of the universe. As sketched in Figure 1, we suppose that we can divide the universe into two mutually exclusive pieces: the ***system*** that we are studying and the ***surroundings***, which we take to encompass everything else.

We imagine the system to be enclosed by a boundary, which may or may not correspond to a material barrier surrounding the collection of matter that we designate as the system. (For our purposes, a system will always contain a macroscopic quantity of matter. However, this is not necessary; thermodynamic principles can be applied to a volume that is occupied only by radiant energy.) Everything inside the boundary is part of the system. Everything outside the boundary is part of the surroundings. Every increment of energy that the system receives, as either heat or work, is passed to it from the surroundings, and conversely.

An ***open system*** can exchange both matter and energy with its surroundings. A ***closed system*** can exchange energy but not matter with its surroundings. An ***isolated system*** can exchange neither matter nor energy. Together, system and surroundings comprise the ***universe***.

If we are too literal-minded, this reference to "the universe" can start us off on unnecessary ruminations about cosmological implications. All we really have in mind is an energy-accounting scheme, much like the accountants' system of double-entry bookkeeping, in which every debit to one account is a credit to another. When we talk about "the universe," we are really just calling attention to the fact that our scheme involves only two accounts. One is labeled "system," and the other is labeled "surroundings." Since we do our bookkeeping one system at a time, the combination of system and surroundings encompasses the universe of things affected by the change.

Figure 2 schematically depicts a closed system that can exchange heat and work with its surroundings. The surroundings comprise a heat reservoir and a device that can convert potential energy in the surroundings into

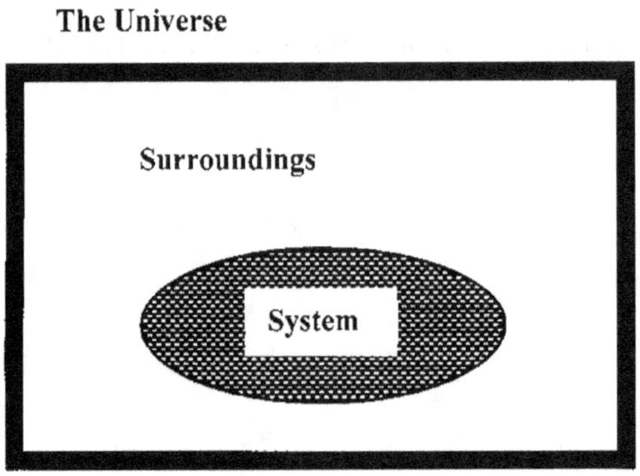

Figure 1. The thermodynamic universe.

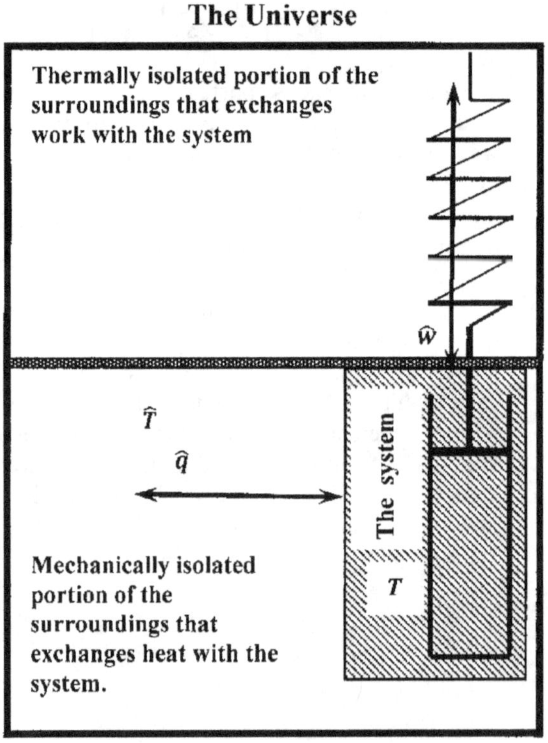

Figure 2. Transferring heat and work in a thermodynamic universe.

work exchanged with the system. The heat reservoir can exchange heat but not work with the system. In this sketch, the heat reservoir is at a constant temperature, $\hat{T}\left(=T_{surroundings}\right)$. It might comprise, for example, a large quantity of ice and water in phase equilibrium. The work-generating device cannot exchange heat, but it can exchange work with the system. A partially extended spring represents the potential energy available in the surroundings. The system can do work on the device and increase the potential energy of the spring. Alternatively, the surroundings can transfer energy to the system at the expense of the potential energy of the spring. Since nothing else in the rest of the universe is affected by these exchanges, our sketch encompasses the entire universe insofar as these changes are concerned. A system that cannot interact with anything external to itself is isolated. The combination of system and surroundings depicted in Figure 2 is itself an isolated system.

When we deal with the entropy change that accompanies some change in the state of the system, the properties of the surroundings become important. We develop the reason for this in Chapter 9. It is useful to introduce notation to distinguish properties of the surroundings from properties of the system. In Figure 2, we indicate the temperatures of system and surroundings by T and $\hat{T}$, respectively. We adopt this general rule: ***When a thermodynamic quantity appears with a superscripted caret, the quantity is that of the surroundings. If there is no superscripted caret, the quantity is that of the system.*** Thus, $\hat{T}$, $\hat{E}$, and $\hat{S}$ are the temperature, the energy, and the entropy of the surroundings, respectively, whereas T, E, and S are the corresponding quantities for the system.

We develop thermodynamics by reasoning about closed chemical systems that consist of one or more ***homogeneous phases***. A phase can be a solid, a liquid, or a gas. A phase can consist of a single chemical substance, or it can be a homogeneous solution containing two or more chemical substances. When we say that a phase is homogeneous, we mean that the pressure, temperature, and composition of the phase are the same in every part of the phase. Since gases are always miscible, a system cannot contain two gas phases that are in contact with one another. However, multiple solid and immiscible-liquid phases can coexist.

§2 Thermodynamic systems and variables

We characterize the system by specifying the values of enough variables so that the system can be exactly replicated. By "exactly replicated" we mean, of course, that we are not able to distinguish the system from its replicate by any experimental measurement. Any variable that can be used to characterize the system in this way is called a ***variable of state***, a ***state variable***, or a ***state function***.

We can say the same thing in slightly different words by saying that the state of a system is completely specified when the values of all of its state variables are specified. If we initially have a system in some equilibrium state and change one or more of the variables that characterize it, the system will eventually reach a new equilibrium state, in which some state variables will have values different from those that characterized the original state. If we want to return the system to its original state, we must arrange matters so that the value of every state variable is the same as it was originally.

The variables that are associated with a chemical system include ***pressure***, ***volume***, ***temperature***, and the number of ***moles*** of each substance present. All of these variables can be measured directly; that is, every equilibrium state of a system is associated with a specific value of each of these variables, and this value can be determined without reference to any other state of the system. ***Energy*** and ***entropy*** are also variables that are associated with a thermodynamic system. We can only measure changes in energy and entropy; that is, we can only measure energy and entropy for a process in which a system passes from one state to another.

Other important thermodynamic variables are defined as functions of pressure, volume, temperature, energy and entropy. These include ***enthalpy***, the ***Gibbs free energy***, the ***Helmholtz free energy***, ***chemical activity***, and the ***chemical potential***. Our goals in developing the subject of chemical thermodynamics are to define each of these state functions, learn how to measure each of them, and provide a theory that relates the change that occurs in any one of them to the changes that occur in the others when a chemical system changes from one state to another.

Any interaction through which a chemical system can exchange work with its surroundings can affect its behavior. Work-producing forces can involve many phenomena, including gravitational, electric, and magnetic fields; surface properties; and sound (pressure) waves. In Chapter 17, we discuss the work done when an electric current passes through an electrochemical cell. Otherwise, this book focuses on pressure–volume work and gives only passing attention to the job of incorporating other forms of work into the general theory. We include pressure–volume work because it occurs whenever the volume of a system changes. A thermodynamic theory that did not include volume as a variable would be of limited utility.

Thermodynamic variables can be sorted into two classes in another way. Consider the pressure, temperature, and volume of an equilibrium system. We can imagine inserting a barrier that divides this original system into two subsystems—without changing anything else. Each of the subsystems then has the temperature and pressure of the original system; however, the volume of each subsystem is different from the volume of the original system. We say that temperature and pressure are ***intensive variables***, by which we mean that the temperature or pressure of an equilibrium system is independent of the size of the system and the same at any location within the system. Intensive variables stand in contrast to ***extensive variables***. The magnitude of an extensive variable is directly proportional to the size of the system. Thus, volume is an extensive variable. Energy is an extensive variable. We shall see that entropy, enthalpy, the Helmholtz free energy, and the Gibbs free energy are extensive variables also.

For any extensive variable, we can create a companion intensive variable by dividing by the size of the system. For example, we can convert the mass of a homogeneous system into a companion variable, the density, by dividing by the system's volume. We will discover that it is useful to define certain ***partial molar quantities***, which have units like energy per mole. Partial molar quantities are intensive variables. We will find a partial molar quantity that is particularly important in describing chemical equilibrium. It is called the ***chemical potential***, and since it is a partial molar quantity, the chemical potential is an intensive thermodynamic variable.

We think of a system as a specific collection of matter containing specified phases. Our goal is to develop mathematical models (equations) that relate a system's state functions to one another. A system can be at equilibrium under a great many different circumstances. We say that the system can have many equilibrium positions. A complete description of all of these equilibrium positions requires models that can specify how much of each of the substances that make up the system is present in each phase.

However, if a system is at equilibrium, a half-size copy of it is also at equilibrium; whether a system is at equilibrium can be specified without specifying the sizes of the phases that make it up. This means that we can characterize the equilibrium states of any system that contains specified substances and phases by specifying the values of the system's intensive variables. In general, not all of these intensive variables will be independent. The number of intensive variables that are independent is called the number of ***degrees of freedom*** available to the system. This is also the number of intensive variables that can change independently while a given system remains at equilibrium.

To completely define a particular system we must specify the size and composition of each phase. To do so, we must specify the values of some number of extensive variables. These extensive variables can change while all of the intensive variables remain constant and the system remains at equilibrium. In the next section, we review the phase equilibria of water. A system comprised of liquid and gaseous water in phase equilibrium illustrates these points. Specifying either the pressure or temperature specifies the equilibrium state to within the sizes of the two phases. For a complete description, we must specify the number of moles of water in each phase. By adding or removing heat, while maintaining the original pressure and temperature, we can change the distribution of the water between the two phases.

In 1875, J. Willard Gibbs developed an equation, called ***Gibbs' phase rule***, from which we can calculate the number of degrees of freedom available to any particular system. We introduce Gibbs' phase rule in §8. The perspective and analysis that underlie Gibbs' phase rule have a significance that transcends use of the rule to find the number of degrees of freedom available to a system. In essence, ***the conditions assumed in deriving Gibbs' phase rule define what we mean by equilibrium in chemical systems***. From experience, we are usually confident that we know when a system is at equilibrium and when it is not. One of our goals is to relate thermodynamic functions to our experience-based ideas about what equilibrium is and is not. To do so, we need to introduce the idea of a ***reversible process***, in which the system undergoes a ***reversible change***.

We will see that the states that are accessible to a system that is at equilibrium in terms of Gibbs' phase rule are identically the states that the system can be in while undergoing a reversible change. A principal goal of the remainder of this chapter is to clarify this equivalence between the range of states accessible to the system at equilibrium and the possible paths along which the system can undergo reversible change.

The thermodynamic theory that we develop predicts quantitatively how a system's equilibrium position changes in response to a change that we impose on one or more of its state functions. ***The principle of Le Chatelier*** makes qualitative predictions about such changes. We introduce the principle of Le Chatelier and its applications later in this chapter. In Chapter 12, we revisit this principle to understand it as a restatement, in qualitative terms, of the thermodynamic criteria for equilibrium.

§3 Equilibrium and reversibility: phase equilibria

To review the general characteristics of phase equilibria, let us consider a familiar system. Suppose that we have a transparent but very strong cylinder, sealed with a frictionless piston, within which we have trapped a quantity of pure liquid water at some high pressure. We can fix the pressure of the liquid water at any value we choose by applying an appropriate force to the piston. Suppose that we hold the temperature constant and force the volume to increase by withdrawing the piston in very small increments. Because pure water is not compressed easily, we find initially that the pressure of the water decreases and does so in very large increments.

However, after some small increase in the volume, we find that imposing a further volume increase changes the system's behavior abruptly. The system undergoes a profound change. What was formerly pure liquid becomes a mixture of liquid and gas. As we impose still further volume increases, the pressure of the system remains constant, additional liquid passes from the liquid to the gas phase, and we find that we must supply substantial amounts of heat in order to keep the temperature of the system constant. If we continue to force volume increases in this manner, vaporization continues until all of the liquid evaporates.

If we impose a decrease in the volume of the two-phase system, we see the process reverse. The pressure of the system remains constant, some of the gas condenses to liquid, and the system gives up heat to the surroundings. For any given temperature, these conversions are precisely balanced at some particular pressure, and these conditions characterize a state of liquid–vapor equilibrium. At any given pressure, the equilibrium temperature is called the ***boiling point*** of the liquid. The equilibrium pressure and temperature completely specify the state of the system, except for the exact amounts of liquid and gaseous water present.

If we begin with this system in a state of liquid–vapor equilibrium, we can increase the amount of vapor by imposing a small volume increase. Conversely, we can decrease the amount of vapor by imposing a very small volume decrease. At the equilibrium temperature and pressure, changing the imposed volume by an arbitrarily small amount (from V to $V \pm dV$) is sufficient to reverse the direction of the change that occurs in the system. ***We call any process whose direction can be reversed by an arbitrarily small change in a thermodynamic state function a reversible process***. Evidently, there is a close connection between reversible processes and equilibrium states. ***If a process is to occur reversibly, the system must pass continuously from one equilibrium state to another.***

In this description, the reversible, constant-temperature vaporization of water is driven by arbitrarily small volume changes. The system responds to these imposed volume changes so as to maintain a constant equilibrium vapor pressure at the specified temperature. We say that the reversible process "takes place at constant pressure and temperature." We can also describe this process as being driven by arbitrarily small changes in the applied pressure: If the applied pressure exceeds the equilibrium vapor pressure by an arbitrarily small increment, $dP > 0$, condensation occurs; if the applied pressure is less than the equilibrium vapor pressure by an arbitrarily small increment, $dP < 0$, vaporization occurs. To describe this tersely, we introduce a figure of speech and say that the reversible process occurs "while the system pressure and the applied pressure are equal." Literally, of course, there can be no change when these pressures are equal.

To cause water to vaporize at a constant temperature and pressure, we must add heat energy to the system. This heat is called the ***latent heat of vaporization*** or the ***enthalpy of vaporization***, and it must be supplied from some entity in the surroundings. When water vapor condenses, this latent heat must be removed from the system and taken up by the surroundings. (The enthalpy change for vaporizing one mole of a substance is usually denoted $\Delta_{vap}H$. It varies with temperature and pressure. Tables usually give experimental values of the equilibrium boiling temperature at a pressure of 1 bar or 1 atm; then they give the enthalpy of vaporization at this temperature and pressure. We discuss the enthalpy function in Chapter 8.)

Four conditions are sufficient to exactly specify either the initial or the final state: the number of moles of liquid, the number of moles of gas, the pressure, and the temperature. The change is a conversion of some liquid to gas, or *vice versa*. We can represent this change as a transition from an initial state to a final state where n^o_{liquid} and n^o_{gas} are the initial numbers of moles of liquid and gas, respectively, and δn is the incremental number of moles vaporized:

$$\left(P, T, n^o_{liquid}, n^o_{gas}\right) \rightarrow \left(P, T, n^o_{liquid} - \delta n, n^o_{gas} + \delta n\right)$$

The initial pressure and temperature are the same as the final pressure and temperature. Effecting this change requires that a quantity of heat, $\left(\Delta_{vap}H\right)\delta n$, be added to the system, without changing the temperature of the system.

This introduces another requirement that a reversible process must satisfy. If the reversibly vaporizing water is to take up an arbitrarily small amount of heat, the system must be in contact with surroundings that are hotter than the system. The temperature difference between the system and its surroundings must be arbitrarily small, because we can describe exactly the same process as being driven by contacting the system, at temperature T, with surroundings at temperature $\hat{T} + \delta\hat{T}$. If we keep the applied pressure constant at the temperature-T equilibrium vapor pressure, the system volume increases. We can reverse the direction of change by changing the temperature of the surroundings from $\hat{T} + \delta\hat{T}$ to $\hat{T} - \delta\hat{T}$. If the process is to satisfy our criterion for reversibility, the difference between these two temperatures must be arbitrarily small. To describe this requirement tersely, we again introduce a figure of speech and say that the reversible process occurs "while the system temperature and the surroundings temperature are equal."

If we repeat the water-in-cylinder experiment with the temperature held constant at a slightly different value, we get similar results. There is again a pressure at which the process of converting liquid to vapor is at equilibrium. At this temperature and pressure, both liquid and gaseous water can be present in the system, and, so long as no heat is added or removed from the system, the amount of each remains constant. When we hold the pressure of the system constant at the equilibrium value and supply a quantity of heat to the system, a quantity of liquid is again converted to gaseous water. (The quantity of heat required to convert one mole of liquid to gaseous water is slightly different from the quantity required in the previous experiment. This is what we mean when we say that the enthalpy of vaporization varies with temperature.)

This experiment can be repeated for many temperatures. So long as the temperature is in the range $273.16 < T < 647.1$ K, we find a pressure at which liquid and gaseous water are in equilibrium. If we plot the results, they lie on a smooth curve, which is sketched in Figure 3. This curve represents the combinations of pressure and temperature at which liquid water and gaseous water are in equilibrium.

Below 273.16 K, an equilibrium system containing only liquid and gaseous water cannot exist. At high pressures, a two-phase equilibrium system contains solid and liquid; at sufficiently low pressures, it contains solid and gas. Above 647.1 K, the distinction between liquid and gaseous water vanishes. The water exists as a single dense phase. This is the ***critical temperature***. Above the critical temperature, there is a single fluid phase at any pressure.

If we keep the pressure constant and remove heat

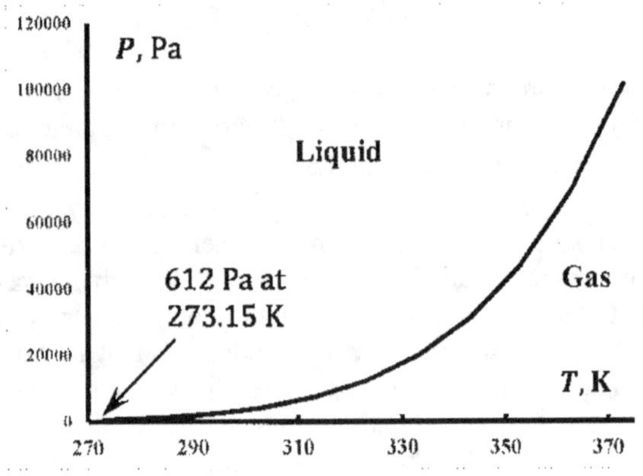

Figure 3. Water liquid–vapor equilibrium.

from a quantity of liquid water, the temperature decreases until we eventually reach a temperature at which the water begins to freeze to ice. At this point, water and ice are in equilibrium. Further removal of heat does not decrease the temperature of the water–ice system; rather, the temperature remains constant and additional water freezes into ice. Only when all of the liquid has frozen does further removal of heat cause a further decrease in the temperature of the system. When we repeat this experiment at a series of temperatures, we find a continuous line of pressure–temperature points that are liquid–ice equilibrium points.

As sketched in Figure 4, the liquid–ice equilibrium line intersects the liquid–vapor equilibrium line. At this intersection, liquid water, ice, and water vapor are all in equilibrium with one another. There is only one such point. It is called the ***triple point*** of water. The ***ice point*** or ***melting point*** of water is the temperature at which solid and liquid water are in equilibrium at one atmosphere in the presence of air. The water contains dissolved air. The triple point occurs in a pure-water system; it is the temperature and pressure at which gaseous, liquid,

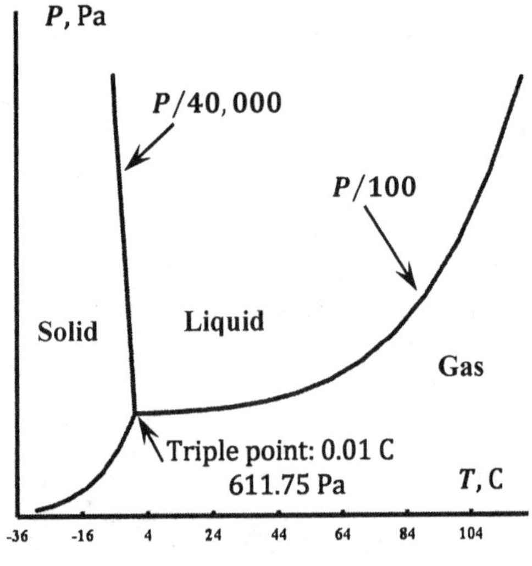

Figure 4. The phase diagram for water. Note that the ordinate values for the solid–liquid and liquid–gas equilibria are severely compressed[1].

and solid water are in equilibrium. By definition, the triple point temperature is 273.16 K. Experimentally, the pressure at the triple point is 611 Pa. Experimentally, the melting point is 273.15 K.

To freeze a liquid, we must remove heat. To fuse (melt) the same amount of the solid at the same temperature and pressure, we must add the same amount of heat. This heat is called the ***latent heat of fusion*** or the ***enthalpy of fusion***. The enthalpy of fusion for one mole of a substance is usually denoted $\Delta_{fus}H$. It varies slightly with temperature and pressure. Tables usually give experimental values of the equilibrium melting temperature at a pressure of 1 bar or 1 atm; then they give the enthalpy of fusion at this temperature and pressure.

At low pressures and temperatures, ice is in equilibrium with gaseous water. A continuous line of pressure–temperature points represents the conditions under which the system contains only ice and water vapor. As the temperature increases, the ice–vapor equilibrium line ends at the triple point. The conversion of a solid directly into its vapor is called sublimation. To sublime a solid to its vapor requires the addition of heat. This heat is called the ***latent heat of sublimation*** or the ***enthalpy of sublimation***. The enthalpy of sublimation for one mole of a substance is usually denoted $\Delta_{sub}H$. It varies slightly with temperature and pressure.

§4 Distribution equilibria

A system can contain more than one phase, and more than one chemical substance can be present in each phase. If one of the substances is present in two phases, we say that the substance is ***distributed*** between the two phases. We can describe the equilibrium distribution quantitatively by specifying the concentration of the substance in each phase. At constant temperature, we find experimentally that the ratio of these concentrations is approximately constant. Letting A be the substance that is distributed, we find for the distribution equilibrium

$$A(phase\ 1) \rightleftharpoons A(phase\ 2)$$

the ***equilibrium constant***

$$K = \frac{[A]_{phase\ 2}}{[A]_{phase\ 1}}$$

where K varies with temperature and pressure.

For example, iodine is slightly soluble in water and much more soluble in chloroform. Since water and chloroform are essentially immiscible, a system containing water, chloroform, and iodine will contain two liquid phases. If there is not enough iodine present to make a saturated solution with both liquids, the system will reach equilibrium with all of the iodine dissolved in the two immiscible solvents. Experimentally, the equilibrium concentration ratio

$$K = \frac{[I_2]_{water}}{[I_2]_{chloroform}}$$

is approximately constant, whatever amounts of the three substances are mixed.

We begin our development of physical chemistry by reasoning about the effects of concentrations on the properties of chemical systems. In Chapter 5, we find that rate laws expressed using concentration variables are adequate for the analysis of reaction mechanisms. Consideration of these rate laws leads us to the equilibrium constant for a chemical reaction expressed as a function of concentrations. Eventually, however, we discover that an adequately accurate theory of chemical equilibrium must be expressed using new quantities, which we call ***chemical activities*** [2]. We can think of a chemical activity as an "effective concentration" or a "corrected centration," where the correction is for the effects of intermolecular interactions. When we allow for the effects of intermolecular interactions, we find that we must replace the concentration terms by chemical activities. For the distribution equilibrium constant, we have

$$K = \frac{\tilde{a}_{A,phase\ 2}}{\tilde{a}_{A,phase\ 1}}$$

where $\tilde{a}_{A,phase\ 1}$ denotes the chemical activity of species A, in phase 1, at equilibrium.

§5 Equilibria in chemical reactions

Equilibria involving chemical reactions share important characteristics with phase and distribution equilibria. In Chapter 5, we develop the equilibrium constant expression from ideas about reaction rates. For the present comparison, let us consider the equilibrium between the gases nitrogen dioxide, NO_2, and dinitrogen tetroxide, N_2O_4:

$$N_2O_4\ (g) \rightleftharpoons 2NO_2\ (g)$$

Suppose that we trap a quantity of pure N_2O_4 in a cylinder closed with a piston. If we fix the temperature and volume of this system, the dissociation reaction occurs until equilibrium is achieved at some system pressure. For present purposes, let us assume that both N_2O_4 and NO_2 behave as ideal gases. The equilibrium system pressure will be equal to the sum of the partial pressures: $P = P_{N_2O_4} + P_{NO_2}$. If we now do a series of experiments, in which we hold the volume constant while allowing the temperature to change, we find a continuous series of pressure–temperature combinations at which the system is at equilibrium. This curve is sketched in Figure 5. It is much like the curve describing the dependence of the water–water-vapor equilibrium on pressure and temperature.

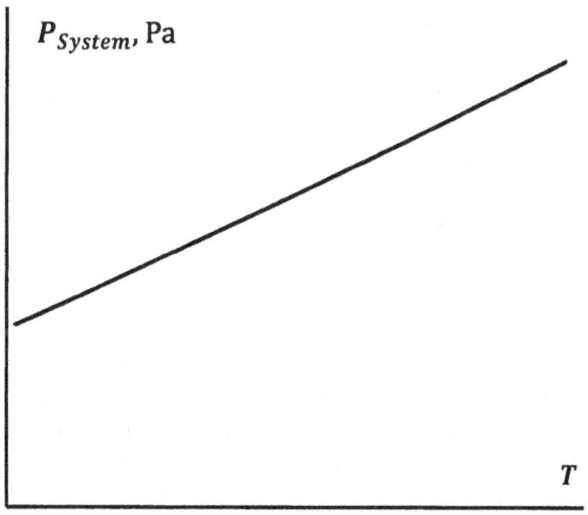

Figure 5. System pressure *versus* temperature for N_2O_4 dissociation.

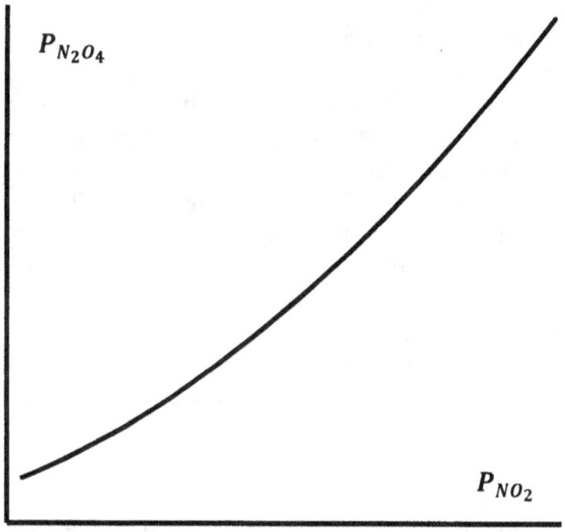

Figure 6. Equilibrium compositions for N_2O_4 dissociation at a fixed temperature.

If we hold the temperature constant and allow the volume to vary, we can change the force on the piston to keep the total pressure constant at a new value, P^*_{total}. The position of the chemical equilibrium will change. At the new equilibrium position, the new N_2O_4 and NO_2 partial pressures will satisfy the total pressure relationship. When we repeat this experiment, we find that, whatever the total pressure, the equilibrium partial pressures are related to one another as sketched in Figure 6. We find that the experimental data fit the equation $K_P = P^2_{NO_2}/P_{N_2O_4}$, where K_P is the equilibrium constant for the reaction. (Pressure is a measure of gas concentration. Later, we see that the equilibrium constant can be expressed more rigorously as a ratio of fugacities—or activities.)

§6 Le Chatelier's principle

In Chapter 1, we describe a very general goal: given that we create a system in some arbitrary initial state (by some "change of conditions" or "removal of some constraint"), we want to predict how the system will respond as it changes on its own ("spontaneously") to some new equilibrium state. Under these circumstances, we need a lot of information about the system before we can make any useful prediction about the spontaneous change. In this case, we have said nothing about the condition of the system before we effect the change of conditions that creates the arbitrary "initial state".

Our ability to make useful predictions is much greater if the system and the change of conditions have a particular character. If we start with a system that is at equilibrium, and we impose a change in conditions on it, the "initial" state of the system after the imposed change of conditions will generally not be an equilibrium state. Experience shows that the system will undergo some spontaneous change to arrive at a new equilibrium state. In these particular circumstances, Le Chatelier's principle enables us to predict the spontaneous change that occurs.

Le Chatelier's principle

If a change is *imposed* on the state of a system at equilibrium, the system adjusts to reach a new equilibrium state. In doing so, the system undergoes a spontaneous change that *opposes* the *imposed* change.

Le Chatelier's principle is useful, and it is worthwhile to learn to apply it. The principle places no limitations on the nature of the imposed change or on the number of thermodynamic variables that might change as the system responds. However, since our reasoning based on the principle is qualitative, it is frequently useful to suppose that the imposed change is made in just one variable and that the opposing change involves just one other variable. That is, we ask how changing one of the variables that characterizes the equilibrated system changes a second such variable, "all else being equal." Successful use of the principle often requires careful thinking about the variable on which change is imposed and the one whose value changes in response. Let us consider some applications.

Vapor–liquid equilibrium. Suppose that we have a sealed vial that contains only the liquid and vapor phases of a pure compound. We suppose that the vial and its contents are at a single temperature and that the liquid and the vapor are in equilibrium with one another at this temperature. What will happen if we now thermostat the vial at some new and greater temperature?

We see that the imposed change is an increase in temperature or, equivalently, an addition of heat to the system. The system cannot respond by decreasing its temperature, because the temperature change is the imposed change. Similarly, it cannot respond by changing its volume, because the system volume is fixed. Evidently, the observable consequence of increasing temperature—adding heat— must be a change in the pressure of the system. The principle asserts that the system will respond so as to consume heat. Converting liquid to vapor consumes the latent heat of vaporization, so the system can oppose the imposed addition of heat by converting liquid to vapor. This increases the pressure of the vapor. We can conclude from Le Chatelier's principle that increasing the temperature of a system at liquid-vapor equilibrium increases the equilibrium vapor pressure.

Now suppose that we have the liquid and vapor phases of the same pure compound in a thermally isolated cylinder that is closed by a piston. We ask what will happen if we decrease the volume. That is, the imposed change is a step decrease in volume, accompanied by an increase in pressure. The new volume is fixed, but the pressure is free to adjust to a new value at the new equilibrium position. The principle asserts that the system will respond so as to decrease its pressure. Decreasing the system pressure is accomplished by condensing vapor to liquid, which is accompanied by the release of the latent heat of vaporization. Since we suppose that the system is thermally isolated during this process, the heat released must result in an increase in the temperature of the system. While the pressure can decrease from the initial non-equilibrium value, it cannot decrease to its original-equilibrium value; evidently, the new equilibrium pressure must be greater than the original pressure.

We again conclude that an increase in the equilibrium vapor pressure requires an increase in the temperature of the system. (If the volume decrease were imposed with the system immersed in a constant temperature bath, the heat evolved would be transferred from the system to the bath. The system would return to its original pressure and original temperature, albeit with fewer moles of the substance present in the gas phase.)

Ice–water equilibrium. Suppose that we have a closed system consisting of ice in equilibrium with liquid water at some temperature and pressure. What will happen if we impose an increase in the temperature this system? We suppose that the system occupies a container of fixed volume. Initially it is at equilibrium with a constant-temperature bath. We impose the change by moving the container to a new bath whose temperature is higher—but not high enough to melt all of the ice. The imposed change is a temperature increase or, equivalently, an addition of heat. The principle asserts that the system will respond by consuming heat, which it can do by converting ice to liquid. Since liquid water occupies less volume than the same mass of ice, the system pressure will decrease. We conclude that the pressure at which ice and water are at equilibrium decreases when the temperature increases. That is, the melting point increases as the pressure decreases.

Again, we can imagine that the equilibrium mixture of ice and water is contained in a thermally isolated cylinder that is closed by a piston and ask how the system must respond if we impose a step decrease in its volume. We impose the volume decrease by applying additional force to the piston. The imposed step change in the volume is accompanied by an increase in the system pressure; the new volume is fixed, but the system pressure can adjust. The principle asserts that the system will respond by decreasing its pressure. The system pressure will decrease if some of the ice melts. Melting ice consumes heat. Since we are now assuming that the system is thermally isolated, this heat cannot come from outside the system, which means that the temperature of the system must decrease. While the pressure can decrease from its initial non-equilibrium value, it cannot decrease to the value that it had in the original equilibrium position. We again conclude that increasing the pressure results in a decrease in temperature; that is, the melting point of ice increases as the pressure decreases.

Chemical reaction between gases. Finally, suppose that we have a chemical equilibrium involving gaseous reagents. To be specific, let us again consider the reaction

$$N_2O_4\ (g) \rightleftharpoons 2\ NO_2\ (g)$$

We suppose that this system is initially at equilibrium at some temperature and that we seek to increase the pressure while maintaining the temperature constant. (We can imagine that the system is contained in a cylinder that is closed by a piston. The cylinder is immersed in a constant-temperature bath. We increase the pressure by applying additional force to the piston. As in the examples above, we view this as a step change in volume that is accompanied by an increase of the pressure to a transitory non-equilibrium value.) The principle asserts that the system will respond by undergoing a change that opposes this pressure increase. The system can reduce its pressure by decreasing the number of moles of gas present, and it can do this by converting NO_2 molecules to N_2O_4 molecules. We conclude that there will be less NO_2 present at equilibrium at the higher pressure.

When we first encounter it, Le Chatelier's principle seems to embody a remarkable insight. As, indeed, it does. However, as we think about it, we come to see it as a logical necessity. Suppose that the response of an equilibrium system to an imposed change were to augment the change rather than oppose it. Then an imposed change would reinforce itself. The slightest perturbation of any equilibrium system would cause the system to "run away" from that original position. Since no real system can be maintained at a perfectly constant set of conditions, any real system could undergo spontaneous change. Equilibrium would be unattainable. If we assume that a system must behave oppositely to the way that is predicted by Le Chatelier's principle, we arrive at a prediction that contradicts our experience.

Le Chatelier's principle is inherently qualitative. We will discuss it further after we develop the thermodynamic criteria for equilibrium. We will find that the thermodynamic criteria for equilibrium tell us quantitatively how two (or more) thermodynamic variables must change in concert if a system is to remain at equilibrium while also undergoing some change of condition.

§7 The number of variables required to specify some familiar systems

If we are to model a physical system mathematically, we must abstract measurable properties from it—properties that we can treat as variables in our model. In §2 we found that the size of the system does not matter when we consider the variables that specify an equilibrium state. A half-size version of an equilibrium system has the same equilibrium properties. We can say that only intensive properties are relevant to the question of whether a system is at equilibrium.

The idea that we can subdivide a system without changing its equilibrium properties is subject to an important qualification. We intend that both subsystems be qualitatively equivalent to the original. For example, if we divide a system at vapor–liquid equilibrium into subsystems, each subsystem must contain some liquid and some vapor. If we subdivide it into one subsystem that is all liquid and another that is all vapor, the subsystems are not qualitatively equivalent to the original.

We can be more precise about the criterion we have in mind: An equilibrium system consists of one or more homogenous phases. Two systems can be in the same equilibrium condition only if all of the phases present in one are also present in the other. If a process changes the number of phases present in a system, we consider that the system changes from one kind of equilibrium system to a second one. We can describe one kind of equilibrium system by specifying a sufficient number of intensive variables. This description will be complete to within a specification of the exact amount of each phase present.

If we apply these ideas to a macroscopic sample of a pure gas, we know that we need four variables to completely describe the state of the gas: the number of moles of the gas, its pressure, its volume, and its temperature. This assumes that we are not interested in the motion of the container that contains the gas. It assumes also that no other extrinsic factors—like gravitational, electric, or magnetic fields— affect the behavior that we propose to model

When we do experiments in which the amount, pressure, volume, and temperature of a pure gas vary, we find that we can develop an equation that relates the values that we measure. We call this an ***equation of state***, because it is a mathematical model that describes the state of the system. In Chapter 2, we reviewed the ideal gas equation, van der Waals equation, and the virial equation; however, we can devise many others. Whatever equation of state we develop, we know that it must have a particular property: At constant pressure and temperature, the volume must be directly proportional to the number of moles. This means that any equation of state can be rewritten as a function of concentration. For the case of an ideal gas, we have $P = (n/V)RT$, where n/V, the number of moles per unit volume, is the gas concentration. We see that any equation of state can be expressed as a function of three intensive variables: pressure, temperature, and concentration.

The existence of an equation of state means that only two of the three intensive variables that describe the gas sample are independent of one another. At equilibrium, a sample of pure gas has two degrees of freedom. Viewed as a statement about the mathematical model, this is true because knowledge of the equation of state and any two of the intensive variables enables us to calculate the third variable. Viewed as a statement about our experimental observations, this is true because, so long as the changes are consistent with the system remaining a gas, we can change any two of these variables independently. That only two are independent is shown experimentally by the observation that we can start with a fixed quantity of gas at any pressure, temperature, and concentration and find, after taking the system through any sequence of changes whatsoever, that returning to the original pressure and temperature also restores the original concentration.

In the experiment or in the mathematical model, fixing two of the three intensive variables is sufficient to fix the equilibrium properties of the system. Fixing the equilibrium properties means, of course, that the state of the system is fixed to within an arbitrary factor, which can be specified either as the number of moles present or as the system volume.

Similar results are obtained when we study the pressure–volume–temperature behavior of pure substances in condensed phases. At equilibrium, a pure liquid or a pure solid has two degrees of freedom.

If we consider a homogeneous mixture of two non-reacting gases, we discover that three variables are necessary to fix the equilibrium properties of the system. We must know the pressure and temperature of the system and the concentration of each gas. Because the mixture must obey an equation of state, determination of any three of these variables is sufficient to fix the value of the fourth. Note that we can conclude that three intensive variables are sufficient to determine the equilibrium properties of the system even if we do not have a mathematical model for the equation of state.

If we experiment with a system in which the liquid and vapor of a pure substance are in phase equilibrium with one another, we find that there is only one independent intensive variable. (Figure 4 illustrates this for water.) To maintain phase equilibrium, the system pressure must be the equilibrium vapor pressure of the substance at the system temperature. If we keep the pressure and temperature constant at equilibrium values, we can increase or decrease the concentration (moles per unit system volume) by removing or adding heat. In this process, we change one variable, concentration, while maintaining phase equilibrium.

If we keep the pressure constant and impose a temperature increase, vaporization continues (the concentration decreases) until the liquid phase is completely consumed. In this process, two variables change, and phase equilibrium cannot be maintained. To reach a new equilibrium state in which both liquid and gas are present at a higher temperature, we must increase the pressure to the new equilibrium vapor pressure; the magnitude of the temperature increase completely determines the required pressure increase. Two intensive variables change, but the changes are not independent.

If we have pure gas, there are two independent intensive variables. If we have pure liquid, there are two independent intensive variables. However, if we have liquid and gas in equilibrium with one another, there is only one independent intensive variable. In the liquid region of the water phase diagram, we can vary pressure and temperature and the system remains liquid water. Along the liquid-gas equilibrium line, we can vary the temperature and remain at equilibrium only if we simultaneously vary the pressure so as to remain on the liquid-gas equilibrium line.

Similar statements apply if we contrast varying pressure and temperature for the pure solid to varying the pressure and temperature along the solid-liquid or the solid-gas equilibrium line. At the triple point, nothing is variable. For a fixed quantity of water, the requirement that the system be at equilibrium at the triple point fixes the system pressure, temperature, and concentration. Evidently, maintaining a phase equilibrium in a system imposes a constraint that reduces the number of intensive variables that we can control independently.

The equilibrium between water and ice is completely unaffected by the state of subdivision of the ice. The ice can be present in a single lump or as a large number of very small pieces; from experience, we know that the equilibrium behavior of the system is the same so long as some ice and some water are both present. A system contains as many phases as there are ***kinds*** of macroscopic, homogeneous, bounded portions that are either solid, liquid, or gas.

If we add a lump of pure aluminum to our ice-water system, the new system contains three phases: water, ice, and aluminum. The equilibrium properties of the new system are the same if the aluminum is added as a ground-up powder. The powder contains many macroscopic, homogeneous, bounded portions that are aluminum, but each of these portions has the same composition; there is only one kind of aluminum particle. (Molecules on the surface of a substance can behave differently from those in the bulk. When a substance is very finely divided, the fraction of the molecules that is on the surface can become large enough to have a significant effect on the behavior of the system. In this book, we do not consider systems whose behavior is surface-area dependent.)

§8 Gibbs' phase rule

Gibbs found an important relationship among the number of chemical constituents, the number of phases present, and the number of intensive variables that must be specified in order to characterize an equilibrium system. This number is called the number of ***degrees of freedom*** available to the system and is given the symbol F. By specifying F intensive variables, we can specify the state of the system—except for the amount of each phase. The number of chemical constituents is called the ***number of components*** and is given the symbol C. The number of components is the smallest number of pure chemical compounds that we can use to prepare the equilibrium system so that it contains an arbitrary amount of each phase. The ***number of phases*** is given the symbol P. The relationship that Gibbs found between C, P, and F is called ***Gibbs' phase rule*** or just ***the phase rule***. The phase rule applies to equilibrium systems in which any component can move freely between any two phases in which that component is present.

We suppose that the state of the system is a

continuous function of its state functions. If F, intensive, independent variables, $X_1, X_2, \ldots, X_F$, are sufficient to specify the state of an equilibrium system, then $X_1 + dX_1, X_2 + dX_2, \ldots, X_F + dX_F$ specify an incrementally different equilibrium state of the same system. This means that the number of degrees of freedom is also the number of intensive variables that can be varied independently while the system changes reversibly—subject to the condition that there is no change in either the number or kinds of phases present. Moreover, if we keep the system's intensive variables constant, we can change the size of any phase without changing the nature of the system. This means that Gibb's phase rule applies to any equilibrium system, whether it is open or closed.

A system containing only liquid water contains one component and one phase. By adjusting the temperature and pressure of this system, we can arrive at a state in which both liquid and solid are present. For present purposes, we think of this as a second system. Since the second system can be prepared using only liquid water (or, for that matter, only ice) it too contains only one component. However, since it contains both liquid and solid phases, the second system contains two phases. We see that the number of components required to prepare a system in such a way that it contains an arbitrary amount of each phase is not affected by phase equilibria. However, the number of components is affected by chemical equilibria and by any other stoichiometric constraints that we impose on the system. The number of components is equal to the number of chemical substances present in the system, less the number of stoichiometric relationships among these substances.

Let us consider an aqueous system containing dissolved acetic acid, ethanol, and ethyl acetate. For this system to be at equilibrium, the esterification reaction

$$CH_3CO_2H + CH_3CH_2OH \rightleftharpoons CH_3CO_2CH_2CH_3 + H_2O$$

must be at equilibrium. In general we can prepare a system like this by mixing any three substances chosen from the set: acetic acid, ethanol, ethyl acetate, and water. Hence, there are three components. The esterification reaction, or its reverse, then produces an equilibrium concentration of the fourth substance. However, there is a special case with only two components. Suppose that we require that the equilibrium concentrations of ethanol and acetic acid be exactly equal. In this case, we can prepare the system by mixing ethyl acetate and water. Then the stoichiometry of the reaction assures that the concentration condition will be met; indeed, this is the only way that the equal-concentration condition can be met exactly.

In this example, there are four chemical substances. The esterification reaction places one stoichiometric constraint on the amounts of these substances that can be present at equilibrium, which means that we can change only three concentrations independently. The existence of this constraint reduces the number of components from four to three. An additional stipulation that the product concentrations be equal is a second stoichiometric constraint that reduces the number of independent components to two.

If we have a one-phase system at equilibrium, we see that the pressure, the temperature, and the C component-concentrations constitute a set of variables that must be related by an equation of state. If we specify all but one of these variables, the remaining variable is determined, and can be calculated from the equation of state. There are $C + 2$ variables, but the existence of the equation of state means that only $C + 1$ of them can be changed independently. Evidently, the number of degrees of freedom for a one-phase system is $F = C + 1$.

To find the number of degrees of freedom when P such phases are in equilibrium with one another requires a similar but more extensive analysis. We first consider the number of intensive variables that are required to describe completely a system that contains C components and P phases, if the phases are not at equilibrium with one another. (Remember that the description we seek is complete except for a specification of the absolute amount of each phase present. For the characterization of equilibrium that we seek, these amounts are arbitrary.) In this case, each phase is a subsystem in its own right. Each phase can have a pressure, a temperature, and a concentration for each component. Each of these properties can have a value that is independent of its value in any other phase. There are $C + 2$ variables for each phase or $P(C + 2)$ variables for all P phases. Table 1 displays these variables.

Table 1. The $P(C + 2)$ variables for P subsystems

	Phase number			
	1	2	...	P
Pressure	P_1	P_2	...	P_P
Temperature	T_1	T_2	...	T_P
Component 1	$n_{1,1}/V_1$	$n_{1,2}/V_2$	...	$n_{1,P}/V_P$
Component 2	$n_{2,1}/V_1$	$n_{2,2}/V_2$	...	$n_{2,P}/V_P$
...	...	...	...	...
Component C	$n_{C,1}/V_1$	$n_{C,2}/V_2$	...	$n_{C,P}/V_P$

If the system is at equilibrium, there are numerous relationships among these $P(C + 2)$ variables. We want to know how many independent relationships there are among them. Each such relationship decreases by one the number of independent intensive variables that are needed to specify the state of the system when all of the phases are at equilibrium. Let us count these relationships.

- The pressure must be the same in each phase. That is, $P_1 = P_2, P_1 = P_3, \ldots, P_1 = P_P, P_2 = P_3, \ldots, P_2 = P_P$, *etc.* Since $P_1 = P_2$ and $P_1 = P_3$ implies that $P_2 = P_3$, *etc.*, there are only $P - 1$ independent equations that relate these pressures to one another.
- The temperature must be the same in each phase. As for the pressure, there are $P - 1$ independent relationships among the temperature values.

- The concentration of species A in phase 1 must be in equilibrium with the concentration of species A in phase 2, and so forth. We can write an equation for phase equilibrium involving the concentration of A in any two phases; for example,

$$K = \frac{(n_{A,2}/V_{A,2})}{(n_{A,1}/V_{A,2})}$$

(In Chapter 14, we will find that this requirement can be stated more rigorously using a thermodynamic function that we call the chemical potential. At equilibrium, the chemical potential of species A must be the same in each phase.) For the P phases, there are again $P-1$ independent relationships among the component-A concentration values. This is true for each of the C components, so the total number of independent relationships among the concentrations is $C(P-1)$.
- While every component need not be present in each phase, there must be a finite amount of each phase present. Each phase must have a non-zero volume. To express this requirement using intensive variables, we can say that the sum of the concentrations in each phase must be greater than zero. For phase 1, we must have

$$(n_{A,1}/V_1) + (n_{B,1}/V_1) + \cdots + (n_{Z,1}/V_1) > 0$$

and so on for each of the P phases. There are P such relationships that are independent of one another.

If we subtract, from the total number of relevant relationships, the number of independent relationships that must be satisfied at equilibrium, we find ***Gibbs' phase rule***: There are

$$\begin{aligned} F &= P(C+2) - (P-1) - (P-1) - C(P-1) - P \\ &= 2 + C - P \end{aligned}$$

independent relationships or degrees of freedom needed to describe the equilibrium system containing C components and P phases.

A component may not be present in some particular phase. If this is the case, the total number of relationships is one less than the number that we used above to derive the phase rule. The number of equilibrium constraints is also one less than the number we used. Consequently, the absence of a component from any particular phase has no effect on the number of degrees of freedom available to the system at equilibrium.

§9 Reversible *vs.* irreversible processes

When we think about a physical system that is undergoing a reversible change, we imagine that the system passes through a series of states. In each of these states, every thermodynamic variable has a well-defined value in every phase of the system. We suppose that successive states of the changing system are arbitrarily close to one another in the sense that the successive values of every thermodynamic are arbitrarily close to one another. These suppositions are equivalent to assuming that the state of the system and the value of every thermodynamic variable are continuous functions of time. Then every thermodynamic variable is either constant or a continuous function of other thermodynamic variables. When we talk about a reversible process, we have in mind a physical system that behaves in this way and in which an arbitrarily small change in one of the thermodynamic variables can reverse the direction in which other thermodynamic variables change.

A process that is not reversible is said to be ***irreversible***. We distinguish between two kinds of irreversible processes. A process that cannot occur under a given set of conditions is said to be an ***impossible*** process. A process that can occur, but does not do so reversibly, is called a ***possible*** process or a ***spontaneous*** process.

Another essential characteristic of a reversible process is that changes in the system are driven by conditions that are imposed on the system by the surroundings. In our discussion of the phase equilibria of water, we note that the surroundings can transfer heat to the system only when the temperature of the surroundings is greater than that of the system. However, if the process is to be reversible, this temperature difference must be arbitrarily small, so that heat can be made to flow from the system to the surroundings by an arbitrarily small decrease in the temperature of the surroundings.

Similar considerations apply when the process involves the exchange of work between system and surroundings. We focus on changes in which the work exchanged between system and surroundings is pressure–volume work. A process can occur reversibly only if the pressure of the system and the pressure applied to the system by the surroundings differ by an arbitrarily small amount. To abbreviate these statements, we customarily introduce a figure of speech and say that, for a reversible process, $T_{system} = T_{surroundings}$ (or $T = \hat{T}$) and that $P_{system} = P_{surroundings}$ (or $P = P_{applied}$).

Since a reversible process involves a complementary exchange of energy increments between system and surroundings, it is evident that an isolated system cannot undergo a reversible change. Any change that occurs in an isolated system must be spontaneous. By the contrapositive, an isolated system that cannot undergo change must be at equilibrium.

While $T = \hat{T}$ and $P = P_{applied}$ are necessary conditions for a reversible process, they are not sufficient. A spontaneous process can occur under conditions in which the system temperature is arbitrarily close to the temperature of the surroundings and the system pressure is arbitrarily close to the applied pressure. Consider a mixture of hydrogen and oxygen in a cylinder closed by a frictionless piston. We suppose that the surroundings are maintained at a constant temperature and that the surroundings apply a constant pressure to the piston. We suppose that the system contains a small quantity of a poorly effective catalyst. By controlling the activity of the catalyst, we can arrange for the formation of water to occur at an arbitrarily slow rate—a rate so slow that the temperature and pressure gradients that occur in the neighborhood of the catalyst are arbitrarily small. Nevertheless, the reaction is a spontaneous process, not a reversible one. If the process were reversible, an arbitrarily small increase in the applied pressure would be sufficient to reverse the direction of reaction, causing water to decompose to hydrogen and oxygen.

§10 Duhem's Theorem: Specifying reversible change in a closed system

We view a chemical system as a collection of substances that occupies some volume. Let us consider a closed system whose volume is variable, and in which no work other than pressure–volume work is possible. If this system is undergoing a reversible change, it is at equilibrium, and it is in contact with its surroundings. Because the system is at equilibrium, all points inside the system have the same pressure and the same temperature. Since the change is reversible, the interior pressure is arbitrarily close to the pressure applied to the system by the surroundings. If the reversibly changing system can exchange heat with its surroundings, the temperature of the surroundings is arbitrarily close to the temperature of the system. (If a process takes place in a system that cannot exchange heat with its surroundings, we say that the process is ***adiabatic***.)

We can measure the pressure, temperature, and volume of such a system without knowing anything about its composition. For a system composed of a known amount of a single phase of a pure substance, we know from experience that any cyclic change in pressure or temperature restores the initial volume. That is, for a pure phase, there is an equation of state that we can rearrange as $V = V(P, T)$, meaning that specifying P and T is sufficient to specify V uniquely.

For other reversible systems, the function $V = V(P, T)$ may not exist. For example, consider a system that consists of a known amount of water at liquid–vapor equilibrium and whose pressure and temperature are known. For this system, the volume can have any value between that of the pure liquid and that of the pure gas. Specifying the pressure and temperature of this system is not sufficient to specify its state. However, if we specify the temperature of this system, the pressure is fixed by the equilibrium condition; and if we specify the volume of the system, we can find how much water is in each phase from the known molar volumes of the pure substances at the system pressure and temperature. For the water–water-vapor equilibrium system, we can write $P = P(V, T)$.

In each of these cases, we can view one of the variables as a function of the other two and represent it as a surface in a three dimensional space. The two independent variables define a plane. Projecting the system's location in this independent-variable plane onto the surface establishes the value of the dependent variable. The two independent-variable values determine the point on the surface that specifies the state of the system. In the liquid–vapor equilibrium system, the pressure is a surface above the volume–temperature plane.

A complete description of the state of the system must also include the number of moles of liquid and the number of mole of vapor present. Each of these quantities can also be described as a surface in a three dimensional space in which the other two dimensions are volume and temperature. ***Duhem's theorem*** asserts that these observations are special cases of a more general truth:

Duhem's theorem
For a closed, reversible system in which only pressure–volume work is possible, specifying how some pair of state functions changes is sufficient to specify how the state of the system changes.

Duhem's theorem asserts that two variables are sufficient to specify the state of the system in the following sense: Given the values of the system's thermodynamic variables in some initial state, say $\{X_1, Y_1, Z_1, W_1, ...\}$, specifying the change in some pair of variables, say ΔX and ΔY, is sufficient to determine the change in the remaining variables, ΔZ, ΔW,... so that the system's thermodynamic variables in the final state are $\{X_2, Y_2, Z_2, W_2, ...\}$, where $W_2 = W_1 + \Delta W$, etc. The theorem does not specify which pair of variables is sufficient. In fact, from the discussion above of the variables that can be used to specify the state of a system containing only water, it is evident that a particular pair may not remain sufficient if there is a change in the number of phases present.

In Chapter 10, we see that Duhem's theorem follows from the first and second laws of thermodynamics, and we consider the particular pairs of variables that can be used. For now, let us consider a proof of Duhem's theorem for a system in which the pressure, temperature, volume, and composition can vary. We consider systems in which only pressure–volume work is possible. Let the number of chemical species present be C' and the number of phases be P. (C, the number of components in the phase rule, and C' differ by the number of stoichiometric constraints that apply to the system: C is C' less the number of stoichiometric constraints.) We want to know how many variables can be changed independently while the system remains at equilibrium.

This is similar to the question we answered when we developed Gibbs' phase rule. However, there are important differences. The phase rule is independent of the size of the system; it specifies the number of intensive variables required to prescribe an equilibrium state in which specified phases are present. The size of the system is not fixed; we can add or remove matter to change the size of any phase without changing the number of degrees of freedom. In the present problem, the system cannot exchange matter with its surroundings. Moreover, the number of phases present can change. We require only that any change be reversible, and a reversible process can change the number of phases. (For example, reversible vaporization can convert a two-phase system to a gaseous, one-phase system.)

We want to impose a change on an initial state of a closed system. This initial state is an equilibrium state, and we want to impose a change that produces a new Gibbsian equilibrium state of the same system. This means that the change we impose can neither eliminate an existing chemical species nor introduce a new one. A given phase can appear or disappear, but a given chemical species cannot.

We can find the number of independent variables for this system by an argument similar to the one we used to find the phase rule. To completely specify this system, we must specify the pressure, temperature, and volume of each phase. We must also specify the number of moles of each of C' chemical species in each phase. This means that $P(C'+3)$ variables must be specified. Every relationship that exists among these variables decreases by one the number that are independent. The following relationships exist:

- The pressure is the same in each phase. There are $P-1$ pressure constraints.
- The temperature is the same in each phase. There are $P-1$ temperature constraints.
- The volume of each phase is determined by the pressure, the temperature, and the number of moles of each species present in that phase. (In Chapter 14, we find that the volume of a phase, V, is given rigorously by the equation $V=\sum_k n_k \overline{V}_k$, where n_k and $\overline{V}_k$ are the number of moles and the partial molar volume of the kth species in that phase. The $\overline{V}_k$ depend only on pressure, temperature, and composition.) For P phases, there are P constraints, one for the volume of each phase.
- To completely specify the system, the concentration of each species must be specified in each phase. This condition creates $C'P$ constraints. (We can also reach this conclusion by a slightly different argument. To specify the concentrations of C' species in some one phase requires C' constraints. A distribution equilibrium relates the concentrations of each species in every pair of phases. There are $P-1$ independent pairs of phases. For C' chemical species, there are $C'(P-1)$ such constraints. This is equivalent to the requirement in our phase rule analysis that there are $C(P-1)$ equilibrium relationships among C components in P phases. In the present problem, the total number of concentration constraints is $C'+C'(P-1)=C'$P.

Subtracting the number of constraints from the number of variables, we find that there are

$$P(C'+3)-(P-1)-(P-1)-P-C'P=2$$

independent variables for a reversible process in a closed system, if all work is pressure–volume work. The number of independent variables is constant; it is independent of the species that are present and the number of phases.

It is important to appreciate that there is no conflict between Duhem's theorem and the phase-rule conclusion that F degrees of freedom are required to specify an equilibrium state of a system containing specified phases. When we say that specifying some pair of variables is sufficient to specify the state of a particular closed system undergoing reversible change, we are describing a system that is continuously at equilibrium as it goes from a first equilibrium state to a second one. Because it is closed and continuously in an equilibrium state, the range of variation available to the system is circumscribed in such a way that specifying two variables is sufficient to specify its state. On the other hand, when we say that F degrees of freedom are required to specify an equilibrium state of a system containing specified phases, we mean that we must know the values of F intensive variables in order to establish that the state of the system is an equilibrium state.

To illustrate the compatibility of these ideas and the distinction between them, let us consider a closed system that contains nitrogen, hydrogen, and ammonia gases. In the presence of a catalyst, the reaction $N_2+3H_2 \rightleftharpoons 2NH_3$ occurs. For simplicity, let us assume that these gases behave ideally. (If the gases do not behave ideally, the argument remains the same, but more complex equations are required to express the equilibrium constant and the system pressure as functions of the molar composition.) This system has two components and three degrees of freedom. When we say that the system is closed, we mean that the total number of moles of the elements nitrogen and hydrogen are known and constant. Let these be n_N and n_H, respectively. Letting the moles of ammonia present be $n_{NH_3}=x$, the number of moles of dihydrogen and dinitrogen are $n_{H_2}=(n_H-3x)/2$ and $n_{N_2}=(n_N-x)/2$, respectively.

If we know that this system is at equilibrium, we know that the equilibrium constant relationship is satisfied. We have

$$\begin{aligned} K_P &= \frac{P_{NH_3}^2}{P_{H_2}^3 P_{N_2}} \\ &= \frac{n_{NH_3}^2}{n_{H_2}^3 n_{N_2}}\left(\frac{RT}{V}\right)^{-2} \\ &= \frac{16x^2}{(n_H-3x)^3(n_N-x)}\left(\frac{RT}{V}\right)^{-2} \end{aligned}$$

where V is the volume of the system. The ideal-gas equilibrium constant is a function only of temperature. We assume that we know this function; therefore, if we know the temperature, we know the value of the equilibrium constant. The pressure of the system can also be expressed as a function of x and V. We have

$$P = P_{H_2} + P_{N_2} + P_{NH_3}$$
$$= \left[\frac{(n_H + n_N)}{2} - x\right]\left(\frac{RT}{V}\right)$$

If we know the system pressure and we know that the system is at equilibrium, we can solve the equations for K and P simultaneously to find the unknowns x and V. From these, we can calculate the molar composition of the system and the partial pressure of each of the gases. (We discuss ideal-gas equilibrium calculations in detail in Chapter 13.) Thus, if we know that the system is at equilibrium, knowledge of the pressure and temperature is sufficient to determine its composition and all of its other properties.

If we do not know that this system is at equilibrium, but instead want to collect sufficient experimental data to prove that it is, the phase rule asserts that we must find the values of some set of three intensive variables. Two are not sufficient. From the perspective provided by the equations developed above, we can no longer use the equilibrium constant relationship to find x and V. Instead, our problem is to find the composition of the system by other means, so that we can test for equilibrium by comparing the value of the quantity

$$\frac{P_{NH_3}^2}{P_{H_2}^3 P_{N_2}} = \frac{16x^2}{(n_H - 3x)^3(n_N - x)}\left(\frac{RT}{V}\right)^{-2}$$

to the value of the equilibrium constant. We could accomplish this goal by measuring the values of several different combinations of three intensive variables. A convenient combination is pressure, temperature, and ammonia concentration, x/V. When we rearrange the equation for the system pressure to

$$P = \left[\frac{(n_H + n_N)}{2V} - \left(\frac{x}{V}\right)\right]RT$$

it is easy to see that knowing P, T, and x/V enables us to find the volume of the system. Given the volume, we can find the molar composition of the system and the partial pressure of each of the gases. With these quantities in hand, we can determine whether the equilibrium condition is satisfied.

§11 Reversible motion of a mass in a constant gravitational field

Let us explore our ideas about reversibility further by considering the familiar case of a bowling ball that can move vertically in the effectively constant gravitational field near the surface of the earth.

We begin by observing that we develop our description by abstracting from reality. We consider idealized models because we want to develop theories that capture the most important features of real systems. We ignore less important features. In the present example, we know that the behavior of the bowling ball will be slightly influenced by its frictional interaction with the surrounding atmosphere. (We attribute these interactions to a property of air that we call viscosity.) We assume that this effect can be ignored. This causes no difficulty so long as our experiments are too insensitive to observe the effects of this atmospheric drag. If necessary, of course, we could do our experiments inside a vacuum chamber, so that the system we study experimentally better meets the assumptions we make in our analysis. Alternatively, we could expand our theory to include the effects of atmospheric drag.

To raise an initially stationary bowling ball to a greater height requires that we apply a vertical upward force that exceeds the downward gravitational force on the ball. Let height increase in the upward direction, and let $h(t)$ and $v(t)$ be the height and (vertical) velocity of the ball at time t. Let the mass of the ball be m, and let the ball be at rest at time zero. Representing the initial velocity and height as v_0 and h_0, we have $v_0 = v(0) = 0$ and $h_0 = h(0) = 0$. Letting the gravitational acceleration be g, the gravitational force on the ball is $f_{gravitation} = -mg$. To raise the ball, we must apply a vertical force, $f_{applied} > 0$, that makes the net force on the ball greater than zero. That is, we require

$$f_{net} = f_{applied} - mg > 0$$

so that

$$m\frac{d^2h}{dt^2} = f_{net}$$

If $f_{applied}$ is constant, f_{net} is constant; we find for the height and velocity of the ball at any later time t,

$$v(t) = \left(\frac{f_{net}}{m}\right)t$$

and

$$h(t) = \left(\frac{f_{net}}{m}\right)\frac{t^2}{2}$$

Let us consider the state of the system when the ball reaches a particular height, h_S. Let the corresponding time, velocity, kinetic energy, and potential energy at h_S, be t_S, v_S, τ_S, and υ_S, respectively. Since

$$v_S = \left(\frac{f_{net}}{m}\right)t_S$$

and

$$h_S = \left(\frac{f_{net}}{m}\right)\frac{t_S^2}{2}$$

we have

$$\tau(h_S) = \frac{mv_S^2}{2}$$
$$= \frac{m}{2}\left(\frac{f_{net}}{m}\right)^2 t_S^2$$
$$= f_{net}h_S$$

The energy we must supply to move the ball from height

zero to h_S is equal to the work done by the surroundings on the ball. The increase in the energy of the ball is $-\widehat{w}$. At h_S this input energy is present as the kinetic and potential energy of the ball. We have

$$-\widehat{w} = \int_{h=o}^{h_S} f_{applied}\,dh$$

$$= \int_{h=0}^{h_S} (f_{net} + mg)dh$$

$$= f_{net}h_S + mgh_S$$

$$= \tau(h_S) + \upsilon(h_S)$$

where the kinetic and potential energies are $\tau(h_S) = f_{net}h_S$ and $\upsilon(h_S) = mgh_S$, respectively.

The ball rises only if the net upward force is positive: $f_{net} = f_{applied} - mg > 0$. Then the ball arrives at h_S with a non-zero velocity and kinetic energy. If we make f_{net} smaller and smaller, it takes the ball longer and longer to reach h_S; when it arrives, its velocity and kinetic energy are smaller and smaller. However, no matter how long it takes the ball to reach h_S, when it arrives, its potential energy is $\upsilon(h_S) = mgh_S$,

Now, let us consider the energy change in a process in which the ball begins at rest at height zero and ends at rest at h_S. At the end, we have $\tau(h_S) = 0$. To effect this change in a real system, we must apply a net upward force to the ball to get it moving; later we must apply a net downward force to slow the ball in such a way that its velocity becomes zero at exactly the time that it reaches h_S. There are infinitely many ways we could apply forces to meet these conditions. The net change in the ball's energy is the same for all of them.

We find it useful to use a hypothetical process to calculate this energy change. In this hypothetical process, the upward force is always just sufficient to oppose the gravitational force on the ball. That is, $f_{net} = 0$ so that $f_{applied} = mg$, and from the development above $v_S = 0$ and $\tau(h_S) = 0$. Of course, $t_\infty = \infty$. This is a hypothetical process, because the ball would not actually move under these conditions. We see that the hypothetical process is the limiting case in a series of real processes in which we make $f_{net} > 0$ smaller and smaller. In all of these processes, the potential energy change is

$$\upsilon(h_S) = \int_{h=0}^{h_S} mgdh = mgh_S$$

If the ball is stationary and $f_{applied} = mg$, the ball remains at rest, whatever its height. If we make $f_{applied} > mg$, the ball rises. If we make $f_{applied} < mg$, the ball falls. If $f_{net} \approx 0$ and the ball is moving only slowly in either direction, a very small change in $f_{net} = f_{applied} - mg$ can be enough to reverse the direction of motion. These are the characteristics of a reversible process: an arbitrarily small change in the applied force changes the direction of motion.

The advantage of working with the hypothetical reversible process is that the integral of the applied force over the distance through which it acts is the change in the potential energy of the system. While we cannot actually carry out a reversible process, we can compute the work that must be done if we know the limiting force that is required in order to effect the change. This is true because the velocity and kinetic energy of the ball are zero throughout the process. When the process is reversible, the change in the potential energy of the ball is equal to the work done on the ball; we have $-\widehat{w}(h_S) = w(h_S) = \upsilon(h_S)$

Gravitational potential energy is an important factor in some problems of interest in chemistry. Other forms of potential energy are important much more often. Typically, our principal interest is in the potential energy change associated with a change in the chemical composition of a system. We are seldom interested in the kinetic energy associated with the motion of a macroscopic system as a whole. We can include effects that arise from gravitational forces or from the motion of the whole system in our thermodynamic models, but we seldom find a need to do so. For systems in which the motion of the whole system is important, the laws of mechanics are usually sufficient; we find out what we want to know about such systems by solving their equations of motion.

When we discuss the first law of thermodynamics, we write $E = q + w$ (or $dE = dq + dw$) for the energy change that accompanies some physical change in a system. Since chemical applications rarely require that we consider the location of the system or the speed with which it may be moving, "w" usually encompasses only work that changes the energy of the system itself. Then, E designates the energy of the macroscopic system itself. As noted earlier, we often recognize this by calling the energy of the system its ***internal energy***. Some writers use the symbol U to represent the internal energy, intending thereby to make it explicit that the energy under discussion is independent of the system's location and motion.

§12 Equilibria and reversible processes

The distinction between a system at equilibrium and a system undergoing reversible change is razor-thin. What we have in mind goes to the way we choose to define the system and centers on the origin of the forces that affect its energy. For a system at equilibrium, the forces are fixed. For a system undergoing reversible change, some of the forces originate in the surroundings, and those that do are potentially variable.

To raise a bowling ball reversibly, we apply an upward force, $+mg$, exactly equal and opposite to the downward force, $-mg$, due to gravity. At any point in this reversible motion, the ball is stationary, which is the reason we say that a reversible process is a hypothetical change. If we were to change the system slightly, by adding a shelf to support the ball at exactly the same height, the forces on the ball would be the same; however, the forces would be fixed and we would say that the ball is at equilibrium.

We can further illustrate this distinction by returning to the water–water-vapor system. If an unchanging water–water-vapor mixture is enclosed in a container whose dimensions are fixed (like a sealed glass bulb) we say that the system is at equilibrium. If a piston encloses the same collection of matter, and the surroundings apply a force on the piston that balances the pressure exerted by the mixture, we can say that the system is changing reversibly.

In §1-6, we used the term "primitive equilibrium" to refer to an equilibrium state in which all of the state functions are fixed. A system that can undergo reversible change without changing the number or kinds of phases present can be in an infinite number of such states. Since the set of such primitive equilibrium states encompasses the accessible equilibrium conditions in the sense of Gibb's phase rule, we can call this set a ***Gibbsian equilibrium manifold***.

§13 The laws of thermodynamics

We usually consider that the first, second, and third laws of thermodynamics are basic postulates. One of our primary objectives is to understand the ideas that are embodied in these laws. We introduce these ideas here, using statements of the laws of thermodynamics that are immediately applicable to chemical systems. In the next three chapters, we develop some of the most important consequences of these ideas. In the course of doing so, we examine other ways that these laws have been stated.

The first law deals with the definition and properties of energy. The second and third laws deal with the definition and properties of entropy. The laws of thermodynamics assert that energy and entropy are state functions. In the next chapter, we discuss the mathematical properties of state functions. Energy and entropy changes are defined in terms of the heat and work exchanged between a system and its surroundings. ***We adopt the convention that heat and work are positive if they increase the energy of the system.*** In a process in which a closed system accepts increments of heat, dq , and work, dw, from its surroundings, we define the changes in the energy, dE, and the entropy, dS, of the system in terms of dq, dw, and the temperature.

The meaning of the first law is intimately related to a crucial distinction between the character of energy on the one hand and that of the variables heat and work on the other. When we say that energy is a state function, we mean that the energy is a property of the system. In contrast, heat and work are not properties of the system; rather they describe a process in which the system changes. When we say that the heat exchanged in a process is q, we mean that q units of thermal energy are transferred from the surroundings to the system. If $q > 0$, the energy of the system increases by this amount, and the energy of the surroundings decreases by the same amount. q has meaning only as a description of one aspect of the process.

When the process is finished, the system has an energy, but q exists only as an accounting record. Like the amount on a cancelled check that records how much we paid for something, q is just a datum about a past event. Likewise, w is the record of the amount of non-thermal energy that is transferred. Because we can effect the same change in the energy of a system in many different ways, we have to measure q and w for a particular process as the process is taking place. We cannot find them by making measurements on the system after the process has gone to completion.

In §1, we introduce a superscripted caret to denote a property (state function) of the surroundings. Thus, E is the energy of the system; $\hat{E}$ is the energy of the surroundings; dE is an incremental change in the energy of the system; and $d\hat{E}$ is an incremental change in the energy of the surroundings. If we are careful to remember that heat and work are not state functions, it is useful to extend this notation to increments of heat and work. If q units of energy are transmitted to the system as heat, we let $\hat{q}$ be the thermal energy transferred to the surroundings in the same process. Then $\hat{q} = -q$, and $\hat{q} + q = 0$. Likewise, we let w be the work done on the system and $\hat{w}$ be the work done on the surroundings in the same process, so that $\hat{w} = -w$, and $\hat{w} + w = 0$. Unlike E and $\hat{E}$, which are properties of different systems, q and $\hat{q}$ (or w and $\hat{w}$) are merely alternative expressions of the same thing—the quantity of energy transferred as heat (or work).

We define the incremental change in the energy of a closed system as $dE = dq + dw$. The accompanying change in the energy of the surroundings is $d\hat{E} = d\hat{q} + d\hat{w}$, so that $dE + d\hat{E} = 0$. Whereas $\hat{q} + q = 0$ (or $d\hat{q} + dq = 0$) is a tautology, because it merely defines $\hat{q}$ as $-q$, the first law asserts that $dE_{universe} = dE + d\hat{E}$ is a fundamental property of nature. Any increase in the energy of the system is accompanied by a decrease in the energy of the surroundings, and conversely. Energy is conserved; heat is not; work is not.

The first law of thermodynamics

In a process in which a closed system accepts increments of heat, dq, and work, dw, from its surroundings, the change in the energy of the system, dE, is $dE = dq + dw$. Energy is a state function. For any process, $dE_{universe} = 0$.

For a reversible process in which a system passes from state A to state B, the amount by which the energy of the system changes is the line integral of dE along the path followed. Denoting an incremental energy change along this path as $d_{AB}E$, we have $\Delta_{AB}E = \int_A^B d_{AB}E$. (We review line integrals in the next chapter.) The energy change for the surroundings is the line integral of $d\hat{E}$ along the path followed by the surroundings during the same process: $\Delta_{AB}\hat{E} = \int_A^B d_{AB}\hat{E}$.

For any process in which energy is exchanged with

the surroundings, the change in the system's energy is $\Delta E = q + w$, where q and w are the amounts of thermal and non-thermal energy delivered to the system. We can compute ΔE from q and w whether the process is reversible or irreversible.

In contrast, the definition of entropy change applies only to reversible processes. In a process in which a system reversibly accepts an increment of heat, dq^{rev}, from its surroundings, the entropy change is defined by $dS = dq^{rev}/T$. (We introduce the superscript, "rev", to distinguish heat and work exchanged in reversible processes from heat and work exchanged in irreversible, "irrev", or spontaneous, "spon", processes.) When a system passes reversibly from state A to state B, the entropy change for the system is the line integral of $d_{AB}S = q^{rev}/T$ along the path followed: $\Delta_{AB}S = \int_A^B d_{AB}q^{rev}/T$. The entropy change for the surroundings is defined by the same relationship, $d_{AB}\hat{S} = d_{AB}\,\hat{q}^{rev}/T$. Every system has an entropy. The entropies of the system and of its surroundings can change whenever a system undergoes a change. If the change is reversible, $\Delta S = -\Delta\hat{S}$.

The second law of thermodynamics

In a reversible process in which a closed system accepts an increment of heat, dq^{rev}, from its surroundings, the change in the entropy of the system, dS, is $dS = dq^{rev}/T$. Entropy is a state function. For any reversible process, $dS_{universe} = 0$, and conversely. For any spontaneous process, $dS_{universe} > 0$, and conversely.

We define the entropy change of the universe by $dS_{universe} = dS + d\hat{S}$; it follows that $\Delta_{AB}S_{universe} = \Delta_{AB}S + \Delta_{AB}\hat{S}$ for any process in which a system passes from a state A to a state B, whether the process is reversible or not. Since $dS_{universe} = 0$ for every part of a reversible process, we have $\Delta S_{universe} = 0$ for any reversible process. Likewise, since $dS_{universe} > 0$ for every part of a spontaneous process, we have $\Delta S_{universe} > 0$ for any spontaneous process.

The third law deals with the properties of entropy at temperatures in the neighborhood of absolute zero. It is possible to view the third law as a statement about the properties of the temperature function. It is also possible to view it as a statement about the properties of heat capacities. A statement in which the third law attributes particular properties to the entropy of pure substances is directly applicable to chemical systems. This statement is that of Lewis and Randall[3]:

The third law of thermodynamics

If the entropy of each element in some crystalline state be taken as zero at the absolute zero of temperature, every substance has a positive finite entropy; but at the absolute zero of temperature the entropy may become zero, and does so become in the case of perfect crystalline substances.

The Lewis and Randall statement focuses on the role that the third law plays in our efforts to express the thermodynamic properties of pure substances in useful ways. To do so, it incorporates a matter of definition when it stipulates that "the entropy of each element be taken as zero at the absolute zero of temperature." The third law enables us to find thermodynamic properties ("absolute entropies" and Gibbs free energies of formation) from which we can make useful predictions about the equilibrium positions of reactions. The third law can be inferred from experimental observations on macroscopic systems. It also arises in a natural way when we develop the theory of statistical thermodynamics. In both developments, the choice of zero for the entropy of "each element in some crystalline state" at absolute zero is—while arbitrary—logical, natural, and compellingly convenient.

§14 Thermodynamic criteria for change

When the state of an isolated system can change, we say that the system is capable of spontaneous change. When an isolated system is incapable of spontaneous change, we say that it is at equilibrium. Ultimately, this statement defines what we mean by (primitive) equilibrium. From our statement of the second law of thermodynamics, we have criteria for spontaneous change and for equilibrium in any macroscopic system:

An isolated system can undergo any change that results in an increase in the entropy of the system. The converse is also true; an isolated system whose entropy can increase can undergo change. Any such change is said to be spontaneous. If an isolated system cannot change in such a way that its entropy increases, the system cannot change at all and is said to be at equilibrium.

A system that is not isolated can undergo any change that results in an increase in the entropy of the universe, and conversely. Such changes are also said to be spontaneous.

If a system that is not isolated undergoes a change, but the entropy of the universe remains constant, the change is not spontaneous. The entropy changes for the system and the surroundings are equal in magnitude and opposite in sign. The change is said to be reversible.

Although the first statement applies to isolated systems and the second applies to systems that are not isolated, we usually consider that both are statements of the same criterion, because the second statement follows from the first when we view the universe as an isolated system. We can restate these criteria for spontaneous change and equilibrium using the compact notation that we introduce in §13.

From our definitions, any change that occurs in an isolated system must be spontaneous. From our statement of the second law, the entropy of the universe must increase in any such process. To indicate this, we write

$\Delta S_{universe} > 0$. The surroundings must be unaffected by any change in an isolated system; hence, none of the surroundings' state functions can change. Thus, $\Delta\hat{S} = 0$, and since $\Delta S + \Delta\hat{S} = \Delta S_{universe} > 0$, we have $\Delta S > 0$.

For a spontaneous change in a system that is not isolated, ΔS can be greater or less than zero. However, ΔS and $\Delta\hat{S}$ must satisfy $\Delta S + \Delta\hat{S} = \Delta S_{universe} > 0$.

In a system that is not isolated, reversible change may be possible. A system that undergoes a reversible change is at—or is arbitrarily close to—one of its equilibrium states during every part of the process. For a reversible change, it is always true that $\Delta S = -\Delta\hat{S}$, so that $\Delta S + \Delta\hat{S} = \Delta S_{universe} = 0$.

Our criteria for change are admirably terse, but to appreciate them we need to understand precisely what is meant by "entropy". To use the criteria to make predictions about a particular system, we need to find the entropy changes that occur when the system changes. To use these ideas to understand chemistry, we need to relate these statements about macroscopic systems to the properties of the molecules that comprise the system.

Since an isolated system does not interact with its surroundings in any way, no change in an isolated system can cause a change in its surroundings. If an isolated system is at equilibrium, no change is possible, and hence there is no system change for which the entropy of the universe can increase. Evidently, the entropy of the universe is at a maximum when the system is at equilibrium.

Typically, we are interested in what happens when the interaction between the system and surroundings serves to impose conditions on the final state of a system. A common example of such conditions is that the surroundings maintain the system at a constant pressure, while providing a constant-temperature heat reservoir, with which the system can exchange heat. In such cases, the system is not isolated. It turns out that we can use the entropy criterion to develop supplemental criteria based on other thermodynamic functions. These supplemental criteria provide the most straightforward means to discuss equilibria and spontaneous change in systems that are not isolated. Which thermodynamic function is most convenient depends upon the conditions that we impose on the system.

§15 State functions in systems undergoing spontaneous change

In this chapter, we introduce ideas that underlie classical thermodynamics. Because the development of classical thermodynamics relies on the properties of reversible processes, we have devoted considerable attention to specifying what we mean by a reversible process and to the relationship between reversible processes and the equilibrium states available to a system. In Chapters 7-17, we develop this theory. The development assumes that we can measure the heat and work exchanged between a system and its surroundings. It assumes that we can measure the state functions volume, pressure, temperature, and the amounts (moles) of chemical substances in a system. We define other state functions, whose values can be computed from these measurable quantities. From observations that we make on systems that are undergoing reversible change, we develop numerous relationships among these state functions.

No real system can change in exactly the manner we have in mind when we talk about a reversible process. Strictly speaking, any process that actually occurs must be spontaneous. The idea of reversible change is clearly an abstraction from reality. Nevertheless, we can determine—to a good approximation—the way in which one variable depends on another in a reversible process. We accomplish this by making measurements on a real system whose behavior approximates the ideal of reversibility as closely as possible. We express the—approximate—result of any such measurement as a number. Normally, we view the approximate character of the number to be a consequence of experimental error. When we say that we make measurements on a system that is undergoing a reversible change, we mean that we are making the measurements on a process that satisfies our definition of reversibility closely enough for the purpose at hand.

There are two reasons for the fact that reversible processes play an essential role in the development of the equations of thermodynamics. The first is that we can measure the entropy change for a process only if the process is reversible. The second and subtler reason is that an intensive variable may not have a unique value in a system that is undergoing a spontaneous change. If the temperature, the pressure, or the concentration of a component varies from point to point within the system, then that state function does not have a unique value, and we cannot use it to model the change. This occurs, for example, when gasoline explodes in a cylinder of a piston engine. The system consists of the contents of the cylinder. At any given instant, the pressure, temperature, and component concentrations vary from place to place within the cylinder. In general, no single value of any of these intensive variables is an adequate approximation for use in the thermodynamic equations that characterize the system as a whole.

To explore this idea further, let us think about measuring changes in extensive state functions during a spontaneous process. Since we are free to define the system as we please, we can choose a definition that makes the volume readily measurable. In the piston-engine example, there is no ambiguity about the volume of the system at any instant. While point-to-point variability means that the concentrations of the chemical components are not defined for the system as a whole, we are confident that there is some specific number of moles of each component present in the system at every instant. We can reach this conclusion by imagining that we can instantaneously freeze the composition by stopping all reactions. We can then find the number of moles of each component at our leisure.

If the system is not too inhomogeneous, we can devise an alternative procedure for making—in concept—such composition measurements. We imagine dividing the system into a large number of macroscopic subsystems. Each of these subsystems has a well-defined volume. We suppose also that each of them has well-defined thermodynamic functions at any given instant; that is, we assume that the pressure, temperature, and concentrations are approximately homogeneous within each of these subsystems. If this condition is satisfied, we can sum up the number of moles of a component in each of the sub-volumes to obtain the number of moles of that component in the whole system.

We can make a similar argument for any extensive thermodynamic function, so it applies to the energy, entropy, enthalpy, and the Helmholtz and Gibbs free energies. As long as the point-to-point variability within the system is small enough so that a division of the system into macroscopic subsystems produces subsystems that are approximately homogeneous, we can find the value for any extensive thermodynamic function in each individual sub-system and for the system as a whole. The measurement we propose has the character of a *gedanken* experiment. We can describe a procedure for making the measurement, whether we can actually perform the procedure or not.

This argument does not work for intensive thermodynamic functions. It is true that we could produce a weighted-average value for the temperature by multiplying the temperature of each subsystem by the subsystem volume, adding up the products, and dividing the sum by the volume of the whole system; however, the result would not be an intensive property of the whole system. For one thing, we could produce a different average temperature for every extensive variable by using it rather than the volume as the weighting factor in the average-temperature computation. Moreover, no such weighted-average temperature can reflect the fact that different temperatures in different subsystems result in grossly different reaction rates. No single temperature represents the state of the whole system, and we can make the same statement about any other intensive thermodynamic function. For a non-homogeneous system that can be subdivided into approximately homogeneous macroscopic subsystems, we can measure, in principle, the values of the system's extensive state functions; however, its intensive state functions are essentially undefined.

On the other hand, we may be able to assume that an effectively homogeneous subsystem of macroscopic proportions does have well-defined extensive and intensive state functions, even if it is not in an equilibrium state. While a spontaneously changing system need not be homogenous, we commonly encounter systems that are homogeneous to within some arbitrarily small deviation. Consider a closed and well-stirred system in which some chemical reaction is occurring slowly. We immerse this system in a constant-temperature bath, and arrange for the applied pressure to be constant. From experience, we know that the temperature and pressure within such a system will be essentially constant, equal to the bath temperature and the applied pressure, respectively, and homogeneous throughout the system. In such a system, the temperature and the pressure of the system are at equilibrium with those imposed by the surroundings. The chemical process is not at equilibrium, but the component concentrations are homogeneous.

An important question now arises: Are all of the equations of equilibrium thermodynamics applicable to a system in which some processes occur spontaneously? That they are not is evident from the fact that we can calculate an entropy change from its defining equation, $dS = dq^{rev}/T$, only if the behavior of the system is reversible.

Nevertheless, we will find that the relationships among state functions that we derive for reversible processes can be augmented to describe spontaneous processes that occur in homogeneous systems. The necessary augmentation consists of the addition of terms that express the effects of changing composition. (In §9-14, we develop the ***fundamental equation***,

$$dE = TdS - PdV + dw_{NPV}$$

which applies to any reversible process in a closed system. In §14-1, we infer that the fundamental equation becomes

$$dE = TdS - PdV + dw_{NPV} + \sum_i \mu_i dn_i$$

for a spontaneous process in which μ_i and dn_i are the chemical potential and the change in the number of moles of component i, respectively.)

The distinction between reversible and spontaneous processes plays a central role in our theory. We find a group of relationships that express this distinction, and we call these relationships ***criteria for change***. (In §9-19, we find that $(dE)_{SV} = dw_{NPV}$ if and only if the process is reversible, while $(dE)_{SV} < dw_{NPV}$ if and only if the process is spontaneous. We find a close connection between the criteria for change and the composition-dependent terms that are needed to model the thermodynamic functions during spontaneous processes. We find that $\sum_i \mu_i dn_i = 0$ if and only if the process is reversible, while $\sum_i \mu_i dn_i < 0$ if and only if the process is spontaneous.)

In thinking about spontaneous processes, we should also keep in mind that the validity of our general relationships among state functions does not depend on our ability to measure the state functions of any particular state of a system. For example, we can find relationships among the molar volume and other thermodynamic properties of liquid water. Liquid water does not exist at 200 C and 1 bar, so we cannot undertake to measure its thermodynamic properties. However, by using our relationships among state functions and properties that we measure for liquid water where it does exist, we can estimate the thermodynamic properties of liquid water at 200 C and 1 bar. The results are two steps removed from reality; they are the estimated properties of a hypothetical substance. Nevertheless, they have predictive value; for example, we can use them to predict that liquid water at

200 C and 1 bar will spontaneously vaporize to form gaseous water at 1 bar. The equations of thermodynamics are creatures of theory. We should not expect every circumstance that is described by the theory to exist in reality. What we require is that the theory accurately describe every circumstance that actually occurs.

To develop the equations of classical thermodynamics, we consider reversible processes. We then find general criteria for change that apply to any sort of change in any system. Later, we devise criteria based on the changes that occur in the composition of the system. In this book, we consider such composition-based criteria only for homogeneous systems. An extensive theory[4.5] has been developed to model spontaneous processes in systems that are not necessarily homogeneous. This theory is often called ***irreversible thermodynamics*** or ***non-equilibrium thermodynamics***. Development of this theory has led to a wide variety of useful insights about various molecular processes. However, much of what we are calling classical thermodynamics also describes irreversible processes. Even as we develop our theory of reversible thermodynamics, we use arguments that apply the equations we infer from reversible processes to describe closely related systems that are not at equilibrium.

Problems

Use Le Chatelier's principle to answer questions 1-8.

1. One gram of iodine just dissolves in 2950 mL of water at ambient temperatures. One gram of iodine is added to 1000 mL of pure water and the resulting system is allowed to come to equilibrium. When equilibrium is reached, will all of the iodine have dissolved? What will happen if a small amount of water is added to the equilibrated system?

2. A saturated solution of barium sulfate is in contact with excess solid barium sulfate.

$$BaSO_4(s) \rightleftharpoons Ba^{2+} + SO_4^{2-}$$

A small amount of a concentrated solution of $BaCl_2$ is added. How does the system respond?

3. In the presence of a catalyst, oxygen reacts with sulfur dioxide to produce sulfur trioxide.

$$SO_2(g) + ½O_2(g) \rightleftharpoons SO_3(g)$$

A particular system consists of an equilibrated mixture of these three gases. A small amount of oxygen is added. How does the system respond?

4. A system containing these three gases is at equilibrium.

$$SO_2(g) + ½O_2(g) \rightleftharpoons SO_3(g)$$

We suddenly decrease the volume of this system. How does the system respond?

5. In the presence of a catalyst, oxygen reacts with nitrogen to produce nitric oxide.

$$N_2(g) + O_2(g) \rightleftharpoons 2NO(g)$$

A particular system consists of an equilibrated mixture of these three gases. While keeping the temperature constant, we suddenly increase the volume of this system. How does the system respond?

6. Nitric oxide formation

$$N_2(g) + O_2(g) \rightleftharpoons 2NO(g)$$

is endothermic. (At constant temperature, the system absorbs heat as reaction occurs from left to right.) How does the position of equilibrium change when we increase the temperature of this system?

7. Pure water dissociates to a slight extent, producing hydronium, H_3O^+, and hydroxide, OH^-, ions. This reaction is called the autoprotolysis of water.

$$2H_2O \rightleftharpoons H_3O^+ + OH^-$$

Is the autoprotolysis reaction endothermic or exothermic? (What happens to the temperature when we mix an acid with a base?) How does the autoprotolysis equilibrium change when we increase the temperature of pure water?

8. At the melting point, most substances are more dense in their solid state than they are in their liquid state. Such a substance is at its melting point at a particular pressure. Suppose that we now increase the pressure on this system. Does the melting point of the substance increase or decrease?

For each of the systems 9 – 20, specify
(a) what phases are present,
(b) the number of phases, P,
(c) the substances that are present,
(d) the number of components*, C, and
(e) the number of degrees of freedom, F.
Assume that the temperature and pressure of each system is constant and that all relevant chemical reactions are at equilibrium.

9. Pure helium gas, He, sealed in a glass bulb.

10. A mixture of helium gas, He, and neon gas, Ne, sealed in a glass bulb.

11. A mixture of N_2O_4 gas and NO_2 gas sealed in a glass bulb. These compounds react according to the equation

$$N_2O_4(g) \rightleftharpoons 2NO_2(g)$$

12. A mixture of N_2O_4 gas, NO_2 gas, and He gas sealed in a glass bulb.

13. A mixture of PCl_5 gas, PCl_3 gas, and Cl_2 gas, sealed in a glass bulb. The proportions of PCl_3 and Cl_2 are arbitrary. These compounds react according to the equation

$$PCl_5(g) \rightleftharpoons PCl_3(g) + Cl_2(g)$$

14. A mixture of PCl_5 gas, PCl_3 gas, Cl_2 gas, and He gas sealed in a glass bulb. The proportions of PCl_3 and Cl_2 are arbitrary.

15. A mixture of PCl_5 gas, PCl_3 gas, and Cl_2 gas sealed in a glass bulb. In this particular system, the number of moles of PCl_3 is the same as the number of moles of Cl_2.

16. A saturated aqueous solution of iodine, I_2. The solution is in contact with a quantity of solid I_2.

17. An aqueous solution that contains potassium ion, K^+, iodide ion, I^-, triiodide ion, I_3^- , and dissolved I_2. The solution is in contact with a quantity of solid I_2. Recall that triiodide ion is formed by the reaction

$$I^- + I_2 \rightleftharpoons I_3^-$$

18. An aqueous solution that contains K^+, and I^-. This solution is in contact with a quantity of solid silver iodide, AgI. Recall that AgI is quite insoluble. Neutral molecules of AgI do not exist as such in aqueous solution. The solid substance equilibrates with its dissolved ions according to the reaction

$$AgI(s) \rightleftharpoons Ag^+ + I^-$$

19. An aqueous solution that contains K^+, I^-, and chloride ions, Cl^-. This solution is in contact with a mixture of (pure) solid AgI and (pure) solid $AgCl$.

20. An aqueous solution that contains K^+, I^-, Cl^-, and nitrate ions, NO_3^-. This solution is in contact with a mixture of (pure) solid AgI and (pure) solid $AgCl$.

21. A large vat contains oil and water. The oil floats as a layer on top of the water. Orville has another tank with a reserve supply of oil. He also has pipes and pumps that enable him to pump oil between his tank and the vat. Wilbur has a third tank with a reserve supply of water. Wilbur has pipes and pumps that enable him to pump water between his tank and the vat. Their pumps are calibrated to show the volume of oil or water added to or removed from the vat. Normally, Orville and Wilbur work as a team to keep the total mass of liquid in the vat constant. The oil and water have densities of 0.80 kg L^{-1} and 1.00 kg L^{-1}, respectively. Let M_{vat} be the total mass of the liquids in the vat.

(a) If Orville pumps a small volume of oil, dV_{oil}, into or out of the vat, while Wilbur does nothing, what is the change in the mass of liquid in the vat? (i.e., dM_{vat} =?)

(b) If Wilbur pumps a small volume of water, dV_{water}, into or out of the vat, while Orville does nothing, what is the change in the mass of liquid in the vat? (i.e., dM_{vat} =?)

(c) Suppose that Orville and Wilbur make adjustments, dV_{oil} and dV_{water} at the same time, but contrary to their customary practice, they do not coordinate their adjustments with one another. What would be the change in the mass of liquid in the vat? (i.e., dM_{vat} =?)

(d) If Orville and Wilbur make adjustments, dV_{oil} and dV_{water}, at the same time, in such a way as to keep the mass of liquid in the vat constant, what value of dM_{vat} results from this combination of adjustments? (i.e., dM_{vat} =?)

(e) From your answers to (c) and (d), what relationship between dV_{oil} and dV_{water} must Orville and Wilbur maintain in order to keep dM_{vat} constant?

(f) One day, the boss, Mr. Le Chatelier, instructs Orville to add 1.00 L of oil to the vat. Qualitatively, what change does Mr. Le Chatelier impose on the mass of the vat's contents? (That is, what is the direction of the imposed change?)

(g) Quantitatively, what is the change in mass that Mr. Le Chatelier imposes? (That is, what is the equation for the change in the mass in the vat in kg?

(h) Qualitatively, how must Wilbur respond?

(i) Quantitatively, how must Wilbur respond? (That is, what is the equation for the change in the mass in the vat in kg?)

22. Which of the following processes can be carried out reversibly?

(a) Melting an ice cube.

(b) Melting an ice cube at 273.15 K and 1.0 bar.

(c) Melting an ice cube at 275.00 K.

(d) Melting an ice cube at 272.00 K and 1.0 bar.

(e) Frying an egg.

(f) Riding a roller coaster.

(g) Riding a roller coaster and completing the ride in 10 minutes.

(h) Separating pure water from a salt solution at 1 bar and 280.0 K.

(i) Dissolving NaCl in an aqueous solution that is saturated with NaCl.

(j) Compressing a gas.

(k) Squeezing juice from a lemon.

(l) Growing a bacterial culture.

(m) Bending (flexing) a piece of paper.

(n) Folding (creasing) a piece of paper.

Notes

[1] The ordinate (pressure) values for the solid–liquid and liquid–gas equilibrium lines are severely compressed. The ranges of pressure values are so different that the three equilibrium lines cannot otherwise be usefully exhibited on the same graph.

[2] We also use closely related quantities that we call ***fugacities***. We think of a fugacity as a "corrected pressure." For present purposes, we can consider a fugacity to be a particular type of activity.

[3] G. N. Lewis and M. Randall, ***Thermodynamics and the Free Energy of Chemical Substances***, 1st Ed., McGraw-Hill, Inc., New York, 1923, p. 448.

[4] Ilya Prigogine, ***Introduction to the Thermodynamics of Irreversible Processes***, Second Edition, Interscience Publishers, 1961.

[5] S. R. de Groor and P. Mazur, ***Non-Equilibrium Thermodynamics***, Dover Publications, New York, 1984. (Published originally by North Holland Publishing Company, Amsterdam, 1962.)

7

State Functions and the First Law

§1 Changes in a state function are independent of path

We can specify an equilibrium state of a physical system by giving the values of a sufficient number of the system's measurable properties. We call any measurable property that can be used in this way a state function or a state variable. If a system undergoes a series of changes that return it to its original state, any state function must have the same value at the end as it had at the beginning. The relationship between our definition of a physical state and our definition of a state function is tautological. A system can return to its initial state only if every state variable returns to its original value.

It is evident that the change in a state function when the system goes from an initial state, I, to some other state, II, must always be the same. Consider the state functions X, Y, Z, and W. Suppose that functions Y, Z, and W are sufficient to specify the state of a particular system. Let their values in state I be X_I, Y_I, Z_I, and W_I. We can express their interdependence by saying that X_I is a function of the other state functions Y_I, Z_I, and W_I: $X_I = f(Y_I, Z_I, W_I)$. In state II, this relationship becomes $X_{II} = f(Y_{II}, Z_{II}, W_{II})$. The difference

$$X_{II} - X_I = f(Y_{II}, Z_{II}, W_{II}) - f(Y_I, Z_I, W_I)$$

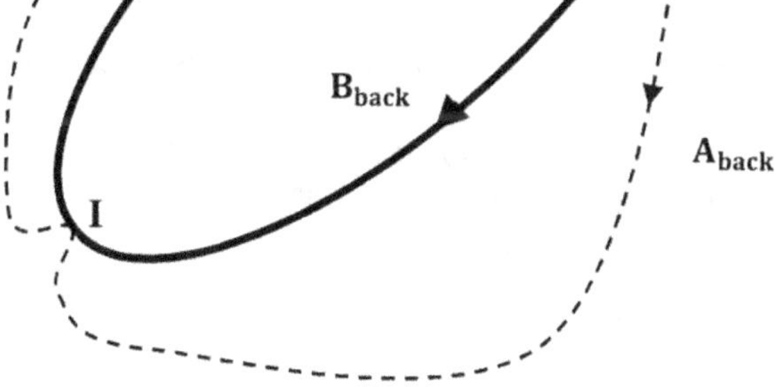

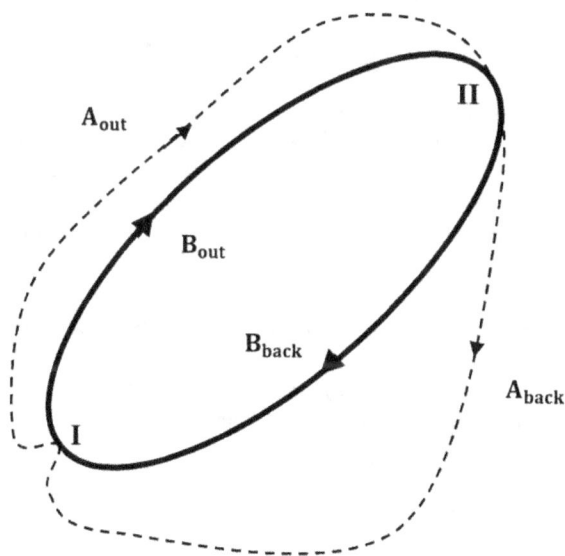

Figure 1. Paths between states I and II.

depends only on the states I and II. In particular, $X_{II} - X_I$ is independent of the values of Y, Z, and W in any intermediate states that the system passes through as it undergoes the change from state I to state II. We say that the change in the value of a state function depends only on the initial and final states of the system; the change in the value of a state function does not depend on the path along which the change is effected.

We can also develop this conclusion by a more explicit argument about the path. Suppose that the system goes from state I to state II by path A_{out} and then returns to state I by path A_{back}, as sketched in Figure 1. Let X be some state function. If the change in X as the system traverses path A_{out} is $X_{II} - X_I = \Delta X(A_{out})$, and the change in X as the system traverses A_{back} is $X_I - X_{II} = \Delta X(A_{back})$, we must have $\Delta X(A_{out}) + \Delta X(A_{back}) = 0$, so that

$$\Delta X(A_{out}) = -\Delta X(A_{back})$$

For some second path comprising B_{out} followed by B_{back}, the same must be true:

$$\Delta X(B_{out}) + \Delta X(B_{back}) = 0$$

and

$$\Delta X(B_{out}) = -\Delta X(B_{back})$$

The same is true for any other path. In particular, in must be true for the path A_{out} followed by B_{back}, so that $\Delta X(A_{out}) + \Delta X(B_{back}) = 0$, and hence

$$\Delta X(A_{out}) = -\Delta X(B_{back})$$

But this means that

$$\Delta X(A_{out}) = \Delta X(B_{out})$$

Since the paths A and B are arbitrary, the change in X in going from state I to state II must have the same value for any path.

§2 The total differential

If $f(x, y)$ is a continuous function of the variables x and y, we can think of $f(x, y)$ as a surface in a three-dimensional space. $f(x, y)$ is the height of the surface above the xy-plane at the point (x, y) in the plane. If we consider points (x_1, y_1) and (x_2, y_2) in the xy-plane, the vertical separation between the corresponding points on the surface, $f(x_1, y_1)$ and $f(x_2, y_2)$, is

$$\Delta f = f(x_2, y_2) - f(x_1, y_1)$$

We can add $f(x_1, y_2) - f(x_1, y_2)$ to Δf without changing its value. Then

$$\Delta f = [f(x_2, y_2) - f(x_1, y_2)] + [f(x_1, y_2) - f(x_1, y_1)]$$

If we consider a small change, such that $x_2 = x_1 + \Delta x$ and $y_2 = y_1 + \Delta y$, we have

$$\Delta f = \frac{[f(x_1 + \Delta x, y_1 + \Delta y_1) - f(x_1, y_1 + \Delta y_1)]\Delta x}{\Delta x} + \frac{[f(x_1, y_1 + \Delta y_1) - f(x_1, y_1)]\Delta y}{\Delta y}$$

Letting $df = \lim_{\substack{\Delta x \to 0 \\ \Delta y \to 0}} \Delta f$, we have

$$df = \lim_{\substack{\Delta x \to 0 \\ \Delta y \to 0}} \left\{ \frac{[f(x_1 + \Delta x, y_1 + \Delta y_1) - f(x_1, y_1 + \Delta y_1)]\Delta x}{\Delta x} \right\} + \lim_{\Delta y \to 0} \left\{ \frac{[f(x_1, y_1 + \Delta y_1) - f(x_1, y_1)]\Delta y}{\Delta y} \right\}$$

$$= \lim_{\Delta y \to 0} \left\{ \left(\frac{\partial f(x_1, y_1 + \Delta y)}{\partial x} \right)_y dx \right\} + \left(\frac{\partial f(x_1, y_1)}{\partial y} \right)_x dy$$

$$= \left(\frac{\partial f(x_1, y_1)}{\partial x} \right)_y dx + \left(\frac{\partial f(x_1, y_1)}{\partial y} \right)_x dy$$

We call df the ***total differential*** of the function $f(x, y)$:

$$df = \left(\frac{\partial f}{\partial x} \right)_y dx + \left(\frac{\partial f}{\partial y} \right)_x dy$$

where df is the amount by which $f(x, y)$ changes when x changes by an arbitrarily small increment, dx, and y changes by an arbitrarily small increment, dy. We use the notation

$$f_x(x, y) = \left(\frac{\partial f}{\partial x} \right)_y$$

and

$$f_y(x, y) = \left(\frac{\partial f}{\partial y} \right)_x$$

to represent the partial derivatives more compactly. In this notation, $df = f_x(x, y)dx + f_y(x, y)dy$. We indicate the partial derivative with respect to x with y held constant at the particular value $y = y_0$ by writing $f_x(x, y_0)$.

We can also write the total differential of $f(x, y)$ as

$$df = M(x, y)dx + N(x, y)dy$$

in which case $M(x, y)$ and $N(x, y)$ are merely new names for $(\partial f / \partial x)_y$ and $(\partial f / \partial y)_x$, respectively. To express the fact that there exists a function, $f(x, y)$, such that $M(x, y) = (\partial f / \partial x)_y$ and $N(x, y) = (\partial f / \partial y)_x$, we say that df is an exact differential.

It is important to recognize that a differential expression, $df = M(x, y)dx + N(x, y)dy$, may not be exact. In our efforts to model physical systems, **we encounter differential expressions that have this form, but for which there is no function, $f(x, y)$,** such that $M(x, y) = (\partial f / \partial x)_y$ and $N(x, y) = (\partial f / \partial y)_x$. We call a differential expression, $df(x, y)$, for which there is no corresponding function, $f(x, y)$, an **inexact differential**. Heat and work are important examples. We will develop differential expressions that describe the amount of heat, dq, and work, dw, exchanged between a system and its surroundings. We will find that these differential expressions are not necessarily exact. (We develop examples in §17 to §20.) It follows that heat and work are not state functions.

§3 Line integrals

The significance of the distinction between exact and inexact differential expressions comes into focus when we use the differential, df, to find how the quantity, f, changes when the system passes from the state defined by (x_1, y_1) to the state defined by (x_2, y_2). We suppose that the system undergoes this change along some continuous path in the xy-plane. We can specify such a path as a function, $c = g(x, y)$, where c is a constant, or as $y = h(x)$. Whether the differential is exact or inexact, we can sum up increments of change, Δf, along short segments of the path to find the change in f between (x_1, y_1) and (x_2, y_2). Let (x_i, y_i) and $(x_i + \Delta x, y_i + \Delta y)$ be two neighboring points on the curve $c = g(x, y)$. As the system traverses $c = g(x, y)$ between these points, the change in f is

$$\Delta f \approx M(x_i, y_i)\Delta x + N(x_i, y_i)\Delta y$$

If we sum up such increments of Δf, along the curve $c = g(x, y)$, from (x_1, y_1) to (x_2, y_2), the sum approximates the change in f along this path. In the limit that all of the incremental Δx and Δy become arbitrarily small, the approximation becomes exact. The limit of this sum is called the ***line integral*** of df along the path $c = g(x, y)$, between (x_1, y_1) and (x_2, y_2).

Whether df is exact or inexact, the line integral of df is defined along any continuous path in the xy-plane. If the path is $c = g(x, y)$ and it connects the points (x_1, y_1) and (x_2, y_2) in the xy-plane, we designate the value of the line integral as

$$\Delta f = \int_g df = \int_{c=g(x_1, y_1)}^{c=g(x_2, y_2)} df$$

(any differential expression)

However, if df is exact, we know that $\Delta f = f(x_2, y_2) - f(x_1, y_1)$. In this case, the line integral of df along curve $c = g(x, y)$ between these points has the value

$$\Delta f = f(x_2, y_2) - f(x_1, y_1) = \int_{c=g(x_1,y_1)}^{c=g(x_2,y_2)} df$$

(for exact differential df)

Because the value of the line integral depends only on the values of $f(x, y)$ at the end points of the integration path, the line integral of the total differential, df, is independent of the path, $c = g(x, y)$. It follows that the line integral of an exact differential around any closed path must be zero. A circle in the middle of the integral sign is often used to indicate that the line integral is being taken around a closed path. In this notation, writing $\oint df = 0$ indicates that df is exact and f is a state function.

In concept, the evaluation of line integrals is straightforward. Since the path of integration is a line, the integrand involves only one dimension. A line integral can always be expressed using a single variable of integration. Three approaches to the evaluation of line integrals are noteworthy.

If we are free to choose an arbitrary path, we can choose the two-segment path $(x_1, y_1) \rightarrow (x_2, y_1) \rightarrow (x_2, y_2)$. Along the first segment, y is constant at y_1, so we can evaluate the change in f as

$$\Delta f_I = \int_{x_1}^{x_2} M(x, y_1)dx$$

Along the second segment, x is constant at x_2, so we can evaluate the change in f as

$$\Delta f_{II} = \int_{y_1}^{y_2} N(x_2, y)dy$$

Then $\Delta f = \Delta f_I + \Delta f_{II}$.

If the path, $c = g(x, y)$, , is readily solved for y as a function of x, say $y = h(x)$, substitution converts the differential expression into a function of only x:

$$df = M(x, h(x))dx + N(x, h(x))\left(\frac{dh}{dx}\right)dx$$

Integration of this expression from x_1 to x_2 gives Δf.

The path, $c = g(x, y)$, can always be expressed as a parametric function of a dummy variable, t. That is, we can always find functions $x = x(t)$ and $y = y(t)$ such that $c = g(x(t), y(t)) = g(t)$, $x_1 = x(t_1)$, $y_1 = y(t_1)$, $x_2 = x(t_2)$, and $y_2 = y(t_2)$. Then substitution converts the differential expression into a function of t:

$$df = M(x(t), y(t))dt + N(x(t), y(t))dt$$

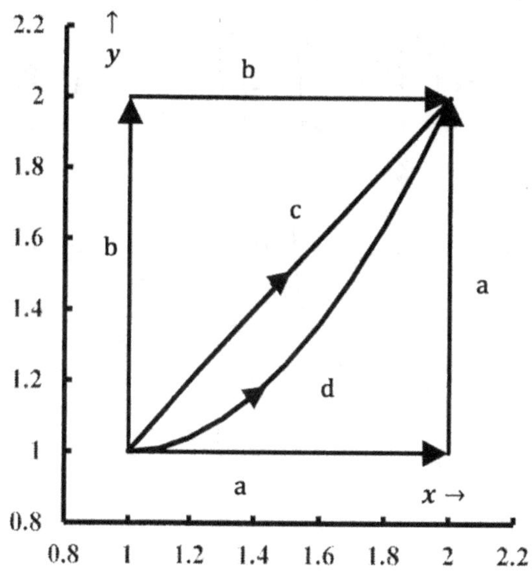

Figure 2. Paths a, b, c, and d.

Integration of this expression from t_1 to t_2 gives Δf.

While the line integral of an exact differential between two points is independent of the path of integration, this not the case for an inexact differential. For an inexact differential, the integral between two points depends on the path of integration. To illustrate these ideas, let us consider some examples. These examples illustrate methods for finding the integral of a differential along a particular path. They illustrate also the path-independence of the integral of an exact differential and the path-dependence of the integral of an inexact differential.

We begin by considering the function $f(x, y) = xy^2$, for which $df = y^2 dx + 2xy\, dy$. Since $f(x, y)$ exists, df must be exact. Let us integrate df between the points $(1, 1)$ and $(2, 2)$ along four different paths, sketched in Figure 2, that we denote as paths a, b, c, and d.

Path a has two linear segments. The first segment is the portion of the line $y = 1$ from $x = 1$ to $x = 2$. Along this segment, $dy = 0$. The second segment is portion of the line $x = 2$ from $y = 1$ to $y = 2$. Along the second segment, $dx = 0$.

Path b has two linear segments also. The first segment is the portion of the line $x = 1$ from $y = 1$ to $y = 2$. Along the first segment, $dx = 0$. The second segment is portion of the line $y = 2$ from $x = 1$ to $x = 2$. Along the second segment, $dy = 0$.

Path c is the line $y = x$, from $x = 1$ to $x = 2$, and for which $dy = dx$.

Path d is the line $y = x^2 - 2x + 2$, which we can express in parametric form as $y = t^2 + 1$ and $x = t + 1$. At $(1, 1)$, $t = 0$. At $(2, 2)$, $t = 1$. Also, $dx = dt$ and $dy = 2t\, dt$.

The integrals along these paths are

Path a:

$$\int_a df = \int_{x=1}^{x=2} 1^2 dx + \int_{y=1}^{y=2} (2)(2)\,y\,dy = [x]_1^2 + [2y^2]_1^2 = 7$$

Path b:

$$\int_b df = \int_{x=1}^{x=2} 2^2 dx + \int_{y=1}^{y=2} (2)(1)\,y\,dy = [4x]_1^2 + [y^2]_1^2 = 7$$

Path c:

$$\int_c df = \int_{x=1}^{x=2} 3x^2\,dx = [x^3]_1^2 \qquad = 7$$

Path d:

$$\int_d df = \int_{t=0}^{t=1} \{(t^2+1)^2 + 2(t+1)(t^2+1)(2t)\}\,dt$$

$$= \int_{t=0}^{t=1} \{5t^4 + 4t^3 + 6t^2 + 4t + 1\}dt$$

$$= [t^5 + t^4 + 2t^3 + 2t^2 + t]_0^1 \qquad = 7$$

The integrals along all four paths are the same. The value is 7, which, as required, is the difference $f(2,2) - f(1,1) = 7$.

Now, let us consider the differential expression $dh = y\,dx + 2xy\,dy$. This expression has the form of a total differential, but we will see that there is no function, $h(x,y)$, for which this expression is the total differential. That is, dh is an inexact differential. If we integrate dh over the same four paths, we find

Path a:

$$\int_a dh = \int_{x=1}^{x=2} (1)dx + \int_{y=1}^{y=2} (2)(2)\,y\,dy = [x]_1^2 + [2y^2]_1^2 = 7$$

Path b:

$$\int_b dh = \int_{x=1}^{x=2} (2)dx + \int_{y=1}^{y=2} (2)(1)\,y\,dy = [2x]_1^2 + [y^2]_1^2 = 5$$

Path c:

$$\int_c dh = \int_{x=1}^{x=2} (x + 2x^2)\,dx = \left[\frac{x^2}{2} + \frac{2x^3}{3}\right]_1^2 \qquad = 6\frac{1}{6}$$

Path d:

$$\int_d dh = \int_{t=0}^{t=1} \{(t^2+1) + 2(t+1)(t^2+1)(2t)\}\,dt$$

$$= \int_{t=0}^{t=1} \{4t^4 + 4t^3 + 5t^2 + 4t + 1\}dt$$

$$= [4t^5/5 + t^4 + 5t^3/3 + 2t^2 + t]_0^1 \qquad = 6\frac{7}{15}$$

For $dh(x,y)$, the value of the integral depends on the path of integration, confirming that $dh(x,y)$ is an inexact differential: Since the value of the integral depends on path, there can be no $h(x,y)$ for which

$$\Delta h = h(x_2,y_2) - h(x_1,y_1) = \int_{(x_1,y_1)}^{(x_2,y_2)} dh$$

That is, $h(x_2,y_2) - h(x_1,y_1)$ cannot have four different values.

§4 Exact differentials and state functions

Now, let us consider the general case of a continuous function $f(x,y)$, for which the exact differential is $df = f_x(x,y)dx + f_y(x,y)dy$. We want to integrate the exact differential over very short paths like paths a and b in §3. Let us evaluate the integral between (x_0,y_0) and $(x_0+\Delta x, y_0+\Delta y)$ over the paths a* and b* sketched in Figure 3.

Path a* has two linear segments. The first segment is the portion of the line $y = y_0$ as x goes from x_0 to $x_0 + \Delta x$. Along the first segment $\Delta y = 0$. The second segment is the portion of the line $x = x_0 + \Delta x$ as y goes from y_0 to $y_0 + \Delta y$. Along the second segment, $\Delta x = 0$.

Path b* has two linear segments also. The first segment is the portion of the line $x = x_0$ as y goes from y_0 to $y_0 + \Delta y$. Along the first segment, $\Delta x = 0$. The second segment is the portion of the line $y = y_0 + \Delta y$ as x goes from x_0 to $x_0 + \Delta x$. Along the second segment, $\Delta y = 0$.

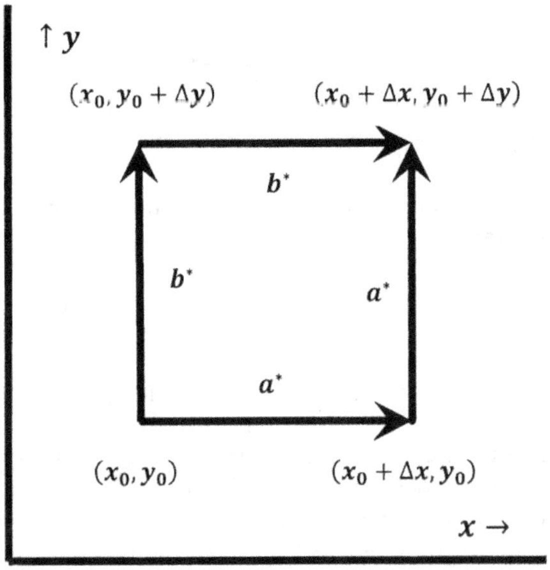

Figure 3. Alternative paths from (x_0, y_0) to $(x_0 + \Delta x, y_0 + \Delta y)$

Along path a*, we have

$$\Delta_{a^*}f = f_x(x_0, y_0)\Delta x + f_y(x_0 + \Delta x, y_0)\Delta y$$

Along path b*,

$$\Delta_{b^*}f = f_x(x_0, y_0 + \Delta y)\Delta x + f_y(x_0, y_0)\Delta y$$

In the limit as Δx and Δy become arbitrarily small, we must have $\Delta_{a^*}f = \Delta_{b^*}f$, so that

$$f_x(x_0, y_0)\Delta x + f_y(x_0 + \Delta x, y_0)\Delta y = f_x(x_0, y_0 + \Delta y)\Delta x + f_y(x_0, y_0)\Delta y$$

Rearranging this equation so that terms in f_x are on one side and terms in f_y are on the other side, dividing both sides by $\Delta x\Delta y$, and taking the limit as $\Delta x \to 0$ and $\Delta y \to 0$, we have

$$\lim_{\Delta x\to 0}\left[\frac{f_y(x_0 + \Delta x, y_0) - f_y(x_0, y_0)}{\Delta x}\right] = \lim_{\Delta y\to 0}\left[\frac{f_x(x_0, y_0 + \Delta y) - f_x(x_0, y_0)}{\Delta y}\right]$$

These limits are the partial derivative of $f_y(x_0, y_0)$ with respect to x and of $f_x(x_0, y_0)$ with respect to y. That is

$$\left[\frac{\partial}{\partial x} f_y(x_0, y_0)\right]_y = \left[\frac{\partial}{\partial x}\left(\frac{\partial f(x_0, y_0)}{\partial y}\right)\right]_y = \frac{\partial^2 f(x_0, y_0)}{\partial y\partial x}$$

and

$$\left[\frac{\partial}{\partial y} f_x(x_0, y_0)\right]_x = \left[\frac{\partial}{\partial y}\left(\frac{\partial f(x_0, y_0)}{\partial x}\right)\right]_x = \frac{\partial^2 f(x_0, y_0)}{\partial x\partial y}$$

This shows that, if $f(x, y)$ is a continuous function of x and y whose partial derivatives exist, then

$$\frac{\partial^2 f(x_0, y_0)}{\partial y\partial x} = \frac{\partial^2 f(x_0, y_0)}{\partial x\partial y}$$

The mixed second partial derivative of $f(x, y)$ is independent of the order of differentiation. We also write these second partial derivatives as $f_{xy}(x_0, y_0)$ and $f_{yx}(x_0, y_0)$.

To summarize these points, if $f(x, y)$ is a continuous function of x and y, all of the following are true:

- $f(x, y)$ represents a surface in a three-dimensional space.
- $f(x, y)$ is a state function.
- The total differential is
  $$df = (\partial f/\partial x)_y dx + (\partial f/\partial y)_x dy.$$
- The total differential is exact.
- The line integral of df between two points is independent of the path of integration.
- The line integral of df around any closed path is zero: $\oint df = 0$.
- The mixed second-partial derivatives are equal; that is,
  $$\frac{\partial^2 f}{\partial y\partial x} = \frac{\partial^2 f}{\partial x\partial y}$$

§5 Determining whether an expression is an exact differential

Since exact differentials have these important characteristics, it is valuable to know whether a given differential expression is exact or not. That is, given a differential expression of the form $df = M(x, y)dx + N(x, y)dy$, we would like to be able to determine whether df is exact or inexact. It turns out that there is a simple test for exactness:

The differential $df = M(x, y)dx + N(x, y)dy$ is exact if and only if $\partial M/\partial y = \partial N/\partial x$.

That is, this condition is necessary and sufficient for the existence of a function, $f(x, y)$, for which $M(x, y) = f_x(x, y)$ and $N(x, y) = f_y(x, y)$.

In §4 we demonstrate that the condition is necessary. Now we want to show that it is sufficient. That is, we want to demonstrate: If $\partial M/\partial y = \partial N/\partial x$, then there exists a $f(x, y)$, such that $M(x, y) = f_x(x, y)$ and $N(x, y) = f_y(x, y)$. To do this, we show how to find a function, $f(x, y)$, that satisfies the given differential relationship. If we integrate $M(x, y)$ with respect to x, we have

$$f(x, y) = \int M(x, y)dx + h(y)$$

where $h(y)$ is a function only of y; it is the arbitrary constant in the integration with respect to x, which we carry out with y held constant.

To complete the proof, we must find a function $h(y)$ such that this $f(x, y)$ satisfies the conditions:

(1) $M(x, y) = f_x(x, y) \quad \Leftrightarrow$

$$M(x, y) = \frac{\partial}{\partial x}\left[\int M(x, y)dx + h(y)\right]$$

(2) $N(x, y) = f_y(x, y) \quad \Leftrightarrow$

$$N(x, y) = \frac{\partial}{\partial y}\left[\int M(x, y)dx + h(y)\right]$$

The validity of condition (1) follows immediately from the facts that the order of differentiation and integration can be interchanged for a continuous function and that $h(y)$is a function only of y, so that $\partial h/\partial x = 0$.

To find $h(y)$ such that condition (2) is satisfied, we observe that

$$\frac{\partial}{\partial y}\left[\int M(x, y)dx + h(y)\right] = \int\left(\frac{\partial M(x, y)}{\partial y}\right)dx + \frac{dh(y)}{dy}$$

But since

$$\frac{\partial M(x,y)}{\partial y} = \frac{\partial N(x,y)}{\partial x}$$

this becomes

$$\frac{\partial}{\partial y}\left[\int M(x,y)dx + h(y)\right]$$
$$= \int \left(\frac{\partial N(x,y)}{\partial x}\right) dx + \frac{dh(y)}{dy}$$
$$= N(x,y) + \frac{dh(y)}{dy}$$

Hence, condition (2) is satisfied if and only if $dh(y)/dy = 0$, so that $h(y)$ is simply an arbitrary constant.

§6 The chain rule and the divide-through rule

If we have $f(x,y)$ while x and y are functions of another variable, u, the chain rule states that

$$\frac{df}{du} = \left(\frac{\partial f}{\partial x}\right)_y \frac{dx}{du} + \left(\frac{\partial f}{\partial y}\right)_x \frac{dy}{du}$$

If x and y are functions of variables u and v; that is, $x = x(u,v)$ and $y = y(u,v)$, the chain rule for partial derivatives is

$$\left(\frac{\partial f}{\partial u}\right)_v = \left(\frac{\partial f}{\partial x}\right)_y \left(\frac{\partial x}{\partial u}\right)_v + \left(\frac{\partial f}{\partial y}\right)_x \left(\frac{\partial y}{\partial u}\right)_v$$

A useful mnemonic recognizes that these equations can be generated from the total differential by "dividing through" by du. We must specify that the "new" partial derivatives are taken with v held constant. This is sometimes called the ***divide-through rule.***

The divide-through rule is a reliable expedient for generating new relationships among partial derivatives. As a further example, dividing by dx and specifying that any other variable is to be held constant produces a valid equation. Letting w be the variable held constant, we obtain

$$\left(\frac{\partial f}{\partial x}\right)_w = \left(\frac{\partial f}{\partial x}\right)_y \left(\frac{\partial x}{\partial x}\right)_w + \left(\frac{\partial f}{\partial y}\right)_x \left(\frac{\partial y}{\partial x}\right)_w$$
$$= \left(\frac{\partial f}{\partial x}\right)_y + \left(\frac{\partial f}{\partial y}\right)_x \left(\frac{\partial y}{\partial x}\right)_w$$

where we recognize that $(\partial x/\partial x)_w = 1$. The result is just the chain rule for$(\partial f/\partial x)_w$ when $f = f(x,y)$ and $y = y(x,w)$; that is, when $f = f\big(x, y(x,w)\big)$.

If we require that $f(x,y)$ remain constant while x and y vary, we can use the divide-though rule to obtain another useful relationship from the total differential. If $f(x,y)$ is constant, $df(x,y) = 0$. This can only be true if there is a relationship between x and y. To find this relationship we use the divide-through rule to find $(\partial f/\partial y)_f$ when $f = f(x(y),y)$. Dividing

$$df = \left(\frac{\partial f}{\partial x}\right)_y dx + \left(\frac{\partial f}{\partial y}\right)_x dy$$

by dy, and stipulating that f is constant, we find

$$\left(\frac{\partial f}{\partial y}\right)_f = \left(\frac{\partial f}{\partial x}\right)_y \left(\frac{\partial x}{\partial y}\right)_f + \left(\frac{\partial f}{\partial y}\right)_x \left(\frac{\partial y}{\partial y}\right)_f$$

Since $(\partial f/\partial y)_f = 0$ and $(\partial y/\partial y)_f = 1$, we have

$$\left(\frac{\partial f}{\partial y}\right)_x = -\left(\frac{\partial f}{\partial x}\right)_y \left(\frac{\partial x}{\partial y}\right)_f$$

In Chapter 10, we find that the divide-through rule is a convenient way to generate thermodynamic relationships.

§7 Measuring work: pressure–volume work

By definition, the energy of a system can be exploited to produce a mechanical change in the surroundings. The energy of the surroundings increases; the energy of the system decreases. Raising a weight against the earth's gravitational force is the classical example of a mechanical change in the surroundings. When we say that work is done on a system, we mean that the energy of the system increases because of some non-thermal interaction between the system and its surroundings. The amount of work done on a system is determined by the non-thermal energy change in its surroundings. We define work as the scalar product of a vector representing an applied force, $\vec{F}_{applied}$, and a second vector, $\vec{r}$, representing the displacement of the object to which the force is applied. The definition is independent of whether the process is reversible or not. If the force is a function of the displacement, we have

$$dw = \vec{F}(\vec{r})_{applied} \bullet d\vec{r}$$

Pressure–volume work is done whenever a force in the surroundings applies pressure on the system while the volume of the system changes. Because chemical changes typically do involve volume changes, pressure–volume work often plays a significant role. Perhaps the

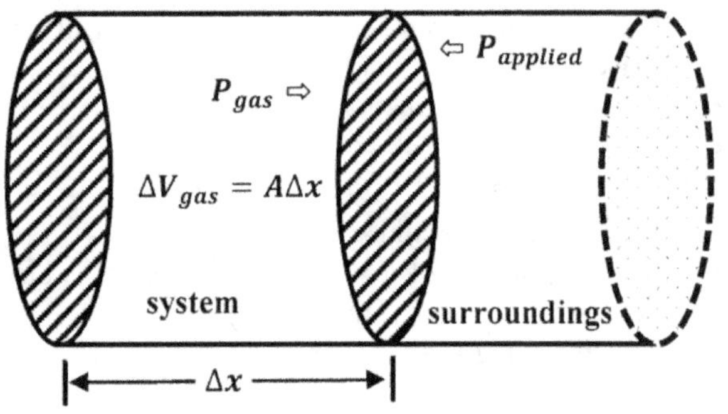

Figure 4. Pressure—volume work.

most typical chemical experiment is one in which we carry out a chemical reaction at the constant pressure imposed by the earth's atmosphere. When the volume of such a system increases, the system pushes aside the surrounding atmosphere and thereby does work on the surroundings.

When a pressure, $P_{applied}$, is applied to a surface of area A, the force normal to the area is $F = P_{applied}A$. For a displacement, dx, normal to the area, the work is $Fdx = dw = P_{applied}A\ dx$. We can find the general relationship between work and the change in the volume of a system by supposing that the system is confined within a cylinder closed by a piston. (See Figure 4.) The surroundings apply pressure to the system by applying force to the piston. We suppose that the motion of the piston is frictionless.

The system occupies the volume enclosed by the piston. If the cross-sectional area of the cylinder is A, and the system occupies a length x, the magnitude of the system's volume is $V = Ax$. If an applied pressure moves the piston a distance dx, the volume of the system changes by $dV_{system} = A\ dx$. The magnitude of the work done in this process is therefore

$$\left|dw_{system}\right| = \left|P_{applied}Adx\right| = \left|P_{applied}dV_{system}\right|$$

We are using the convention that work is positive if it is done on the system. This means that a compression of the system, for which $dx < 0$ and $dV_{system} < 0$, does a positive quantity of work on the system. Therefore, the work done on the system is $dw_{system} = -P_{applied}dV_{system}$ or, using our convention that unlabeled variables always characterize the system,

$$dw = -P_{applied}dV$$

§8 Measuring work: non-pressure–volume work

For chemical systems, pressure–volume work is usually important. Many other kinds of work are possible. From our vector definition of work, any force that originates in the surroundings can do work on a system. The force drives a displacement in space of the system or some part of the system. Stretching a strip of rubber is a one-dimensional analog of pressure–volume work. Changing the surface area of a liquid is a two-dimensional analog of pressure–volume work. When only internal forces act, a liquid system minimizes its surface area. We can model this property by attributing a surface-area minimizing force, which we call the surface tension, to the surface of the liquid. We can think of the layer of molecules at the surface as a film that separates the bulk liquid from its surroundings. To increase the area of a liquid system requires an expenditure of work by the surroundings against the surface tension of the film. Gravitational, electrical, and magnetic forces can all do work on particular systems.

In this book, we give little attention to the details of the various kinds of non-pressure–volume work that can be important. (There are two exceptions: Electrical work is important in electrochemistry, which we discuss in Chapter 17. We discuss gravitational work in examples that illustrate reversible processes and some aspects of the criteria for change.) Nevertheless, no development of the basic concepts can be complete without including the effects of non-pressure–volume work. For this reason, we include non-pressure–volume work in our discussions frequently. For the most part, however, we do so in a generalized or abstract way. To do so, we must identify some essential features of any process that does work on a system.

Whenever a particular kind of work is done on a system, some change occurs in a thermodynamic variable that is characteristic of that kind of work. For pressure–volume work this is the volume change. For stretching a strip of rubber, it is the change in length. For gravitational work, it is the displacement of a mass in a gravitational field. For changing the shape of a liquid, it is the change in surface area. For electrical work, it is the displacement of a charge in an electrical field. For magnetic work, it is the displacement of a magnetic moment in a magnetic field. For an arbitrary form of non-pressure–volume work, let us use θ to represent this variable. We can think of θ as a generalized displacement. When there is an incremental change, $d\theta$, in this variable, there is a corresponding change, dE, in the energy of the system.

For a displacement, $d\theta$, let the increase in the energy of the system be dw_θ. The energy increase also depends on the magnitude of the force that must be applied to the system, parallel to the displacement $d\theta$. Let this force be f_θ. Then, for this arbitrary abstract process, we have $dw_\theta = f_\theta d\theta$, or $f_\theta = dw_\theta/d\theta$. Since dw_θ is the contribution to the incremental change in the energy of the system associated with the displacement $d\theta$, we can also write this as

$$f_\theta = \frac{dw_\theta}{d\theta} = \frac{\partial E}{\partial \theta}$$

We can generalize this perspective. θ need not be a vector, and $\partial E/\partial\theta$ need not be a mechanical force. So long as $d\theta$ determines the energy change, dE, we have

$$dw_\theta = \left(\frac{\partial E}{\partial \theta}\right) d\theta$$

We call $\partial E/\partial\theta$ a potential. If we let

$$\Phi_\theta = \left(\frac{\partial E}{\partial \theta}\right)$$

the energy increment becomes $dw_\theta = \Phi_\theta d\theta$. If multiple forms of work are possible, we can distinguish them by their characteristic variables, which we label θ_1, θ_2,..., θ_k,..., θ_ω. For each of these characteristic variables, there is a corresponding potential, Φ_1, Φ_2,..., Φ_k,..., Φ_ω. The total energy increment, which we also call the non-pressure–volume work, dw_{NPV}, becomes

$$dw_{NPV} = \sum_{k=1}^{\omega} \Phi_k d\theta_k$$

For pressure–volume work, $dw_{PV} = -PdV$. The characteristic variable is volume, $d\theta = dV$, and the potential is the negative of the pressure, $\Phi_V = -P$. For gravitational work, the characteristic variable is elevation, $d\theta = dh$; for a given system, the potential depends on the gravitational acceleration, g, and the mass of the system: $\Phi_h = mg$.

When a process changes the composition of a system, it is often important to relate the work done on the system to the composition change. Formally, we express the incremental work resulting from the k-th generalized displacement as

$$dw_k = \left(\frac{\partial E}{\partial \theta_k}\right)\left(\frac{\partial \theta_k}{\partial n}\right) dn$$

where dw_k and dn are the incremental changes in the work done on the system and the number of moles of the substance in the system. To see how this works out in practice, let us consider the particular case of electrical work. The electrodes of an electrochemical cell can be at different electric potentials. We usually designate the potential difference between the electrodes as $\mathcal{E}_{cell}$. (We can also write $\Phi_{cell} = \mathcal{E}_{cell}$ when we want to keep our notation uniform. The unit of electrical potential is the ***volt***, V. One volt is one joule per coulomb, 1 V = 1 J C^{-1}.) We are usually interested in cases in which we can assume that $\mathcal{E}_{cell}$ is constant.

Whenever a current flows in an electrochemical cell, electrons flow through an external circuit from one electrode to the other. By our definition of electrical potential, the energy change that occurs when a charge dq passes through a potential difference, $\mathcal{E}_{cell}$, is

$$dE = \mathcal{E}_{cell} dq = dw_{elect}$$

We have $\mathcal{E}_{cell} = (\partial E/\partial q)$. Evidently, charge is the characteristic variable for electrical work; we have $\theta_{elect} = q$, and

$$\left(\frac{\partial E}{\partial \theta_{elect}}\right) = \left(\frac{\partial E}{\partial q}\right) = \mathcal{E}_{cell}$$

Letting the magnitude of the electron charge be e, dN electrons carry charge $dq = -e\, dN$. Then, $dq = (-e\overline{N})(dN/\overline{N})$. The magnitude of the charge carried by one mole of electrons is the faraday, $\mathcal{F}$. That is, $1\ \mathcal{F} = |e\overline{N}| = 96{,}485\ \text{C mol}^{-1}$. (See §17-8.) Letting dn be the number of moles of electrons, we have $dn = dN/\overline{N}$ and $dq = -\mathcal{F}dn$, so that

$$\left(\frac{\partial \theta_{elect}}{\partial n}\right) = \left(\frac{\partial q}{\partial n}\right) = -\mathcal{F}$$

The work done when dn moles of electrons pass through the potential difference $\mathcal{E}_{cell}$ becomes

$$\begin{aligned} dw_{elect} &= \left(\frac{\partial E}{\partial \theta_{cell}}\right)\left(\frac{\partial \theta_{cell}}{\partial n}\right) dn \\ &= \left(\frac{\partial E}{\partial q}\right)\left(\frac{\partial q}{\partial n}\right) dn \\ &= -\mathcal{F}\mathcal{E}_{cell}\, dn \end{aligned}$$

We find the work done when ions pass through a potential difference $\mathcal{E}$ by essentially the same argument. If ions of species j carry charge $z_j e$, then dN_j ions carry charge $dq = z_j e\, dN_j = z_j \mathcal{F} dn_j$, and the electrical work is $(dw_{elect})_j = z_j \mathcal{F}\mathcal{E} dn_j$. If ω different species pass through the potential difference, the total electrical work becomes

$$dw_{elect} = \sum_{j=1}^{\omega} z_j \mathcal{F}\mathcal{E} dn_j$$

§9 Measuring heat

As the idea of heat as a form of energy was first being developed, a unit amount of heat was taken to be the amount that was needed to increase the temperature of a reference material by one degree. Water was the reference material of choice, and the calorie was defined as the quantity of heat that raised the temperature of one gram of water one degree kelvin. The amount of heat exchanged by a known amount of water could then be calculated from the amount by which the temperature of the water changed. If, for example, introducing 63.55 g (1 mole) of copper metal, initially at 274.0 K, into 100 g of water, initially at 373.0 K, resulted in thermal equilibrium at 288.5 K, the water surrendered

$$100\ \text{g} \times 1\ \text{cal g}^{-1}\ \text{K}^{-1} \times 84.5\ \text{K} = 8450\ \text{cal}$$

This amount of heat was taken up by the copper, so that 0.092 cal was required to increase the temperature of one gram of copper by one degree K. Given this information, the amount of heat gained or lost by a known mass of copper in any subsequent experiment can be calculated from the change in its temperature.

Joule developed the idea that mechanical work can be converted entirely into heat. The quantity of heat that could be produced from one unit of mechanical work was called the ***mechanical equivalent of heat***. Today we ***define*** the unit of heat in mechanical units. That is, we define the unit of energy, the joule (J), in terms of the mechanical units mass (kg), distance (m), and time (s). One joule is one newton-meter or one kg m^2 s^{-2}. ***One calorie is now defined as 4.184 J, exactly.*** This definition assumes that heat and work are both forms of energy. This assumption is an intrinsic element of the first law of thermodynamics. This aspect of the first law is, of course, just a restatement of Joule's original idea.

When we want to measure the heat added to a system, measuring the temperature increase that occurs is often the most convenient method. If we know the temperature increase in the system, and we know the temperature increase that accompanies the addition of one unit of

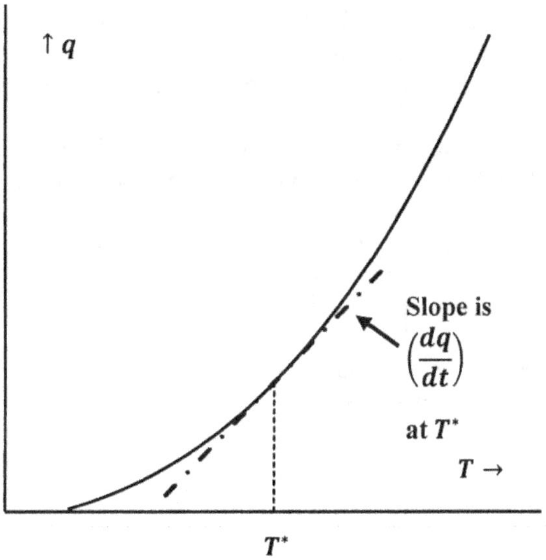

Figure 5. Heat capacity is the slope of *q versus T*.

heat, we can calculate the heat input to the system. Evidently, it is useful to know how much the temperature increases when one unit of heat is added to various substances. Let us consider a general procedure for accumulating such information.

First, we need to choose some standard amount of the substance in question. After all, if we double the amount, it takes twice as much heat to effect the same temperature change. One mole is a natural choice for this standard amount. If we add small increments of heat to one mole of a pure substance, we can measure the temperature after each addition and plot heat *versus* temperature. Figure 5 shows such a plot. (In experiments like this, it is often convenient to introduce the heat by passing a known electrical current, I, through a known resistance, R, immersed in the substance. The rate at which heat is produced is I^2R. Except for the usually negligible amount that goes into warming the resistor, all of it is transferred to the substance.) At any particular temperature, the slope of the graph is the increment of heat input divided by the incremental temperature increase. This slope is so useful, it is given a name; it is the molar heat capacity of the substance, C. Since this slope is also the derivative of the q-*versus*-T curve, we have

$$C = \frac{dq}{dT}$$

The temperature increase accompanying a given heat input varies with the particular conditions under which the experiment is done. In particular, the temperature increase will be less if some of the added heat is converted to work, as is the case if the volume of the system increases. If the volume increases, the system does work on the surroundings. For a given q, ΔT will be less when the system is allowed to expand, which means that $q/\Delta T$ will be greater. Heat capacity measurements are most conveniently done with the system at a constant pressure. However, the heat capacity at constant volume plays an important role in our theoretical development. The heat capacity is denoted C_P when the pressure is constant and C_V when the volume is constant. We have the important definitions

$$C_P = \left(\frac{\partial q}{\partial T}\right)_P$$

and

$$C_V = \left(\frac{\partial q}{\partial T}\right)_V$$

Since no pressure–volume work can be done when the volume is constant, less heat is required to effect a given temperature change, and we have $C_P > C_V$, as a general result. (In §14, we consider this point further.) If the system contains a gas, the effect of the volume increase can be substantial. For a monatomic ideal gas, the temperature increase at constant pressure is only 60% of the temperature increase at constant volume for the same input of heat.

§10 The first law of thermodynamics

A state function must return to its original value if a system is taken through a series of changes and finally returned to its original state. We say that the change in a state function must be zero if the system is taken through a cyclic process, or somewhat more picturesquely, if the system traverses a cyclic path. While we can measure the heat and work that a system exchanges with its surroundings, neither the heat nor the work is necessarily zero when the system traverses a cycle. Heat and work are not state functions. Nevertheless, adding heat to a system increases its energy. Likewise, doing work on a system increases its energy. If the system surrenders heat to the surroundings or does work on the surroundings, the energy of the system is decreased. In any change that a closed system undergoes, the total energy change is $E = q + w$, where q and w can be either positive or negative. For very small changes, we write $dE = dq + dw$. Anything we do to increase the energy of a closed system can be classified as either adding heat to the system or doing work on the system.

Heat, work, and energy are all extensive variables. They are additive. If a system acquires an increment of heat q_1 from one source and an increment q_2 from another source, the total heat acquired by the system is $q_1 + q_2$. If work, w_1, of one kind, and work, w_2, of a second kind are done on the system, the total work is $w_1 + w_2$.

In keeping with the thermodynamic perspective that we can partition the universe into system and surroundings, we assume that any energy lost by the system is taken up by the surroundings, and *vice versa*. By definition, we have $q = -\hat{q}$, $w = -\hat{w}$, and $E = -\hat{E}$; for any process, $E_{universe} = E + \hat{E}$. This is the ***principle of conservation of energy***, which is usually stated: ***For any change in any system, the energy of the universe remains constant.***

So, conservation of energy is built into our energy accounting scheme. It is a consequence of the thermodynamic perspective and our rules for keeping track of exchanges of heat and work between system and surroundings. Conservation of energy is an "accounting convention," but it is not arbitrary. That is, we are not free to choose another convention for this "energy accounting." Ample experimental evidence supports our assumption that energy conservation is a fundamental property of nature.

In summary, we postulate that for any change whatsoever that a closed system may undergo, we can identify energy inputs either as heat or as one or more forms of work such that

$$E = q + w$$

and, if a system undergoes a series of changes that ultimately return it to its original state, the energy change for the entire series of changes will be zero. There are two components to this postulate. The first component is an operational definition of energy. An important aspect of this definition is that the principle of conservation of energy is embedded in it. The second is an assertion that energy is a state function. Our operational definition of energy is open-ended. An essential element of the postulate is that we can always identify work inputs that make the energy, whose changes we compute as $q + w$, a state function. The facts that energy is conserved and that energy is a state function are related properties of a single aspect (energy) of nature. The relationship between these facts is a characteristic property of physical reality; it is not a matter of logic in the sense that one fact implies the other.

All of these ideas are essential components of the concept of energy. We roll them all together and assert them as a postulate that we call the first law of thermodynamics. We introduce the first law in Chapter 6. We repeat it here.

The first law of thermodynamics

In a process in which a closed system accepts increments of heat, dq, and work, dw, from its surroundings, the change in the energy of the system, dE, is $dE = dq + dw$. Energy is a state function. For any process, $dE_{universe} = 0$.

This statement of the first law does not deal explicitly with the mechanical energy of the system as a whole or with the energy effects of a transport of matter across the boundary of an open system. Because dw can include work that changes the position or motion of a system relative to an external reference frame, increments of mechanical energy can be included in dE. In chemical applications, we seldom need to consider the mechanical energy of the system as a whole; we can assume that the system has no kinetic or potential energy associated with the movement or location of its mass. When this is case, the total incremental energy change, dE, is the same thing as the incremental change in the internal energy of the system. When we need to distinguish the internal energy of a system from its total energy, we write U for the internal energy and E for the total energy. Letting incremental changes in the kinetic and potential energy of the whole system be $d\tau$ and $d\upsilon$, respectively, we have $dU = dq + dw$ and $dE = dU + d\tau + d\upsilon$. For processes of interest in chemical systems, we normally have $d\tau = d\upsilon = 0$. Then the total energy and the internal energy are the same thing: $dE = dU$.

In §8, we introduce characteristic variables, θ_k, to represent changes in the system that result from various forms of non-pressure–volume work done on the system. We let $\Phi_k = (\partial E/\partial \theta_k)$, so that we can represent the incremental energy change that results from the non-pressure–volume work of all kinds as

$$dw_{NPV} = \sum_{k=1}^{\omega} \Phi_k d\theta_k$$

When both pressure–volume and non-pressure–volume work occur, we have

$$dw = dw_{PV} + dw_{NPV} = -PdV + \sum_{k=1}^{\omega} \Phi_k d\theta_k$$

When a non-thermal process changes the energy of a closed, constant-volume system, we have $dE = dw_{NPV}$.

We state the first law for a closed system. Extending the first law to open systems is straightforward. The energy of a system depends on the substances that are present, their amounts, and their states. At any specified conditions, a given amount of a particular substance makes a fixed contribution to the energy of the system. If we transfer matter across the boundary of a system, we change the energy of the system. We can always alter the original system to include the matter that is to be transferred. The altered system is closed; and so, by the first law, its energy is the same after the transfer as it was before. In §14-2 we develop an explicit mathematical function to model the contribution made to the energy of a system by a specified quantity of matter in a specified state. If matter crosses the boundary of a system, the energy models for the separate collections of substances pre-transfer must equate to that for the new system post transfer.

Finally, we make a further simple but important observation: We imagine that we can always identify an energy increment that crosses a system boundary as work, dw, or heat, dq. However, the essence of the first law is that these increments lose their identities—so to speak—in the system. The effect of a work input, dw, doesn't necessarily appear as an increase in the mechanical energy of the system; a heat input, dq, doesn't necessarily appear as in increase in the thermal energy of the system.

To illustrate this point, let us consider a reversible process and an irreversible process, each of which increases the temperature of one gram of water by one degree K. The initial and final states of the system are the same for both processes. In the reversible process, we bring the water, whose temperature is T, into contact with a thermal reservoir at an incrementally higher temperature $T + dT$ and allow 4.184 J of heat to transfer to the system by convection[1]. No work is done. We have $q = -\hat{q} = 4.184$ J, $w = -\hat{w} = 0$, and $E = q = 4.184$ J. In this case, there is no inter-conversion of heat and work.

In the irreversible process, we stir one gram of water that is thermally isolated. The stirring generates heat in the system. We supply the energy to drive the stirrer from the surroundings, perhaps by allowing a spring to uncoil. When 4.184 J of work from the surroundings has been frictionally dissipated in the system, the state of the one-gram system is the same as it was at the end of the reversible process. In this irreversible process, all of the energy traverses the system boundary as work (non-thermal energy). No heat traverses the system boundary. We have $w = -\hat{w} = 4.184$ J, $q = -\hat{q} = 0$, and $\Delta E = w = 4.184$ J. Uncoiling the spring generates bulk motion within the water. Within a short time, the energy of this bulk motion is completely dissipated as molecular-level kinetic energy, or heat.

Beginning in §18, we consider the reversible isothermal expansion of an ideal gas. This simple process provides a further illustration of the inter-conversion of heat and work within a system. For this process, we find that thermal energy, $q > 0$, crosses the system boundary and is converted entirely into work that appears in the surroundings, $\hat{w} = -w > 0$. The energy of the system is unchanged. The energy of an ideal gas depends only on temperature. In an isothermal expansion, the temperature is constant, so the transport of heat across the system boundary has no effect on the energy of the system.

§11 Other statements of the first law

The first law has been stated in many ways. Some are intended to be humorous or evocative rather than precise statements; for example, "You can't get something (useful work in some system) for nothing (no decrease in the energy of some other system)." Others are potentially ambiguous, because we construct them to be as terse as possible. To make them terse, we omit ideas that we deem to be implicit.

A compact and often used statement is, "$E = q + w$, and E is a state function." In this statement, the fact that energy is conserved is taken to be implicit in the operational definition, $E = q + w$. We can give an equally valid statement by saying, "Energy is conserved ($E_{universe} = E + \hat{E} = 0$) in all processes." In making this statement, we assume that the definition of energy ($E = q + w$) is understood and that the state-function postulate is implicit in this definition.

To see that the postulate that energy is conserved and the postulate that energy is a state function are logically independent, let us consider a system that undergoes a particular cyclic process, which we call "Cycle A." In Cycle A, the final state of the system is the same as its initial state; the postulate that energy is a state function is then equivalent to the statement that $E_{Cycle\ A} = 0$. The postulate that energy is conserved is equivalent to the statement that $E_{Cycle\ A} + \hat{E}_{Cycle\ A} = 0$. Now, what can we say about $\hat{E}_{Cycle\ A}$? Obviously, if we combine the information from the two postulates, it follows that $\hat{E}_{Cycle\ A} = 0$. The essential point, however, is that $\hat{E}_{Cycle\ A} = 0$ is not required by either postulate alone.

- $\hat{E}_{Cycle\ A} = 0$ is not required by the postulate that energy is a state function, because the surroundings do not necessarily traverse a cycle whenever the system does.

- $\hat{E}_{Cycle\ A} = 0$ is not required by conservation of energy, which merely requires $\hat{E}_{Cycle\ A} = -E_{Cycle\ A}$, and absent the requirement that E be a state function, $E_{Cycle\ A}$ could be anything.

In Chapter 9, we explore a statement of the second law that denies the possibility of constructing a "perpetual motion machine of the second kind." Such a perpetual motion machine converts heat from a constant-temperature reservoir into work. This statement is: "It is impossible to construct a machine that operates in a cycle, exchanges heat with its surroundings at only one temperature, and produces work in the surroundings."

A parallel statement is sometimes taken as a statement of the first law. This statement denies the possibility of constructing a "perpetual motion machine of the first kind." This statement is, "It is impossible to construct a machine that operates in a cycle and converts heat into a greater amount of work." The shared perspective and phrasing of these statements is esthetically pleasing. Let us consider the relationship between this statement of the first law and the statement given in §10. (For brevity, let us denote this impossibility statement as the "machine-based" statement of the first law and refer to it as proposition "MFL." We refer to the statement of the first law given in §10 as proposition "FL.")

In the machine-based statement (MFL), we mean by "a machine" a system that accepts heat from its surroundings and produces a greater amount of work, which appears in the surroundings. If such a machine exists, the machine-based statement of the first law is false, and proposition ~MFL is true. For one cycle of this first-law violating machine, we have $\hat{w} > q > 0$. Since $q = -\hat{q}$, we have $\hat{w} > -\hat{q} > 0$. It follows that $\hat{E} = \hat{q} + \hat{w} > 0$. Our statement of the principle of conservation of energy ($E + \hat{E} = 0$) then requires that, for one cycle of this perpetual motion machine, $E < 0$. Our statement of the first law, FL, requires that, since energy is a state function, $E = 0$. Since this is a contradiction, the existence of a perpetual motion machine of the first kind (proposition ~MFL) implies that the first law (energy is a state function, proposition FL) is false (~MFL ⇒ ~FL).

From this result, we can validly conclude: If the first law is true, the existence of a perpetual motion machine of the first kind is impossible:

$$(\sim\text{MFL} \Rightarrow \sim\text{FL}) \Rightarrow (\sim\sim\text{FL} \Rightarrow \sim\sim\text{MFL}) \Rightarrow (\text{FL} \Rightarrow \text{MFL})$$

We cannot conclude that the impossibility of perpetual

motion of the first kind implies that energy is a state function

(~MFL ⇒ ~FL) does not imply (MFL⇒FL)

That is, the impossibility of perpetual motion of the first kind, as we have interpreted it, is not shown (by this argument) to be equivalent to the first law, as we have stated it. (It remains possible, of course, that this equivalence could be proved by some other argument.)

Evidently, when we take the impossibility of constructing a perpetual motion machine of the first kind as a statement of the first law, we have a different interpretation in mind. The difference is this: When we specify a machine that operates in a cycle, we intend that everything about the machine shall be the same at the end of the cycle as at the beginning—including its energy. That is, we intend the statement to be understood as requiring that, for one cycle of the perpetual motion machine $E = 0$. Equivalently, we intend the statement to be understood to include the stipulation that energy is a state function.

Now, for one cycle of the perpetual motion machine, we have $E = 0$ and $\hat{E} > 0$. Given the basic idea that energy is additive, so that $E_{universe} = E + \hat{E}$, we have that $E_{universe} > 0$. The impossibility statement asserts that this is false; equivalently, the impossibility statement asserts that energy cannot be created. This conclusion is a weak form of the principle of conservation of energy; it says less than we want the first law to say. We postulate that energy can be neither created nor destroyed. That is, $E_{universe} > 0$ and $E_{universe} < 0$ are both false. When we consider the impossibility statement to assert the principle of energy conservation, we implicitly stipulate that the machine can also be run in reverse. (See problem 10.)

We intend the first law to assert the existence of energy and to summarize its properties. However we express the first law, we recognize that the concept of energy encompasses several closely interrelated ideas.

§12 Notation for changes in thermodynamic quantities: *E* *versus* Δ*E*

From the outset of our study of energy, we recognize that we are always dealing with energy changes. Even when we write $E = E(P, T, h)$ to indicate that energy is a function of P, T, and h, we recognize that E represents the energy difference between the state of the system characterized by P, T, and h and the state of the system when the independent variables correspond to a reference state in which, by definition, $E = 0$. As we observe in §6-2, we can sort thermodynamic variables into two classes. Some, like P, V, and T, can be measured only for a system. Others, like q, w, E, S, H, G, and A, can be measured only for a process. To say that the volume of a system is one cubic meter has absolute significance. To say that the energy of a system is one joule means nothing unless we know the reference state.

When we intend to specify that the reference state for energy is the particular state specified by $P = P_0$, $T = T_0$, and $h = h_0$, we write "$E(P_0, T_0, h_0) = 0$." Otherwise, when we write "$E = E(P, T, h)$," we could equally well write "$\Delta E = E(P, T, h)$." We intend either of these formulations to mean the same thing as "$E = E(P, T, h) - E(P_0, T_0, h_0)$ and $E(P_0, T_0, h_0) = 0$."

Whether we write E or ΔE, the quantity represented is the difference in energy between some initial and some final state. When we focus on very small changes, we can write dE or $d(\Delta E)$. If our perspective is that we are describing a process, we may prefer to write "E"; if our perspective is that we are describing a change in the system, we may prefer to write "ΔE." In practice, our choice depends primarily on what we have grown accustomed to in the context at hand. In the discussion above, we write $E = q + w$. We could equally well write $\Delta E = q + w$. The meaning is the same. We can make similar statements about most thermodynamic functions. Often there is no particular reason to prefer X over ΔX, or *vice versa*.

However, there are circumstances in which the delta notation serves particular purposes. If a system undergoes a change in which some thermodynamic variables remain constant, the delta notation provides a convenient way to indicate that a particular variable is not constant. For example, if the volume of a system changes while the applied pressure remains constant, we write $w = -P_{applied}\Delta V$.

Similarly, we often want to describe processes in which some state functions are different in the final state than they are in the initial state, while other state functions are the same in both states, but not necessarily constant throughout the process. In the next few chapters, we develop properties of the state functions entropy, S, enthalpy, H, and Gibbs free energy, G. We define the Gibbs free energy by the relationship $G = H - TS$. To specify the relationship among the changes in these state functions when the final temperature is the same as the initial temperature, we write $\Delta G = \Delta H - T\Delta S$. Here too, we often say that this relationship relates the changes in ΔG, ΔH, and ΔS when "the temperature is constant." This is another useful, but potentially misleading, figure of speech. It is important to remember that the equation is valid for any path between the same two states, even if the temperature varies wildly along that path, so long as the initial and final states are at the same temperature.

Finally, we find it convenient to use subscripted versions of the delta notation to specify particular kinds of processes. For a process in which one mole of a pure substance vaporizes to its gas at a particular temperature, we write $\Delta_{vap}H$ and $\Delta_{vap}G$ to denote the changes in enthalpy and Gibbs free energy, respectively. (We can write $\Delta_{vap}E$ to denote the change in the energy; however, $\Delta_{vap}E$ is not a quantity that we find useful very often.) Similarly for the fusion and sublimation of one mole of a pure substance at a particular temperature, we write $\Delta_{fus}G$, $\Delta_{sub}G$, $\Delta_{fus}H$, and $\Delta_{sub}H$. We also find it convenient to write $\Delta_r H$ and $\Delta_r G$ to denote the changes in these quantities when a chemical reaction occurs. When we do so, it is essential that we specify the corresponding stoichiometric equation.

§13 Heat capacities for gases: C_V, C_P

If we heat or do work on any gas—real or ideal—the energy change is $E = q + w$. When we investigate the energy change that accompanies a temperature change, we can obtain reproducible results by holding either the pressure or the volume constant. With volume held constant, we measure C_V. With pressure held constant, the energy change we measure depends on both C_P and the relationship among the pressure, volume, and temperature of the gas. If we know an equation of state for the gas and the values of both C_V and C_P, we can find the energy change between any two states of the gas, because the same change of state can be achieved in two steps, one at constant pressure and one at constant volume.

To see this, we recognize that the state of any pure gas is completely specified by specifying its pressure, temperature, and volume. Any change of state necessarily involves changing at least two of these state functions. Any change of state that changes all three of them can be achieved in an alternate way that involves two changes, each of which occurs with one variable held constant. For example, the change

$$(P_1, V_1, T_1) \to (P_2, V_2, T_2)$$

can be achieved by the constant-pressure sequence

$$(P_1, V_1, T_1) \to (P_1, V_2, T_i)$$

followed by the constant-volume sequence

$$(P_1, V_2, T_i) \to (P_2, V_2, T_2)$$

where T_i is some intermediate temperature. Note that this sequence has to be possible: with P held constant, specifying a change in T is sufficient to determine the change in V; with V held constant, specifying a change in T is sufficient to determine the change in P.

Let us consider how the energy of one mole of any pure substance changes with temperature at constant volume. The rate of change of E with T is

$$\left(\frac{\partial E}{\partial T}\right)_V = \left(\frac{\partial q}{\partial T}\right)_V + \left(\frac{\partial w}{\partial T}\right)_V = C_V + \left(\frac{\partial w}{\partial T}\right)_V$$

where we use the definition of C_V. For any system, and hence for any substance, the pressure–volume work is zero for any process in which the volume remains constant throughout; therefore, we have $(\partial w/\partial T)_V = 0$ and

$$\left(\frac{\partial E}{\partial T}\right)_V = C_V$$

(one mole of any substance, only PV work possible)

When we develop the properties of ideal gases by treating them as point mass molecules, we find that their average translational kinetic energy is $3RT/2$ per mole or $3kT/2$ per molecule, which clearly depends only on temperature. Translational kinetic energy is the only form of energy available to a point-mass molecule, so these relationships describe all of the energy of any point-mass molecule. In particular, they describe all of the energy of a monatomic ideal gas. Since the energy of a monatomic ideal gas is independent of pressure and volume, the temperature derivative must be independent of pressure and volume. The ordinary derivative and the partial derivatives at constant pressure and constant volume all describe the same thing, which, we have just seen, is C_V.

$$\frac{dE}{dT} = \left(\frac{\partial E}{\partial T}\right)_P = \left(\frac{\partial E}{\partial T}\right)_V = C_V = \frac{3}{2}R$$

(one mole of a monatomic ideal gas)

It is useful to extend the idea of an ideal gas to molecules that are not monatomic. When we do so, we have in mind molecules that do not interact significantly with one another. Another way of saying this is that the energy of the collection of molecules is not affected by any interactions among the molecules; we can get the energy of the collection by adding up the energies that the individual molecules would have if they were isolated from one another. In our development of statistical thermodynamics, we find that the energy of a collection of non-interacting molecules depends only on the molecules' energy levels and the temperature. The molecules' energy levels are fixed. This means that if we extend our idea of ideal gases to include non-interacting polyatomic compounds, the energies of such gases still depend only on temperature. For any ideal gas, we have

$$\frac{dE}{dT} = \left(\frac{\partial E}{\partial T}\right)_P = \left(\frac{\partial E}{\partial T}\right)_V = C_V$$

(one mole of any ideal gas)

However, for polyatomic molecules it will no longer be true that $C_V = 3R/2$. Let us see why. Recall that we construct our absolute temperature scale by extrapolating the Charles' law graph of volume versus temperature to zero volume. (Figure 2-2.) By experiment, we find that this graph is the same for one mole of a polyatomic ideal gas as it is for one mole of a monatomic ideal gas. Evidently, our definition of temperature depends only on the translational energy of ideal gas molecules and *vice-versa.* At a fixed temperature, the average translational kinetic energy is the same for any ideal gas; it is independent of the mass of the molecule and of the kinds of atoms in it. To increase the temperature by one degree requires that the translational kinetic energy increase by $3R/2$, and *vice versa.*

Consider what happens when we add energy to a polyatomic ideal gas. Polyatomic gas molecules have energy in rotational and vibrational modes of motion. When we add energy to such molecules, some of the added energy goes into these rotational and vibrational modes. To achieve the same increase in translational kinetic energy, the total amount of energy added must be greater. We find that we need a larger ΔE to achieve the same ΔT, which means that the heat capacity (either C_V or C_P) of the polyatomic ideal gas is greater than that of a monatomic ideal gas.

Now let us consider the rate of change of E with T at constant pressure. For one mole of any substance, we have

$$\left(\frac{\partial E}{\partial T}\right)_P = \left(\frac{\partial q}{\partial T}\right)_P + \left(\frac{\partial w}{\partial T}\right)_P$$
$$= C_P + \left(\frac{\partial w}{\partial T}\right)_P$$

This equation is as far as we can go, unless we can focus on a particular situation for which we know how work varies with temperature at constant pressure.

For one mole of an ideal gas, we have this information. From $PV = RT$ at constant P, we have $PdV = RdT$. If reversible work is done on the ideal gas, $w = \int -P_{applied}dV = \int -PdV$ and

$$\left(\frac{\partial w}{\partial T}\right)_P = \left[\frac{\partial}{\partial T}\int -PdV\right]_P$$
$$= \left[\frac{\partial}{\partial T}\int -RdT\right]_P$$
$$= -R$$

(any ideal gas)

That is, when enough heat is added to increase the temperature of one mole of ideal gas by one degree kelvin at constant pressure, $-R$ units of work are done on the gas. This is the energy change that occurs because of the increase in volume that accompanies the one-degree temperature increase. Since, for any ideal gas,

$$C_V = \left(\frac{\partial E}{\partial T}\right)_P$$
$$= \left(\frac{\partial q}{\partial T}\right)_P + \left(\frac{\partial w}{\partial T}\right)_P$$
$$= C_P - R$$

we have

$$C_P = C_V + R$$

(one mole of any ideal gas)

For a monatomic ideal gas,

$$C_P = C_V + R$$
$$= \frac{3}{2}R + R$$
$$= \frac{5}{2}R$$

(one mole of a monatomic ideal gas)

The heat capacity functions have a pivotal role in thermodynamics. We consider many of their properties further in the next section and in later chapters (particularly §10-9 and §10-10.) Because we want to use these properties before we get around to justifying them all, let us summarize them now:

- For monatomic ideal gases, C_V and C_P are independent of temperature.
- For polyatomic gases, real or ideal, C_V and C_P are functions of temperature.
- C_P is always greater than C_V, but as the temperature decreases, their values converge, and both vanish at absolute zero.
- At ordinary temperatures, C_V and C_P increase only slowly as temperature increases. For many purposes they can be taken to be constant over rather wide temperature ranges.
- For real substances, C_V is a weak function of volume, and C_P is a weak function of pressure. These dependencies are so small that they can be neglected for many purposes.
- For ideal gases, C_V is independent of volume, and C_P is independent of pressure.

§14 Heat capacities of solids: The law of Dulong and Petit

It is easy to maintain a constant pressure on a solid while varying its temperature. To keep its volume rigorously constant over a range of temperatures is difficult. Because the direct measurement of C_P is straightforward, most heat-capacity experiments on solids measure C_P. In §10-9, we derive a general relationship between C_P, C_V, and other measurable properties of a substance. This relationship makes it possible to evaluate C_V indirectly. For a solid, this relationship shows that C_P and C_V are usually about the same.

Heat capacities of solids have been investigated over wide temperature ranges. For most solids, C_P is approximately constant at room temperature and above. For any of the heavier elements, this constant has about the same value. This observation was first made in 1819. It is called ***the law of Dulong and Petit***, in honor of the discoverers. It played an important role in the establishment of correct atomic weights for the elements. The value of the constant found by Dulong and Petit is about

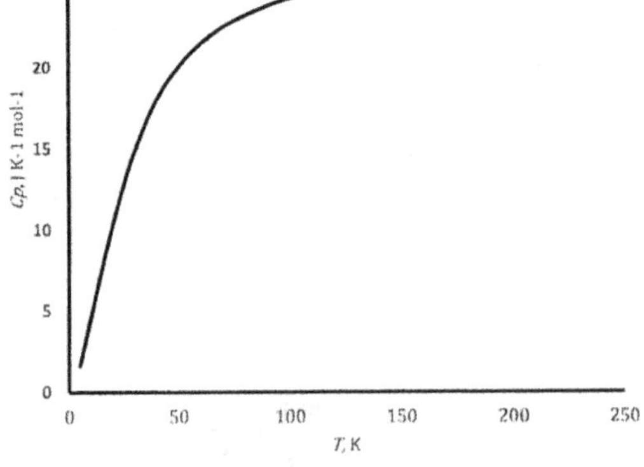

Figure 6. Heat capacity of solid rhombohedral mercury. Data from D. R. Linde, Editor, ***The Handbook of Chemistry and Physics***, CRC Press, 79th Edition (1998-1999), p 6-134.

$3R$. Remarkably, the law can be extended to polyatomic molecules containing only the heavier elements. Often the solid-state heat capacity of such molecules is about $3R$ per mole of atoms in the molecule. Correlations that are more detailed have been developed. These relate the heat capacity of a mole of a molecular solid to its molecular formula. In such correlations, the heat capacity per mole increases by a fixed increment for each atom of, say, carbon in the molecule; by a different fixed increment for each atom of nitrogen in the molecule; *etc.* For the lighter elements, the increments are less than $3R$. For the heavier elements, the increment is approximately $3R$, as observed by Dulong and Petit.

As the temperature of any solid decreases, its heat capacity eventually begins to decrease. At temperatures near absolute zero, the heat capacity approaches zero. The graph in Figure 6 shows the shape of the heat capacity versus temperature curve for solid mercury. The shape of this curve can be predicted from a very simple model for the energy modes available to the atoms in a solid. Albert Einstein developed this model in 1907. Einstein's model for the heat capacity of a solid was an important milestone in the development of quantum mechanics. Since then, the basic ideas have been extended and refined to create more detailed theories that achieve good quantitative agreement with the experimental results for particular substances. We discuss Einstein's treatment in §22-6.

§15 Defining enthalpy, H

Any mathematical expression that involves only state functions must itself be a state function. We could define $\Gamma = E^2P^2VT$, and Γ would be a state function. However, it is not a useful state function. We can define several state functions that have the units of energy and that turn out to be particularly useful. One of them is named enthalpy and is customarily represented by the symbol H.

We define ***enthalpy***: $H = E + PV$.

One reason that enthalpy is a useful state function emerges if we examine the change in H when the system pressure is equal to the applied pressure, and both are constant. (When these conditions are satisfied, we usually denote the heat accepted by the system as "q_P.") If all of the work is pressure–volume work, we have

$$\begin{aligned}\Delta H &= \Delta E + P\Delta V \\ &= q_P + w + P\Delta V \\ &= q_P - P_{applied}\int dV + P\Delta V \\ &= q_P - \left(P_{applied} - P\right)\Delta V \\ &= q_P\end{aligned}$$

If these conditions are satisfied, the enthalpy change is the same thing as the heat added to the system. When we want to express the requirement that the system and applied pressures are equal and constant, we often just say that the process "occurs at a constant pressure." This is another convenient figure of speech. It also reflects our expectation that the system pressure and the applied pressure will equilibrate rapidly in most circumstances.

For an ideal gas, the molar energy depends only on temperature. Since $PV = RT$ for an ideal gas, PV depends only on temperature. Hence, the molar enthalpy of an ideal gas also depends only on temperature. For an ideal gas, we have the parallel relationships:

$$\left(\frac{\partial E}{\partial T}\right)_V = \left(\frac{\partial E}{\partial T}\right)_P = \left(\frac{\partial q}{\partial T}\right)_V = C_V$$

and

$$\left(\frac{\partial H}{\partial T}\right)_V = \left(\frac{\partial H}{\partial T}\right)_P = \left(\frac{\partial q}{\partial T}\right)_P = C_P$$

Earlier we asserted that, while energy is a state function, heat and work are not. ***Hess's law***, as originally formulated in 1840, says that the heat changes for a series of chemical reactions can be summed to get the heat change for the overall process described by the sum of the chemical reactions. This amounts to saying that heat is a state function. As it stands, this is a contradiction. The resolution is, of course, that Hess's law was formulated for a series of chemical reactions that occur at the same constant pressure. Then the heat involved in each step is the enthalpy change for that step, and since enthalpy is a state function, there is no contradiction. Modern statements of Hess's law frequently forego historical accuracy in favor of scientific accuracy to assert that the enthalpy change for a series of reactions can be summed to get the enthalpy change for the overall process. Thus revised, Hess's law ceases to be a seminal but imperfect conjecture and becomes merely a special case of the principle that enthalpy is a state function.

§16 Heat transfer in reversible processes

If a system is in thermal contact with its surroundings, a reversible change can involve the exchange of heat between the system and the surroundings. In Chapter 6, we make a number of important observations about the nature of any heat transfer that occurs during a reversible process. Let us review these ideas.

A system can undergo a change in which it accepts (or liberates) heat while its temperature remains constant. If we boil a liquid at constant pressure, a thermometer immersed in the liquid continues to show the same temperature even though we add more and more heat energy. The added heat is used within the system to convert the liquid to its vapor. If the liquid is stirred well, any localized temperature excursions away from the equilibrium temperature are small; it is a good approximation to say that the temperature of the system is homogenous throughout the system and that it has a constant value.

Nevertheless, for a finite boiling rate we recognize that the idea of an isothermal process is indeed an approximation. For heat transfer to occur from the surroundings to the system, the surroundings must be at a higher temperature than the system. The portion of the system in immediate contact with the wall of the vessel must be at a higher temperature than the portion in the interior of the vessel.

When we think about a constant-temperature system undergoing a reversible change while in thermal contact with its surroundings, we imagine that heat can be transferred in either direction with equal facility. If the system is taking up heat as the process proceeds, we imagine that we can reverse the direction of the change simply by changing the direction of heat transfer. Heat will flow from the system to the surroundings, and the process will run backwards. We can reverse the direction of heat flow by changing the temperature of the surroundings. Initially the surroundings must be hotter than the system. To reverse the direction of heat flow, we must make the temperature of the surroundings less than that of the system. Since a reversible process is one whose direction can be reversed by an arbitrarily small change in some state function, the original temperatures must be arbitrarily close to one another.

For a system that exchanges heat with its surroundings, ***a process can be reversible only if the temperatures of the system and the surroundings are arbitrarily close to one another.*** In a reversible process, net heat transfer occurs between two entities—the system and its surroundings—that are arbitrarily close to thermal equilibrium. Such a process is an idealization. As we have noted several times, a reversible process is a creature of theory that is merely approximated in real systems. A reversible process does not have to be a constant-temperature process. If the temperatures of system and surroundings change simultaneously, they can remain arbitrarily close to one another throughout the process. Nor must a system undergoing reversible change be in thermal contact with its surroundings. A system can undergo a reversible change adiabatically.

Finally, we have noted that the term "isothermal process" is often intended to mean a constant-temperature thermally-reversible process. However, the same words are frequently intended to indicate only that the final temperature of the system is the same as the initial temperature. This is the case whenever the "isothermal process" is a spontaneous process. The intended meaning is usually clear from the context.

§17 Free expansion of a gas

To develop the theory of thermodynamics, we must be able to model the thermodynamic properties of gases as functions of pressure, temperature, and volume. To do so, we consider processes in which the volume of a gas changes. For the expansion (or compression) of a gas to be a reproducible process, the exchange of heat between the system and its surroundings must be controlled. There are two straightforward ways to do this. We can immerse the system in a constant temperature bath whose temperature is the same as that of the system; in this case, $\Delta T = 0$, and we can say that the process is isothermal.

Alternatively, we can isolate the system so that it cannot exchange heat with the surroundings; in this case $q = 0$, and the process is said to be adiabatic. In §7 we find that the work done on a system when its volume changes by dV under the influence of an applied pressure, $P_{applied}$, is $dw = -P_{applied}dV$. Any expansion of a system in which the applied pressure is less than the system pressure can be called a ***free expansion***. In §10-14 we consider the adiabatic expansion of a real gas against a constant applied pressure—a process known as a Joule-Thomson expansion. We find that we must introduce a new parameter—the Joule-Thomson coefficient—in order to describe the behavior of a real gas in a free expansion. The Joule-Thomson coefficient varies with pressure and temperature.

Literally, an ***isothermal process*** is one in which the temperature of the system remains the same throughout the process. However, we often use the term to mean merely that the process occurs while the system is in thermal contact with constant-temperature surroundings. The free expansion of a gas is an irreversible process; in principle, the temperature of a gas undergoing a free expansion is not a meaningful quantity. When we talk about an isothermal free expansion of a gas, we mean that the final temperature is the same as the initial temperature.

Here we consider the behavior of ideal gases, and we begin by considering the limiting case of a free expansion in which the applied pressure is zero. Physically, this corresponds to the expansion of a system into a (very large) evacuated container. Under this condition, $dw = 0$, and the energy change is $dE = dq$. For one mole of any substance, $C_V = (\partial E/\partial T)_V$. If only pressure–volume work is possible and the applied pressure is zero, we have $dE = dq = C_V dT$, and

$$\Delta E = q = \int_{T_1}^{T_2} C_V dT$$

where T_1 and T_2 are the temperatures of the substance before and after the expansion, respectively.

At ordinary temperatures, C_V changes only slowly as the temperature changes. Over a short temperature range, it is usually a good approximation to assume that C_V is constant. We have

$$\Delta E = q = C_V(T_2 - T_1)$$

(one mole of any gas or other substance)

For a monatomic ideal gas, the energy change is exactly

$$\Delta E = q = \frac{3}{2}R(T_2 - T_1)$$

(one mole of a monatomic ideal gas)

The enthalpy change for any process is $\Delta H = \Delta E + \Delta(PV)$. If the system is one mole of an ideal gas, we have, because $\Delta(PV) = R\Delta T = R(T_2 - T_1)$,

$$\begin{aligned}\Delta H &= C_V(T_2 - T_1) + R(T_2 - T_1)\\ &= (C_V + R)(T_2 - T_1)\end{aligned}$$

$$= C_P(T_2 - T_1)$$
(one mole of any ideal gas)

For an isothermal free expansion against an applied pressure of zero, we have $\Delta T = 0$, and so neither the energy nor the enthalpy of the gas changes. Since also $dw = 0$, there can be no exchange of heat with the surroundings. We have

$$w = q = \Delta T = \Delta E = \Delta H = 0$$
(free expansion, ideal gas)

For an adiabatic free expansion, we have $dq = 0$ and $dw = 0$, and it follows again that $w = q = \Delta T = \Delta E = \Delta H = 0$. We see that the isothermal and adiabatic expansions of an ideal gas into a vacuum are equivalent processes. If the expansion is opposed by a non-zero applied pressure, the two processes cease to be equivalent.

§18 Reversible *versus* irreversible pressure–volume work

In §16 we consider heat transfer in reversible processes. Similar considerations apply to the exchange of work between a system and its surroundings. When we use a piston to compress a gas in a cylinder, we must apply sufficient inward force on the piston to overcome the outward force applied by the gas. In any real system, it is necessary also to overcome the force of friction in order to slide the piston into the cylinder. We ignore friction, imagining that we can make its effects arbitrarily small.

The gas can be compressed only if the applied pressure exceeds the gas pressure. If the applied pressure equals the gas pressure, the piston remains stationary. If the applied pressure is greater than the gas pressure by any ever-so-small amount, the gas will be compressed. Conversely, if the applied pressure is infinitesimally less than the gas pressure, the gas will expand. The work done under such conditions is reversible work; an arbitrarily small change in the relative pressures can reverse the direction in which the piston moves. We summarize these conditions by saying that reversible pressure–volume work can occur only if the system and its surroundings are at mechanical equilibrium.

Now, let us think about calculating the reversible work for isothermally compressing a gas by sliding a piston into a cylinder. In any real experiment, we must have $P_{applied} > P_{gas}$, and any real experiment is necessarily irreversible. In a reversible experiment, we have $P_{applied} = P_{gas} = P$, and the reversible work, w^{rev}, is

$$w^{rev} = \int_{V_1}^{V_2} -P_{applied} dV = \int_{V_1}^{V_2} -P_{gas} dV$$

For one mole of an ideal gas, we have $P = RT/V$. Since the temperature is constant, the reversible isothermal work becomes

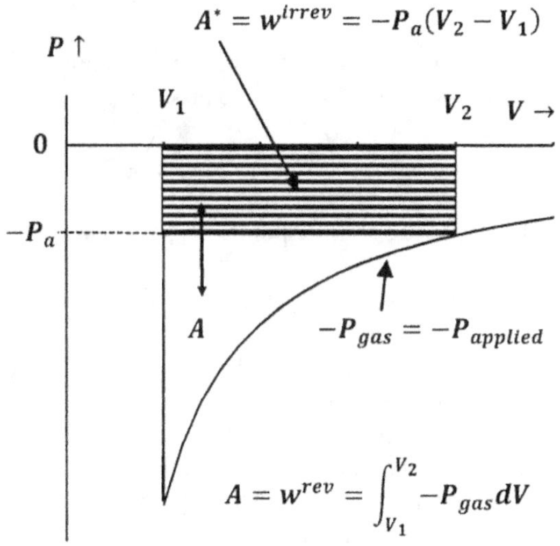

Figure 7. Reversible *versus* irreversible expansion of an ideal gas.

$$w^{rev} = \int_{V_1}^{V_2} -\frac{RT}{V} dV = -RT \ln \frac{V_2}{V_1}$$

where V_1 and V_2 are the initial and final volumes of the gas, respectively. This has a straightforward graphical interpretation. For an ideal gas at constant temperature, $-P_{gas}$ is inversely proportional to V_{gas}. As sketched in Figure 7, the reversible work corresponds to the area between this curve and the abscissa and between the initial, V_1, and the final, V_2, gas volumes.

In contrast, an irreversible expansion corresponds to movement of the piston when $P_{gas} > P_{applied}$, or equivalently, $-P_{gas} < -P_{applied}$. Therefore, the work done on the gas is less in the reversible case than it is in the irreversible case. (Both work terms are less than zero. The absolute value of the reversible work is greater than the absolute value of the irreversible work.) From our definitions of reversible and irreversible pressure–volume work, we have $dw^{rev} < dw^{irrev}$ and $w^{rev} < w^{irrev}$, so long as the initial and final states are the same in the irreversible process as they are in the reversible constant-temperature process. The shaded area in Figure 7 represents the work done on the gas when the applied pressure is instantaneously decreased to the final pressure, P_a, attained by the gas in the reversible process.

For the reversible process, the pressure–volume curve accurately depicts the state of the gas as the volume increase takes place. The temperature of the gas is constant along this curve. While we can trace a similar line of pressure–volume points for the irreversible expansion, this line does not define a set of intermediate states that the system occupies during the irreversible expansion. The state of the gas is well defined only in the equilibrium state that precedes the irreversible pressure drop and in the equilibrium state that the system ultimately attains. It is convenient to describe these two processes as a reversible process and a spontaneous process that "take the system from the same initial state to the same final state." However, this language obscures a significant point. In the initial state for the reversible process, we have

$$P_{gas} = RT/V_1 = P_{applied}$$

In the initial state for the spontaneous process, we have

$$P_{gas} = RT/V_1 \quad \text{and} \quad P_{applied} = RT/V_2$$

What we mean, of course, is that the values of all of the state functions for the hypothetical initial state of the spontaneous process are the same as those for the equilibrium initial state of the reversible process.

So long as we can say that the process takes the system from the same initial state to the same final state, a similar argument can be made for reversible and irreversible work of any kind. Whatever the force, the isothermal reversible work done on the system is always less than the irreversible work for taking the system between the same two states. This is an important result. In Chapter 9, we find that it is a logical consequence of the second law of thermodynamics.

Finally, let us consider a reversible process in which a system completes a pressure–volume cycle. The system traverses a closed path in the pressure–volume plane. Such a path is depicted in Figure 8. We let the smallest and largest volumes reached during the cycle be V_ℓ and V_h, respectively. The closed path is composed of a high-pressure segment and a low-pressure segment that meet at V_ℓ and V_h. On each of these segments, the pressure is a function of volume. We let pressures on the high- and low-pressure segments be $P_h(V)$ and $P_\ell(V)$, respectively. (In the interval $V_\ell < V < V_h$, we have $P_h(V) > P_\ell(V) > 0$. At the limiting volumes, we have $P_h(V_h) = P_\ell(V_h)$ and $P_h(V_\ell) = P_\ell(V_\ell)$.) The system temperature varies continuously around the closed path. The work done on the system as it traverses the high-pressure segment from V_ℓ to V_h is represented in Figure 8 by area A_h. We have

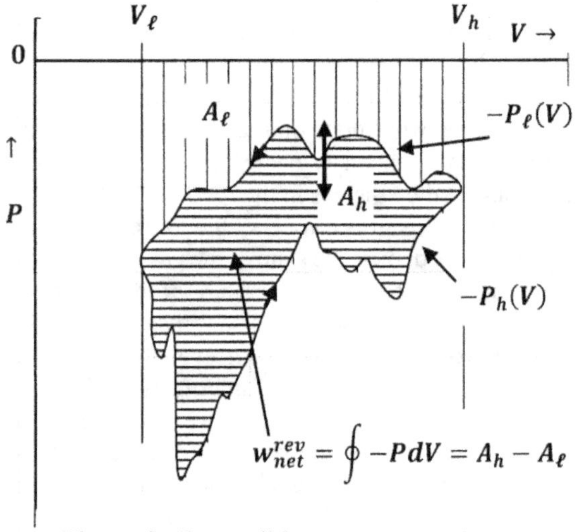

Figure 8. Reversible pressure—volume work in a cycle.

$$\begin{aligned} w_h(V_\ell \rightarrow V_h) &= \int_{V_\ell}^{V_h} -P_h(V)dV \\ &= A_h \\ &< 0 \end{aligned}$$

The work done on the system as it traverses the low-pressure segment from V_ℓ to V_h is represented by area A_ℓ. We have

$$\begin{aligned} w_\ell(V_\ell \rightarrow V_h) &= \int_{V_\ell}^{V_h} -P_\ell(V)dV \\ &= A_\ell \\ &< 0 \end{aligned}$$

When the low-pressure segment is traversed in the opposite direction, we have $w_\ell(V_h \rightarrow V_\ell) = -A_\ell > 0$. When the system traverses the cycle in the counterclockwise direction, the net work done on the system is

$$\begin{aligned} w_{net} &= w_h(V_\ell \rightarrow V_h) + w_\ell(V_h \rightarrow V_\ell) \\ &= \int_{V_\ell}^{V_h} -P_h(V)dV + \int_{V_h}^{V_\ell} -P_\ell(V)dV \\ &= \oint -P(V)dV \\ &= A_h - A_\ell < 0 \end{aligned}$$

Thus the net work done on the system is represented on the graph by the area $A_h - A_\ell$, which is just the (negative) area in the pressure–volume plane that is bounded by the closed path.

§19 Isothermal expansions of an ideal gas

For an isothermal reversible expansion of an ideal gas, we have by definition that $\Delta T = 0$. Since the energy of an ideal gas depends only on the temperature, a constant temperature implies constant energy, so that $\Delta E = 0 = q^{rev} + w^{rev}$. Using the equation we find for w^{rev} in the previous section, we have

$$-q^{rev} = w^{rev} = -RT \ln \frac{V_2}{V_1}$$

(ideal gas, isothermal reversible expansion)

where V_1 and V_2 are the initial and final volumes, respectively. Since enthalpy is defined as $H = E + PV$, we have $\Delta H = \Delta E + \Delta(PV) = \Delta E + \Delta(RT) = 0$.

For the spontaneous isothermal expansion of an ideal gas from V_1 to V_2 against a constant applied pressure, we again have $\Delta T = \Delta E = \Delta H = 0$. These are state functions, and the amounts by which they change in this spontaneous process must be the same as those for the reversible process between the same two states. The heat and work exchanged in the spontaneous process are different, demonstrating that heat and work are not state functions. We have

$$
\begin{aligned}
-q &= w \\
&= \int_{V_1}^{V_2} -P_{applied} dV \\
&= -P_{applied}(V_2 - V_1) \\
&= RT\left(\frac{P_{applied}}{P_1} - 1\right)
\end{aligned}
$$

(one mole ideal gas, isothermal free expansion, $P_{applied} > 0$)

§20 Adiabatic expansions of an ideal gas

Consider an ideal gas that undergoes a reversible adiabatic expansion from an initial state, specified by known values V_1 and T_1, to a new state in which the value of the volume, V_2, is known but the value of the temperature, T_2, is not known. For an adiabatic reversible process, $q = 0$, and $w = \Delta E$. Since $(\partial E/\partial T)_V = C_V$, we have $dE = C_V dT$, so that

$$
w = \Delta E = \int_{T_1}^{T_2} C_V \, dT
$$

For any gas, we can assume that C_V is approximately constant over a small temperature range. Taking C_V to be constant in the interval $T_1 < T < T_2$, we have $w = \Delta E = C_V(T_2 - T_1)$. We obtain the enthalpy change from

$$
\begin{aligned}
\Delta H &= \Delta E + \Delta(PV) \\
&= \Delta E + \Delta(RT) \\
&= C_V(T_2 - T_1) + R(T_2 - T_1) \\
&= C_P(T_2 - T_1)
\end{aligned}
$$

where we use our ideal-gas result from §16, $C_P = C_V + R$.

While these relationships yield the values of the various thermodynamic quantities in terms of the temperature difference, $T_2 - T_1$, we have yet to find the final temperature, T_2. To find T_2, we return to the first law: $dE = dq + dw$. Substituting for dE, dq, and dw, and making use of the ideal gas equation, we have

$$
\begin{aligned}
C_V dT &= -PdV \\
&= -\frac{RT}{V} dV
\end{aligned}
$$

from which, by separation of variables, we have

$$
\int_{T_1}^{T_2} C_V \frac{dT}{T} = -R \int_{V_1}^{V_2} \frac{dV}{V}
$$

(one mole ideal gas, reversible adiabatic expansion)

If we know C_V as a function of temperature, we can integrate to find a relationship among T_1, T_2, V_1, and V_2. Given any three of these quantities, we can use this relationship to find the fourth. If C_V is independent of temperature, as it is for a monatomic ideal gas, we have

$$
\begin{aligned}
\ln\frac{T_2}{T_1} &= -\frac{R}{C_V}\ln\frac{V_2}{V_1} \\
&= \frac{R}{C_V}\ln\frac{V_1}{V_2} \\
&= \ln\left(\frac{V_1}{V_2}\right)^{R/C_V}
\end{aligned}
$$

so that

$$
\frac{T_2}{T_1} = \left(\frac{V_1}{V_2}\right)^{R/C_V}
$$

(monatomic ideal gas, reversible adiabatic expansion)

For the spontaneous adiabatic expansion of an ideal gas against a constant applied pressure, we have $dq = 0$, so that $dE = dw$, and $C_V dT = -P_{applied} dV$. Given the initial conditions, we can find the final temperature from

$$
\begin{aligned}
\int_{T_1}^{T_2} C_V dT &= \int_{V_1}^{V_2} -P_{applied} dV \\
&= -P_{applied}\left(\frac{RT_2}{P_{applied}} - \frac{RT_1}{P_1}\right) \\
&= R\left(\frac{P_{applied} T_1}{P_1} - T_2\right)
\end{aligned}
$$

(spontaneous adiabatic process)

The changes in the remaining state functions can then be calculated from the relationships above. In this spontaneous adiabatic process, all of the other thermodynamic quantities are different from those of a reversible adiabatic process that reaches the same final volume.

Problems

1. Which of the following differential expressions are exact?

a. $df = ydx + xdy$
b. $df = 2xy^2 dx + 2x^2 y dy$
c. $df = 2xydx + 2x^2 ydy$
d. $df = [(1 - xy)e^{-xy}]dx - [x^2 e^{-xy}]dy$
e. $df = (\cos x \cos y)dx - (\sin x \sin y)dy$
f. $df = (\cos x \cos y)dx - (\sin y)dy$

2. Show that $df = e^{-y}dx - xe^{-y}dy$ is exact. Find $f(x,y)$ by integrating the dx term. Find $f(x,y)$ by integrating the dy term.

3. A marble of mass m is free to move on a surface whose height above the x,y-plane is $h = ax^2 + by^2$.

a. What is the gravitational potential energy of the marble expressed as a function of x and y, $E(x,y)$?

b. The force experienced by the marble due to gravity is the vector function

$$\vec{f}(x,y) = -\nabla E(x,y) = -\left(\frac{\partial E}{\partial x}\right)_y \vec{\imath} - \left(\frac{\partial E}{\partial y}\right)_x \vec{\jmath}$$

What is $\vec{f}(x,y)$ on this surface?

c. What is the differential of E? Is dE exact or inexact?

d. The vector description of a general path, $\{(x,y)\}$, is the position vector, $\vec{r} = x\vec{\imath} + y\vec{\jmath}$, and so $d\vec{r} = dx\,\vec{\imath} + dy\,\vec{\jmath}$. If we push the marble up the surface from point $(0,0,E(0,0))$ to point $(2,2,E(2,2))$ along the path $y = x$, express $d\vec{r}$ as a vector function of dx.

e. If we push the marble along the path in part d with a force just large enough to overcome the force of gravity, what is the increment of work, dw, associated with an increment of motion, $d\vec{r}$?

f. How much work must we do if we are to move the marble from $(0,0,E(0,0))$ to point $(2,2,E(2,2))$ along the path in part d, using the force in part e? What is the relationship between this amount of work and the change in the energy of the marble during this process?

g. Suppose that we push the marble up the surface from point $(0,0,E(0,0))$ to point $(2,2,E(2,2))$ along the path $y = x^2/2$. What is the vector description of this path?

h. How much work must we do if we are to move the marble from point $(0,0,E(0,0))$ to point $(2,2,E(2,2))$ along the path in part g using the force in part b? Compare this result to your result in part f. Explain.

4. Consider the plane, $f(x,y) = 1 - 2x - 3y$. What is df for this surface? Evaluate $\Delta f = f(1,1) - f(-1,-1)$by integrating df along each of the following paths:

a. $y = x$

b. $y = x^3$

c. $y = 1 + x - x^2$

d. $y = \sin(\pi x/2)$

5. A 2.00 mole sample of a monatomic ideal gas is expanded reversibly and isothermally at 350 K from 5.82 L to 58.20 L. How much work is done on the gas? What are q, ΔE, and ΔH for the gas in this process?

6. A 2.00 mole sample of a monatomic ideal gas is expanded irreversibly from 5.82 L to 58.20 L at a constant applied pressure equal to the final pressure of the gas. The initial and final temperatures are 350 K. How much work is done on the gas? What are q, ΔE, and ΔH for the gas in this process? Compare w, q, ΔE, and ΔH for this process to the corresponding quantities for the process in problem 5. Compare the initial and final states of the gas to the corresponding states in problem 5.

7. A 2.00 mole sample of a monatomic ideal gas is expanded reversibly and adiabatically from 5.82 L to 58.20 L. The initial temperature is 350 K. What is the final temperature? What are the initial and final pressures? How much work is done on the gas? What are q, ΔE, and ΔH for the gas in this process?

8. The equation of state for a "hard-sphere gas" is $P(V - nb) = nRT$, where n is the number of moles and b is the molar volume of the hard spheres. How much work is done on this gas when n moles of it expand reversibly and isothermally from V_1 to V_2?

9. Strictly speaking, can the spontaneous expansion of a real gas be isothermal? Can it be free? Can it be adiabatic? Can the reversible expansion of a gas be isothermal? Can it be free? Can it be adiabatic?

10. Consider a machine that operates in a cycle and converts heat into a greater amount of work. What would happen to the energy of the universe if this machine could be operated in reverse?

11. Show that the product of pressure and volume has the units of energy.

12. Give a counter-example to prove that each of the following propositions is false:

a. If X is a state function, X is conserved.

b. If X is an extensive quantity that satisfies $X + \hat{X} = 0$, X is a state function.

Notes

[1] Since the temperature of the water increases and the process is to be reversible, we must keep the temperature of the thermal reservoir just dT greater than that of the water throughout the process. We can accomplish this by using a quantity of ideal gas as the heat reservoir. By reversibly compressing the ideal gas, we can reversibly deliver the required heat while maintaining the required temperature. We consider this operation further in §12-5.

8

Enthalpy and Thermochemical Cycles

§1 Enthalpy

In Chapter 7, we introduce the enthalpy function, which we define as

$$H = E + PV$$

When the only form of work possible is pressure–volume work and a system change occurs at constant pressure, the enthalpy change is synonymous with the heat added to the system.

$$H = q_P \qquad \text{(only PV work)}$$

Since we define $C_P = (\partial q/\partial T)_P$, it follows that $(\partial H/\partial T)_P = C_P$. Recalling our earlier discovery that $(\partial E/\partial T)_V = C_V$, we have the important parallel relationships:

$$C_P = \left(\frac{\partial q}{\partial T}\right)_P = \left(\frac{\partial H}{\partial T}\right)_P$$

and

$$C_V = \left(\frac{\partial q}{\partial T}\right)_V = \left(\frac{\partial E}{\partial T}\right)_V$$

We can find the enthalpy change for heating a substance at constant pressure by integrating its constant-pressure heat capacity, C_P, over the change in temperature. That is, $C_P = (\partial H/\partial T)_P$ implies that

$$\Delta H = \int_{T_1}^{T_2} C_P dT$$

Similarly, we have

$$\Delta E = \int_{T_1}^{T_2} C_V dT$$

for a process in which a substance is heated at constant volume.

One reason that the enthalpy function is useful in chemistry is that many processes are carried out at conditions (constant pressure, only PV work) where the enthalpy change is synonymous with the heat exchanged. The heat exchanged in a process is frequently an important consideration. If we want to carry out an endothermic process, we must provide means to add sufficient heat. If we want to carry out an exothermic process, we may have to make special arrangements to safely transfer the heat evolved from the system to its surroundings.

One of our principal objectives is to predict whether a given process can occur spontaneously. We will see that the heat evolved in a process is ***not*** a generally valid predictor of whether or not the process can occur spontaneously; however, it is true that a very exothermic process is usually one that can occur spontaneously. (We will see that $\Delta H < 0$ is a rigorous criterion for whether the process can occur spontaneously if and only if the process is one for which both the entropy and the pressure remain constant.)

§2 Using thermochemical cycles to find enthalpy changes

Because enthalpy is a state function, the enthalpy change in going between any two states of a system is independent of the path. For a series of changes that restore a system to its original state, the sum of all the enthalpy changes must be zero. This fact enables us to find the enthalpy changes for many processes for which it is difficult to measure heat and work directly. It is easiest to see what is involved by considering a specific example. Figure 1 shows a cyclic path, A→A*→B→…→A,

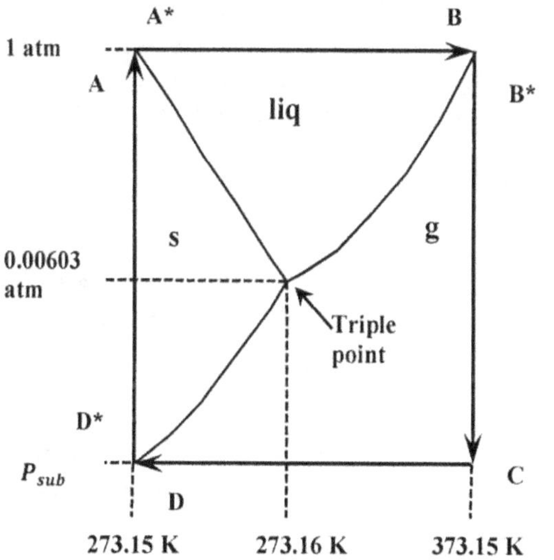

Figure 1. A cyclic path on the phase diagram for water.

superimposed on a not-to-scale presentation of the phase diagram for water. Let us look at the sublimation of ice at the melting point of pure water. The sublimation of ice is the conversion of pure ice to pure water vapor. (The melting point of pure water is the temperature at which pure ice is at equilibrium with pure liquid water at a pressure of one atmosphere; it is represented by points A and A* on the diagram. We want to find the enthalpy of sublimation at the temperature and pressure represented by points D and D*.)

Points A, A*, D, and D* are all at the same temperature; this temperature is about 273.153 K or 0.003 C. (This temperature is very slightly greater than 273.15 K or 0 C—which is the temperature at which ice and water are at equilibrium in the presence of air at a total pressure of one atmosphere.) We want to calculate the enthalpy change for the equilibrium conversion of one mole of ice to gaseous water at the pressure where the solid–gas equilibrium line intersects the line $T = 273.153\text{ K} \approx 0\text{ C}$.

On the diagram, this sublimation pressure is represented as P_{sub} and the sublimation process is represented as the transition from D* to D. P_{sub} is less than the triple-point pressure of 611 Pa or 6.03×10^{-3} atm. However, the difference is less than 1.4×10^{-5} atm or 1.4 Pa. In equation form, the successive states traversed in this cycle are:

A (ice at 0 C and 1 atm) →
A* (water at 0 C and 1 atm) →
B (water at 100 C and 1 atm) →
B* (water vapor at 100 C and 1 atm) →
C (water vapor at 100 C and P_{sub}) →
D (water vapor at 0 C and P_{sub}) →
D*(ice at 0 C and P_{sub}) →
A (ice at 0 C and 1 atm)

We select these steps because it is experimentally straightforward to find the enthalpy change for all of them except the sublimation step (D*→D). All of these steps can be carried out reversibly. This strategy is useful in general. We make extensive use of reversible cycles to find thermodynamic information for chemical systems. The enthalpy changes for these steps are

H_2O (s, 0 C, 1 atm) → H_2O (liq, 0 C, 1 atm)

$$\Delta H(\mathrm{A}\to\mathrm{A}^*) = \Delta_{fus}H$$

H_2O (liq, 0 C, 1 atm) → H_2O (liq, 100 C, 1 atm)

$$\Delta H(\mathrm{A}^*\to\mathrm{B}) = \int_{273.15\text{ K}}^{372.15\text{ K}} C_P(H_2O,\text{liq})\,dT$$

H_2O (liq, 100 C, 1 atm) → H_2O (g, 100 C, 1 atm)

$$\Delta H(\mathrm{B}\to\mathrm{B}^*) = \Delta_{vap}H$$

H_2O (g, 100 C, 1 atm) → H_2O (g, 100 C, P_{sub})

$$\Delta H(\mathrm{B}^*\to\mathrm{C}) = \int_{P=1}^{P=P_{sub}} \left(\frac{\partial H(H_2O,\text{g})}{\partial P}\right)_T dP \approx 0$$

H_2O (g, 100 C, P_{sub}) →H_2O (g, 0 C, P_{sub})

$$\Delta H(\mathrm{C}\to\mathrm{D}) = \int_{373.15\text{ K}}^{272.15\text{ K}} C_P(H_2O,\text{g})\,dT$$

H_2O (g, 0 C, P_{sub}) → H_2O (s, 0 C, P_{sub})

$$\Delta H(\mathrm{D}\to\mathrm{D}^*) = -\Delta_{sub}H$$

H_2O (s, 0 C, P_{sub}) →H_2O (s, 0 C, 1 atm)

$$\Delta H(\mathrm{D}^*\to\mathrm{A}) = \int_{P=P_{sub}}^{P=1} \left(\frac{\partial H(H_2O,\text{s})}{\partial P}\right)_T dP \approx 0$$

Summing the enthalpy changes around the cycle gives

$$0 = \Delta_{fus}H + \int_{273.15\text{ K}}^{372.15\text{ K}} C_P(H_2O,\text{liq})\,dT + \Delta_{vap}H$$
$$+\Delta H(\mathrm{B}^*\to\mathrm{C}) + \int_{373.15\text{ K}}^{272.15\text{ K}} C_P(H_2O,\text{g})\,dT - \Delta_{sub}H$$
$$+\Delta H(\mathrm{D}^*\to\mathrm{A})$$

Using results that we find in the next section, $H(\mathrm{B}^*\to\mathrm{C}) \approx 0$ and $\Delta H(\mathrm{D}^*\to\mathrm{A}) \approx 0$, we have

$$0 = \Delta_{fus}H + \int_{273.15\text{ K}}^{372.15\text{ K}} C_P(H_2O,\text{liq})\,dT + \Delta_{vap}H$$
$$+ \int_{373.15\text{ K}}^{272.15\text{ K}} C_P(H_2O,\text{g})\,dT - \Delta_{sub}H$$

The enthalpy of fusion, the enthalpy of vaporization, and the heat capacities are measurable in straightforward experiments. Their values are given in standard compilations, so we are now able to evaluate $\Delta_{sub}H$, a quantity that is not susceptible to direct measurement, from other thermodynamic quantities that are. (See Problem 8.)

§3 How enthalpy depends on pressure

Let us look briefly at the approximations $\Delta H(\mathrm{B}^*\to\mathrm{C}) \approx 0$ and $\Delta H(\mathrm{D}^*\to\mathrm{A}) \approx 0$ that we used in §2. In these steps, the pressure changes while the temperature remains constant. In Chapter 10, we find a general relationship for the pressure-dependence of a system's enthalpy:

$$\left(\frac{\partial H}{\partial P}\right)_T = -T\left(\frac{\partial V}{\partial T}\right)_P + V$$

This evaluates to zero for an ideal gas and to a negligible quantity for many other systems.

For liquids and solids, information on the variation of volume with temperature is collected in tables as the ***coefficient of thermal expansion***, α, where

$$\alpha = \frac{1}{V}\left(\frac{\partial V}{\partial T}\right)_P$$

Consequently, the dependence of enthalpy on pressure is given by

$$\left(\frac{\partial H}{\partial P}\right)_T = V(1 - \alpha T)$$

For ice, $\alpha \approx 50 \times 10^{-6}\ \mathrm{K}^{-1}$ and the molar volume near 0 C is 19.65 $\mathrm{cm^3\ mol^{-1}}$. The enthalpy change for compressing one mole of ice from the sublimation pressure to 1 atm is $\Delta H(\mathrm{D}^* \rightarrow \mathrm{A}) = 2\ \mathrm{J\ mol^{-1}}$.

To find the enthalpy change for expanding one mole of water vapor at 100 C from 1 atm to the sublimation pressure, we use the virial equation and tabulated coefficients for water vapor to calculate $(\partial H/\partial P)_{398\ \mathrm{K}}$. We find $\Delta H(\mathrm{B}^* \rightarrow \mathrm{C}) = 220\ \mathrm{J\ mol^{-1}}$. (See problem 9.)

§4 Standard states and enthalpies of formation

A useful convention makes it possible to tabulate enthalpy data for individual compounds in such a way that the enthalpy change for any chemical reaction can be calculated from the tabulated information for the reaction's reactants and products. The convention comprises the following rules:

I. At any particular temperature, **we define the standard state of any liquid or solid substance to be the most stable form of that substance at a pressure of one bar**. For example, for water at - 10 C, the standard state is ice at a pressure of one bar; at +10 C, it is liquid water at a pressure of one bar.

II. At any particular temperature, we define the standard state of a gas to be the ***ideal gas standard state*** at that temperature. By the ideal gas standard state, we mean a finite low pressure at which the real gas behaves as an ideal gas. We know that it is possible to find such a pressure, because any gas behaves as an ideal gas at a sufficiently low pressure. Since the enthalpy of an ideal gas is independent of pressure, **we can also think of a substance in its ideal gas standard state as a hypothetical substance whose pressure is one bar but whose molar enthalpy is that of the real gas at an arbitrarily low pressure**.

III. For any substance at any particular temperature, **we define the standard enthalpy of formation as the enthalpy change for a reaction in which the product is one mole of the substance and the reactants are the compound's constituent elements in their standard states.**

For water at **–10 C**, this reaction is

$$H_2(\mathrm{g}, -10\ \mathrm{C}, 1\ \mathrm{bar}) + \tfrac{1}{2}O_2(\mathrm{g}, -10\ \mathrm{C}, 1\ \mathrm{bar}) \rightarrow H_2O(\mathrm{s}, -10\ \mathrm{C}, 1\ \mathrm{bar})$$

For water at **+10 C**, it is

$$H_2(\mathrm{g}, +10\ \mathrm{C}, 1\ \mathrm{bar}) + \tfrac{1}{2}O_2(\mathrm{g}, +10\ \mathrm{C}, 1\ \mathrm{bar}) \rightarrow H_2O(\mathrm{liq}, +10\ \mathrm{C}, 1\ \mathrm{bar})$$

For water at **+110 C**, it is

$$H_2(g, +110\ \mathrm{C}, 1\ \mathrm{bar}) + \tfrac{1}{2}O_2(\mathrm{g}, +110\ \mathrm{C}, 1\ \mathrm{bar}) \rightarrow H_2O(\mathrm{g}, +110\ \mathrm{C}, 1\ \mathrm{bar})$$

IV. The standard enthalpy of formation is given the symbol $\Delta_f H^o$, where the superscript degree sign indicates that the reactants and products are all in their standard states. The subscript, f, indicates that the enthalpy change is for the formation of the indicated compound from its elements. Frequently, the compound and other conditions are specified in parentheses following the symbol. The solid, liquid, and gas states are usually indicated by the letters "s", " ℓ " (or "liq"), and "g", respectively. The letter "c" is sometimes used to indicate that the substance is in a crystalline state. In this context, specification of the gas state normally means the ideal gas standard state.

Thermochemical-data tables that include standard enthalpies of formation can be found in a number of publications or on the internet. For some substances, values are available at a number of temperatures. For substances for which less data is available, these tables usually give the value of the standard enthalpy of formation at 298.15 K. (In this context, 298.15 K is frequently abbreviated to 298 K.)

V. For any element at any particular temperature, we define the standard enthalpy of formation to be zero. When we define standard enthalpies of formation, we choose the elements in their standard states as a common reference state for the enthalpies of all substances at a given temperature. While we could choose any arbitrary value for the enthalpy of an element in its standard state, choosing it to be zero is particularly convenient.

§5 The ideal gas standard state

The ideal gas standard state is a useful invention, which has additional advantages that emerge as our development proceeds. For permanent gases—gases whose behavior is approximately ideal anyway—there is a negligible difference between the enthalpy in the ideal gas state and the enthalpy at 1 bar.

For volatile substances that are normally liquid or solid at 1 bar, the ideal gas standard state becomes a second standard state. For such substances, data tables frequently give the standard enthalpy of formation for both the condensed phase (designated $\Delta_f H^o(\mathrm{liq})$ or $\Delta_f H^o(\mathrm{s})$) and the ideal gas standard state (designated $\Delta_f H^o(\mathrm{g})$). For example, the CODATA[1] values for the standard enthalpies of formation for liquid and ideal-gas methanol are - 239.2 and - 201.0 $\mathrm{kJ\ mol^{-1}}$, respectively, at

298.15 K. The difference between these values is the enthalpy change in vaporizing one mole of liquid methanol to its ideal gas standard state at 298.15 K:

$$CH_3OH(\text{liq}, 298.15\text{ K}, 1\text{ bar}) \rightarrow CH_3OH(\text{ideal gas}, 298.15\text{ K}, \sim 0\text{ bar})$$

Since this is the difference between the enthalpy of methanol in its standard state as an ideal gas and methanol in its standard state as a liquid, we can call this difference the standard enthalpy of vaporization for methanol:

$$\Delta_{vap}H^o = \Delta_f H^o(\text{g}, 298.15\text{ K}, \sim 0\text{ bar}) - \Delta_f H^o(\text{g}, 298.15\text{ K}, 1\text{ bar}) = 37.40\text{ kJ mol}^{-1}$$

This is not a reversible process, because liquid methanol at 1 bar is not at equilibrium with its vapor at an arbitrarily low pressure at 298.15 K.

Note that $\Delta_{vap}H^o$ is not the same as the ordinary enthalpy of vaporization, $\Delta_{vap}H$. The ordinary enthalpy of vaporization is the enthalpy change for the reversible vaporization of liquid methanol to real methanol vapor at a pressure of 1 atm and the normal boiling temperature. We write it without the superscript degree sign because methanol vapor is not produced in its standard state. For methanol, the normal boiling point and enthalpy of vaporization[2] are 337.8 K and 35.21 kJ mol^{-1}, respectively.

We can devise a cycle that relates these two vaporization processes to one another: Summing the steps below yields the process for vaporizing liquid methanol in its standard state to methanol vapor in its standard state.

1) CH_3OH (liq, 298.15 K, 1 bar) → CH_3OH (liq, 337.8 K, 1 bar) $\quad \Delta_{(1)}H$

2) CH_3OH (liq, 337.8 K, 1 bar) → CH_3OH (liq, 337.8 K, 1 atm) $\quad \Delta_{(2)}H$

3) CH_3OH (liq, 337.8 K, 1 atm) → CH_3OH (g, 337.8 K, 1 atm) $\quad \Delta_{(3)}H = \Delta_{vap}H$

4) CH_3OH (g, 337.8 K, 1 atm) → CH_3OH (g, 337.8 K, ~0 bar) $\quad \Delta_{(4)}H$

5) CH_3OH (g, 337.8 K, ~0 bar) → CH_3OH (g, 298.15 K, ~0 bar) $\quad \Delta_{(5)}H$

Thus, we have

$$\Delta_{vap}H^o = \Delta_{(1)}H + \Delta_{(2)}H + \Delta_{vap}H + \Delta_{(4)}H + \Delta_{(5)}H$$

$\Delta_{(1)}H$ and $\Delta_{(5)}H$ can be evaluated by integrating the heat capacities for the liquid and gas, respectively. $\Delta_{(2)}H$ and $\Delta_{(4)}H$ can be evaluated by integrating $(\partial H/\partial P)_T$ for the liquid and gas, respectively. $\Delta_{(2)}H$ is negligible. (For the evaluation of these quantities, see problem 10.)

§6 Standard enthalpies of reaction

The benefit of these conventions is that, **at any particular temperature**, the standard enthalpy change for a reaction

$$aA + bB + \cdots \rightarrow cC + dD + \cdots$$

which we designate as $\Delta_r H^o$, is given by

$$\Delta_r H^o = c\Delta_f H^o(C) + d\Delta_f H^o(D) + \cdots - a\Delta_f H^o(A) - b\Delta_f H^o(B) - \cdots$$

If we have the enthalpies of formation, we can compute the enthalpy change for the reaction. We can demonstrate this by writing out the chemical equations corresponding to the formation of A, B, C, and D from their elements. When we multiply these chemical equations by the appropriately signed stoichiometric coefficient and add them, we obtain the chemical equation for the indicated reaction of A and B to give C and D. (See below.) Because enthalpy is a state function, the enthalpy change that we calculate this way will be valid for any process that converts the specified reactants into the specified products.

The oxidation of methane to methanol is a reaction that illustrates the value of this approach. The normal products in the oxidation of methane are, of course, carbon dioxide and water. If the reaction is done with an excess of methane, a portion of the carbon-containing product will be carbon monoxide rather than carbon dioxide. In any circumstance, methanol is, at best, a trace product. Nevertheless, it would be very desirable to devise a catalyst that quantitatively—or nearly quantitatively—converted methane to methanol according to the equation

$$CH_4 + \tfrac{1}{2}O_2 \rightarrow CH_3OH$$

(This is frequently called a ***selective oxidation***, to distinguish it from the ***non-selective oxidation*** that produces carbon dioxide and water.)

If the catalyst were not inordinately expensive or short-lived, and the operating pressure were sufficiently low, this would be an economical method for the manufacture of methanol. (Methanol is currently manufactured from methane. However, the process involves two steps and requires a substantial capital investment.) If the cost of manufacturing methanol could be decreased sufficiently, it would become economically feasible to convert natural gas, which cannot be transported economically unless it is feasible to build a pipeline for the purpose, into liquid methanol, which is readily transported by ship. (At present, the economic feasibility of marine transport of liquefied natural gas, LNG, is marginal, but it appears to be improving.) This technology would make it possible to utilize the fuel value of known natural gas resources that are presently useless because they are located too far from population centers.

When we contemplate trying to develop a catalyst and a manufacturing plant to carry out this reaction, we soon discover reasons for wanting to know the enthalpy change. One is that the oxidative manufacture of methanol will be exothermic, so burning the methanol produced will yield less heat than would be produced by burning the methane from which it was produced. We want to know how much heat energy is lost in this way.

Another reason is that a manufacturing plant will have to control the temperature of the oxidation reaction in order to maintain optimal performance. (If the temperature is too low, the reaction rate will be too slow. If the temperature is too high, the catalyst may be deactivated in a short time, and the production of carbon oxides will probably be excessive.) A chemical engineer designing a plant will need to know how much heat is produced so that he can provide adequate cooling equipment.

Because we do not know how to carry out this reaction, we cannot measure its enthalpy change directly. However, if we have the enthalpies of formation for methane and methanol, we can compute this enthalpy change:

$$C(s) + 2\,H_2(g) + \tfrac{1}{2}O_2(g) \rightarrow CH_3OH(g) \qquad \Delta H = \Delta_f H^o(CH_3OH, g)$$

$$CH_4(g) \rightarrow C(s) + 2\,H_2(g) \qquad \Delta H = -\Delta_f H^o(CH_4, g)$$

$$\tfrac{1}{2}O_2(g) \rightarrow \tfrac{1}{2}O_2(g) \qquad \Delta H = -\tfrac{1}{2}\,\Delta_f H^o(O_2, g) = 0$$

Summing the reactions gives

$$CH_4(g) + \tfrac{1}{2}O_2(g) \rightarrow CH_3OH(g) \qquad \Delta H = \Delta_r H^o$$

and summing the enthalpy changes gives

$$\Delta_r H^o = \Delta_f H^o(CH_3OH, g) - \Delta_f H^o(CH_4, g) - \tfrac{1}{2}\,\Delta_f H^o(O_2, g)$$

The diagram in Figure 2 shows how these conventions, and the fact that enthalpy is a state function, work together to produce, for the reaction $aA + bB + \cdots \rightarrow cC + dD + \cdots$, the result that the standard reaction enthalpy is given by

$$\Delta_r H^o = c\,\Delta_f H^o(C) + d\,\Delta_f H^o(D) + \cdots - a\,\Delta_f H^o(A) - b\,\Delta_f H^o(B) - \cdots$$

This cycle highlights another aspect of the conventions that we have developed. Note that $\Delta_r H^o$ is the difference between the enthalpies of formation of the ***separated*** products and the enthalpies of formation of the ***separated*** reactants. We often talk about $\Delta_r H^o$ as if it were the enthalpy change that would occur if we mixed a moles of A with b moles of B and the reaction proceeded quantitatively to yield a mixture containing c moles of C and d moles of D. This is usually a good approximation. However, to relate rigorously the standard enthalpy of reaction to the enthalpy change that would occur in a real system in which this reaction took place, it is necessary to recognize that there can be enthalpy changes associated with the pressure–volume changes and with the processes of ***mixing*** the reactants and ***separating*** the products.

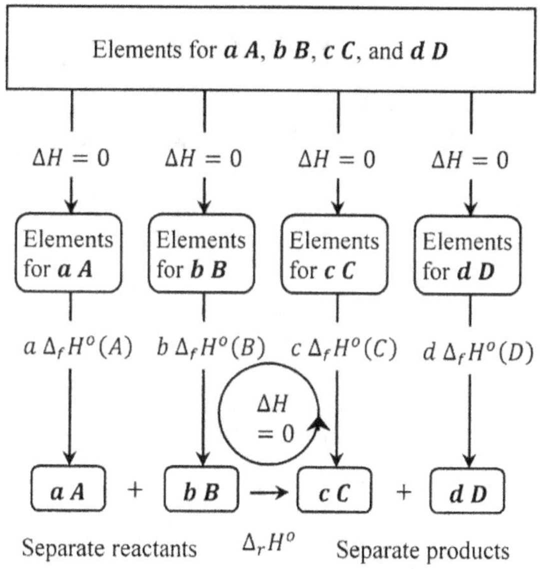

Figure 2. A thermochemical cycle to find $\Delta_r H^o$.

Let us suppose that the reactants and products are gases in their hypothetical ideal-gas states at 1 bar, and that we carry out the reaction by mixing the reactants in a sealed pressure vessel. We suppose that the reaction is then initiated and that the products are formed rapidly, reaching some new pressure and an elevated temperature. (To be specific, we could imagine the reaction be the combustion of methane. We would mix known amounts of methane and oxygen in a pressure vessel and initiate the reaction using an electrical spark.) We allow the temperature to return to the original temperature of the reactants; there is an accompanying pressure change.

Experimentally, we measure the heat evolved as the mixed reactants are converted to the mixed products, at the original temperature. To complete the process corresponding to the standard enthalpy change, however, we must also separate the products and bring them to a pressure of 1 bar. That is, the standard enthalpy of reaction and the enthalpy change we would measure are related by the following sequence of changes, where the middle equation corresponds to the process whose enthalpy change we actually measure.

$$(aA + bB)_{\text{separate reactants at } P=1\text{ bar}} \rightarrow (aA + bB)_{\text{separate reactants at } P} \qquad \Delta H_{\text{compression}}$$

$$(aA + bB)_{\text{separate reactants at } P} \rightarrow (aA + bB)_{\text{homogeneous mixture at } P} \qquad \Delta H_{\text{mixing}}$$

$(aA + bB)_{\text{homogeneous mixture at } P}$
$\rightarrow (cC + dD)_{\text{homogeneous mixture at } P^*}$
$\Delta H_{\text{measured}}$

$(cC + dD)_{\text{homogeneous mixture at } P^*}$
$\rightarrow (cC + dD)_{\text{separate products at } P^*}$
$\Delta H_{\text{separation}}$

$(cC + dD)_{\text{separate products at } P^*}$
$\rightarrow (cC + dD)_{\text{separate products at } P=1 \text{ bar}}$
$\Delta H_{\text{expansion}}$

Summing the reaction equations gives

$(aA + bB)_{\text{separate reactants at } P=1 \text{ bar}}$
$\rightarrow (cC + dD)_{\text{separate products at } P=1 \text{ bar}}$
$\Delta_r H^o$

and summing the enthalpy changes for the series of steps gives the standard enthalpy change for the reaction:

$$\Delta_r H^o = \Delta H_{\text{compression}} + \Delta H_{\text{mixing}} + \Delta H_{\text{measured}} + \Delta H_{\text{separation}} + \Delta H_{\text{expansion}}$$

It turns out that the enthalpy changes for the compression, mixing, separation, and expansion processes are usually small compared to $\Delta_r H^o$. This is the principal justification for our frequent failure to consider them explicitly. For ideal gases, these enthalpy changes are ***identically*** zero. (In Chapter 13, we see that the ***entropy*** changes for the mixing and separation processes are important.)

When we call $\Delta_r H^o$ the standard enthalpy change "for the reaction," we are indulging in a degree of poetic license. Since $\Delta_r H^o$ is a computed difference between the enthalpies of the pure products and those of the pure reactants, the corresponding "reaction" is a purely formal change, which is a distinctly different thing from the real-world process that actually occurs.

§7 Standard state heat capacities

We have observed that C_V depends on volume and temperature, while C_P depends on pressure and temperature. Compilations of heat capacity data usually give values for C_P, rather than C_V. When the temperature-dependence of C_P is known, such compilations usually express it as an empirical polynomial function of temperature. In Chapter 10, we find an explicit function for the dependence of C_P on pressure:

$$\left(\frac{\partial C_P}{\partial P}\right)_T = -T\left(\frac{\partial^2 V}{\partial T^2}\right)_P$$

If we have an equation of state for a substance, we can find this pressure dependence immediately. It is usually negligible. For ideal gases, it is zero, and C_P is independent of pressure.

Compilations often give data for the ***standard state heat capacity***, C_P^o, at a specified temperature. For condensed phases, this is the heat capacity for the substance at one bar. For gases, this is the heat capacity of the substance in its ideal gas standard state.

Table 1. Some Thermochemical Data

	$\Delta_f H^o$ (kJ mol^{-1})		$C_P = a + bT$	
	300 K	400 K	a(J)	b(J K^{-1})
$C(s)$	0	0	−1.482	0.03364
$H_2(g)$	0	0	27.853	0.00332
$O_2(g)$	0	0	27.221	0.00722
$CH_4(g)$	−74.656	−77.703	21.167	0.04866
$CH_3OH(g)$	−201.068	−204.622	21.737	0.07494

Data from reference 1. The heat capacity parameters are calculated from values given at 300 and 400 K.

§8 How the enthalpy change for a reaction depends on temperature

In §6, we see how to use tabulated enthalpies of formation to calculate the enthalpy change for a particular chemical reaction. Such tables typically give enthalpies

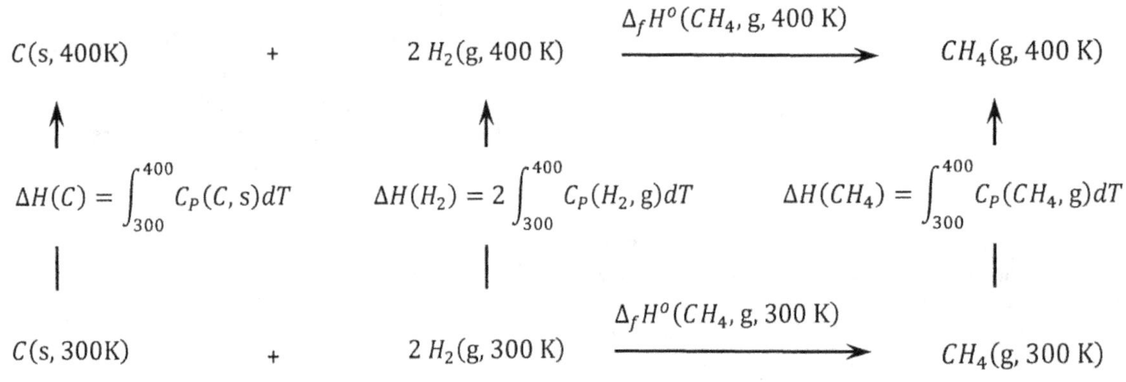

$\Delta_f H^o(CH_4, \text{g}, 400 \text{ K})$
$C(\text{s}, 400\text{K})$ + $2\,H_2(\text{g}, 400 \text{ K})$ → $CH_4(\text{g}, 400 \text{ K})$
$\Delta H(C) = \int_{300}^{400} C_P(C, \text{s})dT$
$\Delta H(H_2) = 2\int_{300}^{400} C_P(H_2, \text{g})dT$
$\Delta H(CH_4) = \int_{300}^{400} C_P(CH_4, \text{g})dT$
$\Delta_f H^o(CH_4, \text{g}, 300 \text{ K})$
$C(\text{s}, 300\text{K})$ + $2\,H_2(\text{g}, 300 \text{ K})$ → $CH_4(\text{g}, 300 \text{ K})$

Figure 3. A thermochemical cycle relating $\Delta_f H^o(CH_4)$ at two temperatures.

of formation at a number of different temperatures, so that the enthalpy change for a given reaction can also be calculated at these different temperatures; it is just a matter of repeating the same calculation at each temperature.

We often need to find the enthalpy change associated with increasing the temperature of a substance at constant pressure. As we observe in §1, this enthalpy change is readily calculated by integrating the heat capacity over the temperature change. We may want to know, for example, the enthalpy change for increasing the temperature of one mole of methane from 300 K to 400 K, with the pressure held constant at one bar. In Table 1, we find

$$\Delta_f H^o(CH_4,\mathrm{g},300\ \mathrm{K}) = -74.656\ \mathrm{kJ\ mol^{-1}}$$

and

$$\Delta_f H^o(CH_4,\mathrm{g},400\ \mathrm{K}) = -77.703\ \mathrm{kJ\ mol^{-1}}$$

We might be tempted to think that the difference represents the enthalpy change associated with heating the methane. This is not so! The reason becomes immediately apparent if we consider a cycle in which we go from the elements to a compound at two different temperatures. For methane, this cycle is shown in Figure 3.

The difference between the standard enthalpies of formation of methane at 300 K and 400 K reflects the enthalpy change for increasing the temperatures of all of the reactants and products from 300 K to 400 K. That is,

$$\Delta_f H^o(CH_4,\mathrm{g},400\ \mathrm{K}) - \Delta_f H^o(CH_4,\mathrm{g},300\ \mathrm{K}) = \int_{300}^{400} C_P(CH_4,\mathrm{g})dT - \int_{300}^{400} C_P(\mathrm{C,s})dT - 2\int_{300}^{400} C_P(H_2,\mathrm{g})dT$$

Over the temperature range from 300 K to 400 K, the heat capacities of carbon, hydrogen, and methane are approximated by $C_P = a + bT$, with values of a and b given in Table 1. From this information, we calculate the enthalpy change for increasing the temperature of one mole of each substance from 300 K to 400 K at 1 bar: $\Delta H(C) = 1{,}029\ \mathrm{J\ mol^{-1}}$, $\Delta H(H_2) = 2{,}902\ \mathrm{J\ mol^{-1}}$, and $\Delta H(CH_4) = 3{,}819\ \mathrm{J\ mol^{-1}}$. Thus, from the cycle, we calculate:

$$\begin{aligned}\Delta_f H^o(CH_4,\mathrm{g},400\ \mathrm{K}) &= -74{,}656 + 3{,}819 - 1{,}029 - 2(2{,}902)\ \mathrm{J\ mol^{-1}}\\ &= -77{,}670\ \mathrm{J\ mol^{-1}}\end{aligned}$$

The tabulated value is $-77{,}703\ \mathrm{J\ mol^{-1}}$. The two values differ by $33\ \mathrm{J\ mol^{-1}}$, or about 0.04%. This difference arises from the limitations of the two-parameter heat-capacity equations.

As another example of a thermochemical cycle, let us consider the selective oxidation of methane to methanol at 300 K and 400 K. From the enthalpies of formation in Table 1, we calculate the enthalpies for the reaction to be $\Delta_r H^o(300\ \mathrm{K}) = -126.412\ \mathrm{kJ\ mol^{-1}}$ and $\Delta_r H^o(400\ \mathrm{K}) = -126.919\ \mathrm{kJ\ mol^{-1}}$. As in the previous example, we use the tabulated heat-capacity parameters to calculate the enthalpy change for increasing the temperature of one mole of each of these gases from 300 K to 400 K at 1 bar. We find: $\Delta H(CH_3OH) = 4{,}797\ \mathrm{J\ mol^{-1}}$, $\Delta H(CH_4) = 3{,}819\ \mathrm{J\ mol^{-1}}$, and $\Delta H(O_2) = 2{,}975\ \mathrm{J\ mol^{-1}}$.

The cycle is shown in Figure 4. Inspecting this cycle, we see that we can calculate the enthalpy change for warming one mole of methanol from 300 K to 400 K by summing the enthalpy changes around the bottom, left side, and top of the cycle; that is,

$$\Delta H(CH_3OH) = 126{,}412 + 3{,}819 + \left(\tfrac{1}{2}\right)2{,}975 - 126{,}919\ \mathrm{J\ mol^{-1}} = 4{,}800\ \mathrm{J\ mol^{-1}}$$

This is 3 J or about 0.06 % larger than the value obtained (4,797 J) by integrating the heat capacity for methanol.

§9 Calorimetry

Calorimetry is the experimental science of measuring the heat changes that accompany chemical or physical changes. The accurate measurement of small amounts of heat is experimentally challenging. Nevertheless, calorimetry is an area in which great

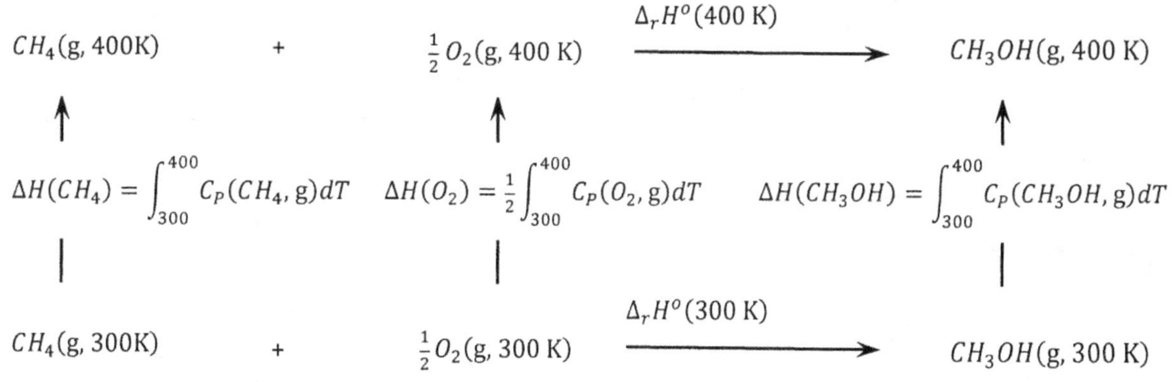

Figure 4. A thermochemical cycle relating $\Delta_r H^o$ at two temperatures.

experimental sophistication has been achieved and remarkably accurate measurements can be made. Numerous devices have been developed to measure heat changes.

Some of these devices measure a (usually small) temperature change. Such devices are calibrated by measuring how much their temperature increases when a known amount of heat is introduced. This is usually accomplished by passing a known electric current through a known resistance for a known time. Other calorimeters measure the amount of some substance that undergoes a phase change. The ice calorimeter is an important example of the latter method. In an ice calorimeter, the heat of the process is transferred to a mixture of ice and water. The amount of ice that melts is a direct measure of the amount of heat released by the process. The amount of ice melted can be determined either by direct measurement of the increase in the amount of water present or by measuring the change in the volume of the ice–water mixture. (Since ice occupies a greater volume than the same mass of water, melting is accompanied by a decrease in the total volume occupied by the mixture of ice and water.)

The processes that can be investigated accurately using calorimetry are limited by two important considerations. One is that the process must go to completion within a relatively short time. No matter how carefully it is constructed, any calorimeter will exchange thermal energy with its environment at some rate. If this rate is not negligibly small compared to the rate at which the process evolves heat, the accuracy of the measurement is degraded. The second limitation is that the process must involve complete conversion of the system from a known initial state to a known final state. When the processes of interest are chemical reactions, these considerations mean that the reactions must be quantitative and fast.

Combustion reactions and catalytic hydrogenation reactions usually satisfy these requirements, and they are the most commonly investigated. However, even in these cases, there can be complications. For a compound containing only carbon, hydrogen, and oxygen, combustion using excess oxygen produces only carbon dioxide and water. For compounds containing heteroatoms like nitrogen, sulfur, or phosphorus, there may be more than one heteroatom-containing product. For example, combustion of an organosulfur compound might produce both sulfur dioxide and sulfur trioxide. To utilize the thermochemical data obtained in such an experiments, a chemical analysis must be done to determine the amount of each oxide present.

Problems

1. One mole of an ideal gas reversibly traverses Cycle I below. Step a is isothermal. Step b is isochoric (constant volume). Step c is isobaric (constant pressure). Assume C_V and C_P are constant. Find q, w, ΔE, and ΔH for each step and for the cycle. Prove $C_P = C_V + R$.

2. One mole of an ideal gas reversibly traverses Cycle II below. Step a is the same isothermal process as in problem 1. Step d is adiabatic. Step e is isobaric. Assume C_V and C_P are constant. Find q, w, ΔE, and ΔH for each step and for the cycle.

3. One mole of an ideal gas reversibly traverses Cycle III below. Step a is the same isothermal process as in problem 1. Step f is adiabatic. Step g is isochoric. Assume C_V and C_P are constant. Find q, w, ΔE, and ΔH for each step and for the cycle.

4. One mole of an ideal gas reversibly traverses Cycle IV. Step h is isobaric. Step f is the same adiabatic process as in problem 3. Step i is isochoric. Assume C_V and C_P are constant. Find q, w, ΔE, and ΔH for each step and for the cycle.

5. Prove that the work done on the system is positive when the system traverses Cycle I. Note that Cycle I traverses the region of the PV plane that it encloses in a counter-clockwise direction. Hint: Note that $T_2 < T_1$. Show that $V_2/V_1 = T_2/T_1$.

6. Cycles III and IV share a common adiabatic step. Express the work done in each of these cycles in terms of V_1, V_2, and T_1. Prove that the work done in Cycle IV is greater than the work done in Cycle III.

7. Cycles I, II, and III share a common first step, a. Express V_3, T_3, and T_4 in terms of V_1, V_2, and T_1. For $V_1 = 10$ L, $V_2 = 2$ L, and $T_1 = 400$ K, show that the work done decreases in the order Cycle I > Cycle III > Cycle II.

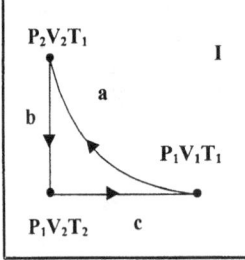

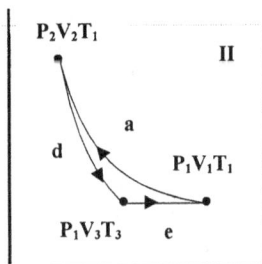

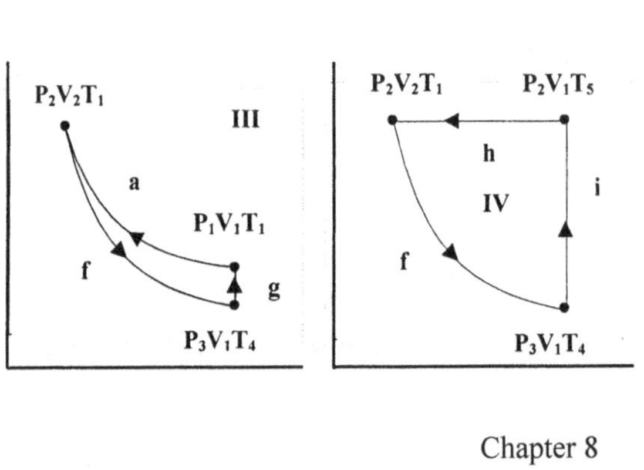

8. For water, the enthalpies of fusion and vaporization are 6.009 and 40.657 kJ mol^{-1}, respectively. The heat capacity of liquid water varies only weakly with temperature and can be taken as 75.49 J mol^{-1} K^{-1}. The heat capacity of water vapor varies with temperature:

$$C_P(H_2O, g) = 30.51 + (1.03 \times 10^{-2})T$$

where T is in degrees K and the heat capacity is in J mol^{-1} K^{-1}. Estimate the enthalpy of sublimation of water.

9. If we truncate the virial equation ($Z = 1 + B^*(T)P + \cdots$) and make use of $B(T) = RTB^*(T)$, where $B(T)$ is the "second virial coefficient" most often given in data tables, the molar volume is

$$\overline{V} = \frac{RT}{P} + B(T)$$

Show that

$$\left(\frac{\partial H}{\partial P}\right)_T = B(T) - T\left(\frac{dB}{dT}\right)$$

The Handbook of Chemistry and Physics (CRC Press, 79th Ed., 1999, p. **6**–25) gives the temperature dependence of B for water vapor as

$$B = -1158 - 5157t - 10301t^2 - 10597t^3 - 4415t^4$$

where $t = (298.15/T) - 1$, T is in degrees kelvin, and the units of B are cm^{-3} mol^{-1}. Estimate the enthalpy change when one mole of water vapor at 1 atm and 100 C is expanded to the equilibrium sublimation pressure, which for this purpose we can approximate as the triple-point pressure, 610 Pa. How does this value compare to the result of problem 8?

10. The heat capacities of methanol liquid and gas are 81.1 and 44.1 J mol^{-1} K^{-1}, respectively. The second virial coefficient for methanol vapor is

$$B = -1752 - 4694t$$

where $t = (298.15/T) - 1$, T is in degrees kelvin, and the units of B are cm^{-3} mol^{-1}. Referring to the discussion of methanol vaporization in §5, calculate $\Delta_{(1)}H$, $\Delta_{(4)}H$, $\Delta_{(5)}H$, $\Delta_{(vap)}H^o$. Compare this value of $\Delta_{(vap)}H^o$ to the value given in the text. [Data from the ***Handbook of Chemistry and Physics***, CRC Press, 79th Ed., 1999, p. **5**-27 and p. **6**-31.]

Some Enthalpies of Formation at 298 K

Data from the ***Handbook of Chemistry and Physics***, CRC Press, 79th Ed., 1999, pp. 5-5 and following.

Molecular formula	Name	$\Delta_f H^o$ (kJ mol^{-1})
H_2O (liq)	Water	−285.8
CO (g)	Carbon monoxide	−110.5
CO_2 (g)	Carbon dioxide	−393.5
CH_4(g)	Methane	−74.6
C_2H_4(g)	Ethylene	52.4
C_2H_6(g)	Ethane	−84.0
CH_3CH_2OH (liq)	Ethanol	−277.6
CH_3CHO (liq)	Acetaldehyde	−192.2
CH_3CO_2H (liq)	Acetic acid	−484.3
CH_3CH_2CHO (liq)	Propanal	−215.6
C_6H_6 (liq)	Benzene	49.1
$C_6H_5CO_2H$ (s)	Benzoic acid	−385.2

11. Using data from the table above, find the enthalpy change for each of the following reactions at 298 K.

(a) $C_2H_6(g) + \frac{1}{2}O_2(g) \rightarrow CH_3CH_2OH(liq)$

(b) $C_2H_4(g) + \frac{1}{2}O_2(g) \rightarrow CH_3CHO(liq)$

(c) $C_2H_6(g) + \frac{1}{2}O_2(g) \rightarrow CH_3CHO(liq) + H_2O(liq)$

(d) $C_6H_6(liq) + CO_2(g) \rightarrow C_6H_5CO_2H(s)$

(e) $CH_3CHO(liq) + \frac{1}{2}O_2(g) \rightarrow CH_3CO_2H(liq)$

(f) $CH_4(g) + H_2O(liq) \rightarrow CO(g) + 3\,H_2(g)$

(g) $CH_4(g) + H_2O(liq) + \frac{1}{2}O_2(g) \rightarrow CO_2(g) + 3\,H_2(g)$

(h) $C_2H_4(g) + CO(g) + H_2(g) \rightarrow CH_3CH_2CHO(liq)$

Notes

[1] Data compiled by The Committee on Data for Science and Technology (CODATA) and reprinted in D. R. Linde, Editor, ***The Handbook of Chemistry and Physics***, 79th Edition (1998-1999), CRC Press, §5.

[2] D. R. Linde, *op. cit.*, p. **6**-104.

9

The Second Law: Entropy and Spontaneous Change

§1 The second law of thermodynamics

The first law of thermodynamics is concerned with energy and its properties. As we saw in Chapter 7, the first law arose from the observation that the dissipation of mechanical work through friction creates heat. In a synthesis that was partly definition and partly a generalization from experience, it was proposed that mechanical energy and heat are manifestations of a common quantity, energy. Later, by further definition and generalization, the concept was expanded to include other forms of energy. The energy concept evolved into the prescript that there exists a quantity (state function) that is conserved through any manner of change whatsoever.

The element of definition arises from the fact that we recognize new forms of energy whenever necessary in order to ensure that the conservation condition is satisfied. The element of experience arises from the fact that this prescript has resulted in a body of theory and a body of experimental results that are mutually compatible. When we define and measure energy "correctly" we do indeed find that energy is a state function and that it is conserved.

The theory of relativity introduced a significant expansion of the energy concept. For chemical processes, we can view mass and energy conservation as independent postulates. For processes in which fundamental particles undergo changes and for systems moving at velocities near that of light, we cannot. Relativity asserts that the energy of a particle is given by Einstein's equation, $E^2 = p^2c^2 + m_0^2c^4$. In this equation, E is the particle energy, p is its momentum, m_0 is its rest mass, and c is the speed of light. In transformations of fundamental particles in which the sum of the rest masses of the product particles is less than that of the reactant particles, conservation of energy requires that the sum of the momenta of the product particles exceed that of the reactant particles. The momentum increase means that the product particles have high velocities, corresponding to a high temperature for the product system. The most famous expression of this result is that $E = m_0c^2$, meaning that we can associate this quantity of energy with the mass, m_0, of a stationary particle, for which $p = 0$.

The situation with respect to the second law is similar. From experience with devices that convert heat into work, the idea evolved that such devices must have particular properties. Consideration of these properties led to the discovery of a new state function, which we call entropy, and to which we customarily assign the symbol "S". We introduce the laws of thermodynamics in §6-13. We repeat our statement of the second law here:

The second law of thermodynamics
In a reversible process in which a closed system accepts an increment of heat, dq^{rev}, from its surroundings, the change in the entropy of the system, dS, is $dS = dq^{rev}/T$. Entropy is a state function. For any reversible process, $dS_{universe} = 0$, and conversely. For any spontaneous process, $dS_{universe} > 0$, and conversely.

If a spontaneous process takes a system from state A to state B, state B may or may not be an equilibrium state. State A cannot be an equilibrium state. Since we cannot use the defining equation to find the entropy change for a spontaneous process, we must use some other method if we are to estimate the value of the entropy change. This means that we must have either an empirical mathematical model from which we can estimate the entropy of a non-equilibrium state or an equilibrium system that is a good model for the initial state of the spontaneous process.

We can usually find an equilibrium system that is a good model for the initial state of a spontaneous process. Typically, some alteration of an equilibrium system makes the spontaneous change possible. The change-enabled state is the initial state for a spontaneous process, but its thermodynamic state functions are essentially identical to those of the pre-alteration equilibrium state. For example, suppose that a solution contains the reactants and products for some reaction that occurs only in the presence of a catalyst. In this case, the solution can be effectively at equilibrium even when the composition does not correspond to an equilibrium position of the reaction. (In an effort to be more precise, we can term this a quasi-equilibrium state, by which we mean that the system is unchanging even though a spontaneous change is possible.) If we introduce a very small quantity of catalyst, and consider the state of the system before any reaction occurs, all of the state functions that characterize the system must be essentially unchanged. Nevertheless, as soon as the catalyst is introduced, the system can no longer be considered to be in an equilibrium state. The spontaneous reaction proceeds until it reaches equilibrium. We can find the entropy change for the spontaneous process by finding the entropy change for a reversible process that takes the initial, pre-catalyst, quasi-equilibrium state to the final, post-catalyst, equilibrium state.

Our statement of the second law establishes the properties of entropy by postulate. While this approach is rigorously logical, it does not help us understand the ideas involved. Like the first law, the second law can be stated several ways. To develop our understanding of entropy and its properties, it is useful to again consider a more traditional statement of the second law:

A traditional statement of the second law
It is impossible to construct a machine that operates in a cycle, exchanges heat with its surroundings at only one temperature, and produces work in the surroundings.

When we introduce the qualification that the machine "exchanges heat with its surroundings at only one temperature," we mean that the temperature of the surroundings has a particular value whenever the machine and surroundings exchange heat. The statement does not place any conditions on the temperature of the machine at any time.

In this chapter, we have frequent occasion to refer to each of these statements. To avoid confusing them, we will refer to our statement of the second law as the ***entropy-based*** statement. We will refer to the statement above as the ***machine-based*** statement of the second law.

By "a machine", we mean a heat engine—a device that accepts heat and produces mechanical work. This statement asserts that a "perpetual motion machine of the second kind" cannot exist. Such a machine accepts heat energy and converts all of it into work, while itself returning to the same state at the end of each cycle. (In §7-11, we note that a "perpetual motion machine of the first kind" is one whose operation violates the principle of conservation of energy.) Normally, we view this statement as a postulate. We consider that we infer it from experience. Unlike our statements about entropy, which are entirely abstract, this statement makes an assertion about real machines of the sort that we encounter in daily life. We can understand the assertion that it makes in concrete terms: A machine that could convert heat from a constant-temperature source into work could extract heat from ice water, producing ice cubes in the water and an equivalent amount of work elsewhere in the surroundings. This machine would not exchange heat with any other heat reservoir. Our machine-based statement of the second law postulates that no such machine can exist.

Our entropy-based statement of the second law arose from thinking about the properties of machines that do convert heat into work. We trace this thinking to see how our entropy-based statement of the second law was developed. Understanding this development gives us a better appreciation for the meaning of entropy. We find that we must supplement the machine-based statement of the second law with additional assumptions in order to arrive at all of the properties of the entropy function that are asserted in the entropy-based statement.

However, before we undertake to develop the entropy-based statement of the second law from the machine-based statement, let us develop the converse; that is, let us show that the machine-based statement is a logical consequence of the entropy-based statement. To do so, we assume that a perpetual motion machine of the second kind is possible. To help keep our argument clear, let proposition MSL be the machine-based statement. We are assuming that proposition MSL is false, so that proposition ~MSL is true. We let SL be the entropy-based statement of the second law.

The sketch in Figure 1 describes the interaction of this perpetual motion machine, PPM, with its surroundings. From our entropy-based statement of the second law, we can assert some important facts about the entropy changes that accompany operation of the machine. Since entropy is a state function, $\Delta S = 0$ for one cycle of the machine. If the machine works (that is, ~MSL is true), then the entropy-based statement requires that $\Delta S_{universe} = \Delta S + \Delta \hat{S} \geq 0$. Since $\Delta S = 0$, it follows that $\Delta \hat{S} \geq 0$. We can make this more explicit by writing: (SL and ~MSL) $\Rightarrow \Delta \hat{S} \geq 0$.

The machine-based statement of the second law also enables us to determine the entropy change in the surroundings from our second-law definition of entropy. In one cycle, this machine (system) delivers net work, $\hat{w} > 0$, to the surroundings; it accepts a net quantity of heat, $q > 0$, from the surroundings, which are at temperature, $\hat{T}$. Simultaneously, the surroundings surrender a quantity of heat, $\hat{q}$, where $\hat{q} = -q$, and $\hat{q} < 0$. The change that occurs in one cycle of the machine need not be reversible. However, whether the change is reversible or not, the entire thermal change in the surroundings consists in the exchange of an amount of heat, $\hat{q} < 0$, by a constant temperature reservoir at $\hat{T}$. We can effect identically the same change in the surroundings using some other process to reversibly extract this amount of heat. The entropy change in the surroundings in this reversible process will be $\hat{q}/\hat{T}$, and this will be the same as the entropy change for the surroundings in one cycle of the machine. (We consider this conclusion further in §15.) It follows that $\Delta \hat{S} = \hat{q}/\hat{T}$, and since $\hat{q} < 0$, while $\hat{T} > 0$, we have $\Delta \hat{S} < 0$. We can write this conclusion more explicitly: (SL and ~MSL) $\Rightarrow \Delta \hat{S} < 0$.

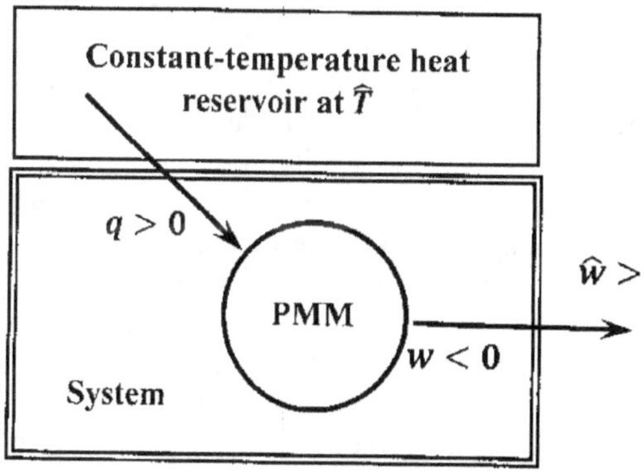

Figure 1. A perpetual motion machine (PPM) that violates the second law.

By assuming a perpetual motion machine of the second kind is possible—that is, by assuming ~MSL is true—we derive the contradiction that both $\Delta\hat{S} \geq 0$ and $\Delta\hat{S} < 0$. Therefore, proposition ~MSL must be false. Proposition MSL must be true. The entropy-based second law of thermodynamics implies that a perpetual motion machine of the second kind is not possible. That is, the entropy-based statement of the second law implies the machine-based statement. (We prove that ~(SL and ~MSL); it follows that SL ⇒MSL. For a more detailed argument, see problem 2.)

§2 The Carnot cycle for an ideal gas and the entropy concept

Historically, the steam engine was the first machine for converting heat into work that could be exploited on a large scale. The steam engine played a major role in the industrial revolution and thus in the development of today's technology-intensive economy. It was important also in the development of the basic concepts of thermodynamics. A steam engine produces work when hot steam under pressure is introduced into a cylinder, driving a piston outward. A shaft connects the piston to a flywheel. When the connecting shaft reaches its greatest extension, the spent steam is vented to the atmosphere. Thereafter the flywheel drives the piston inward.

The economic viability of the steam engine derives, in part, from the fact that the spent steam can be vented to the atmosphere at the end of each cycle. However, this is not a necessary feature of heat engines. We can devise engines that alternately heat and cool a captive working fluid to convert heat energy into mechanical work. Stirling engines are practical devices of this type. A Carnot engine is a conceptual engine that exploits the response of a closed system to temperature changes. A ***Carnot engine*** extracts heat from one reservoir at a fixed high temperature and discharges a lesser amount of heat into a second reservoir at a fixed lower temperature. An amount of energy equal to the difference between these increments of heat energy appears in the surroundings as work.

For one cycle of the Carnot engine, let the heat transferred to the system from the hot and cold reservoirs be q_h and q_ℓ respectively. We have $q_h > 0$ and $q_\ell < 0$. Let the net work done on the system be w_{net} and the net work that appears in the surroundings be $\hat{w}_{net}$. We have $\hat{w}_{net} > 0$, $\hat{w}_{net} = -w_{net}$, and $w_{net} < 0$. For one cycle of the engine, $\Delta E = 0$, and since $\Delta E = q_h + q_\ell + w_{net} = q_h + q_\ell - \hat{w}_{net}$, it follows that $\hat{w}_{net} = q_h + q_\ell$. The energy input to the Carnot engine is q_h, and the useful work that appears in the surroundings is $\hat{w}_{net}$. (The heat accepted by the low-temperature reservoir, $\hat{q}_\ell = -q_\ell > 0$, is a waste product, in the sense that it represents energy that cannot be converted to mechanical work using this cycle. All feasible heat engines share this feature of the Carnot engine. In contrast, a perpetual motion machine of the second kind converts its entire heat intake to work; no portion of its heat intake goes unused.) The efficiency, ϵ, with which the Carnot engine converts the input energy, q_h, to *useful* output energy, $\hat{w}_{net}$, is therefore,

$$\epsilon = \frac{\hat{w}_{net}}{q_h} = \frac{q_h + q_\ell}{q_h} = 1 + \frac{q_\ell}{q_h}$$

We can generalize our consideration of heat engines to include any series of changes in which a closed system exchanges heat with its surroundings at more than one temperature, delivers a positive quantity of work to the surroundings, and returns to its original state. We use the Carnot cycle and the machine-based statement of the second law to analyze systems that deliver pressure–volume work to the surroundings. We consider both reversible and irreversible systems. We begin by considering reversible Carnot cycles. If any system reversibly traverses any closed path on a pressure–volume diagram, the area enclosed by the path represents the pressure–volume work exchanged between the system and its surroundings. If the area is not zero, the system temperature changes during the cycle. If the cycle is reversible, all of the heat transfers that occur must occur reversibly. **We can apply our reasoning about reversible cycles to any closed system containing any collection of chemical substances, so long as any phase changes or chemical reactions that occur do so reversibly.** This means that all phase and chemical changes that occur in the system must adjust rapidly to the new equilibrium positions that are imposed on them as a system traverses a Carnot cycle reversibly.

In Figure 2, we describe the operation of a reversible Carnot engine in which the working fluid is an ideal gas. We designate the system's initial pressure, volume, and temperature by P_1, V_1, and T_h. From this initial state, we cause the ideal gas to undergo a reversible isothermal expansion in which it absorbs a quantity of heat, q_h, from a high-temperature heat reservoir at $\hat{T}_h$. We designate the

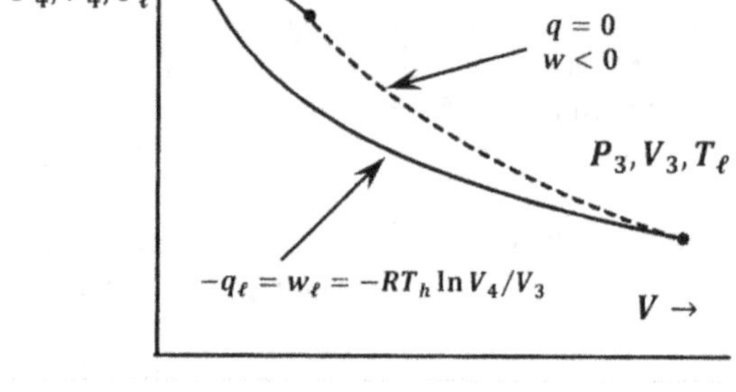

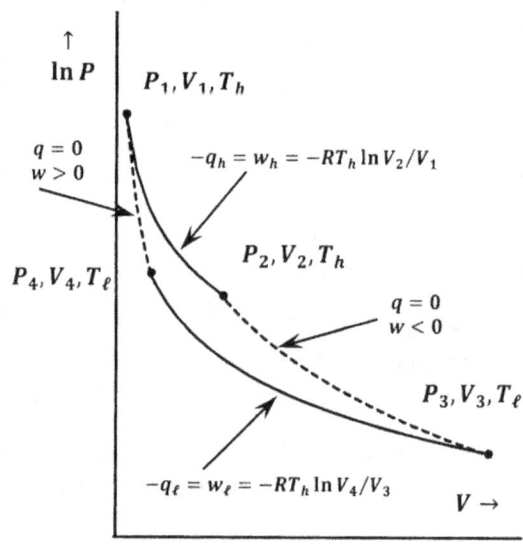

Figure 2. An ideal gas Carnot cycle. Note that the pressure axis is compressed: ln *P* is plotted vs. *V*.

pressure, volume, and temperature at the end of this isothermal expansion as P_2, V_2, and T_h. In a second step, we reversibly and adiabatically expand the ideal gas until its temperature falls to that of the second, low-temperature, heat reservoir. We designate the pressure, volume, and temperature at the end of this adiabatic expansion as P_3, V_3, and T_ℓ. We begin the return portion of the cycle by reversibly and isothermally compressing the ideal gas at the temperature of the cold reservoir. We continue this reversible isothermal compression until the ideal gas reaches the pressure and volume from which an adiabatic compression will just return it to the initial state. We designate the pressure, volume, and temperature at the end of this isothermal compression by P_4, V_4, and T_ℓ. During this step, the ideal gas gives up a quantity of heat, $q_\ell < 0$, to the low-temperature reservoir. Finally, we reversibly and adiabatically compress the ideal gas to its original pressure, volume, and temperature.

For the high-temperature isothermal step, we have

$$-q_h = w_h = -RT_h \ln\left(\frac{V_2}{V_1}\right)$$

and for the low-temperature isothermal step, we have

$$-q_\ell = w_\ell = -RT_\ell \ln\left(\frac{V_4}{V_3}\right)$$

For the adiabatic expansion and compression, we have

$$q_{exp} = q_{comp} = 0$$

The corresponding energy and work terms are

$$\Delta_{exp}E = w_{exp} = \int_{T_h}^{T_\ell} C_V dT$$

for the adiabatic expansion and

$$\Delta_{comp}E = w_{comp} = \int_{T_\ell}^{T_h} C_V dT$$

for the adiabatic compression. The heat-capacity integrals are the same except for the direction of integration; they sum to zero, and we have $w_{exp} + w_{comp} = 0$. The net work done on the system is the sum of the work for these four steps, $w_{net} = w_h + w_{exp} + w_\ell + w_{comp} = w_h + w_\ell$. The heat input occurs at the high-temperature reservoir, so that $q_h > 0$. The heat discharge occurs at the low-temperature reservoir, so that $q_\ell < 0$.

For one cycle of the reversible, ideal-gas Carnot engine,

$$\epsilon = 1 + \frac{q_\ell}{q_h} = 1 + \frac{RT_\ell \ln(V_4/V_3)}{RT_h \ln(V_2/V_1)}$$

Because the two adiabatic steps involve the same limiting temperatures, the energy of an ideal gas depends only on temperature, and $dE = dw$ for both steps, we see from §7-20 that

$$\int_{T_h}^{T_\ell} \frac{C_V}{T} dT = -\int_{V_2}^{V_3} \frac{R}{V} dV = -R \ln\left(\frac{V_3}{V_2}\right)$$

and

$$\int_{T_\ell}^{T_h} \frac{C_V}{T} dT = -\int_{V_4}^{V_1} \frac{R}{V} dV = -R \ln\left(\frac{V_1}{V_4}\right)$$

The integrals over T are the same except for the direction of integration. They sum to zero, so that $-R\ln(V_3/V_2) - R\ln(V_1/V_4) = 0$ and

$$\frac{V_2}{V_1} = \frac{V_3}{V_4}$$

Using this result, the second equation for the reversible Carnot engine efficiency becomes

$$\epsilon = 1 - \frac{T_\ell}{T_h}$$

Equating our expressions for the efficiency of the reversible Carnot engine, we find

$$\epsilon = 1 + \frac{q_\ell}{q_h} = 1 - \frac{T_\ell}{T_h}$$

from which we have

$$\frac{q_h}{T_h} + \frac{q_\ell}{T_\ell} = 0$$

Since there is no heat transfer in the adiabatic steps, $q_{exp} = q_{comp} = 0$, and we can write this sum as

$$\sum_{cycle} \frac{q_i}{T_i} = 0$$

If we divide the path around the cycle into a large number of very short segments, the limit of this sum as the q_i become very small is

$$\oint \frac{dq^{rev}}{T} = 0$$

where the superscript "rev" serves as a reminder that the cycle must be traversed reversibly. Now, we can define a new function, S, by the differential expression

$$dS = \frac{dq^{rev}}{T}$$

In this expression, dS is the incremental change in S that occurs when the system reversibly absorbs a small of increment of heat, dq^{rev}, at a particular temperature,

T. For an ideal gas traversing a Carnot cycle, we have shown that

$$\Delta S = \oint dS = \oint \frac{dq^{rev}}{T} = 0$$

S is, of course, the entropy function described in our entropy-based statement of the second law.

We now want to see what the machine-based statement of the second law enables us to deduce about the properties of S. Since the change in S is zero when an ideal gas goes around a complete Carnot cycle, we can conjecture that S is a state function. Of course, the fact that $\Delta S = 0$ around one particular cycle does not prove that S is a state function. If S is a state function, it must be true that $\Delta S = 0$ around any cycle whatsoever. We now prove this for any reversible cycle.

The proof has two steps. In the first, we show that $\oint dq^{rev}/T = 0$ for a machine that uses any reversible system operating between two constant-temperature heat reservoirs to convert heat to work. In the second step, we show that $\oint dq^{rev}/T = 0$ for any system that reversibly traverses any closed path.

§3 The Carnot cycle for any reversible system

To show that $\oint dq^{rev}/T = 0$ for any reversible system taken around a Carnot cycle, we first observe that the Carnot cycle can be traversed in the opposite direction. In this case, work is delivered to the engine and a quantity of heat is transferred from the low-temperature reservoir to the high-temperature reservoir. Operated in reverse, the Carnot engine is a refrigerator. Suppose that we have two identical ideal-gas Carnot machines, one of which we operate as an engine while we operate the other as a refrigerator. If we configure them so that the work output of the engine drives the refrigerator, the effects of operating them together cancel completely. The refrigerator exactly consumes the work output of the engine. The heat transfers to and from the heat reservoirs offset exactly.

Now, let us consider an ideal-gas Carnot engine and any other reversible engine that extracts heat from a high-temperature reservoir and rejects a portion of it to a low-temperature reservoir. Let us call these engines A and B. We suppose that one is operated to produce work in its surroundings ($w < 0$); the other is operated to consume this work and transfer net heat energy from the low-temperature to the high-temperature reservoir. Let the net work done in one cycle on machines A and B be w_{netA} and w_{netB}, respectively. We can choose to make these engines any size that we please. Let us size them so that one complete cycle of either engine exchanges the same quantity of heat with the high-temperature reservoir. That is, if the high-temperature reservoir delivers heat q_{hA} to engine A, then it delivers heat $q_{hB} = q_{hA}$ to engine B. Figure 3 diagrams these engines. With one operating as an engine and the other operating as a refrigerator, we have $q_{hA} + q_{hB} = 0$. When both engine and refrigerator have completed a cycle, the high temperature reservoir has returned to its original state.

We can create a combined device that consists of A running as an engine, B running as a refrigerator, and the high-temperature reservoir. Figure 3 also diagrams this combination. When it executes one complete cycle, the initial condition of the combined device is restored. Therefore, since E is a state function, we have

$$\begin{aligned}\Delta E &= w_{netA} + q_{hA} + q_{\ell A} + w_{netB} + q_{hB} + q_{\ell B} \\ &= w_{netA} + w_{netB} + q_{\ell A} + q_{\ell B} \\ &= 0.\end{aligned}$$

where we use the constraint $q_{hA} + q_{hB} = 0$. Let us consider the possibility that $w_{netA} + w_{netB} < 0$; that is, the combined device does net work on the surroundings. Then, $\Delta E = 0$ implies that $q_{\ell A} + q_{\ell B} > 0$.

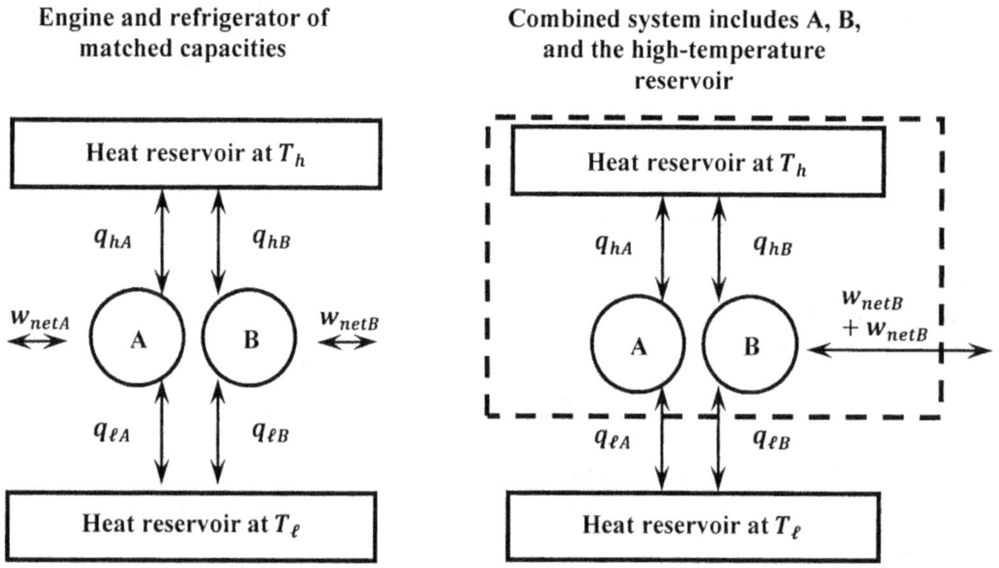

Figure 3. Matched heat engine and refrigerator and a system that combines them.

In this cyclic process, the combined device takes up a positive quantity of heat from a constant-temperature reservoir and delivers a positive quantity of work to the surroundings. There is no other change in either the system or the surroundings. This violates the machine-based statement of the second law. Evidently, it is not possible for the combined device to operate in the manner we have hypothesized. We conclude that any such machine must always operate such that $w_{netA} + w_{netB} \geq 0$; that is, the net work done on the combined machine during any complete cycle must be either zero or some positive quantity.

In concluding that $w_{netA} + w_{netB} \geq 0$, we specify that the combined machine has A running as a heat engine and B running as a refrigerator. Now, suppose that we reverse their roles, and let w^*_{netA} and w^*_{netB} represent the net work for the reversed combination. Applying the same argument as previously, we conclude that $w^*_{netA} + w^*_{netB} \geq 0$. But, since the direction of operation is reversed for both machines, we must also have $w^*_{netA} = -w_{netA}$ and $w^*_{netB} = -w_{netB}$. Hence we have $-w_{netA} - w_{netB} \geq 0$ or $w_{netA} + w_{netB} \leq 0$. We conclude, therefore, that

$$w_{netA} + w_{netB} = 0$$

for any two, matched, reversible engines operating around a Carnot cycle.

This conclusion can be restated as a condition on the efficiencies of the two machines. The individual efficiencies are $\epsilon_A = -w_{netA}/q_{hA}$ and $\epsilon_B = -w_{netB}/q_{hB}$. (The efficiency equation is unaffected by the direction of operation, because changing the direction changes the sign of every energy term in the cycle. Changing the direction of operation is equivalent to multiplying both the numerator and denominator by minus one.) Then, from $w_{netA} + w_{netB} = 0$, it follows that

$$\epsilon_A q_{hA} + \epsilon_B q_{hB} = 0$$

Since we sized A and B so that $q_{hA} + q_{hB} = 0$, we have

$$\epsilon_A q_{hA} - \epsilon_B q_{hA} = 0$$

so that

$$\epsilon_A = \epsilon_B$$

for any reversible Carnot engines A and B operating between the same two heat reservoirs.

For the ideal gas engine, we found $\epsilon = 1 - T_\ell/T_h$. For any reversible Carnot engine, we have $\Delta E = 0 = w_{net} + q_h + q_\ell$, so that $-w_{net} = q_h + q_\ell$, and

$$\epsilon = \frac{-w_{net}}{q_h} = 1 + \frac{q_\ell}{q_h}$$

This means that the efficiency relationship

$$\epsilon = 1 - \frac{T_\ell}{T_h} = 1 + \frac{q_\ell}{q_h}$$

applies to any reversible Carnot engine. It follows that the integral of dq^{rev}/T around a Carnot cycle is zero for any reversible system.

The validity of these conclusions is independent of type of work that the engine produces; if engine A is an ideal-gas engine, engine B can be comprised of any system and can produce any kind of work. In obtaining this result from the machine-based statement of the second law, we make the additional assumption that pressure–volume work can be converted entirely to any other form of work, and *vice versa.* That is, we assume that the work produced by engine A can reversibly drive engine B as a refrigerator, whether engines A and B produce the same or different kinds of work.

§4 The entropy change around any cycle for any reversible system

Any system reversibly traversing any closed curve on a pressure–volume diagram exchanges work with its surroundings, and the area enclosed by the curve represents the amount of this work. In the previous section, we found $\oint dq^{rev}/T = 0$ for any system that traverses a Carnot cycle reversibly. We now show that this is true for any system that traverses any closed path reversibly. This establishes that ΔS is zero for any system traversing any closed path reversibly and proves that S, defined by $dS = dq^{rev}/T$, is a state function.

To do so, we introduce an experience-based theorem: The pressure–volume diagram for any reversible system can be tiled by intersecting lines that represent isothermal and adiabatic paths. These lines can be packed as densely as we please, so that the tiling of the pressure–volume diagram can be made as closely spaced as we please. The perimeter of any one of the resulting tiles corresponds to a path around a Carnot cycle. Given any arbitrary closed curve on the pressure–volume diagram, we can select a set of tiles that just encloses it. See Figure 4. The perimeter of this set of tiles approximates

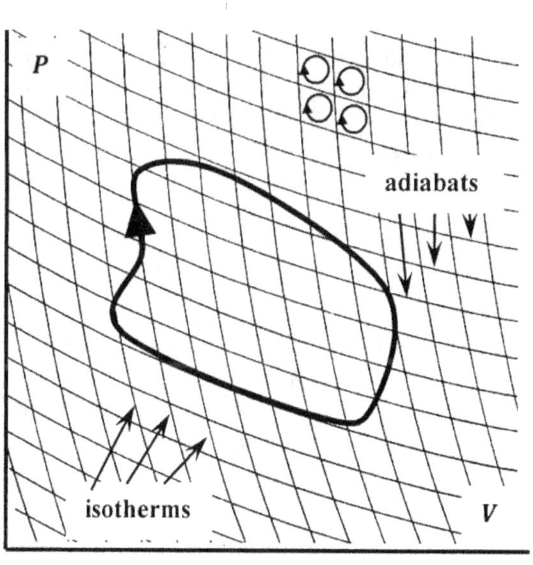

Figure 4. Tiling the PV-plane with isotherms and adiabats.

the path of the arbitrary curve. Since the tiling can be made as fine as we please, the perimeter of the set of tiles can be made to approximate the path of the arbitrary curve as closely as we please.

Suppose that we traverse the perimeter of each of the individual tiles in a clockwise direction, adding up q^{rev}/T as we go. Segments of these perimeters fall into two groups. One group consists of segments that are on the perimeter of the enclosing set of tiles. The other group consists of segments that are common to two tiles. When we traverse both of these tiles in a clockwise direction, the shared segment is traversed once in one direction and once in the other. When we add up q^{rev}/T for these two traverses of the same segment, we find that the sum is zero, because we have q^{rev}/T in one direction and $-q^{rev}/T$ in the other. This means that the sum of q^{rev}/T around all of the tiles will just be equal to the sum of q^{rev}/T around those segments that lie on the perimeter of the enclosing set. That is, we have

$$\sum_{\substack{\text{cycle}\\\text{perimeter}}} \frac{q^{rev}}{T} + \sum_{\substack{\text{interior}\\\text{segments}}} \frac{q^{rev}}{T} = \sum_{\substack{\text{all}\\\text{tiles}}} \left\{ \sum_{\substack{\text{tile}\\\text{perimeter}}} \frac{q^{rev}}{T} \right\}$$

where

$$\sum_{\substack{\text{interior}\\\text{segments}}} \frac{q^{rev}}{T} = 0$$

because each interior segment is traversed twice, and the two contributions cancel exactly.

This set of tiles has another important property. Since each individual tile represents a reversible Carnot cycle, we know that

$$\sum_{\substack{\text{tile}\\\text{perimeter}}} \frac{q^{rev}}{T} = 0$$

around each individual tile. Since the sum around each tile is zero, the sum of all these sums is zero. It follows that the sum of q^{rev}/T around the perimeter of the enclosing set is zero:

$$\sum_{\substack{\text{cycle}\\\text{perimeter}}} \frac{q^{rev}}{T} = 0$$

By tiling the pressure–volume plane as densely as necessary, we can make the perimeter of the enclosing set as close as we like to any closed curve. The heat increments become arbitrarily small, and

$$\lim_{q^{rev} \to dq^{rev}} \left[\sum_{\substack{cycle\\perimeter}} \frac{q^{rev}}{T} \right] = \oint \frac{dq^{rev}}{T} = 0$$

For any reversible engine producing pressure–volume work, we have $\oint dS = 0$ around any cycle.

We can extend this analysis to reach the same conclusion for a reversible engine that produces any form of work. To see this, let us consider the tiling theorem more carefully. When we say that the adiabats and isotherms tile the pressure–volume plane, we mean that each point in the pressure–volume plane is intersected by one and only one adiabat and by one and only one isotherm. When only pressure–volume work is possible, every point in the pressure–volume plane represents a unique state of the system. Therefore, the tiling theorem asserts that every state of the variable-pressure system can be reached along one and only one adiabat and one and only one isotherm.

From experience, we infer that this statement remains true for any form of work. That is, every state of any reversible system can be reached by one and only one isotherm and by one and only one adiabat when any form of work is done. If more than one form of work is possible, there is an adiabat for each form of work. If changing θ_1 and changing θ_2 change the energy of the system, the effects on the energy of the system are not necessarily the same. In general, Φ_1 is not the same as Φ_2, where

$$\Phi_i = \left(\frac{\partial E}{\partial \theta_i} \right)_{V,\theta_{m \neq i}}$$

From §3, we know that a reversible Carnot engine doing any form of work can be matched with a reversible ideal-gas Carnot engine in such a way that the engines complete the successive isothermal and adiabatic steps in parallel. At each step, each engine experiences the same heat, work, energy, and entropy changes as the other. Just as we can plot the reversible ideal-gas Carnot cycle as a closed path in pressure–volume space, we can plot a Carnot cycle producing any other form of work as a closed path with successive isothermal and adiabatic steps in $\Phi_i - \theta_i$ space. Just as any closed path in pressure–volume space can be tiled (or built up from) arbitrarily small reversible Carnot cycles, so any closed path in $\Phi_i - \theta_i$ space can be tiled by such cycles. Therefore, the argument we use to show that $\oint dS = 0$ for any closed reversible cycle in pressure–volume space applies equally well to a closed reversible cycle in which heat is used to produce any other form of work.

§5 The tiling theorem and the paths of cyclic processes in other spaces

We view the tiling theorem as a generalization from experience, just as the machine-based statement of the second law is such a generalization. Let us consider the kinds of familiar observations from which we infer that every equilibrium state of any system is intersected by one and only one adiabat and by one and only one isotherm.

When only pressure–volume work is possible, each pressure–volume point specifies a unique equilibrium state of the system. Since temperature is a state function, the temperature of this state has one and only one value. When another form of work is possible, every $\Phi_i - \theta_i$ point specifies a unique state for which the temperature has one and only one value. From experience, we know that we can produce a new state of the system, at the same temperature, by exchanging heat and work with it in a concerted fashion. We can make this change of state arbitrarily small, so that successive equilibrium states with the same temperature are arbitrarily close to one another. This succession of arbitrarily close equilibrium states is an isotherm. Therefore, at least one isotherm intersects any equilibrium state. There cannot be two such isotherms. If there were two isotherms, the system would have two temperatures, violating the principle that temperature is a state function.

In an adiabatic process, the system exchanges energy as work but not as heat. From experience, we know that we can effect such a change with any reversible system. The result is a new equilibrium state. When we make the increment of work arbitrarily small, the new equilibrium state is arbitrarily close to the original state. Successive exchanges of arbitrarily small work increments produce successive equilibrium states that are arbitrarily close to one another. This succession of arbitrarily close equilibrium states is an adiabat.

If the same state of a system could be reached by two reversible adiabats involving the same form of work, the effect of doing a given amount of this work on an equilibrium system would not be unique. From the same initial state, two reversible adiabatic experiments could do the same amount of the same kind of work and reach different final states of the system. For example, in two different experiments, we could raise a weight reversibly from the same initial elevation, do the same amount of work in each experiment, and find that the final elevation of the weight is different. Any such outcome conflicts with the observations that underlie our ideas about reversible processes.

More specifically, the existence of two adiabats through a given point, in any $\Phi_i - \theta_i$ space, violates the machine-based statement of the second law. Two such adiabats would necessarily intersect a common isotherm. A path along one adiabat, the isotherm, and the second adiabat would be a cycle that restored the system to its original state. This path would enclose a finite area. Traversed in the appropriate direction, the cycle would produce work in the surroundings. By the first law, the system would then accept heat as it traverses the isotherm. The system would exchange heat with surroundings at a single temperature and produce positive work in the surroundings, thus violating the machine-based statement.

If an adiabatic process that connects two states A and B is reversible, we see that the system follows the same path, in opposite directions, when it does work going from A to B as it does when work is done on it as it goes from B to A.

From another perspective, we can say that the tiling theorem is a consequence of our assumptions about reversible processes. Our conception of a reversible process is that the energy, pressure, temperature, and volume are continuous functions of state, with continuous derivatives. That there is one and only one isotherm for every state is equivalent to the assumption that temperature is a continuous (single-valued) function of the state of the system. That there is one and only one adiabat for every state is equivalent to the assumption that $(\partial E/\partial V)_{T,\theta_1}$, or generally, $(\partial E/\partial \theta_i)_{T,V,\theta_{m\neq i}}$, is a continuous, single-valued function of the state of the system.

With these ideas in mind, let us now observe that any reversible cycle can be described by a closed path in a space whose coordinates are T and q^{rev}/T (entropy). In Figure 5, we sketch this space with q^{rev}/T on the abscissa; then an isotherm is a horizontal line, and line of constant entropy (an ***isentrope***) is vertical. A reversible Carnot cycle is a closed rectangle, and the area of this rectangle corresponds to the reversible work done by the system on its surroundings in one cycle. Any equilibrium state of the system corresponds to a particular point in this space. Any closed path can be tiled arbitrarily densely by isotherms and isentropes. Any reversible cycle involving any form of work is represented by a closed path in this space. Figure 5 is an alternative illustration of the argument that we make in §4. The path in this space is independent of the kind of work done, reinforcing the conclusion that $\oint dq^{rev}/T = 0$ for a reversible Carnot cycle producing any form of work. The fact that a cyclic process corresponds to a closed path in this space is equivalent to the fact that entropy is a state function.

To appreciate this aspect of the path of a cyclic process in $T - q^{rev}/T$ space, let us describe the path of the same process in a space whose coordinates are T and q^{rev}. With q^{rev} on the abscissa, isotherms are again horizontal lines and adiabats are vertical lines. In this space, a reversible Carnot cycle does not begin and end at the

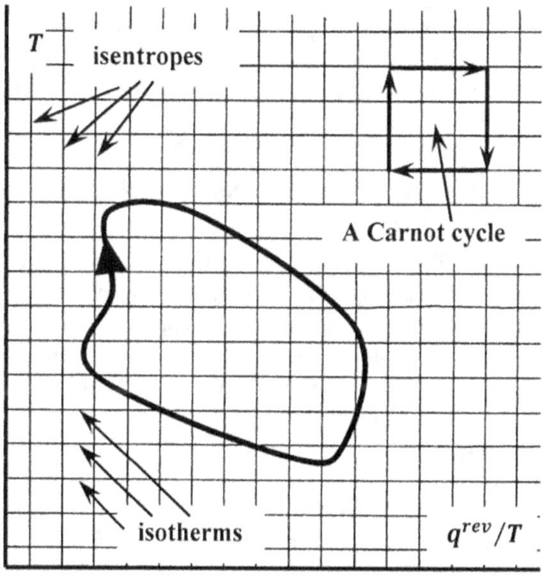

Figure 5. A reversible cycle described using coordinates T and q^{rev}/T.

same point. The path is not closed. Similarly, the representation of an arbitrary reversible cycle is not a closed figure. See Figure 6. The difference between the representations of a reversible cyclic process in these two spaces illustrates graphically the fact that entropy is a state function while heat is not.

§6 Entropy changes for a reversible process

Let us consider a closed system that undergoes a reversible change while in contact with its surroundings. Since the change is reversible, the portion of the surroundings that exchanges heat with the system is at the same temperature as the system: $T = \hat{T}$. From $q^{rev} = -\hat{q}^{rev}$ and the definition, $dS = dq^{rev}/T$, the entropy changes are

$$\Delta S = q^{rev}/T$$

and

$$\Delta\hat{S} = \hat{q}^{rev}/T = -q^{rev}/T = -\Delta S$$

Evidently, for any reversible process, we have

$$\Delta S_{universe} = \Delta S + \Delta\hat{S} = 0$$

Note that these ideas are not sufficient to prove that the converse is true. From only these ideas, we cannot prove that $\Delta S_{universe} = 0$ for a process means that the process is reversible; it remains possible that there could be a spontaneous process for which $\Delta S_{universe} = 0$. However, our entropy-based statement of the second law does assert that the converse is true, that $\Delta S_{universe} = 0$ is necessary and sufficient for a process to be reversible.

In the next section, we use the machine-based statement of the second law to show that $\Delta S \geq 0$ for any spontaneous process in an isolated system. We introduce heuristic arguments to infer that $\Delta S = 0$ is not possible for a spontaneous process in an isolated system. From this, we show that $\Delta S_{universe} > 0$ for any spontaneous process and hence that $\Delta S_{universe} = 0$ is not possible for any spontaneous process. We conclude that $\Delta S_{universe} = 0$ is sufficient to establish that the corresponding process is reversible.

§7 Entropy changes for a spontaneous process in an isolated system

In §6, we consider the entropy changes in a system and its surroundings when the process is reversible. We consider now the diametrically opposite situation in which an isolated system undergoes a spontaneous change. From the entropy-based statement of the second law, we know how the entropy of this system and its surroundings change. Since the system is isolated, no change occurs in the surroundings. Thus, $\Delta\hat{S} = 0$; and since $\Delta S + \Delta\hat{S} > 0$, we have $\Delta S > 0$.

Let us attempt to develop these conclusions from the machine-based statement of the second law. Since the process occurs irreversibly, we cannot use the heat of the process to find the entropy change for the system. We

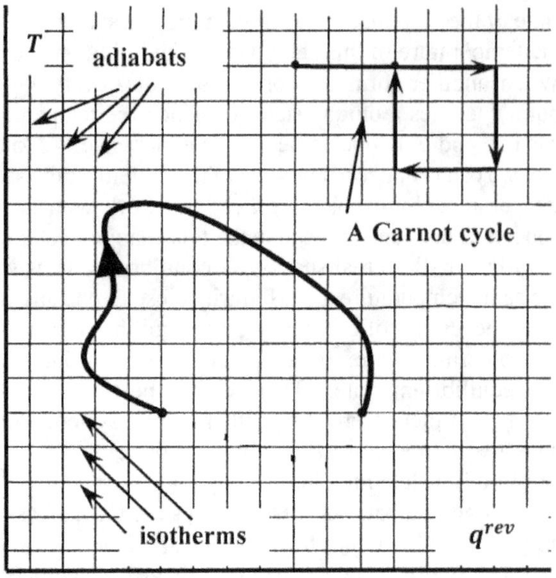

Figure 6. A reversible cycle described using coordinates T and q^{rev}

can calculate the entropy change for a process from the defining equation only if the process is reversible. However, entropy is a state function; using the figure of speech that we introduce in §7-21, we can find the entropy change for the spontaneous process by evaluating ΔS along a second and reversible path that connects the same initial and final states.

In Figure 7, these paths are diagrammed in temperature–entropy space. The transition from state A to state B occurs irreversibly, and therefore it does not necessarily correspond to a path that we can specify on this diagram. The dashed line drawn for this transition is supposed to remind us of this fact. We can readily devise a reversible path from B back to A. First, we reversibly and adiabatically return the temperature of the system to its original value $\hat{T}$. In this step, the system does work on

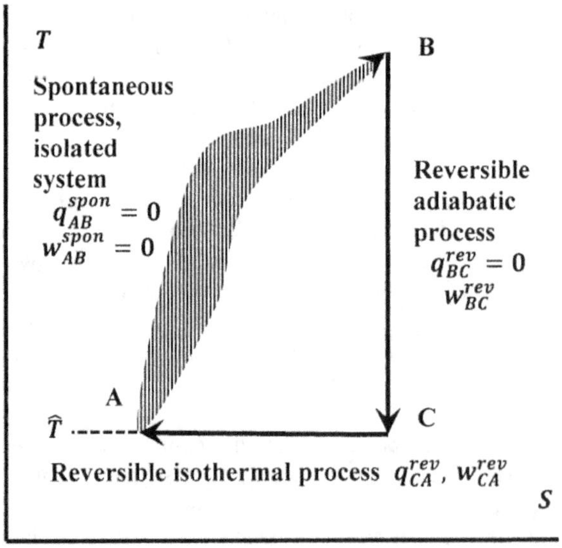

Figure 7. A cycle that includes a spontaneous process in an isolated system.

the surroundings, or *vice versa*. The system reaches point C on the diagram. Then we reversibly and isothermally add or remove heat from the system to return to the original state at point A. For the transfer of heat to be reversible, we must have $T = \hat{T}$ for this step. Hence, the final (and original) temperature of the system at point A is equal to the temperature of the surroundings. The reversible path $B \to C \to A$ must exist, because the tiling theorem asserts that adiabats (vertical lines) and isotherms (horizontal lines) tile the T-S-plane arbitrarily densely.

Taken literally, this description of state A is inconsistent. We suppose that the initial state A is capable of spontaneous change; therefore, it cannot be an equilibrium state. We suppose that the final state A is reached by a reversible process; therefore, it must be an equilibrium state. We bridge this contradiction by refining our definition of the initial state. The final state A is an equilibrium state with well-defined state functions. What we have in mind is that these final equilibrium-state values also characterize the initial non-equilibrium state. Evidently, the initial state A that we have in mind is a hypothetical state. This hypothetical state approximates the state of a real system that undergoes spontaneous change. By invoking this hypothetical initial state, we eliminate the contradiction between our descriptions of initial state A and final state A. Given a real system that undergoes spontaneous change, we must find approximate values for the real system's state functions by finding an equilibrium—or quasi-equilibrium—system that adequately models the initial state of the spontaneously changing system.

In the development below, we place no constraints on the nature of the system or the spontaneous process. We assume that the state functions of any hypothetical initial state A can be adequately approximated by some equilibrium-state model. However, before we consider the general argument, let us show how these conditions can be met for another specific system. Consider a vessel whose interior is divided by a partition. The real gas of a pure substance occupies the space on one side of the partition. The space on the other side of the partition is evacuated. We suppose that this vessel is isolated. The real gas is at equilibrium. We can measure its state functions, including its pressure, volume, and temperature. Now suppose that we puncture the partition. As soon as we do so, the gas expands spontaneously to fill the entire vessel, reaching a new equilibrium position, at a new pressure, volume, and temperature. The gas undergoes a free expansion, as defined in §7-17.

At the instant the partition is punctured, the system becomes able to undergo spontaneous change. In this hypothetical initial state, before any significant quantity of gas passes through the opening, neither the actual condition of the gas nor the values of its state functions have changed. After the expansion to the new equilibrium state, the original state can be restored by reversible processes of adiabatic compression and isothermal volume adjustment. (Problems 13 and 14 in Chapter 10 deal with the energy and entropy changes for ideal and real gases around a cycle in which spontaneous expansion in an isolated system is followed by reversible restoration of the initial state.)

Returning to the general cycle depicted in Figure 7, we see that there are some important conditions on the heat and work terms in the individual steps. Since the system is isolated while it undergoes the transition from A to B, it exchanges no heat or work with the surroundings in this step: $q_{AB}^{spon} = w_{AB}^{spon} = 0$. For the reversible adiabatic transition from B to C, $d_{BC}q^{rev} = 0$ in every incremental part of the path. The transition from C to A occurs reversibly and isothermally; letting the heat of this step be q_{CA}^{rev}, the entropy changes for these reversible steps are, from the defining equation,

$$\Delta_{BC}S = \int_{T_B}^{T_C} \frac{d_{BC}q^{rev}}{T} = 0$$

and

$$\Delta_{CA}S = \frac{q_{CA}^{rev}}{\hat{T}}$$

The energy and entropy changes around this cycle must be zero, whether the individual steps occur reversibly or irreversibly. We have

$$\begin{aligned}\Delta E &= q_{AB}^{spon} + w_{AB}^{spon} + q_{BC}^{rev} + w_{BC}^{rev} + q_{CA}^{rev} + w_{CA}^{rev} \\ &= q_{CA}^{rev} + w_{BC}^{rev} + w_{CA}^{rev} \\ &= 0\end{aligned}$$

and

$$\begin{aligned}\Delta S &= \Delta_{AB}S + \Delta_{BC}S + \Delta_{CA}S \\ &= \Delta_{AB}S + q_{CA}^{rev}/\hat{T} = 0\end{aligned}$$

We want to analyze this cycle using the machine-based statement of the second law. We have $w_{BC}^{rev} = -\hat{w}_{BC}^{rev}$, $w_{CA}^{rev} = -\hat{w}_{CA}^{rev}$, and $q_{CA}^{rev} = -\hat{q}_{CA}^{rev}$. Let us assume that the system does net work on the surroundings as this cycle is traversed so that $\hat{w}_{BC}^{rev} + \hat{w}_{CA}^{rev} > 0$. Then,

$$-(\hat{w}_{BC}^{rev} + \hat{w}_{CA}^{rev}) = w_{BC}^{rev} + w_{CA}^{rev} < 0$$

and it follows that $q_{CA}^{rev} > 0$. The system exchanges heat with the surroundings in only one step of this process. In this step, the system extracts a quantity of heat from a reservoir in the surroundings. The temperature of this reservoir remains constant at $\hat{T}$ throughout the process. The heat extracted by the system is converted entirely into work. This result contradicts the machine-based statement of the second law. Hence, $w_{BC}^{rev} + w_{CA}^{rev} < 0$ is false; it follows that

$$w_{BC}^{rev} + w_{CA}^{rev} \geq 0$$

and that

$$q_{CA}^{rev} \leq 0$$

For the entropy change in the spontaneous process in the isolated system, we have

$$\Delta_{AB}S = -\,q_{CA}^{rev}/\hat{T} \geq 0$$

Now, we introduce the premise that $q_{CA}^{rev} \neq 0$. If this is true, the entropy change in the spontaneous process in the isolated system becomes

$$\Delta_{AB}S > 0$$

(The converse is also true; that is, $\Delta_{AB}S > 0$ implies that $q_{CA}^{rev} \neq 0$.) The premise that $q_{CA}^{rev} \neq 0$ is independent of the machine-based statement of the second law, which requires only that $q_{CA}^{rev} \leq 0$, as we just demonstrated. It is also independent of the first law, which requires only that $q_{CA}^{rev} = -w_{BC}^{rev} - w_{CA}^{rev}$. If $q_{CA}^{rev} \neq 0$, we can conclude that, for a spontaneous process in an isolated system, we must have $w_{BC} + w_{CA} > 0$ and $q_{CA}^{rev} < 0$. These conditions correspond to doing work on the system and finding that heat is liberated by the system. There is no objection to this; it is possible to convert mechanical energy into heat quantitatively. The conclusions that $q_{CA}^{rev} < 0$ and $\Delta_{AB}S > 0$ have important consequences; we consider them below. First, however, we consider a line of thought that leads us to infer that $q_{CA}^{rev} \neq 0$ and hence that $q_{CA}^{rev} < 0$ must be true.

Because $q_{AB} = 0$ and $w_{AB} = 0$, we have $E_A = E_B$. The system can be taken from state A to state B by the reversible process A$\rightarrow C \rightarrow B$. Above we see that if $q_{CA}^{rev} = 0$, we have $S_A = S_B$. In §6-10, we introduce Duhem's theorem, which asserts that two thermodynamic variables are sufficient to specify the state of a closed reversible system in which only pressure–volume work is possible. We gave a proof of Duhem's theorem when the two variables are chosen from among the pressure, temperature, and composition variables that describe the system. We avoided specifying whether other pairs of variables can be used. If we assume now that specifying the variables energy and entropy is always sufficient to specify the state of such a system, it follows that states A and B must in fact be the same state. (In §14, and in greater detail in Chapter 10, we see that the first law and our entropy-based statement of the second law do indeed imply that specifying the energy and entropy specifies the state of a closed reversible system in which only pressure–volume work is possible.)

If state A and state B are the same state; that is, if the state functions of state A are the same as those of state B, it is meaningless to say that there is a spontaneous process that converts state A to state B. Therefore, if A can be converted to B in a spontaneous process in an isolated system, it must be that $q_{CA}^{rev} \neq 0$. That is,

$$[(q_{CA}^{rev} = 0) \Rightarrow \sim(\text{A can go to B spontaneously})]$$

$$\Rightarrow [(A \text{ can go to B spontaneously}) \Rightarrow (q_{CA}^{rev} \neq 0)]$$

From the machine-based statement of the second law, we find $\Delta_{AB}S = -\,q_{CA}^{rev}/\hat{T} \geq 0$. When we supplement this conclusion with our Duhem's theorem-based inference that $q_{CA}^{rev} \neq 0$, we can conclude that $\Delta S > 0$ for any spontaneous process in any isolated system. Because the system is isolated, we have $\hat{q} = 0$, and $\Delta\hat{S} = 0$. ***For any spontaneous process in any isolated system*** we have

$$\Delta S_{universe} = \Delta S + \Delta\hat{S} > 0.$$

We can also conclude that the converse is true; that is, if $\Delta_{AB}S = S_B - S_A > 0$ for a process in which an isolated system goes from state A to state B, the process must be spontaneous. Since any process that occurs in an isolated system must be a spontaneous process, it is only necessary to show that $\Delta_{AB}S > 0$ implies that state B is different from state A. This is trivial. Because entropy is a state function, $S_B - S_A > 0$ requires that state B be different from state A.

None of our arguments depends on the magnitude of the change that occurs. Evidently, the same inequality must describe every incremental portion of any spontaneous process; otherwise, we could define an incremental spontaneous change for which the machine-based statement of the second law would be violated. For every incremental part of any spontaneous change in any isolated system we have $dS > 0$ and

$$dS_{universe} = dS + d\hat{S} > 0.$$

These are pivotally important results; we explore their ramifications below. Before doing so, however, let us again consider a system in which only pressure–volume work is possible. There is an alternative way to express the idea that such a system is isolated. Since an isolated system cannot interact with its surroundings in any way, it cannot exchange energy with its surroundings. Its energy must be constant. Since it cannot exchange pressure–volume work, its volume must be constant. Hence, isolation implies constant E and V. If only pressure–volume work is possible, the converse must be true; that is, if only pressure–volume work is possible, constant energy and volume imply that there are no interactions between the system and its surroundings. Therefore, constant E and V imply that the system is isolated, and it must be true that $\Delta\hat{S} = 0$. In this case, a spontaneous process in which E and V are constant must be accompanied by an increase in the entropy of the system. (If V is constant and only pressure–volume work is possible, the process involves no work.) We have a criterion for spontaneous change:

$$(\Delta S)_{EV} > 0$$

(spontaneous process, only pressure–volume work)

where the subscripts indicate that the energy and volume of the system are constant. (In §21, we arrive at this conclusion by a different argument.)

§8 The entropy of the universe

In §7, we conclude that the entropy change is positive for any spontaneous change in an isolated system. Since we can consider the universe to be an isolated system, it follows that $\Delta S_{universe} > 0$ for any spontaneous process.

To reach this conclusion by a more detailed argument, let us consider an arbitrary system that is in contact with its surroundings. We can subdivide these surroundings into subsystems. As diagrammed in Figure 8, we define a surroundings subsystem (Surroundings 1) that interacts with the system and a more remote surroundings subsystem (Surroundings 2) that does not. That is, we assume that we can define Surroundings 2 so that it is unaffected by the process. Then we define an augmented system consisting of the original system plus Surroundings 1. The augmented system is isolated from the remote portion of the surroundings, so that the entropy change for the augmented system is positive by the argument in the previous section. Denoting entropy changes for the system, Surroundings 1, Surroundings 2, and the augmented system by ΔS, $\Delta \hat{S}_1$, $\Delta \hat{S}_2$, and $\Delta S_{augmented}$, respectively, we have $\Delta S_{augmented} = \Delta S + \Delta \hat{S}_1 > 0$, and $\Delta \hat{S} = \Delta \hat{S}_1 + \Delta \hat{S}_2 > 0$. Since the remote portion of the surroundings is unaffected by the change, we have $\Delta \hat{S}_2 = 0$. For any spontaneous change, whether the system is isolated or not, we have

$$\begin{aligned}\Delta S_{universe} &= \Delta S + \Delta \hat{S}_1 + \Delta \hat{S}_2 \\ &= \Delta S_{augmented} + \Delta \hat{S}_2 \\ &= \Delta S_{augmented} > 0\end{aligned}$$

(any spontaneous change)

This statement is an essential part of the entropy-based statement of the second law. We have now developed it from the machine-based statement of the second law by convincing, but not entirely rigorous arguments. In §6 we find that $\Delta S_{universe} = \Delta S + \Delta \hat{S} = 0$ for any reversible process. Thus, for any possible process, we have

$$\Delta S_{universe} = \Delta S + \Delta \hat{S} \geq 0$$

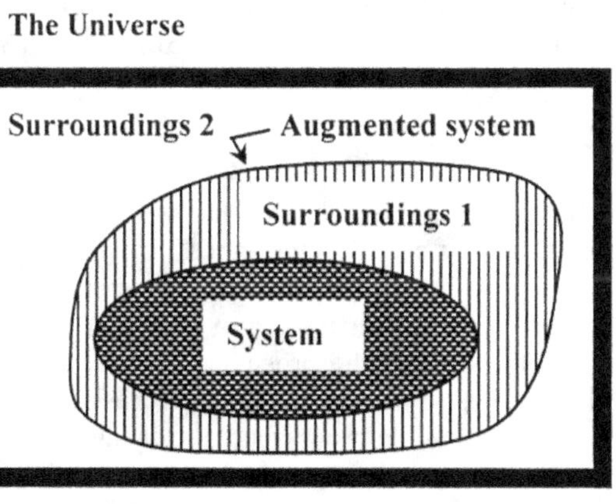

Figure 8. Expanding a system to create a new, augmented, isolated system.

The equality applies when the process is reversible; the inequality applies when it is spontaneous.

Because entropy is a state function, ΔS and $\Delta \hat{S}$ change sign when the direction of a process is reversed. We say that a process for which $\Delta S + \Delta \hat{S} < 0$ is an impossible process. Our definitions mean that these classifications—reversible, spontaneous, and impossible—are exhaustive and mutually exclusive. We conclude that $\Delta S_{universe} = \Delta S + \Delta \hat{S} = 0$ is necessary and sufficient for a process to be reversible; $\Delta S_{universe} = \Delta S + \Delta \hat{S} > 0$ is necessary and sufficient for a process to be spontaneous. (See problem 19.)

§9 The significance of the machine-based statement of the second law

Our entropy-based statement of the second law asserts the definition and basic properties of entropy that we need in order to make predictions about natural processes. The ultimate justification for these assertions is that the predictions they make agree with experimental observations. We have devoted considerable attention to arguments that develop the definition and properties of entropy from the machine-based statement of the second law. These arguments parallel those that were made historically as these concepts were developed. Understanding these arguments greatly enhances our appreciation for the relationship between the properties of the entropy function and the changes that can occur in various physical systems.

While these arguments demonstrate that our machine-based statement implies the entropy-based statement, we introduce additional postulates in order to make them. These include: the premise that the pressure, temperature, volume, and energy of a reversible system are continuous functions of one another; Duhem's theorem; the tiling theorem; and the presumption that the conclusions we develop for pressure–volume work are valid for any form of work. We can sum up this situation by saying that our machine-based statement serves a valuable heuristic purpose. The entropy-based statement of the second law is a postulate that we infer by reasoning about the consequences of the machine-based statement. When we want to apply the second law to physical systems, the entropy-based statement and other statements that we introduce below are much more useful.

Finally, we note that our machine-based statement of the second law is not the only statement of this type. Other similar statements have been given. The logical relationships among them are interesting, and they can be used to develop the entropy-based statement of the second law by arguments similar to those we make in §2 to §8.

§10 A slightly philosophical digression on energy and entropy

The content of the first law of thermodynamics is that there is a state function, which we call energy, which has the property that $\Delta E_{universe} = 0$ for any process that can occur. The content of the second law is that there is a state function, which we call entropy, which has the property that $\Delta S_{universe} > 0$ for any spontaneous process.

These two state functions exhaust the range of independent possibilities: Suppose that we aspire to find a new and independent state function, call it B, which further characterizes the possibilities open to the universe. What other condition could B impose on the universe—or *vice versa*? The only available candidate might appear to be $\Delta B_{universe} < 0$. However, this does not represent an independent condition, since its role is already filled by the quantity $-\Delta S_{universe}$.

Of course, we can imagine a state function, B, which is not simply a function of S, but for which $\Delta B_{universe} > 0$, $\Delta B_{universe} = 0$, or $\Delta B_{universe} < 0$, according as the process is spontaneous, reversible, or impossible, respectively. For any given change, ΔB would not be the same as ΔS; however, ΔB and ΔS would make exactly the same predictions. If $\Delta B_{universe}$ were more easily evaluated than $\Delta S_{universe}$, we would prefer to use B rather than S. Nevertheless, if there were such a function B, its role in our description of nature would duplicate the role played by S.

§11 A third statement of the second law

Let us consider another frequently cited alternative statement of the second law, which, for easy reference, we call the **temperature-based** statement of the second law:

The temperature-based statement of the second law
The spontaneous transfer of heat from a colder body to a warmer body is impossible.

In the discussion below, we refer to this statement as proposition TSL. By "body", we simply mean any system or object. By the "spontaneous transfer of heat," we mean that the transfer of heat energy can be initiated by bringing the two bodies into contact with one another or by enabling the transmission of radiant energy between them. The surroundings do no work and exchange no heat with either reservoir; there is no change of any sort in the surroundings.

We can show that the entropy-based statement and the temperature-based statement of the second law are equivalent: Given the definition of entropy, one implies the other.

Let us begin by showing that the entropy-based statement implies the temperature-based statement of the second law. That is, we prove SL ⇒ TSL. To do so, we prove ~TSL ⇒ ~SL. That is, we assume that spontaneous transfer of heat from a colder to a warmer body is possible and show that this leads to a contradiction of the entropy-based statement of the second law. Let the quantity of heat received by the warmer body be $dq_{warmer} > 0$, and let the temperatures of the warmer and colder bodies be T_{warmer} and T_{colder}, respectively. We have $T_{warmer} - T_{colder} > 0$. The colder body receives heat $dq_{colder} = -dq_{warmer} < 0$. We make the heat increment so small that there is no significant change in the temperature of either body. No other changes occur. The two bodies are the only portions of the universe that are affected. Let the entropy changes for the warmer and colder bodies be dS_{colder} and dS_{warmer}, respectively.

To find dS_{colder} and dS_{warmer} we must find a reversible path to effect the same changes. This is straightforward. We can effect identically the same change in the warmer body by transferring heat, $q_{warmer} > 0$, to it through contact with some third body, whose temperature is infinitesimally greater than T_{warmer}. This process is reversible, and the entropy change is $dS_{warmer} = dq_{warmer}/T_{warmer}$. Similarly, the entropy change for the colder body is $dS_{colder} = dq_{colder}/T_{colder} = -dq_{warmer}/T_{colder}$. It follows that

$$\begin{aligned} dS_{universe} &= dS_{warmer} + dS_{colder} \\ &= \frac{dq_{warmer}}{T_{warmer}} - \frac{dq_{warmer}}{T_{colder}} \\ &= -dq_{warmer}\left(\frac{T_{warmer} - T_{colder}}{T_{warmer}T_{colder}}\right) \\ &< 0 \end{aligned}$$

However, if $dS_{universe} < 0$ for a spontaneous process, the second law (SL) must be false. We have shown that a violation of the temperature-based statement implies a violation of the entropy-based statement of the second law: ~TSL ⇒ ~SL, so that SL ⇒ TSL.

It is equally easy to show that the temperature-based statement implies the entropy-based statement of the second law. To do so, we assume that the entropy-based statement is false and show that this implies that the temperature-based statement must be false. By the arguments above, the entropy change that the universe experiences during the exchange of the heat increment is

$$dS_{universe} = -dq_{warmer}\left(\frac{T_{warmer} - T_{colder}}{T_{warmer}T_{colder}}\right)$$

If the entropy-based statement of the second law is false, then $dS_{universe} < 0$. It follows that $dq_{warmer} > 0$; that is, the spontaneous process transfers heat from the colder to the warmer body. This contradicts the temperature-based statement. That is, ~SL ⇒ ~TSL, so that TSL ⇒ SL.

§12 Entropy and predicting change

The entropy-based criteria that we develop in §2 through §8 are of central importance. If we are able to evaluate the change in the entropy of the universe for a prospective process and find that it is greater than zero, we can conclude that the process can occur spontaneously. The reverse of a spontaneous process cannot occur; it is an impossible process and the change in the entropy of the universe for such a process must be less than zero. Since an equilibrium process is a reversible process, the entropy of the universe must remain unchanged when a system goes from an initial state to a final state along a path whose every point is an equilibrium state. Using another figure of speech, we often say that a change that occurs along a reversible path is a change that "occurs at equilibrium."

These conclusions are what make the entropy function useful: If we can calculate $\Delta S_{universe}$ for a prospective process, we know whether the system is at equilibrium with respect to that process; whether the process is possible; or whether the process cannot occur. If we find $\Delta S_{universe} > 0$ for a process, we can conclude that the process is possible; however, we cannot conclude that the process will occur. Indeed, many processes can occur spontaneously but do not do so. For example, hydrocarbons can react spontaneously with oxygen; most do so only at elevated temperatures or in the presence of a catalyst.

The criteria $\Delta S_{universe} = \Delta S + \Delta \hat{S} \geq 0$ are completely general. They apply to any process occurring under any conditions. To apply them we must determine both ΔS and $\Delta \hat{S}$. By definition, the system comprises the part of the universe that is of interest to us; the need to determine $\Delta \hat{S}$ would appear to be a nuisance. This proves not to be the case. So long as the surroundings have a well-defined temperature, we can develop additional criteria for equilibrium and spontaneous change in which $\Delta \hat{S}$ does not occur explicitly. In §14, we develop criteria that apply to reversible processes. In §15, we find a general relationship for $\Delta \hat{S}$ that enables us to develop criteria for spontaneous processes.

To develop the criteria for spontaneous change, we must define what we mean by spontaneous change more precisely. To define a spontaneous process in an isolated system as one that can take place on its own is reasonably unambiguous. However, when a system is in contact with its surroundings, the properties of the surroundings affect the change that occurs in the system. To specify a particular spontaneous process we must specify some properties of the surroundings or—more precisely—properties of the system that the surroundings act to establish. The ideas that we develop in §15 lead to criteria for changes that occur while one or more thermodynamic functions remain constant. These criteria supplement the second-law criteria $\Delta S + \Delta \hat{S} \geq 0$. In using these criteria, we can say that the change occurs subject to one or more constraints.

Some of these criteria depend on the magnitudes of ΔE and ΔH in the prospective process. We also find criteria that are expressed using new state functions that we call the Helmholtz and Gibbs free energies. In the next section, we introduce these functions.

§13 Defining the Helmholtz and Gibbs free energies

The first and second laws of thermodynamics define energy and entropy. Energy and entropy are fundamental state functions that we use to define other state functions. In Chapter 8, we use the energy function to define enthalpy. We use the energy and entropy functions to define two more state functions that also prove to have useful properties. These are the Helmholtz and Gibbs free energies. The ***Helmholtz free energy*** is usually given the symbol A, and the ***Gibbs free energy*** is usually given the symbol G. We define them by

$$A = E - TS \qquad \text{(Helmholtz free energy)}$$

and

$$G = H - TS \qquad \text{(Gibbs free energy)}$$

Note that PV, TS, H, A, and G all have the units of energy, E.

The sense of the name "free energy" is that a constant-temperature process in which a system experiences an entropy increase ($\Delta S > 0$) is one in which the system's ability to do work in the surroundings is increased by an energy increment $T\Delta S$. Then, adding $T\Delta S$ to the internal energy lost by the system yields the amount of energy that the process actually has available (energy that is "free") to do work in the surroundings. When we consider how ΔA and ΔG depend on the conditions under which system changes, we find that this idea leads to useful results.

The rest of this chapter develops important equations for ΔE, ΔH, ΔS, ΔA, and ΔG that result when we require that a system change occur under particular sets of conditions.

§14 The fundamental equation and other criteria for reversible change

To begin exploring the possibilities for stating the criteria for change using only the properties of the system, let us consider how some thermodynamic functions change when a process is reversible. We consider a closed system and focus on making incremental changes in the state of the system. For a reversible process, we have $dq^{rev} = TdS$. The reversible pressure–volume work is $dw_{PV}^{rev} = -PdV$. If non-pressure–volume work is also possible, the reversible work becomes $dw^{rev} = -PdV + dw_{NPV}^{rev}$, where dw_{NPV}^{rev} is the increment of reversible, non-pressure–volume work. The energy change is

$$\begin{aligned} dE &= dq^{rev} + dw^{rev} \\ &= TdS - PdV + dw_{NPV}^{rev} \end{aligned}$$

(any reversible process)

This equation is of central importance. It is sometimes called ***the combined first and second laws of thermodynamics*** or the ***fundamental equation***. It applies to any closed system that is undergoing reversible change. It specifies a relationship among the changes in energy, entropy, and volume that must occur if the system is to remain at equilibrium while an increment of non-pressure–volume work, dw_{NPV}^{rev}, is done on it. The burden of our entire development is that any reversible process must satisfy this equation. Conversely, any process that satisfies this equation must be reversible.

For a reversible process at constant entropy, we have $dS = 0$, so that $(dE)_S = -PdV + dw_{NPV}^{rev}$. Since $-PdV$ is the reversible pressure–volume work, dw_{PV}^{rev}, and the sum $dw_{net}^{rev} = -PdV + dw_{NPV}^{rev}$ is the net work, we have

$$(dE)_S = dw_{net}^{rev}$$

(reversible process, constant S)

where the subscript "S" specifies that the entropy is constant. For a reversible process in which all of the work is pressure–volume work, we have $dw_{NPV}^{rev} = 0$, and the fundamental equation becomes

$$dE = TdS - PdV$$

(reversible process, only pressure–volume work)

For a reversible process in which only pressure–volume work is possible, this equation gives the amount, dE, by which the energy must change when the entropy changes by dS and the volume changes by dV.

Now, let us apply the fundamental equation to an arbitrary process that occurs reversibly and at constant entropy and constant volume. Under these conditions, $dS = 0$ and $dV = 0$. Therefore, at ***constant entropy and volume***, a necessary and sufficient condition for the process to be reversible—and hence to be continuously in an equilibrium state as the process takes place—is that

$$(dE)_{SV} = dw_{NPV}^{rev}$$

(reversible process)

and if only pressure–volume work is possible,

$$(dE)_{SV} = 0$$

(reversible process, only pressure–volume work)

where the subscripts indicate that entropy and volume are constant.

If we consider an arbitrary reversible process that occurs at ***constant energy and volume***, we have $dE = 0$ and $dV = 0$, and the fundamental equation reduces to

$$(dS)_{EV} = -\frac{dw_{NPV}^{rev}}{T}$$

(reversible process)

and if only pressure–volume work is possible,

$$(dS)_{EV} = 0$$

(reversible process, only pressure–volume work)

In this case, as noted in §7, the system is isolated. In §1-6, we note that an isolated system in an equilibrium state can undergo no further change. Thus, the condition $(dS)_{EV} = 0$ defines a unique or primitive equilibrium state.

If a closed system behaves reversibly, any composition changes that occur in the system must be reversible. For chemical applications, composition changes are of paramount importance. We return to these considerations in Chapter 14, where we relate the properties of chemical substances—their chemical potentials—to the behavior of systems undergoing both reversible and spontaneous composition changes.

If a closed system behaves reversibly and only pressure–volume work is possible, we see from the fundamental equation that specifying the changes in any two of the three variables, E, S, and V, is sufficient to specify the change in the system. In particular, if energy and entropy are constant, $dE = dS = 0$, the volume is also constant, $dV = 0$, and the system is isolated. Thus, the state of an equilibrium system whose energy and entropy are fixed is unique; $dE = dS = 0$ specifies a primitive equilibrium state. We see that the internal consistency of our model passes a significant test: From the entropy-based statement of the second law, we deduce the same proposition that we introduce in §7 as a heuristic conjecture. In Chapter 10, we expand on this idea.

Starting from the fundamental equation, we can find similar sets of relationships for enthalpy, the Helmholtz free energy, and the Gibbs free energy. We define $H = E + PV$. For an incremental change in a system we, have

$$dH = dE + PdV + VdP$$

Using the fundamental equation to substitute for dE, this becomes

$$\begin{aligned} dH &= TdS - PdV + dw_{NPV}^{rev} + PdV + VdP \\ &= TdS + VdP + dw_{NPV}^{rev} \end{aligned}$$

For a reversible process in which all of the work is pressure–volume work, we have

$$dH = TdS + VdP$$

(reversible process, only pressure–volume work)

For a reversible process in which only pressure–volume work is possible, this equation gives the amount, dH, by which the enthalpy must change when the entropy changes by dS and the pressure changes by dP. If a reversible process occurs at constant entropy and pressure, then $dS = 0$ and $dP = 0$. At ***constant entropy and pressure***, the process is reversible if and only if

$$(dH)_{SP} = dw_{NPV}^{rev}$$

(reversible process)

If only pressure–volume work is possible,

$$(dH)_{SP} = 0$$

(reversible process, only pressure–volume work)

where the subscripts indicate that entropy and pressure are constant.

If we consider an arbitrary reversible process that occurs at ***constant enthalpy and pressure***, we have $dH = 0$ and $dP = 0$, and the total differential for dH reduces to

$$(dS)_{HP} = -\frac{dw_{NPV}^{rev}}{T}$$

(reversible process)

and if only pressure–volume work is possible,

$$(dS)_{HP} = 0$$
(reversible process, only pressure–volume work)

From $A = E - TS$, we have $dA = dE - TdS - SdT$. Using the fundamental equation to substitute for dE, we have

$$\begin{aligned} dA &= TdS - PdV + dw_{NPV}^{rev} - TdS - SdT \\ &= -PdV - SdT + dw_{NPV}^{rev} \end{aligned}$$

For a reversible process in which all of the work is pressure–volume work,

$$dA = -SdT - PdV$$
(reversible process, only pressure–volume work)

For a reversible process in which only pressure–volume work is possible, this equation gives the amount, dA, by which the Helmholtz free energy must change when the temperature changes by dT and the volume changes by dV. For a reversible isothermal process, we have $dT = 0$, and from $dA = -PdV - SdT + dw_{NPV}^{rev}$ we have

$$\begin{aligned} (dA)_T &= -PdV + dw_{NPV}^{rev} \\ &= dw_{PV}^{rev} + dw_{NPV}^{rev} \\ &= dw_{net}^{rev} \end{aligned}$$
(reversible isothermal process)

where we recognize that the reversible pressure–volume work is $dw_{PV}^{rev} = -PdV$, and the work of all kinds is $dw_{net}^{rev} = dw_{PV}^{rev} + dw_{NPV}^{rev}$. We see that $(dA)_T$ is the total of all the work done on the system in a reversible process at constant temperature. This is the reason that "A" is used as the symbol for the Helmholtz free energy: "A" is the initial letter in "Arbeit," a German noun whose meaning is equivalent to that of the English noun "work."

If a reversible process occurs at constant temperature and volume, we have $dT = 0$ and $dV = 0$. At ***constant temperature and volume***, a process is reversible if and only if

$$(dA)_{TV} = dw_{NPV}^{rev}$$
(reversible process)

If only pressure–volume work is possible,

$$(dA)_{TV} = 0$$
(reversible process, only pressure–volume work)

where the subscripts indicate that volume and temperature are constant. (Of course, these conditions exclude all work, because constant volume implies that there is no pressure–volume work.)

From

$$\begin{aligned} G &= H - TS \\ &= E + PV - TS \end{aligned}$$

and the fundamental equation, we have

$$\begin{aligned} dG &= dE + PdV + VdP - SdT - TdS \\ &= TdS - PdV + dw_{NPV}^{rev} \\ &\qquad - TdS + PdV + VdP - SdT \\ &= -SdT + VdP + dw_{NPV}^{rev} \end{aligned}$$

For a reversible process in which all of the work is pressure–volume work,

$$dG = -SdT + VdP$$
(reversible process, only pressure–volume work)

For a reversible process in which only pressure–volume work is possible, this equation gives the amount, dG, by which the Gibbs free energy must change when the temperature changes by dT and the pressure changes by dP. For a reversible process that occurs at constant temperature and pressure, $dT = 0$ and $dP = 0$. At ***constant temperature and pressure***, the process will be reversible if and only if

$$(dG)_{TP} = dw_{NPV}^{rev}$$
(any reversible process)

If only pressure–volume work is possible,

$$(dG)_{TP} = 0$$
(reversible process, only pressure–volume work)

where the subscripts indicate that temperature and pressure are constant.

In this section, we develop several criteria for reversible change, stating these criteria as differential expressions. Since each of these expressions applies to every incremental part of a reversible change that falls within its scope, corresponding expressions apply to finite changes. For example, we find $(dE)_S = dw_{net}^{rev}$ for every incremental part of a reversible process in which the entropy has a constant value. Since we can find the energy change for a finite amount of the process by summing up the energy changes in every incremental portion, it follows that

$$(\Delta E)_S = w_{net}^{rev}$$
(reversible process)

Each of the other differential-expression criteria for reversible change also gives rise to a corresponding criterion for a finite reversible change. These criteria are summarized in §25.

In developing the criteria in this section, we stipulate that various combinations of the thermodynamic functions that characterize the system are constant. We develop these criteria for systems undergoing reversible change; consequently, the requirements imposed by reversibility must be satisfied also. In particular, the system must be composed of homogeneous phases and its temperature must be the same as that of the surroundings. The pressure of the system must be equal to the pressure applied to it by the surroundings. When we specify that a reversible process occurs at constant temperature, we mean that $T = \hat{T} =$ constant. When we specify that a reversible process occur at constant pressure, we mean that $P = P_{applied} =$ constant.

§15 Entropy and spontaneous change

In a reversible process, the changes that occur in the system are imposed by the surroundings; reversible change occurs only because the system responds to changes in the conditions imposed on it by its surroundings. A reversible process is driven by the surroundings. In contrast, a spontaneous process is driven by the system. Nevertheless, when a spontaneous process occurs under some specific set of imposed conditions (specific values of the temperature and pressure, for example) the system's equilibrium state depends on these conditions. To specify a particular spontaneous change, we must specify enough constraints to fix the final state of the system.

To see these points from a slightly different perspective, let us consider a closed reversible system in which only pressure–volume work is possible. Duhem's theorem asserts that a change in the state of this system can be specified by specifying the changes in some pair of state functions, say X and Y. If the imposed values of X and Y are constant at their eventual equilibrium values, but the system is changing, the system cannot be on a Gibbsian equilibrium manifold. We say that the system is undergoing a spontaneous change at constant X and Y.

This description is a figure of speech in that the system's X and Y values do not necessarily attain the imposed values and become constant until equilibrium is reached. An example is in order: A system whose original pressure and temperature are P_i and T_i can undergo a spontaneous change while the surroundings impose a constant pressure, $P_{applied} = P_f$, and the system is immersed in constant temperature bath at $T = T_f$. The pressure and temperature of the system may be indeterminate as the process occurs, but the equilibrium pressure and temperature must be P_f and T_f.

If the surroundings operate to impose particular values of X and Y on the system, then the position at which the system eventually reaches equilibrium is determined by these values. The same equilibrium state is reached for any choice of surroundings that imposes the same values of X and Y on the system at the time that the system reaches equilibrium. For every additional form of non-pressure–volume work that affects the system, we must specify the value of one additional variable in order to specify a unique equilibrium state.

The entropy changes that occur in the system and its surroundings during a spontaneous process have predictive value. However, our definitions do not enable us to find the entropy change for a spontaneous process, and the temperature of the system may not have a meaningful value. On the other hand, we can always carry out the process so that the temperature of the surroundings is known at every point in the process. Indeed, if the system is in thermal contact with its surroundings as the process occurs, we cannot specify the conditions under which the process occurs without specifying the temperature of the surroundings along this path.

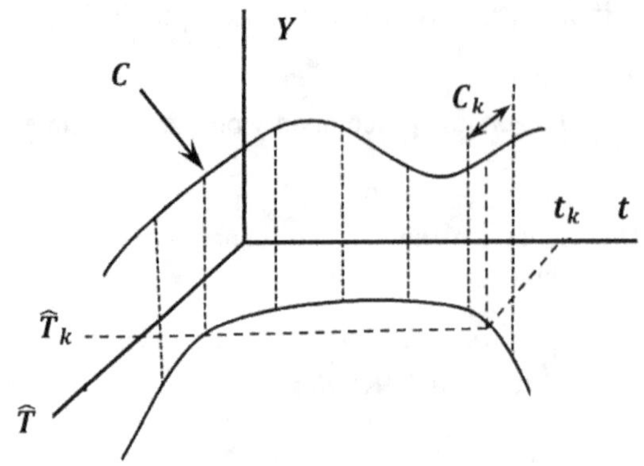

Figure 9. The path of a spontaneously changing system.

Figure 9 describes a spontaneous process whose path can be specified by the values of thermodynamic variable Y and the temperature of the surroundings, $\hat{T}$, as a function of time, t. Let us denote the curve that describes this path as C. We can divide this path into short intervals. Let C_k denote a short segment of this path along which the temperature of the surroundings is approximately constant. For our present purposes, the temperature of the system, T, is irrelevant; since the process is spontaneous, the temperature of the system may have no meaningful value within the interval C_k. As the system traverses segment C_k, it accepts a quantity of heat, q_k, from the surroundings, which are at temperature $\hat{T}_k$. The heat exchanged by the surroundings within C_k is $\hat{q}_k = -q_k$. Below, we show that it is always possible to carry out the process in such a way that the change in the surroundings occurs reversibly. Then

$$\Delta\hat{S}_k = \frac{\hat{q}_k}{\hat{T}_k} = -\frac{q_k}{\hat{T}_k}$$

and since $\Delta S_k + \Delta\hat{S}_k > 0$, it follows that

$$\Delta S_k > \frac{q_k}{\hat{T}_k}$$

This is the ***Clausius inequality***. It plays a central role in the thermodynamics of spontaneous processes. When we make the intervals C_k arbitrarily short, we have

$$dS_k > \frac{dq_k}{\hat{T}_k}$$

To demonstrate that we can measure the entropy change in the surroundings during a spontaneous process, let us use a conceptual device to transfer the heat, q_k, that must be exchanged from the surroundings at temperature, $\hat{T}_k$ to the system. As sketched in Figure 10, we imagine a very small, reversible, ideal-gas Carnot engine, whose

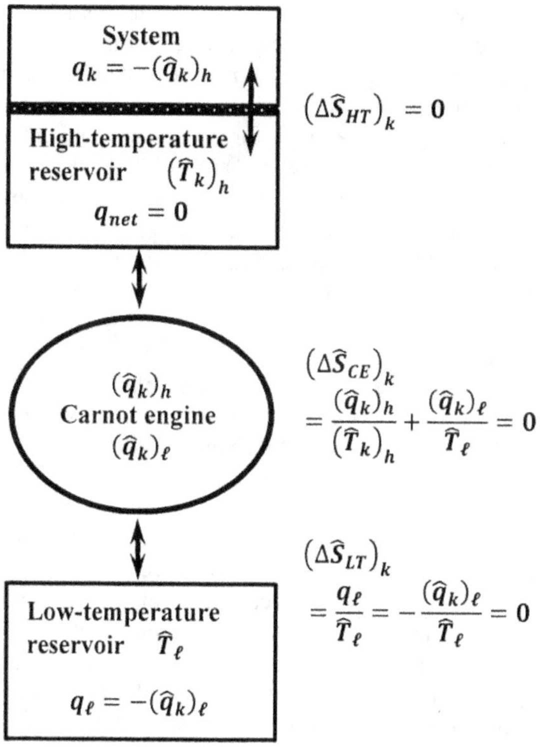

Figure 10. Using a reversible Carnot engine to exchange heat with a spontaneously changing system.

high-temperature reservoir is also very small. We suppose that the Carnot engine delivers a very small heat increment δq to the high temperature reservoir in every cycle. While the system is within C_k, we maintain the Carnot engine's high temperature reservoir at $\hat{T}_k$, and allow heat q_k to pass from the high temperature reservoir to the system. The high temperature reservoir is the only part of the surroundings that is in thermal contact with the system; q_k is the only heat exchanged by the system while it is within C_k.

To maintain the high temperature reservoir at $\hat{T}_k$ we operate the Carnot engine for a large integral number of cycles, n, such that $q_k \approx n \times \delta q$, and do so at a rate that just matches the rate at which heat passes from the high-temperature reservoir to the system. When the system passes from path-segment C_k to path-segment C_{k+1}, we alter the steps in the reversible Carnot cycle to maintain the high-temperature reservoir at the new surroundings temperature, $\hat{T}_{k+1}$. The low-temperature heat reservoir for this Carnot engine is always at the constant temperature $\hat{T}_\ell$. Let the heat delivered from the high-temperature reservoir to the Carnot engine within C_k be $(\hat{q}_k)_h$. We have $q_k = -(\hat{q}_k)_h$. Let the heat delivered from the low-temperature reservoir to the Carnot engine within C_k be $(\hat{q}_k)_\ell$. Let the heat delivered to the low-temperature reservoir within C_k be q_ℓ. We have $q_\ell = -(\hat{q}_k)_\ell$. Since the Carnot engine is reversible, we have

$$\frac{(\hat{q}_k)_h}{\hat{T}_k} + \frac{(\hat{q}_k)_\ell}{\hat{T}_\ell} = 0$$

and

$$-\frac{q_k}{\hat{T}_k} - \frac{q_\ell}{\hat{T}_\ell} = 0$$

so that

$$\frac{q_\ell}{\hat{T}_\ell} = -\frac{q_k}{\hat{T}_k}$$

While the system is within C_k, it receives an increment of heat q_k from the high temperature reservoir. Simultaneously, three components in the surroundings also exchange heat. Let the entropy changes in the high-temperature reservoir, the Carnot engine, and the low-temperature reservoir be $(\Delta\hat{S}_{HT})_k$, $(\Delta\hat{S}_{CE})_k$, and $(\Delta\hat{S}_{LT})_k$, respectively. The high temperature reservoir receives heat q_k from the Carnot engine and delivers the same quantity of heat to the system. The net heat accepted by the high temperature reservoir is zero. No change occurs in the high-temperature reservoir. We have $(\Delta\hat{S}_{HT})_k = 0$. The reversible Carnot engine completes an integral number of cycles, so that $(\Delta\hat{S}_{CE})_k = 0$. The low temperature reservoir accepts heat $-(\hat{q}_k)_\ell = q_\ell$, at the fixed temperature $\hat{T}_\ell$, during the reversible operation of the Carnot engine, so that

$$(\Delta\hat{S}_{LT})_k = \frac{q_\ell}{\hat{T}_\ell} = -\frac{q_k}{\hat{T}_k}$$

The entropy change in the surroundings as the system passes through C_k is

$$\Delta\hat{S}_k = (\Delta\hat{S}_{HT})_k + (\Delta\hat{S}_{CE})_k + (\Delta\hat{S}_{LT})_k = \frac{q_\ell}{\hat{T}_\ell} = -\frac{q_k}{\hat{T}_k}$$

so that, as we observed above,

$$\Delta S_k > -\Delta\hat{S}_k = \frac{q_k}{\hat{T}_k}$$

Since C_k can be any part of path C, and C_k can be made arbitrarily short, we have for every increment of any spontaneous process occurring in a closed system that can exchange heat with its surroundings, $d\hat{S} = -dq/\hat{T}$, and

$$dS > \frac{dq}{\hat{T}}$$

If the temperature of the surroundings is constant between any two points A and B on curve C, we can integrate over this interval to obtain $\Delta_{AB}\hat{S} = -q_{AB}/\hat{T}$ and

$$\Delta_{AB}S > \frac{q_{AB}}{\hat{T}}$$

For an adiabatic process, $q = 0$. For any arbitrarily small increment of an adiabatic process, $dq = 0$. It follows that $\Delta S > 0$ and $dS > 0$ for any spontaneous adiabatic process.

§16 Internal entropy and the second law

For every incremental part of any process, we have $dS + d\hat{S} \geq 0$. Let us define a new quantity, the ***external entropy*** change, as $d_eS = -d\hat{S}$. The change criteria become $dS - d_eS \geq 0$. Now, let us define the ***internal entropy***[1] change as $d_iS = dS - d_eS$. The entropy change for a system is the sum of its internal and external entropy changes, $dS = d_iS + d_eS$. We use d_iS and d_eS to represent incremental changes. To represent macroscopic changes, we use Δ_iS and Δ_eS. Since two processes can effect different changes in the surroundings while the change that occurs in the system is the same, $\Delta\hat{S}$ and Δ_eS are not completely determined by the change in the state of the system. Neither the internal nor the external entropy change depends solely on the change in the state of the system. Nevertheless, we see that $d_iS \geq 0$ or $\Delta_iS \geq 0$ is an alternative expression of the thermodynamic criteria.

The external entropy change is that part of the entropy change that results from the interaction between the system and its surroundings. The internal entropy is that part of the entropy change that results from processes occurring entirely within the system. (We also use the term "internal energy." The fact that the word "internal" appears in both of these terms does not reflect any underlying relationship of material significance.) The criterion $d_iS > 0$ makes it explicit that a process is spontaneous if and only if the events occurring within the system act to increase the entropy of the system. In one common figure of speech, we say "entropy is produced" in the system in a spontaneous process. (It is, of course, possible for a spontaneous process to have $d_iS > 0$ while $d_eS < 0$, and $dS < 0$.)

In §14-1, we introduce a quantity,

$$\sum_{j=1}^{\omega} \mu_j \, dn_j$$

that we can think of as a change in the chemical potential energy of a system. The internal entropy change is closely related to this quantity: We find

$$d_iS = -\frac{1}{T}\sum_{j=1}^{\omega} \mu_j \, dn_j$$

As required by the properties of d_iS, we find that $\sum_{j=1}^{\omega} \mu_j \, dn_j \leq 0$ is an expression of the thermodynamic criteria for change. Internal entropy is a useful concept that is applied to particular advantage in the analysis of many different kinds of spontaneous processes in non-homogeneous systems.

§17 Notation and terminology: conventions for spontaneous processes

We now want to consider criteria for a spontaneous process in which a closed system passes from state A to state B. State B can be an equilibrium state, but state A is not. We can denote the energy change for this process as $\Delta_{AB}E$, and we can find it by measuring the heat and work exchanged with the surroundings as the process takes place, $\Delta_{AB}E = q + w$, or for a process in which the increments of heat and work are arbitrarily small, $d_{AB}E = dq + dw$. Likewise, we can denote the entropy change for the spontaneous process as $\Delta_{AB}S$ or $d_{AB}S$, but we cannot find the entropy change by measuring q^{spon} or dq^{spon}. If we cannot find the entropy change, we cannot find the Helmholtz or Gibbs free energy changes from their defining relationships, $A = E - TS$ and $G = H - TS$. Moreover, intensive variables—pressure, temperature, and concentrations—may not have well-defined values in a spontaneously changing system.

When we say that a reversible process occurs with some thermodynamic variable held constant, we mean what we say: The thermodynamic variable has the same value at every point along the path of reversible change. In the remainder of this chapter, we develop criteria for spontaneous change. These criteria are statements about the values of ΔE, ΔH, ΔA, and ΔG for a system that can undergo spontaneous change under particular conditions. In stating some of these criteria, we specify the conditions by saying that the pressure or the temperature is constant. As we develop these criteria, we will see that these stipulations have specific meanings. When we say that a process occurs "at constant volume" (isochorically), we mean that the volume of the system remains the same throughout the process. When we say that a spontaneous process occurs "at constant pressure" (isobarically or isopiestically), we mean that the pressure applied to the system by the surroundings is constant throughout the spontaneous process and that the system pressure is equal to the applied pressure, $P = P_{applied}$, at all times. When we say that a spontaneous process occurs "at constant temperature", we may mean only that

- the system is continuously in thermal contact with its surroundings
- the temperature of the surroundings is constant
- in the initial and final states, the system temperature is equal to the surroundings temperature.

§18 The heat exchanged by a spontaneous process at constant S

To continue our effort to find change criteria that use only properties of the system, let us consider a spontaneous process, during which the system is in contact with its surroundings and the entropy of the system is constant. For every incremental part of this process, we have $dS = 0$ and $dS + d\hat{S} > 0$. Hence, $d\hat{S} > 0$. It follows that $\Delta S = 0$, $\Delta S + \Delta\hat{S} > 0$, and $\Delta\hat{S} > 0$. (Earlier, we found that the entropy changes for a spontaneous process in an isolated system are $\Delta S > 0$ and $\Delta\hat{S} = 0$. The present system is not isolated.) Since the change that occurs in the system is irreversible, $dS = 0$ does not mean that $dq = 0$. The requirement that $dS = 0$ places no constraints on the temperature of the system or of the surroundings at any time before, during, or after the process occurs.

In §15, we find $dS > dq^{spon}/\hat{T}$ for any spontaneous process in a closed system. If the entropy of the system is constant, we have

$$dq^{spon} < 0$$
(spontaneous process, constant entropy)

for every incremental part of the process. For any finite change, it follows that the overall heat must satisfy the same inequality:

$$q^{spon} < 0$$
(spontaneous process, constant entropy)

For a spontaneous process that occurs with the system in contact with its surroundings, but in which the entropy of the system is constant, the system must give up heat to the surroundings. $dq < 0$ and $q < 0$ are criteria for spontaneous change at constant system entropy.

In §14, we develop criteria for reversible processes. The criteria relate changes in the system's state functions to the reversible non-pressure–volume work that is done on the system during the process. Now we can develop parallel criteria for spontaneous processes.

§19 The energy change for a spontaneous process at constant S and V

From the fundamental equation, $dE - TdS + PdV = dw_{NPV}^{rev}$ for a reversible process. We find that the criterion for reversible change at constant entropy is $(dE)_S = dw_{net}^{rev}$. For a reversible process at constant entropy and volume, we find $(dE)_{SV} = dw_{NPV}^{rev}$

To consider the energy change for a spontaneous process, we begin with $dE = dq + dw$, which is independent of whether the change is spontaneous or reversible. For a spontaneous process in which both pressure–volume, dw_{PV}^{spon}, and non-pressure–volume work, dw_{NPV}^{spon}, are possible, we have $dE = dq^{spon} + dw_{PV}^{spon} + dw_{NPV}^{spon}$, which we can rearrange to

$$dE - dw_{PV}^{spon} - dw_{NPV}^{spon} = dq^{spon}$$

For a spontaneous, constant-entropy change that occurs while the system is in contact with its surroundings, we have $dq^{spon} < 0$. Hence, we have $(dE)_S - dw_{PV}^{spon} - dw_{NPV}^{spon} < 0$. Lettting $dw_{net}^{spon} = dw_{PV}^{spon} + dw_{NPV}^{spon}$, we can express this as

$$(dE)_S < dw_{net}^{spon}$$
(spontaneous process, constant S)

If we introduce the further condition that the spontaneous process occurs while the volume of the system remains constant, we have $dw_{PV}^{spon} = 0$. Making this substitution and repeating our earlier result for a reversible process, we have the parallel relationships

$$(dE)_{SV} < dw_{NPV}^{spon}$$
(spontaneous process, constant S and V)

$$(dE)_{SV} = dw_{NPV}^{rev}$$
(reversible process, constant S and V)

If we introduce the still further requirement that only pressure–volume work is possible, we have $dw_{NPV}^{spon} = 0$. The parallel relationships become

$$(dE)_{SV} < 0$$
(spontaneous process, constant S and V, only PV work)

$$(dE)_{SV} = 0$$
(reversible process, constant S and V, only PV work)

These equations state the criteria for change under conditions in which the entropy and volume of the system remain constant. If the process is reversible, the energy change must be equal to the non-pressure–volume work. If the process is spontaneous, the energy change must be less than the non-pressure volume work. If only pressure–volume work is possible, the energy of the system must decrease in a spontaneous process and remain constant in a reversible process. Each of these differential-expression criteria applies to every incremental part of a change that falls within its scope. In consequence, corresponding criteria apply to finite spontaneous changes. These criteria are listed in the summary in §25.

Now the question arises: What sort of system can undergo a change at constant entropy? If the process is reversible and involves no heat, the entropy change will be zero. If we have a system consisting of a collection of solid objects at rest, we can rearrange the objects without transferring heat between the objects and their surroundings. For such a process, the change in the energy of the system is equal to the net work done on the system. Evidently, reversible changes in mechanical systems occur at constant entropy and satisfy the criterion

$$(dE)_S = dw_{net}^{rev}$$

For a change that occurs reversibly and in which the entropy of the system is constant, the energy change is equal to the net work (of all kinds) done on the system. A spontaneous change in a mechanical system dissipates mechanical energy as heat by friction. If this heat appears in the surroundings and the thermal state of the system remains unchanged, such a spontaneous processes satisfies the criterion

$$(dE)_S < dw_{net}^{spon}$$

We have arrived at the criterion for change that we are accustomed to using when we deal with a change in the potential energy of a constant-temperature mechanical system: A spontaneous change can occur in such a system if and only if the change in the system's energy is less than the net work done on it. The excess work is degraded to heat that appears in the surroundings. This convergence notwithstanding, the principles of mechanics and those of thermodynamics, while consistent with one another, are substantially independent. We address this issue briefly in §12-2.

In the next section, we develop spontaneous-change criteria based on the enthalpy change for a constant-entropy process. In subsequent sections, we consider other constraints and find other criteria. We find that the Helmholtz and Gibbs free energy functions are useful because they provide criteria for spontaneous change when the process is constrained to occur isothermally.

§20 The enthalpy change for a spontaneous process at constant S and P

From $H = E + PV$, we have $dH = dE + d(PV)$. For a spontaneous process in which both pressure–volume and non-pressure–volume work are possible, we can write this as $dH = dq^{spon} + dw_{PV}^{spon} + dw_{NPV}^{spon} + d(PV)$, which we can rearrange to $dH - dw_{PV}^{spon} - dw_{NPV}^{spon} - d(PV) = dq^{spon}$. For a spontaneous constant-entropy change that occurs while the system is in contact with its surroundings, we have $dq^{spon} < 0$, so that

$$(dH)_S - dw_{PV}^{spon} - dw_{NPV}^{spon} - d(PV) < 0.$$

Now, let us introduce the additional constraint that the system is subjected to a constant applied pressure, $P_{applied}$, throughout the process. Thus $P_{applied}$ is a well-defined property that can be measured at any stage of the process. The incremental pressure–volume work done by the surroundings on the system is $dw_{PV}^{spon} = -P_{applied}dV$. In principle, the system can undergo spontaneous change so rapidly that there can be a transitory difference between the system pressure and the applied pressure. In practice, pressure adjustments occur very rapidly. Except in extreme cases, we find that $P = P_{applied}$ is a good approximation at all times. Then the change in the pressure volume product is $d(PV) = P_{applied}dV$. Making these substitutions, the enthalpy inequality becomes

$$(dH)_{SP} < dw_{NPV}^{spon}$$

(spontaneous process, constant S and $P_{applied}$)

From our earlier discussion of reversible processes, we have the parallel relationship

$$(dH)_{SP} = dw_{NPV}^{rev}$$

(any reversible process, constant S and $P_{applied}$)

If we introduce the still further requirement that only pressure–volume work is possible, we have $dw_{NPV} = 0$. The parallel relationships become

$$(dH)_{SP} < 0$$

(spontaneous process, constant S and P, only PV work)

$$(dH)_{SP} = 0$$

(reversible process, constant S and P, only PV work)

These equations state the criteria for change under conditions in which the entropy and pressure of the system remain constant. If the process is reversible, the enthalpy change must be equal to the non-pressure–volume work. If the process is spontaneous, the enthalpy change must be less than the non-pressure–volume work. If only pressure–volume work is possible, the enthalpy of the system must decrease in a spontaneous process and remain constant in a reversible process. Since each of these differential criteria applies to every incremental part of a reversible change that falls within its scope, corresponding criteria apply to finite spontaneous changes. These criteria are listed in the summary in §25.

§21 The entropy change for a spontaneous process at constant E and V

For any spontaneous process, we have $dE = dq^{spon} + dw^{spon}$, which we can rearrange to $dq^{spon} = dE - dw^{spon}$. Substituting our result from §15, we have

$$\hat{T}dS > dE - dw^{spon}$$

(spontaneous process)

If the energy of the system is constant throughout the process, we have $dE = 0$ and

$$\hat{T}(dS)_E > -dw^{spon}$$

(spontaneous process, constant energy)

The spontaneous work is the sum of the pressure–volume work and the non-pressure–volume work, $dw^{spon} = dw_{PV}^{spon} + dw_{NPV}^{spon}$. If we introduce the further condition that the spontaneous process occurs while the volume of the system remains constant, we have $dw_{PV}^{spon} = 0$. Making this substitution and repeating our earlier result for a reversible process, we have the parallel relationships

$$(dS)_{EV} > \frac{-dw_{NPV}^{spon}}{\hat{T}}$$

(spontaneous process, constant E and V)

$$(dS)_{EV} = \frac{-dw_{NPV}^{spon}}{\hat{T}}$$

(reversible process, constant E and V)

(For a reversible process, $T = \hat{T}$.) If the spontaneous process occurs while $\hat{T}$ is constant, summing the incremental contributions to a finite change of state produces the parallel relationships

$$(\Delta S)_{EV} > \frac{-w_{NPV}^{spon}}{\hat{T}}$$

(spontaneous process, constant E, V, and $\hat{T}$)

$$(\Delta S)_{EV} = \frac{-w_{NPV}^{spon}}{\hat{T}}$$

(reversible process, constant E, V, and $\hat{T}$)

Constant $\hat{T}$ corresponds to the common situation in chemical experimentation in which we place a reaction vessel in a constant-temperature bath. If we introduce the further condition that only pressure–volume work is possible, we have $dw_{NPV}^{spon} = 0$. The parallel relationships become

$$(dS)_{EV} > 0$$
(spontaneous process, constant E and V, only PV work)

$$(dS)_{EV} = 0$$
(reversible process, constant E and V, only PV work)

If the energy and volume are constant for a system in which only pressure–volume work is possible, the system is isolated. The conditions we have just derived are entirely equivalent to our earlier conclusions that $dS = 0$ and $dS > 0$ for an isolated system that is at equilibrium or undergoing a spontaneous change, respectively. Summing the incremental contributions to a finite change of state produces the parallel relationships

$$(\Delta S)_{EV} > 0$$
(spontaneous process, only PV work)

$$(\Delta S)_{EV} = 0$$
(reversible process, only PV work)

The validity of these expressions is independent of any variation in either T or $\hat{T}$.

§22 The entropy change for a spontaneous process at constant *H* and *P*

For any spontaneous process, we have

$$\begin{aligned} dH &= dE + PdV + VdP \\ &= dq^{spon} - P_{applied}dV + dw_{NPV}^{spon} + PdV + VdP \end{aligned}$$

If the pressure is constant ($P = P_{applied} =$ constant), this becomes $dq^{spon} = dH - dw_{NPV}^{spon}$. Substituting our result from §15, we have

$$\hat{T}(dS)_P > dH - dw_{NPV}^{spon}$$
(spontaneous process, constant P)

If the enthalpy of the system is also constant throughout the process, we have

$$\hat{T}(dS)_{HP} > -dw_{NPV}^{spon}$$
(spontaneous process, constant H and P)

Dividing by $\hat{T}$ and repeating our earlier result for a reversible process, we have the parallel relationships

$$(dS)_{HP} > \frac{-dw_{NPV}^{spon}}{\hat{T}}$$
(spontaneous process, constant H and P)

$$(dS)_{HP} = \frac{-dw_{NPV}^{rev}}{\hat{T}}$$
(reversible process, constant H and P)

If it is also true that the temperature of the surroundings is constant, summing the incremental contributions to a finite change of state produces the parallel relationships

$$(\Delta S)_{HP} > \frac{-w_{NPV}^{spon}}{\hat{T}}$$
(spontaneous process, constant H, P, and $\hat{T}$)

$$(\Delta S)_{HP} > \frac{-w_{NPV}^{rev}}{\hat{T}}$$
(reversible process, constant H, P, and $\hat{T} = T$)

If only pressure–volume work is possible, we have $dw_{NPV}^{spon} = 0$, and

$$(dS)_{HP} > 0$$
(spontaneous process, constant H, P, only PV work)

$$(dS)_{HP} = 0$$
(reversible process, constant H and P, only PV work)

and for a finite change of state,

$$(\Delta S)_{HP} > 0$$
(spontaneous process, only PV work)

$$(\Delta S)_{HP} = 0$$
(reversible process, only PV work)

In this and earlier sections, we develop criteria for spontaneous change that are based on dE and dH. We are now able to develop similar criteria for a spontaneous change in a system that is in thermal contact with constant-temperature surroundings. These criteria are based on dA and dG. However, before doing so, we develop a general relationship between the isothermal work in a spontaneous process and the isothermal work in a reversible process, when these processes take a system from a common initial state to a common final state.

§23 The reversible work is the minimum work at constant $\hat{T}$

The Clausius inequality leads to an important constraint on the work that can be done on a system during a spontaneous process in which the temperature of the surroundings is constant. As we discuss in §9-7, the initial state of the spontaneous process cannot be a true equilibrium state. In our present considerations, we assume that the initial values of all the state functions of the spontaneously changing system are the same as those of a true equilibrium system. Likewise, we assume that the final state of the spontaneously changing system is either a true equilibrium state or a state whose thermodynamic functions have the same values as those of a true equilibrium system.

From the first law applied to any spontaneous process in a closed system, we have $\Delta E^{rev} = \Delta E^{spon}$ and $q^{rev} + w^{rev} = q^{spon} + w^{spon}$. Since the temperature of the system and its surroundings are equal and constant for the reversible process, we have $q^{rev} = T\Delta S = \hat{T}\Delta S$. So long as the temperature of the surroundings is constant, we have $q^{spon} < \hat{T}\Delta S$ for the spontaneous process. It follows that

$$\hat{T}\Delta S + w^{rev} - w^{spon} = q^{spon} < \hat{T}\Delta S$$

so that

$$w^{rev} < w^{spon} \qquad (\hat{T} \text{ constant})$$

A given isothermal process does the minimum possible amount of work on the system when it is carried out reversibly. (In §7-20, we find this result for the special case in which the only work is the exchange of pressure–volume work between an ideal gas and its surroundings.) Equivalently, a given isothermal process produces the maximum amount of work in the surroundings when it is carried out reversibly: Since $w^{rev} = -\hat{w}^{rev}$ and $w^{spon} = -\hat{w}^{spon}$, we have $-\hat{w}^{rev} < -\hat{w}^{spon}$ or

$$\hat{w}^{rev} > \hat{w}^{spon}$$

§24 The free energy changes for a spontaneous process at constant $\hat{T}$

Now let us consider the change in the Helmholtz free energy when a system undergoes a spontaneous change while in thermal contact with surroundings whose temperature remains constant at $\hat{T}$. We begin by considering an arbitrarily small increment of change in a process in which the temperature of the system remains constant at $T = \hat{T}$. The change in the Helmholtz free energy for this process is $(dA)_T = dE - TdS$. Substituting $dE = dq^{spon} + dw^{spon}$ gives

$$(dA)_T = dq^{spon} + dw^{spon} - TdS$$
(spontaneous process, constant T)

Rearranging, we have $(dA)_T - dw^{spon} + TdS = dq^{spon}$. Using the inequality $dq^{spon} < \hat{T}dS$, we have

$$(dA)_T - dw^{spon} + TdS < \hat{T}dS$$

When we stipulate that $T = \hat{T} =$ constant, this becomes

$$(dA)_T < dw^{spon}$$
(spontaneous process, constant T)

where dw^{spon} is all of the work of any kind done on the system during a small increment of the spontaneous process. If we introduce the still further requirement that the volume is constant, we have $dw_{PV}^{spon} = 0$ and $dw^{spon} = dw_{NPV}^{spon}$. Then

$$(dA)_{TV} < dw_{NPV}^{spon}$$
(spontaneous process, constant T and V)

and if only pressure–volume work is possible,

$$(dA)_{TV} < 0$$
(spontaneous process, constant T and V, only PV work)

From our earlier discussion of reversible processes, we have the parallel relationships

$$(dA)_T = dw_{net}^{rev}$$
(reversible isothermal process)

$$(dA)_{TV} = dw_{NPV}^{rev}$$
(reversible process at constant T and V)

$$(dA)_{TV} = 0$$
(reversible process at constant T and V, only PV work)

Similarly, under these conditions, the change in the Gibbs free energy for a spontaneous isothermal process is

$$\begin{aligned}(dG)_T &= dH - TdS \\ &= dE + d(PV) - TdS \\ &= dq^{spon} + dw_{PV}^{spon} + dw_{NPV}^{spon} + d(PV) - TdS\end{aligned}$$

Rearranging, we have

$$(dG)_T - dw_{PV}^{spon} - dw_{NPV}^{spon} - d(PV) + TdS = dq^{spon} < \hat{T}dS$$

and since $T = \hat{T} =$ constant,

$$(dG)_T < dw_{PV}^{spon} + dw_{NPV}^{spon} + d(PV)$$
(spontaneous process, constant T)

As we did when considering the enthalpy change for a spontaneous process, we introduce the additional constraints that the system is subjected to a constant applied pressure, $P_{applied}$, and that $P = P_{applied}$ throughout the process. The irreversible pressure–volume work done by the surroundings on the system becomes $dw_{PV}^{spon} = -P_{applied}dV$, and the change in the pressure volume product becomes $d(PV) = P_{applied}dV$. The Gibbs free energy inequality becomes

$$(dG)_{TP} < dw_{NPV}^{spon}$$
(spontaneous process, constant $P_{applied}$ and T)

If only pressure–volume work is possible, this becomes

$$(dG)_{TP} < 0$$
(spontaneous process, constant $P_{applied}$ and T, only PV work)

From our earlier discussion of reversible processes, we have the parallel relationships

$$(dG)_{TP} = dw_{NPV}^{rev}$$
(reversible process, constant P and T)

$$(dG)_{TP} = 0$$

(reversible process, constant P and T, only PV work)

Since each of these differential-expression criteria applies to every incremental part of a reversible change that falls within its scope, we have the following criteria for finite spontaneous changes when the temperature of the system is constant:

$$(\Delta A)_T < w^{spon}$$

(spontaneous process, constant T)

$$(\Delta A)_{TV} < w_{NPV}^{spon}$$

(spontaneous process, constant T and V)

$$(\Delta A)_{TV} < 0$$

(spontaneous process, constant T and V, only PV work)

$$(\Delta G)_{TP} < w_{NPV}^{spon}$$

(spontaneous process, constant $P_{applied}$ and T)

$$(\Delta G)_{TP} = 0$$

(spontaneous process, constant $P_{applied}$ and T, only PV work)

While the development we have just made assumes that the system temperature is strictly constant, the validity of these finite-change inequalities is not restricted to the condition of strictly constant system temperature. We can derive these finite-change inequalities by essentially the same argument from less restrictive conditions.

Let us consider a spontaneous process in which a system goes from state B to state C while in contact with surroundings whose temperature remains constant at $\hat{T}$. We suppose that in both state B and state C the system temperature is equal to the surroundings temperature; that is, $T_B = T_C = \hat{T} = \text{constant}$. However, at any intermediate point in the process, the system can have any temperature whatsoever. In states B and C, the Helmholtz free energies are $A_B = E_B - \hat{T}S_B$ and $A_C = E_C - \hat{T}S_C$. The change in the Helmholtz free energy is $(A_C - A_B) = (E_C - E_B) - \hat{T}(S_C - S_B)$ or $(\Delta A)_{\hat{T}} = \Delta E - \hat{T}\Delta S = q^{spon} + w^{spon} - \hat{T}\Delta S$. Rearranging, and using $q^{spon} < \hat{T}\Delta S$, we have
$(\Delta A)_{\hat{T}} - w^{spon} + \hat{T}\Delta S = q^{spon} < \hat{T}\Delta S$, so that

$$(\Delta A)_{\hat{T}} < w^{spon}$$

(spontaneous process, constant $\hat{T}$)

If we require further that the system volume remain constant, there is no pressure–volume work, and we have

$$(\Delta A)_{V\hat{T}} < w_{NPV}^{spon}$$

(spontaneous process, constant $\hat{T}$ and V)

If only pressure–volume work is possible, $w_{NPV}^{spon} = 0$, and

$$(\Delta A)_{V\hat{T}} < 0$$

(spontaneous process, constant $\hat{T}$ and V, only PV work)

Under the same temperature assumptions, and assuming that $P_B = P_C = P_{applied} = \text{constant}$, the Gibbs free energies are $G_B = E_B + P_{applied}V_B - \hat{T}S_B$ and $G_C = E_C + P_{applied}V_C - \hat{T}S_C$. So that $(G_C - G_B) = (E_C - E_B) + P_{applied}(V_C - V_B) - \hat{T}(S_C - S_B)$ or

$$\begin{aligned}(\Delta G)_{P\hat{T}} &= \Delta E + P_{applied}\Delta V - \hat{T}\Delta S \\ &= q^{spon} + w_{PV}^{spon} + w_{NPV}^{spon} + P_{applied}\Delta V - \hat{T}\Delta S\end{aligned}$$

The pressure–volume work is $w_{PV}^{spon} = -P_{applied}\Delta V$. Cancelling and rearranging, we have

$$(\Delta G)_{P\hat{T}} - w_{NPV}^{spon} + \hat{T}\Delta S = q^{spon} < \hat{T}\Delta S$$

and

$$(\Delta G)_{P\hat{T}} < w_{NPV}^{spon}$$

(spontaneous process, constant $\hat{T}$ and P)

If only pressure–volume work is possible,

$$(\Delta G)_{P\hat{T}} < 0$$

(spontaneous process, constant $\hat{T}$ and P, only PV work)

We find $(\Delta G)_{P\hat{T}} < w_{NPV}^{spon}$ for any spontaneous process that occurs at constant pressure, while the system is in contact with surroundings at the constant temperature $\hat{T}$, and in which the initial and final system temperatures are equal to $\hat{T}$. These are the most common conditions for carrying out a chemical reaction. Consider the situation after we mix non-volatile reactants in an open vessel in a constant-temperature bath. We suppose that the initial temperature of the mixture is the same as that of the bath. The atmosphere applies a constant pressure to the system. The reaction is an irreversible process. It proceeds spontaneously until its equilibrium position is reached. Until equilibrium is reached, the reaction cannot be reversed by an arbitrarily small change in the applied pressure or the temperature of the surroundings. $(\Delta G)_{P\hat{T}} < w_{NPV}^{spon}$ and $(\Delta G)_{P\hat{T}} < 0$ are criteria for spontaneous change that apply to this situation whatever the temperature of the system might be during any intermediate part of the process.

§25 Summary: thermodynamic functions as criteria for change

For a spontaneous process, we conclude that the entropy change of the system must satisfy the inequality $\Delta S + \Delta\hat{S} > 0$. For any process that occurs reversibly, we conclude that $\Delta S + \Delta\hat{S} = 0$. For every incremental part of a reversible process that occurs in a closed system, we have the following relationships:

$$dE = TdS - PdV + dw_{NPV}^{rev}$$
$$dH = TdS + VdP + dw_{NPV}^{rev}$$
$$dA = -SdT - PdV + dw_{NPV}^{rev}$$
$$dG = -SdT + VdP + dw_{NPV}^{rev}$$

At constant entropy, the energy relationship becomes:

$$(dE)_S = dw_{net}^{rev}$$
$$(\Delta E)_S = w_{net}^{rev}$$

At constant temperature, the Helmholtz free energy relationship becomes:

$$(dA)_T = dw_{net}^{rev}$$
$$(\Delta A)_T = w_{net}^{rev}$$

For reversible processes in which all work is pressure–volume work:

$$dE = TdS - PdV$$
$$dH = TdS + VdP$$
$$dA = -SdT - PdV$$
$$dG = -SdT + VdP$$

From these general equations, we find the following relationships for reversible processes when various pairs of variables are held constant:

$(dS)_{EV} = -dw_{NPV}^{rev}/T$	$(\Delta S)_{EV} = -w_{NPV}^{rev}/T$
$(dS)_{HP} = -dw_{NPV}^{rev}/T$	$(\Delta S)_{HP} = -w_{NPV}^{rev}/T$
$(dE)_{SV} = dw_{NPV}^{rev}$	$(\Delta E)_{SV} = w_{NPV}^{rev}$
$(dH)_{SP} = dw_{NPV}^{rev}$	$(\Delta H)_{SP} = w_{NPV}^{rev}$
$(dA)_{TV} = dw_{NPV}^{rev}$	$(\Delta A)_{TV} = w_{NPV}^{rev}$
$(dG)_{TP} = dw_{NPV}^{rev}$	$(\Delta G)_{TP} = w_{NPV}^{rev}$

If the only work is pressure–volume work, then $dw_{NPV}^{rev} = 0$, $w_{NPV}^{rev} = 0$, and these relationships become:

$(dS)_{EV} = 0$	$(\Delta S)_{EV} = 0$
$(dS)_{HP} = 0$	$(\Delta S)_{HP} = 0$
$(dE)_{SV} = 0$	$(\Delta E)_{SV} = 0$
$(dH)_{SP} = 0$	$(\Delta H)_{SP} = 0$
$(dA)_{TV} = 0$	$(\Delta A)_{TV} = 0$
$(dG)_{TP} = 0$	$(\Delta G)_{TP} = 0$

For every incremental part of an irreversible process that occurs in a closed system at constant entropy:

$dq^{spon} < 0$ and $(dE)_S < dw_{net}^{spon}$

and

$q^{spon} < 0$ and $(\Delta E)_S < w_{net}^{spon}$

For an irreversible process at constant temperature:

$dq^{spon} < \hat{T}dS$ and $(dA)_{\hat{T}} < dw_{net}^{spon}$

and

$q^{spon} < \hat{T}\Delta S$ and $(\Delta A)_{\hat{T}} < w_{net}^{spon}$

When an irreversible process occurs with various pairs of variables held constant, we find:

$(dS)_{EV} > -dw_{NPV}^{spon}/\hat{T}$	$(\Delta S)_{EV} = -w_{NPV}^{spon}/\hat{T}$
$(dS)_{HP} > -dw_{NPV}^{spon}/\hat{T}$	$(\Delta S)_{HP} > -w_{NPV}^{spon}/\hat{T}$
$(dE)_{SV} < dw_{NPV}^{spon}$	$(\Delta E)_{SV} < w_{NPV}^{spon}$
$(dH)_{SP} < dw_{NPV}^{spon}$	$(\Delta H)_{SP} < w_{NPV}^{spon}$
$(dA)_{\hat{T}V} < dw_{NPV}^{spon}$	$(\Delta A)_{\hat{T}V} < w_{NPV}^{spon}$
$(dG)_{\hat{T}P} < dw_{NPV}^{spon}$	$(\Delta G)_{\hat{T}P} < w_{NPV}^{spon}$

For irreversible processes in which the only work is pressure–volume work, these inequalities become:

$(dS)_{EV} > 0$	$(\Delta S)_{EV} > 0$
$(dS)_{HP} > 0$	$(\Delta S)_{HP} > 0$
$(dE)_{SV} < 0$	$(\Delta E)_{SV} < 0$
$(dH)_{SP} < 0$	$(\Delta H)_{SP} < 0$
$(dA)_{\hat{T}V} < 0$	$(\Delta A)_{\hat{T}V} < 0$
$(dG)_{\hat{T}P} < 0$	$(\Delta G)_{\hat{T}P} < 0$

Problems

1. Does a perpetual motion machine of the second kind violate the principle of conservation of energy?

2. What is the contrapositive of (SL and ~MSL) ⇒ $(\Delta\hat{S} < 0)$? It is a theorem of logic that ~(B and C) ⇒ (~B and/or ~C). Interpret this theorem. Given that SL is true and that ~(SL and ~MSL) is true, prove that ~MSL is true.

3. Max Planck introduced the following statement of the second law:

"It is impossible to construct an engine which will work in a complete cycle, and produce no effect except the raising of a weight and the cooling of a heat-reservoir."

(M. Planck, ***Treatise on Thermodynamics***, 3rd Edition, translated from the seventh German Edition, Dover Publications, Inc., p 89.) Since we take "raising a weight" to be equivalent to "produces work in the surroundings," the Planck statement differs from our machine-based statement only in that it allows the temperature of the heat source to decrease as the production of work proceeds. We can now ask whether this difference has any material consequences. In particular, can we prove that the Planck statement implies our machine-based statement, or vice versa? (Suggestion: Suppose that we have identical Planck-type machines, each with its own heat reservoir. We dissipate by friction the work produced by one machine in the heat reservoir of the other.)

4. Our statements of the first and second laws have a common format: Assertion that a state function exists; operational definition by which the state function can be measured; statement of a property exhibited by this state function. Express the zero-th law of thermodynamics (Chapter 1) in this format.

5. A 0.400 mol sample of N_2 is compressed from 5.00 L to 2.00 L, while the temperature is maintained constant at 350 K. Assume that N_2 is an ideal gas. Calculate the change in the Helmholtz free energy, ΔA.

6. Show that $\Delta G = \Delta A$ when an ideal gas undergoes a change at constant temperature.

7. Calculate ΔE, ΔH, and ΔG for the process in problem 5.

8. A sample of 0.200 mol of an ideal gas, initially at 5.00 bar, expands reversibly and isothermally from 1.00 L to 10.00 L. Calculate ΔE, ΔH, and ΔG for this process.

9. A 100.0 g sample of carbon tetrachloride is compressed from 1.00 bar to 10.00 bar at a constant temperature of 20 C. At 20 C, carbon tetrachloride is a liquid whose density is 1.5940 g mL^{-1}. Assume that the density does not change significantly with pressure. What is ΔG for this process?

10. Calculate the Helmholtz free energy change (ΔA) in problem 9.

11. If C_V is constant, show that the initial and final temperatures and volumes for an adiabatic ideal-gas expansion are related by the equation

$$\left(\frac{T_f}{T_i}\right) = \left(\frac{V_i}{V_f}\right)^{R/C_V}$$

12. At 25 C, the initial volume of a monatomic ideal gas is 5 L at 10 bar. This gas expands to 20 L against a constant applied pressure of 1 bar.
(a) Is this process impossible, spontaneous, or reversible?
(b) What is the final temperature?
(c) Find q, w, ΔE, and ΔH for this process.

13. The same change of state experience by the monatomic ideal gas in problem 12 can be effected in two steps. Let step A be the reversible cooling of the gas to its final temperature while the pressure is maintained constant at 10 bar. Let step B be the reversible isothermal expansion of the resulting gas to a pressure of 1 bar.
(a) Find q, w, ΔE, and ΔH for step A.
(b) Find q, w, ΔE, and ΔH for step B.
(c) From your results in (a) and (b), find q, w, ΔE, and ΔH for the overall process of step A followed by step B.
(d) Compare the values of q, w, ΔE, and ΔH that you find in (c) to the values for the same overall process that you found in problem 12.
(e) Find ΔS and $\Delta \hat{S}$ for step A.
(f) Find ΔS and $\Delta \hat{S}$ for step B.
(g) Find ΔS, $\Delta \hat{S}$, and $\Delta S_{universe}$ for the overall process.

14. Assume that the process in problem 12 occurs while the gas is in thermal contact with its surroundings and that the temperature of the surroundings is always equal to the final temperature of the gas. Find $\Delta \hat{S}$ and $\Delta S_{universe}$ for this process.

15. At 25 C, the initial volume of a monatomic ideal gas is 5 L at 10 bar. The gas expands to 20 L while in thermal contact with surroundings at 125 C. During the expansion, the applied pressure is constant and equal to the equilibrium pressure at the final volume and temperature.
(a) Is this process impossible, spontaneous, or reversible?
(b) Find q, w, ΔE, ΔH, and $\Delta \hat{S}$ for this process.
(c) Find ΔS and $\Delta S_{universe}$ for this process. To find ΔS, it is necessary to find a reversible alternative path that effects the same change in the system's state functions.

16. At 60 C, the density of water is 0.98320 g cm^{-3}, the vapor pressure is 19,932 Pa, and the enthalpy of vaporization is 42,482 J mol^{-1}. Assume that gaseous water behaves as an ideal gas. A vessel containing liquid and gaseous water is placed in a constant 60 C bath, and the applied pressure is maintained at 19,932 Pa while 100 g of water vaporizes.
(a) Is this process impossible, spontaneous, or reversible?
(b) Find q, w, ΔE, ΔH, ΔS, ΔA, and ΔG for this process.
(c) Is $(\Delta S)_{EV} = 0$ a criterion for equilibrium that applies to this system? Why or why not? $(\Delta H)_{SP} = 0$? $(\Delta A)_{VT} = 0$? $(\Delta G)_{PT} = 0$?

17. This problem compares the efficiency and $\sum q/T$ for one mole of a monatomic ideal gas taken around a reversible Carnot cycle to the same quantities for the same gas taken around an irreversible cycle using the same two heat reservoirs.

(i) Let the successive step of the reversible Carnot cycle be a, b, c, and d. Isothermal step a begins with the gas occupying 5.00 L at 600 K and ends with the gas occupying 20.00 L. Adiabatic expansion step b ends with the gas at 300 K. After the isothermal compression step c, the gas is adiabatically compressed in step d to the original state. Find P, V, and T for the gas at the end of each step of this reversible cycle. Find $\sum q/\hat{T}$, ΔS, and $\Delta\hat{S}$ for the cycle a, b, c, d. What is the efficiency of this cycle?

(ii) Suppose that following step b, the ideal gas is warmed at constant volume to 400 K by exchanging heat with the 600 K reservoir. Call this step e. Following step e, the gas is cooled at constant pressure to 300 K by contact with the 300 K reservoir. Call this step f. Following step f, the gas is isothermally and reversibly compressed at 300 K to the same P, V, and T as the gas reaches at the end of step c. Call this step g. Find P, V, and T for the gas at the ends of steps e, f, and g. Although steps e and f are not reversible, the same changes can be effected reversibly by keeping $T = \hat{T}$ as the gas is warmed at constant volume (step e) and cooled at constant pressure (step f). (We discuss this point in further in §12-4.) Consequently, $\Delta_e S = \int_{300\ K}^{400\ K} \frac{C_V}{T} dT$ and $\Delta_f S = \int_{300\ K}^{400\ K} \frac{C_P}{T} dT$. Find $\sum q/\hat{T}$, ΔS, and $\Delta\hat{S}$, and $\Delta S_{universe}$ for the cycle a, b, e, f, g, d. What is the efficiency of this cycle?

(iii) Compare the value of $\sum q/\hat{T}$ that you obtained in part (ii) to value of $\sum q/\hat{T}$ that you obtained in part (i).

(iv) Clausius' theorem states that $\sum q/\hat{T} = 0$ for a cycle traversed reversibly, and $\sum q/\hat{T} < 0$ for a cycle traversed spontaneously. Comment.

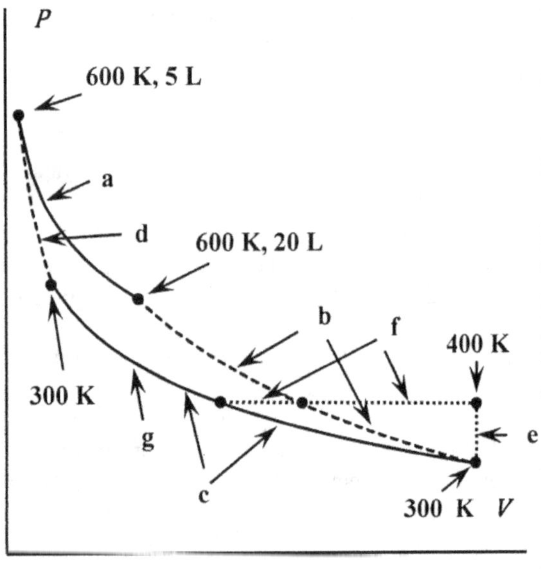

18. For a spontaneous cycle traversed while the temperature changes continuously, Clausius' theorem asserts that $\oint dq/\hat{T} < 0$. Show that this inequality follows from the result, $dS > dq/\hat{T}$, that we obtained in §15 for any spontaneous process in a closed system.

19. In § 6 through §8, we conclude that $\Delta S + \Delta\hat{S} = 0$ is necessary for a reversible process, $\Delta S + \Delta\hat{S} > 0$ is necessary for a spontaneous process, and $\Delta S + \Delta\hat{S} < 0$ is necessary for an impossible process. That is:

(Process is reversible) $\Rightarrow (\Delta S + \Delta\hat{S} = 0)$

(Process is spontaneous) $\Rightarrow (\Delta S + \Delta\hat{S} > 0)$, and

(Process is impossible) $\Rightarrow (\Delta S + \Delta\hat{S} < 0)$.

Since we have defined the categories reversible, spontaneous, and impossible so that they are exhaustive and mutually exclusive, the following proposition is true:

~(Process is spontaneous) and ~(Process is impossible) $\Rightarrow$ (Process is reversible)

(a) Prove that $\Delta S + \Delta\hat{S} = 0$ is sufficient for the process to be reversible; that is, prove: $(\Delta S + \Delta\hat{S} = 0) \Rightarrow$ (Process is reversible)

(b) Prove that $\Delta S + \Delta\hat{S} > 0$ is sufficient for the process to be spontaneous; that is, prove: $(\Delta S + \Delta\hat{S} > 0) \Rightarrow$ (Process is spontaneous)

(c) Prove that $\Delta S + \Delta\hat{S} < 0$ is sufficient for the process to be impossible; that is, prove: $(\Delta S + \Delta\hat{S} < 0) \Rightarrow$ (Process is impossible)

20. Label the successive steps in a reversible Carnot cycle A, B, C, and D, where A is the point at which the pressure is greatest.

(a) Sketch the path ABCD in P–V space.
(b) Sketch the path ABCD in T–dq^{rev}/T space.
(c) Sketch the path ABCE in T– q^{rev} space.
(d) Sketch the path BCDA in T– q^{rev} space.
(e) Sketch the path CDAB in T– q^{rev} space.
(d) Sketch the path DABC in T– q^{rev} space.

21. Assume that the earth's atmosphere is pure nitrogen and that it behaves as an ideal gas. Assume that the molar energy of this nitrogen is constant and that its molar entropy changes are adequately modeled by $d\overline{S} = (C_V/T)dT + (R/\overline{V})d\overline{V}$. For this atmosphere, show that

$$\left(\frac{\partial T}{\partial h}\right)_E = \frac{-\overline{M}g}{C_V}$$

where h is the height above the earth's surface, $\overline{M}$ is the molar mass of dinitrogen (0.0280 kg mol^{-1}), g is the acceleration due to gravity (9.80 m s^{-1}), and C_V is the constant-volume heat capacity (20.8 J K^{-1} mol^{-1}). [Suggestion: Write the total differential for $\overline{E} = \overline{E}(\overline{S}, \overline{V}, h)$. What are $(\partial\overline{E}/\partial\overline{S})_{\overline{V},h}$, $(\partial\overline{E}/\partial\overline{V})_{\overline{S},h}$, and $(\partial\overline{E}/\partial h)_{\overline{S},\overline{V}}$?] If the temperature at sea level is 300 K, what is the temperature on the top of a 3000 m mountain?

22. Assume that the earth's atmosphere is pure nitrogen and that it behaves as an ideal gas. Assume that the molar enthalpy of this nitrogen is constant and that its molar entropy changes are adequately modeled by $d\overline{S} = (C_P/T)dT - (R/P)dP$. For this atmosphere, show that

$$\left(\frac{\partial T}{\partial h}\right)_S = \frac{-\overline{M}g}{C_P}$$

where h is the height above the earth's surface, $\overline{M}$ is the molar mass of dinitrogen (0.0280 kg mol^{-1}), g is the acceleration due to gravity (9.80 m s^{-1}), and C_P is the constant-pressure heat capacity (29.1 J K^{-1} mol^{-1}). [Suggestion: Write the total differential for $\overline{H} = \overline{H}(\overline{S}, P, h)$. What are $(\partial \overline{H}/\partial \overline{S})_{P,h}$, $(\partial \overline{H}/\partial P)_{\overline{V},h}$, and $(\partial \overline{H}/\partial h)_{\overline{S},\overline{V}}$?] Use this approximation to calculate the temperature on the top of a 3000 m mountain when the temperature at sea level is 300 K.

23. Hikers often say that, as a rule-of-thumb, the temperature on a mountain decreases by 1 C for every 100 m increase in elevation. Is this rule in accord with the relationships developed in problems 21 and 22? In these problems, we assume that the temperature of an ideal-gas atmosphere varies with altitude but that the molar energy or enthalpy does not. Does this assumption contradict the principle that the energy and enthalpy of an ideal gas depend only on temperature?

24. Derive the barometric formula (§2-10) from the assumptions that the earth's atmosphere is an ideal gas whose molar mass is $\overline{\boldsymbol{M}}$ and whose temperature and Gibbs free energy are independent of altitude.

25. Run in reverse, a Carnot engine consumes work $(w > 0)$ and transfers heat $(q_\ell > 0)$ from a low-temperature reservoir to a high temperature reservoir $(q_h < 0)$. The work done by the machine is converted to heat that is discharged to the high-temperature reservoir. In one cycle of the machine, $\Delta E = q_\ell + q_h = 0$. For a refrigerator—or for a heat pump operating in air-conditioning mode—we are interested in the quantity of heat removed (q_ℓ) per unit of energy expended (w). We define the coefficient of performance as $COP(cooling) = q_\ell/w$. This is at a maximum for the reversible Carnot engine. Show that the theoretical maximum is

$$COP(cooling) = \frac{1-\epsilon}{\epsilon} = \frac{T_\ell}{T_h - T_\ell}$$

where ϵ is the reversible Carnot-engine efficiency,

$$\epsilon = 1 - \frac{T_\ell}{T_h}$$

26. For a heat pump operating in heating mode—as a "furnace"—we are interested in the quantity of heat delivered to the space being heated $(-q_h)$ per unit of energy expended (w). We define the coefficient of performance as $COP(heating) = -q_h/w$. Show that the theoretical maximum is

$$COP(heating) = \frac{1}{\epsilon} = \frac{T_h}{T_h - T_\ell}$$

27. For $T_\ell = 300$ K and $T_h = 500$ K, calculate the theoretical maxima for $COP(cooling)$ and $COP(heating)$.

28. Find the theoretical maximum $COP(cooling)$ for a refrigerator at 40 F in a room at 72 F.

29. Find the theoretical maximum $COP(cooling)$ for a heat pump that keeps a room at 72 F when the outside temperature is 100 F.

30. Find the theoretical maximum $COP(heating)$ for a heat pump that keeps a room at 72 F when the outside temperature is 32 F.

Notes

[1] For an introduction to the concept of internal entropy and its applications, see Ilya Prigogine, ***Introduction to the Thermodynamics of Irreversible Processes***, Interscience Publishers, 1961.

10

Some Mathematical Consequences of the Fundamental Equation

§1 Thermodynamic relationships from $dE, dH, dA,$ and dG

In Chapter 9, we substitute $dq^{rev} = TdS$, from the second law, into $dE = dq - PdV$, from the first law, to obtain, for any closed system undergoing a reversible change in which the only work is pressure–volume work, the fundamental equation, $dE = TdS - PdV$. In view of the mathematical properties of state functions that we develop in Chapter 7, this result means that we can express the energy of the system as a function of entropy and volume, $E = E(S,V)$. With this choice of independent variables, the total differential of E is $dE = (\partial E/\partial S)_V dS + (\partial E/\partial V)_S dV$. Equating these expressions for dE, we find

$$\left[\left(\frac{\partial E}{\partial S}\right)_V - T\right] dS + \left[\left(\frac{\partial E}{\partial V}\right)_S + P\right] dV = 0$$

for any such system. Since S and V are independent variables, this equation can be true for any arbitrary state of the system only if the coefficients of dS and dV are each identically equal to zero. It follows that

$$\left(\frac{\partial E}{\partial S}\right)_V = T$$

and

$$\left(\frac{\partial E}{\partial V}\right)_S = -P$$

Moreover, because dE is an exact differential, we have

$$\frac{\partial}{\partial V}\left(\frac{\partial E}{\partial S}\right)_V = \frac{\partial}{\partial S}\left(\frac{\partial E}{\partial V}\right)_S$$

so that

$$\left(\frac{\partial T}{\partial V}\right)_S = -\left(\frac{\partial P}{\partial S}\right)_V$$

Using the result $dH = TdS + VdP$, parallel arguments show that enthalpy can be expressed as a function of entropy and pressure, $H = H(S,P)$, so that

$$\left(\frac{\partial H}{\partial S}\right)_P = T$$

and

$$\left(\frac{\partial H}{\partial P}\right)_S = V$$

and

$$\left(\frac{\partial T}{\partial P}\right)_S = \left(\frac{\partial V}{\partial S}\right)_P$$

Since $dA = -SdT - PdV$, the Helmholtz free energy must be a function of temperature and volume, $A = A(T,V)$, and we have

$$\left(\frac{\partial A}{\partial T}\right)_V = -S$$

and

$$\left(\frac{\partial A}{\partial V}\right)_T = -P$$

and

$$\left(\frac{\partial S}{\partial V}\right)_T = \left(\frac{\partial P}{\partial T}\right)_V$$

Likewise, $dG = -SdT + VdP$ implies that the Gibbs free energy is a function of temperature and pressure, $G = G(P,T)$, so that

$$\left(\frac{\partial G}{\partial T}\right)_P = -S$$

and

$$\left(\frac{\partial G}{\partial P}\right)_T = V$$

and

$$\left(\frac{\partial V}{\partial T}\right)_P = -\left(\frac{\partial S}{\partial P}\right)_T$$

§2 $dE = TdS - PdV$ and internal consistency

In Chapter 1, we observe that the business of science is the creation of models that are internally consistent and that accurately describe reality. Logical deduction from tentative hypotheses is a valuable tool in our effort to create new models. Such logical arguments often take the form of "*Gedanken* (thought) experiments," as exemplified by our various arguments about the properties of hypothetical, friction-free, piston engines. Nevertheless, the route by which we arrive at a theory is irrelevant; what counts are the theory's internal consistency and predictive capability. Let us pause, therefore, to note that we have arrived at mathematical expressions of ideas that we initially introduced as principles inferred from experience.

In Chapter 6, we prove Duhem's theorem when the variables are chosen from the set pressure, temperature, volume, and component concentrations. However, the theorem is more general. It asserts that two variables are sufficient to specify changes in the state of a closed, reversible system, in which only pressure–volume work is possible. Our derivations have now led us to the conclusion that the energy of such a system can be expressed as a function of entropy and volume. Given the entropy, the volume, and the function $E = E(S,V)$, the relationships developed above mean that we know S, V, E, P, and T for the system. Given these, we can calculate H, A, and G. That is, specifying the changes in the two variables S and V is sufficient to specify the change in the state of the system. Moreover, we can rearrange the fundamental equation to

$$dV = \left(\frac{T}{P}\right) dS - \left(\frac{1}{P}\right) dE$$

so that the volume can be expressed as a function of entropy and energy. Given S, E, and the function $V = V(S,E)$, we can find P and T. Specifying the changes in S and E is sufficient to specify the change in the system. Finally, we can rearrange the fundamental equation to

$$dS = \left(\frac{1}{T}\right) dE + \left(\frac{P}{T}\right) dV$$

so that $S = S(V,E)$ and specifying changes in E and V is sufficient to specify the change in the system.

Now, let us return to our discussion in §9-7 of the entropy change for an isolated system undergoing a spontaneous change. That discussion explores the use of the machine-based statement of the second law to establish that the entropy of an isolated system must increase during any spontaneous process. To infer that the system's entropy must increase in such a process, we consider the special case in which only pressure–volume work is possible and argue that a change in which $\Delta E = \Delta S = 0$ is no change at all. That is, we assume that specifying the change in E and S is sufficient to specify the change in the state of such a system. It is, therefore, a significant check on the internal consistency of our thermodynamic model to see that $dE = TdS - PdV$ implies that E and S are indeed a sufficient pair.

Finally, let us consider the relationship of a spontaneous process in a closed system to the surface that describes reversible processes in the same system. The energy of the system undergoing reversible change is expressed as $E = E(S,V)$. An energy surface in V-S-E-space is sketched in Figure1. At any point on this surface, the system is at equilibrium. The point (V_0, S_0, E_0) is such a point. The tangent to this surface at (V_0, S_0, E_0) and in the plane $V = V_0$ is the partial derivative $T = (\partial E/\partial S)_V$. The tangent to the surface at (V_0, S_0, E_0) and in the plane $S = S_0$ is the partial derivative $P = -(\partial E/\partial V)_S$.

No point in V-S-E-space that is off the $E = E(S,V)$ surface can describe an equilibrium state of the closed system. As a practical matter, some of these points may represent states that the system can attain. If so, they are transient states of a spontaneously changing

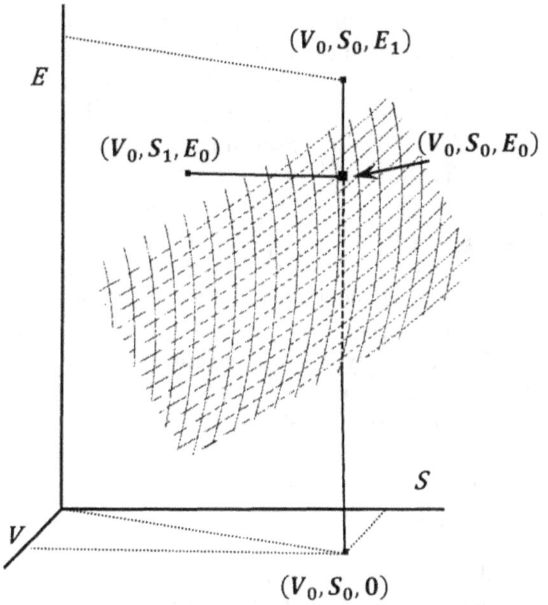

Figure 1. An energy surface in volume–entropy–energy—space.

system. Let us suppose, for example, that we are able to maintain $S = S_0$ and $V = V_0$ while a slow chemical reaction occurs in the system. At every instant, such a state must have a non-equilibrium composition. It must have an energy, and this energy must exceed E_0; since we must have $(\Delta E)_{SV} < 0$ for a spontaneous process, a point (V_0, S_0, E_1) can represent a state of the system during a spontaneous change only if $E_1 > E_0$. If $E_2 < E_0$, the point (V_0, S_0, E_2) cannot represent a state of the system that can spontaneously go to equilibrium at (V_0, S_0, E_0). Similarly, if $S_1 < S_0$ and the system is isolated with $E = E_0$ and $V = V_0$, the point (V_0, S_1, E_0) represents a state of the system that can go to equilibrium at (V_0, S_0, E_0) spontaneously. This process would satisfy the entropy criterion, $(\Delta S)_{EV} > 0$.

§3 Expressing thermodynamic functions with other independent variables

We have found simple differential expressions for E, S, H, A, and G that apply to closed, reversible systems in which only pressure–volume work is possible. From $dE = TdS - PdV$, we infer that $E = E(S,V)$. From $dH = TdS + VdP$, we infer that $H = H(S,P)$. An argument parallel to that above leads us to the conclusion that specifying the changes in S and P is sufficient to specify the change in the state of the system. Similarly, from $dA = -SdT - PdV$ and $dG = -SdT + VdP$, we see that it is sufficient to specify the changes in either V and T or P and T. These total differentials show that that specifying the change in two state functions is sufficient to specify the change that occurs in the state of a closed system, when the change is reversible and all of the work is pressure–volume work. We have now found seven pairs of state functions that are sufficient; they are $\{S,V\}$, $\{S,E\}$, $\{E,V\}$, and the four pairs in which we choose one variable from the set $\{S,T\}$ and one from the set $\{P,V\}$.

However, each of the equations we have obtained so far uses a different pair of independent variables.

Evidently, we should be able to express any thermodynamic function using various pairs of state functions. We can do this by transforming the equations that we have already derived. We are particularly interested in P, V, and T as independent variables, because these quantities are readily measured for most systems. In the sections below, we find exact differentials for dS, dE, dH, dA, and dG with V and T and with P and T as the independent variables.

While specifying the change in some pair of variables is always sufficient to specify the change in the state of a closed reversible system, we should note that it is not always necessary. If the system has only one degree of freedom, specifying some single variable is sufficient. For example, so long as both phases remain present, the change in the state of a pure substance at liquid–vapor equilibrium can be specified by specifying the change in the temperature, the pressure, the volume, or the number of moles of either phase. We discuss this further in §7.

At present, we are developing relationships among state functions that are valid for any closed reversible system in which all work is pressure–volume work. The next several chapters explore the implications of these results. If the composition of the closed reversible system changes during these processes, this composition change does not affect the relationships we develop here. Of course, any composition change that occurs during a reversible process must be reversible; if the components of the system can react, this reaction must be at equilibrium throughout the process. In Chapter 14, we extend the relationships that we develop here to explicitly include molar compositions as independent variables. This enables us to express our theory for equilibrium using composition variables.

In §6-10, we assume—infer from experience—that specifying the changes in P and T is sufficient to specify a change in the state of a closed equilibrium system whose phase composition is fixed and in which only pressure–volume work is possible. We use this assumption to give a partial proof of Duhem's theorem. In §5, we see that this assumption is also a consequence of the theory we have developed. This is another check on the internal consistency of the theory.

Finally, it is time to consider a question we have thus far avoided: Is any arbitrary pair of state functions a sufficient set? The answer is no. In §8, we find that neither $\{P, V\}$ nor $\{S, T\}$ is a sufficient pair in all cases.

§4 Expressing thermodynamic functions with independent variables V and T

If we choose V and T as the independent variables, we can express the differential of E as a function of V and T. We also have the differential relationship $dE = TdS - PdV$. These expressions for dE must be equal:

$$dE = \left(\frac{\partial E}{\partial V}\right)_T dV + \left(\frac{\partial E}{\partial T}\right)_V dT = TdS - PdV$$

Rearranging, we find a total differential for dS with V and T as the independent variables:

$$dS = \frac{1}{T}\left(\frac{\partial E}{\partial T}\right)_V dT + \frac{1}{T}\left[\left(\frac{\partial E}{\partial V}\right)_T + P\right] dV$$

From the coefficient of dT, we have

$$\left(\frac{\partial S}{\partial T}\right)_V = \frac{1}{T}\left(\frac{\partial E}{\partial T}\right)_V = \frac{C_V}{T}$$

where we use the definition $(\partial E/\partial T)_V = C_V$. (When we write "$C_V$," we usually think of it as a property of a pure substance. The relationship above is valid for any reversible system. When we are describing a system that is not a pure substance, C_V is just an abbreviation for $(\partial E/\partial T)_V$.) From the coefficient of dV, we have

$$\left(\frac{\partial S}{\partial V}\right)_T = \frac{1}{T}\left[\left(\frac{\partial E}{\partial V}\right)_T + P\right] = \left(\frac{\partial P}{\partial T}\right)_V$$

where we use the relationship $(\partial S/\partial V)_T = (\partial P/\partial T)_V$ that we find in §1. Substituting into the expression for dS, we find

$$dS = \frac{C_V}{T} dT + \left(\frac{\partial P}{\partial T}\right)_V dV$$

Now, from $dE = TdS - PdV$, we have

$$dE = C_V dT + \left[T\left(\frac{\partial P}{\partial T}\right)_V - P\right] dV$$

From $H = E + PV$, we have

$$\begin{aligned} dH &= dE + d(PV) \\ &= dE + \left(\frac{\partial (PV)}{\partial T}\right)_V dT + \left(\frac{\partial (PV)}{\partial V}\right)_T dV \\ &= dE + V\left(\frac{\partial P}{\partial T}\right)_V dT + \left[P + V\left(\frac{\partial P}{\partial V}\right)_T\right] dV \\ &= \left[C_V + V\left(\frac{\partial P}{\partial T}\right)_V\right] dT + \left[T\left(\frac{\partial P}{\partial T}\right)_V + V\left(\frac{\partial P}{\partial V}\right)_T\right] dV \end{aligned}$$

Of course, we already have

$$dA = -SdT - PdV$$

From $G = H - TS$, by an argument that parallels the above derivation of dH, we obtain

$$dG = \left[V\left(\frac{\partial P}{\partial T}\right)_V - S\right] dT + V\left(\frac{\partial P}{\partial V}\right)_T dV$$

Finally, we can write $P = P(T, V)$ to find

$$dP = \left(\frac{\partial P}{\partial T}\right)_V dT + \left(\frac{\partial P}{\partial V}\right)_T dV$$

P, T, V, C_V, $(\partial P/\partial T)_V$, and $(\partial P/\partial V)_T$ are all experimentally accessible for any reversible system. If we have this information for a system that undergoes a change from a state specified by T_1 and V_1 to a second state specified by T_2 and V_2, we can use these relationships to calculate ΔE, ΔS, and ΔH. To do so, we calculate the appropriate line integral along a reversible path. One such path is an isothermal reversible change, at T_1, from V_1 to V_2, followed by a constant-volume change, at V_2, from T_1 to T_2. In principle, the same procedure can used to calculate ΔA and ΔG. However, because S appears in the differentials dA and dG, this requires that we first find S as a function of V and T.

If the system is a pure substance for which we have an equation of state, we can find $(\partial P/\partial T)_V$, and $(\partial P/\partial V)_T$ by straightforward differentiation. When the substance is a gas, an equation of state may be available in the literature. When the substance is a liquid or a solid, these partial derivatives can still be related to experimentally accessible quantities. The compressibility of a substance is the change in its volume that results from a change in the applied pressure, at a constant temperature. The thermal expansion of a substance is the change in its volume that results from a change in its temperature, at a constant applied pressure. It is convenient to convert measurements of these properties into intensive functions of the state of the substance by expressing the volume change as a fraction of the original volume. That is, we define

The coefficient of thermal expansion:

$$\alpha = \frac{1}{V}\left(\frac{\partial V}{\partial T}\right)_P$$

The coefficient of isothermal compressibility:

$$\beta = -\frac{1}{V}\left(\frac{\partial V}{\partial P}\right)_T$$

Coefficients of thermal expansion and isothermal compressibility are available in compilations of thermodynamic data for many liquids and solids. In general, both coefficients are weak functions of temperature. We have

$$\left(\frac{\partial P}{\partial V}\right)_T = -\frac{1}{\beta V}$$

and

$$\left(\frac{\partial P}{\partial T}\right)_V = -\left(\frac{\partial V}{\partial T}\right)_P \Big/ \left(\frac{\partial V}{\partial P}\right)_T = \frac{\alpha}{\beta}$$

Using these coefficients, we can estimate a pressure change, for example, as a line integral of

$$dP = \left(\frac{\alpha}{\beta}\right) dT - \left(\frac{1}{\beta V}\right) dV$$

§5 Expressing thermodynamic functions with independent variables P and T

We can follow a parallel development to express these thermodynamic functions with P and T as the independent variables. We have the differential relationship $dH = TdS + VdP$. We expand dH with P and T as the independent variables. Equating these, we obtain

$$dH = \left(\frac{\partial H}{\partial P}\right)_T dP + \left(\frac{\partial H}{\partial T}\right)_P dT = TdS + VdP$$

so that we have

$$dS = \frac{1}{T}\left(\frac{\partial H}{\partial T}\right)_P dT + \frac{1}{T}\left[\left(\frac{\partial H}{\partial P}\right)_T - V\right] dP$$

From the coefficient of dT and the definition $(\partial H/\partial T)_P = C_P$, we have

$$\left(\frac{\partial S}{\partial T}\right)_P = \frac{1}{T}\left(\frac{\partial H}{\partial T}\right)_P = \frac{C_P}{T}$$

(When we are describing a reversible system that is not a pure substance, C_P is just an abbreviation for $(\partial H/\partial T)_P$.) From the coefficient of dP and the relationship $(\partial S/\partial P)_T = -(\partial V/\partial T)_P$ that we find in §1, we have

$$\left(\frac{\partial S}{\partial P}\right)_T = \frac{1}{T}\left[\left(\frac{\partial H}{\partial P}\right)_T - V\right] = -\left(\frac{\partial V}{\partial T}\right)_P$$

Substituting into the expression for dS, we find

$$dS = \frac{C_P}{T} dT - \left(\frac{\partial V}{\partial T}\right)_P dP$$

Using the same approach as in the previous section, we can now obtain

$$dE = \left[C_P - P\left(\frac{\partial V}{\partial T}\right)_P\right] dT - \left[P\left(\frac{\partial V}{\partial P}\right)_T + T\left(\frac{\partial V}{\partial T}\right)_P\right] dP$$

$$dH = C_P dT + \left[V - T\left(\frac{\partial V}{\partial T}\right)_P\right] dP$$

$$dA = -\left[S + P\left(\frac{\partial V}{\partial T}\right)_P\right] dT - P\left(\frac{\partial V}{\partial P}\right)_T dP$$

and, we already have

$$dG = VdP - SdT$$

Finally, we can write $V = V(P, T)$ to find

$$dV = \left(\frac{\partial V}{\partial P}\right)_T dP + \left(\frac{\partial V}{\partial T}\right)_P dT$$

so that we have total differentials for all of the principal thermodynamic functions when they are expressed as functions of P and T. If an equation of state is not known but the coefficients of thermal expansion and isothermal compressibility are available, we have $(\partial V/\partial T)_P = \alpha V$ and $(\partial V/\partial P)_T = -\beta V$. Then we can estimate a volume change, for example, as a line integral of

$$dV = \alpha V dT - \beta V dP$$

§6 The transformation of thermodynamic variables in general

Let us suppose that M, Q, R, X, and Y are state functions and that we know the total differentials

$$dM = \left(\frac{\partial M}{\partial X}\right)_Y dX + \left(\frac{\partial M}{\partial Y}\right)_X dY$$
$$dQ = \left(\frac{\partial Q}{\partial X}\right)_Y dX + \left(\frac{\partial Q}{\partial Y}\right)_X dY$$
$$dR = \left(\frac{\partial R}{\partial X}\right)_Y dX + \left(\frac{\partial R}{\partial Y}\right)_X dY$$

To find the total differential of $M(Q,R)$,

$$dM = \left(\frac{\partial M}{\partial Q}\right)_R dQ + \left(\frac{\partial M}{\partial R}\right)_Q dR$$

we solve the total differentials of $Q(X,Y)$ and $R(X,Y)$ to find dX and dY in terms of dQ and dR. Since dQ and dR are simultaneous equations in the variables dX and dY, we can apply Cramer's rule to obtain

$$dX = \frac{\begin{vmatrix} dQ & (\partial Q/\partial Y)_X \\ dR & (\partial R/\partial Y)_X \end{vmatrix}}{J\left(\frac{Q,R}{X,Y}\right)} = \frac{\left(\frac{\partial R}{\partial Y}\right)_X dQ - \left(\frac{\partial Q}{\partial Y}\right)_X dR}{J\left(\frac{Q,R}{X,Y}\right)}$$

and

$$dY = \frac{\begin{vmatrix} (\partial Q/\partial X)_Y & dQ \\ (\partial R/\partial X)_Y & dR \end{vmatrix}}{J\left(\frac{Q,R}{X,Y}\right)} = \frac{-\left(\frac{\partial R}{\partial X}\right)_Y dQ + \left(\frac{\partial Q}{\partial X}\right)_Y dR}{J\left(\frac{Q,R}{X,Y}\right)}$$

where $J((Q,R)/(X,Y))$ is the Jacobian of the transformation of variables X and Y to variables Q and R:

$$J\left(\frac{Q,R}{X,Y}\right) = \begin{vmatrix} (\partial Q/\partial X)_Y & (\partial Q/\partial Y)_X \\ (\partial R/\partial X)_Y & (\partial R/\partial Y)_X \end{vmatrix} = \left(\frac{\partial Q}{\partial X}\right)_Y \left(\frac{\partial R}{\partial Y}\right)_X - \left(\frac{\partial Q}{\partial Y}\right)_X \left(\frac{\partial R}{\partial X}\right)_Y$$

To find

$$dM = \left(\frac{\partial M}{\partial Q}\right)_R dQ + \left(\frac{\partial M}{\partial R}\right)_Q dR$$

We substitute these results for dX and dY into the total differential of $M = M(X,Y)$:

$$\begin{aligned} dM &= \left(\frac{\partial M}{\partial X}\right)_Y dX + \left(\frac{\partial M}{\partial Y}\right)_X dY \\ &= \frac{\left(\frac{\partial M}{\partial X}\right)_Y \left[\left(\frac{\partial R}{\partial Y}\right)_X dQ - \left(\frac{\partial Q}{\partial Y}\right)_X dR\right]}{J\left(\frac{Q,R}{X,Y}\right)} + \frac{\left(\frac{\partial M}{\partial Y}\right)_X \left[-\left(\frac{\partial R}{\partial X}\right)_Y dQ + \left(\frac{\partial Q}{\partial X}\right)_Y dR\right]}{J\left(\frac{Q,R}{X,Y}\right)} \\ &= \left[\frac{\left(\frac{\partial M}{\partial X}\right)_Y \left(\frac{\partial R}{\partial Y}\right)_X - \left(\frac{\partial M}{\partial Y}\right)_X \left(\frac{\partial R}{\partial X}\right)_Y}{J\left(\frac{Q,R}{X,Y}\right)}\right] dQ \\ &\quad + \left[\frac{-\left(\frac{\partial M}{\partial X}\right)_Y \left(\frac{\partial Q}{\partial Y}\right)_X + \left(\frac{\partial M}{\partial Y}\right)_X \left(\frac{\partial Q}{\partial X}\right)_Y}{J\left(\frac{Q,R}{X,Y}\right)}\right] dR \end{aligned}$$

where the coefficients of dQ and dR are $(\partial M/\partial Q)_R$ and $(\partial M/\partial R)_Q$, respectively. In §5, we find the other total differentials in terms of dP and dT. If we set $X = T$ and $Y = P$, we can use these relationships to find the total differential for any state function expressed in terms of any two other state functions.

To illustrate this point, let us use these relationships to find the total differential of S expressed as a function of P and V, $S = S(P,V)$. In this case, we are transforming from the variables (P,T) to the variables (P,V). This is a one-variable transformation. To effect it, we make the additional substitutions $M = S$, $Q = V$, and $R = P$. Since we have $Y = R = P$, the transformation equations simplify substantially. We have

$$\left(\frac{\partial R}{\partial Y}\right)_X = \left(\frac{\partial P}{\partial P}\right)_X = 1$$
$$\left(\frac{\partial R}{\partial X}\right)_Y = \left(\frac{\partial P}{\partial T}\right)_P = 0$$
$$\left(\frac{\partial Q}{\partial Y}\right)_X = \left(\frac{\partial V}{\partial P}\right)_T$$
$$\left(\frac{\partial Q}{\partial X}\right)_Y = \left(\frac{\partial V}{\partial T}\right)_P$$

The Jacobian becomes

$$J\left(\frac{Q,R}{X,Y}\right) = \left(\frac{\partial V}{\partial T}\right)_P$$

and the partial derivatives of S become

$$\left(\frac{\partial S}{\partial V}\right)_P = \left(\frac{\partial S}{\partial T}\right)_P \Big/ \left(\frac{\partial V}{\partial T}\right)_P = \frac{C_P}{T}\left(\frac{\partial T}{\partial V}\right)_P$$

and

$$\left(\frac{\partial S}{\partial P}\right)_V = \frac{-\left(\frac{\partial S}{\partial T}\right)_P \left(\frac{\partial V}{\partial P}\right)_T + \left(\frac{\partial S}{\partial P}\right)_T \left(\frac{\partial V}{\partial T}\right)_P}{\left(\frac{\partial V}{\partial T}\right)_P} = \frac{C_P}{T}\left(\frac{\partial T}{\partial P}\right)_V + \left(\frac{\partial S}{\partial P}\right)_T$$

§7 Reversibility and thermodynamic surfaces

We have found total differentials for the principal thermodynamic functions with V and T and with P and T as the independent variables. In §6, we see how to find such differentials, $dM = M_X dX + M_Y dY$, for any pair of independent variables, X and Y. These equations express our physical theory as a mathematical structure. Because E, T, S, P, V, H, A, and G are state functions, the mathematical properties of state functions enable us to obtain the relationships $M_X = (\partial M/\partial X)_Y$, $M_Y = (\partial M/\partial Y)_X$, and $(\partial M_Y/\partial X) = (\partial M_X/\partial Y)$ that we find in §1. These equations apply to reversible processes in closed systems when only pressure–volume work is possible. Using these equations to describe reversible processes involves a number of important ideas. Let us consider four cases: (1) None of M, X, and Y are constant; (2) M is constant, but X and Y are not; (3) X is constant, but M and Y are not; (4) X and Y are constant, but M is not.

None of *M*, *X*, and *Y* is constant.– The properties of state functions and the existence of the exact differential, $dM = M_X dX + M_Y dY$ imply that M is a function of X and Y, $M = M(X, Y)$. If M_X and M_Y are single-valued and continuous along some path in the XY-plane, we can evaluate the line integral of dM along this path. Given M at a first point, $M(X_1, Y_1)$, we can find M at a second point, $M(X_2, Y_2)$, by evaluating this line integral along some path in the XY-plane between the points (X_1, Y_1) and (X_2, Y_2). Given the values of X and Y, the value of M is uniquely determined; $M(X, Y)$ is a two-dimensional surface in the three-dimensional space whose dimensions are M, X, and Y.

The $M(X, Y)$ surface is just the set of points that are accessible to the system in reversible processes in which X and Y change. Conversely, the only values of M that are accessible to the system in reversible processes in which X and Y change correspond to those points that lie on the surface. In §5, we find the total differentials for dE, dS, dH, dA, and dG using P and T as the independent variables. Evidently, for a reversible process in a closed system, there is a surface representing each of E, S, H, A, and G over the P–T-plane. Since the system is also characterized by an equation of state that relates the values of P, V, and T, there is also a surface representing V over the P–T-plane.

In general, a given system can also undergo spontaneous changes. Suppose that a system is originally at equilibrium at temperature T_1 and pressure P_1. If we contact this system with surroundings at some arbitrary $\hat{T}$ and we arrange for the pressure applied to the system to have some arbitrary value, $P_{applied}$, the system will respond, eventually reaching equilibrium with the system temperature equal to $\hat{T}$ and the system pressure equal to $P_{applied}$. (The change-enabled state in which $P_{system} = P_1$ and $T_{system} = T_1$, while the applied pressure is $P_{applied}$ and the surroundings temperature is $\hat{T}$, is a hypothetical state. It is not an equilibrium state, because $P_{system} \neq P_{applied}$ and $T_{system} \neq \hat{T}$. The change-enabled state can undergo spontaneous change; however, its thermodynamic functions have the same values as they have in the original equilibrium state, in which $P_{system} = P_1$ and $T_{system} = T_1$.) Since this change is spontaneous, it may not be possible to trace the path of the system in the P–T-plane as the change occurs. If we can trace the path in the P–T-plane, the energy of the system can be described as a line in E–P–T-space, but this line will not lie on the reversible-process surface specified by the function $E = E(P, T)$. Nevertheless, we can select paths in the P–T-plane that connect the initial point (P_1, T_1) to the final point $(P_{applied}, \hat{T})$. There are reversible processes that correspond to these paths. By evaluating the line integral for any state function along any of these paths, we can find the change that occurs in the state function during the spontaneous process.

Cases arise in which M is not single-valued or continuous along some or all of the paths that connect points (X_1, Y_1) and (X_2, Y_2). Then M_X or M_Y may not exist for some points (X, Y). In this case, it may not be possible to evaluate the line integral to find the change $M(X_2, Y_2) - M(X_1, Y_1)$. This can occur when there is a phase change. If (P_{vp}, T_{bp}) specifies a state of liquid–vapor equilibrium, the enthalpy of the system is not single-valued. Below, we consider the thermodynamic surfaces of water when a phase change occurs. In §8, we see that M may not be single-valued when V and P or when T and S are the independent variables.

***M* is constant, *X* and *Y* are not.**– If M is constant, we have $dM = 0$. If X and Y are not constant, if M_X and M_Y are defined, and if $M_Y \neq 0$, we can apply the divide-through rule to obtain

$$\left(\frac{\partial Y}{\partial X}\right)_M = -\frac{M_X}{M_Y}$$

Such relationships are useful. In Chapter 12, we discuss the Clapeyron and Clausius-Clapeyron equations, which we obtain from $dG = -SdT + VdP$ using this argument.

***X* is constant, *M* and *Y* are not.**– If X is constant, we have $dX = 0$. Instead of, $dM = M_X dX + M_Y dY$, we have , $dM = M_X dX$, and we must ask whether M is indeed expressible as a function of Y only. If M can be expressed as a function of Y only, so that M_Y is single-valued and continuous, we can integrate to find

$$M_2 - M_1 = \int_{Y_1}^{Y_2} M_Y \, dY$$

***X* and *Y* are constant, *M* is not constant.**– An interesting and important case arises when X and Y are constant, but M is not constant. When X and Y are constant, $dX = dY = 0$, and from $dM = M_X dX + M_Y dY$, it follows that $dM = 0$. Nevertheless, we can readily identify processes in which some state function, M, changes while two (or more) others, X and Y, remain constant. In this case, it is clear that X and Y are not sufficient to model the change in M. Recall from our discussion of Duhem's Theorem that, while two independent variables are sufficient to describe a reversible process in which pressure-volume work is the only work, which pair of variables is adequate depends on the system.

Below we discuss the reversible vaporization of water at constant P and T. For this process we have $dG = 0$. However, we know that $dS > 0$. For independent variables P and T, our differential expressions for dG and dS are

$$dG = VdP - SdT$$

and

$$dS = \frac{1}{T}\left(\frac{\partial H}{\partial T}\right)_P dT - \left(\frac{\partial V}{\partial T}\right)_T dP$$

Setting $dP = dT = 0$ in these equations correctly gives $dG = 0$; however, $dS = 0$ is false. Evidently, variables P and T are not sufficient to model the entropy change in this process.

However, at constant P and T, variables V and T are adequate. We have

$$dS = \frac{1}{T}\left(\frac{\partial E}{\partial T}\right)_V dT + \left(\frac{\partial P}{\partial T}\right)_V dV$$

In §12-10 we develop the Clausius-Clapeyron equation for this vaporization process; we find

$$\left(\frac{\partial P}{\partial T}\right)_V = \frac{P\Delta_{vap}\overline{H}}{RT^2}$$

($\Delta_{vap}\overline{H}$ is the enthalpy of vaporization per mole.) Since the volume of the system is essentially the volume of the gas phase, we have, assuming the vapor behaves as an ideal gas, $V = n_g RT/P$, and

$$dV = \frac{RT}{P} dn_g$$

The entropy equation then becomes

$$\begin{aligned} dS &= \left(\frac{\partial P}{\partial T}\right)_V dV \\ &= \frac{P\Delta_{vap}\overline{H}}{RT^2} \cdot \frac{RT}{P} dn_g \\ &= \frac{\Delta_{vap}\overline{H}}{T} dn_g \end{aligned}$$

so that the entropy change for this reversible process is directly proportional to the number of moles of vapor produced.

We see that we must introduce an extensive variable to model the entropy change in the vaporization process. The system volume serves this purpose, although we wind up expressing this volume in terms of the number of moles of vapor in the system.

From another perspective, we can write the entropy as a function of P, T, and n_g: $S = S(P, T, n_g)$. Then

$$dS = \frac{1}{T}\left(\frac{\partial H}{\partial T}\right)_P dT - \left(\frac{\partial V}{\partial T}\right)_T dP + \left(\frac{\partial S}{\partial n_g}\right)_{PT} dn_g$$

which becomes

$$(dS)_{PT} = \left(\frac{\partial S}{\partial n_g}\right)_{PT} dn_g$$

with

$$\left(\frac{\partial S}{\partial n_g}\right)_{PT} = \frac{\Delta_{vap}\overline{H}}{T}$$

In Chapter 14, we extend all of our thermodynamic models to include variables that specify the system composition—the number of moles of the substances present in the system.

Thermodynamic surfaces in the reversible vaporization of water.– To illustrate the fact that $dT = dP = 0$ has different implications for dS than it does for dG, let us consider the reversible vaporization of one mole of water at constant P and T. ΔG for this process is zero, but ΔS and ΔH are not. We can describe a system comprised of one mole of liquid water using pressure and temperature as the independent variables. There is a wide range of pressure and temperature values that is consistent with the system remaining entirely liquid. Every combination of pressure and temperature at which the system remains entirely liquid can be reached by a reversible process from any other such combination of pressure and temperature. For every combination of pressure and temperature within this range, there is one and only one value for every other thermodynamic function. Choosing the enthalpy function to be specific, we can say that the set of enthalpy–pressure–temperature-points for which the system remains entirely liquid is a thermodynamic surface on which reversible change is possible.

We can say all of these same things about a system that consists of one mole of gaseous water. Of course, the enthalpy surface for gaseous water is a different surface from the enthalpy surface for liquid water. At any given temperature, there is a pressure at which liquid and gaseous water are at equilibrium. Above this pressure, the system is entirely liquid; below it, the system is entirely gas. The enthalpy surface of the liquid lies over a different part of the pressure–temperature-plane than does that of the gas. (If the liquid can be superheated or the gas can be supercooled, a given pressure and temperature may be represented by a point on the enthalpy surfaces for both the pure liquid and the pure gas.) The enthalpy surface for the gaseous system lies at higher energies than that for the liquid system; the two enthalpy surfaces do not intersect.

To reversibly transform pure liquid water to pure gaseous water, we must move on the enthalpy surface of the liquid to a pressure and temperature at which the liquid and the gas are at equilibrium. At this pressure and temperature, we can reversibly increase the volume of the system, causing the reversible vaporization of liquid water, and we can continue this process until all of the liquid has been vaporized. When all of the liquid has been vaporized, the system is on the enthalpy surface of the gas. Thereafter, we can change the pressure and temperature of the system to reversibly change the state of the pure gas. While we can describe this process in terms of the successive changes that we impose on the state functions of a system that consists of one mole of water, we are considering three different systems when we describe the overall process from the perspective afforded by Gibbs' phase rule.

- The first system is one mole of pure liquid. This system has one phase. There are two degrees of freedom, which we take to be pressure and temperature.
- The second system is one mole of water, of which x_ℓ mole is liquid and $1 - x_\ell$ mole is gas, at equilibrium, at a fixed pressure and temperature. In the vaporization process, the pressure and temperature are constant while the volume of the system increases (x_ℓ decreases) reversibly. The one mole of water is described by this system from the time the first bubble of gas appears to the time the last drop of liquid vaporizes. There are two phases and one degree of freedom. When we reversibly vaporize water at a fixed pressure and temperature, one variable must describe the composition: We can take it to be the volume of the system or the liquid mole fraction, x_ℓ. (Of course, we can reversibly vaporize water in a process in which the pressure, temperature, and composition all change; however, because there is only one degree of freedom, specifying a temperature change uniquely determines the pressure change, and conversely. In H-P-T-space, a reversibly vaporizing system traces a path on a vertical plane between the enthalpy surfaces of the liquid and the gas. If pressure and temperature are constant, this path is a vertical line. If reversibility is achieved through synchronous variation of pressure and temperature, the path is not vertical, but it remains in a vertical plane.)
- The third system describes the mole of water after all of the water has been converted to the gas. This system has one phase and two degrees of freedom, which we again take to be pressure and temperature.

We can say that this description of the reversible conversion of liquid water to gaseous water involves three Gibbsian H-P-T-manifolds. Two of these are the enthalpy surfaces for the gas and the liquid. The third is a line of enthalpies at constant pressure and temperature; successive points on this line represent different mole fractions of liquid water.

We can track the reversible conversion of one mole of pure liquid water to pure gas on other thermodynamic surfaces. For example, if we consider enthalpy as a function of volume and temperature, the entire process can be traced on a single $H(V,T)$ surface; that is, every volume–temperature point, (V,T), specifies a unique state of the system, and conversely.

When we use pressure and temperature as the independent variables, the Gibbs free energy provides the criterion for reversibility. Unlike the corresponding enthalpy surfaces, which never meet, the Gibbs free energy surfaces for the pure liquid and the pure gas intersect along a line of pressure and temperature values. At an equilibrium pressure and temperature, the Gibbs free energy change for the reversible vaporization of water is zero, which means that the Gibbs free energy for a mole of liquid water is the same as the Gibbs free energy for a mole of gaseous water at that pressure and temperature.

When we trace the reversible conversion of a mole of liquid water to a mole of gaseous water on the Gibbs free energy surfaces, the point representing the state of the mole of water moves on the Gibbs free energy surface of the liquid from the initial pressure and temperature to the pressure–temperature equilibrium line. The pressure–temperature equilibrium line is formed by the intersection of the Gibbs free energy surface of the liquid with the Gibbs free energy surface of the gas. (The projection of this line of intersection onto the P-T-plane is a line of points in the P-T-plane that satisfies the differential relationship $0 = -\Delta_{vap}SdT + \Delta_{vap}VdP$. The point $P = 1$ atm, $T = 373.15$ K, lies on this line.)

The conversion of liquid to gas can occur while the mole of water remains at the same point in pressure–temperature–Gibbs free energy space, and the mole fractions, x_ℓ and $1 - x_\ell$, vary continuously over the range $0 < x_\ell < 1$. During this reversible vaporization process, $dG = dT = dP = 0$, while $\Delta_{vap}V > 0$, $\Delta_{vap}H > 0$, and $\Delta_{vap}S > 0$. (We find $\Delta_{vap}H$ and $\Delta_{vap}S$ by measuring the heat required to vaporize a mole of water at P and T. Then $q_P = \Delta_{vap}H$, and $q_P/T = \Delta_{vap}S$. Since the process is reversible, we have $\Delta_{vap}S = -\Delta_{vap}\hat{S}$, and $\Delta_{vap}S + \Delta_{vap}\hat{S} = 0$.) When the conversion of liquid to gas is complete, reversible changes to the one mole of gaseous water correspond to motion of a point on the Gibbs free energy surface of the gas.

§8 Using the pair (V,P) or the pair (T,S) as independent variables

When only pressure–volume work is possible, various pairs of state functions can specify the state of any closed equilibrium system. A given pair may be sufficient to specify the state on a given Gibbsian manifold but not to specify the state of the same system on a different Gibbsian manifold. (In §7, we see that the set $\{T,P\}$ is sufficient to specify the state of a closed system containing only liquid water or only gaseous water. However, specifying T and P is not sufficient to establish the amount of water vapor in a closed system in which the liquid is vaporizing reversibly.)

It can also happen that a given pair of state functions specifies the state of a closed system over part but not all of a given Gibbsian manifold. Specifying the values of such a pair is not sufficient to describe the entire manifold. In particular, we can show that neither the set $\{P,V\}$ nor the set $\{T,S\}$ is always a sufficient pair in this sense.

When we say that specifying P and V is sufficient to specify the state of a system on a particular Gibbsian manifold, we mean that any state function, M, must be uniquely specified when P and V are specified; a single-valued function, $M(P,V)$, must exist. Conversely, if for any choice of M in any system, $M(P,V)$ is not single-valued, P and V are not always a sufficient set. In §6, we see how to find the total differential,

$$dM = \left(\frac{\partial M}{\partial P}\right)_V dP + \left(\frac{\partial M}{\partial V}\right)_P dV$$

It might seem that this is sufficient to ensure that specifying P and V always enables us to find $M(P,V)$ relative to its value in an initial reference state $M(P_1,V_1)$. To do so, we need only evaluate dM as a line integral along some reversible path that leads from (P_1,V_1) to (P_2,V_2). However, we can evaluate this line integral only if both of the partial derivatives can be integrated. If one of the partial derivatives is undefined along any path that connects (P_1,V_1) to (P_2,V_2), we cannot find $M(P,V)$ by this method.

Let us consider a closed reversible system that consists of one mole of liquid water. At ordinary pressures, the density of liquid water is not a monotonic function of temperature. At one atmosphere, the density of liquid water reaches a maximum at 4 C. Therefore, at a pressure of one atmosphere, the molar volume of water is a minimum at 4 C, as indicated in Figure 2. This means that, at one atmosphere and a range of volumes, liquid water can be at either of two temperatures for specified values of P and $\overline{V}$. Therefore, specifying P and $\overline{V}$ does not specify T; temperature is not a single-valued function of pressure and volume; we cannot uniquely express the temperature as the required function $T = T(P,\overline{V})$. Moreover, because the density has a maximum, we have

$$\left(\frac{\partial \overline{V}}{\partial T}\right)_P = 0$$

at this maximum, and it follows that

$$\left(\frac{\partial T}{\partial \overline{V}}\right)_P$$

is not defined at this temperature and pressure. In §6, we find

$$\left(\frac{\partial \overline{S}}{\partial \overline{V}}\right)_P = \frac{C_P}{T}\left(\frac{\partial T}{\partial \overline{V}}\right)_P$$

so that $\left(\partial \overline{S}/\partial \overline{V}\right)_P$ is also undefined. Hence, we cannot evaluate $\Delta\overline{S}$ by evaluating the line integral of $d\overline{S}(P,\overline{V})$ along any path that includes a point of maximum density. These examples show that pressure and volume are not sufficient to describe the entire Gibbsian manifold for liquid water.

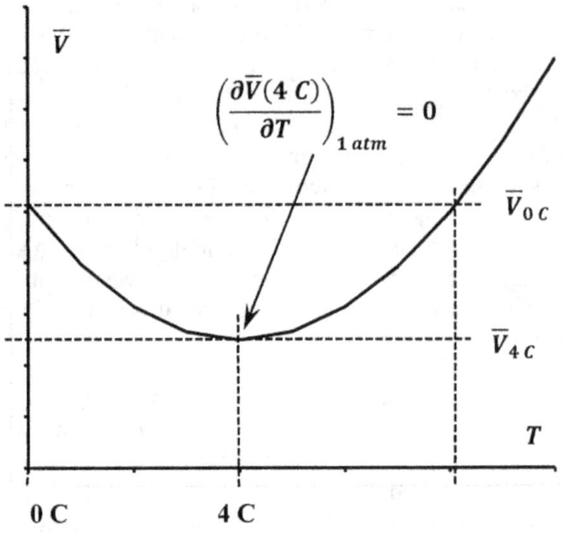

Figure 2. Molar volume of water *versus* temperature.

Temperature and entropy are likewise not sufficient. Since we have

$$d\overline{S} = \frac{C_P}{T}dT - \left(\frac{\partial \overline{V}}{\partial T}\right)_P dP$$

the total differential for pressure as a function of entropy and temperature is

$$dP = \left[\frac{C_P}{T}dT - d\overline{S}\right]\left(\frac{\partial T}{\partial \overline{V}}\right)_P$$

so that dP is not defined at pressures and temperatures of maximum density. Consequently, we cannot express the pressure as $P = P(\overline{S},T)$ over the entire liquid region.

§9 The relationship between C_P and C_V for any substance

In Chapter 7, we derive the relationship between C_P and C_V for an ideal gas. It is useful to have a relationship between these quantities that is valid for any substance. We can derive this relationship from the equations for dS that we develop in §4 and §5. If we apply the divide-through rule to dS expressed as a function of dT and dV, at constant pressure, we have

$$\left(\frac{\partial S}{\partial T}\right)_P = \frac{C_V}{T} + \left(\frac{\partial P}{\partial T}\right)_V \left(\frac{\partial V}{\partial T}\right)_P$$

From dS expressed as a function of T and P,

$$dS = \frac{C_P}{T}dT - \left(\frac{\partial V}{\partial T}\right)_P dP$$

we have

$$\left(\frac{\partial S}{\partial T}\right)_P = \frac{C_P}{T}$$

so that

$$\frac{C_P}{T} = \frac{C_V}{T} + \left(\frac{\partial P}{\partial T}\right)_V \left(\frac{\partial V}{\partial T}\right)_P$$

and

$$C_P - C_V = T\left(\frac{\partial P}{\partial T}\right)_V \left(\frac{\partial V}{\partial T}\right)_P$$

For an ideal gas, the right side of the latter equation reduces to R, in agreement with our previous result. Note also that, for any substance, C_P and C_V become equal when the temperature goes to zero.

The partial derivatives on the right hand side can be related to the coefficients of thermal expansion, α, and isothermal compressibility, β. Using

$$\frac{\alpha}{\beta} = \left(\frac{\partial P}{\partial T}\right)_V$$

we can write the relationship between C_P and C_V as

$$C_P - C_V = \frac{VT\alpha^2}{\beta}$$

§10 The dependence of C_V on volume and of C_P on pressure

The heat capacities of a substance increase with temperature. The rate of increase decreases as the temperature increases. To achieve adequate accuracy in calculations, we often need to know how heat capacities depend on temperature. In contrast, the dependence of heat capacities on pressure and volume is usually negligible; that is, the dependence of C_V on V and the dependence of C_P on P can usually be ignored. Nevertheless, we need to know how to find them.

An exact equation for the dependence of C_V on V follows readily from dS expressed as a function of dT and dV

$$dS = \frac{C_V}{T}dT + \left(\frac{\partial P}{\partial T}\right)_V dV$$

Since the mixed second-partial derivatives must be equal, we have

$$\left[\frac{\partial}{\partial V}\left(\frac{C_V}{T}\right)\right]_T = \left[\frac{\partial}{\partial T}\left(\frac{\partial P}{\partial T}\right)_V\right]_V$$

and thus

$$\left(\frac{\partial C_V}{\partial V}\right)_T = T\left(\frac{\partial^2 P}{\partial T^2}\right)_V$$

Similarly, the dependence of C_P on P follows from dS expressed as a function of dT and dP,

$$dS = \frac{C_P}{T}dT - \left(\frac{\partial V}{\partial T}\right)_P dP$$

Equating the mixed second-partial derivatives, we have

$$\left[\frac{\partial}{\partial P}\left(\frac{C_P}{T}\right)\right]_T = \left[-\frac{\partial}{\partial T}\left(\frac{\partial V}{\partial T}\right)_P\right]_P$$

and thus

$$\left(\frac{\partial C_P}{\partial P}\right)_T = -T\left(\frac{\partial^2 V}{\partial T^2}\right)_P$$

For an ideal gas, it follows that C_V is independent of V, and C_P is independent of P.

When we use the coefficient of thermal expansion to describe the variation of volume with temperature, we have

$$\left(\frac{\partial V}{\partial T}\right)_P = \alpha V$$

When it is adequate to approximate α as a constant, another partial differentiation with respect to temperature gives

$$\left(\frac{\partial C_P}{\partial P}\right)_T = -T\left(\frac{\partial(\alpha V)}{\partial T}\right)_P = -\alpha^2 TV$$

Since α is normally small, this result predicts weak dependence of C_P on P. If α and β are both adequately approximated as constants, we have from

$$\left(\frac{\partial P}{\partial T}\right)_V = \frac{\alpha}{\beta}$$

that

$$\left(\frac{\partial C_V}{\partial V}\right)_T = T\left(\frac{\partial(\alpha/\beta)}{\partial T}\right)_V = 0$$

§11 The Gibbs-Helmholtz equation

When temperature and pressure are the independent variables, the Gibbs free energy is the change criterion that takes the most simple form: $dG = -SdT + VdP$. In chemical applications, temperature and pressure are often the most convenient choice of independent variables, making the Gibbs free energy a particularly useful function. Constant Gibbs free energy is the criterion for equilibrium at constant pressure and temperature. The Gibbs free energy of the system does not change when the system goes from one equilibrium state to another at the same temperature and pressure, $(dG)_{PT} = 0$. An equilibrium system of ice and water is one example; we can melt a portion of the ice, changing the composition of the system, while maintaining equilibrium at constant pressure and temperature. Similarly, we may be able to change the equilibrium composition of an equilibrium system that consists of reacting gases by changing the volume of the system while maintaining constant pressure and temperature.

Consider a system that undergoes some arbitrary change from a state A, in which its Gibbs free energy is G_A, to a second state B, in which its Gibbs free energy is G_B. In general, $G_A \neq G_B$; in the most general case, the pressures and temperatures of states A and B are different. (For example, state A might be a mole of ice at - 10 C and 0.5 bar, while state B is a mole of water at +10 C and 2.0 bar. Either of these states can be converted to the other; however, they are not at equilibrium with one another, and their Gibbs free energies are not equal.) Representing the pressures and temperatures as P_1, T_1, P_2, T_2, we can express the Gibbs free energies of these two state as $G_A = G_A(P_1, T_1)$ and $G_B = G_B(P_2, T_2)$, respectively. The difference is the change in the Gibbs free energy when the system passes from state A to state B:

$$\Delta_{AB}G = G_B(P_2, T_2) - G_A(P_1, T_1)$$

Often, we are interested in Gibbs free energy differences between states that are at the same pressure and temperature, say P_1 and T_1. Then the Gibbs free energy difference is

$$\Delta_{AB}G(P_1, T_!) = G_B(P_1, T_!) - G_A(P_1, T_1)$$

(For example, state A might be a mole of ice at - 10 C and 0.5 bar, while state B is a mole of water, also at - 10 C and 0.5 bar. This would be a super-cooled state of liquid water. These states are not at equilibrium with one another, and their Gibbs free energies are not equal. The difference between the Gibbs free energies of these states is the change in the Gibbs free energy when ice goes to super-cooled water at - 10 C and 0.5 bar.)

Similar considerations apply to expressing differences between the enthalpies and the entropies of two states that are available to a system. The Gibbs free energy is defined by $G = H - TS$. When we are interested in a process that converts some state A to a second state B at constant pressure and temperature, we usually write $\Delta G = \Delta H - T\Delta S$, relying on the context for the information about the pressure and temperature and the initial and final states. To explicitly denote that the change is one that occurs at a constant temperature, T_0, we can write $\Delta G(T_0) = \Delta H(T_0) - T_0\Delta S(T_0)$.

Frequently we are interested in the way that ΔG, ΔH, and ΔS vary with temperature at constant pressure. If we know how G, H, and S vary with temperature for each of the two states of interest, we can find the temperature dependence of ΔG, ΔH, and ΔS. The Gibbs-Helmholtz equation is a frequently useful expression of the temperature dependence of G or ΔG. Since it is a mathematical consequence of our thermodynamic relationships, we derive it here.

At constant pressure, the temperature derivative of the Gibbs free energy is $-S$; that is,

$$\left(\frac{\partial G}{\partial T}\right)_P = -S$$

Using this result and the definition, $G = H - TS$, we obtain the temperature dependence of G as

$$\left(\frac{\partial G}{\partial T}\right)_P = \frac{G - H}{T}$$

However, the Gibbs-Helmholtz equation can be expressed most compactly as the temperature derivative of G/T. As a matter of calculus, we have

$$\left(\frac{\partial (G/T)}{\partial T}\right)_P = -\frac{G}{T^2} + \frac{1}{T}\left(\frac{\partial G}{\partial T}\right)_P$$

Using the relationships above, this becomes

$$\left(\frac{\partial (G/T)}{\partial T}\right)_P = -\frac{H - TS}{T^2} - \frac{S}{T} = -\frac{H}{T^2}$$

Since $\Delta_{AB}G = G_B - G_A$, we have

$$(\partial \Delta_{AB}G/\partial T)_P = -(S_B - S_A) = -\Delta_{AB}S$$

and

$$\begin{aligned}\left(\frac{\partial (\Delta_{AB}\, G/T)}{\partial T}\right)_P &= \left(\frac{\partial (G_B/T)}{\partial T}\right)_P - \left(\frac{\partial (G_A/T)}{\partial T}\right)_P \\ &= -\left(\frac{H_B}{T^2} - \frac{H_A}{T^2}\right) \\ &= -\frac{\Delta_{AB}H}{T^2}\end{aligned}$$

(The Gibbs-Helmholtz equation)

If we know the temperature dependence of ΔS or ΔH, we can find the temperature dependence of ΔG by integrating the relationships above. That is, given ΔG at T_1, we can find ΔG at T_2. Thus, from $(\partial G/\partial T)_P = -S$, we have

$$\int_{\Delta G(T_1)}^{\Delta G(T_2)} \left(\frac{\partial \Delta G}{\partial T}\right)_P dT = \Delta G(T_2) - \Delta G(T_1) = -\int_{T_1}^{T_2} \Delta S\, dT$$

and from $(\partial(\Delta G/T)/\partial T)_P = -\Delta H/T^2$, we have

$$\begin{aligned}\int_{\Delta G(T_1)/T_1}^{\Delta G(T_2)/T_2} \left(\frac{\partial (\Delta G/T)}{\partial T}\right)_P dT &= \frac{\Delta G(T_2)}{T_2} - \frac{\Delta G(T_1)}{T_1} \\ &= -\int_{T_1}^{T_2} \frac{\Delta H}{T^2} dT\end{aligned}$$

For small temperature differences, ΔH is often approximately constant. Then, we can evaluate the change in ΔG from

$$\frac{\Delta G(T_2)}{T_2} - \frac{\Delta G(T_1)}{T_1} = \Delta H\left(\frac{1}{T_2} - \frac{1}{T_1}\right)$$

Another common application arises when we know ΔG at several temperatures. A plot of $\Delta G/T$ versus $1/T$ is then approximately linear with a slope that approximates the average value of ΔH in the temperature interval.

§12 The second law and the properties of ideal gases

We make extensive use of the principle that the energy of an ideal gas depends only on temperature when only pressure–volume work is possible. In Chapter 2, we consider the Joule experiment, which provides weak evidence that this principle is correct. In the Joule experiment, no temperature change is observed during the adiabatic free expansion of a gas whose behavior is approximately ideal at the initial temperature and pressure. While this observation supports the principle, the accuracy attainable in the Joule experiment is poor. Otherwise, the most compelling evidence for this principle that we have developed is the theoretical relationship between the pressure–volume product of an ideal gas model and the mean-square velocity of its molecules. We derive this relationship from the Maxwell-Boltzmann distribution law for gas velocities and use the ideal gas equation to find that the mean squared velocity depends only on temperature.

To appreciate the importance of this principle, let us review some of the important steps in our development of the second law. In Chapter 7, we observe that, since its energy depends only on temperature, C_V for an ideal gas must also depend only on temperature; this follows immediately from the definition, $C_V = (\partial E/\partial T)_V$. In Chapter 9, we use the conclusion that C_V depends only on temperature in our development of the relationships among the heat, work, volume, and temperature changes for an ideal gas traversing a Carnot cycle. In considering these relationships, we observe that the values of the terms q^{rev}/T for the steps in this cycle sum to zero, as required for a state function. This leads us to define entropy by the differential expression $dS = dq^{rev}/T$ and to infer that the entropy so defined is a state function. Reasoning from the machine-based statement of the second law, we conclude that this inference is correct.

That the energy of an ideal gas depends only on temperature is therefore of central importance to the internal consistency of the thermodynamic theory we have developed. It is easy to demonstrate this internal consistency. From the ideal gas equation and the relationships developed earlier in this chapter, we can show that the quantities $(\partial E/\partial V)_T$, $(\partial E/\partial P)_T$, $(\partial H/\partial V)_T$, $(\partial H/\partial P)_T$, $(\partial C_V/\partial V)_T$, and $(\partial C_P/\partial P)_T$ are all identically zero.

The fact that our theory passes this test of internal consistency is independent of the properties of real gases. However, since we want to make predictions about the behavior of real gases, we need to be able to measure these quantities for real gases. Moreover, because we want to understand the properties of real gases in terms of their molecular characteristics, we want to be able to interpret these quantities for real gases using real-gas models that explain the differences between real gas molecules and ideal gas molecules. The van der Waals equation of state provides a simple model for the effects of attractive and repulsive molecular interactions. In the next section, we first consider simple qualitative arguments about the effects of intermolecular interactions on the energy of a real gas. We then investigate these effects for a van der Waals gas. We see that the van der Waals model and our qualitative arguments are consistent.

§13 The pressure-dependence of the energy and enthalpy of a real gas

Let us consider the effects that intermolecular forces of attraction and repulsion must have in the adiabatic free expansion $\left(P_{applied} = 0\right)$ of a real gas. In such an expansion, no energy can be exchanged between the gas and its surroundings.

Suppose that the molecules of the gas are attracted to one another. Then energy must be expended to separate the molecules as the expansion takes place. (To achieve the expansion, work must be done against the intermolecular attractive forces.) Since the system cannot obtain this energy from its surroundings, it must be obtained by decreasing the translational kinetic energy (and the rotational and vibrational energy) of the gas molecules themselves. This means that the temperature of the gas must decrease during the expansion.

Conversely, if the molecules repel one another, energy is released as the expansion takes place, and the temperature of the gas increases during the expansion. The temperature can remain unchanged after the adiabatic free expansion only if the effects of the intermolecular forces of attraction and repulsion offset one another exactly.

We can express these conclusions more precisely by saying that we expect $(\partial T/\partial P)_E > 0$ if forces of attraction dominate the intermolecular interactions. We expect $(\partial T/\partial P)_E < 0$ if forces of repulsion dominate. Now, as a matter of mathematics, we have

$$\left(\frac{\partial E}{\partial P}\right)_T = -\left(\frac{\partial E}{\partial T}\right)_P \left(\frac{\partial T}{\partial P}\right)_E$$

As a matter of experience, increasing the temperature of any gas at constant pressure always increases the energy of the gas; that is, we observe $(\partial E/\partial T)_P > 0$. It follows that we can expect

$$\left(\frac{\partial E}{\partial P}\right)_T < 0$$

(attraction dominates)

when intermolecular forces of attraction dominate and

$$\left(\frac{\partial E}{\partial P}\right)_T > 0$$

(repulsion dominates)

when forces of repulsion dominate.

In the Joule experiment, a gas is allowed to expand into an initially evacuated container. The Joule experiment is a direct test of these ideas; however, as we have noted, it is difficult to carry out accurately. Fortunately, a simple modification of the Joule experiment produces an experiment that is much more sensitive. Instead of allowing the gas to expand freely into a fixed volume, we allow it to expand adiabatically against a constant applied pressure. This is the Joule-Thomson experiment. In the next section, we show that the enthalpy of the gas does not change in such a process. We measure the temperature change as the gas expands adiabatically from an initial, constant, higher pressure to a final, constant, lower pressure. Since this process occurs at constant enthalpy, the Joule-Thomson experiment measures

$$\left(\frac{\partial T}{\partial P}\right)_H$$

from which we can obtain

$$\left(\frac{\partial H}{\partial P}\right)_T$$

To interpret the Joule-Thomson experiment in terms of intermolecular forces, we need to show that

$$\left(\frac{\partial H}{\partial P}\right)_T < 0$$

(attraction dominates)

at pressures and temperatures where intermolecular forces of attraction dominate and

$$\left(\frac{\partial H}{\partial P}\right)_T > 0$$

(repulsion dominates)

where forces of repulsion dominate. To do this using an explicit mathematical model for a real gas, let us find

$$\left(\frac{\partial E}{\partial P}\right)_T$$

and

$$\left(\frac{\partial H}{\partial P}\right)_T$$

for a van der Waals gas. Writing van der Waals equation in terms of the molar volume, $\left(P + a/\overline{V}^2\right)\left(\overline{V} - b\right) = RT$, and introducing

$$\gamma\left(P,\overline{V}\right) = P - \frac{a}{\overline{V}^2} + \frac{2ab}{\overline{V}^3}$$

so that we can express the results more compactly, we find

$$\left(\frac{\partial P}{\partial \overline{V}}\right)_T = -\frac{\gamma\left(P,\overline{V}\right)}{\overline{V} - b}$$

so that

$$\left(\frac{\partial \overline{V}}{\partial P}\right)_T = -\frac{\overline{V} - b}{\gamma\left(P,\overline{V}\right)}$$

and

$$\left(\frac{\partial \overline{V}}{\partial T}\right)_P = \frac{R}{\gamma\left(P,\overline{V}\right)}$$

Substituting into results we develop in §5, we have

$$\left(\frac{\partial \overline{E}}{\partial P}\right)_T = -P\left(\frac{\partial \overline{V}}{\partial P}\right)_T - T\left(\frac{\partial \overline{V}}{\partial T}\right)_P = -\frac{a\left(\overline{V} - b\right)}{\overline{V}^2\gamma\left(P,\overline{V}\right)}$$

and

$$\left(\frac{\partial \overline{H}}{\partial P}\right)_T = \overline{V} - T\left(\frac{\partial \overline{V}}{\partial T}\right)_P = \overline{V} - \frac{RT}{\gamma\left(P,\overline{V}\right)}$$

We introduce van der Waals equation in §2-12. By the argument we make there, the $\left(P + a/\overline{V}^2\right)$ term models the effects of attractive intermolecular interactions when $a > 0$. By a parallel argument, we can see that it models the effects of repulsive interactions when $a < 0$. Parameter b models the effects of intermolecular repulsive interactions that come into play when the molecules come into contact with one another. For present purposes, we can consider molecules for which $b = 0$; this simplifies our equations without affecting the description they give of the phenomena that are of current interest.

This gives us a model in which the effects of intermolecular interactions are described by the values of a single parameter that has a straightforward physical interpretation. Thus, we can write

$$\left(P + a/\overline{V}^2\right)\overline{V} = RT$$

to describe a gas of point-mass molecules that experience intermolecular forces. When $a > 0$, these forces are attractive; when $a < 0$ they are repulsive. (For any given real gas, our equation can only be an approximation that is valid over a limited range of conditions. In some ranges, $a > 0$; in others, $a < 0$.) With $b = 0$ we must have

$$\gamma\left(P,\overline{V}\right) = P - \frac{a}{\overline{V}^2} > 0$$

(If $\gamma\left(P,\overline{V}\right) \le 0$, we have $\left(\partial P/\partial \overline{V}\right)_T \ge 0$. As a matter of experience, the pressure of a gas always decreases with increasing volume at constant temperature. It follows that van der Waals equation with $b = 0$ and $\left(P - a/\overline{V}^2\right) \le 0$ cannot describe any gas.) With $b = 0$ we have

$$\left(\frac{\partial \overline{E}}{\partial P}\right)_T = \frac{-a}{\overline{V}\left(P - a/\overline{V}^2\right)}$$

and

$$\left(\frac{\partial \overline{H}}{\partial P}\right)_T = RT\left[\frac{1}{P + a/\overline{V}^2} - \frac{1}{P - a/\overline{V}^2}\right]$$

For a gas at conditions in which forces of attraction dominate, we have $a > 0$, so that

$$\left(\partial \overline{E}/\partial P\right)_T < 0 \qquad \text{and} \qquad \left(\partial \overline{H}/\partial P\right)_T < 0$$
(attraction dominates)

Conversely, at conditions in which forces of repulsion dominate, we have $a < 0$, and

$$\left(\partial \overline{E}/\partial P\right)_T > 0 \qquad \text{and} \qquad \left(\partial \overline{H}/\partial P\right)_T > 0$$
(repulsion dominates)

§14 The Joule-Thomson effect

In practice, the Joule-Thomson experiment is done by allowing gas from a pressure vessel to pass through an insulated tube. The tube contains a throttling valve or a porous plug through which gas flows slowly enough so that the gas upstream from the plug is at a uniform pressure P_1, and the gas downstream is at a uniform pressure P_2. In general, the temperature of the downstream gas is different from that of the upstream gas. Depending on the initial temperature and pressure, the pressure drop, and the gas, the temperature of the gas can either decrease or increase as it passes through the plug. (We see below that it must be constant if the gas is ideal.)

The temperature change is called the Joule-Thomson effect. The enthalpy of the gas remains constant. If the measured temperature and pressure changes are ΔT and ΔP, their ratio is called the Joule-Thomson coefficient, μ_{JT}. We define

$$\mu_{JT} = \left(\frac{\partial T}{\partial P}\right)_H \approx \frac{\Delta T}{\Delta P}$$

To see that the enthalpy of the gas is the same on both sides of the plug, we consider an idealized version of the experiment, in which the flow of gas through the plug is controlled by the coordinated movement of two pistons. (See Figure 3.) We suppose that the gas is pushed through the plug in such a way that the upstream pressure remains constant at P_1 and the downstream pressure remains constant at P_2. Let us consider the changes that result when one mole of gas passes through the plug under these conditions. Initially, there are $n_1 + 1$ moles of gas on the upstream side at a pressure P_1, occupying a volume $(n_1 + 1)\overline{V}_1$, at a temperature T_1, and having an energy per mole of $\overline{E}_1$. On the downstream side, there are n_2 moles of gas at a pressure P_2, occupying a volume $n_2\overline{V}_2$, but having a temperature T_2 and an energy per mole of $\overline{E}_2$.

When the process is complete, there are n_1 moles of gas on the upstream side, still at a pressure P_1 and temperature T_1, but occupying a volume $n_1\overline{V}_1$. On the downstream side, there are $n_2 + 1$ moles of gas at pressure P_2, occupying volume $(n_2 + 1)\overline{V}_2$ at a temperature T_2 and with an energy per mole of $\overline{E}_2$. On the upstream side, $\Delta E_1 = -\overline{E}_1$ and

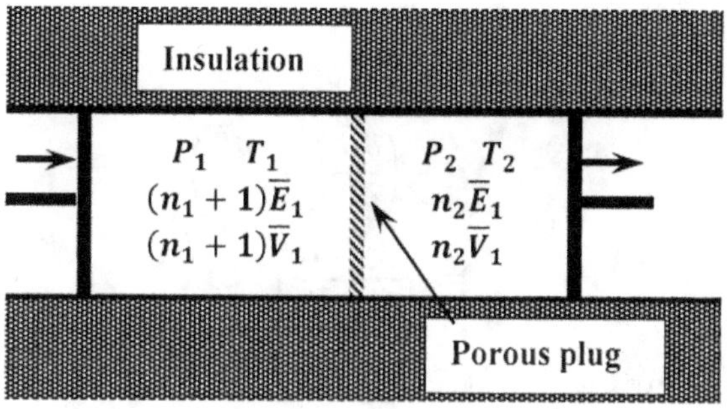

Figure 3. The idealized Joule-Thomson experiment.

$$w_1 = -P_1\left[n_1\overline{V}_1 - (n_1 + 1)\overline{V}_1\right] = P_1\overline{V}_1$$

On the downstream side, $\Delta E_2 = \overline{E}_2$, and

$$w_2 = -P_2\left[(n_2 + 1)\overline{V}_1 - n_2\overline{V}_1\right] = -P_2\overline{V}_2$$

Since the process is adiabatic, any heat taken up by the upstream gas must be surrendered by the downstream gas, so that $q_1 + q_2 = 0$. For the process of moving the mole of gas across the plug,

$$\begin{aligned} \Delta E &= \Delta E_1 + \Delta E_2 \\ &= -\overline{E}_1 + \overline{E}_2 \\ &= q_1 + q_2 + w_1 + w_2 \\ &= P_1\overline{V}_1 - P_2\overline{V}_2 \end{aligned}$$

from which

$$\overline{E}_1 + P_1\overline{V}_1 = \overline{E}_2 + P_2\overline{V}_2$$

or

$$\overline{H}_1 = \overline{H}_2$$

so that we have $\Delta H = 0$ for the expansion.

In practice, it is convenient to measure downstream pressures and temperatures, P_2 and T_2, in a series of experiments in which the upstream pressure and temperature, P_1 and T_1, are constant. The enthalpy of the gas is the same at each of these pressure-temperature points. A graph of these points is an isenthalpic (constant enthalpy) curve. At any given pressure and temperature, the Joule-Thomson coefficient, μ_{JT}, is the slope of this curve.

We can also express μ_{JT} as a function of the heat capacity, C_P, and the coefficient of thermal expansion, α, where $\alpha = V^{-1}(\partial V/\partial T)_P$. We begin by expressing $d\overline{H}$ as a function of temperature and pressure:

$$d\overline{H} = \left(\frac{\partial \overline{H}}{\partial T}\right)_P dT + \left(\frac{\partial \overline{H}}{\partial P}\right)_T dP$$

If we divide through by dP and hold $\overline{H}$ constant, we obtain

$$0 = \left(\frac{\partial \overline{H}}{\partial T}\right)_P \left(\frac{\partial T}{\partial P}\right)_{\overline{H}} + \left(\frac{\partial \overline{H}}{\partial P}\right)_T$$

so that

$$\mu_{JT} = \left(\frac{\partial T}{\partial P}\right)_{\overline{H}}$$
$$= -\left(\frac{\partial \overline{H}}{\partial P}\right)_T \Big/ \left(\frac{\partial \overline{H}}{\partial T}\right)_P$$
$$= -\frac{1}{C_P}\left(\frac{\partial \overline{H}}{\partial P}\right)_T$$

If we substitute the coefficient of thermal expansion into the expression for $\left(\partial\overline{H}/\partial P\right)_T$ that we develop in §5, we have

$$\left(\frac{\partial \overline{H}}{\partial P}\right)_T = \overline{V} - T\left(\frac{\partial \overline{V}}{\partial T}\right)_P$$
$$= \overline{V} - \alpha\overline{V}T$$
$$= \overline{V}(1-\alpha T)$$

For an ideal gas, $\left(\partial\overline{V}/\partial T\right)_P = \overline{V}/T$, so that both $\left(\partial\overline{H}/\partial P\right)_T$ and μ_{JT} are zero. For real gases, we substitute into the expression for μ_{JT} to find

$$\mu_{JT} = \left(\frac{\partial T}{\partial P}\right)_{\overline{H}}$$
$$= -\frac{1}{C_P}\left(\frac{\partial \overline{H}}{\partial P}\right)_T$$
$$= -\frac{\overline{V}}{C_P}(1-\alpha T)$$

Given $\overline{V}$ and any two of μ_{JT}, C_P, or α, we can find the third from this relationship.

Making the same substitutions using the partial derivatives we found above for a van der Waals gas, we find

$$\mu_{JT} = -\frac{1}{C_P}\left(\overline{V} - \frac{RT}{\gamma(P,\overline{V})}\right)$$

Given that the van der Waals equation oversimplifies the effects of intermolecular forces, we can anticipate that calculation of the Joule-Thomson coefficient from the van der Waals parameters is likely to be qualitatively correct, but in poor quantitative agreement with experimental results. Figure 4 compares calculated and experimental curves for the Joule-Thomson coefficient of nitrogen gas at 0 C from 1 to 200 bar. (Calculated values take $a = 0.137$ Pa m^6 mol^{-2} and $b = 3.81 \times 10^{-5}$ m^3 mol^{-1}. The experimental data are from reference 1.)

We anticipate that the Joule-Thomson coefficient becomes zero at pressures and temperatures where the effects of intermolecular attractions and repulsions exactly offset one another. For interactions between molecules, attractive forces have the dominant effect at long distances, while repulsive forces dominate at short distances. The lower the pressure, the greater the average distance between gas molecules. Therefore, at any given

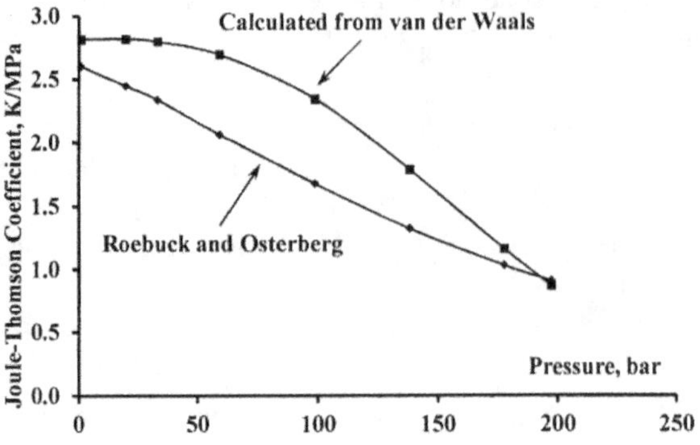

Figure 4. The Joule-Thomson coefficient for N_2 at 0 C.

temperature and a sufficiently low pressure, the effects of intermolecular attractive forces are more important than those of intermolecular repulsive forces. At low pressures, the Joule-Thomson coefficient should be positive. As the pressure increases, the effects of both attractive and repulsive forces must both increase, but at a sufficiently high pressure, the average intermolecular distance becomes so small that the effects of intermolecular repulsive forces become dominant. Therefore, we anticipate that the Joule-Thomson coefficient decreases as the pressure increases, eventually becoming negative.

Experiments confirm these expectations. A temperature and pressure at which the Joule-Thomson coefficient becomes zero is called a Joule-Thomson inversion point. The experimentally determined curve for nitrogen gas[1] is graphed in Figure 5. The van der Waals model also exhibits this effect. The inversion curve can be found from the expression for μ_{JT} developed above for a van der Waals gas. The inversion curve for nitrogen that is found in this way is also graphed in Figure 5. Qualitatively, the agreement is a satisfying confirmation of the basic interpretation that we have given for the role of intermolecular forces. Quantitatively, the agreement is poor, as we expect given the overly simple character of the van der Waals model.

The Joule-Thomson coefficient for an ideal gas is zero, and we normally expect the properties of real gases

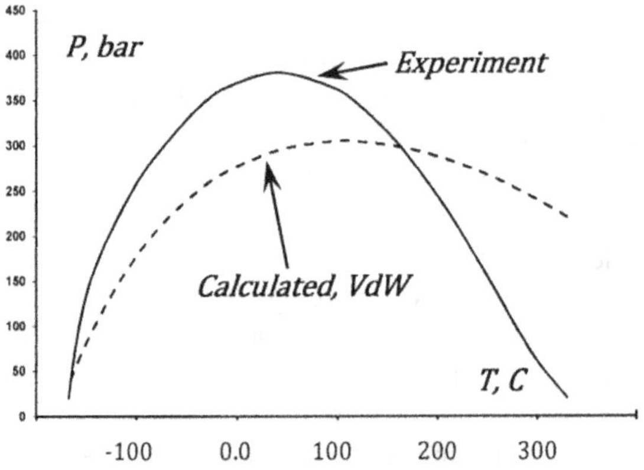

Figure 5. The Joule-Thomson inversion temperature for N_2 (see reference 1).

to approach those of an ideal gas as the pressure falls to zero. However, both experiment and the van der Waals model indicate that the Joule-Thomson coefficient converges to a finite value as the pressure decreases to zero at a fixed temperature. A statistical thermodynamic model[2] also predicts this outcome. This model calculates the coefficients in the virial equation of state. In it, the second virial coefficient reflects the net effect of attractive and repulsive forces between a pair of molecules, and it is the second virial coefficient and its temperature derivative determine that the value of $\left(\partial\overline{H}/\partial P\right)_T$. (Higher-order virial coefficients reflect interactions among larger numbers of molecules.)

Problems

1. Show that

$$\left(\frac{\partial(A/T)}{\partial T}\right)_V = -\frac{E}{T^2}$$

2. At 60 C, the vapor pressure of water is 19,932 Pa, and the enthalpy of vaporization is 42.482 kJ mol^{-1}.

(a) Is the vaporization of water at these conditions impossible, spontaneous, or reversible? What is ΔG for this process?

(b) Estimate ΔG for the vaporization of liquid water at 19,932 Pa and 70 C. Is this process impossible, spontaneous, or reversible?

(c) Estimate ΔG for the vaporization of liquid water at 19,932 Pa and 50 C. Is this process impossible, spontaneous, or reversible?

3. At 298.15 K and 1 bar, the Gibbs free energy of one mole of N_2O_4 is 4.729 kJ less than the Gibbs free energy of two moles of NO_2. The enthalpy of one mole of N_2O_4 is 57.111 kJ less than the enthalpy of two moles of NO_2. We customarily express these facts by saying that the Gibbs free energy and the enthalpy changes for the reaction $2\,NO_2 \rightarrow N_2O_4$ are $\Delta_r G^o(298.15\text{ K}) = -4.729$ kJ and $\Delta_r H^o(298.15\text{ K}) = -57.111$ kJ. Assume that the enthalpy change for this process is independent of temperature. Estimate the Gibbs free energy change for this reaction at 500 K and 1 bar, $\Delta_r G^o(500\text{ K})$.

4. Over the temperature range $300\text{ K} < T < 1000\text{ K}$, the Gibbs free energy change for the formation of ammonia from the elements, $\frac{1}{2}N_2 + \frac{3}{2}H_2 \rightarrow NH_3$, is well approximated by

$$\Delta_f G^o(NH_3) = a + b(T-600) + c(T-600)^2 + d(T-600)^3$$

where $a = 15.824$ kJ, $b = 0.1120$ kJ K^{-1}, $c = 1.316 \times 10^{-5}$ kJ K^{-2}, and $d = -1.324 \times 10^{-8}$ kJ K^{-3}. Estimate the enthalpy change for this process, $\Delta_f H^o(NH_3)$, at 600 K.

5. Consider the total differentials for $S = S(P,T)$, $E = E(P,T)$, $H = H(P,T)$, $A = A(P,T)$, and $G = G(P,T)$. Can we ever encounter an undefined integrand when we evaluate the line integral of one of these total differentials between any two points (P_1T_1) and (P_2T_2)? (In the next chapter, we find that, because of the third law of thermodynamics, no real system can ever reach the absolute zero of temperature.)

6. Consider the total differentials for $S = S(V,T)$, $E = E(V,T)$, $H = H(V,T)$, $A = A(V,T)$, and $G = G(V,T)$. Can we ever encounter an undefined integrand when we evaluate the line integral of one of these total differentials between any two points (V_1T_1) and (V_2T_2)?

7. The normal boiling point of methanol is 337.8 K at 1 atm. The enthalpy of vaporization at the normal boiling point is $\Delta_{vap}H = 35.21$ kJ mol^{-1}. Is the process impossible, spontaneous, or reversible? Find q, w, $\Delta_{vap}E$, $\Delta_{vap}S$, $\Delta_{vap}A$, $\Delta_{vap}G$ for the vaporization of one mole of methanol at the normal boiling point. Assume that methanol vapor behaves as an ideal gas.

8. For $S = S(P,V)$, we obtain

$$\left(\frac{\partial S}{\partial V}\right)_P = \frac{C_P}{T}\left(\frac{\partial T}{\partial V}\right)_P$$

For $S = S(P,T)$, we obtain

$$\left(\frac{\partial S}{\partial T}\right)_P = \frac{C_P}{T}$$

For temperatures near 4 C and at a pressure of 1 atm, the molar volume of water is given by

$$\overline{V} = \overline{V}_4 + a(T - 277.15)^2$$

where $\overline{V}_4 = 1.801575 \times 10^{-6}$ m^3 mol^{-1} and $a = 1.45 \times 10^{-11}$ m^3 K^{-1}. The heat capacity of liquid water is 75.49 J mol^{-1}.

(a) Using $(\partial S/\partial T)_P$, calculate the entropy change when one mole of water is warmed from 2 C to 6 C while the pressure is constant at 1 atm.

(b) Repeat the calculation in (a), for warming the water from 4 C to 6 C.

(c) Can we calculate the entropy change when one mole of water is warmed from 2 C to 6 C using $(\partial S/\partial V)_P$? Why, or why not? The required integral can be transformed to

$$\int \frac{A\,du}{u^{1/2} + Bu} = \left(\frac{2A}{B}\right)\ln\left(1 + \beta u^{1/2}\right) + C$$

where C is an arbitrary constant.

(d) Using $(\partial S/\partial V)_P$, calculate the entropy change when one mole of water is warmed from 4 C to 6 C. Compare this result to the value obtained in (b).

9. For an ideal gas, show that $(\partial E/\partial V)_T$, $(\partial E/\partial P)_T$, $(\partial H/\partial V)_T$, $(\partial H/\partial P)_T$, $(\partial C_V/\partial V)_T$, and $(\partial C_P/\partial P)_T$ are all zero.

10. Find $(\partial E/\partial P)_T$ for a gas that obeys the virial equation of state $P[\overline{V} - B(T)] = RT$, in which $B(T)$ is a function of temperature.

11. Derive the following relationships for an ideal gas:
(a) $dE = C_V dT$
(b) $dS = (C_V/T)dT + (R/V)dV$
(c) $dS = (C_P/T)dT - (R/P)dP$

12. Derive the following relationships for a gas that obeys the virial equation, $P[\overline{V} - B(T)] = RT$, where $B(T)$ is a function of temperature:
(a)

$$d\overline{E} = C_V dT - \left[\frac{RT}{\overline{V} - B} + \frac{RT}{\overline{V} - B}\left(\frac{dB}{dT}\right) - \frac{RT^2}{\left(\overline{V} - B\right)^2}\left(\frac{dB}{dT}\right)\right] d\overline{V}$$

(b)

$$d\overline{E} = \left[C_P - R - P\left(\frac{dB}{dT}\right)\right] dT - T\left(\frac{dB}{dT}\right) dP$$

(c)

$$d\overline{S} = \frac{C_V}{T} dT + \left[\frac{R}{\overline{V} - B} + \frac{RT}{\left(\overline{V} - B\right)^2}\left(\frac{dB}{dT}\right)\right] d\overline{V}$$

(d)

$$d\overline{S} = \frac{C_P}{T} dT - \left[\frac{R}{P} + \left(\frac{dB}{dT}\right)\right] dP$$

13. One mole of a monatomic ideal gas ($C_V = 3R/2$), originally at 10 bar and 300 K (state A), undergoes an adiabatic free expansion against a constant applied pressure of 1 bar to reach state B. Thereafter the gas is warmed reversibly at constant volume back to 300 K, reaching state C. Finally, the warmed gas is compressed reversibly and isothermally to the original pressure. What is are the temperature and volume in state B, following the original adiabatic free expansion? Find q, w, ΔE, ΔH, and ΔS for each of the steps and for the cycle A→B→C→A.

14. As in problem 13, one mole of a monatomic ideal gas ($C_V = 3R/2$), originally at 10 bar and 300 K (state A), undergoes an adiabatic free expansion against a constant applied pressure of 1 bar to reach state B. The gas is then returned to its original state in a different two-step process. From state A a reversible constant-pressure warming takes the gas to state D at the original temperature of 300 K. The gas is then returned to state A by an isothermal compression to the original volume. What are the temperature and volume after the constant-pressure warming step? Find q, w, ΔE, ΔH, and ΔS for each of the steps and for the cycle A→B→D→A.

15. As in problem 13, one mole of a monatomic ideal gas ($C_V = 3R/2$), originally at 10 bar and 300 K (state A), undergoes an adiabatic free expansion against a constant applied pressure of 1 bar to reach state B. Now consider a reversible adiabatic expansion from the same initial state, A, that reaches the same temperature as the gas in state B. Call this state F. Find q, w, ΔE, ΔH, and ΔS for the step A→F. Find q, w, ΔE, ΔH, and ΔS for reversible isothermal expansion from state F to state B. What are q, w, ΔE, ΔH, ΔS, and $\Delta \hat{S}$ for the cycle A→F→B→A. Does this cycle violate the machine-based statement of the second law?

16. One mole of carbon dioxide, originally at 10 bar and 300 K, is taken around the cycle in problem 13. Find the energy and entropy changes for the steps in this cycle using the ideal gas equation and the temperature-dependent heat capacity. The constant-volume heat capacity is $C_V = 14.7 + 0.046 \times T$. Find q, w, ΔE, ΔH, and ΔS for each of the steps and for the cycle when CO_2 is taken around the cycle A→B→C→A.

17. Ten moles of a monatomic ideal gas, initially occupying a volume of 30 L at 25 C, is expanded against a constant applied pressure of 2 bar. The final temperature is 25 C.
(a) What is the initial pressure? The final volume?
(b) Is this process impossible, spontaneous, or reversible?
(c) Find q, w, ΔE, ΔH, ΔA, ΔS, and ΔG for this process.

18. One mole of CO_2, originally at 1.00 bar and 300 K, expands adiabatically against a constant applied pressure of 0.200 bar. Assume that CO_2 behaves as an ideal gas with constant heat capacity, $C_V = 28.5$ J mol^{-1} K^{-1}.
(a) For the spontaneous expansion, we have $dE = C_V dT - P_{applied} dV$. Find the final temperature and volume for this spontaneous expansion. What is ΔE for this process?
(b) Find the volume and pressure after the gas is compressed adiabatically and reversibly to the original temperature of 300 K. What are ΔS and ΔE for this step?
(c) Find ΔE when the gas in the final state of part (b) is compressed isothermally to the original volume. What is ΔS for this step?
(d) What are ΔE and ΔS for the cycle comprised of the spontaneous expansion of part (a), the adiabatic compression of part (b), and the isothermal compression of part (c)?
(e) What are ΔS, $\Delta \hat{S}$, and $\Delta S_{universe}$ for the spontaneous expansion?

19. Consider the energy surface depicted in Figure 1. As sketched, E increases monotonically as S increases. E decreases monotonically as V increases. Could the energy surface decrease as S increases or increase as V increases?

20. At 298.15 K, the vapor pressure of water is 3.169×10^{-3} Pa. Some thermodynamic properties for liquid and gaseous water at this temperature and pressure are given in the table below.

Thermodynamic properties of water at 298.15 K and 3.169 x 10³ Pa		
	liquid	gas
$\overline{G}$, kJ mol^{-1}	– 237.1	– 237.1
$\overline{S}$, J mol^{-1} K^{-1}	70.0	217.5
$\overline{E}$, kJ mol^{-1}	– 285.5	– 245.1
C_P, J mol^{-1} K^{-1}	75.3	33.6
C_V, J mol^{-1} K^{-1}	67.0	25.3

(a) Find $\Delta_{vap}\overline{G}$, $\Delta_{vap}\overline{S}$, $\Delta_{vap}\overline{E}$ for water at this temperature and pressure. Is this process reversible, spontaneous, or impossible?

(b) Sketch $\overline{G}(\ell)$ and $\overline{G}(g)$ *vs.* T for $288.15 < T < 308.15$ K. What path is followed when one mole of water at 288.15 K and 3.169×10^3 Pa goes reversibly to 308.15 K at the same pressure?

(c) On the graph of part (b), indicate the transition in which superheated liquid water at 300 K and 3.169×10^3 Pa goes to gaseous water at 300 K and the same pressure. Is this process spontaneous, reversible, or impossible? Is $\Delta\overline{G}$ for this process positive, zero, or negative?

(d) Sketch $\overline{E}(\ell)$ and $\overline{E}(g)$ *vs.* T for $288.15 < T < 308.15$ K. What path is followed when one mole of water at 288.15 K and 3.169×10^3 Pa goes reversibly to 308.15 K at the same pressure?

(e) On the graph of part (b), indicate the transition in which superheated liquid water at 300 K and 3.169×10^3 Pa goes to gaseous water at 300 K and the same pressure. Is $\Delta\overline{E}$ for this process positive, zero, or negative?

21. At 273.15 K and 1 bar, the enthalpy of fusion of ice is 6010 J mol^{-1}. Estimate the Gibbs free energy change for the fusion of ice at 283.15 K and 1 bar.

Notes

[1] J. R. Roebuck and H. Osterberg, ***The Joule-Thomson Effect in Nitrogen***, **Phys. Rev.**, Vol. 48, pp 450-457 (1935).

[2] See T. L. Hill, ***An Introduction to Statistical Thermodynamics***, Addison-Wesley Publishing Co., Reeding, MA, 1960, pp 266-268.

11

The Third Law, Absolute Entropy, and the Gibbs Free Energy of Formation

§1 Heat capacity as a function of temperature

It is relatively easy to measure heat capacities as a function of temperature. If we measure the constant-pressure heat capacity of a pure substance over a wide temperature range, we typically observe a curve like that in Figure 1. The heat capacity is a smooth, continuous function of temperature except for a small number of discontinuities. These occur at temperatures where the substance undergoes phase changes. These can be changes from one solid phase to another, melting to convert a solid phase to the liquid, or vaporization to convert the liquid to the gas. The details of the curve are pressure dependent; for example, at a low pressure, we might observe sublimation of the material from a solid phase directly into its gas phase.

Another general feature of these curves is that the heat capacity of the solid substance decreases to zero as the absolute temperature decreases to zero; the curve meets the abscissa at the zero of temperature and does so asymptotically. That this is true for all substances seems like an odd sort of coincidence. Why should all solid substances exhibit essentially the same heat capacity (zero) at one temperature (absolute zero)?

As it turns out, this result has a straightforward molecular interpretation in the theory of statistical thermodynamics. In §22-6, we consider a theory of low-temperature heat capacity developed by Einstein. Einstein's theory explains all of the qualitative features that are observed when we measure heat capacities at low temperatures, but its predictions are not quantitatively exact.

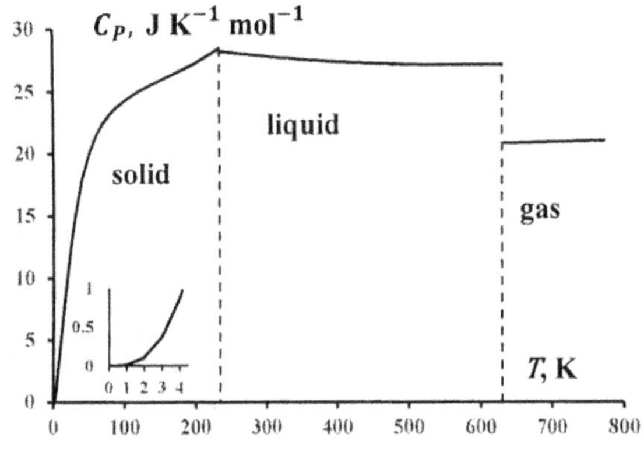

Figure 1. Heat capacity of mercury *versus* temperature.

Debye extended the Einstein model and developed a theory that gives generally excellent quantitative predictions. The Debye theory predicts that, at temperatures near absolute zero, the heat capacity varies as the cube of temperature: $C_P = AT^3$, where A is a constant. If we have heat capacity data down to a temperature near absolute zero, we can estimate the value of A from the value of C_P at the lowest available temperature.

Anticipating results that we develop in Chapter 22, we can characterize the statistical interpretation as follows: When a system of molecules gives up heat to its surroundings, some of the molecules move from higher energy levels to lower ones. Statistical thermodynamics posits that the fraction of the molecules that are in the lowest energy level approaches one as the temperature goes to zero. If nearly all of the molecules are already in the lowest energy level, decreasing the temperature still further has a negligible effect on the energy and enthalpy of the system.

Given such heat capacity data, we can find the enthalpy or entropy change that occurs as we change the temperature of a quantity of the substance from some reference temperature to any other value. When we use pressure and temperature as the independent variables, we have

$$dH = C_P dT + \left[V - T\left(\frac{\partial V}{\partial T}\right)_P\right] dP$$

and

$$dS = \frac{C_P}{T} dT - \left(\frac{\partial V}{\partial T}\right)_P dP$$

At constant pressure, we have

$$(dH)_P = C_P dT$$

so that

$$H(T) - H(T_{ref}) = \int_{T_{ref}}^{T} C_P dT$$

and

$$(dS)_P = \frac{C_P}{T} dT$$

so that

$$S(T) - S(T_{ref}) = \int_{T_{ref}}^{T} \frac{C_P}{T} dT$$

If phase transitions occur as the temperature goes

from the reference temperature to the temperature of interest, these integrations must be carried out in steps. Also, we must include the enthalpy and entropy changes that occur during these phase changes.

§2 Enthalpy as a function of temperature

The fact that C_P goes to zero asymptotically as the temperature goes to zero has no practical ramifications for the measurement or use of enthalpy. We can only measure changes in energy and enthalpy; no particular state of any system is a uniquely useful reference state for the enthalpy function. Experimental convenience is the only consideration that makes one reference state a better choice than another. In Chapter 8, we define standard enthalpies of formation for elements and compounds. For this purpose, we choose to define the standard enthalpy of formation of each element to be zero at every temperature. For standard enthalpies of formation, the reference state is different at every temperature.

Compilations of thermodynamic data often choose 298.15 K and one bar as the zero of enthalpy for pure substances. (In citing data from such compilations, 298.15 K is frequently abbreviated to "298 K.") Because it is near the ambient temperature of most laboratories, much thermochemical data has been collected at or near this temperature. Choosing a reference temperature near ambient laboratory temperatures helps minimize the errors introduced when we extrapolate experimental thermodynamic data to the reference temperature. Expressed relative to a reference temperature, the substance's enthalpy at any other temperature is the change in enthalpy that occurs when the substance is taken from the reference temperature to that temperature. Such enthalpy changes are often called ***absolute enthalpies***. Enthalpy-data tables frequently include values for "$H_T^o - H_{298\text{K}}^o$" at a series of temperatures. These data should not be confused with enthalpies of formation.

§3 The Third Law

For entropy on the other hand, the fact that the heat capacity goes to zero as the temperature decreases has important consequences. Consider the change in the entropy of a pure substance whose heat capacity approaches some finite limiting value as its temperature decreases to absolute zero. For such a substance, C_P/T becomes arbitrarily large as the temperature decreases, and the entropy integral

$$\int_T^0 \frac{C_P}{T}\,dT$$

approaches minus infinity as the temperature approaches zero. For real substances, this does not occur. In the neighborhood of absolute zero, heat capacities decrease more rapidly than temperature. The entropy change approaches zero as the temperature approaches zero.

The idea that the entropy change for a pure substance goes to zero as the temperature goes to zero finds expression as the third law of thermodynamics. In 1923, Lewis and Randall[1] gave a statement of the third law that is particularly convenient in chemical applications:

If the entropy of each element in some crystalline state be taken as zero at the absolute zero of temperature, every substance has a positive finite entropy; but at the absolute zero of temperature the entropy may become zero, and does so become in the case of perfect crystalline substances.

Implicitly, the Lewis and Randall statement defines the entropy of any substance, at any temperature, T, to be the difference between the entropy of the constituent elements, at absolute zero, and the entropy of the substance at temperature T. Equivalently, we can say that it is the entropy change when the substance is formed at temperature T from its constituent elements at absolute zero. Arbitrarily, but very conveniently, the statement sets the entropy of an element to zero at absolute zero.

The distinction between perfect crystalline substances and less-than-perfect crystalline substances lies in the regularity of the arrangement of the molecules within the crystal lattice. In any lattice, each molecule of the substance is localized at a specific site in the lattice. In a perfect crystal, all of the molecules are in oriented the same way with respect to the lattice. Some substances form crystals in which the molecules are not all oriented the same way. This can happen when the molecule can fit into a lattice site of the same shape in more than one way. For example, in solid carbon monoxide, the individual molecules occupy well-defined lattice sites. If the carbon monoxide crystal were perfect, all of the molecules would point in the same direction, as diagrammed in Figure 2. Instead, they point randomly in either of two possible directions.

CO CO CO CO CO CO CO CO
CO CO CO CO CO CO CO CO
CO CO CO CO CO CO CO CO
CO CO CO CO CO CO CO CO
CO CO CO CO CO CO CO CO
CO CO CO CO CO CO CO CO
CO CO CO CO CO CO CO CO
CO CO CO CO CO CO CO CO

Figure 2. A two-dimensional representation of a perfect CO crystal.

§4 Genesis of the third law: The Nernst heat theorem

The third law arises in a natural way in the development of statistical thermodynamics. It is probably fair to say that the classical thermodynamic treatment of the third law was shaped to a significant degree by the statistical thermodynamic treatment that developed about the same time. Nevertheless, we can view the third law as an inference from thermochemical observations.

Walther Nernst was the first to recognize the principle that underlies the third law. From published experimental results, Nernst inferred a postulate known as the ***Nernst heat theorem***. The experimental results that inspired Nernst were measurements of enthalpy and Gibbs free energy differences, $\Delta_r H$ and $\Delta_r G$, for particular reactions at a series of temperatures. (We define $\Delta_r H^o$ in §8-6. We define $\Delta_r H$ the same way, except that the reactants and products are not all in their standard states. Likewise, $\Delta_r G$ and $\Delta_r S$ are differences between Gibbs free energies and entropies of reactants and products. We give a more precise definition for $\Delta_r G^o$ in §13-2.) As the temperature decreased to a low value, the values of $\Delta_r H$ and $\Delta_r G$ converged. Since $T\Delta_r S = \Delta_r H - \Delta_r G$, this observation was consistent with the fact that the temperature was going to zero. However, Nernst concluded that the temperature factor in $T\Delta_r S$ was not, by itself, adequate to explain the observed dependence of $\Delta_r H - \Delta_r G$ on temperature. He inferred that the entropy change for these reactions decreased to zero as the temperature decreased to absolute zero and postulated that this observation would prove to be generally valid. The Nernst heat theorem asserts that the entropy change for any reaction of pure crystalline substances goes to zero as the temperature goes to zero.

Subsequently, Max Planck suggested that the entropy of reaction goes to zero because of a still more basic phenomenon: the entropy of every crystalline substance goes to zero as the temperature goes to zero. Further investigation then showed that Planck's formulation fails for substances like carbon monoxide, in which the crystalline solid does not become perfectly ordered at the temperature goes to zero. The Lewis and Randall statement refines the Planck formulation by recognizing that non-zero entropies will be observed at absolute zero for solids that are not crystalline and for crystalline solids that are not perfectly ordered. The Lewis and Randall statement also makes a choice (implicit also in the Planck formulation) of the zero point for the entropies of chemical substances—namely, "some crystalline state" of each element at absolute zero. This choice ensures that, at any temperature greater than zero, the entropy of every substance will be greater than zero.

§5 Absolute entropy

By the Lewis and Randall statement of the third law, the entropy of a substance that forms a perfect crystal is identically equal to zero at absolute zero. Much as the ideal gas temperature scale has a natural zero at the temperature at which the volume extrapolates to zero, a perfect crystalline substance has a natural zero of entropy at this same temperature. We can choose a non-zero value for the absolute zero of temperature. The Centigrade scale is based on such a choice. However, for thermodynamic purposes, any such choice is much less convenient. Similarly, we could choose arbitrary values for the entropies of the elements at the absolute zero of temperature. The entropy of a perfect crystalline substance at absolute zero would then be the sum of the entropies of its constituent elements. (See problem 5.) However, choosing non-zero values proves to be much less convenient.

Given the entropy of a substance at absolute zero, its entropy at any higher temperature can be calculated from the entropy changes that occur as the substance is warmed to the new temperature. At the very lowest temperatures, this entropy change is calculated by integrating C_P/T, using Debye's theoretical relationship, $C_P = AT^3$; A is obtained from the value of C_P at the lowest temperature for which an experimental value of C_P is available. In temperature ranges where experimental heat capacity data are available, the entropy change is obtained by integration using these data.

Phase changes are isothermal and reversible. Where the substance undergoes phase changes, the contribution that the phase change makes to the entropy of the substance is equal to the enthalpy change for the phase change divided by the temperature at which it occurs.

At any given temperature, the entropy value that is obtained in this way is called the substance's ***absolute entropy*** or its ***third-law entropy***. When the entropy value is calculated for one mole of the substance in its standard state, the resulting absolute entropy is called the ***standard entropy***. The standard entropy is usually given the symbol S^o. It is usually included in compilations of thermodynamic data for chemical substances.

We write $S_A^o(T)$ to indicate the absolute entropy of substance A in its standard state at temperature T. $S_A^o(T)$ is the entropy of the substance in its standard state at absolute zero plus the entropy increase that occurs as the substance changes reversibly to its standard state at T. So long as substance A forms a perfect crystal at absolute zero, $S_A^o(T)$ is the difference between its molar entropy at T and its molar entropy at absolute zero—as calculated from heat capacity and phase-change enthalpy data.

If substance A does not form a perfect crystal at absolute zero, the true value of $S_A^o(T)$ exceeds the calculated value. The excess is the molar entropy of the imperfect crystal at absolute zero. We observe the discrepancy when measured values, at T, of entropies of reactions that involve A fail to agree with those calculated using the incorrect value of $S_A^o(T)$.

In §2 we note that many tables of thermochemical properties present "absolute enthalpy" data for chemical substances. An absolute enthalpy is the difference between the enthalpies of a substance at two different temperatures, but the reference temperature is not absolute zero. In §6-4 and §6-5, we introduce enthalpy standard states and the standard enthalpy of formation of substance A at temperature T, which we designate as $\Delta_f H_A^o(T)$. We define the standard enthalpy of formation of any element at any temperature to be zero. In §8-6, we find that the enthalpy difference between reactants in their standard states and products in their standard states is readily calculated from the standard enthalpies of formation of the participating substances. As illustrated in Figure 8-2, this calculation is successful because it utilizes an isothermal cycle which begins and ends in a common set of elements, all of which are at the same temperature.

We can also define the standard entropy of formation of any substance to be the difference between its standard entropy, $S_A^o(T)$, and those of its pure constituent elements in their standard states at the same temperature. This definition is embedded in the Lewis and Randall statement of the third law. For example, the standard entropy of formation of water at 400 K is the difference

$$\Delta_f S^o(H_2O, 400\ \text{K}) = \Delta S^o(H_2O, 400\ \text{K}) - \Delta S^o(H_2, 400\ \text{K}) - \tfrac{1}{2}\Delta S^o(O_2, 400\ \text{K})$$

Because of this definition, the standard entropy of formation of an element in its standard state is zero. We can calculate the standard entropy change for any reaction, $\Delta_r S^o(T)$, either as the difference between the standard entropies of formation (the $\Delta_f S^o(T)$ values) of the reactants and products or as the difference between their standard entropies (the $S_A^o(T)$ values). Either calculation is successful because it begins and ends with a common set of elements, all of which are at the same temperature. When we compute $\Delta_r S^o(T)$ using values of $S_A^o(T)$ for the reactants and products the reference temperature for the elements is absolute zero. When we compute $\Delta_r S^o(T)$ using values of $\Delta_f S^o(T)$ for the reactants and products, the reference temperature is T.

Given $\Delta_f H^o(T)$ and $\Delta_f S^o(T)$, the standard Gibbs free energy of formation is immediately obtained from $\Delta_f G^o(T) = \Delta_f H^o(T) - T\Delta_f S^o(T)$. For any element at any temperature, we have $\Delta_f H^o = 0$ and $\Delta_f S^o = 0$; it follows that the standard Gibbs free energy of formation of an element in its standard state is zero. Tables of thermodynamic data usually give values for$\Delta_f H^o$, $\Delta_f G^o$, and S^o. (A set of standard entropies contains the same information as the corresponding set of entropies of formation. Entropies of formation are seldom tabulated. If $\Delta_f S^o$ is needed, it can be calculated either from $\Delta_f H^o$ and $\Delta_f G^o$ or from the absolute entropies of the substance and the elements from which it is formed.)

§6 The standard state for third-law entropies

The standard state for entropies is essentially the same as the standard state for enthalpies. For liquids and solids, the standard state for entropies is identical to that for enthalpies: At any given temperature, the standard state is the most stable form of the substance at that temperature and a pressure of 1 bar.

For gases, the ideal gas standard state for entropy tabulations is the hypothetical ideal gas state at a pressure of 1 bar. A substance in a hypothetical ideal gas state is a creature of theory, and we obtain its thermodynamic properties from calculations that use the experimentally determined properties of the real gas. The idea behind these calculations is that we can—by calculation—remove the effects of intermolecular interactions from the measured properties. When we want the properties of the real gas, we put these effects back. From this perspective, we can say that a substance in its hypothetical ideal gas standard state has the entropy that the real gas would have if it behaved as an ideal gas at pressures of 1 bar and below.

In principle, this differs from the standard state chosen for enthalpy, because the enthalpy standard state is defined to be an arbitrary low pressure at which the substance behaves as an ideal gas. However, because the enthalpy of an ideal gas is independent of pressure, we can consider the standard state for either enthalpy or entropy to be the hypothetical ideal gas at a pressure of 1 bar. (Below, we write "HIG^o" to designate this state.) Since the Gibbs free energy is defined by $G = H - TS$, we can also describe the standard state for the Gibbs free energy of gases as the hypothetical ideal gas at 1 bar.

To see precisely what we mean by the hypothetical ideal gas standard state, let us consider the conversion of one mole of a real gas, initially at pressure P and temperature T, to its hypothetical ideal gas state at 1 bar and the same temperature. We accomplish this in a three-step process. In the first step, we reversibly and isothermally expand the real gas to some arbitrary low pressure at which the real gas behaves ideally. The second step is purely conceptual: We suppose that the real gas changes into a hypothetical gas that behaves ideally. The third step is the reversible isothermal compression of the hypothetical ideal gas to a pressure of 1 bar. Letting A denote the gas, these steps are

(1) $A(\text{real gas}, P) \rightarrow A(\text{real gas}, P^* \approx 0)$
(2) $A(\text{real gas}, P^* \approx 0) \rightarrow A(\text{ideal gas}, P^* \approx 0)$
(3) $A(\text{ideal gas}, P^* \approx 0) \rightarrow A(HIG^o, P^o = 1\ \text{bar})$

We use the symbol "P^o" to designate the standard-state pressure. P^o is a constant, whose value is 1 bar.

Let the enthalpy, entropy, and Gibbs free energy changes for the ith step be $\Delta_i\overline{H}$, $\Delta_i\overline{S}$, and $\Delta_i\overline{G}$. (We use the overbar to emphasize that the system consists of one mole of a pure substance. Since the superscript zero implies that $H_A^o(HIG^o)$, for example, is a property of one mole of A in its standard state, we omit the overbar from standard state properties.) From our expressions for dH and dS as functions of pressure and temperature, we have, at constant temperature,

$$\left(d\overline{H}\right)_T = \left[\overline{V} - T\left(\frac{\partial \overline{V}}{\partial T}\right)_P\right] dP$$

and

$$\left(d\overline{S}\right)_T = -\left(\frac{\partial \overline{V}}{\partial T}\right)_P dP$$

For an ideal gas,

$$\left(d\overline{H}\right)_T = 0$$

and

$$\left(d\overline{S}\right)_T = -R/P$$

For step (1), the enthalpy change is

$$\Delta_1\overline{H} = \int_P^{P^*} \left[\overline{V} - T\left(\frac{\partial \overline{V}}{\partial T}\right)_P\right] dP$$

the entropy change is

$$\Delta_1\overline{S} = \int_P^{P^*} -\left(\frac{\partial \overline{V}}{\partial T}\right)_P dP$$

We must evaluate the partial derivative using an equation of state that describes the real gas.

Step (2) is merely a change in our perspective; nothing actually happens to the gas. There is no enthalpy or entropy change: $\Delta_2\overline{S} = 0$ and $\Delta_2\overline{H} = 0$.

For step (3), the enthalpy change is zero because the gas is ideal, $\Delta_3\overline{H} = 0$. We evaluate the entropy change using the ideal gas equation; then

$$\Delta_3\overline{S} = \int_{P^*}^{P^o} -\frac{R}{P} dP$$

Let $\overline{H}_A(P)$, $\overline{S}_A(P)$, and $\overline{G}_A(P)$ be the molar enthalpy, entropy, and Gibbs free energy of real gas A at pressure P. Let us express the enthalpy, entropy, and Gibbs free energy of the standard state substance relative to the corresponding properties of the constituent elements in their standard states at temperature T. Then the molar enthalpy, entropy, and Gibbs free energy of A in its hypothetical ideal gas standard state are $\Delta_f H_A^o(HIG^o)$, $\Delta_f S_A^o(HIG^o)$, and $\Delta_f G_A^o(HIG^o)$. Since these are state functions, we have

$$\begin{aligned}\Delta_f H_A^o(HIG^o) &= \overline{H}_A(P) + \Delta_1\overline{H} + \Delta_2\overline{H} + \Delta_3\overline{H} \\ &= \overline{H}_A(P) + \int_P^{P^*} \left[\overline{V} - T\left(\frac{\partial \overline{V}}{\partial T}\right)_P\right] dP\end{aligned}$$

and

$$\begin{aligned}\Delta_f S_A^o(HIG^o) &= \overline{S}_A(P) + \Delta_1\overline{S} + \Delta_2\overline{S} + \Delta_3\overline{S} \\ &= \overline{S}_A(P) - \int_P^{P^*} \left(\frac{\partial \overline{V}}{\partial T}\right)_P dP - \int_{P^*}^{P^o} \frac{R}{P} dP\end{aligned}$$

These equations relate the enthalpy and entropy of the hypothetical ideal gas standard state to the enthalpy and entropy of the real gas at pressure P, at the same temperature.

To evaluate the enthalpy change, we can use the virial equation for the volume of a real gas, set $P^* = 0$, and evaluate the resulting integral. (See problem 1a.) For the entropy, there is a complication. If we set $P^* = 0$, neither integral is finite. We can overcome this difficulty by choosing a small positive value for P^*. Then both integrals are finite, and the value of their sum remains finite in the limit as $P^* \to 0$. This occurs because the molar volume of any gas approaches the molar volume of an ideal gas at a sufficiently low pressure. (See problem 1.)

§7 The fugacity of a gas

As it turns out, the pressure-dependence of the Gibbs free energy is useful more often than that of the enthalpy or entropy. From the defining relationship, $G = H - TS$, and the results above, we find

$$\begin{aligned}\Delta_f G_A^o(HIG^o) &= \Delta_f H_A^o(HIG^o) - T\Delta_f S_A^o(HIG^o) \\ &= \overline{H}_A(P) + \int_P^{P^*} \left[\overline{V} - T\left(\frac{\partial \overline{V}}{\partial T}\right)_P\right] dP \\ &\quad -T\overline{S}_A(P) + \int_P^{P^*} T\left(\frac{\partial \overline{V}}{\partial T}\right)_P dP + \int_{P^*}^{P^o} \frac{RT}{P} dP \\ &= \overline{G}_A(P) - \int_{P^*}^{P} \overline{V} dP - \int_{P^o}^{P^*} \frac{RT}{P} dP\end{aligned}$$

where we have reversed the direction of integration in the remaining integrals. Again, we cannot evaluate the integrals with P* = 0. To overcome this difficulty, we introduce a clever expedient: we add and subtract the same integral, obtaining

$$\begin{aligned}\Delta_f G_A^o(HIG^o) &= \overline{G}_A(P) - \int_{P^*}^{P} \overline{V} dP - \int_{P^o}^{P^*} \frac{RT}{P} dP \\ &\quad - \int_{P^*}^{P} \frac{RT}{P} dP + \int_{P^*}^{P} \frac{RT}{P} dP \\ &= \overline{G}_A(P) - \int_{P^*}^{P} \left[\overline{V} - \frac{RT}{P}\right] dP - \int_{P^o}^{P} \frac{RT}{P} dP\end{aligned}$$

This result is pivotal. We will derive it again later by a slightly different argument. For now, we evaluate the last integral and rearrange the equation to the form in which we write it most often:

$$\begin{aligned}\overline{G}_A(P) &= \Delta_f G_A^o(HIG^o) + RT\ln\left(\frac{P}{P^o}\right) \\ &\quad + RT\int_0^P \left[\frac{\overline{V}}{RT} - \frac{1}{P}\right] dP\end{aligned}$$

In doing so, we use the fact that

$$\frac{\overline{V}}{RT} - \frac{1}{P}$$

vanishes for any gas when the pressure is sufficiently low, which means that the integral from $P^*$to P remains finite when we let $P^* \to 0$.

Recapitulating: The last equation relates the Gibbs free energy of one mole of a real gas at pressure P and temperature T, $\overline{G}_A(P)$, to the Gibbs free energy of formation of the gas in its standard state, $\Delta_f G_A^o(HIG^o)$, where the standard state is the hypothetical ideal gas at $P = P^o = 1$ bar and the same temperature. (We write "$RT \ln(P/P^o)$" rather than "$RT \ln P$" to emphasize that the argument of the logarithm is a dimensionless quantity.)

If the gas is ideal, the integrand is zero at any pressure. For an ideal gas, the Gibbs free energy, at pressure P and temperature T, is related to the Gibbs free energy in the standard state by

$$\overline{G}_A(P) = G_A^o(1 \text{ bar}) + RT \ln\left(\frac{P}{P^o}\right)$$

(ideal gas)

This observation enables us to give a physical interpretation to the integral

$$\int_0^P \left[\overline{V} - \frac{RT}{P}\right] dP = RT \int_0^P \left[\frac{\overline{V}}{RT} - \frac{1}{P}\right] dP$$

Evidently, this integral is the difference between the Gibbs free energy of one mole of the real gas and the Gibbs free energy that it would have if it behaved as an ideal gas. When we add it to the Gibbs free energy of the hypothetical ideal gas, we get the Gibbs free energy of the real gas.

For an ideal gas, the Gibbs free energy is a simple function of its pressure. It turns out to be useful to view the integral as a contribution to a "corrected pressure."

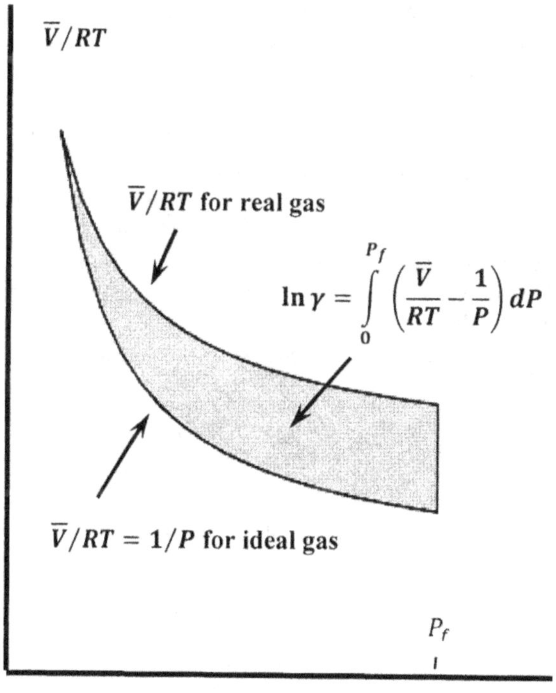

Figure 3. Graphical description of $\ln \gamma$.

The "correction" is an adjustment to the pressure that, in our calculations, makes the real gas behave as an ideal gas. The idea is that we can express the Gibbs free energy as a function of this corrected pressure, which we call the ***fugacity***, and to which we give the symbol "f". Fugacity is therefore a function of pressure. Fugacity also has the units of pressure, which we always take to be bars. We define the fugacity of A at pressure P, $f_A(P)$, by

$$\overline{G}_A(P) = \Delta_f G_A^o(HIG^o) + RT \ln\left(\frac{f_A(P)}{f_A(HIG^o)}\right)$$

(real gas)

The fugacity, $f_A(HIG^o)$, and the standard Gibbs free energy of formation, $\Delta_f G_A^o(HIG^o)$, describe the same state. We define the fugacity in this standard state to be one bar. That is, the fugacity of a substance in its hypothetical ideal-gas standard state, $f_A(HIG^o)$, is a constant whose value is 1 bar. (If we want to express fugacity in units of, say, pascals, then $P^o = f_A(HIG^o) = 10^5$ Pa.)

We can calculate the fugacity of a real gas, at pressure P and temperature T, from

$$RT \ln\left(\frac{f_A(P)}{f_A(HIG^o)}\right) = RT \ln\left(\frac{P}{P_0}\right) + RT \int_0^P \left[\frac{\overline{V}}{RT} - \frac{1}{P}\right] dP$$

We find it useful to introduce a function of pressure that we call the ***fugacity coefficient***, $\gamma_A(P)$. We define it as

$$\gamma_A(P) = \frac{f_A(P)}{P}$$

Since $f_A(HIG^o) = P^o = 1$ bar, we have

$$\ln \gamma_A(P) = \int_0^P \left[\frac{\overline{V}}{RT} - \frac{1}{P}\right] dP$$

Figure 3 exhibits these quantities graphically. The shaded area is $\ln \gamma_A(P)$.

§8 A general strategy for expressing the thermodynamic properties of a substance

The equations we develop in §7 and §8 express the differences $\Delta_f H_A^o(HIG^o) - \overline{H}_A(P)$, $\Delta_f S_A^o(HIG^o) - \overline{S}_A(P)$, and $\Delta_f G_A^o(HIG^o) - \overline{G}_A(P)$ between thermodynamic functions for one mole of a gas at two different pressures and the same temperature. They follow from the properties of gases and the relationships that result when we express the enthalpy, entropy, and Gibbs free energy as functions of temperature and pressure. Because $\Delta_f H_A^o(HIG^o)$, $\Delta_f S_A^o(HIG^o)$, and $\Delta_f G_A^o(HIG^o)$ are measured relative to the constituent elements of substance A at the same temperature, $\overline{H}_A(P)$, $\overline{S}_A(P)$, and $\overline{G}_A(P)$ are the differences in these properties between the real substance at pressure P and temperature T and the real constituent elements in their standard states at the same temperature.

We have just found a way to express the thermodynamic functions of a pure real gas at any pressure and temperature. This development shows us the way toward a broader goal. Ultimately, we want to be able to express—and to find the values of—the thermodynamic functions of any substance in any system at any temperature and pressure. ***Our goal is to create a scheme in which the enthalpy, the entropy, or the Gibbs free energy of any substance in any arbitrary state is equal to the change in that thermodynamic property when the substance is produced, in that state, from its pure, separate, constituent elements, in their standard states at the same temperature.***

The scheme we create uses two steps to convert the constituent elements into the substance in the arbitrary state. The elements are first converted into the pure substance in its standard state at the same temperature. The substance is then taken from its standard state to the state it occupies in the arbitrary system, at the same temperature. While straightforward in principle, finding changes in thermodynamic properties for this last step is often difficult in practice.

The value of the Gibbs free energy of substance A in any arbitrary system, $\overline{G}_A$, becomes equal to the sum of its Gibbs free energy in the standard state, $\Delta_f G_A^o$, and the Gibbs free energy change when the substance passes from its standard state into its state in the arbitrary system, $\Delta\overline{G}_A$. That is,

$$\overline{G}_A = \Delta_f G_A^o + \Delta\overline{G}_A$$

and $\overline{G}_A$ is the same thing as the Gibbs free energy change when the substance is formed, as it is found in the arbitrary system, from its constituent elements. The same relationships apply to the enthalpies and entropies of these states:

$$\overline{H}_A = \Delta_f H_A^o + \Delta\overline{H}_A$$

and

$$\overline{S}_A = \Delta_f S_A^o + \Delta\overline{S}_A$$

Measuring the differences $\Delta\overline{H}_A$, $\Delta\overline{S}_A$, and $\Delta\overline{G}_A$ become important objectives—and major challenges—in the study of the thermodynamics of chemical systems.

In §6, we find the changes that occur in the enthalpy, entropy, and Gibbs free energy when one mole of a pure gas is taken from its real gas state at any pressure to its hypothetical ideal-gas standard state, at the same temperature. In Chapter 13, we extend this development to express the thermodynamic properties of any mixture of ideal gases in terms of the properties of the individual pure gases. As a result, we can find the equilibrium position for any reaction of ideal gases from the thermodynamic properties of the individual pure gases. This application is successful because we can find both $\Delta_f G_A^o$ and $\Delta\overline{G}_A$ for an ideal gas at any pressure. Beginning in Chapter 14, we extend this success by finding ways to measure $\Delta\overline{G}_A$ for the process of taking A from its standard state as a pure substance to any arbitrary state in which A may be only one component of a solution or mixture.

§9 The standard entropy and the Gibbs free energy of formation

Given $\Delta_f G^o$ for all of the species involved in a reaction, we can calculate the difference between the Gibbs free energies of formation of the pure separate products and those of the pure separate reactants. We call this difference the standard Gibbs free energy change for the reaction, $\Delta_r G^o$. A standard Gibbs free energy of formation is the standard Gibbs free energy change for the reaction that forms a substance from its elements. Likewise, from the absolute entropies, S^o, of the reactants and products, we can calculate the standard entropy change for the reaction, $\Delta_r S^o$. In doing so, we utilize the thermochemical cycle that we introduced to calculate $\Delta_r H^o$ from the values of $\Delta_f H^o$ for the reacting species. For $a\,A + b\,B \rightarrow c\,C + d\,D$, we have

$$\Delta_r G^o = c\,\Delta_f G^o(C) + d\,\Delta_f G^o(D) - a\,\Delta_f G^o(A) - b\,\Delta_f G^o(B)$$

and

$$\Delta_r S^o = c\,S^o(C) + d\,S^o(D) - a\,S^o(A) - b\,S^o(B)$$

We use $\Delta_f G_A^o(HIG^o)$ to denote the Gibbs free energy of one mole of a gas in its hypothetical ideal-gas standard state. Because the fugacity of the ideal gas standard state is 1 bar, $f_A(HIG^o) = P^o$, the Gibbs free energy of a gas at unit fugacity becomes the Gibbs free energy change for the formation of the substance in its hypothetical ideal gas standard state. For an ideal gas, unit fugacity occurs at a pressure of one bar. For real gases, the standard state of unit fugacity occurs at a real-gas pressure that is, in general, different from one bar.

The Gibbs free energy of the gas at any other pressure, $\overline{G}_A(P)$, becomes identical to the difference between the Gibbs free energy of the gas in that state and the Gibbs free energy of its constituent elements in their standard states at the same temperature. This convention makes the Gibbs free energy of the elements the "zero point" for the Gibbs free energy of the gas. As indicated in Figure 4, the Gibbs free energy of the gas at any pressure, P, becomes

$$\overline{G}_A(P) = \Delta_f G_A^o(HIG^o) + RT\ln\left(\frac{f_A(P)}{f_A(HIG^o)}\right) = \Delta_f G_A^o(HIG^o) + RT\ln\left(\frac{f_A(P)}{P^o}\right)$$

§10 The nature of hypothetical states

It is worthwhile to call attention to some important aspects of this development. The hypothetical ideal gas standard state is a wholly theoretical construct. We create this "substance" only because it is convenient to have a name for the "unreal" state of substance A, whose Gibbs free energy we have denoted as $G_A^o(HIG^o) = \Delta_f G_A^o(HIG^o)$. We have developed procedures for calculating $\Delta_f G_A^o(HIG^o)$ from the properties of the corresponding real gas. Given the properties of real gas A, these procedures determine $\Delta_f G_A^o(HIG^o)$ uniquely. $\Delta_f G_A^o(HIG^o)$ is a useful quantity; the calculation of $\Delta_f G_A^o(HIG^o)$ is "real" even though the substance it putatively describes is not.

Other hypothetical states are frequently useful. Problem 8 in this chapter considers a hypothetical liquid state of methanol at 500 K and 1 atm—conditions at which the real substance is a gas. Alternative approximations enable us to calculate the Gibbs free energy of this hypothetical state in different ways. The results have predictive value. Not surprisingly, however, the alternative approximations produce Gibbs free energy values whose quantitative agreement is poor. Were it useful to do so, we could select one particular approximation and define the Gibbs free energy of the hypothetical superheated liquid methanol to be the value produced using that approximation. This would not make the superheated liquid methanol any more real, but it would uniquely define the Gibbs free energy of the hypothetical substance.

Later in our development, we create other hypothetical reference states. As for the hypothetical ideal gas standard state, we specify unique ways to calculate the properties of these hypothetical states from measurements that we can make on real systems.

§11 The fugacity and Gibbs free energy of a substance in any system

We can find the Gibbs free energy of formation for substances whose standard states are condensed phases. As indicated in §9, we adopt the same rule for any substance; we set $G_A^o = \Delta_f G^o(A)$ for any substance, whether its standard state is a gas, liquid, or solid. The Gibbs free energy of the elements becomes the "zero point" for the Gibbs free energy of any substance.

In Chapters 14 and 15, we see that we can also define the fugacity of any substance in any system; that is, we can define the fugacity for a pure liquid, a pure solid, or for one component in any mixture. When we do so, the Gibbs free energy of one mole of the substance in the system, $\overline{G}_A(\text{system}, P)$, is given by the same relationship we developed for the molar Gibbs free energy of a pure gas. We find

$$\overline{G}_A(\text{system}, P) = \Delta_f G_A^o(HIG^o) + RT \ln\left(\frac{f_A(\text{system}, P)}{f_A(HIG^o)}\right)$$

To obtain this result and to see how to find the fugacity of A in any system, $f_A(\text{system}, P)$, we must introduce a number of additional ideas. For now, let us note some of the consequences.

The essential consequence is that the difference

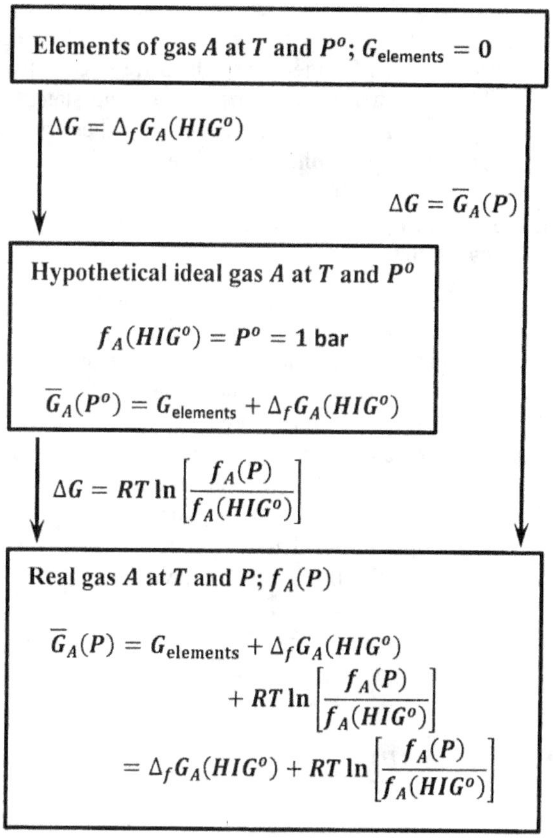

Figure 4. Fugacity and Gibbs free energy.

between the Gibbs free energy of one mole of a substance in two different systems, say system X and system Y, can be expressed using the ratio of the corresponding fugacities. That is,

$$\overline{G}_A(\text{system X}, P_X) - \overline{G}_A(\text{system Y}, P_Y) = RT \ln\left(\frac{f_A(\text{system}, P_X)}{f_A(\text{system Y}, P_Y)}\right)$$

where P_X and P_Y are the pressures of systems X and Y, respectively, and both systems are at the same temperature.

For liquids and solids, the standard state is the pure substance in its most stable form at one bar and the temperature of interest. The fugacity in the standard state must be determined experimentally. If the liquid or solid has negligible vapor pressure, this may not be possible. Since we intend "any system" to include all manner of mixtures and solutions, it can be very difficult to find the Gibbs free energy change for taking the substance from its standard state to the arbitrary system in which its fugacity is $f_A(\text{system}, P)$. In Chapter 14, we introduce the chemical activity of the substance to cope with such cases.

When we define the chemical activity of a substance in a particular system, we also introduce a new standard state. The primary criterion for our choice of this activity standard state is that we be able to measure how much the Gibbs free energy of the substance differs between the activity standard state and other states of the system. A principal object of the next seven chapters is to introduce ideas for measuring the difference between the Gibbs free energy of a substance in two states of a given system. Even so, our treatment of the issues involved in this step is quite incomplete.

§12 Evaluating entropy changes using thermochemical cycles

As for the standard enthalpy of reaction, we can obtain the standard entropy of reaction at a new temperature by evaluating entropy changes around a suitable thermochemical cycle. To do so, we need the standard entropy change at one temperature. We also need heat capacity data for all of the reactants and products. For the reaction $a A + b B \rightarrow c C + d D$, we can evaluate the entropy change at a second temperature by summing the individual contributions to the change in entropy around the cycle in Figure 5. For this cycle, we have

$$\Delta_r S^o(T_2) = \Delta_r S^o(T_1) + c \int_{T_1}^{T_2} \frac{C_P(C)}{T} dT$$

$$+ d \int_{T_1}^{T_2} \frac{C_P(D)}{T} dT - a \int_{T_1}^{T_2} \frac{C_P(A)}{T} dT - b \int_{T_1}^{T_2} \frac{C_P(B)}{T} dT$$

§13 Absolute zero is unattainable

The third law postulates that the entropy of a substance is always finite and that it approaches a constant as the temperature approaches zero. The value of this constant is independent of the values of any other state functions that characterize the substance. For any given substance, we are free to assign an arbitrarily selected value to the zero-temperature limiting value. However, we cannot assign arbitrary zero-temperature entropies to all substances. The set of assignments we make must be consistent with the experimentally observed zero-temperature limiting values of the entropy changes of reactions among different substances. For perfectly crystalline substances, these reaction entropies are all zero. We can satisfy this condition by assigning an arbitrary value to the zero-temperature molar entropy of each element

and stipulating that the zero-temperature entropy of any compound is the sum of the zero-temperature entropies of its constituent elements. This calculation is greatly simplified if we let the zero-temperature entropy of every element be zero. This is the essential content of the third law.

The Lewis and Randall statement incorporates this selection of the zero-entropy reference state for entropies, specifying it as "a crystalline state" of each element at zero degrees. As a result, the entropy of any substance at zero degrees is greater than or equal to zero. That is, the Lewis and Randall statement includes a convention that fixes the zero-temperature limiting value of the entropy of any substance. In this respect, the Lewis and Randall statement makes an essentially arbitrary choice that is not an intrinsic property of nature. We see, however, that it is an overwhelmingly convenient choice.

We have discussed alternative statements of the first and second laws. A number of alternative statements of the third law are also possible. We consider the following:

It is impossible to achieve a temperature of absolute zero.

This statement is more general than the Lewis and Randall statement. If we consider the application of this statement to the temperatures attainable in processes involving a single substance, we can show that it implies, and is implied by, the Lewis and Randall statement.

The properties of the heat capacity, C_P, play a central role in these arguments. We have seen that C_P is a function of temperature. While it is not useful to do so, we can apply the defining relationship for C_P to a substance undergoing a phase transition and find $C_P = \infty$. If we think about a substance whose heat capacity is less than zero, we encounter a contradiction of our basic ideas about heat and temperature: If $q > 0$ and $q/\Delta T < 0$, we must have $\Delta T < 0$; that is, heating the substance causes its temperature to decrease. In short, the theory we have developed embeds premises that require $C_P > 0$ for any system on which we can make measurements.

Let us characterize a pure-substance system by its pressure and temperature and consider reversible constant-pressure processes in which only pressure–volume

$$a A(T_2) \quad + \quad b B(T_2) \quad \xrightarrow{\Delta_r S(T_2)} \quad c C(T_2) \quad + \quad d D(T_2)$$

$$\Delta S = a \int_{T_1}^{T_2} \frac{C_P(A)}{T} dT \quad \Delta S = b \int_{T_1}^{T_2} \frac{C_P(B)}{T} dT \quad \Delta S = c \int_{T_1}^{T_2} \frac{C_P(C)}{T} dT \quad \Delta S = d \int_{T_1}^{T_2} \frac{C_P(D)}{T} dT$$

$$a A(T_1) \quad + \quad b B(T_1) \quad \xrightarrow{\Delta_r S(T_1)} \quad c C(T_1) \quad + \quad d D(T_1)$$

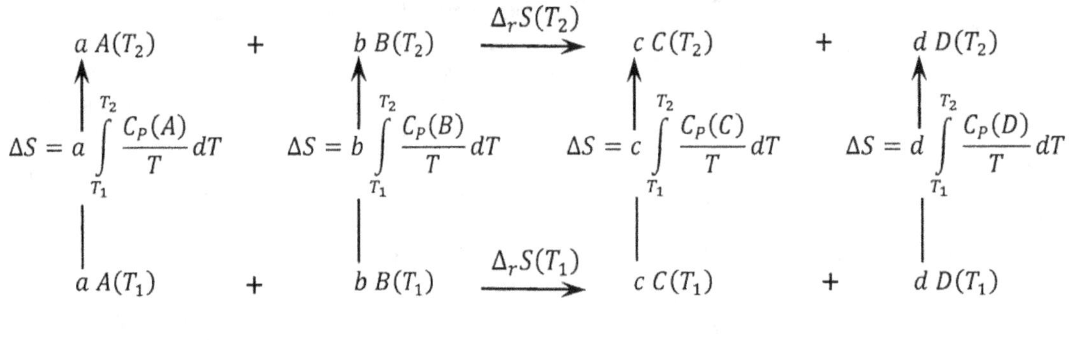

Figure 5. Cycle relating the entropy changes for a reaction at two temperatures.

work is possible. Then $(\partial S/\partial T)_P = C_P/T$ and $dS = C_P dT/T$. We now want to show: the Lewis and Randall stipulation that the entropy is always finite requires that the heat capacity go to zero when the temperature goes to zero. (Since we are going to show that the third law prohibits measurements at absolute zero, this conclusion is consistent with our conclusion in the previous paragraph.) That the heat capacity goes to zero when the temperature goes to zero is evident from $S = C_P dT/T$. If C_P does not go to zero when the temperature goes to zero, dS becomes arbitrarily large as the temperature goes to zero, which contradicts the Lewis and Randall statement.

To develop this result more explicitly, we let the heat capacities at temperatures T and zero be $C_P(T)$ and $C_P(0)$, respectively. Since $C_P(T) > 0$ for any $T > 0$, we have $S(T) - S(T^*) > 0$ for any $T > T^* > 0$. Since the entropy is always finite, $\infty > S(T) - S(T^*) > 0$, so that

$$\infty > \lim_{T^* \to 0} [S(T) - S(T^*)] > 0$$

and

$$\infty > \lim_{T^* \to 0} \int_{T^*}^{T} \frac{C_P}{T} dT > 0$$

For temperatures in the neighborhood of zero, we can expand the heat capacity, to arbitrary accuracy, as a Taylor series polynomial in T:

$$C_P(T) = C_P(0) + \left(\frac{\partial C_P(0)}{\partial T}\right)_P T + \frac{1}{2}\left(\frac{\partial^2 C_P(0)}{\partial T^2}\right)_P T^2 + \cdots$$

The inequalities become

$$\infty > \lim_{T^* \to 0} \left\{ C_P(0) \ln \frac{T}{T^*} + \left(\frac{\partial C_P(0)}{\partial T}\right)_P (T - T^*) + \frac{1}{4}\left(\frac{\partial^2 C_P(0)}{\partial T^2}\right)_P (T - T^*)^2 + \cdots \right\} > 0$$

The condition on the left requires $C_P(0) = 0$.

We could view the third law as a statement about the heat capacities of pure substances. We infer not only that $C_P > 0$ for all $T > 0$, but also that

$$\lim_{T \to 0} \left(\frac{C_P}{T}\right) = 0$$

More generally, we can infer corresponding assertions for closed reversible systems that are not pure substances: $(\partial H/\partial T)_P > 0$ for all $T > 0$, and $\lim_{T \to 0} T^{-1}(\partial H/\partial T)_P = 0$. (The zero-temperature entropies of such systems are not zero, however.) In the discussion below, we describe the system as a pure substance. We can make essentially the same arguments for any system; we need only replace C_P by $(\partial H/\partial T)_P$. The Lewis and Randall statement asserts that the entropy goes to a constant at absolute zero, irrespective of the values of any other thermodynamic functions. It follows that the entropy at zero degrees is independent of the value of the pressure. For any two pressures, P_1 and P_2, we have $S(P_2, 0) - S(P_1, 0) = 0$. Letting $P = P_1$ and $P_2 = P + \Delta P$ and, we have

$$\frac{S(P + \Delta P, 0) - S(P, 0)}{\Delta P} = 0$$

for any ΔP. Hence, we have

$$\left(\frac{\partial S}{\partial P}\right)_{T=0} = 0$$

In Chapter 10, we find $(\partial S/\partial P)_T = -(\partial V/\partial T)_P$, so both the entropy and the volume approach their zero-temperature values asymptotically.

When we say that absolute zero is unattainable, we mean that no system can undergo any change in which its final temperature is zero. To see why absolute zero must be unattainable, let us consider processes that can decrease the temperature of a system. In general, we have heat reservoirs available at various temperatures. We can select the available reservoir whose temperature is lowest, and bring the system to this temperature by simple thermal contact. This much is trivial; clearly, the challenge is to decrease the temperature further. To do so, we must effect some other change. Whatever this change may be, it cannot be aided by an exchange of heat with the surroundings. Once we have brought the system to the temperature of the coldest available portion of the surroundings, any further exchange of heat with the surroundings can only be counter-productive. We conclude that any process suited to our purpose must be adiabatic. Since an adiabatic process exchanges no heat with the surroundings, $\Delta \hat{S} = 0$.

The process must also be a possible process, so that $\Delta S + \Delta \hat{S} \geq 0$, and since it is adiabatic, $\Delta S \geq 0$. Let us consider a reversible process and an irreversible process in which the same system[2] goes from the state specified by P_1 and T_1 to a second state in which the pressure is P_2. The final temperatures and the entropy changes of these processes are different. For the reversible process, $\Delta S = 0$; we designate the final temperature as T_2. For the irreversible process, $\Delta S > 0$; we designate the final temperature as T_2^*. As it turns out, the temperature change is less for the irreversible process than for the reversible process; that is, $T_2 - T_1 < T_2^* - T_1$. Equivalently, the reversible process reaches a lower temperature: $T_2 < T_2^*$. From

$$dS = \frac{C_P}{T} dT - \left(\frac{\partial V}{\partial T}\right)_P dP$$

we can calculate the entropy changes for these processes. For the reversible process, we calculate

$$\Delta S^{rev} = S(P_2,T_2) - S(P_1,T_1)$$

To do so, we first calculate

$$(\Delta S)_T = S(P_2,T_1) - S(P_1,T_1)$$

for the isothermal reversible transformation from state P_1, T_1 to the state specified by P_2 and T_1. For this step, dT is zero, and so

$$(\Delta S)_T = \int_{P_1}^{P_2} \left(\frac{\partial V}{\partial T}\right)_P dP$$

We then calculate

$$(\Delta S)_P = S(P_2,T_2) - S(P_2,T_1)$$

for the isobaric reversible transformation from state P_2, T_1 to state P_2, T_2. For this transformation, dP is zero, and

$$(\Delta S)_P = -\int_{T_1}^{T_2} \frac{C_P}{T} dT$$

Then,

$$\begin{aligned}\Delta S^{rev} &= S(P_2,T_2) - S(P_1,T_1)\\ &= \int_{T_1}^{T_2} \frac{C_P}{T} dT - \int_{P_1}^{P_2} \left(\frac{\partial V}{\partial T}\right)_P dP\\ &= 0\end{aligned}$$

Because $\Delta S^{rev} = 0$, the reversible process is unique; that is, given P_1, T_1, and P_2, the final temperature of the system is determined. We find T_2 from

$$\int_{T_1}^{T_2} \frac{C_P}{T} dT = \int_{P_1}^{P_2} \left(\frac{\partial V}{\partial T}\right)_P dP$$

To understand the entropy change for the irreversible process, we note first that there are an infinite number of such processes. There is nothing unique about the final temperature. Given P_1, T_1, and P_2, the final temperature, T_2^*, can have any value consistent with the properties of the substance. To specify a particular irreversible process, we must specify all four of the quantities P_1, T_1, P_2, and T_2^*. Having done so, however, we can calculate the entropy change for the *irreversible process*,

$$\Delta S^{irrev} = S(P_2,T_2^*) - S(P_1,T_1) > 0$$

by computing the entropy changes as we *reversibly* carry the system along the isothermal two-step path from P_1, T_1 to P_2, T_1 and then along the isobaric path from P_2, T_1 to P_2, T_2^*. The calculation of ΔS^{irrev} for this reversible path from P_1, T_1 to P_2, T_2^* employs the same logic as the calculation, in the previous paragraph, of ΔS for the reversible path from P_1, T_1 to P_2, T_2. The difference is that T_2^* replaces T_2 as the upper limit in the temperature integral. The pressure integral is the same. We have

$$\begin{aligned}\Delta S^{irrev} &= S(P_2,T_2^*) - S(P_1,T_1)\\ &= \int_{T_1}^{T_2^*} \frac{C_P}{T} dT - \int_{P_1}^{P_2} \left(\frac{\partial V}{\partial T}\right)_P dP\\ &> 0\end{aligned}$$

From $\Delta S^{irrev} > \Delta S^{rev}$, we have

$$\int_{T_1}^{T_2^*} \frac{C_P}{T} dT > \int_{T_1}^{T_2} \frac{C_P}{T} dT$$

Since the integrands are the same and positive, it follows that $T_2^* > T_2$, as asserted above.

Figure 6 shows the relationships among the various quantities discussed in this argument. In the first instance, Figure 6 shows a plot of two of the system's isobars in temperature—entropy space. That is, the line labeled $P = P_1$ depicts the set of temperature—entropy points at which the equilibrated system has pressure P_1; the line labeled $P = P_2$, depicts the equilibrium positions at pressure P_2. Other lines in this sketch represent paths along which the system can undergo reversible changes at constant entropy or constant temperature. The dotted line represents the irreversible process in which the system goes from the state specified by P_1, T_1 to the state specified by P_2, T_2^*. This line is dotted to represent the fact that the system's temperature may not be well defined during the irreversible process.

Effective cooling can be achieved using pressure changes if the system is a gas. However, for liquids and solids, $(\partial V/\partial T)_P$ is small; consequently, the temperature change for a reversible pressure change is also small. At temperatures near absolute zero, nearly all substances are

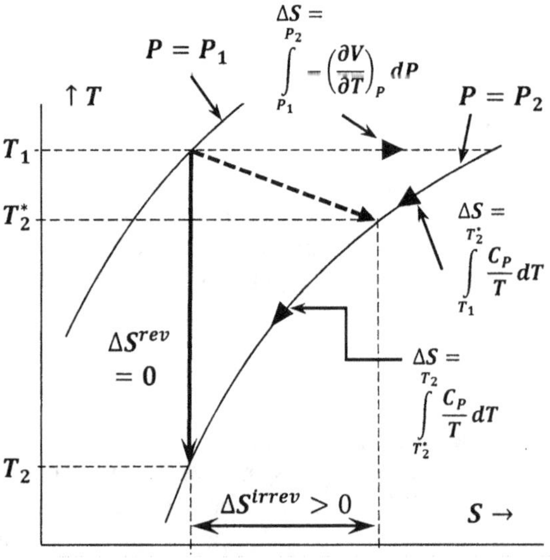

Figure 6. Temperature *versus* entropy for spontaneous and reversible processes.

solid; to achieve effective cooling we must change a thermodynamic variable for which a solid's temperature coefficient is as large as possible. To consider the general problem of decreasing the temperature of a system by varying something other than pressure, we must consider a system in which some form of non-pressure–volume work is possible. Such a system is subject to an additional force, and its energy changes as this force changes.

The practical method by which extremely low temperatures are achieved is called ***adiabatic demagnetization***. This method exploits the properties of paramagnetic solids. In such solids, unpaired electrons localized on individual atoms give rise to a magnetic moment. Quantum mechanics leads to important conclusions about the interaction between such magnetic moments and an applied magnetic field: In an applied magnetic field, the magnetic moment of an individual atom is quantized. In the simplest case, it can be aligned in only two directions; it must be either parallel or anti-parallel to the applied magnetic field. When an atom's magnetic moment is parallel to the magnetic field the energy of the system is less than when the alignment is anti-parallel. The applied magnetic field exerts a force on the magnetic moments associated with individual atoms. The energy of the system depends on the magnitude of the applied magnetic field.

Rather than focus on the particular case of adiabatic demagnetization, let us consider the energy and entropy changes associated with changes in a generalized potential, Φ_θ, and its generalized displacement, θ. (For adiabatic demagnetization, θ would be the applied magnetic field.) Three variables are required to describe reversible changes in this system. We can express the energy and entropy as functions of temperature, pressure, and θ: $E = E(T,P,\theta)$ and $S = S(T,P,\theta)$. The total differential of the entropy includes a term that specifies the dependence of entropy on θ. We have

$$\begin{aligned} dS &= \left(\frac{\partial S}{\partial T}\right)_{P,\theta} dT + \left(\frac{\partial S}{\partial P}\right)_{T,\theta} dP + \left(\frac{\partial S}{\partial \theta}\right)_{T,P} d\theta \\ &= \frac{C(T,P,\theta)}{T} dT - \left(\frac{\partial V}{\partial T}\right)_{P,\theta} dP + \left(\frac{\partial S}{\partial \theta}\right)_{T,P} d\theta \end{aligned}$$

where we write $C(T,P,\theta)$ to emphasize that our present purposes now require that we measure the heat capacity at constant pressure and constant θ.

For constant pressure, P, and constant displacement, θ, the entropy depends on temperature as

$$\begin{aligned} S(T,P,\theta) &= S(0,P,\theta) + \int_0^T \left(\frac{\partial S}{\partial T}\right)_{P,\theta} dT \\ &= S(0,P,\theta) + \int_0^T \frac{C(T,P,\theta)}{T} dT \end{aligned}$$

The postulate that the entropy be finite at any temperature implies that the pressure- and θ-dependent heat capacity becomes zero at absolute zero. That is, at absolute zero, the heat capacity vanishes whatever the values of P and θ. The argument is exactly the same as before. Earlier, we wrote $C_P(0) = 0$; for the present generalized case, we write $C(0,P,\theta) = 0$.

Similarly, from the postulate that the entropy goes to a constant at absolute zero for all values of the other thermodynamics variables, it follows that, for any two pressures P_1 and P_2, and for any two values of the generalized displacement, θ_1 and θ_2,

$$\begin{aligned} S(0,P_1,\theta_1) &= S(0,P_2,\theta_1) \\ &= S(0,P_1,\theta_2) \\ &= S(0,P_2,\theta_2) \\ &= 0 \end{aligned}$$

and hence that

$$\left(\frac{\partial S}{\partial P}\right)_{T=0,\theta} = \left(\frac{\partial S(0,P,\theta)}{\partial P}\right)_{T,\theta} = 0$$

and

$$\left(\frac{\partial S}{\partial \theta}\right)_{T=0,P} = \left(\frac{\partial S(0,P,\theta)}{\partial \theta}\right)_{T,P} = 0$$

We want to consider a process in which a system goes from the lowest temperature available in the surroundings to a still lower temperature. To minimize the final temperature, this process must be carried out adiabatically. It must also be a possible process, so that $dS \geq 0$. For simplicity, let us now assume that we carry out this process at a constant pressure, P, and that the system goes from the state specified by P, T_1, θ_1 to the state specified by P, T_2, θ_2 where $T_1 > T_2$. The entropies of these two states are

$$S(T_1,P,\theta_1) = S(0,P,\theta_1) + \int_0^{T_1} \frac{C(T,P,\theta_1)}{T} dT$$

and

$$S(T_2,P,\theta_2) = S(0,P,\theta_2) + \int_0^{T_2} \frac{C(T,P,\theta_2)}{T} dT$$

The entropy change for this process is

$$\begin{aligned} S(T_2,P,\theta_2) - S(T_1,P,\theta_1) &= S(0,P,\theta_2) - S(0,P,\theta_1) \\ &+ \int_0^{T_2} \frac{C(T,P,\theta_2)}{T} dT - \int_0^{T_1} \frac{C(T,P,\theta_1)}{T} dT \geq 0 \end{aligned}$$

Now, let us suppose that the final temperature is zero; that is, $T_2 = 0$, so that

$$\int_0^{T_2} \frac{C(T,P,\theta_2)}{T} dT = 0$$

It follows that

$$S(0,P,\theta_2) - S(0,P,\theta_1) \geq \int_0^{T_1} \frac{C(T,P,\theta_1)}{T} dT > 0$$

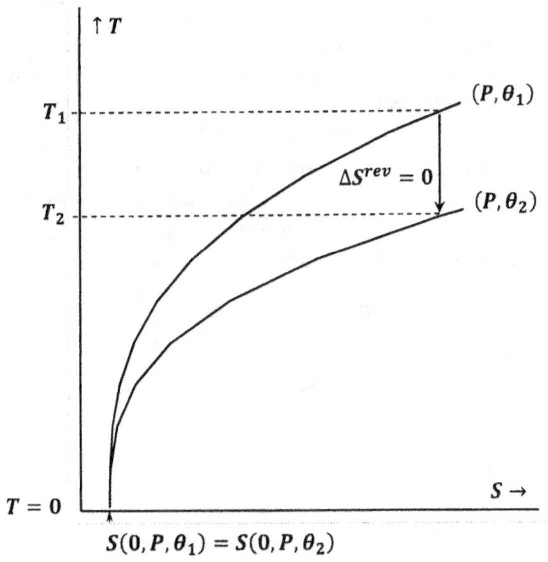

Figure 7. Reversible change in a system that satisfies the Lewis and Randall statement.

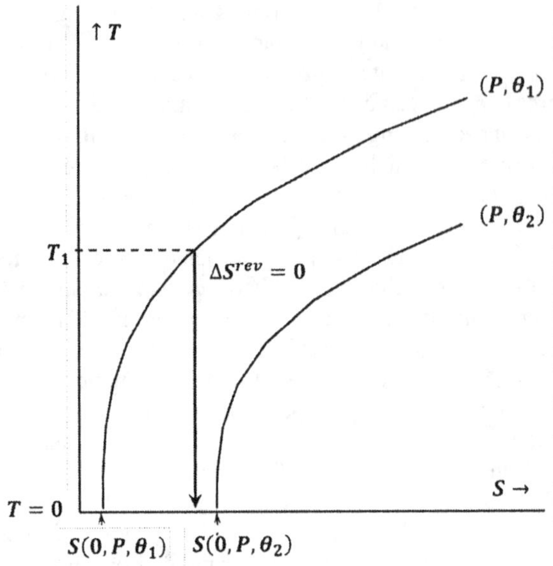

Figure 8. Reversible change in a system that does not satisfy the Lewis and Randall statement.

where the inequality on the right follows from the fact that that $C(T,P,\theta_1) > 0$. Then, it follows that

$$S(0,P,\theta_2) - S(0,P,\theta_1) > 0$$

which contradicts the Lewis and Randall statement of the third law. The assumption that the system can reach absolute zero leads to a contradiction of the Lewis and Randall statement of the third law. Therefore, if the Lewis and Randall statement is true, absolute zero is unattainable.

The converse applies also; that is, from the proposition that absolute zero is unattainable, we can show that the Lewis and Randall statement is true. To do so, we rearrange the above equation for ΔS,

$$\int_0^{T_2} \frac{C(T,P,\theta_2)}{T} dT \geq \int_0^{T_1} \frac{C(T,P,\theta_1)}{T} dT - S(0,P,\theta_2) + S(0,P,\theta_1)$$

If we now assume that the Lewis and Randall statement is false, the expression on the right can be less than or equal to zero. The integral on the left can then be zero, in which case the system can reach absolute zero. If the Lewis and Randall statement is false, it is true that the system can reach absolute zero. Therefore: If the system cannot reach absolute zero, the Lewis and Randall statement is true.

Figures 7 and 8 depict these ideas using contour plots in temperature–entropy space. Each figure shows two contour lines. One of these contour lines is a set of temperature and entropy values along which the pressure is constant at P and θ is constant at θ_1. The other contour line is a set of temperature and entropy values along which the pressure is constant at P and θ is constant at θ_2. The slope of a contour line is

$$\left(\frac{\partial T}{\partial S}\right)_{P,\theta} = \frac{T}{C(T,P,\theta)}$$

Because the heat capacity is always positive, this slope is always positive.

In Figure 7, the Lewis and Randall statement is satisfied. When the temperature goes to zero, the contour lines meet at the same value of the entropy; these contours satisfy the relationship

$$S(0,P,\theta_1) = S(0,P,\theta_2)$$

An adiabatic (vertical) path from the contour for P and θ_1 meets the contour for P and θ_2 at a positive temperature, $T_2 > 0$. Since this is evidently true for any P and any θ_2, the final state for any adiabatic process will have $T_2 > 0$. Because the Lewis and Randall statement is satisfied, the system cannot reach absolute zero, and *vice versa*.

In Figure 8, the Lewis and Randall statement is violated, because we have $S(0,P,\theta_1) < S(0,P,\theta_2)$. In this case, an adiabatic process initiated from a low enough initial temperature, T_1, will reach absolute zero without intersecting the contour for constant P and θ_2. Because the Lewis and Randal statement is violated, the system can reach absolute zero, and *vice versa*.

Problems

1. The relationships between H^o, S^o, and G^o for the standard state of a gas and the molar enthalpy, $\overline{H}(P)$, entropy, $\overline{S}(P)$, and Gibbs free energy, $\overline{G}(P)$, of the real gas at pressure P and temperature T involve several integrals. Given the virial equation for a real gas,

$$Z = \frac{P\overline{V}}{RT} = 1 + B^*(T)P + C^*(T)P^2 + D^*(T)P^3 + \cdots$$

evaluate the following:

(a)

$$\int_0^P \left[\overline{V} - T\left(\frac{\partial \overline{V}}{\partial T}\right)_P\right] dP$$

(b)

$$\int_0^P \left[\frac{\overline{V}}{RT} - \frac{1}{P}\right] dP$$

(c)

$$\int_{P^*}^{P} \left(\frac{\partial \overline{V}}{\partial T}\right)_P dP - \int_{P^*}^{1\text{ bar}} \frac{R}{P} dP$$

2. Why does $(\partial S/\partial P)_{T=0} = -(\partial V/\partial T)_{P,T=0}$ imply that both the entropy and the volume approach their zero-temperature values asymptotically? Is this consistent with defining absolute zero to be the temperature at which the volume of an ideal gas extrapolates to zero—at constant pressure?

3. Prove that

$$\overline{S}_A(P) - S_A^0 = \int_{P}^{P^*} \left(\frac{\partial \overline{V}}{\partial T}\right)_P dP + \int_{P^*}^{1\text{ bar}} \frac{R}{P} dP$$

remains finite for any gas in the limit as $P^* \to 0$. Hint: Express the integral from P to P^* as the sum of integrals from P^* to 1 bar and from 1 bar to P.

4. Let A, B, C, and D be elements, whose absolute entropies at 1 bar and temperature T are $S^o(A,T)$, $S^o(B,T)$, $S^o(C,T)$, and $S^o(D,T)$, respectively. Let AB, CD, AC, and BD be binary compounds of these elements, and represent their absolute entropies at these conditions by $S^o(AB,T)$, $S^o(CD,T)$, $S^o(AC,T)$, and $S^o(BD,T)$.

(a) What is the entropy change, $\Delta_r S^o(T)$, for the reaction $AB + CD \to AC + BD$?

(b) What are $\Delta_f S^o(AB,T)$, $\Delta_f S^o(CD,T)$, $\Delta_f S^o(AC,T)$, and $\Delta_f S^o(BD,T)$?

(c) Show that

$$\Delta_r S^o(T) = \Delta_f S^o(AC,T) + \Delta_f S^o(BD,T) - \Delta_f S^o(AB,T) - \Delta_f S^o(CD,T)$$

5. Let A and B be elements; let A_aB_b be a binary compound of these elements. At temperature T and 1 bar, let the entropy of these substances be $S^o(A,T)$, $S^o(B,T)$, and $S^o(A_aB_b,T)$, respectively. At absolute zero and 1 bar, let these entropies be $S^o(A,0)$, $S^o(B,0)$, and $S^o(A_aB_b,0)$. Represent the change in entropy when these substances are warmed from 0 K to T, at a constant pressure of 1 bar, as $\Delta_{0\to T}S^o(A)$, $\Delta_{0\to T}S^o(B)$, and $\Delta_{0\to T}S^o(A_aB_b)$. These quantities are related by the following equations:

$$S^o(A,T) = S^o(A,0) + \Delta_{0\to T}S^o(A)$$
$$S^o(B,T) = S^o(B,0) + \Delta_{0\to T}S^o(B)$$
$$S^o(A_aB_b,T) = S^o(A_aB_b,0) + \Delta_{0\to T}S^o(A_aB_b)$$

By the Nernst Heat Theorem, the entropy change for formation of A_aB_b at absolute zero must be zero:

$$\Delta_f S^o(A_aB_b,0) = S^o(A_aB_b,0) - a\,S^o(A,0) - b\,S^o(B,0)$$

When, following Planck and Lewis and Randall, we choose to let the entropies of the elements be zero at absolute zero, the entropy of A_aB_b and its entropy of formation also become zero at absolute zero:

$$\Delta_f S^o(A_aB_b,0) = S^o(A_aB_b,0) = 0$$

Suppose that we decide to create an alternative set of absolute entropies by assigning non-zero values to the entropies of the elements at absolute zero. Let us distinguish entropy values in this new scheme with a tilde. Then, the non-zero values that we assign to the elements at absolute zero are $\tilde{S}^o(A,0) \neq 0$ and $\tilde{S}^o(B,0) \neq 0$. By the Nernst Heat Theorem, we have

$$\Delta_f \tilde{S}^o(A_aB_b,0) = \tilde{S}^o(A_aB_b,0) - a\,\tilde{S}^o(A,0) - b\,\tilde{S}^o(B,0)$$

Evidently, we have

$$\Delta_f \tilde{S}^o(A_aB_b,0) = \Delta_f S^o(A_aB_b,0) = 0$$

and, since the values of $\tilde{S}^o(A,0)$ and $\tilde{S}^o(B,0)$ are arbitrary, we can choose them so that $\tilde{S}^o(A_aB_b,0)$ is non-zero also

$$\tilde{S}^o(A_aB_b,0) = a\,\tilde{S}^o(A,0) + b\,\tilde{S}^o(B,0) \neq 0$$

(a) What are $\tilde{S}^o(A,T)$, $\tilde{S}^o(B,T)$, and $\tilde{S}^o(A_aB_b,T)$?

(b) What is $\Delta_f \tilde{S}^o(A_aB_b,T)$?

(c) Show that $\Delta_f \tilde{S}^o(A_aB_b,T) = \Delta_f S^o(A_aB_b,T)$.

(d) Consider a reaction in which A_aB_b is a reactant or a product. How will the alternative choice of values for the entropies of A and B at absolute zero affect the values we calculate for $\Delta_r S^o$, $\Delta_r H^o$, $\Delta_r G^o$?

(e) Can you think of any circumstance in which there would be an advantage to choosing $\tilde{S}^o(A,0) \neq 0$ and $\tilde{S}^o(B,0) \neq 0$?

6. Find $\ln \gamma$ for a gas that obeys the equation of state $P(\overline{V} - b) = RT$. For CO_2 at 300 K, the value of b (the second virial coefficient) is -1.26×10^{-4} m^3 mol^{-1}. Calculate the fugacity coefficient and the fugacity of CO_2 at 300 K and pressures of 1, 10, and 100 bar.

7. Consider the following sequence of steps that convert a van der Waals gas, vdwg, at an arbitrary pressure to the corresponding hypothetical ideal gas in its standard state.

(I) $A(\text{vdwg}, P, T) \xrightarrow{\Delta_I G} A(\text{vdwg}, P^* \approx 0, T)$

(II) $A(\text{vdwg}, P^* \approx 0, T) \xrightarrow{\Delta_{II} G} A(\text{ideal gas}, P^* \approx 0, T)$

(III) $A(\text{ideal gas}, P^* \approx 0, T) \xrightarrow{\Delta_{III} G} A(HIG^o, T)$

Show that the fugacity of the van der Waals gas is given by

$$\ln f_{vdw} = \frac{b}{\overline{V} - b} - \frac{2a}{RT\overline{V}} + \ln \frac{RT}{\overline{V} - b}$$

Hint: Find $\Delta G = \Delta_I G + \Delta_{II} G + \Delta_{III} G$. To calculate $\Delta_I G$, use integration by parts:

$$\Delta_I G = \int_P^{P^*} \overline{V}_{vdw} dP = \left[P\overline{V}\right]_P^{P^*} - \int_{\overline{V}}^{\overline{V}^*} P d\overline{V}$$

When $P^* \to 0, \overline{V}^* \to \infty$.

8. The normal boiling point of methanol is 337.8 K at 1 atm. The enthalpy of vaporization at the normal boiling point is $\Delta_{vap} H = 35.21$ kJ mol^{-1}.

(a) What are $\Delta_{vap} G$ and $\Delta_{vap} S$ for methanol at its normal boiling point?

(b) At 1 bar, the heat capacity of gaseous methanol depends on temperature as

$$C_P(CH_3OH, \text{g}, 1 \text{ bar}) = 21.737 + 0.07494\, T \quad \left[\text{J K}^{-1} \text{ mol}^{-1}\right]$$

Assume that the heat capacity at 1 atm is the same as it is at 1 bar. Calculate the enthalpy change and the entropy change when one mole of gaseous methanol is heated from the normal boiling point to 500 K at 1 atm.

(c) At 1 bar, the absolute entropy of gaseous methanol depends on temperature as

$$S^o(CH_3OH, \text{g}, 1 \text{ bar}) = 192.8 + 0.1738\, T - (5.367 \times 10^{-5}) T^2 \quad \left[\text{J K}^{-1} \text{ mol}^{-1}\right]$$

Assume that the absolute entropy at 1 atm is the same as it is at 1 bar. Calculate the Gibbs Free Energy change, ΔG, when one mole of gaseous methanol is heated from the normal boiling point to 500 K at a constant pressure of 1 atm.

(d) The heat capacity of liquid methanol,

$$C_P(CH_3OH, \text{liq}, 298.15 \text{ K}, 1 \text{ bar})$$

is 1.1 J K^{-1} mol^{-1}. Assume that this heat capacity remains constant for superheated liquid methanol. Calculate the enthalpy and entropy changes when liquid methanol is heated from the normal boiling point to 500 K.

(e) The molar entropy of liquid methanol at 1 bar and 298.15 K is 126.8 J K^{-1} mol^{-1}. We can estimate the molar entropy of superheated liquid methanol by using the heat capacity at 298.15 K to estimate the entropy at higher temperatures:

$$S^o(T) \approx S^o(298.15 \text{ K}) + \int_{298.15}^{T} \frac{C_P}{T} dT$$

Find $S^o(T)$ and use this equation to calculate ΔG for heating liquid methanol from the normal boiling point to 500 K.

(f) Devise a cycle that enables you to use the results you obtain in parts (a)–(e) to calculate $\Delta_{vap} H(500 \text{ K})$, $\Delta_{vap} S(500 \text{ K})$, and $\Delta_{vap} G(500 \text{ K})$ when one mole of methanol vaporizes at 1 atm and 500 K.

(g) Use the values you obtain for $\Delta_{vap} H(500 \text{ K})$ and $\Delta_{vap} S(500 \text{ K})$ in part (f) to calculate $\Delta_{vap} G(500 \text{ K})$.

(h) Use the Gibbs-Helmholtz equation to calculate the Gibbs free energy change, $\Delta_{vap} G(500 \text{ K})$, when one mole of methanol vaporizes at 1 atm and 500 K. Compare this result to those you obtained in parts (f) and (g). Is this process impossible, spontaneous, or reversible?

(i) How much heat is taken up by the system when one mole of methanol is vaporized reversibly at its normal boiling point and the resulting vapor is heated reversibly at 1 atm to 500 K? How much work is done on the system in this process? What are ΔE and ΔS for this process?

(j) Suppose that the change of state in part (i) is effected irreversibly by contacting the liquid methanol with surroundings at 500 K, while maintaining the applied pressure constant at 1 atm. How much work is done on the system in this spontaneous process? How much heat is taken up by the system? What are $\Delta \hat{S}$ and $\Delta S_{universe}$ for this spontaneous process?

9. To maximize the temperature change for a given change in system pressure, the value of $|(\partial T/\partial P)_S|$ should be as large as possible. If only pressure–volume work is possible, we have

$$dS = \frac{C_P}{T}dT - \left(\frac{\partial V}{\partial T}\right)_P dP$$

Show that

$$\left(\frac{\partial T}{\partial P}\right)_S = \frac{T}{C_P}\left(\frac{\partial V}{\partial T}\right)_P$$

What happens to T/C_P as the temperature approaches absolute zero? For condensed phases, we find $(\partial V/\partial T)_P \ll V$. Consider the behavior of the molecules in a lattice as the temperature approaches absolute zero. Is it reasonable to expect

$$\lim_{T\to 0}\left(\frac{\partial V}{\partial T}\right)_P = 0$$

Why?

Notes

[1] Lewis, G.N., Randall, M., K. S. Pitzer, and L. Brewer, ***Thermodynamics***, 2nd Edition, McGraw-Hill, New York, 1981, p 130.

[2] More precisely, we consider a reversible process and a spontaneous process whose initial state is a "change-enabled" modification of the reversible-process initial state. The state functions are the same in both initial states.

12

Applications of the Thermodynamic Criteria for Change

§1 Processes in which the composition of the system changes

The equations we derive in Chapters 9 and 10 are the core of chemical thermodynamics. However, we have yet to deal with the effects of changing the concentrations of the substances present in the system. To apply our theory to chemical changes, we must extend our theory so that it can model these effects. In this chapter, we consider some basic applications that do not involve chemical reactions and in which both intermolecular interactions and the effects of mixing can be ignored. In Chapters 13-16, we develop the application of thermodynamic concepts to processes in which a chemical reaction occurs. We do so in two steps. In Chapter 13, we consider an approximation in which the properties of a multicomponent system are determined by the effects of mixing pure substances whose molecules neither attract nor repel one another. In Chapter 14, we begin to consider the general case in which intermolecular interactions can be important.

§2 Mechanical processes

When we talk about a purely mechanical process, we have in mind a system in which one or more unchanging objects can move relative to some reference frame. Their movements are described completely by Newton's laws of motion. The objects are characterized by their masses, locations, velocities, and accelerations. They may be subject to the effects of force fields, whose magnitudes can vary with location and time. We stipulate that the volume, pressure, temperature, entropy, composition, and the internal energy, U, of an individual object remain constant. Since entropy and volume are extensive state functions, we can obtain the entropy and volume of the objects in aggregate by summing up those for the individual objects. Moreover, the total entropy and the total volume are constants.

The energy of a purely mechanical system is the sum of its internal energy, U, its kinetic energy, τ, and its mechanical potential energy, υ; that is, $E = U + \tau + \upsilon$. U is the sum of the internal energies of the unchanging constituent objects. Since it is constant, the internal energy can be given an arbitrary value that we usually choose to be zero. When we do so, the energy of the system becomes the sum of its kinetic and potential energies. Noting explicitly that the entropy and volume are constant, we have $(\Delta E)_{SV} = \tau + \upsilon$.

The essential distinction between a purely mechanical system and a thermodynamic system is that our models for mechanical systems focus on the motions of unchanging objects; our models for thermodynamic systems focus on the internal changes of stationary objects.

An important aspect of this distinction is that our definitions of equilibrium, reversibility, and spontaneous change in mechanical systems are not wholly congruent with the definitions we use in developing the principles of thermodynamics. Thus, an equilibrium state of a mechanical system is one in which the objects comprising the system are stationary with respect to some reference frame. For a mechanical system at equilibrium, the kinetic energy is constant and can usually be taken to be zero, $\tau = 0$. The motion of a system generally has no bearing on whether the substances that comprise the system are at equilibrium in a thermodynamic sense.

We normally consider that a sufficient condition for a mechanical system to be reversible is that—following some excursion to other states—the initial conditions of both the system and the surroundings can be restored exactly[1]. In our thermodynamic view of reversibility, this condition is necessary but not sufficient: We say that a thermodynamic system is reversible only if the direction in which it is changing can be reversed at any time by an arbitrarily small change in its interaction with its surroundings. The initial conditions can be restored after any excursion.

Our treatments of the frictionless harmonic oscillator illustrate this imperfect congruence. Viewed as a mechanical system, a frictionless harmonic oscillator is a reversible system. If we adopt the view that it is continuously undergoing spontaneous change, our thermodynamic principles mean that its entropy is continuously increasing. However, since its state and that of its surroundings reproduce themselves exactly after every period of oscillation, our thermodynamic principles mean also that there is zero net change in the entropy over any complete oscillation. Clearly, this is a contradiction. If we attempt to salvage the situation by supposing that the entropy of an isolated freely moving harmonic oscillator is constant, our thermodynamic principles require us to say that it is at equilibrium. This contradicts the view that a mechanical system is at equilibrium only if its kinetic energy is zero. Since neither formulation is satisfactory, we recognize that we cannot expect to describe every mechanical system in purely thermodynamic terms.

Nor can we expect to describe every

thermodynamic system in mechanical terms. This becomes obvious when we observe that the second law of thermodynamics is essential to our description of thermodynamic systems, but it is not among the principles of mechanics. Beginning in Chapter 20, we find that we can model the thermodynamic properties of a system that is itself a collection of a large number of subsystems by focusing on the average values of the properties of the subsystems. The laws of motion model the movements of the individual particles of a system. The laws of thermodynamics model the average properties of the particles in a system that contains a very large number of particles. While we cannot usefully describe an individual harmonic oscillator as a thermodynamic system, we see in §22-6 that the thermodynamic properties of a system composed of many identical harmonic oscillators can be modeled very successfully.

In short, mechanics and thermodynamics model different kinds of systems from fundamentally different perspectives. Nevertheless, when we limit our consideration of mechanical systems to the prediction of spontaneous change from one equilibrium state to another, we can recognize that the criteria we are accustomed to apply to mechanical systems are analogous to our thermodynamic criteria as they apply to such processes.

From our thermodynamic perspective, a purely mechanical process involves no change in the entropy or volume of the system, and the criterion for irreversible change is $(\Delta E)_{SV} < w_{NPV}$. From our mechanical perspective, an irreversible transition between equilibrium states is one in which various objects interact with one another or with various force fields. We consider that the process can occur if the change in the potential energy of the system is less than the work done on it; that is $\Delta \upsilon < w_{NPV}$. In the mechanical system, some of the work done on the system is dissipated by frictional forces as heat that appears in the surroundings. (If the process involves no exchange of work with the surroundings, the criterion becomes $\Delta \upsilon < 0$.) If the mechanical system begins and ends at rest, we have $(\Delta E)_{SV} < \Delta \upsilon$, so that again $(\Delta E)_{SV} < w_{NPV}$.

If no work is exchanged with the surroundings, potential energy minimization, $\Delta \upsilon < 0$, is a sufficient condition for spontaneous change to be possible in mechanical systems under the circumscribed conditions we have outlined. In §14-2, we find that minimization of the chemical potential energy,

$$\sum_{j=1}^{\omega} \mu_j \, dn_j < 0$$

is a necessary and sufficient condition for spontaneous change to be possible in a thermodynamic system. These conditions are parallel, but they are not equivalent to one another.

§3 The direction of spontaneous heat transfer

The idea that thermal energy can be transferred from a warmer body to a colder one, but not in the opposite direction, is a fundamental assumption in our development of the thermodynamic criteria for change. Therefore, if our theory is to be internally consistent, we must be able to deduce this principle from the criteria we have developed. Let us consider one way in which this can be done: We consider an isolated system comprised of two subsystems, A and B, that are in thermal contact with one another. We suppose that the temperatures are T_A and T_B and that $T_A \neq T_B$. If the energy of subsystem A increases, heat transfers from subsystem B to subsystem A. In this case, we know that $q_A > 0$ and $T_B > T_A$.

When we seek to analyze this process using our thermodynamic theory, we encounter a problem that arises for any spontaneous process: Since the process is not reversible, we must introduce approximating assumptions. For the present analysis, we want to estimate the entropy change that occurs in each subsystem. To do so, we suppose that an increment of heat, dq, can pass from one subsystem to the other without significantly changing the temperature of either one. It is evident that we could—by some other process—effect this change in either subsystem as nearly reversibly as we wish. (In §5, we consider such a process.) Even though the present process is not reversible, we have good reason to assume that the entropy changes in the subsystems are well approximated as dq_A/T_A and $dq_B/T_B = -\,dq_A/T_B$. Since the system is isolated, the process can be spontaneous only if its entropy change is positive; that is, the relevant thermodynamic criterion is $dq_A/T_A - dq_A/T_B > 0$. With $dq_A > 0$, we find $1/T_A - 1/T_B > 0$, or $T_B > T_A$. When heat is spontaneously transferred from B to A, our thermodynamic criterion also requires that subsystem B be warmer than subsystem A.

§4 Phase changes: The fusion of ice

Let us consider processes in which transfer of heat from the surroundings melts one mole of ice. We suppose that the ice is initially at 0 C and one bar. At these conditions, the enthalpy change for melting a mole of ice is 6010 J. If the ice melts reversibly at these conditions, the temperature of the surroundings is also 0 C. As it melts, the ice takes up 6010 J of heat, which is given up by the surroundings. For this process, we have $q_P^{rev} = \Delta_{fus}H = 6010$ J. The temperature is constant, and the entropy change for the system is

$$\Delta S = \frac{q_P^{rev}}{T} = \frac{\Delta_{fus}H}{T}$$

Since $\hat{q} = -q_P^{rev}$, we have

$$\Delta \hat{S} = \frac{-q_P^{rev}}{\hat{T}} = \frac{-\Delta_{fus}H}{\hat{T}}$$

so that

$$\Delta S = 6010 \text{ J}/273.15 \text{ K} = 22.00 \text{ J K}^{-1}$$

and

$$\Delta \hat{S} = -\,6010 \text{ J}/273.15 \text{ K} = -22.00 \text{ J K}^{-1}$$

As required for a reversible process, we have $\Delta S + \Delta \hat{S} = 0$. The Gibbs free energy change is

$$\begin{aligned}(\Delta G)_{PT} &= (\Delta H)_{PT} - T(\Delta S)_{PT} \\ &= q_P^{rev} - T(q_P^{rev}/T) \\ &= 0\end{aligned}$$

which is also as required for a reversible process.

Now let us consider a spontaneous process, in which the ice melts while in thermal contact with surroundings at 10 C. To reach equilibrium, the system must reach the temperature of the surroundings, which we assume to be constant. In this process, the ice melts and the melt water warms to 10 C. To find the entropy change, we must find a reversible process that effects the same change. A two-step process effects this conveniently. The first step is the one we have just considered: Surroundings at 0 C transfer 6010 J of heat to the system, reversibly melting the ice to water at 0 C. We have $q_1 = 6010$ J and

$$\Delta S_1 = 22.00 \text{ J K}^{-1}$$

In the second reversible step, surroundings that are always at the same temperature as the system transfer heat to the system as the temperature increases from 273.15 K to 283.15 K. The heat capacity of liquid water is 75.3 J mol^{-1}. For this step,

$$\begin{aligned}q_2 &= \int_{273.15}^{283.15} C_P dT \\ &= (75.3 \text{ J K}^{-1})(10 \text{ K}) \\ &= 753 \text{ J}\end{aligned}$$

and

$$\begin{aligned}\Delta S_2 &= \int_{273.15}^{283.15} \frac{C_P}{T} dT \\ &= (75.3 \text{ J K}^{-1}) \ln \frac{283.15}{273.15} \\ &= 2.71 \text{ J K}^{-1}\end{aligned}$$

For these reversible changes in the system, we have $\Delta S = \Delta S_1 + \Delta S_2 = 24.71 \text{ J K}^{-1}$. This is also the value of ΔS for the spontaneous process. The heat taken up by the system in the two-step reversible process is $q = q_1 + q_2 = 6763$ J. This heat is surrendered by the surroundings, and we could effect identically the same change in the surroundings by exchanging this quantity of heat reversibly. For the spontaneous process, therefore, we have $\hat{q} = -6763$ J and

$$\Delta \hat{S} = -6763 \text{ J}/283.15 \text{ K} = -23.88 \text{ J K}^{-1}$$

For the universe, we have

$$\Delta S_{universe} = \Delta S + \Delta \hat{S} = +0.83 \text{ J K}^{-1}$$

which is greater than zero, as required for a spontaneous process.

Because this reversible two-step process does not occur at a constant temperature, its Gibbs free energy change is not zero. However, we can use the Gibbs-Helmholtz equation to estimate the Gibbs free energy change for the related process in which ice at 10 C and 1 bar (a hypothetical substance) melts to form liquid water at the same temperature and pressure. For this process, we estimate

$$\Delta G = -220 \text{ J mol}^{-1} \text{ K}^{-1}$$

(See problem 10-21.) Since we have $\Delta G < 0$ for the process, our change criterion asserts that, in agreement with our experience, superheated ice melts spontaneously at 10 C.

§5 Measuring the entropy change for any reversible process

We define entropy in terms of its differential as $dS = dq^{rev}/T$. To measure an entropy change using this definition requires that the process be reversible, which means that the system and surroundings must be at the same temperature as the heat transfer occurs. We understand reversible heat transfer to be the limiting case in which the temperature difference between system and surroundings becomes arbitrarily small. Here we want to consider the conceptual problems associated with keeping the temperature of the surroundings arbitrarily close to the temperature of the system while the system undergoes an arbitrary reversible change, which may include a temperature change.

We can illuminate one necessary aspect by posing a trivial dilemma: Since system and surroundings jointly comprise the universe, the requirement that system and surroundings be at the same temperature might seem to require that the entire universe be at a single temperature. Plainly, this condition is not met; the temperature of the universe varies from place to place. In fact, the requirement is only that the system and that portion of the surroundings with which the system exchanges heat be at the same temperature. We can satisfy this requirement by permitting the exchange of heat between the system and a portion of the surroundings, under conditions in which the combination of the two is thermally isolated from the rest of the universe.

The non-trivial aspect of this problem arises from the requirement that the temperature of the surroundings remain arbitrarily close to the temperature of the system, while both temperatures change and heat is exchanged between the system and the surroundings. A clumsy solution to this problem is to suppose that we exchange one set of surroundings (at temperature T) to a new set (at temperature $T + \Delta T$) whenever the temperature of the system changes by ΔT. A more elegant solution is to use a machine that can measure the entropy change associated with an arbitrary reversible change in any closed system. This is a conceptual device, not a practical machine. We can use it in *gedanken* experiments to make arbitrarily small changes in the temperature and pressure of the system along any reversible path. At every step along this path, the entropy change is

$$dS = \frac{1}{T}\left(\frac{\partial H}{\partial T}\right)_P dT - \left(\frac{\partial V}{\partial T}\right)_P dP$$

The entropy-measuring machine is sketched in Figure 1. In this device, the portion of the surroundings with which the system can exchange heat is a quantity of ideal gas, which functions as a heat reservoir. This heat reservoir is in thermal contact with the system. The combination of system and ideal-gas heat reservoir is thermally isolated from the rest of the universe. We consider the case in which only pressure–volume work can be done on either the system or the ideal-gas heat reservoir. In this device, all changes are driven by changes in the pressures applied to the surroundings (the ideal-gas heat reservoir) and the system. The pressure applied to the system and the pressure applied to the ideal-gas heat reservoir can be varied independently. We suppose that the system is initially at equilibrium and that changes in the applied pressures are effected in such a manner that all changes in the system and in the ideal-gas heat reservoir occur reversibly. For any change effected in the entropy-measuring machine, the heat and entropy changes in the heat reservoir are $\hat{q}$ and $\Delta\hat{S}$.

Isothermal process.– Let us consider first a process in which a quantity of heat must be transferred from the surroundings to the system while both are at the constant temperature T_P. To be specific, let this be a process in which a mole of pure liquid vaporizes at constant pressure, taking up a quantity of heat equal to the molar enthalpy of vaporization. We can supply this heat by reversibly and isothermally compressing the ideal-gas heat reservoir. To keep the temperature of the ideal-gas heat reservoir constant, we reversibly withdraw the piston that controls the pressure of the system, causing the vaporization of liquid in the system and absorption by the system of the heat given up by the ideal-gas heat reservoir. Overall, we apply forces to the two pistons to achieve reversible isothermal compression of the ideal gas in the heat reservoir and reversible isothermal vaporization of a quantity of liquid in the system. Because $q = -\hat{q}$, the magnitude of the entropy change for the surroundings is equal to that for the ideal-gas heat reservoir. The entire process is reversible, the entropy change for the system and the entropy change for the surroundings sum to zero: $\Delta S + \Delta\hat{S} = 0$.

We can calculate the entropy change for the ideal-gas heat reservoir. Overall, the $\hat{n}$ moles of ideal gas in the heat reservoir go from the state specified by $\hat{P}_1$ and T_P to the state specified by $\hat{P}_2$ and T_P. From the general relationship, $dS = T^{-1}(\partial H/\partial T)_P dT - (\partial V/\partial T)_P dP$, with $dT = 0$, we have $d\hat{S} = -(\hat{n}R/P)dP$ and

$$\Delta\hat{S} = -\hat{n}R\ln\frac{\hat{P}_2}{\hat{P}_1}$$
$$= \hat{n}R\ln\frac{\hat{V}_2}{\hat{V}_1}$$

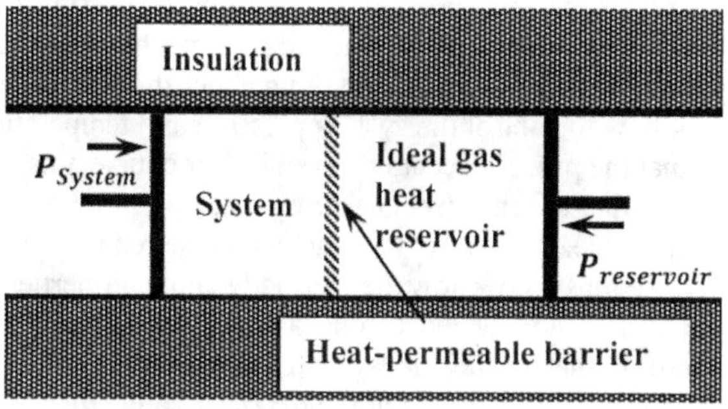

Figure 1. An entropy-measuring machine.

The entropy change for the system is $\Delta S = -\Delta\hat{S}$. So long as we carry out the process isothermally and reversibly, we can determine the entropy change for the system simply by measuring the initial and final pressures (or volumes) of the ideal-gas heat reservoir.

Any reversible process.– If the system undergoes a reversible change in which the temperature of the system is not constant, we can operate the entropy-measuring machine in essentially the same manner as before. The only difference is that we must adjust the pressure applied to the system so that the system temperature changes in the manner required to keep the process reversible—that is, to maintain the system at equilibrium. Then the change to the system and the change to the ideal-gas heat reservoir both take place reversibly. Even though the temperatures change, appropriate control of the applied pressures assures that the system is always in an equilibrium state and that the temperature of the system is always arbitrarily close to the temperature of the ideal-gas heat reservoir.

We can calculate the entropy change for the ideal-gas heat reservoir. Overall, the $\hat{n}$ moles of ideal gas in the heat reservoir go from the state specified by $\hat{P}_1$ and $\hat{T}_1$ to the state specified by $\hat{P}_2$ and $\hat{T}_2$. We can evaluate the entropy change for taking the ideal gas from state 1 to state 2 by a two-step path. We first compress the gas isothermally at $\hat{T}_1$ from $\hat{P}_1$ to $\hat{P}_2$. We then warm the gas at constant pressure $\hat{P}_2$ from $\hat{T}_1$ to $\hat{T}_2$. For the first step, $d\hat{T} = 0$, and, as before, we find

$$\Delta\hat{S}_{isothermal} = -\hat{n}R\ln\frac{\hat{P}_2}{\hat{P}_1} = \hat{n}R\ln\frac{\hat{V}_2}{\hat{V}_1}$$

For the second step, $d\hat{P} = 0$, and

$$\Delta\hat{S}_{isobaric} = \int_{\hat{T}_1}^{\hat{T}_2}\frac{\hat{n}C_P}{T}dT = \hat{n}C_P\ln\frac{\hat{T}_2}{\hat{T}_1}$$

The entropy change for the ideal-gas heat reservoir is thus

$$\Delta\hat{S} = \Delta\hat{S}_{isothermal} + \Delta\hat{S}_{isobaric}$$
$$= -\hat{n}R\ln\frac{\hat{P}_2}{\hat{P}_1} + \hat{n}C_P\ln\frac{\hat{T}_2}{\hat{T}_1}$$

and we have $\Delta S = -\Delta\hat{S}$.

The essential point of the entropy-measuring machine is that we can determine the entropy change for any process without knowing anything about the process except how to control the system pressure and temperature so that the process occurs reversibly. Of course, this one reason that the entropy-measuring machine is not a practical device. To control the machine in the required manner, we must know how the thermodynamic properties of the system are related to one another on the Gibbsian manifold that defines the system's equilibrium states. If we know this, then we know $(\partial H/\partial T)_P$ and $(\partial V/\partial T)_P$ for the system along any reversible path, and we can calculate the entropy change for the system in the same way that we calculate the entropy change for the ideal-gas heat reservoir. If we have the information needed to perform the measurement, we can calculate the entropy change without using the machine.

§6 Another perspective on the principle of Le Chatelier

When we apply Le Chatelier's principle, we imagine an equilibrium system on which we impose some step-wise change. Immediately following the imposition of this change, we isolate the system from all further interactions with its surroundings. This isolated system is a hypothetical construct, which can be only approximated in any real experiment. It has peculiar features: While the changed and isolated system has the properties of the original system, it is also free to undergo a further change that the original system could not. The hypothetical isolated system is no longer at equilibrium; it can undergo a spontaneous process of further change until it reaches a new position of equilibrium. The principle asserts that this further change opposes the imposed change.

The principle is inherently qualitative. This contributes to its utility in that we do not have to have quantitative data in order to use it. However, a qualitative prediction is less useful than a quantitative one. Let us now attempt to apply our second-law based quantitative models to the sequence of changes envisioned by Le Chatelier's principle. We begin by restating the principle in more mathematical language. We then illustrate these ideas for the specific case of vapor–liquid equilibrium with temperature and pressure as the independent variables.

Let us suppose that W, X, Y, and Z are a set of thermodynamic variables that is adequate to specify the state of the system. In any equilibrium state, the entropy of the system is then a function of these variables; we have $S = S(W,X,Y,Z)$. For present purposes, we assume that we know the function $S(W,X,Y,Z)$. Given small changes, dW, dX, dY, and dZ, in the independent variables, we can find the change in dS for a reversible transition from (W,X,Y,Z) to $(W+dW, X+dX, Y+dY, Z+dZ)$:

$$dS = \left(\frac{\partial S}{\partial W}\right)_{XYZ} dW + \left(\frac{\partial S}{\partial X}\right)_{WYZ} dX + \left(\frac{\partial S}{\partial Y}\right)_{WXZ} dY + \left(\frac{\partial S}{\partial Z}\right)_{WXY} dZ$$

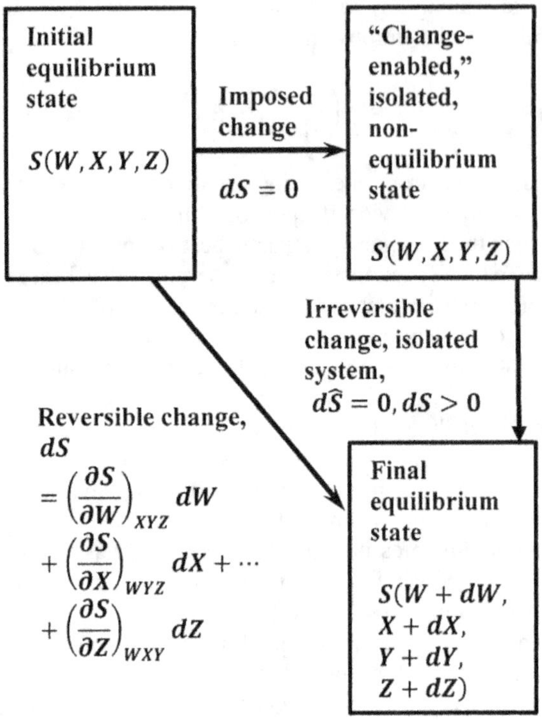

Figure 2. Spontaneous and reversible processes connecting the same equilibrium states.

When we impose the change creating the hypothetical isolated system, we imagine that some characteristic of the system changes instantaneously, and that it does so without changing the other properties of the system. Since we suppose that nothing about the system changes in the perturbation and isolation step, the entropy of the perturbed, isolated, hypothetical system remains the same as that of the original equilibrium system.

Figure 2 shows the entropies for three states in the cycle that comprises the Le Chatelier model for change. The entropy of the original equilibrium system is $S(W,X,Y,Z)$ and that of the final equilibrium system is $S(W+dW, X+dX, Y+dY, Z+dZ)$. The same final equilibrium state is reached by both the irreversible transition from the change-enabled hypothetical state and by a reversible transition from the initial equilibrium state. Since entropy is a state function, its change around this cycle must be zero. Hence, the incremental changes dW, dX, dY, and dZ that occur in the thermodynamic variables must satisfy the inequality

$$\left(\frac{\partial S}{\partial W}\right)_{XYZ} dW + \left(\frac{\partial S}{\partial X}\right)_{WYZ} dX + \left(\frac{\partial S}{\partial Y}\right)_{WXZ} dY + \left(\frac{\partial S}{\partial Z}\right)_{WXY} dZ > 0$$

We can view the application of this inequality to the hypothetical, change-enabled, isolated system as a mathematical expression of Le Chatelier's principle. To see this more clearly, let us suppose that we are able to keep W and Z constant. We suppose that the imposed change requires that the final value of X be $X + dX$. For the system to remain at equilibrium, the remaining variable, Y, must change by an amount, dY, that satisfies this inequality. That is, to reach the new equilibrium state, the change in Y must satisfy

$$\left(\frac{\partial S}{\partial X}\right)_{WYZ} dX + \left(\frac{\partial S}{\partial Y}\right)_{WXZ} dY > 0$$

In this model, variables X and Y drive the entropy change as the hypothetical system moves toward its new equilibrium position. The imposed change in X changes the entropy of the system by

$$dS_{imposed} = \left(\frac{\partial S}{\partial X}\right)_{WYZ} dX$$

Since the effect of the imposed change is to drive the system away from its original equilibrium position, we have $dS_{imposed} < 0$. The system's response changes the entropy of the system by

$$dS_{response} = \left(\frac{\partial S}{\partial Y}\right)_{WXZ} dY$$

We have $dS_{response} > -dS_{imposed} > 0$, so that we can reasonably describe the response, dY, that makes $dS_{response} > 0$, as a change that opposes the imposed change, dX, that makes $dS_{imposed} < 0$.

Applying Le Chatelier's principle is something of an art. Central to this art is an ability to devise a hypothetical, change-enabled, isolated, non-equilibrium state that is a good model for the initial state of the spontaneous process. In §6-6, we use qualitative arguments to apply Le Chatelier's principle to vapor–liquid equilibrium. To relate these qualitative arguments to the mathematical model we have developed, let us consider the *gedanken* experiment depicted in Figure 3. We suppose that the initial equilibrium system contains the liquid and vapor of a pure substance at pressure, P, and temperature, T. We imagine that we can create the hypothetical isolated system by imposing a step change to the applied pressure without changing the pressure of the system.

To model the pressure perturbation, let us view the liquid–vapor mixture as a subsystem, which is enclosed in a vertical cylinder that is sealed by a frictionless piston. A mass, M, rests on top of the piston. For present purposes, we consider this mass to be a portion of a second subsystem. The gravitational force on this mass creates the pressure applied to the liquid–vapor mixture. Since this is an equilibrium state of the substance, this applied pressure is equal to the pressure, P, of the liquid–vapor subsystem. A small mass, dM, is also a part of the second subsystem. In this original equilibrium state of the system, this smaller mass is supported in some manner, so that it does not contribute to the applied pressure. We assume that the piston is a perfect thermal insulator, so that no heat can be exchanged between the two subsystems.

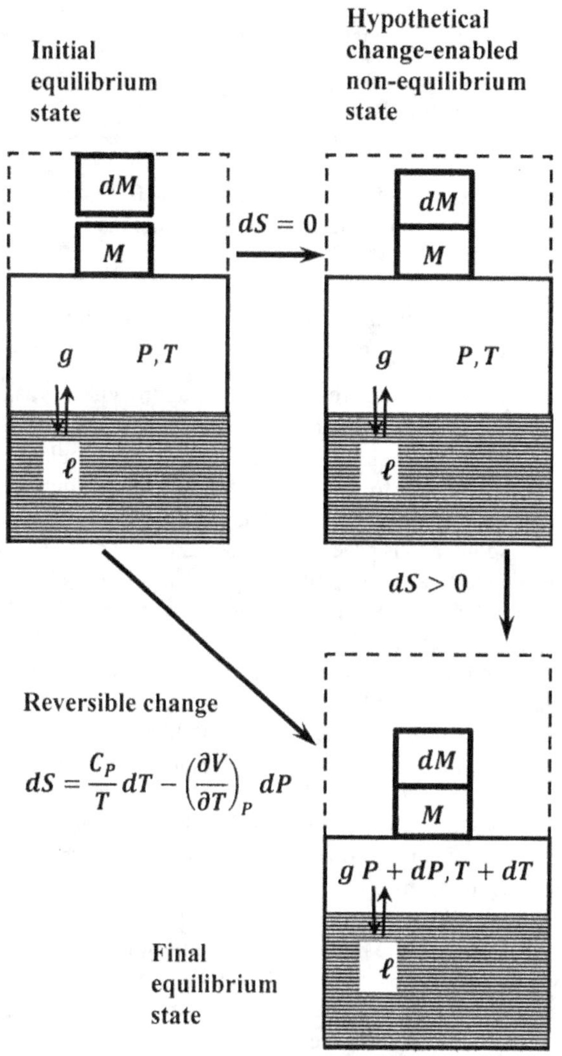

Figure 3. Spontaneous and reversible processes connecting the same liquid–vapor equilibrium states.

As sketched in Figure 3, we create the hypothetical change-enabled system by moving the smaller mass so that it too rests on top of the piston. Immediately thereafter, we completely isolate the system from the rest of the universe. We suppose that the applied pressure instantaneously increases to $P + dP$. However, since the liquid–vapor subsystem is unchanged, we suppose that the pressure, entropy, and all other thermodynamic properties of the liquid–vapor subsystem remain unchanged in this hypothetical state. The system is not at equilibrium in this hypothetical state, because the applied pressure is not equal to that of the liquid–vapor subsystem. Spontaneous change to a new equilibrium state can occur. Because the system is isolated, $\Delta\hat{S} = 0$. Therefore, we have $\Delta S > 0$. The final equilibrium temperature is $T + dT$.

With pressure and temperature as the independent variables, this model for Le Chatelier's principle gives rise to the following mathematical requirement: $dS = T^{-1}(\partial H/\partial T)_P dT - (\partial V/\partial T)_P dP > 0$. We know that

T, $(\partial H/\partial T)_P$, and $(\partial V/\partial T)_P$ are positive. Therefore we can rearrange the inequality to find

$$dT > T\left[\left(\frac{\partial V}{\partial T}\right)_P \Big/ \left(\frac{\partial H}{\partial T}\right)_P\right] dP$$

If we have $dP > 0$, it follows that $dT > 0$; that is, the liquid–vapor equilibrium temperature increases with pressure.

§7 Phase equilibria: Temperature dependence of the boiling point

In §4 and §6, we explore two approaches to using the entropy-based criterion for spontaneous change. In discussing the melting of ice at +10 C, we calculate the entropy changes for both the system and the surroundings to show that $\Delta S_{universe} > 0$, as the second law requires for a spontaneous process. In discussing the pressure-dependence of a liquid's boiling point in §6, we relate the second law criterion for spontaneous change to Le Chatelier's principle. We turn now to specifying the pressures and temperatures at which two phases of a pure substance are in equilibrium. When we choose pressure and temperature as the independent variables, the Gibbs free energy criteria specify the equilibrium state and the direction of spontaneous change.

For a reversible process in which all the work is pressure–volume work and in which the pressure and temperature change by dP and dT, the change in the Gibbs free energy is $dG = VdP - SdT$. Let us apply this relationship to the liquid–vapor equilibrium problem that we discuss in §6. To do so, we view the process from a slightly different perspective. We suppose that we have two systems. These are identical to the initial and final states of the system in our discussion above. One of these systems is at liquid–vapor equilibrium at a particular pressure, P, and temperature, T. The other is at liquid–vapor equilibrium at $P + dP$ and $T + dT$. We consider the change in the Gibbs free energy of a mole of the substance as it reversibly traverses the cycle sketched in Figure 4.

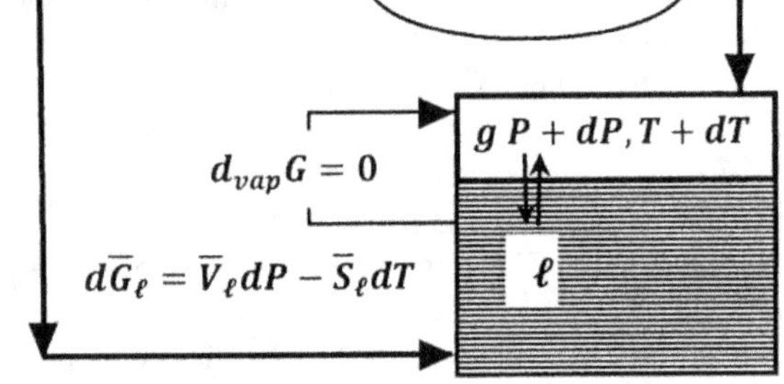

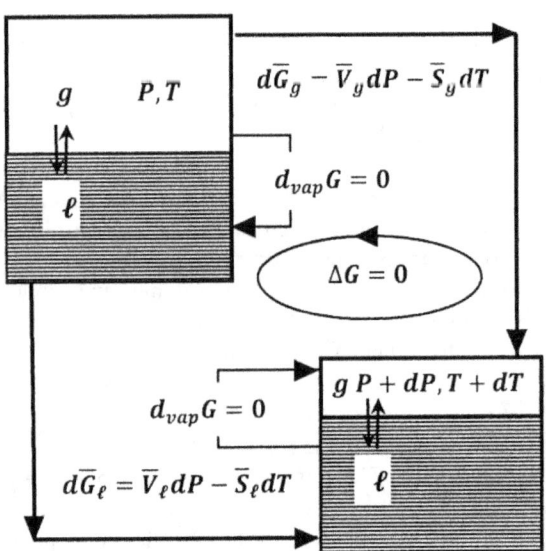

Figure 4. Molar Gibbs free energy differences between two liquid–vapor equilibrium states.

The pressure and temperature are constant in each of the two equilibrium states. In either of these equilibrium states, the Gibbs free energy does not change when a mole of liquid is converted to its gas, $\Delta_{vap}G(P,T) = 0$ and $\Delta_{vap}G(P + dP, T + dP) = 0$. When the pressure and temperature of one mole of liquid change from P and T to $P + dP$ and $T + dT$, the Gibbs free energy change is $d\overline{G}(\ell) = \overline{V}_\ell dP - \overline{S}_\ell dT$. For a mole of gas, this change in the pressure and temperature change the Gibbs free energy by $d\overline{G}(g) = \overline{V}_g dP - \overline{S}_g dT$. ($\overline{V}_\ell, \overline{V}_g, \overline{S}_\ell$. and $\overline{S}_g$ are evaluated at P and T. However, since dP and dT are small, these quantities are essentially constant over the pressure and temperature ranges involved.) For the individual steps in this cycle, we have

$(\ell, P, T) \rightarrow (\ell, P + dP, T + dT)$

$$d\overline{G}(\ell) = \overline{V}_\ell dP - \overline{S}_\ell dT$$

$(\ell, P + dP, T + dT) \rightarrow (g, P + dP, T + dT)$

$$\Delta_{vap}G(P + dP, T + dT) = 0$$

$(g, P + dP, T + dT) \rightarrow (g, P, T)$

$$-d\overline{G}(g) = -\left(\overline{V}_g dP - \overline{S}_g dT\right)$$

$(g, P, T) \rightarrow (\ell, P, T)$

$$-\Delta_{vap}G(P,T) = 0$$

Since the Gibbs free energy is a state function, the sum of these terms is zero. We have

$$\begin{aligned}\Delta_{vap}G(P + dP, T + dT) - \Delta_{vap}G(P,T)&\\ +d\overline{G}(\ell) - d\overline{G}(g)&\\ = d\overline{G}(\ell) - d\overline{G}(g)&\\ = 0&\end{aligned}$$

so that $d\overline{G}(\ell) = d\overline{G}(g)$. That is, the Gibbs free energy of the liquid changes by the same amount as the Gibbs free energy of the gas when a mole of either is taken from one liquid–vapor equilibrium state to another. Substituting, we find a condition that the pressure and temperature changes must satisfy when the system goes from the liquid–vapor equilibrium state at (P, T) to the liquid–vapor equilibrium state at $(P + dP, T + dT)$:

$$\left(\overline{V}_g - \overline{V}_\ell\right)dP - \left(\overline{S}_g - \overline{S}_\ell\right)dT = 0$$

We let $\Delta_{vap}\overline{V} = \overline{V}_g - \overline{V}_\ell$ and $\Delta_{vap}\overline{S} = \overline{S}_g - \overline{S}_\ell$, where $\Delta_{vap}\overline{V}$ and $\Delta_{vap}\overline{S}$ are the volume and entropy changes that accompany the vaporization of one mole of the liquid at P and T. $\Delta_{vap}\overline{V}$ and $\Delta_{vap}\overline{S}$ are essentially constant over the small pressure and temperature ranges involved. Substituting, we have $\Delta_{vap}\overline{V}dP - \Delta_{vap}\overline{S}dT = 0$, which we can rearrange to give

$$\frac{dP}{dT} = \frac{\Delta_{vap}\overline{S}}{\Delta_{vap}\overline{V}}$$

As one mole of liquid vaporizes reversibly at P and T, the system accepts heat $q_P^{rev} = \Delta_{vap}\overline{H}$. Hence, the entropy of vaporization at P and T is $\Delta_{vap}\overline{S} = \Delta_{vap}\overline{H}/T$, and the relationship between dP and dT becomes

$$\frac{dP}{dT} = \frac{\Delta_{vap}\overline{H}}{T\,\Delta_{vap}\overline{V}}$$

Below we see that such a relationship holds for any equilibrium between two pure phases. The general relationship is called the ***Clapeyron equation***.

This analysis is successful because the constituents are pure phases; the properties of the liquid are independent of how much vapor is present and *vice versa*. When we analyze the equilibrium between a liquid solution and a gas of the solution's components, the problem is more complex, because the properties of the phases depend on their compositions.

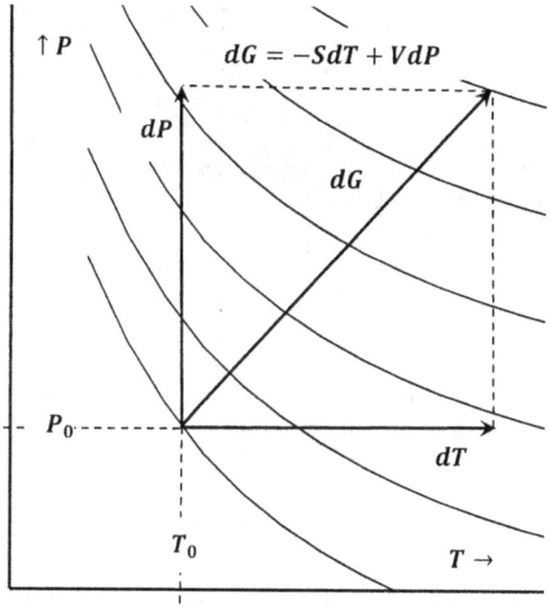

Figure 5. The path of a changing system depicted on a map of constant Gibbs free energy contours.

§8 Phase equilibria: Temperature dependence of the melting point

We can also represent reversible changes by paths on contour maps. In Figure 5, a Gibbs free energy surface is represented as a contour map. For small changes in T and P, we can evaluate

$$dG = G(P_0 + dP, T_0 + dT) - G(P_0, T_0)$$

from

$$dG = -S(P_0, T_0)dT + V(P_0, T_0)dP$$

For larger changes, we can integrate along the paths $P = P_0$ and $T = T_0 + \Delta T$ to find

$$\Delta G = \int_{T_0}^{T_0+\Delta T} -S(P_0, T)dT + \int_{P_0}^{P_0+\Delta P} V(P, T_0 + \Delta T)dP$$

The calculation of ΔS in §5 could be similarly represented as a path in the temperature–pressure plane that connects two constant-entropy contours.

Analysis of solid–liquid equilibrium parallels that of liquid–vapor equilibrium. Let us again consider the equilibrium between ice and water. Given that ice and water are at equilibrium at a particular temperature and pressure, and supposing that we increase the pressure from this equilibrium value, how must the temperature change in order that the system remain at equilibrium? In §6-6, we use Le Chatelier's principle to answer this question qualitatively. Now, we find a quantitative answer by an argument that closely parallels that in §7.

Figure 6 depicts the line of pressures and temperatures along which ice and water are in equilibrium. We can view this as a contour map. In this case, the contours are sets of pressures and temperatures for which $\Delta_{fus}\overline{G}$ is constant. Only the contour for $\Delta_{fus}\overline{G} = 0$ is shown. The figure also depicts paths along which ice and water can individually be taken from their equilibrium state at P and T to their equilibrium state at $P + dP$ and $T + dT$. The Gibbs free energy change for the ice must equal that for water. Letting $\overline{G}_\ell$, $\overline{S}_\ell$, and $\overline{V}_\ell$ be the Gibbs free energy, the entropy, and the volume of one mole of water at temperature T and pressure P, the equation

$$d\overline{G}_\ell = \overline{V}_\ell dP - \overline{S}_\ell dT$$

specifies the change in the Gibbs free energy of one mole of water when the pressure changes P to $P + dP$ and the

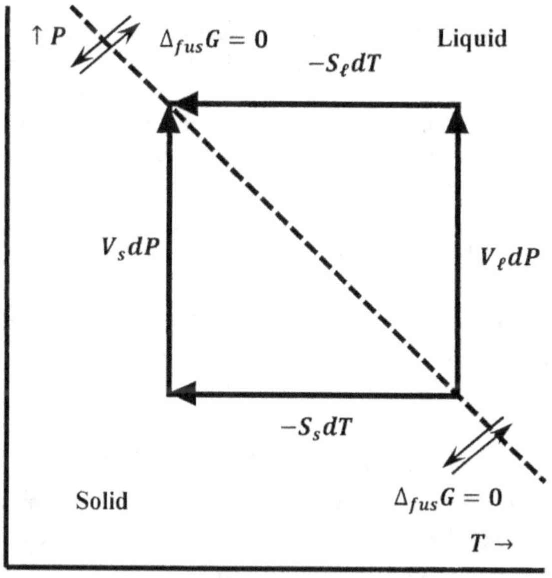

Figure 6. Gibbs free energy changes in a cycle connecting two solid–liquid equilibrium states.

temperature changes from T to $T + dT$. Similarly, using the subscript "s" to denote ice, we have

$$d\overline{G}_s = \overline{V}_s dP - \overline{S}_s dT$$

Since these Gibbs free energy changes connect states of ice–water equilibrium, they must be equal, and we have

$$\begin{aligned} d\overline{G}_\ell - d\overline{G}_s &= (\overline{V}_\ell - \overline{V}_s)dP - (\overline{S}_\ell - \overline{S}_s)dT \\ &= \Delta_{fus}\overline{V}dP - \Delta_{fus}\overline{S}dT \\ &= 0 \end{aligned}$$

where we introduce $\Delta_{fus}\overline{S}$ and $\Delta_{fus}\overline{V}$ to represent the entropy and volume changes that occur when one mole of ice melts reversibly at P and T. Rearranging gives

$$\frac{dP}{dT} = \frac{\Delta_{fus}\overline{S}}{\Delta_{fus}\overline{V}}$$

Since $\Delta_{fus}\overline{S} = \Delta_{fus}\overline{H}/T$, the Clapeyron equation becomes

$$\frac{dP}{dT} = \frac{\Delta_{fus}\overline{H}}{T\,\Delta_{fus}\overline{V}}$$

At a pressure of one bar and a temperature of 273.15 K, the enthalpy of fusion is 6010 J mol^{-1} K^{-1}. The enthalpy value changes only slowly as the equilibrium temperature changes. The volumes of one mole of ice and one mole of water are 19.651 and 18.019 cm^3, respectively. At 273.15 K, we obtain

$$\frac{dP}{dT} = -143.7 \text{ bar K}^{-1}$$

If the pressure increases to 1000 bar, the change in the melting point is about –6.96 K, so that $T_{mp}(1000 \text{ bar}) =$ 266.2 K.

Again, this analysis is successful because the constituents are pure phases; the properties of the ice are independent of how much water is present and *vice versa.* When we analyze the equilibrium between ice and salt water, the properties of the salt water depend on the kind of salt present and on its concentration.

§9 The Clapeyron equation

The analysis in the two previous sections can be repeated for any phase change of a pure substance. Let α and β denote the two phases that are at equilibrium.

$$\alpha \rightleftharpoons \beta$$

Let $\overline{G}_\alpha$, $\overline{S}_\alpha$, and $\overline{V}_\alpha$ represent the Gibbs free energy, the entropy, and the volume of one mole of pure phase α at pressure P and temperature T. Let $\overline{G}_\beta$, $\overline{S}_\beta$, and $\overline{V}_\beta$ represent the corresponding properties of one mole of pure phase β. The equations $d\overline{G}(\alpha) = \overline{V}_\alpha dP - \overline{S}_\alpha dT$ and $d\overline{G}(\beta) = \overline{V}_\beta dP - \overline{S}_\beta dT$ describe the changes in the Gibbs free energy of a mole of α and a mole of β when they go from one α-β-equilibrium state at P and T to a second α-β-equilibrium state at $P + dP$ and $T + dT$. Since these Gibbs free energy changes must be equal, we have

$$\begin{aligned} d\overline{G}(\beta) - d\overline{G}(\alpha) &= (\overline{V}_\beta - \overline{V}_\alpha)dP - (\overline{S}_\beta - \overline{S}_\alpha)dT \\ &= \Delta\overline{V}dP - \Delta\overline{S}dT \\ &= 0 \end{aligned}$$

and

$$\frac{dP}{dT} = \frac{\Delta\overline{S}}{\Delta\overline{V}}$$

where $\Delta\overline{S}$ and $\Delta\overline{V}$ are the entropy and volume changes that occur when one mole of the substance goes from phase α to phase β. Since $\Delta\overline{S} = \Delta\overline{H}/T$, the condition for equilibrium between phases α and β becomes

$$\frac{dP}{dT} = \frac{\Delta\overline{H}}{T\,\Delta\overline{V}}$$

(The Clapeyron equation)

§10 The Clausius-Clapeyron equation

To use the Clapeyron equation we must know the enthalpy and volume differences at one equilibrium temperature and pressure. In general, these properties are readily measured. If we fix the pressure, we can measure the corresponding equilibrium temperature. We can obtain the enthalpy change at this pressure by measuring the heat required to convert a mole of the substance from one phase to the other. We can obtain the volume change from the molar volumes, which we can obtain by measuring the density of each phase. The enthalpy of the phase change varies only weakly as the equilibrium pressure and temperature vary. Similarly, for condensed phases, the densities are weak functions of temperature. This means that, for transitions between condensed phases, $\Delta\overline{H}/\Delta\overline{V}$ is approximately constant over a modest temperature range.

For a sublimation or vaporization process, the product is a gas. Then the molar volume of the product is a sensitive function of both pressure and temperature. However, the molar volume of the product phase is much greater than the molar volume of the initial solid or liquid phase. To a good approximation, the volume change for the process equals the volume of the gas produced. If we have an equation of state for the gas, the volume calculated from the equation of state is a good approximation to ΔV for the phase change. The ideal gas equation is usually adequate for this purpose. Then, $\Delta\overline{V} \approx RT/P$, and

$$\frac{dP}{dT} = \frac{P\Delta\overline{H}}{RT^2}$$

(The Clausius-Clapeyron equation)

This equation for the pressure–temperature relationship for a phase equilibrium involving a gas is called the Clausius-Clapeyron equation. Dividing both sides by the pressure, we can put the Clausius-Clapeyron equation into an alternative and often-useful form:

$$\frac{d\ln P}{dT} = \frac{\Delta\overline{H}}{RT^2}$$

If we can assume that $\Delta\overline{H}$ is independent of pressure, we can separate variables and integrate to obtain the Clausius-Clapeyron equation in integrated form. If we can assume further that $\Delta\overline{H}$ is constant, the integration yields

$$\int_{P_0}^{P} \frac{dP}{P} = \frac{\Delta\overline{H}}{R}\int_{T_0}^{T}\frac{dT}{T^2}$$

and

$$\ln\frac{P}{P_0} = -\frac{\Delta\overline{H}}{R}\left(\frac{1}{T}-\frac{1}{T_0}\right)$$

where P_0 and T_0 are the initial equilibrium position.

Problems

1. For any change in a reversible system, we have $dG = -SdT + VdP$. Consider two systems, α and β, where system α can be converted to system β. (Below, we will let α and β be the solid and liquid phases of the same pure substance, but this is not a necessary restriction.) For incremental changes in temperature and pressure, represented by dT and dP, we have

$$dG_\alpha = V_\alpha dP - S_\alpha dT$$

and

$$dG_\beta = V_\beta dP - S_\beta dT$$

We can subtract to find

$$d(G_\beta - G_\alpha) = (V_\beta - V_\alpha)dP - (S_\beta - S_\alpha)dT$$

or

$$d(\Delta_{\alpha\to\beta}G) = (\Delta_{\alpha\to\beta}V)dP - (\Delta_{\alpha\to\beta}S)dT$$

which we usually write as

$$d(\Delta G) = \Delta V dP - \Delta S dT$$

Here $\Delta_{\alpha\to\beta}X$ (or ΔX) is the change in the state function X that occurs when system α is converted to system β. For many inter-convertible systems, it is a good approximation to say that $\Delta_{\alpha\to\beta}S$ (or ΔS) and $\Delta_{\alpha\to\beta}V$ (or ΔV) are constant for modest changes in temperature or pressure. Then, representing the pressure and temperature in the initial and final states as (P_1, T_1) and (P_2, T_2), respectively, the ***change*** in $\Delta_{\alpha\to\beta}G$ (or ΔG) can be obtained by integration:

$$\int_{P_1,T_1}^{P_2,T_2} d(\Delta G) = -\Delta S\int_{T_1}^{T_2} dT + \Delta V\int_{P_1}^{P_2} dP$$

or

$$\begin{aligned}\Delta(\Delta G) &= \Delta G(P_2,T_2) - \Delta G(P_1,T_1)\\ &= -\Delta S(T_2 - T_1) + \Delta V(P_2 - P_1)\end{aligned}$$

Note that α and β need not be in equilibrium with one another at either the condition specified by (P_1, T_1) or that specified by (P_2, T_2).

However, in the important special case that α and β are in equilibrium at (P_1, T_1), we have $\Delta_{\alpha\to\beta}G(P_1,T_1) = \Delta G(P_1,T_1) = 0$. Then

$$\begin{aligned}\Delta(\Delta G) &= \Delta G(P_2,T_2)\\ &= -\Delta S(T_2 - T_1) + \Delta V(P_2 - P_1)\end{aligned}$$

Consider the application of these observations to the case where α and β are solid and liquid aluminum metal, respectively. At one bar, aluminum melts at 933.47 K. At the melting point, the enthalpy of fusion is $\Delta_{fus}\overline{H} = +10.71\ \text{kJ mol}^{-1}$. The atomic weight of aluminum is $26.9815\ \text{g mol}^{-1}$. The density of liquid aluminum at the melting point is $2.375\ \text{g cm}^{-1}$; the molar volume of the liquid is therefore $1.1361 \times 10^{-5}\ \text{m}^3\ \text{mol}^{-1}$. At 20 C, the density of solid aluminum is $2.70\ \text{g cm}^{-1}$, and the molar volume is $0.9993 \times 10^{-5}\ \text{m}^3\ \text{mol}^{-1}$. Assuming the molar volume of the solid to be independent of temperature, the change in the molar volume that occurs when aluminum melts is $\Delta_{fus}\overline{V} = 1.368 \times 10^{-6}\ \text{m}^3\ \text{mol}^{-1}$.

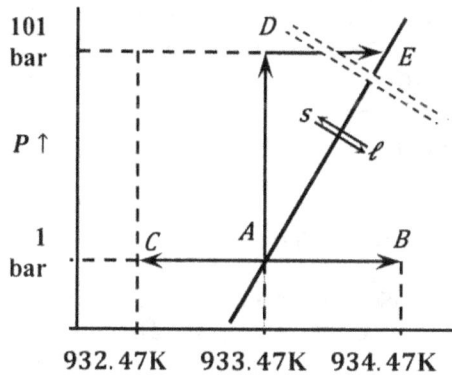

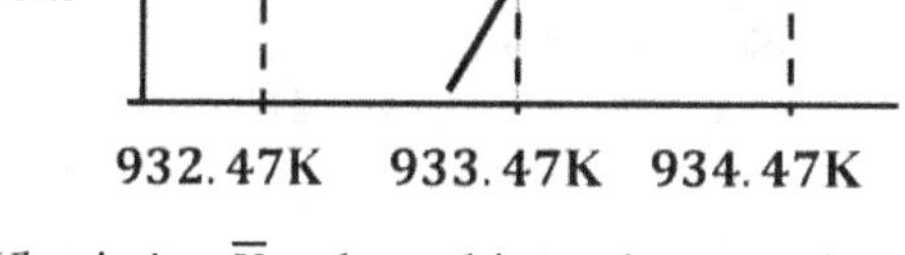

(a) What is $\Delta_{fus}\overline{V}$ at the melting point at one bar (point A on the diagram)?

(b) Is the conversion of solid aluminum to liquid aluminum a reversible, spontaneous, or impossible process at (933.47 K, 1 bar)—that is, at point A?

(c) What is $\Delta_{fus}\overline{S}$ at the melting point at one bar—that is, at point A?

(d) What is $\Delta(\Delta_{fus}\overline{G})$ when the pressure and temperature go from (933.47 K, 1 bar) to (934.47 K, 1 bar)? (That is from point A to point B on the diagram.) What is $\Delta_{fus}\overline{G}$ at (934.47 K, 1 bar)?

(e) Is the conversion of solid aluminum to liquid aluminum a reversible, spontaneous, or impossible process at (934.47 K, 1 bar)—that is, at point B?

(f) What is $\Delta(\Delta_{fus}\overline{G})$ when the pressure and temperature go from (933.47 K, 1 bar) to (932.47 K, 1 bar)? (That is from point A to point C on the diagram.) What is $\Delta_{fus}\overline{G}$ at (932.47 K, 1 bar)?

(g) Is the conversion of solid aluminum to liquid aluminum a reversible, spontaneous, or impossible process at (932.47 K, 1 bar)—that is, at point C?

(h) What is $\Delta(\Delta_{fus}\overline{G})$ when the pressure and temperature go from (933.47 K, 1 bar) to (933.47 K, 101 bar)? (That is from point A to point D on the diagram.) What is $\Delta_{fus}\overline{G}$ at (933.47 K, 101 bar)?

(i) Is the conversion of solid aluminum to liquid aluminum a reversible, spontaneous, or impossible process at (933.47 K, 101 bar)—that is, at point D?

(j) If we maintain the pressure constant at 101 bar, how much would we have to change the temperature to just offset the change in $\Delta(\Delta_{fus}\overline{G})$ that occurred in part h? Note that this change will reach the conditions represented by point E on the diagram

(k) Is the conversion of solid aluminum to liquid aluminum a reversible, spontaneous, or impossible process at point E?

(l) What is $\Delta(\Delta_{fus}\overline{G})$ in going from point D (933.47 K, 101 bar) to point B (934.47 K, 1 bar)? Is this value equal to the difference between the Gibbs free energy of a mole of liquid aluminum at point B and a mole of solid aluminum at point D?

2. In the temperature interval $0.01 > t > -10$ C, the vapor pressure of water (in Pa) above pure ice is aproximated by $\ln P = a + bt + ct^2$, where $a = 6.41532$, $b = 8.229 \times 10^{-2}$, $c = -3.2 \times 10^{-4}$, $t = T - 273.15$, and T is the temperature in degrees kelvin. Estimate the enthalpy of sublimation of ice at 273.15 K.

3. In the temperature interval $373.15 > T > 273.15$ K, the vapor pressure of water (in Pa) is approximated by $\ln P = a + bt + ct^2 + dt^3 + et^4$, where $a = 9.42095$, $b = 4.960 \times 10^{-2}$, $c = -1.7536 \times 10^{-4}$, $d = 6.02 \times 10^{-7}$, $e = -2.0 \times 10^{-9}$, $t = T - 323.15$, and T is the temperature in degrees kelvin. Estimate the enthalpy of vaporization of water at 323.15 K and 373.15 K.

4. The normal (1 atm) boiling point of acetone is 56.05 C. The enthalpy of vaporization at the normal boiling point is 29.10 kJ mol^{-1}. What is the entropy of vaporization of acetone at the normal boiling point? Estimate the vapor pressure of acetone at 25.0 C.

5. Two allotropic forms of tin, gray tin and white tin, are at equilibrium at 13.2 C and 1 atm. The density of gray tin is 5.769 g cm^{-1}; the density of white tin is 7.265 g cm^{-1}. Assume that the densities are independent of temperature. At 298.15 K and 1 bar, S^o and C_P for gray tin are 44.1 J mol^{-1} K^{-1} and 25.8 J mol^{-1} K^{-1}, respectively. For white tin, S^o and C_P are 51.2 J mol^{-1} K^{-1} and 27.0 J mol^{-1} K^{-1}, respectively. Estimate $\Delta\overline{S}$ and $\Delta\overline{V}$ for the conversion of gray tin to white tin at 13.2 C. At what temperature are gray tin and white tin at equilibrium at a pressure of 100 atm?

6. At 1000K, the standard Gibbs free energies of formation of graphite and diamond are 0.000 and +5.905 kJ mol^{-1}, respectively. At 298.15 K and 1.000 bar, the molar volumes of graphite and diamond are 5.46×10^{-6} m^3 mol^{-1} and 3.42×10^{-6} m^3 mol^{-1}, respectively. Let $P^\#$ be the pressure at which graphite and diamond are at equilibrium at 1000 K.

(a) What is the value of $\Delta\overline{G}$ for
$$C(\text{graphite}, 1\text{ bar}, 1000\text{K}) \rightarrow C(\text{diamond}, 1\text{ bar}, 1000\text{K})?$$

(b) Express $\Delta\overline{G}$ for
$$C(\text{graphite}, 1\text{ bar}, 1000\text{K}) \rightarrow C(\text{graphite}, P^\#, 1000\text{K})$$
as a function of $P^\#$.

(c) Express $\Delta\overline{G}$ for
$$C(\text{diamond}, 1\text{ bar}, 1000\text{K}) \rightarrow C(\text{diamond}, P^\#, 1000\text{K})$$
as a function of $P^\#$.

(d) What is the value of $\Delta\overline{G}$ for
$$C(\text{graphite}, P^\#, 1000\text{K}) \rightarrow C(\text{diamond}, P^\#, 1000\text{K})?$$

(e) Assume that the molar volumes are independent of pressure and temperature. Estimate the value of the equilibrium pressure, $P^\#$.

Notes

[1] See Robert Bruce Lindsay and Henry Margenau, ***Foundations of Physics***, Dover Publications, Inc., New York, 1963, p 195.

13

Equilibria in Reactions of Ideal Gases

§1 The Gibbs free energy of an ideal gas

In Chapter 11, we find a general equation for the molar Gibbs free energy of a pure gas. We adopt the Gibbs free energy of formation of the hypothetical ideal gas, in its standard state at 1 bar, P^o, as the reference state for the Gibbs free energy of the gas at other pressures and the same temperature. Then, the molar Gibbs free energy of pure gas A, at pressure P, is

$$\overline{G}_A(P) = \Delta_f G^o(A, HIG^o) + RT \ln\left(\frac{P}{P^o}\right) + RT \int_0^P \left(\frac{\overline{V}}{RT} - \frac{1}{P}\right) dP$$

(any pure gas)

$\overline{G}_A(P)$ is the difference between the Gibbs free energy of the gas at pressure P and that of its constituent elements at 1 bar and the same temperature. If gas A is an ideal gas, the integral is zero, and the standard-state Gibbs free energy of formation is that of an "actual" ideal gas, not a "hypothetical state" of a real gas. To recognize this distinction, let us write $\Delta_f G^o(A, P^o)$, rather than $\Delta_f G^o(A, HIG^o)$, when the gas behaves ideally. In a mixture of ideal gases, the partial pressure of gas A is given by $P_A = x_A P$, where x_A is the mole fraction of A and P is the pressure of the mixture. In §3, we find that the Gibbs free energy of one mole of pure ideal gas A at pressure P_A has the same Gibbs free energy as one mole of gas A in a gaseous mixture in which the partial pressure of A is $P_A = x_A P$. Recognizing these properties of an ideal gas, we can express the molar Gibbs free energy of an ideal gas—pure or in a mixture—as

$$\overline{G}_A(P_A) = \Delta_f G^o(A, P^o) + RT \ln\left(\frac{P_A}{P^o}\right)$$

(ideal gas)

Note that we can obtain this result for pure gas A directly from $\left(\partial \overline{G}/\partial P\right)_T = \overline{V} = RT/P$ by evaluating the definite integrals

$$\int_{\Delta_f G^o(A,P^o)}^{\overline{G}_A(P_A)} d\overline{G} = \int_{P^o}^{P_A} \frac{RT}{P} dP$$

Including the constant, P^o, in these relationships is a useful reminder that $RT \ln(P_A/P^o)$ represents a Gibbs free energy difference. Including P^o makes the argument of the natural-log function dimensionless; if we express P in bars, including $P^o = 1$ bar leaves the numerical value of the argument unchanged. If we express P in other units, P^o becomes the conversion factor for converting those units to bars; if we express P in atmospheres, we have $P^o = 1 \text{ bar} = 0.986923 \text{ atm}$.

However, including the "P^o" is frequently a typographical nuisance. Therefore, let us introduce another bit of notation; we use a lower-case "p" to denote the ratio "P/P^o". That is, p_A is a dimensionless quantity whose numerical value is that of the partial pressure of A, expressed in bars. The molar Gibbs free energy becomes

$$\overline{G}_A(P_A) = \Delta_f G^o(A, P^o) + RT \ln p_A$$

(ideal gas)

§2 The Gibbs free energy change for a reaction of ideal gases

Let us consider a system that consists of a mixture of ideal gases A, B, C, and D, at a particular fixed temperature. We suppose that reaction occurs according to

$$a\,A + b\,B \rightleftharpoons c\,C + d\,D$$

We want to think about what happens when a moles of A (at pressure P_A) and b moles of B (at pressure P_B) react to form c moles of C (at pressure P_C) and d moles of D (at pressure P_D) under conditions in which the partial pressures in the mixture remain constant at P_A, P_B, P_C, and P_D. At first encounter, these conditions may appear to be impossible; if A reacts, its partial pressure must change. However, on reflection, we recognize that by making the system very large, a moles of A and b moles of B can disappear without changing P_A or P_B very much. In fact, we can make the change in P_A and P_B as small as we like just by making the original system large enough. The same considerations apply to P_C and P_D. We let $\Delta_r G$ be the Gibbs free energy change under these conditions. We want to know how the Gibbs free energy of this system, G, and the Gibbs free energy change, $\Delta_r G$, depend on the partial pressures of the gases involved in the reaction.

We have observed repeatedly that the temperature and pressure of a system undergoing spontaneous change may not be well defined, and the concentration of a component may vary from point to point within a given phase. If any of these inhomogeneities are substantial, the reaction conditions in the previous paragraph are not met. On the other hand, if any pressure and temperature variations are too small to have observable effects, and there are no point-to-point concentration variations, it is entirely reasonable to suppose that the Gibbs free energies of the system, and of its individual components, are described by the thermodynamic models we have developed for reversible systems.

To see that there can be no objection in principle to measuring Gibbs free energies in a non-equilibrium system, we need only find a hypothetical equilibrium system whose state functions must have the same values. Thus, if the reaction $a\,A + b\,B \rightleftharpoons c\,C + d\,D$ does not occur, mixtures of gases A, B, C, and D in any proportions can be at equilibrium, and the thermodynamic properties of any such mixture are well defined. In concept, we can produce a hypothetical equilibrium state equivalent to any intermediate state in the spontaneous process if we suppose that the reaction occurs only in the presence of a solid catalyst. By introducing and then removing the catalyst from the reaction mixture, we can produce a quasi-equilibrium state whose composition is identical to that of the spontaneous reaction at any particular extent of reaction. In this quasi-equilibrium state, the pressure is equal to the applied pressure, and the temperature is equal to that of the surroundings. If the spontaneous uncatalyzed reaction is slow compared to the rate at which the pressure and temperature of the system can equilibrate with the applied pressure and the surroundings temperature, the state functions of a spontaneously reacting system and a static, quasi-equilibrium system with the same composition must be essentially identical.

It turns out that we can find the dependence of $\Delta_r G$ on concentrations by considering a fundamentally different system—one that is composed of exactly the same amount of each of these gases, but in which the gases are not mixed. Each gas occupies its own container. For the partial pressures P_A, P_B, P_C, and P_D, the Gibbs free energies per mole are

$$\overline{G}_A(P_A) = \Delta_f G^o(A, P^o) + RT \ln p_A$$

$$\overline{G}_B(P_B) = \Delta_f G^o(B, P^o) + RT \ln p_B$$

$$\overline{G}_A(P_C) = \Delta_f G^o(C, P^o) + RT \ln p_C$$

$$\overline{G}_D(P_D) = \Delta_f G^o(D, P^o) + RT \ln p_D$$

The Gibbs free energy of this system, which is just a composite of the separated gases, is

$$G = n_A \overline{G}_A(P_A) + n_B \overline{G}_B(P_B) + n_C \overline{G}_C(P_C) + n_D \overline{G}_D(P_D)$$

If we subtract the Gibbs free energies of a moles of reactant A and b moles of reactant B from the Gibbs free energies of c moles of product C and d moles of product C, wc find

$$\begin{aligned} c\,\overline{G}_C(P_C) + d\,\overline{G}_D(P_D) - a\,\overline{G}_A(P_A) + b\,\overline{G}_B(P_B) \\ = c\,\Delta_f G^o(C, P^o) + d\,\Delta_f G^o(D, P^o) - a\,\Delta_f G^o(A, P^o) \\ - b\,\Delta_f G^o(B, P^o) + RT \ln \frac{p_C p_D}{p_A p_B} \end{aligned}$$

To represent these free energy differences when the gases are in separate containers, we introduce the abbreviations

$$\Delta_{sep} G = c\,\overline{G}_C(P_C) + d\,\overline{G}_D(P_D) - a\,\overline{G}_A(P_A) - b\,\overline{G}_B(P_B)$$

and

$$\begin{aligned} \Delta_r G^o = c\,\Delta_f G^o(C, P^o) + d\,\Delta_f G^o(D, P^o) \\ - a\,\Delta_f G^o(A, P^o) - b\,\Delta_f G^o(B, P^o) \end{aligned}$$

so that the difference between the Gibbs free energies of the separated reactants and products can be written more compactly as

$$\Delta_{sep} G = \Delta_r G^o + RT \ln \frac{p_C p_D}{p_A p_B}$$

$\Delta_{sep} G$ is the difference in the Gibbs free energies when the pressures of the separated gases are fixed at P_A, P_B, P_C, and P_D. Note that, if any of the pressures changes, $\Delta_{sep} G$ changes. When we introduce $\Delta_r G^o$ in §11-10, we emphasize that this quantity is the difference between the standard Gibbs free energies of the separated products

[c moles of C at 1 bar and d moles of D at 1 bar]

and the separated reactants

[a moles of A at 1 bar and b moles of B at 1 bar].

We call $\Delta_r G^o$ the ***standard Gibbs free energy change*** for the reaction. At a given temperature, $\Delta_r G^o$ is a constant. Our choice of standard state for the Gibbs free energy of a compound means that we can calculate the standard free energy change for a reaction from the standard free energies of formation of the products and reactants:

$$\begin{aligned} \Delta_r G^o = c\,\Delta_f G^o(C, P^o) + d\,\Delta_f G^o(D, P^o) \\ - a\,\Delta_f G^o(A, P^o) - b\,\Delta_f G^o(B, P^o) \end{aligned}$$

Let us recapitulate: In the first system, we are interested in the Gibbs free energy change, $\Delta_r G$, for a process in which a moles of A and b moles of B are converted to c moles of C and d moles of D in a (large) mixed system where the partial pressures are constant at the values P_A, P_B, P_C, and P_D. In the second system, there is actually no process at all. The Gibbs free energy change, $\Delta_{sep} G$, is merely a computed difference between the Gibbs free energies of the specified quantities of product and reactant gases, where each gas is in its own container at its specified pressure. When the gases are ideal, the Gibbs free energy differences (changes) for these two systems turn out to be the same. That is, $\Delta_{sep} G = \Delta_r G$.

This relationship is valid for systems in which the

properties of one substance are not affected by the concentrations of other substances present. It is not true in general. On the other hand, $\Delta_r G^o$ is always the computed difference between the standard Gibbs free energies of the pure separated products and reactants. For ideal gases,

$$\Delta_r G = \Delta_r G^o + RT \ln \frac{p_C p_D}{p_A p_B}$$
(reaction of ideal gases)

We have asserted that we can equate $\Delta_{sep} G$ and $\Delta_r G$ for systems composed of ideal gases. Now we need to show that this is true. This is easy if we first understand a related problem—the thermodynamics of mixing ideal gases.

§3 The thermodynamics of mixing ideal gases

When we talk about the ***thermodynamics of mixing***, we have a very particular process in mind. By convention, the process of mixing two gases, call them A and B, is the process in which the two gases initially occupy separate containers, but are both at a common pressure and temperature. (We denote the common initial pressure by "P_0". P_0 is not to be confused with the constant P^o.) The final state after the mixing process is one in which there is a homogeneous mixture of A and B at the same temperature as characterized the initial state. The final volume is the sum of the initial volumes. If the gases are ideal, the final pressure is the same as the initial pressure, and the partial pressures are $P_A = n_A RT/(V_A + V_B)$ and $P_B = n_B RT/(V_A + V_B)$.

The mixing process is represented by the change on the right side of Figure 1. The gases are always in thermal contact with constant-temperature surroundings. We imagine that we bring the initially separate containers together and then remove the overlapping walls. (Or we can imagine connecting the two containers by a tube—whose volume is negligibly small—that allows the molecules to move from one container to the other.) Molecular diffusion eventually causes the concentration of either gas to be the same in any macroscopic portion of the combined volumes. This diffusive mixing begins as soon as we provide a path for the molecules to move between their containers. Isothermal mixing is a spontaneous process. The reverse process does not occur. Isothermal mixing is irreversible.

Since the temperature is constant and the gases are ideal, the energy of the A molecules is constant; likewise, the energy of the B molecules constant. It follows that the energy of a system containing molecules of A and B is independent of their concentrations and that the energy

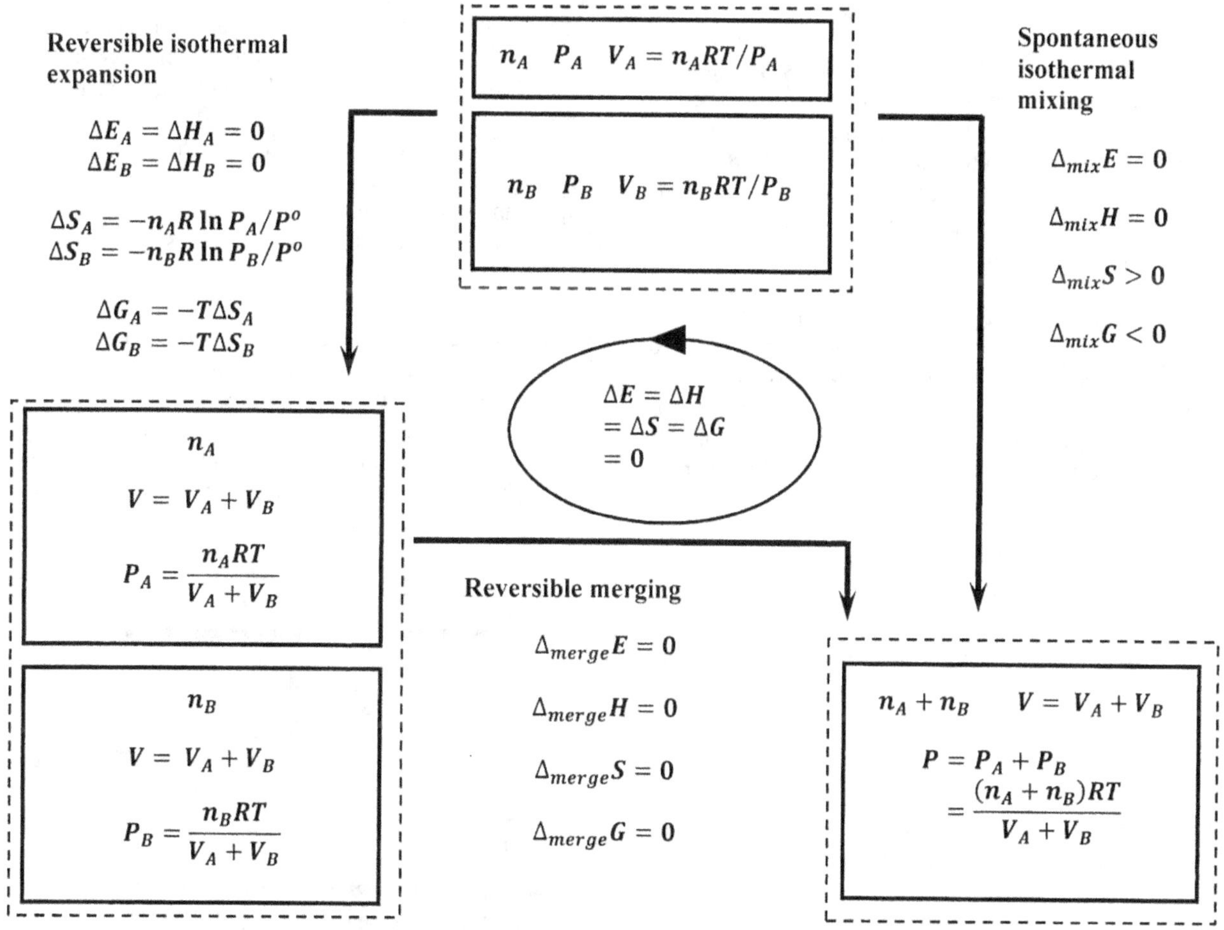

Figure 1. Changes in thermodynamic functions during the mixing of ideal gases.

of the mixture is the sum of the energies of the separated components. That is, we have $\Delta_{mix}E = 0$. Since the volume and the pressure of the two-gas system are constant, it follows that $\Delta_{mix}(PV) = 0$ and $\Delta_{mix}H = 0$. Constant volume also means that $w_{mix} = 0$. It follows that $q_{mix} = 0$, and that $\Delta_{mix}\hat{S} = 0$. Then, because the mixing process is spontaneous, we have $\Delta_{mix}S > 0$. Finally, since $\Delta_{mix}H = 0$, we have $\Delta_{mix}G = -T\Delta_{mix}S < 0$. However, in order to calculate the values of $\Delta_{mix}S = 0$ and $\Delta_{mix}G < 0$, we must find a reversible path for the state change that occurs during mixing.

The two-step path represented by the changes along the left side and the bottom of Figure 1 is such a reversible path. The first step is simply the reversible isothermal expansion of the separate gases from their initial volumes (V_A and V_B, respectively) to the common final volume, $V_A + V_B$. At the end of this step, the two gases are still in separate containers. Reversible isothermal expansion of an ideal gas is a familiar process; for this step, we know:

$$\Delta E_A = \Delta H_A = 0$$
$$\Delta E_B = \Delta H_B = 0$$

$$\Delta S_A = -n_A R\ln(P_A/P_0) > 0$$
$$\Delta S_B = -n_B R\ln(P_B/P_0) > 0$$

$$\Delta G_A = -T\Delta S_A < 0$$
$$\Delta G_B = -T\Delta S_B < 0$$

We call the change along the bottom of the diagram the ***merging process***. The merging process is the isothermal, reversible blending of the separate gas samples, each initially occupying a volume $V_A + V_B$, in such a way that the final state has all of the molecules of the two gases in the same container, whose volume is also $V_A + V_B$. While the final state of the merging process is identical to the final state of the mixing process, the initial state is distinctly different.

For ideal gases, it turns out that all of the thermodynamic functions are unchanged during the merging process. Consequently, the thermodynamic functions for mixing are just the sums of the thermodynamic functions for the reversible expansions of A and B separately. Equating thermodynamic functions for the two paths to the mixed state, we have

$$\Delta_{mix}E = \Delta E_A + \Delta E_B + \Delta_{merge}E = 0$$
$$\Delta_{mix}H = \Delta H_A + \Delta H_B + \Delta_{merge}H = 0$$

$$\begin{aligned}\Delta_{mix}S &= \Delta S_A + \Delta S_B + \Delta_{merge}S \\ &= -n_A R\ln(P_A/P_0) - n_B RT\ln(P_B/P_0) \\ &> 0\end{aligned}$$

$$\Delta_{mix}G = -T\Delta_{mix}S < 0$$

The pressure ratios equal the mole fractions of the compounds in the mixture. Therefore, the entropy of mixing is also given by

$$\Delta_{mix}S = -n_A R\ln x_A - n_B R\ln x_B > 0$$

If we calculate the entropy of mixing per mole of A–B-mixture, $\Delta_{mix}\overline{S}$, we find

$$\begin{aligned}\Delta_{mix}\overline{S} &= \frac{\Delta_{mix}S}{n_A + n_B} \\ &= -\left(\frac{n_A}{n_A + n_B}\right)R\ln x_A - \left(\frac{n_B}{n_A + n_B}\right)R\ln x_B \\ &= -x_A R\ln x_A - x_B R\ln x_B\end{aligned}$$

It remains to prove our assertion that the thermodynamic functions are unchanged during the merging process. That the energy is unchanged follows from the fact that the energy of an ideal gas depends only on temperature; it is therefore independent of pressure and of the presence of any other substance. We can give a qualitative argument for the idea that other thermodynamic quantities are also unchanged. In this argument, the enthalpy, entropy, and free energy functions are unchanged because ideal gas molecules do not interact with one another. If A molecules do not interact with B molecules, it follows that the properties of the A molecules are independent of whether the B molecules share the same container or are present in a separate container of identical volume. At the same temperature, the A molecules generate a pressure P_A, and the B molecules independently generate a pressure P_B. Since these pressures are generated independently, we conclude that the total pressure is the sum of the two partial pressures—which is, of course, just Dalton's law of partial pressures. The same argument applies to the enthalpy, entropy, and free energy functions, so these should also be unchanged during the merging process.

We can also create a device in which we can—in concept—carry out the reversible isothermal merging process and calculate thermodynamic-function changes. (We rely on the argument above to establish Dalton's law of partial pressures and the conclusion that $\Delta_{merge}E = 0$. We use the device to show from these results that there is no change in the other thermodynamic functions.) The device is sketched in Figure 2, at an intermediate stage of the process. It consists of three cylinders, each closed by a frictionless piston. The first contains unmerged gas A, the second contains unmerged gas B, and the third contains the mixture that results from merging them. The head of the A cylinder is shared with

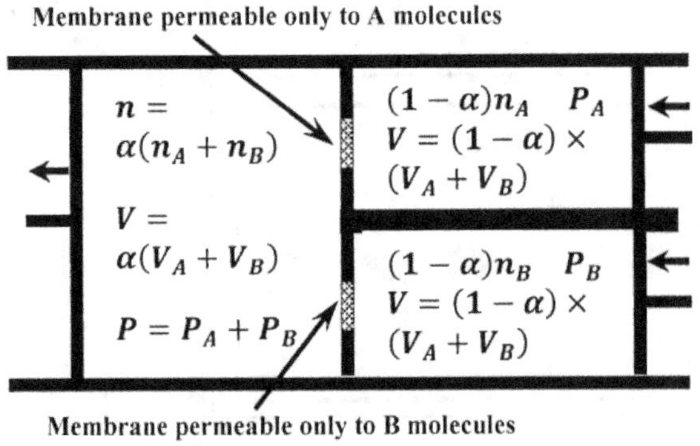

Figure 2. A device to merge gases.

part of the head of the merging cylinder. Likewise, the head of the B cylinder is shared with another part of the head of the merging cylinder. We suppose that the A and B cylinder heads are comprised in part of molecule-selective gas-permeable membranes. The membrane in the head of the A cylinder allows the diffusion of A molecules in either direction, but does not permit B molecules to pass. The membrane in the head of the B cylinder allows the diffusion of B molecules in either direction, but does not permit A molecules to pass.

In the merging cylinder, we reversibly accumulate A molecules from the A cylinder and B molecules from the B cylinder. We accomplish this by controlling the pressures in the three cylinders. The pressure in the A cylinder is always infinitesimally greater than P_A. The pressure in the B cylinder is always infinitesimally greater than P_B. The partial pressure of A in the merging cylinder is infinitesimally less than P_A; the partial pressure of B in the merging cylinder is infinitesimally less than P_B; and the total pressure in the merging cylinder is infinitesimally less than $P_A + P_B$. (There is always a difference in the total pressure across each membrane.) The system consists of the contents of the cylinders. Work is done on the system by the forces acting on the A and B cylinders. Work is done by the system on the surroundings as the merging cylinder fills with the mixture.

As far as the A cylinder is concerned, the process is reversible because the pressure due to the A molecules is just infinitesimally greater in the A cylinder than in the merging cylinder. Therefore, a very small decrease in the pressure of the A cylinder would cause the net flow of A molecules to change direction. The same is true for the B cylinder.

The initial and final states of the apparatus when we use it to merge n_A moles of A at P_A with n_B moles of B at P_A are shown in Figure 3. Let $V_{mix} = V_A + V_B$ be the sum of the initial volumes of cylinders A and B. This is also the final volume of the merging cylinder. The reversible work done in the A cylinder is

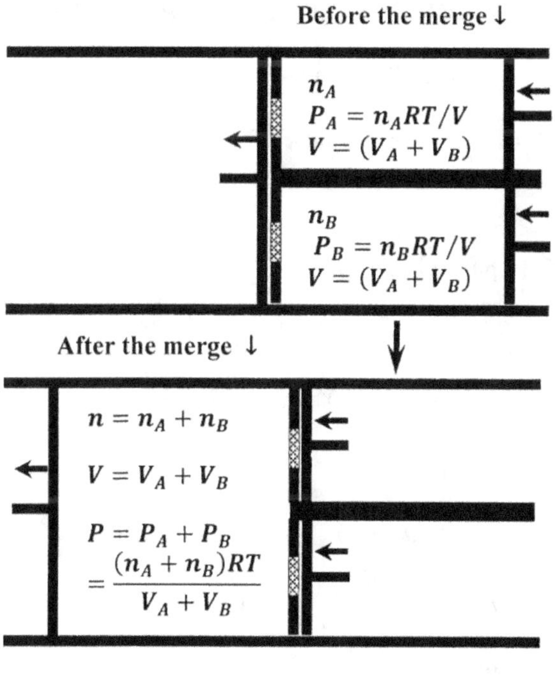

Figure 3. The device before and after merging the gases.

$$\begin{aligned} w_A^{rev} &= -\int_{V_A+V_B}^{0} P_A dV \\ &= -P_A(0 - V_A - V_B) \\ &= P_A(V_A + V_B) \\ &= n_A RT \end{aligned}$$

Similarly, the work in the B cylinder is

$$\begin{aligned} w_B^{rev} &= P_B(V_A + V_B) \\ &= n_B RT \end{aligned}$$

In the merging cylinder it is

$$\begin{aligned} w_{A+B}^{rev} &= -(P_A + P_B)(V_A + V_B) \\ &= -(n_B + n_B)RT \end{aligned}$$

For the merging process the net work is

$$\begin{aligned} w_{merge}^{rev} &= w_A^{rev} + w_B^{rev} + w_{A+B}^{rev} \\ &= 0 \end{aligned}$$

and the net change in the pressure-volume product is

$$\begin{aligned} \Delta_{merge}(PV) &= (P_A + P_B)(V_A + V_B) \\ &\quad -P_A(V_A + V_B) - P_B(V_A + V_B) \\ &= 0 \end{aligned}$$

Since constant temperature ensures that $\Delta_{merge}E = 0$, it follows from $\Delta_{merge}(PV) = 0$ that $\Delta_{merge}H = 0$, and from $w_{merge}^{rev} = 0$ that $q_{merge}^{rev} = 0$. Since the merging process is reversible, we have also that $\Delta_{merge}S = 0$ and $\Delta_{merge}G = 0$.

These arguments can be extended to merging any number of ideal gases. In the initial state for this merging process each gas is at the same temperature, but occupies a separate container; all of these containers have the same volume. Each gas can be at a different pressure. In the final state, all of the gases occupy a common container, whose volume is the same as the common volume of their initial containers. The temperature of the mixture in the final state is equal to the common initial temperature of the separate gases. In the final state, the partial pressure of each gas is equal to its pressure in the initial state. For any number of gases, we have

$$\Delta_{merge}E = \Delta_{merge}H = \Delta_{merge}S = \Delta_{merge}G = 0$$

Likewise, these arguments can be extended to the mixing of multiple ideal gases, all at the same original pressure and temperature, into a final volume that is equal to the sum of the initial volumes—and is at the original temperature. If there are ω such gases,

$$\Delta_{mix}E = \Delta_{mix}H = 0$$

$$\Delta_{mix}S = \sum_{i=1}^{\omega} -n_i R \ln x_i > 0$$

$$\Delta_{mix}G = -T\Delta_{mix}S < 0$$

§4 The Gibbs free energy change for reaction at constant partial pressures

Now we can compare the difference between the Gibbs free energies of reactants and products in the reaction

$$a\,A + b\,B \rightleftharpoons c\,C + d\,D$$

when all of the gases are present in the same system to the same difference when each gas is in its own container.

In the first case, the gases are present in a mixture, and their partial pressures remain constant at P_A, P_B, P_C, and P_D respectively. We called the Gibbs free energy change under these conditions $\Delta_r G$. These conditions can be satisfied if we suppose that the system is very large. That is, $\Delta_r G$ is the limiting Gibbs free energy change for the conversion of an initial mixture into a final mixture. The initial mixture contains $(n_A + a)$ moles of A, $(n_B + b)$ moles of B, n_C moles of C, and n_D moles of D. The final mixture contains n_A moles of A, n_B moles of B, $(n_C + c)$ moles of C, and $(n_D + d)$ moles of D. $\Delta_r G$ is the Gibbs free energy change for this conversion in the limit as n_A, n_B, n_C, and n_D become arbitrarily large. In this limit, we have $n_A \gg a$, $n_B \gg b$, $n_C \gg c$, , and $n_D \gg d$. Since the partial pressures remain (essentially) constant, the n_i must also satisfy

$$P_A = n_A/(n_A + n_B + n_C + n_D)$$
$$P_B = n_B/(n_A + n_B + n_C + n_D)$$
$$P_C = n_C/(n_A + n_B + n_C + n_D)$$
$$P_D = n_D/(n_A + n_B + n_C + n_D)$$

When the ideal gases are separated from one another, the Gibbs free energy difference is the Gibbs free energy of c moles of gas C (at pressure P_C) plus the Gibbs free energy of d moles of gas D (at pressure P_D) minus the Gibbs free energy of a moles of gas A (at pressure P_A) and minus the Gibbs free energy of b moles of gas B (at pressure P_B). In §2, we call the Gibbs free energy change under these conditions $\Delta_{sep}G$.

In §2, we assert that $\Delta_{sep}G = \Delta_r G$. Now we can see why this is so. The cycle shown in Figure 4 relates these two Gibbs free energy differences. We have

$$\Delta_{sep}G(P_A, P_B, P_C, P_D) + \Delta_{merge}G_{final} = \Delta_{merge}G_{initial} + \Delta_r G(P_A, P_B, P_C, P_D)$$

and since

$$\Delta_{merge}G_{final} = \Delta_{merge}G_{initial}$$

we have

$$\Delta_{sep}G(P_A, P_B, P_C, P_D) = \Delta_r G(P_A, P_B, P_C, P_D)$$

Letting $\overline{G}_A(P_A)$ be the Gibbs free energy of one mole of pure ideal gas A at pressure P_A, $\overline{G}_B(P_B)$ be the Gibbs free energy of one mole of pure ideal gas B at pressure P_B, *etc.*, we have

$$\begin{aligned}\Delta_{sep}G(P_A, P_B, P_C, P_D) = &\; n_A\,\overline{G}_A(P_A) + n_B\,\overline{G}_B(P_B) \\ &+(n_C + c)\,\overline{G}_C(P_C) + (n_D + d)\overline{G}_D(P_D) \\ &-(n_A + a)\,\overline{G}_A(P_A) - (n_B + b)\,\overline{G}_B(P_B) \\ &-n_C\,\overline{G}_C(P_C) - n_D\,\overline{G}_D(P_D) \\ = &\; c\,\overline{G}_C(P_C) + d\,\overline{G}_D(P_D) - a\,\overline{G}_A(P_A) - b\,\overline{G}_B(P_B)\end{aligned}$$

Since $\Delta_{sep}G = \Delta_r G$, we have

$$\Delta_r G = c\,\overline{G}_C(P_C) + d\,\overline{G}_D(P_D) - a\,\overline{G}_A(P_A) - b\,\overline{G}_B(P_B)$$

$(n_A + a)A(P_A) + (n_B + b)B(P_B) + n_C C(P_C) + n_D D(P_D)$ **{separate}**

↓ $\Delta_{merge}G_{merge} = 0$

{mixture} $(n_A + a)A(P_A) + (n_B + b)B(P_B) + n_C C(P_C) + n_D D(P_D)$

$\Delta_{sep}G(P_A, P_B, P_C, P_D)$ ↓ $\Delta_r G(P_A, P_B, P_C, P_D)$

{mixture} $n_A A(P_A) + n_B B(P_B) + (n_C + c)C(P_C) + (n_D + d)D(P_D)$

↑ $\Delta_{merge}G_{final} = 0$

$n_A A(P_A) + n_B B(P_B) + (n_C + c)C(P_C) + (n_D + d)D(P_D)$ **{separate}**

Figure 4. Cycle demonstrating that $\Delta_{sep}G = \Delta_r G$.

§5 $\Delta_r G$ is the rate at which the Gibbs free energy changes with the extent of reaction

For the reaction $a\,A + b\,B \rightleftharpoons c\,C + d\,D$, let us call the consumption of a moles of A one "unit of reaction." $\Delta_r G$ corresponds to the actual Gibbs free energy change for one unit of reaction only in the limiting case where the reaction occurs in an arbitrarily large system. For a closed system of specified initial composition, n_A^o, n_B^o, n_C^o, and n_D^o, whose composition at any time is specified by n_A, n_B, n_C, and n_D, the extent of reaction, ξ, is

$$\begin{aligned} \xi &= -\left(\frac{n_A - n_A^o}{a}\right) \\ &= -\left(\frac{n_B - n_B^o}{b}\right) \\ &= \frac{n_C - n_C^o}{c} \\ &= \frac{n_D - n_D^o}{d} \end{aligned}$$

At constant pressure and temperature, every possible state of this system is represented by a point on a plot of $\Delta_r G$ *versus* ξ. Every such state is also represented by a point on a plot of G_{system} *versus* ξ.

From the general result that $\left(dG_{system}\right)_{PT} = 0$ if and only if the system is at equilibrium, it follows that $\Delta_r G\left(\xi_{eq}\right) = 0$ if and only if ξ_{eq} specifies the equilibrium state. (We can arrive at the same conclusion by considering the heat exchanged for one unit of reaction in an infinitely large system at equilibrium. This process is reversible, and it occurs at constant pressure and temperature, so we have $\Delta_r H = q_P^{rev}$, $\Delta_r S = q_P^{rev}/T$, and $\Delta_r G = q_P^{rev} - T(q_P^{rev}/T) = 0$.)

Below, we show that

$$\Delta_r G(\xi) = \left(\frac{\partial G_{system}}{\partial \xi}\right)_{PT}$$

for any value of ξ. (In §15-9, we use essentially the same argument to show that this conclusion is valid for any reaction among any substances.) Given this result, we see that the equilibrium composition corresponds to the extent of reaction, ξ_{eq}, for which the Gibbs free energy change for one unit of the reaction is zero

$$\Delta_r G\left(\xi_{eq}\right) = 0$$

and

$$\left(\frac{\partial G_{system}}{\partial \xi}\right)_{PT} = 0$$

So that the Gibbs free energy of the system is a minimum.

In the next section, we show that the condition $\Delta_r G\left(\xi_{eq}\right) = 0$ makes it easy to calculate the equilibrium extent of reaction, ξ_{eq}. Given the stoichiometry and initial composition, the equation for $\Delta_r G\left(\xi_{eq}\right)$ specifies the equilibrium composition and the partial pressures P_A, P_B, P_C, and P_D. This is the usual application of these results. Setting $\Delta_r G = 0$ enables us to answer the question: If we initiate reaction at a given composition, what will be the equilibrium composition of the system? Usually this is what we want to know. The amount by which the Gibbs free energy changes as the reaction goes to equilibrium is seldom of interest.

To show that $\Delta_r G = (\partial G/\partial \xi)_{PT}$ for any reaction, it is helpful to introduce modified stoichiometric coefficients, ν_j, defined such that $\nu_j > 0$ if the j-th species is a product and $\nu_j < 0$ if the j-th species is a reactant. That is, for the reaction $a\,A + b\,B \rightleftharpoons c\,C + d\,D$, we define $\nu_A = -a$, $\nu_B = -b$, $\nu_C = c$, and $\nu_D = d$. Associating successive integers with the reactants and products, we represent the j-th chemical species as X_j and an arbitrary reaction as

$$|\nu_1|X_1 + |\nu_2|X_2 + \cdots + |\nu_i|X_i \rightleftharpoons \nu_j X_j + \cdots + \nu_\omega X_\omega$$

Let the initial number of moles of ideal gas X_j be n_j^o; then $n_j = n_j^o + \nu_j \xi$. (For species that are present but do not participate in the reaction, we have $\nu_j = 0$.)

We have shown that the Gibbs free energy of a mixture of ideal gases is equal to the sum of the Gibbs free energies of the components. In calculating $\Delta_r G$, we assume that this is as true for a mixture undergoing a spontaneous reaction as it is for a mixture at equilibrium. In doing so, we assume that the reacting system is homogeneous and that its temperature and pressure are well defined. In short, we assume that the Gibbs free energy of the system is the same continuous function of temperature, pressure, and composition, $G = G(T, P, n_1, n_2, \dots)$, whether the system is at equilibrium or undergoing a spontaneous reaction. For the equilibrium system, we have $(\partial G/\partial T)_{P,n_j} = -S$ and $(\partial G/\partial P)_{T,n_j} = V$. When we assume that these functions are the same for a spontaneously changing system as they are for a reversible system, it follows that

$$\begin{aligned} dG &= \left(\frac{\partial G}{\partial T}\right)_{P,n_j} dT + \left(\frac{\partial G}{\partial P}\right)_{T,n_j} dP \\ &\quad + \sum_j \left(\frac{\partial G}{\partial n_j}\right)_{P,T,n_{i\neq j}} dn_j \\ &= -SdT + VdP + \sum_j \left(\frac{\partial G}{\partial n_j}\right)_{P,T,n_{i\neq j}} dn_j \end{aligned}$$

whether the system is at equilibrium or undergoing spontaneous change. At constant temperature and pressure, when pressure–volume work is the only work, the thermodynamic criteria for change, $dG_{TP} \le 0$ become

$$\sum_j \left(\frac{\partial G}{\partial n_j}\right)_{P,T,n_{i\neq j}} dn_j \le 0$$

When a reaction occurs in the system, the

composition is a continuous function of the extent of reaction. We have $G = G(T, P, n_1^o + \nu_1\xi, n_2^o + \nu_2\xi, \ldots)$. At constant temperature and pressure, the dependence of the Gibbs free energy on the extent of reaction is

$$\left(\frac{\partial G}{\partial \xi}\right)_{P,T,n_m} = \sum_j \left(\frac{\partial G}{\partial(n_j^o + \nu_j\xi)}\right)_{P,T,n_{m\neq j}} \left(\frac{\partial(n_j^o + \nu_j\xi)}{\partial \xi}\right)_{P,T,n_{m\neq j}}$$

Since

$$\left(\frac{\partial G}{\partial(n_j^o + \nu_j\xi)}\right)_{P,T,n_{m\neq j}} = \left(\frac{\partial G}{\partial n_j}\right)_{P,T,n_{m\neq j}} = \overline{G}_j$$

and

$$\left(\frac{\partial(n_j^o + \nu_j\xi)}{\partial \xi}\right)_{P,T,n_{m\neq j}} = \nu_j$$

it follows that

$$\left(\frac{\partial G}{\partial \xi}\right)_{P,T,n_m} = \sum_j \nu_j \overline{G}_j = \Delta_r G$$

Moreover, we have

$$(dG)_{PT} = \left(\frac{\partial G}{\partial \xi}\right)_{P,T,n_m} d\xi$$

The criteria for change, $(dG)_{PT} \leq 0$, become

$$\left(\frac{\partial G}{\partial \xi}\right)_{P,T,n_m} d\xi \leq 0$$

From our definition of ξ, we have $d\xi > 0$ for a process that proceeds spontaneously from left to right, so the criteria become

$$\left(\frac{\partial G}{\partial \xi}\right)_{P,T,n_m} \leq 0$$

or, equivalently,

$$\sum_j \nu_j \overline{G}_j = \Delta_r G \leq 0$$

§6 The standard Gibbs free energy change and equilibrium in ideal gas reactions

The relationship between $\Delta_r G$ and $\Delta_r G^o$ is evident from the cycle in Figure 5. Since we have shown that $\Delta_{sep} G(P_A, P_B, P_C, P_D) = \Delta_r G(P_A, P_B, P_C, P_D)$, we can consider the bottom equation in this cycle to represent the reaction occurring in a mixture while calculating its free energy change as the free energy difference between pure products and pure reactants. Since $\Delta_{cycle} G = 0$,

$$\Delta_{cycle} G = \Delta_r G^o - \Delta_r G + RT(\ln p_C^c + \ln p_D^d - \ln p_A^a - \ln p_B^B) = 0$$

which can be rearranged to the result obtained in §2:

$$\Delta_r G = \Delta_r G^o + RT \ln \frac{p_C^c p_D^d}{p_A^a p_B^b}$$

$\Delta_r G$ is the Gibbs free energy change for one unit of the reaction occurring in a system whose composition is specified by P_A, P_B, P_C, and P_D. In this spontaneously reacting system, the molar Gibbs free energy of ideal gas A is

$$\overline{G}_A(P_A) = \Delta_f G^o(A, P^o) + RT \ln p_A$$

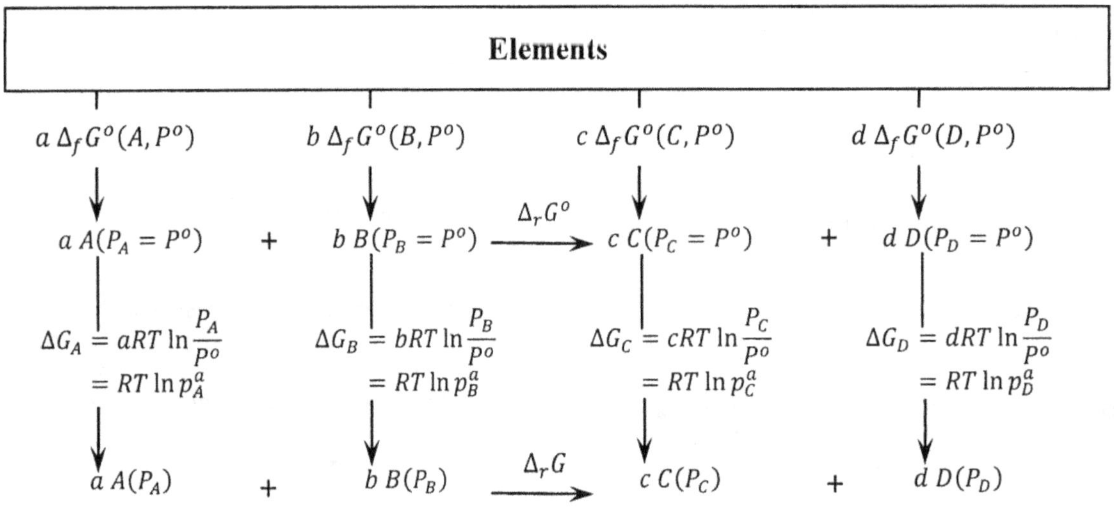

Figure 5. Cycle demonstrating the relationship between $\Delta_r G$ and $\Delta_r G^o$.

If the system is at equilibrium, P_A, P_B, P_C, and P_D are equilibrium pressures; these values characterize an equilibrium state. Then $\Delta_r G$ is the free energy change for a reaction occurring at equilibrium at constant pressure and temperature, and $\Delta_r G$ is zero. The equation

$$0 = \Delta_r G^o + RT \ln \frac{p_C^c p_D^d}{p_A^a p_B^b}$$

is exact. We have, when the partial pressures are those for a system at equilibrium,

$$\Delta_r G^o = -RT \ln \frac{p_C^c p_D^d}{p_A^a p_B^b}$$

Since $\Delta_r G^o$ is a constant, it follows that

$$\frac{p_C^c p_D^d}{p_A^a p_B^b}$$

is a constant. It is, of course, the equilibrium constant. We have

$$K_P = \frac{p_C^c p_D^d}{p_A^a p_B^b}$$

and

$$\Delta_r G^o = -RT \ln K_P$$

or, solving for K_P

$$K_P = \exp\left(-\frac{\Delta_r G^o}{RT}\right)$$

Note that the value of the equilibrium constant is calculated from the Gibbs free energy change at standard conditions, not the Gibbs free energy change at equilibrium, which is zero.

§7 The Gibbs free energy of formation and equilibrium in ideal gas reactions

In §6, we develop a relationship between the standard Gibbs free energy change for a reaction and the equilibrium constant for that reaction. The standard Gibbs free energy change for a reaction is a specific quantity of energy, which depends only on the temperature. In Chapter 11, we develop the Gibbs free energy of formation for a substance in its standard state. We assert that it is convenient to set the Gibbs free energy of every substance in its standard state equal to its Gibbs free energy of formation. Now we can see the reason: When we express the molar Gibbs free energy of an ideal gas as

$$\overline{G}_A(P_A) = \Delta_f G^o(A, P^o) + RT \ln \frac{P_A}{P^o}$$

$\overline{G}_A(P_A)$ is the Gibbs free energy change for producing one mole of ideal gas A, at pressure P_A, from the elements at the same temperature. Because the Gibbs free energies of the reactants and products are measured from a common starting point, we can use them to calculate the Gibbs free energy change for their reaction.

To demonstrate the role of our choice of the elements as the reference state for the Gibbs free energies of chemical substances, we need only expand the Gibbs free energy cycle in Figure 4. In Figure 6, we introduce the Gibbs free energy change for producing the separate

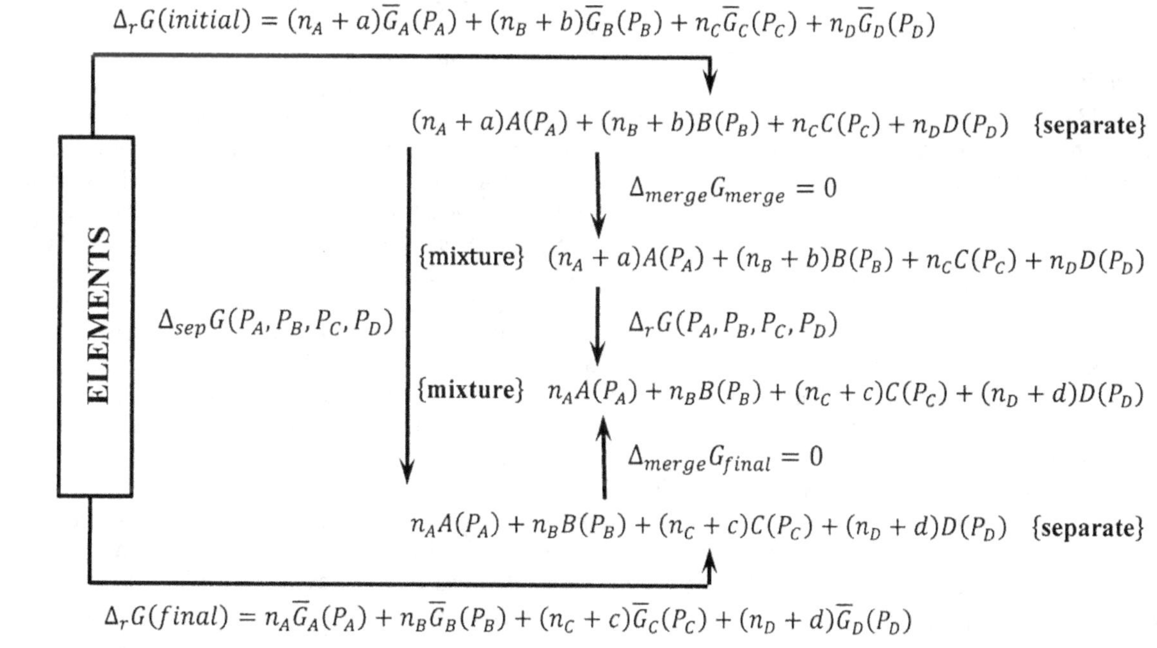

Figure 6. Cycle demonstrating the relationship between $\Delta_{sep} G$ and $\Delta_r G$ for the formation of reactants and products from their constituent elements.

components of the very large system. Computing the Gibbs free energy change in a clockwise direction around the cycle in which the elements are converted first to isolated reactants, then to components of a large equilibrium system, then to separated products, and finally back to the elements, we find

$$0 = \Delta_f G(initial) + \Delta_{merge} G_{initial}$$
$$+\Delta_r G(P_A, P_B, P_C, P_D) - \Delta_{merge} G_{final} - \Delta_f G(final)$$

Since the partial pressures P_A, P_B, P_C, and P_D characterize the system at equilibrium, $\Delta_r G(P_A, P_B, P_C, P_D) = 0$. Also, $\Delta_{merge} G_{initial} = \Delta_{merge} G_{final} = 0$, so the equation for the Gibbs free energy change around the cycle simplifies to

$$\begin{aligned} 0 &= \Delta_f G(initial) - \Delta_f G(final) \\ &= a\,\overline{G}_A(p_A) + b\,\overline{G}_B(p_B) - c\,\overline{G}_C(p_C) - d\,\overline{G}_D(p_D) \\ &= a\,\Delta_f G^o(A, P^o) + b\,\Delta_f G^o(B, P^o) - c\,\Delta_f G^o(C, P^o) \\ &\qquad -d\,\Delta_f G^o(D, P^o) \\ &\qquad +RT \ln p_A^a + RT \ln p_B^b - RT \ln p_C^c - RT \ln p_D^d \\ &= -\Delta_r G^o - RT \ln \frac{p_C^c p_D^d}{p_A^a p_B^b} \end{aligned}$$

So that we have

$$\Delta_r G^o = -RT \ln \frac{p_C^c p_D^d}{p_A^a p_B^b}$$

To illustrate the application of these ideas, let us consider equilibrium in the oxidation of nitric oxide to nitrogen dioxide

$$2\,NO + O_2 \rightleftharpoons 2\,NO_2$$

At 800 K, the standard Gibbs free energy of formation of nitric oxide is +81.298 kJ mol^{-1}, and that of nitrogen dioxide is +83.893 kJ mol^{-1}. Hence, the standard Gibbs free energy change for the oxidation is +5,190 J mol^{-1}, and the equilibrium constant is

$$\begin{aligned} K_P &= \exp\left(-\frac{\Delta_r G^o}{RT}\right) \\ &= 0.458 \end{aligned}$$

Suppose that one mole of NO_2 and one mole of O_2 are mixed, that the mixture is thermostated at 800 K, and that the applied pressure remains constant at 10 bar while the reaction goes to equilibrium.

Let the number of moles of NO present at equilibrium be n_{NO}. Then the moles of NO_2 and O_2 present at equilibrium are $n_{NO_2} = 1 - n_{NO}$ and $n_{O_2} = 1 + n_{NO}/2$. The partial pressures are $P_{NO} = n_{NO}RT/V$, $P_{NO_2} = n_{NO_2}RT/V$, and $P_{O_2} = n_{O_2}RT/V$, so that

$$K_P = \frac{(1 - n_{NO})^2}{n_{NO}^2 (1 + n_{NO}/2)(RT/V)}$$

The value of RT/V depends on the total pressure; we have

$$\begin{aligned} P &= P_{NO} + P_{NO_2} + P_{O_2} \\ &= (2 + n_{NO}/2)(RT/V) \\ &= 10 \text{ bar} \end{aligned}$$

so that

$$\frac{RT}{V} = \frac{20}{4 + n_{NO}}$$

and

$$K_P = \frac{(1 - n_{NO})^2 (4 + n_{NO})}{10 n_{NO}^2 (2 + n_{NO})} = 0.458$$

Solving, we find $n_{NO} = 0.388$.

§8 Equilibrium when a component is present as a condensed phase

Suppose that the very large equilibrium system with ideal gas components at pressures P_A, P_B, P_C, and P_D, also contains a quantity of liquid A. (For the present, we assume that this is pure liquid A; components B, C, and D are insoluble in liquid A.) If this augmented system is at equilibrium, we know that liquid A is in phase equilibrium with ideal gas A at pressure P_A. That is, P_A is the vapor pressure of liquid A at the fixed temperature that we are considering, and the Gibbs free energy change for converting liquid A to its ideal gas at P_A is zero. As long as the liquid is in phase equilibrium with its ideal gas, the relationship between the ideal gas partial pressures and the standard Gibbs free energy change for the reaction is not affected by the presence of the liquid. These same considerations apply when the very large equilibrium system contains both ideal gas A and solid A, so long as the equilibrium sublimation pressure is equal to the partial pressure of ideal gas A, P_A.

Now, let us suppose that substance A is a non-volatile liquid or solid. In this case, we may not be able to measure the standard Gibbs free energy of formation of ideal gas A. From a practical standpoint, this is an important consideration; if we cannot find the standard Gibbs free energy of formation of the ideal gas, we cannot use it to calculate equilibrium constants. From a theoretical standpoint, it is less crucial; we can reasonably imagine that any substance has a finite vapor pressure at any temperature, even if the value is much too small to measure experimentally. We can reason about the relationship of the equilibrium vapor pressure to other quantities whether we can measure it or not.

If substance A is a non-volatile liquid (or solid) at the temperature of interest, it is useful to modify the cycle introduced in the preceding section. We can find the standard Gibbs free energy of formation for the liquid from thermal measurements, and we can use liquid A rather than ideal gas A as the standard state. If A is non-volatile, it is present in the equilibrium system only as the liquid. For present purposes, we again assume that the other components are insoluble in liquid A; then, the only difference between A in its standard state and A in the equilibrium system is that the pressure on A in its standard state is one bar and the pressure on liquid A in the equilibrium system is $P = P_B + P_C + P_D$. Therefore, when a moles of liquid A go from their standard state to the equilibrium system, the Gibbs free energy change is

$$\Delta_{press} G_A = a \int_{P^o}^{P} \overline{V}_A^{\bullet}(\ell) dP$$

where $\overline{V}_A^{\bullet}(\ell)$ is the molar volume of pure liquid A. When a moles of liquid A are produced at the equilibrium pressure from the elements, the Gibbs free energy change is

$$a\overline{G}_A(\ell, P) = a\,\Delta_f G^o(A, \ell) + a \int_{P^o}^{P} \overline{V}_A^{\bullet}(\ell) dP$$

In §15-3, we give further attention to the value of this integral; for now, let us simply note that it is negligible in essentially all circumstances. These considerations mean that we can modify the left side of the cycle in Figure 5 as indicated in Figure 7. When we sum around the modified cycle in the same manner as before, we find

$$\begin{aligned} 0 = a\,\Delta_f G^o(A,\ell) &+ b\,\Delta_f G^o(B,P^o) - c\,\Delta_f G^o(C,P^o) \\ &-d\,\Delta_f G^o(D,P^o) + a \int_{P^o}^{P} \overline{V}_A^{\bullet}(\ell) dP + RT \ln p_B^b \\ &-RT \ln p_C^c - RT \ln p_D^d \end{aligned}$$

where we have supplemented our notation to emphasize that A is a non-volatile liquid while B, C, and D are ideal gases. We continue to use $\Delta_r G^o$ to represent the difference between the standard Gibbs free energies of the products and those of the reactants. In the present circumstances, we have

$$\begin{aligned} \Delta_r G^o = \; & c\,\Delta_f G^o(C,P^o) + d\,\Delta_f G^o(D,P^o) \\ & - a\,\Delta_f G^o(A,\ell) - b\,\Delta_f G^o(B,P^o) \end{aligned}$$

Taking the value of the integral as zero, our result simplifies to

$$\Delta_r G^o = -RT \ln \frac{p_C^c p_D^d}{p_B^b}$$

We have again arrived at the conclusion that the "concentration" of a pure solid or liquid can be set equal to unity in the equilibrium constant expression for a reaction in which it participates. When we do so, we must use the Gibbs free energy of formation of the condensed phase in the calculation of $\Delta_r G^o$.

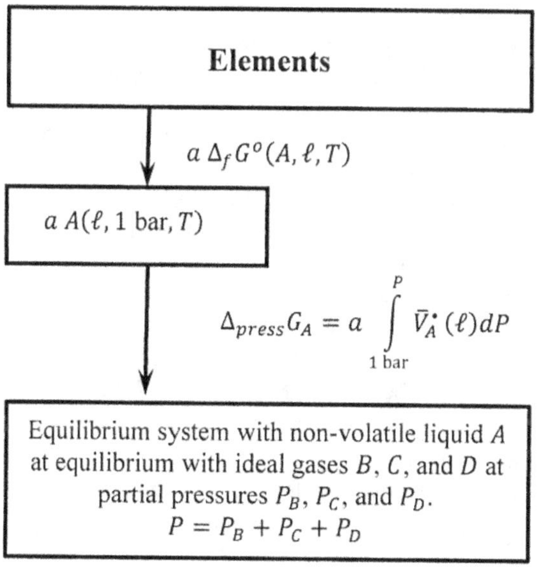

Figure 7. The Gibbs free energy of liquid A in an equilibrium system with ideal gases B, C, and D.

§9 Equilibrium when an ideal gas component is also present as a solute

Finally, let us consider a very large equilibrium system that contains ideal gas components A, B, C, and D, at pressures P_A, P_B, P_C, and P_D. We suppose that this system also contains a very large quantity of an inert solvent. This solvent is a liquid in which components A, B, C, and D are soluble. Let the concentrations of components A, B, C, and D in this solvent be $[A]$, $[B]$, $[C]$, and $[D]$. Since this solvent-containing system is at equilibrium, we know that A dissolved in the solvent at concentration $[A]$ is in equilibrium with ideal gas A at pressure P_A. Since we assume that both the gas and the solution phases are very large, the transport of a moles of A from one to the other does not significantly change any concentration in either phase. The Gibbs free energy change is zero for this phase-transfer process. The Gibbs free energy change for the ideal-gas reaction and its equilibrium position are unaffected by the presence of the solution.

In Chapter 6, we observe that a distribution equilibrium is characterized by an equilibrium constant; the ratio of the concentration of a given chemical species in one phase to its concentration in a second phase is (approximately) a constant. In the present instance, the partial pressure of component A is a measure of its gas-phase concentration and $[A]$ is a measure of its solution-phase concentration. Letting

$$\kappa_A = \left(\frac{P_A/P^o}{[A]}\right) = \frac{p_A}{[A]}$$

be the distribution constant for component A, we have $p_A = \kappa_A [A]$, and parallel relationships for components B, C, and D. Substituting the distribution equilibrium relationships into the equilibrium constant equation, we have

$$\begin{aligned} \Delta_r G^o &= -RT \ln \frac{p_C^c p_D^d}{p_A^a p_B^b} \\ &= -RT \ln \frac{\kappa_C^c \kappa_D^d}{\kappa_A^a \kappa_B^b} - RT \ln \frac{[C]^c [D]^d}{[A]^a [B]^b} \end{aligned}$$

Evidently, we can characterize the position of equilibrium in this system using either the pressure-based constant,

$$K_P = \frac{p_C^c p_D^d}{p_A^a p_B^b}$$

or the concentration based constant,

$$K_C = \frac{[C]^c[D]^d}{[A]^a[B]^b}$$

The relationship between the pressure-based and the concentration-based constants is

$$K_C = \left(\frac{\kappa_C^c \kappa_D^d}{\kappa_A^a \kappa_B^b}\right) K_P$$

In our present discussion, the concentrations can be expressed in any convenient units. The numerical value of the concentration-based constant depends on the units of concentration and the values of the distribution-equilibrium constants as well as the standard Gibbs free energy change for the reaction of the ideal gases.

When all of the reacting species are non-volatile, all of the reacting substances are present in the solution. The partial pressures of these species in any gas phase above the solution is immeasurably small, and the Gibbs free energies of formation of the ideal gases are not accessible by thermal measurements. Since $\Delta_r G^o$ for the ideal-gas reaction is not available, it cannot be used to find K_P. Nevertheless, our thermodynamic model presumes that these parameters have finite—albeit immeasurable—values. The concentration-based constant

$$K_C = \frac{[C]^c[D]^d}{[A]^a[B]^b}$$

characterizes the position of equilibrium in solution even when the data to characterize the gas-phase equilibrium process are immeasurable.

We have again arrived at the same function of concentrations to characterize the position of equilibrium. It is the same whether the equilibrating species are present in the gas phase or in an inert-liquid solvent. To obtain this result, we have used our general thermodynamic model for equilibrium, but we have made special assumptions about the properties of the reacting species. We have assumed that the gases behave ideally and that the distribution-equilibrium constants can be expressed using the species' concentrations. In Chapter 15, we return to this subject and develop a rigorous model for chemical equilibrium that does not require these special assumptions.

Problems

1. Ethylene is the most important feedstock for the organic chemicals industry. The ethylene-production process with the lowest processing cost is the "thermal cracking" of ethane to produce hydrogen and ethylene. The table below gives $\Delta_f G^o$ for ethane and ethylene at 1000 K, 1100 K, and 1200 K.

	$\Delta_f G^o$, J mol^{-1}	
T, K	C_6H_6	C_6H_4
1000	110,750	119,067
1100	132,385	127,198
1200	154,096	135,402

(a) Calculate $\Delta_r G^o$ and K_P at each temperature.

(b) At each temperature, calculate the extent of reaction when one mole of pure ethane reacts to reach equilibrium, while the pressure of the system is maintained constant at one bar.

(c) At each temperature, calculate the extent of reaction when one mole of pure ethane reacts to reach equilibrium, while the pressure of the system is maintained constant at 0.100 bar.

(d) To minimize side reactions, it is desirable to operate a cracking reactor at the lowest possible temperature. The feed to a cracking reactor contains many moles of water (steam) for every mole of ethane. Steam is simply an inexpensive inert gas in this system. Why is steam fed to cracking reactors?

2. A system initially contains one mole of ethane. The cracking reaction occurs while the system is maintained at a constant pressure of 0.1 bar and a constant temperature of 1000 K.

(a) Write the equations for the molar Gibbs free energies of ethane, ethylene, and hydrogen as a function of the extent of reaction, ξ.

(b) Write the equation for $\Delta_r G^o$ as a function of the extent of reaction. For a constant system pressure of 0.1 bar, calculate $\Delta_r G^o$ for ξ = 0, 0.4, 0.7, 0.8, 0.9, and 1.0. Roughly, what extent of reaction corresponds to $\Delta_r G^o = 0$?

(c) Write the equation for the Gibbs free energy of the system, G_{system}, as a function of the extent of reaction, ξ. For a constant system pressure of 0.1 bar, calculate G_{system} for ξ = 0, 0.4, 0.7, 0.8, 0.9, and 1.0. Roughly, at what extent of reaction is G_{system} a minimum?

(d) From the equation for G_{system} in part (c), find $\left(\partial G_{system}/\partial \xi\right)_{T,P,n_j}$. (When the n_j are constant, the composition of the system is fixed: $\overline{G}_{C_6H_6}$, $\overline{G}_{C_6H_4}$, and $\overline{G}_{H_2}$ are constants.)

3. At –56.6 C, the vapor pressure of solid CO_2 is 5.18 bar. (This is the triple point.) At –78.5 C, the vapor pressure of solid CO_2 is 1.01 bar.

(a) A cold bath is prepared by mixing solid CO_2 with methanol in a Dewar flask that is open to the atmosphere. What is the temperature of this cold bath?

(b) Use the Clausius-Clapeyron equation to estimate the enthalpy of sublimation of CO_2.

(c) Assume that we know the Gibbs free energy of formation for both solid and ideal-gas CO_2 at a particular temperature, T. That is, we know $\Delta_f G^o(CO_2,\text{g},T)$ and $\Delta_f G^o(CO_2,\text{s},T)$. What is the Gibbs free energy of gaseous CO_2 as a function of pressure at constant temperature, T?

(d) What is the Gibbs free energy of solid CO_2 at a constant temperature, T?

(e) Assume that the variation of the Gibbs free energy of the solid with pressure is negligible. Using your answers to (c) and (d), write the Gibbs free energy change, ΔG, for the process $CO_2(\text{solid}, P, T) \rightarrow CO_2(\text{gas}, P, T)$. If (P, T) is a point of solid—gas equilibrium, what is the value of ΔG? How is the equilibrium constant for this process related to P_{CO_2}?

(f) Use the Gibbs-Helmholtz equation to estimate the enthalpy of sublimation of CO_2.

(g) Compare the equation you used in part (b) to the equation that you used in part (f).

4. Ethanol is manufactured by the addition of water to ethylene over a solid-acid catalyst. The process is known as "the hydration of ethylene." One mole of ethylene, 10 moles of water, and a quantity of solid catalyst are charged to a reactor. When equilibrium is reached at 400 K, the system pressure is 10.00 bar and the conversion to ethanol is 64.17%. Assume that ethylene, water, and ethanol behave as ideal gases. What is $\Delta_r G^o$ for the hydration of ethylene at 400 K?

5. At 298.15 K, the Gibbs free energies for formation of NO_2 and N_2O_4, in their hypothetical ideal gas standard states, are 51.3 and 99.8 kJ mol^{-1}, respectively.

(a) At this temperature, what is the value of the equilibrium constant for the reaction $N_2O_4 \rightleftharpoons 2\,NO_2$?

(b) A 10 L sample of gas initially contains one mole of N_2O_4. The balance of the sample is N_2. The pressure and temperature of the system are maintained constant at 3 bar and 298.15 K. How many moles of NO_2 are present at equilibrium? What fraction of the N_2O_4 is converted to NO_2?

(c) A 10 L sample of gas initially contains only one mole of N_2O_4 at 298.15 K. The volume and temperature of the system are maintained constant. What fraction of the N_2O_4 is converted to NO_2?

(d) A 10 L sample of gas initially contains only 10^{-1} mole of N_2O_4 at 298.15 K. The pressure and temperature of the system are maintained constant. What fraction of the N_2O_4 is converted to NO_2?

(e) A 10 L sample of gas initially contains only 10^{-1} mole of N_2O_4 at 298.15 K. The volume and temperature of the system are maintained constant. What fraction of the N_2O_4 is converted to NO_2?

6. At 800 K, the Gibbs free energies for formation of methanol and formaldehyde, in their hypothetical ideal gas standard states, are –88.063 and –87.893 kJ mol^{-1}, respectively.

(a) At 800 K, what is the value of the equilibrium constant for the reaction $CH_3OH \rightleftharpoons CH_2O + H_2$?

(b) Initially, one mole of pure methanol occupies 1 m^3 at 800 K. The volume and temperature of this system are maintained constant. Assume that the presence of a catalyst allows this reaction to occur selectively. How many moles of formaldehyde are present at equilibrium? What fraction of the methanol is converted to formaldehyde?

(c) The initial system in (b) is allowed to equilibrate at constant pressure and temperature. What fraction of the methanol is converted to formaldehyde?

(d) Initially, one mole of hydrogen and one mole of methanol occupy 1 m^3 at 800 K. The volume and temperature of this system are maintained constant. What fraction of the methanol is converted to formaldehyde at equilibrium?

7. At 298.15 K, the Gibbs free energies of formation of Br_2 and $BrCl$, in their hypothetical ideal gas standard states, are +3.1 and –1.0 kJ mol^{-1}, respectively. Why is $\Delta_f G^o(Br_2, HIG^o, 298.15\text{ K})$ not zero? A 1 m^3 vessel initially contains 0.100 mole of Br_2 and 0.200 mole of Cl_2 at 298.15 K. The volume and temperature are maintained constant. What are the partial pressures of Br_2, Cl_2, and $BrCl$ at equilibrium?

8. Devise a non-isothermal process for the reversible mixing of two gases. (Suppose that one of the substances can be condensed to a liquid, whose vapor pressure is negligible, at a temperature at which the other substance remains a gas.)

9. The following thermodynamic data are available:

$$\Delta_f G^o(SO_2,\text{g}, 298.15\text{ K}) = -300.1\text{ kJ }mol^{-1}$$
$$\Delta_f G^o(SO_3,\text{g}, 298.15\text{ K}) = -371.1\text{ kJ }mol^{-1}$$
$$\Delta_f H^o(SO_2,\text{g}, 298.15\text{ K}) = -296.8\text{ kJ }mol^{-1}$$
$$\Delta_f H^o(SO_3,\text{g}, 298.15\text{ K}) = -395.7\text{ kJ }mol^{-1}$$

(a) What are $\Delta_r G^o$ and $\Delta_r H^o$ for the reaction $2\,SO_2 + O_2 \rightleftharpoons 2\,SO_3$ at 298.15 K?

(b) Estimate the temperature, T_0, at which the standard Gibbs free energy change for this reaction is zero.

(c) A mixture is prepared to contain 1.00 mole of SO_2 and 1.00 mole of O_2. The mixture is allowed to equilibrate at T_0 and a pressure of 10.0 bar. What fraction of the SO_2 is converted to SO_3?

(d) A mixture is prepared to contain 1.00 mole of SO_2 and 10.0 mole of O_2. The mixture is allowed to equilibrate at T_0 and a pressure of 10.0 bar. What fraction of the SO_2 is converted to SO_3?

10. At 400 K, the Gibbs free energies of formation for carbon monoxide and methanol are -146.341 kJ mol^{-1} and -148.509 kJ mol^{-1}, respectively. What is $\Delta_r G^o$ for the reaction $CO + 2\,H_2 \rightleftharpoons CH_3OH$? A mixture of carbon monoxide and hydrogen in the stoichiometric proportions is prepared. Assume that the presence of a catalyst allows this reaction to occur selectively. When this mixture reaches equilibrium at 400 K and 1.00 bar, what fraction of the carbon monoxide is converted to methanol?

11. At 500 K, the Gibbs free energies of formation for carbon monoxide and methanol are -155.412 kJ mol^{-1} and -134.109 kJ mol^{-1}, respectively. What is $\Delta_r G^o$ for the reaction $CO + 2\,H_2 \rightleftharpoons CH_3OH$? Assume that the presence of a catalyst allows this reaction to occur selectively. What ratio of hydrogen to carbon monoxide must be charged to a reactor in order for 50% of the carbon monoxide to be converted to methanol when the system reaches equilibrium at a pressure of 100.0 bar?

12. The reaction of carbon monoxide with water to produce carbon dioxide and hydrogen, $CO + H_2O \rightleftharpoons CO_2 + H_2$, is known as "the water-gas shift reaction." It is used in the commercial manufacture of hydrogen. The Gibbs free energy change for this reaction becomes less favorable as the temperature increases. At 1100 K, the standard Gibbs free energy change for this reaction is $\Delta_r G^o = +0.152$ kJ mol^{-1}. Equal numbers of moles of carbon monoxide and water are charged to a reactor. What fraction of the water is converted to hydrogen when the system reaches equilibrium at 1100 K and 10.0 bar?

14

Chemical Potential: Extending the Scope of the Fundamental Equation

§1 Dependence of the internal energy on the composition of the system

We develop the thermodynamic criteria for change in a closed system without specifying the composition of the system. It is clear, therefore, that the validity of the results is not restricted in any way by the composition of the system. For the most part, our development proceeds as if the system is a single substance. However, the validity of the thermodynamic criteria for change is independent of whether the system comprises a single substance in a single phase or multiple substances in multiple phases. The criteria for reversible change are independent of whether the change involves interconversions among the substances comprising the system, so long as these interconversions occur reversibly.

While we develop these criteria without specifying the composition of the system undergoing change, we use them to make predictions about processes in which system compositions do change. In Chapter 12, we apply the Gibbs free energy criterion for reversible change to equilibria between two phases of a pure substance. This application is successful because the thermodynamic properties of one pure phase are independent of how much of any other pure phase is present.

In Chapter 13, we find a relationship between the standard Gibbs free energy change and the equilibrium constant for a reaction of ideal gases. We also use the equation for the Gibbs free energy of an ideal gas to find the value of the Gibbs free energy change for a spontaneous reaction of ideal gases at a constant temperature. This is a noteworthy result. Because we can find this value, we can predict the spontaneous transformation of one non-equilibrium state to a second non-equilibrium state. These applications are successful because ideal gas molecules do not interact with one another; the laws describing the behavior of an ideal gas do not depend on the properties of other substances that may be present.

In general, the behavior of a system depends on the molecular characteristics of all of the components and on their concentrations. For example, the pressure at which pure ice is in equilibrium with a salt solution depends not only on the temperature but also on the choice of salt and its concentration. When it is expressed as a function of reagent concentrations, the equilibrium constant for a reaction of real gases varies to some extent with the composition and with the total pressure of the system. Such observations are consistent with the idea that the thermodynamic properties of a system depend on the attractive and repulsive forces among its molecules. Such forces act between any pair of real molecules, whether they are molecules of the same substance or molecules of different substances.

Our treatment of ideal-gas reactions shows that, if we can model the effects of compositional changes on the values of a system's thermodynamic functions, we can use our thermodynamic theory to predict spontaneous changes in chemical composition. To extend to real substances the treatment that we develop for ideal gases in Chapter 13, we must find general relationships between a system's thermodynamic properties and its chemical composition. These relationships must reflect the dependence of the thermodynamic properties of one substance on the molecular characteristics of the other substances present. To find them, we must introduce additional inferences and assumptions. Since we want to describe spontaneous processes, these ideas must apply to non-equilibrium systems.

To introduce these ideas, let us consider an open system that can undergo a spontaneous change. We begin by considering its energy. We want to model any change in the energy, dE, of this system as a function of a sufficient set of independent variables. Thus far, we have developed equalities that relate thermodynamic functions only for closed systems undergoing reversible change. Consequently, when we now seek an equation for dE that is valid for an open, spontaneously changing system, our task is one of scientific inference, not mathematical deduction from the reversible-process equations. Nevertheless, the reversible-process equations are our primary resource.

For a closed, reversible system, we interpret the fundamental equation to mean that dS determines the effect of thermal processes and that dV determines the effect of pressure–volume work on the energy of the system. We infer that entropy and volume changes will play these same roles in the description of open, spontaneously changing systems. If λ different kinds of non-pressure–volume work are possible, the energy of the system is also a function of λ generalized extensive thermodynamic variables, $\theta_1, \theta_2, \ldots, \theta_k, \ldots, \theta_\lambda$. If the system contains ω substances, adding or removing some amount of any of these substances changes the energy of the system. Evidently, the energy of the system is a function of the number of moles of each substance present. Let n_1,

$n_2,\ldots,n_\omega$, represent the number of moles of each component in the system. We can recognize these dependencies formally by writing

$$E = E(S, V, \theta_1, \theta_2, \ldots, \theta_\lambda, n_1, n_2, \ldots, n_\omega)$$

This formalism applies to both open and closed systems. If the system is open, all of these variables are independent; any one of them can be changed independently of the others. If the system is closed, changes in the n_j are not independent, because the mass of the system is constant. (Letting $\overline{M}_j$ be the molar mass of the j-th substance, we have $0 = \sum_1^\omega \overline{M}_j\, dn_j$.) If the energy is a continuous and differentiable function of each of these variables, the total differential becomes

$$dE = \left(\frac{\partial E}{\partial S}\right)_{V,\theta_m,n_j} dS + \left(\frac{\partial E}{\partial V}\right)_{S,\theta_m,n_j} dV + \sum_{k=1}^{\lambda}\left(\frac{\partial E}{\partial \theta_k}\right)_{S,V,\theta_{m\neq k},n_j} d\theta_k + \sum_{j=1}^{\omega}\left(\frac{\partial E}{\partial n_j}\right)_{S,V,\theta_m,n_{p\neq j}} dn_j$$

(Inclusion of "$\theta_{m\neq k}$" and "$n_{p\neq j}$" in the subscripted variable lists indicates that the partial taken with respect to any θ_k is taken with the other θ-values held constant, and the partial taken with respect to any of the n_j is taken with the other n-values held constant.)

We hypothesize that this total differential describes energy changes in any system whose energy, entropy, volume, work variables, θ_k, and composition variables, n_j, vary in a continuous manner. In particular, we hypothesize that it describes the energy change that results from any change in composition, specified by the dn_j, whether the system is open or closed. All of these partial derivatives are intensive variables. Each of them specifies how much the energy of the system changes when its associated extensive variable undergoes an incremental change. We find it convenient to refer to the partial derivatives as ***potentials***. In this book, we restrict our attention to systems in which a given potential has the same value at every location within the system.

Since

$$\left(\frac{\partial E}{\partial n_j}\right)_{S,V,\theta_m,n_{p\neq j}}$$

is the energy change that occurs when the amount of the j-th chemical substance changes by dn_j, we call this partial derivative a ***chemical potential***. We introduce the new symbol μ_j to denote the chemical potential of the j-th substance; that is,

$$\mu_j = \left(\frac{\partial E}{\partial n_j}\right)_{S,V,\theta_m,n_{p\neq j}}$$

The value of μ_j depends exclusively on the properties of the system. When the amount of the j-th substance changes by dn_j, the contribution to the energy change is $dE = \mu_j dn_j$. When the system is open, the amount of the j-th substance can change either because of a process that occurs entirely within the system or because some of the substance moves across the system's boundary.

Reversible processes in closed systems.– In §9-14, we find, for a reversible process in a closed system, that $dE = TdS - PdV + dw_{NPV}$, where

$$dw_{NPV} = \sum_{k=1}^{\lambda} \Phi_k\, d\theta_k$$

In our hypothesized total differential, the contribution that composition changes make to dE is

$$\sum_{j=1}^{\omega}\left(\frac{\partial E}{\partial n_j}\right)_{S,V,\theta_m,n_{p\neq j}} dn_j = \sum_{j=1}^{\omega} \mu_j dn_j$$

From the empirical fact that the energy change for a reversible process in a closed system can be specified by specifying dS, dV, and the $d\theta_k$, it follows that

$$\sum_{j=1}^{\omega} \mu_j dn_j = 0$$

for any such process. Implicitly using this fact in §10-1, we find that $(\partial E/\partial S)_V = T$ and $(\partial E/\partial V)_S = -P$ for reversible processes in which all of the work is pressure–volume work. We can extend the argument to a reversible process involving any form of work in a closed system. It follows that

$$\left(\frac{\partial E}{\partial S}\right)_{V,\theta_m,n_j} = T$$

$$\left(\frac{\partial E}{\partial V}\right)_{S,\theta_m,n_j} = -P$$

$$\left(\frac{\partial E}{\partial \theta_k}\right)_{S,V,\theta_{m\neq k},n_j} = \Phi_k$$

and

$$\sum_{k=1}^{\lambda}\left(\frac{\partial E}{\partial \theta_k}\right)_{S,V,\theta_{m\neq k},n_j} d\theta_k = dw_{NPV}$$

Substituting into

$$dE = \left(\frac{\partial E}{\partial S}\right)_{V,\theta_m,n_j} dS + \left(\frac{\partial E}{\partial V}\right)_{S,\theta_m,n_j} dV + \sum_{k=1}^{\lambda}\left(\frac{\partial E}{\partial \theta_k}\right)_{S,V,\theta_{m\neq k},n_j} d\theta_k + \sum_{j=1}^{\omega}\left(\frac{\partial E}{\partial n_j}\right)_{S,V,\theta_m,n_{p\neq j}} dn_j$$

we have

$$dE = TdS - PdV + dw_{NPV} + \sum_{j=1}^{\omega} \mu_j dn_j$$

with

$$\sum_{j=1}^{\omega} \mu_j dn_j = 0$$

for any reversible process in a closed system.

Reversible processes in open systems.– Let us suppose that an open system is undergoing a reversible change and that we instantaneously stop any further exchange of matter between the system and its surroundings. Before this action, the system is open; thereafter, it is closed. The act of closure does not change the system's state functions. If, after closure, the system continues to change reversibly, $dE = TdS - PdV + dw_{NPV}$ must be valid. Closure of the system introduces a constraint that reduces the number of independent variables. However, S, V, and the θ_k remain independent after closure, and the act of closure cannot alter the laws that describe the interaction of the system with its surroundings during a reversible process. The relationships $(\partial E/\partial S)_V = T$, $(\partial E/\partial V)_S = -P$, and $\sum_{k=1}^{\lambda}(\partial E/\partial \theta_k)_{S,V,\theta_{m\neq k},n_j} d\theta_k = dw_{NPV}$ are valid for the closed system. We infer that they are also valid for the open system, so that the total differential for a system undergoing reversible change is

$$dE = TdS - PdV + dw_{NPV} + \sum_{j=1}^{\omega} \mu_j dn_j$$

whether the system is open or closed. This comparison implies that

$$\sum_{j=1}^{\omega} \mu_j dn_j = 0$$

for any reversible process, whether the system is open or closed. (As noted below, however, our model implicitly assumes that $\mu_j = \hat{\mu}_j$; that is, μ_j has the same value whether the change dn_j is the result of internal transformation, movement of the j-th substance across the system boundary, or a combination of the two.)

Spontaneous processes.– We hypothesize that the same mathematical model describes energy changes for both spontaneous and reversible processes in both open and closed systems, so long as the process occurs in such a way that all of the system's state functions behave as continuous variables. For reversible systems, we conclude that this model is

$$dE = TdS - PdV + dw_{NPV} + \sum_{j=1}^{\omega} \mu_j dn_j$$

By hypothesis, the model also describes spontaneous processes, so long as the state functions behave as continuous variables. We base the development of our thermodynamic theory for spontaneous reactions on this hypothesis. It is, of course, the utility of the resulting theory, not the logical force of the reasoning by which we reach it, that justifies our acceptance of the hypothesis.

In §9-19, we find that $(dE)_{SV} < dw_{NPV}$ for a spontaneous process, and we reach this conclusion by an argument that is independent of the composition of the system. For a spontaneous process at constant entropy and volume, our model becomes

$$(dE)_{SV} = dw_{NPV} + \sum_{j=1}^{\omega} \mu_j dn_j$$

and since $(dE)_{SV} < dw_{NPV}$, it follows that

$$\sum_{j=1}^{\omega} \mu_j dn_j < 0$$

for any spontaneous process.

All of the variables that appear in our model for dE are properties of the system. The surroundings can affect the behavior of the system by imposing particular values on one or more of the system's thermodynamic functions. (For example, immersing a reacting system in a constant-temperature bath ensures that the system reaches equilibrium only when $T = \hat{T}$.) When the surroundings impose particular values on the system's thermodynamic functions, the system's state functions must approach the imposed values as the system approaches equilibrium.

Internal entropy and chemical potential.– Substituting the internal and external entropies that we introduce in §9-16, our model for the incremental energy change during a spontaneous composition change becomes

$$dE = Td_iS + dq + dw_{PV} + dw_{NPV} + \sum_{j=1}^{\omega} \mu_j dn_j$$

Our first-law equation is $dE = dq + dw_{PV} + dw_{NPV}$. It follows that

$$Td_iS + \sum_{j=1}^{\omega} \mu_j dn_j = 0$$

and we find

$$d_iS = -\frac{1}{T}\sum_{j=1}^{\omega} \mu_j dn_j$$

The change criteria

$$d_iS \geq 0$$

and

$$\sum_{j=1}^{\omega} \mu_j dn_j \leq 0$$

are equivalent.

§2 Dependence of other thermodynamic functions on the composition of the system

Using the appropriate sets of independent variables, we can obtain similar expressions for dH, dA, dG, and

dS under the conditions that we assume in §1. Since entropy and pressure are the natural variables for enthalpy in a closed system, we infer that a change in the enthalpy of any system can be expressed as a function of dS, dP, changes in the non-pressure–volume work variables, $d\theta_k$, and changes in the composition, dn_j. That is, from $H = H(S, P, \theta_1, \theta_2, \dots, \theta_\lambda, n_1, n_2, \dots, n_\omega)$, we have

$$dH = \left(\frac{\partial H}{\partial S}\right)_{P,\theta_m,n_j} dS + \left(\frac{\partial H}{\partial P}\right)_{S,\theta_m,n_j} dP + \sum_{k=1}^{\lambda}\left(\frac{\partial H}{\partial \theta_k}\right)_{S,P,\theta_{m\neq k},n_j} d\theta_k + \sum_{j=1}^{\omega}\left(\frac{\partial H}{\partial n_j}\right)_{S,P,\theta_m,n_{p\neq j}} dn_j$$

From $H = E + PV$, we have $dH = dE + PdV + VdP$ for any system. Substituting our result from §1

$$dE = TdS - PdV + dw_{NPV} + \sum_{j=1}^{\omega} \mu_j dn_j$$

where

$$\mu_j = \left(\frac{\partial E}{\partial n_j}\right)_{S,V,\theta_m,n_{p\neq j}}$$

we have

$$dH = TdS + VdP + dw_{NPV} + \sum_{j=1}^{\omega} \mu_j dn_j$$

Since we assume that both of these equations for dH describe open systems, all of the variables are independent, and the corresponding coefficients must be equal. For systems that satisfy our assumptions, we have

$$T = \left(\frac{\partial H}{\partial S}\right)_{P,\theta_m,n_j}$$

$$V = \left(\frac{\partial H}{\partial P}\right)_{S,\theta_m,n_j}$$

$$dw_{NPV} = \sum_{k=1}^{\lambda}\left(\frac{\partial H}{\partial \theta_k}\right)_{S,V,\theta_{m\neq k},n_j} d\theta_k$$

and

$$\mu_j = \left(\frac{\partial H}{\partial n_j}\right)_{S,P,\theta_m,n_{p\neq j}} = \left(\frac{\partial E}{\partial n_j}\right)_{S,V,\theta_m,n_{p\neq j}}$$

By parallel arguments, we find

$$dA = -SdT - PdV + dw_{NPV} + \sum_{j=1}^{\omega} \mu_j dn_j$$

with

$$\mu_j = \left(\frac{\partial A}{\partial n_j}\right)_{T,V,\theta_m,n_{p\neq j}}$$

and

$$dG = -SdT + VdP + dw_{NPV} + \sum_{j=1}^{\omega} \mu_j dn_j$$

with

$$\mu_j = \left(\frac{\partial G}{\partial n_j}\right)_{T,P,\theta_m,n_{p\neq j}}$$

Rearranging the result for dE, we find

$$dS = \frac{1}{T}dE + \frac{P}{T}dV - \frac{dw_{NPV}}{T} - \frac{1}{T}\sum_{j=1}^{\omega} \mu_j dn_j$$

so that

$$\mu_j = -T\left(\frac{\partial S}{\partial n_j}\right)_{E,V,\theta_m,n_{p\neq j}}$$

and from the result for dH, we have

$$dS = \frac{1}{T}dH - \frac{V}{T}dP - \frac{dw_{NPV}}{T} - \frac{1}{T}\sum_{j=1}^{\omega} \mu_j dn_j$$

so that

$$\mu_j = -T\left(\frac{\partial S}{\partial n_j}\right)_{H,P,\theta_m,n_{p\neq j}}$$

Evidently, for any system whose condition corresponds to our assumptions, all of the relationships that we develop in §10-1 remain valid. The non-pressure–volume work can be expressed in terms of partial derivatives in several equivalent ways:

$$\begin{aligned} dw_{NPV} &= \sum_{k=1}^{\lambda}\left(\frac{\partial E}{\partial \theta_k}\right)_{S,V,\theta_{m\neq k},n_j} d\theta_k \\ &= \sum_{k=1}^{\lambda}\left(\frac{\partial H}{\partial \theta_k}\right)_{S,P,\theta_{m\neq k},n_j} d\theta_k \\ &= \sum_{k=1}^{\lambda}\left(\frac{\partial A}{\partial \theta_k}\right)_{V,T,\theta_{m\neq k},n_j} d\theta_k \\ &= \sum_{k=1}^{\lambda}\left(\frac{\partial G}{\partial \theta_k}\right)_{P,T,\theta_{m\neq k},n_j} d\theta_k \\ &= -T\sum_{k=1}^{\lambda}\left(\frac{\partial S}{\partial \theta_k}\right)_{E,V,\theta_{m\neq k},n_j} d\theta_k \\ &= -T\sum_{k=1}^{\lambda}\left(\frac{\partial S}{\partial \theta_k}\right)_{H,P,\theta_{m\neq k},n_j} d\theta_k \end{aligned}$$

Most importantly, the chemical potential and the criteria for change can be expressed in several alternative ways. We have

$$\mu_j = \left(\frac{\partial E}{\partial n_j}\right)_{S,V,\theta_m,n_{p\neq j}} = \left(\frac{\partial H}{\partial n_j}\right)_{S,P,\theta_m,n_{p\neq j}}$$
$$= \left(\frac{\partial A}{\partial n_j}\right)_{V,T,\theta_m,n_{p\neq j}} = \left(\frac{\partial G}{\partial n_j}\right)_{P,T,\theta_m,n_{p\neq j}}$$
$$= -T\left(\frac{\partial S}{\partial n_j}\right)_{E,V,\theta_m,n_{p\neq j}} = -T\left(\frac{\partial S}{\partial n_j}\right)_{H,P,\theta_m,n_{p\neq j}}$$

For any system, open or closed, that satisfies the assumptions we introduce in §1, the criteria for change are

$$\sum_{j=1}^{\omega} \mu_j dn_j \leq 0$$

where the chemical potential is equivalently expressed as any of the partial derivatives above.

Let us review the scope and significance of these results. We develop the fundamental equation, $dE = TdS - PdV + dw_{NPV}$, by reasoning about the behavior of closed, reversible systems. For any process in a closed system, we develop the criteria for change

$$(\Delta E)_{SV} \leq w_{NPV}$$
$$(\Delta H)_{SP} \leq w_{NPV}$$
$$(\Delta A)_{VT} \leq w_{NPV}$$
$$(\Delta G)_{PT} \leq w_{NPV}$$
$$(\Delta S)_{EV} \geq -w_{NPV}/\hat{T}$$
$$(\Delta S)_{HP} \geq -w_{NPV}/\hat{T}$$

and the corresponding relationships among differentials for incremental changes. Now we are extending these conclusions to produce equations for the changes in the various thermodynamic functions when a system undergoes a spontaneous composition change. To do so, we introduce the idea that composition variables must be included in a complete model for a thermodynamic function. When we write

$$E = E(S, V, \theta_1, \theta_2, \dots, \theta_\lambda, n_1, n_2, \dots, n_\omega)$$

we assert that every set of values, $\{S, V, \theta_1, \theta_2, \dots, \theta_\lambda, n_1, n_2, \dots, n_\omega\}$, corresponds to a state of the system in which the system has a definite energy. (Of course, we are not asserting that every such set of values can actually be attained by the real system. Many such sets correspond to hypothetical states. The particular sets of values that do correspond to realizable states of the system lie on a manifold. Given one such set of values—one point on the manifold—our differential expressions specify all of the other states that lie on the same manifold.)

When we assume that the various partial derivatives, $(\partial E/\partial S)_{V,\theta_k,n_j}$, $(\partial E/\partial V)_{S,\theta_k,n_j}$, $(\partial E/\partial \theta_k)_{S,V\theta_{m\neq k},n_j}$, and $\left(\partial E/\partial n_j\right)_{S,V,\theta_k,n_{p\neq j}}$, exist, we are assuming that E is a smooth, continuous function of each of these variables. Since some of these partial derivatives are synonymous with intensive variables, we are assuming that these intensive variables are well defined. Moreover, since these intensive variables characterize the system as a whole, we are assuming that each intensive variable has the same value in every part of the system[1]. When we assume that $\mu_j = \left(\partial E/\partial n_j\right)_{S,V,\theta_m,n_{p\neq j}}$ has a well-defined value, we are assuming that E is a continuous function of n_j; we are assuming that μ_j exists for any arbitrary state of the system and not just for states undergoing reversible change.

When we assume that an arbitrary change is described by the total differential $dE = TdS - PdV + dw_{NPV} + \sum_{j=1}^{\omega} \mu_j dn_j$, we are going beyond our conclusion that this total differential describes paths of reversible change. We are asserting that it describes any process in which a change in composition is the sole source of irreversibility. As a practical matter, we expect it to describe any process that occurs in a system whose potential functions are well defined. By well defined, we mean, of course, that that they can be measured and that the measurements are reproducible and consistent. We expect these conclusions to apply to multiple-phase, open systems, so long as each potential has the same value in every phase. In thus assuming that we can expand the scope of the fundamental equation, we are not modifying the change criteria that we develop in Chapter 9. Our conclusions that $(dE)_{SV} = dw_{NPV}$ for a reversible process and $(dE)_{SV} < dw_{NPV}$ for a spontaneous process in a closed system are not affected.

Again, while our arguments for them are compelling, these results are not rigorously logical consequences of our earlier conclusions about reversible processes. As for any scientific principle, their validity depends on their predictive capability, not their provenance.

§3 Partial molar quantities

Because they are easy to control in typical laboratory experiments, pressure, temperature, and the number of moles of each component are the independent variables that we find useful most often. Partial derivatives of thermodynamic quantities, taken with respect to the number of moles of a component, ***at constant pressure, temperature, and θ_k***, are given a special designation; they are called ***partial molar quantities***. That is,

$$\left(\frac{\partial E}{\partial n_A}\right)_{P,T,\theta_m,n_{p\neq A}}$$

is the partial molar energy of component A,

$$\left(\frac{\partial G}{\partial n_A}\right)_{P,T,\theta_m,n_{p\neq A}}$$

is the partial molar Gibbs free energy, etc. All partial molar quantities are intensive variables.

Because partial molar quantities are particularly useful, it is helpful to have a distinctive symbol to represent them. We use a horizontal bar over a thermodynamic variable to represent a partial molar quantity. (We have been using the horizontal over-bar to mean simply per mole. When we use it to designate a partial molar quantity, it means per mole of a specific component.) Thus, we write

$$\left(\frac{\partial E}{\partial n_A}\right)_{P,T,\theta_m,n_{p\neq A}} = \overline{E}_A$$

$$\left(\frac{\partial V}{\partial n_A}\right)_{P,T,\theta_m,n_{p\neq A}} = \overline{V}_A$$

$$\left(\frac{\partial G}{\partial n_A}\right)_{P,T,\theta_m,n_{p\neq A}} = \overline{G}_A$$

etc.

In §1 and §2, we introduce the chemical potential for substance A, μ_A, and find that the chemical potential of substance A is equivalently expressed by several partial derivatives. In particular, we have

$$\mu_A = \left(\frac{\partial G}{\partial n_A}\right)_{P,T,\theta_m,n_{p\neq A}} = \overline{G}_A$$

that is, ***the chemical potential is also the partial molar Gibbs free energy***.

It is important to recognize that the other partial derivatives that we can use to calculate the chemical potential are not partial molar quantities. Thus,

$$\mu_A = \left(\frac{\partial E}{\partial n_A}\right)_{S,V,\theta_m,n_{p\neq A}} \neq \left(\frac{\partial E}{\partial n_A}\right)_{P,T,\theta_m,n_{p\neq A}}$$

That is, $\mu_A \neq \overline{E}_A$. Similarly, $\mu_A \neq \overline{H}_A$, $\mu_A \neq \overline{A}_A$, and $\mu_A \neq -T\overline{S}_A$.

We can think of a thermodynamic variable as a manifold—a "surface" in a multidimensional space. If there are two independent variables, the dependent thermodynamic variable is a surface in a three-dimensional space. Then we can visualize the partial derivative of the dependent thermodynamic variable with respect to an independent variable as the slope of a line tangent to the surface. This tangent lies in a plane in which the other independent variable is constant. If the independent variables are pressure, temperature, θ_k-values, and compositions, the slope of the tangent line at $(P, T, \theta_1, \theta_2, \dots, \theta_\lambda, n_1, n_2, \dots, n_\omega)$ is the value of a partial molar quantity at that point.

A more concrete way to think of a partial molar quantity for component A is to view it as the change in that quantity when we add one mole of A to a very large system having the specified pressure, temperature, θ_k-values, and composition. When we add one mole of A to this system, the relative change in any of the system's properties is very small; for example, the ratio of the final volume to the initial volume is essentially unity. Nevertheless, the volume of the system changes by a finite amount. This amount approximates the partial molar volume of substance A. This approximation becomes better as the size of the system becomes larger. We expect the change in the volume of the system to be approximately equal to the volume of one mole of pure A, but we know that in general it will be somewhat different because of the effects of attractive and repulsive forces between the additional A molecules and the molecules comprising the original system.

Partial molar quantities can be expressed as functions of other thermodynamic variables. Because pressure and temperature are conveniently controlled variables, functions involving partial molar quantities are particularly useful for describing chemical change in systems that conform to the assumptions that we introduce in §1. Because the chemical potential is the same thing as the partial molar Gibbs free energy, it plays a prominent role in these equations.

To use these equations to describe a real system, we must develop empirical models that relate the partial molar quantities to the composition of the system. In general, these empirical models are non-linear functions of the system composition. However, simple approximations are sometimes adequate. The simplest approximation is a case we have already considered. If we can ignore the attractive and repulsive interactions among the molecules comprising the system, the effect of increasing n_A by a small amount, dn_A, is simply the effect of adding dn_A moles of pure component A to the system. If we let $\overline{E}_A^\bullet$ be the energy per mole of pure component A, the contribution to the energy of the system, at constant temperature and pressure, is

$$\left(\frac{\partial E}{\partial n_A}\right)_{P,T,\theta_m,n_{p\neq A}} dn_A = \overline{E}_A^\bullet\, dn_A$$

In Chapter 12, we apply the thermodynamic criteria for change to the equilibria between phases of a pure substance. To do so, we use the Gibbs free energies of the pure phases. In Chapter 13, we apply these criteria to chemical reactions of ideal gases, using the Gibbs free energies of the pure gases. In these cases, the properties of a phase of a pure substance are independent of the amounts of any other substances that are present. That is, we use the approximation

$$\left(\frac{\partial G}{\partial n_A}\right)_{P,T,\theta_m,n_{p\neq A}} dn_A = \overline{G}_A^\bullet\, dn_A$$

albeit without using the over-bar or the bullet superscript to indicate that we are using the partial molar Gibbs free energy of the pure substance. In §1, we develop the principle that $\sum_{j=1}^{\omega} \mu_j dn_j \leq 0$ are general criteria for change that are applicable not only to closed systems but also to open systems composed of homogeneous phases.

Thus far in this chapter, we have written each partial derivative with a complete list of the variables that are held constant. This is typographically awkward. Clarity seldom requires that we include the work-related variables, θ_k, and composition variables, n_j, in this list. From here on, we usually omit them.

§4 Chemical potentials and stoichiometry

Let us now apply our chemical-potential equilibrium criterion, $\sum_{j=1}^{\omega} \mu_j dn_j = 0$, to a simple, closed system in which phases α and β of a single component are at equilibrium. The total number of moles, n, is fixed; we have $n = n_\alpha + n_\beta$. Reversible change is possible while the system remains at equilibrium. However, there is a stoichiometric constraint; growth in one phase is exactly matched by shrinkage in the other; $dn = dn_\alpha + dn_\beta = 0$, so that $dn_\alpha = -dn_\beta$. The equilibrium criterion becomes $(\mu_\alpha - \mu_\beta)dn_\alpha = 0$. If reversible change occurs, we have $dn_\alpha \neq 0$, and it follows that $\mu_\alpha = \mu_\beta$.

We infer that this result must always be valid. That is, if a substance is present in two phases of an equilibrium system, the chemical potential of the substance is the same in each phase; no matter what other processes involving the substance may occur. This is so because all processes that occur in the system must be at equilibrium in order for the system as a whole to be at equilibrium. In our discussion of chemical kinetics, we saw this conclusion expressed as the principle of microscopic reversibility. In our development of Gibbs phase rule, we asserted that a distribution-equilibrium constant relates the concentrations of a species in any two phases. Our present conclusion that the chemical potential of a species must be the same in each phase is a more rigorous statement of the same principle.

For a process in which dn moles of a substance transfer spontaneously from phase β to phase α, we have $dn = dn_\alpha = -dn_\beta$. The criterion for spontaneous change, $\sum_{j=1}^{\omega} \mu_j dn_j < 0$, becomes

$$\mu_\alpha dn_\alpha + \mu_\beta dn_\beta = \mu_\alpha dn_\alpha - \mu_\beta dn_\alpha < 0$$

and it follows that $\mu_\alpha < \mu_\beta$. We infer that this result, too, must always be valid. That is, if a substance spontaneously transfers between any two phases, the chemical potential of the transferring substance must be greater in the donor phase than it is in the acceptor phase.

While $\sum_{j=1}^{\omega} \mu_j dn_j \leq 0$ summarizes our criteria for change, it does not specify the process that occurs in the system. In order to describe a particular chemical reaction, we must incorporate the constraints that the reaction stoichiometry imposes on the dn_j. For a description of the equilibrium system that does not require the size of the system to be specified, the position of equilibrium must be given as a function of intensive variables. While the μ_j are intensive variables, the dn_j are not. When we include information about the stoichiometry of the equilibrium process, we obtain a criterion for equilibrium that is specific to that process.

Consider the reaction $a\,A + b\,B \rightleftharpoons c\,C + d\,D$ in a closed system. Let $d\xi$ be the incremental extent of reaction. Using the modified stoichiometric coefficients, ν_j, that we introduce in §13-5, the incremental extent of reaction becomes

$$d\xi = \frac{dn_A}{\nu_A} = \frac{dn_B}{\nu_B} = \frac{dn_C}{\nu_C} = \frac{dn_D}{\nu_D}$$

and we can use the stoichiometric coefficients to express the incremental composition changes, dn_j, as $dn_j = \nu_j d\xi$. The criteria for change become

$$\sum_{j=1}^{\omega} \mu_j \nu_j d\xi \leq 0$$

For the reaction to occur left to right, we must have $d\xi > 0$, so that

$$\sum_{j=1}^{\omega} \mu_j \nu_j \leq 0$$

In this form, the criteria involve only intensive variables, and the stoichiometry of the equilibrium process is uniquely specified. In most circumstances, there are additional relationships among the μ_j for which we may have little or no specific information.

To see the nature of these additional relationships, let us consider a chemical reaction involving three species, $A + B \rightleftharpoons C$, at constant pressure and temperature. No other substances are present. There are two components; so there are three degrees of freedom. Pressure and temperature account for two of the degrees of freedom. The chemical potential of any one of the equilibrating species, say A, accounts for the third. Evidently, fixing pressure, temperature, and μ_A is sufficient to fix the values of the remaining chemical potentials, μ_B and μ_C. μ_A must contain information about the other chemical potentials, μ_B and μ_C.

If the same reaction occurs in the presence of a solvent, there are three components and four degrees of freedom. In this case, two of the chemical potentials can vary independently. Fixing pressure, temperature, and, say μ_A and μ_B, fixes μ_C and the chemical potential of the solvent. We can expect the chemical potential of the solvent to be nearly constant, and we often omit it when we write dG for this system. This omission notwithstanding, the chemical potentials of the equilibrating species include information about the chemical potential of the solvent. In general, the chemical potential of any component of an equilibrium system is a function of the chemical potentials of all of the other components.

§5 $\sum \mu_j dn_j = 0$ and primitive versus Gibbsian equilibrium states

We conclude that $\sum_{j=1}^{\omega} \mu_j dn_j = 0$ is a criterion for reversible change in any system. When the change involves equilibria among two or more phases or substances, it alters the number of moles of the components present. An extent of reaction, $\xi = (n_j - n_j^o)/\nu_j$, characterizes the displacement of every such equilibrium. The magnitude of each incremental equilibrium displacement is specified by composition changes, $dn_j = \nu_j d\xi$, and conversely. The criterion for reversible change becomes $\sum_{j=1}^{\omega} \mu_j \nu_j \, d\xi = 0$. When this criterion is satisfied because $\sum_{j=1}^{\omega} \mu_j \nu_j = 0$, $d\xi$ is arbitrary, and the system can reversibly traverse a range of equilibrium states. In other words, $\sum_{j=1}^{\omega} \mu_j \nu_j = 0$ defines a Gibbsian equilibrium manifold.

We can also have a reversible process for which $d\xi = 0$. If the process is reversible, the state of the system corresponds to a point on the Gibbsian manifold, but $d\xi = 0$ stipulates that the system cannot change: it must remain at the specified point on the manifold. This corresponds to what we are calling a primitive equilibrium state. The system is constrained to remain in this state by the nature of its interactions with its surroundings: the system may be isolated, or the surroundings may act to maintain the system in a fixed state.

§6 The change criteria in a system composed of subsystems

Let us now consider a closed, constant temperature, constant pressure system that is composed of open subsystems. Chemical substances can pass from one subsystem to another, but they cannot enter or leave the system. We assume that our model for dG applies in every subsystem. Each subsystem is at the same temperature and pressure. For the r-th subsystem,

$$G_r = G_r(P, T, \theta_{r,1}, \theta_{r,2}, \dots, \theta_{r,\lambda}, n_{r,1}, n_{r,2}, \dots, n_{r,\omega})$$

For a physical system in which all of these assumptions correspond closely to physical reality, we have, for the r-th subsystem,

$$dG_r = -S_r dT + V_r dP + (dw_{NPV})_r + \sum_{j=1}^{\omega} \mu_j dn_{r,j}$$

For the closed system, we have

$$\sum_r dG_r = -dT \sum_r S_r + dP \sum_r V_r + \sum_r (dw_{NPV})_r + \sum_r \sum_{j=1}^{\omega} \mu_j dn_{r,j}$$

Since Gibbs free energy, entropy, volume, and work are extensive variables, we have, for the closed system, $dG = \sum_r dG_r$, $dS = \sum_r S_r$, $V = \sum_r V_r$, and $dw_{NPV} = \sum_r (dw_{NPV})_r$. Therefore,

$$dG = -SdT + VdP + dw_{NPV} + \sum_r \sum_{j=1}^{\omega} \mu_j dn_{r,j}$$

For any process that occurs in this closed system at constant pressure and temperature, we have $(dG)_{PT} \le dw_{NPV}$, and

$$\sum_r \sum_{j=1}^{\omega} \mu_j dn_{r,j} \le 0$$

expresses the criteria for change in the closed system as a sum of conditions on the open subsystems.

Now let us consider the possibility that, for the ρ-th open subsystem, we have

$$\sum_{j=1}^{\omega} \mu_j dn_{\rho,j} > 0$$

If this were true, the sum over all of the subsystems could still be less than or equal to zero. In this case, the energy increase occurring in the ρ-th subsystem would have to be offset by energy decreases occurring in the other subsystems. This is at odds with the way that physical systems are observed to behave. To see this, let us suppose that the process is a chemical reaction. Then the composition changes are related to the extent of reaction as $dn_{\rho,j} = \nu_j d\xi_\rho$. For the open subsystem, we have

$$\sum_{j=1}^{\omega} \mu_j \nu_j d\xi_\rho > 0$$

Now, we can alter the boundary of this subsystem to make it impermeable to matter, while keeping its state functions unchanged. This change converts the open subsystem to a closed system, for which we know that

$$\sum_{j=1}^{\omega} \mu_j \nu_j d\xi_\rho < 0$$

If the criterion for spontaneous change switches from $\sum_{j=1}^{\omega} \mu_j \nu_j d\xi_\rho > 0$ to $\sum_{j=1}^{\omega} \mu_j \nu_j d\xi_\rho < 0$ the sign of $d\xi_\rho$ must change. The supposition that $\sum_{j=1}^{\omega} \mu_j \nu_j d\xi_\rho > 0$ is possible in an open subsystem implies that the direction of a spontaneous change in a closed system can be opposite the direction of a spontaneous change in an otherwise identical open system. No such thing is ever observed. We conclude that the criterion for spontaneous change,

$$\sum_{j=1}^{\omega} \mu_j dn_j < 0$$

must be satisfied in every part of any system in which the various potentials are the same throughout. Since

$$d_i S = -\frac{1}{T} \sum_{j=1}^{\omega} \mu_j dn_j$$

it follows that $d_i S > 0$ must also be satisfied in every part of the system.

§7 At constant T and P, $\Delta_r\mu$ is the change in Gibbs free energy

At constant temperature and pressure, the chemical potential of component A is the contribution that one mole of A makes to the Gibbs free energy of the system. The Gibbs free energy of the system is the sum of the contributions made by all of its components. The chemical potential change that occurs during a reaction,

$$\Delta_r\mu = \sum_{j=1}^{\omega} \mu_j \nu_j$$

is the same thing as the Gibbs free energy change. To establish these points, we introduce Euler's theorem on homogenous functions. For simplicity, we consider systems in which only pressure–volume work occurs.

A function $f(x, y, z)$ is said to be homogeneous of order n if

$$\lambda^n f(x, y, z) = f(\lambda x, \lambda y, \lambda z)$$

Then,

$$\begin{aligned}\frac{d}{d\lambda}[\lambda^n f(x,y,z)] &= n\lambda^{n-1} f(x,y,z) \\ &= \left(\frac{\partial f}{\partial(\lambda x)}\right)_{yz}\frac{d(\lambda x)}{d\lambda} + \left(\frac{\partial f}{\partial(\lambda y)}\right)_{xz}\frac{d(\lambda y)}{d\lambda} \\ &\quad + \left(\frac{\partial f}{\partial(\lambda z)}\right)_{xy}\frac{d(\lambda z)}{d\lambda} \\ &= x\left(\frac{\partial f}{\partial(\lambda x)}\right)_{yz} + y\left(\frac{\partial f}{\partial(\lambda y)}\right)_{xz} + z\left(\frac{\partial f}{\partial(\lambda z)}\right)_{xy}\end{aligned}$$

Since this must be true for any λ, it must be true for $\lambda = 1$. Making this substitution, we have Euler's theorem for order n:

$$nf(x,y,z) = x\left(\frac{\partial f}{\partial x}\right)_{yz} + y\left(\frac{\partial f}{\partial y}\right)_{xz} + z\left(\frac{\partial f}{\partial z}\right)_{xy}$$

An extensive state function is homogeneous of order one in its extensive variables. In particular, we have

$$\lambda G(P,T,n_A,n_B,n_C,n_D) = G(P,T,\lambda n_A,\lambda n_B,\lambda n_C,\lambda n_D)$$

for any λ. (For, say, $\lambda = 1/2$, this says only that, if we divide a homogeneous equilibrium system into two equal portions, all else remaining constant, the Gibbs free energy of each portion will be half of that of the original system. The pressure and temperature are independent of λ.) Taking the derivative with respect to λ, we find

$$\begin{aligned}\frac{d}{d\lambda}[\lambda G(P,T,n_A,n_B,n_C,n_D)] &= G(P,T,n_A,n_B,n_C,n_D) \\ &= n_A\left(\frac{\partial G}{\partial(\lambda n_A)}\right)_{PT} + n_B\left(\frac{\partial G}{\partial(\lambda n_B)}\right)_{PT} \\ &\quad + n_C\left(\frac{\partial G}{\partial(\lambda n_C)}\right)_{PT} + n_D\left(\frac{\partial G}{\partial(\lambda n_D)}\right)_{PT}\end{aligned}$$

Since this must be true for any λ, it must be true for $\lambda = 1$. We find

$$G(P,T,n_A,n_B,n_C,n_D) = \mu_A n_A + \mu_B n_B + \mu_C n_C + \mu_D n_D$$

The same equation follows from applying Euler's theorem to other state functions. For example, viewing the internal energy as a function of entropy, volume, and composition, we have

$$\begin{aligned}\lambda E(S,V,n_A,n_B,n_C,n_D) \\ = E(\lambda S,\lambda V,\lambda n_A,\lambda n_B,\lambda n_C,\lambda n_D)\end{aligned}$$

so that

$$\begin{aligned}\frac{d}{d\lambda}[\lambda E(S,V,n_A,n_B,n_C,n_D)] &= E(S,V,n_A,n_B,n_C,n_D) \\ &= S\left(\frac{\partial E}{\partial(\lambda S)}\right)_V + V\left(\frac{\partial E}{\partial(\lambda V)}\right)_S \\ &\quad + n_A\left(\frac{\partial E}{\partial(\lambda n_A)}\right)_{SV} + n_B\left(\frac{\partial E}{\partial(\lambda n_B)}\right)_{SV} \\ &\quad + n_C\left(\frac{\partial E}{\partial(\lambda n_C)}\right)_{SV} + n_D\left(\frac{\partial E}{\partial(\lambda n_D)}\right)_{SV}\end{aligned}$$

Setting $\lambda = 1$, we have

$$\begin{aligned}E(S,V,n_A,n_B,n_C,n_D) \\ = TS - PV + \mu_A n_A + \mu_B n_B + \mu_C n_C + \mu_D n_D\end{aligned}$$

which we can rearrange to

$$G = E - TS + PV = \mu_A n_A + \mu_B n_B + \mu_C n_C + \mu_D n_D$$

This equation describes any system whose thermodynamic functions are continuous functions of one another. Evidently, we can model the thermodynamic functions of any such system by modeling the chemical potentials of its components. In the remainder of this chapter, we develop this idea.

By definition, $\Delta_r G$ is the Gibbs free energy change for converting a moles of A and b moles of B to c moles of C and d moles of D at constant pressure and temperature and while the composition remains constant at n_A, n_B, n_C, and n_D. That is,

$$\begin{aligned}\Delta_r G = G(n_A,n_B,n_C+c,n_D+d) \\ - G(n_A+a,n_B+b,n_C,n_D)\end{aligned}$$

where $n_A \gg a$, $n_B \gg b$, $n_C \gg c$, and $n_D \gg d$. Using the relationship we have just found between G and the n_i, we have

$$\begin{aligned}\Delta_r G &= [\mu_A n_A + \mu_B n_B + \mu_C(n_C + c) + \mu_D(n_D + d)] \\ &\quad -[\mu_A(n_A + a) + \mu_B(n_B + b) + \mu_C n_C + \mu_D n_D] \\ &= c\mu_C + d\mu_D - a\mu_A - d\mu_d \\ &= \Delta_r\mu\end{aligned}$$

Since any other extensive thermodynamic function is also homogeneous of order one in the composition variables, a similar relationship will exist between the change in the function itself and its partial molar derivatives. For example, expressing the system volume as a function of pressure, temperature, and composition, we have

$$\lambda V(P,T,n_A,n_B,n_C,n_D) = V(P,T,\lambda n_A,\lambda n_B,\lambda n_C,\lambda n_D)$$

and a completely parallel derivation shows that the volume of the system is related to the partial molar volumes of the components by

$$V(P,T,n_A,n_B,n_C,n_D) = \overline{V}_A n_A + \overline{V}_B n_B + \overline{V}_C n_C + \overline{V}_D n_D$$

The volume change for the reaction is given by

$$\Delta_r V = c\overline{V}_C + d\overline{V}_D - a\overline{V}_A - b\overline{V}_B$$

§8 Gibbs-Duhem equations

An important relationship among the differentials of the chemical potentials for a system follows from the relationships we have just developed. From the fact that the Gibbs free energy, $G(P,T,n_A,n_B,n_C,n_D)$, is homogeneous of order one in the composition variables, we find that the Gibbs free energy of the system is related to its partial molar derivatives by

$$G(P,T,n_A,n_B,n_C,n_D) = \mu_A n_A + \mu_B n_B + \mu_C n_C + \mu_D n_D$$

The differential of the left hand side is

$$dG = -SdT + VdP + \mu_A dn_A + \mu_B dn_B + \mu_C dn_C + \mu_D dn_D$$

and the differential of the right hand side is

$$dG = \mu_A dn_A + n_A d\mu_A + \mu_B dn_B + n_B d\mu_B + \mu_C dn_C + n_C d\mu_C + \mu_D dn_D + n_D d\mu_D$$

Since these differential expressions must be equal, we have

$$-SdT + VdP = n_A d\mu_A + n_B d\mu_B + n_C d\mu_C + n_D d\mu_D$$

for any change in this system.

While we have considered the particular case of a system containing the species A, B, C, and D, it is clear that the same arguments apply to any system. For a system that contains ω species, we can write the result in general form as

$$-SdT + VdP = \sum_{j=1}^{\omega} n_j d\mu_j$$

This relationship is known as the chemical-potential ***Gibbs-Duhem equation***. It is a constraint on the $d\mu_j$ that must be satisfied when any change occurs in a system whose thermodynamic functions are continuous functions of its composition variables. If pressure and temperature are constant and this equation is satisfied, the system is at equilibrium—it is on a Gibbsian equilibrium manifold—and the chemical-potential Gibbs-Duhem equation becomes

$$0 = \sum_{j=1}^{\omega} n_j d\mu_j$$

In the next two sections, we develop a particularly useful expression for $d\mu_j$. We can obtain similar relationships for other partial molar quantities. (See problems 2 and 3.) These relationships are also called Gibbs-Duhem equations. Because the derivation requires only that the thermodynamic function be homogeneous of order one, the same relationships exist among the differentials of the partial molar derivatives of any extensive thermodynamic function. For partial molar volumes at constant pressure and temperature, we find

$$0 = \sum_{j=1}^{\omega} n_j d\overline{V}_j$$

§9 The dependence of chemical potential on other variables

The chemical potential of a substance in a particular system is a function of all of the variables that affect the Gibbs free energy of the system. For component A, we can express this by writing

$$\mu_A = \mu_A(P,T,n_1,n_2,\dots,n_A,\dots,n_\omega)$$

for which the total differential is

$$d\mu_A = \left(\frac{\partial \mu_A}{\partial T}\right)_P dT + \left(\frac{\partial \mu_A}{\partial P}\right)_T dP + \sum_{j=1}^{\omega} \left(\frac{\partial \mu_A}{\partial n_j}\right)_{PT} dn_j$$

Recalling the definition of the chemical potential and the fact that the mixed second-partial derivatives of a state function are equal, we have

$$\begin{aligned}\left(\frac{\partial \mu_A}{\partial T}\right)_P &= \left(\frac{\partial}{\partial T}\right)_P \left(\frac{\partial G}{\partial n_A}\right)_{TP} \\ &= \left(\frac{\partial}{\partial n_A}\right)_{TP} \left(\frac{\partial G}{\partial T}\right)_P \\ &= -\left(\frac{\partial S}{\partial n_A}\right)_{TP} \\ &= -\overline{S}_A\end{aligned}$$

Similarly,

$$\begin{aligned}\left(\frac{\partial \mu_A}{\partial P}\right)_T &= \left(\frac{\partial}{\partial P}\right)_T \left(\frac{\partial G}{\partial n_A}\right)_{TP} \\ &= \left(\frac{\partial}{\partial n_A}\right)_{TP} \left(\frac{\partial G}{\partial P}\right)_T \\ &= \left(\frac{\partial V}{\partial n_A}\right)_{TP} \\ &= \overline{V}_A\end{aligned}$$

Thus, the total differential of the chemical potential for species A can be written as

$$d\mu_A = -\overline{S}_A dT + \overline{V}_A dP + \sum_{j=1}^{\omega} \left(\frac{\partial \mu_A}{\partial n_j}\right)_{PT} dn_j$$

To illustrate the utility of this result, we can use it to derive the Clapeyron equation for equilibrium between two phases of a pure substance. In Chapter 12, we derived the Clayeyron equation using a thermochemical cycle. We can now use the total differential of the chemical potential to present essentially the same derivation using a simpler argument. Letting the two phases be α and β, the total differentials for a system that contains both phases becomes

$$d\mu_\alpha = -\overline{S}_\alpha dT + \overline{V}_\alpha dP + \left(\frac{\partial \mu_\alpha}{\partial n_\alpha}\right)_{PT} dn_\alpha + \left(\frac{\partial \mu_\alpha}{\partial n_\beta}\right)_{PT} dn_\beta$$

and

$$d\mu_\beta = -\overline{S}_\beta dT + \overline{V}_\beta dP + \left(\frac{\partial \mu_\beta}{\partial n_\alpha}\right)_{PT} dn_\alpha + \left(\frac{\partial \mu_\beta}{\partial n_\beta}\right)_{PT} dn_\beta$$

Since equilibrium between phases α and β means that $\mu_\alpha = \mu_\beta$, we have also that $d\mu_\alpha = d\mu_\beta$ for any process in which the phase equilibrium is maintained. Moreover, α and β are pure phases, so that μ_α and μ_β are independent of n_α and n_β. Then

$$\left(\frac{\partial \mu_\alpha}{\partial n_\alpha}\right)_{PT} = \left(\frac{\partial \mu_\beta}{\partial n_\alpha}\right)_{PT} = \left(\frac{\partial \mu_\alpha}{\partial n_\beta}\right)_{PT} = \left(\frac{\partial \mu_\beta}{\partial n_\beta}\right)_{PT} = 0$$

Hence,

$$-\overline{S}_\alpha dT + \overline{V}_\alpha dP = -\overline{S}_\beta dT + \overline{V}_\beta dP$$

and the rest of the derivation follows as before.

§10 Chemical activity

In §5-17, we use mathematical models for the rates of chemical reactions to find the equilibrium constant equation, which specifies the equilibrium position of a reacting system. The variables in the reaction-rate models are the concentrations of reactants and products. Consequently, these concentrations are also the variables in the equilibrium constant equation. We develop our reaction-rate models from simple arguments about the dependence of collision frequencies on reagent concentrations. These arguments ignore the possibility that the effects of intermolecular forces on collision frequencies can change as the composition of the system changes. For a great many purposes, the concentration-based models describe experimental results with sufficient accuracy. However, as we note in our introductory discussion in §1-6, the concentration-based equilibrium constant equation proves to be only approximately constant: To obtain rigorously accurate results, we must introduce new quantities, having the character of "corrected concentrations", that we call activities.

We are now in a position to define the ***chemical activity*** of a substance more precisely. We introduce the activity concept in the particular context of chemical equilibria, but its scope is broader. The activity of a component is a wholly new thermodynamic quantity. We view it as the property that determines the chemical potential, μ, of a substance in a particular system at a given temperature. As such, the activity directly links the properties of the substance to the behavior of the system. Introducing the activity makes our equations for μ and $d\mu$ more compact. It also provides an avenue to relate qualitative ideas about the properties of the substance—ideas that we may not be able to express in mathematical models—to the effects that the substance has on the properties of the system.

The heuristic idea that an activity is a corrected concentration arises in the first instance because we use the pressure-dependence of the Gibbs free energy of an ideal gas as the model for our definition of activity. In §15-7, we find that the equilibrium constant for a chemical reaction can be specified using the activities of the reactants and products and that the form of the equilibrium constant is the same for activities as it is for concentrations. In Chapter 16 we see that assuming ideal behavior for solutions enables us to equate the activity and the mole fraction of a component in such solutions.

Our consideration of the chemical reactions of gases in Chapter 13 introduces the basic ideas that we use: The Gibbs free energy of reaction is the same thing as the change in the chemical potential of the system; a reaction reaches equilibrium when the Gibbs free energy of the system cannot decrease further; the Gibbs free energy of reaction is a function of the chemical potentials of the reactants and products; and the chemical potential of a gas is a function of its partial pressure. In ideal gas mixtures, the chemical potential of a component depends on the partial pressure of that component, but not on the partial pressures of other species that may be present. The model we develop for ideal gases becomes unsatisfactory for real gases under the kinds of conditions that cause the experimentally observed properties of gases to depart from ideal gas behavior—conditions in which intermolecular forces cannot be neglected.

In §9, we observe that the chemical potential of substance A can be expressed as a function of temperature, pressure, and its molar composition:

$$\mu_A(P, T, n_1, n_2, \dots, n_A, \dots, n_\omega)$$

The total differential becomes

$$d\mu_A = -\overline{S}_A dT + \overline{V}_A dP + \sum_{j=1}^{\omega} \left(\frac{\partial \mu_A}{\partial n_j}\right)_{PT} dn_j$$

We can write this as

$$d\mu_A = -\overline{S}_A dT + \overline{V}_A dP + (d\mu_A)_{PT}$$

where

$$(d\mu_A)_{PT} = \sum_{j=1}^{\omega} \left(\frac{\partial \mu_A}{\partial n_j}\right)_{PT} dn_j$$

In general, the change in the chemical potential of A depends on partial derivatives with respect to the amount of every substance present. The dependence on the amount of substance j is given by

$$\left(\frac{\partial \mu_A}{\partial n_j}\right)_{P,T,n_{m\neq j}}$$

and the dependence on the amount of A is given by

$$\left(\frac{\partial \mu_A}{\partial n_A}\right)_{P,T,n_{m\neq A}}$$

In a multi-component system whose components do not interact, $(\partial \mu_A / \partial n_A)_{P,T,n_{m\neq A}}$ is not zero, but all of the other partial derivatives $\left(\partial \mu_A / \partial n_j\right)_{P,T,n_{m\neq j}}$ must, by definition, vanish. This is exemplified in an ideal-gas system, in which the chemical potential of component A is given by

$$\mu_A(P_A) = \Delta_f G^o(A, P^o) + RT \ln(P_A/P^o)$$

and the partial pressure of A depends only on the mole fraction of A in the system, $P_A = x_A P_{system}$. For an ideal gas at constant temperature, changes in the chemical potential do not depend on the amounts of other substances that may be present:

$$(d\mu_A)_T = RT(d \ln P_A)_T$$

In the special case that the system pressure is constant, this becomes:

$$(d\mu_A)_T = RT(d \ln x_A)_T$$

foreshadowing our eventual recognition that mole fraction is the most natural concentration unit for theoretical modeling of the chemical potential.

In real systems, the chemical potential of a component depends on its own concentration, but it also depends—more weakly—on the concentrations of the other species present. To treat real systems adequately, we need a general method to express the chemical potential of a component as a function of the component's concentration, in a way that fully accounts for the effects of the other species. Our grand strategy is to develop the activity of a component as an abstract, dimensionless quantity. We do this by choosing a simple function to define the activity of a component in terms of its chemical potential. This makes it possible to express the thermodynamic properties of any system as relatively simple functions of the activities of its components. To apply these results to specific systems, we must then find empirical equations that express the component activities as functions of component concentrations. So, from one perspective, the activity is merely a convenient intermediary in our overall effort to express the chemical potential as a function of composition.

To begin expressing this strategy in mathematical notation, let us represent the activity and concentration of a component, A, as $\tilde{a}_A$ and c_A, respectively. For simplicity, we consider systems in which pressure–volume work is the only work. The chemical potential is a function of pressure, temperature, and composition; that is $\mu_A = \mu_A(P, T, c_A, c_B, c_C, \ldots)$. We suppose that we have a large volume of such data available; that is, we have measured μ_A at many conditions represented by widely varying values of the variables $P, T, c_A, c_B, c_C, \ldots$. Our strategy is to find two empirical functions. The first empirical function expresses the chemical potential of A as a function of its activity and the temperature of the system, $\mu_A = \mu_A(T, \tilde{a}_A)$. The second expresses the activity of A as a function of the concentrations of the species present and of the pressure and temperature of the system, $\tilde{a}_A = \tilde{a}_A(P, T, c_A, c_B, c_C, \ldots)$. The mathematical composition of these two functions expresses the chemical potential as a function of $P, T, c_A, c_B, c_C, \ldots$; that is, the mathematical composition is the function $\mu_A = \mu_A(P, T, c_A, c_B, c_C, \ldots)$.

If we focus only on finding a function, $\mu_A(P, T, c_A, c_B, c_C, \ldots)$, that fits the data, we can let the dependence of chemical potential on activity be anything we please. All we have to do is fix up the activity function, $\tilde{a}_A = \tilde{a}_A(P, T, c_A, c_B, c_C, \ldots)$, so that the mathematical composition of the two accurately reproduces the experimental data. This is just another way of saying that, if all we want is an empirical correlation of the form $\mu_A = \mu_A(P, T, c_A, c_B, c_C, \ldots)$, we do not need to introduce the activity at all.

However, our goal is more exacting. We want to construct the activity function so that it behaves as much like a concentration as possible. In particular, we want the activity of a component to become equal to its concentration in cases where intermolecular interactions vanish—or become independent of the component's concentration for some other reason. To achieve these objectives, we must choose the form of the function $\mu_A = \mu_A(T, \tilde{a}_A)$ carefully. Here the behavior of ideal gases provides a valuable template. Our observation that the chemical potential is related to the partial pressure by the differential expression $(d\mu_A)_T = RT(d \ln P_A)_T$ suggests that it might be possible to develop the activity function using a relationship of the same form. In fact, this strategy proves to be successful.

We relate the chemical potential of a component to its activity by the differential expression

$$(d\mu_A)_T = RT(d \ln \tilde{a}_A)_T$$

The activity becomes a single function that captures the

dependence of $(d\mu_A)_T$ on the number of moles of each of the components of the system:

$$RT(d\ln\tilde{a}_A)_T = \sum_{j=1}^{\omega}\left(\frac{\partial\mu_A}{\partial n_j}\right)_{PT} dn_j$$

If we hold the temperature constant, we can integrate between some arbitrary base state (in which the activity and chemical potential are $\tilde{a}_A^{\#}$ and $\mu_A^{\#}$) and a second state (in which these quantities are $\tilde{a}_A$ and μ_A). That is, at any given temperature, we can select a particular state that we define to be the base state of A for the purpose of creating an activity scale. We denote the activity in the base state as $\tilde{a}_A^{\#}$. We let the chemical potential of A be $\mu_A^{\#}$ when A is in its base state at this temperature. At a fixed temperature, $\mu_A^{\#}$ is fixed. However, even if we keep all the other variables the same, the properties of the base state change when the temperature changes. This means that the chemical potential in the base state is a function of temperature—and only of temperature.

At constant temperature, the relationship between chemical potential and activity in any other state becomes

$$\int_{\mu_A^{\#}}^{\mu_A} d\mu_A = RT\int_{a_A^{\#}}^{a_A} d\ln\tilde{a}_A$$

or

$$\mu_A = \mu_A^{\#} + RT\ln\left(\frac{\tilde{a}_A}{\tilde{a}_A^{\#}}\right)$$

where it is implicit in the last equation that the activity is a function of pressure, temperature, and the chemical composition of the system. The temperature and chemical potential are state functions, so this expression requires that the activity be a state function also.

It is convenient to let the activity be unity in the base state. When we do this, we give the base state another name; we call it the ***activity standard state***. When we stipulate that the activity in the base state is unity, we use a superscript zero to designate the activity and chemical potential. That is, in the standard state, $\tilde{a}_A^o = 1$, and the value of the chemical potential of A is $\tilde{\mu}_A^o$ when A is in this state. (In many cases, it turns out to be convenient to choose a standard state that does not correspond to any physical condition that can actually be achieved. In such cases, the standard state is a hypothetical system, whose properties we establish by mathematical extrapolation of measurements made on a real system in non-standard states.) With this convention, the chemical potential and activity are related by the equation

$$\mu_A = \tilde{\mu}_A^o + RT\ln\tilde{a}_A$$

at the specified temperature. We adopt this equation as our formal definition of the ***activity*** of component A. The activity and the chemical potential depend on the same variables and contain equivalent information.

Note that we have done nothing to restrict the state we choose to designate as the activity standard state. This creates opportunity for confusion, because it allows us to choose an activity standard state for A in a particular system that is different from the standard state for pure A that we define for the tabulation of thermodynamic data for pure substances. This means that we choose to let the meaning of the words "standard state" be context dependent. The practical effect of this ambiguity is that, whenever we are dealing with a chemical activity, we must be careful to understand what state of the substance is being designated as the standard state—and thereby being assigned chemical potential $\tilde{\mu}_A^o$ and unit activity. Because the standard state is a fixed composition, $\tilde{\mu}_A^o$ is a function of temperature, but not of pressure or composition. At a given temperature, changes in the system pressure or the concentration of A affect the chemical potential, μ_A, only by their effect on the activity, $\tilde{a}_A$.

To complete our program, we define a function that we call the ***activity coefficient***, $\gamma_A(P,T,c_A,c_B,c_C,\dots)$, of component A, in a system characterized by variables $P,T,c_A,c_B,c_C,\dots$, by the equation

$$\tilde{a}_A(P,T,c_A,c_B,c_C,\dots) = c_A\gamma_A(P,T,c_A,c_B,c_C,\dots)$$

This definition places only one constraint on the form of γ_A; the function $\gamma_A(P,T,c_A,c_B,c_C,\dots)$ can be anything that adequately accounts for the experimental data, so long as $\gamma_A \to 1$ when the effects of intermolecular interactions are negligible.

Introducing the activity coefficient does not simplify the job of finding a suitable, empirical, activity function; it just imposes a condition on its form. In Chapters 15 and 16, we find that standard states are best described using mole fraction or molality as the concentration unit. (Note that mole fraction is dimensionless and molality is proportional to mole fraction for dilute solutes.)

In §16-6 we consider the use of molality when A is a solute in a solution whose other components are in fixed proportions to one another. At very low concentrations of A, the environment around every A molecule is essentially the same. The chemical potential of A is observed to be a linear function of $\ln \underline{m}_A$, because a small increase in the concentration of A does not significantly change the intermolecular interactions experienced by A molecules. As the concentration of A increases further, intermolecular interactions among A molecules become increasingly important, and the chemical potential ceases to be a simple linear function of $\ln \underline{m}_A$. The activity coefficient is no longer equal to one.

In the remainder of this chapter, we focus on some of the general properties of the activity function. In the next two chapters, we develop a few basic applications of these ideas. First, however, we digress to relate the ideas that we have just developed about activity to the ideas that we developed in Chapter 11 about fugacity.

§11 Back to the fugacity: the fugacity of a component of a gas mixture

In Chapter 11, we introduce the fugacity as an alternative measure of the difference between the Gibbs free energy of one mole of a pure gas in its hypothetical ideal gas standard state and its Gibbs free energy in any other state at the same temperature. This definition makes the fugacity of a gas an intensive function of pressure and temperature. At a fixed temperature, the state of one mole of a pure gas is specified by its pressure, and the fugacity is a function of pressure only. Fugacity has the units of pressure. Giving effect to our decision to let the fugacity of the gas be unity when the gas is in its hypothetical ideal gas standard state (HIG^o) and using the Gibbs free energy of formation for the gas in this state as the standard Gibbs free energy for fugacity, we define the fugacity of a pure gas, A, by the equation

$$\overline{G}_A(P) = \Delta_f G^o(A, HIG^o) + RT \ln\left[\frac{f_A(P)}{f_A(HIG^o)}\right]$$

For substance A in any system, the chemical potential is the partial molar free energy; that is, $\mu_A(P,T) = \overline{G}_A(P,T)$. Since the Gibbs free energy of formation is defined for one mole of pure substance at a specified pressure and temperature, it is a partial molar quantity. When we elect to use the hypothetical ideal gas at a pressure of one bar as the standard state for the Gibbs free energy of formation of the gas, we also establish the Gibbs free energy of formation of the hypothetical ideal gas in its standard state as the standard-state chemical potential; that is, $\mu_A^o = \Delta_f G^o(A, HIG^o)$. Hence, we can also express the fugacity of a gas by the equation

$$\mu_A(P) = \Delta_f G^o(A, HIG^o) + RT \ln\left[\frac{f_A(P)}{f_A(HIG^o)}\right]$$

or

$$\mu_A(P) = \mu_A^o + RT \ln\left[\frac{f_A(P)}{f_A(HIG^o)}\right]$$

(pure real gas)

For a mixture of real gases, we can extend the definition of fugacity in a natural way. We want the fugacity of a component gas to measure the difference between its chemical potential in the mixture, μ, and its chemical potential in its standard state, μ^o, where its standard state is the pure hypothetical ideal gas at one bar pressure. If gas A is a component of a constant-temperature mixed-gas system, we have $d\mu_A = \overline{V}_A dP$, where $\overline{V}_A$ is the partial molar volume of A in the system, and P is the pressure of the system. Let us find μ_A in a binary mixture that contains one mole of A and n_B moles of a second component, B. Let the partial molar volume of B be $\overline{V}_B$. The system volume is $V = \overline{V}_A + n_B\overline{V}_B$. The mole fractions of A and B are $x_A = 1/(n_B + 1)$ and $x_B = n_B/(n_B + 1)$.

To find the change in μ_A, we need a reversible process that takes one mole of A in its standard state to a mixture of one mole A with n_B moles of B, in which the pressure of the mixture is P. The four-step process described in Figure 1 answers our requirements: One mole of A and n_B moles of B are separately expanded from their hypothetical ideal gas standard states, at P^o, to the arbitrary low pressures $x_A P^*$ and $x_B P^*$, respectively. For this expansion, the change in the Gibbs free energy of one mole of A, which remains in its hypothetical ideal gas state, is

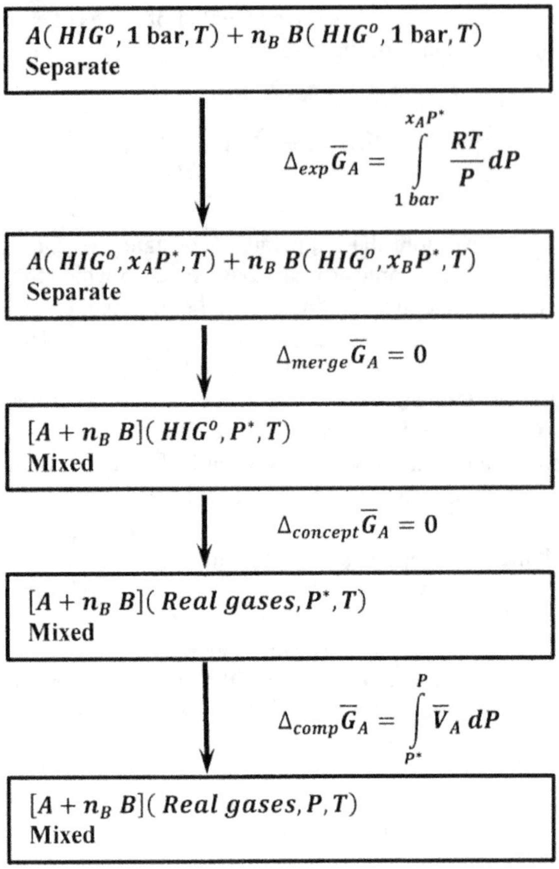

Figure 1. Finding the Gibbs free energy of a real gas in a mixture.

$$\Delta_{exps}\overline{G}_A = \int_{P^o}^{x_A P^*} \overline{V}_A^{\bullet}\, dP$$

$$= \int_{P^o}^{x_A P^*} \frac{RT}{P} dP = RT \ln\left(\frac{x_A P^*}{P^o}\right)$$

Next, these low-pressure ideal gases are merged to form a mixture of one mole of A with n_B moles of B at the total pressure P^*. For this merging process, $\Delta_{merge}\overline{G}_A = 0$. Then, we suppose that the ideal gases become real gases in a mixture whose pressure is P^*. Since this is merely a conceptual change, we have $\Delta_{concept}\overline{G}_A = 0$. Finally, we compress the mixture of real gases from P^* to an arbitrary pressure, P. Since the volume of the mixture is $V = \overline{V}_A + n_B\overline{V}_B$, the Gibbs free energy change for this compression of the mixture is

$$\Delta_{comp}G_{mixture} = \int_{P^*}^{P} V\,dP$$
$$= \int_{P^*}^{P} \overline{V}_A\,dP + n_B \int_{P^*}^{P} \overline{V}_B\,dP$$

We see that the Gibbs free energy change for the real-gas system is the sum of the Gibbs free energy changes for the components; we have

$$\Delta_{comp}\overline{G}_A = \int_{P^*}^{P} \overline{V}_A\,dP$$

For this process, we have

$$\mu_A - \mu_A^o = RT \ln\left[\frac{f_A(P)}{f_A(HIG^o)}\right]$$
$$= \Delta_{exps}\overline{G}_A + \Delta_{merge}\overline{G}_A + \Delta_{concept}\overline{G}_A + \Delta_{comp}\overline{G}_A$$
$$= RT\ln\left[\frac{x_A P^*}{P^o}\right] + \int_{P^*}^{P} \overline{V}_A\,dP - \int_{P^*}^{P} \frac{RT}{P}dP + \int_{P^*}^{P} \frac{RT}{P}dP$$

where we have added and subtracted the quantity

$$\int_{P^*}^{P} \frac{RT}{P}dP$$

Dividing by RT and evaluating the last integral, we find

$$\frac{\mu_A - \mu_A^o}{RT} = \ln\left[\frac{f_A(P)}{f_A(HIG^o)}\right]$$
$$= \ln x_A + \ln P^* - \ln P^o$$
$$+ \int_{P^*}^{P} \left(\frac{\overline{V}_A}{RT} - \frac{1}{P}\right) dP + \ln P - \ln P^*$$

P^* is a finite pressure arbitrarily near zero. At very low pressures, real gas A behaves as an ideal gas; hence, at very low pressures, the partial molar volume of the real gas is well approximated by the partial molar volume of pure gas A. That is, we have $\overline{V}_A \approx \overline{V}_A^{\bullet}$, and

$$\int_{0}^{P^*} \left(\frac{\overline{V}_A}{RT} - \frac{1}{P}\right) dP \approx \int_{0}^{P^*} \left(\frac{\overline{V}_A^{\bullet}}{RT} - \frac{1}{P}\right) dP$$
$$\approx 0$$

where the approximation becomes exact in the limit as $P^* \to 0$. Simplifying the natural logarithm terms and expanding the integral, we obtain

$$\frac{\mu_A - \mu_A^o}{RT} = \ln\left[\frac{f_A(P)}{f_A(HIG^o)}\right]$$
$$= \ln\frac{x_A P}{P^o} + \int_{0}^{P} \left(\frac{\overline{V}_A}{RT} - \frac{1}{P}\right) dP - \int_{0}^{P^*} \left(\frac{\overline{V}_A}{RT} - \frac{1}{P}\right) dP$$
$$= \ln\frac{x_A P}{P^o} + \int_{0}^{P} \left(\frac{\overline{V}_A}{RT} - \frac{1}{P}\right) dP$$

Defining the fugacity coefficient for A in this mixture, γ_A, by

$$\gamma_A = \frac{f_A(P)}{x_A P}$$

and recalling that $f_A(HIG^o) = P^o$, we use this result to find

$$\ln\gamma_A = \ln\left(\frac{f_A(P)}{x_A P}\right) = \int_{0}^{P} \left(\frac{\overline{V}_A}{RT} - \frac{1}{P}\right) dP$$

This differs from the corresponding relationship for the fugacity of a pure gas only in that the partial molar volume is that of gas A in a mixture with other gases. This is a trifling difference in principle, but a major difference in practice. To find the fugacity of pure A, we use the partial molar volume of the pure gas, which is readily calculated from any empirical pure-gas equation of state. However, to experimentally obtain the partial molar volume of gas A in a gas mixture, we must collect pressure–volume–temperature data as a function of the composition of the system. If we contemplate creating a catalog of such data for the mixtures of even a modest number of compounds, we see that an enormous amount of data must be collected. Just the number of systems involving only binary mixtures is large. For N compounds, there are $N\,(N-1)/2$ binary mixtures—each of which would have to be studied at many compositions in order to develop good values for the partial molar volumes.

Fortunately, practical experience shows that a simple approximation often gives satisfactory results. In this approximation, we assume that the partial molar volume of gas A—present at mole fraction x_A in a system whose pressure is P— is equal to the partial molar volume of the pure gas at the same pressure. That is, for a binary mixture of gases A and B, we assume

$$\overline{V}_A(P, x_A, x_B) = \overline{V}_A^{\bullet}(P)$$

In this approximation, we have

$$\frac{\mu_A - \mu_A^o}{RT} = \ln\frac{x_A P}{P^o} + \int_{0}^{P} \left(\frac{\overline{V}_A^{\bullet}}{RT} - \frac{1}{P}\right) dP$$

and

$$\ln\gamma_A = \ln\left(\frac{f_A(P)}{x_A P}\right) = \int_{0}^{P} \left(\frac{\overline{V}_A^{\bullet}}{RT} - \frac{1}{P}\right) dP$$

We make the same assumption for gas B. From

Euler's theorem on homogeneous functions, we have $\overline{V} = x_A\overline{V}_A + x_B\overline{V}_B$. Therefore, in this approximation, we have

$$\overline{V} = x_A\overline{V}_A + x_B\overline{V}_B \approx x_A\overline{V_A^\bullet} + x_B\overline{V_B^\bullet}$$

The last sum is the Amagat's law representation of the molar volume of the gas mixture. We see that our approximation is equivalent to assuming that the system obeys Amagat's law. Physically, this assumes that the gas mixture is an ideal (gaseous) solution. We discuss ideal solutions in Chapter 16. In an ideal solution, the intermolecular interactions between an A molecule and a B molecule are assumed to have the same effect as the interactions between two A molecules or between two B molecules. This differs from the ideal-gas assumption; there is no effect from the interactions between any two ideal-gas molecules.

§12 Relating fugacity and chemical activity

While we develop the fugacity concept by thinking about the Gibbs free energy of a pure real gas, our definition means that the fugacity of a substance in any system depends only on the difference between the chemical potential in that system and its chemical potential in the standard state. There is no reason to confine the definition of fugacity to gaseous systems. We generalize it: ***Our defining relationship specifies the fugacity of any substance in any system as a function of the difference between its chemical potential in that system and its chemical potential in its hypothetical ideal gas standard state.***

Since the Gibbs free energy of an arbitrary system depends on the pressure, temperature, and composition of the system, the fugacity of any component also depends on these variables. However, there is an additional constraint on the fugacity. At a sufficiently low system pressure, the gases in any system behave ideally, and they obey Boyle's law of partial pressures; the integral in the fugacity-coefficient equation becomes zero, and the fugacity coefficient becomes unity. In the limit as $P \to 0$, $f_A \to x_A P$, where x_A is the mole fraction of A in the gas phase. At a sufficiently low pressure, any gas mixture behaves ideally, and the fugacity of a constituent species becomes equal to its gas-phase partial pressure. (For a mixture of ideal gases, the fugacity of a component is always equal to its partial pressure.)

In §10, we define the activity of component A in an arbitrary system by the relationship

$$\mu_A = \tilde{\mu}_A^o + RT\ln[\tilde{a}_A(P,T,c_A,c_B,c_C,\ldots)]$$

where $\tilde{\mu}_A^o$ is the chemical potential of A in an activity standard state in which we stipulate that the activity of A is unity. Since the defining equations for activity and for fugacity are formally identical, the distinction between activity and fugacity lies in our choices of standard states and in the facts that activity is dimensionless while fugacity has the units of pressure. If we use the hypothetical ideal-gas standard state for activity and measure the concentration in bars, the practical distinction between activity and fugacity vanishes. We can view the fugacity as a specialization of the activity concept.

In summary: The fugacity function proves to be a useful way to express the difference between the chemical potential of a substance in two different states. Fugacity is measured in bars with the hypothetical ideal gas standard state as the reference state. We add chemical activity to our list of useful thermodynamic properties because it extends the advantages of the fugacity representation to non-volatile components of systems that contain condensed phases. The standard state for activity can be any particular state of any convenient system that contains the substance. We define the activity of the substance in this reference system to be unity. (As with the hypothetical ideal gas standard state, we often find it useful to define a hypothetical system as the standard state for the activity.) The chemical potential of the substance in the standard-state system is, by definition, the standard chemical potential, $\tilde{\mu}_A^o$, for this particular activity scale.

In the remainder of this chapter and in Chapters 15-17, we consider additional properties and applications of fugacity and chemical activity. Before doing so however, we digress to observe that we can choose an alternate definition of activity: chemical activity can be defined as a ratio of fugacities. Let us consider three systems that contain substance A.

The first is pure gas A in its hypothetical ideal gas standard state at temperature T, $A(HIG^o,T)$. In this state, the fugacity of A is unity, $f_A(HIG^o,T) = 1$ bar. The chemical potential of A is the standard-state chemical potential, which we equate to the Gibbs free energy of formation:

$$\begin{aligned}\overline{G}_A(HIG^o,T) &= \Delta_f G^o(A,HIG^o,T)\\ &= \mu_A^o\end{aligned}$$

The second is a system that we define as the standard state for the activity of A at temperature T. Apart from convenience, there is no reason to prefer any particular system for this role. We denote this system as $A(activity\ standard\ state,T)$ or, for short, $A(ss,T)$. By definition, the activity of A in this state is unity, $\tilde{a}_A^o(ss) = 1$, and the chemical potential of A in this system is the standard chemical potential for this particular activity scale: $\overline{G}_A(ss,T) = \tilde{\mu}_A^o$. We denote the fugacity of A in this state as $f_A(ss)$. By our definition of fugacity, we have

$$\tilde{\mu}_A^o = \mu_A^o + RT\ln\left[\frac{f_A(ss)}{f_A(HIG^o)}\right]$$

The third is an arbitrary system that we denote as $A(P,T,c_A,c_B,c_C,\ldots)$. We denote the chemical potential, fugacity, and activity of A in this system as $\overline{G}_A(P,T,c_A,c_B,c_C,\ldots) = \mu_A(P,T,c_A,c_B,c_C,\ldots)$,

$f_A(P,T,c_A,c_B,c_C,\dots)$, and $\tilde{a}_A(P,T,c_A,c_B,c_C,\dots)$, respectively. By our definition of fugacity, we have

$$\mu_A(P,T,c_A,c_B,c_C,\dots) = \mu_A^o + RT\ln\left[\frac{f_A(P,T,c_A,c_B,c_C,\dots)}{f_A(HIG^o)}\right]$$

and by our definition of activity,

$$\mu_A(P,T,c_A,c_B,c_C,\dots) = \tilde{\mu}_A^o + RT\ln[\tilde{a}_A(P,T,c_A,c_B,c_C,\dots)]$$

Figure 2 summarizes the relationships among the Gibbs free energies of these three states of substance A. From the cycle in Figure 2, we have $\Delta_1 G + \Delta_3 G = \Delta_2 G$, so that

$$RT\ln\left[\frac{f_A(ss,T)}{f_A(HIG^o)}\right] + RT\,ln\left[\frac{f_A(P,T,c_A,c_B,\dots.)}{f_A(ss,T)}\right] = RT\ln\left[\frac{f_A(P,T,c_A,c_B,c_C,\dots)}{f_A(HIG^o)}\right] = RT\ln[\tilde{a}_A(P,T,c_A,c_B,\dots.)]$$

and

$$\tilde{a}_A(P,T,c_A,c_B,c_C,\dots) = \frac{f_A(P,T,c_A,c_B,c_C,\dots)}{f_A(ss,T)}$$

That is, the activity of a substance in a particular system is always equal to its fugacity in that system divided by its fugacity in the standard state for activity. From this relationship, it is evident that activity is always a dimensionless function of concentrations.

§13 Relating the differentials of chemical potential and activity

Let us write $(d\mu_A)_{PT}$ to represent the differential of μ_A at constant pressure and temperature. From the general expression for $d\mu_A$ and the definition of activity, we can write the total differential of the chemical potential of substance A in a particular system in several equivalent ways

$$\begin{aligned} d\mu_A &= \left(\frac{\partial \mu_A}{\partial P}\right)_T dP + \left(\frac{\partial \mu_A}{\partial T}\right)_P dT + (d\mu_A)_{PT} \\ &= \left(\frac{\partial \mu_A}{\partial P}\right)_T dP + \left(\frac{\partial \mu_A}{\partial T}\right)_P dT + \sum_{j=1}^{\omega}\left(\frac{\partial \mu_A}{\partial n_j}\right)_{PT} dn_j \\ &= \overline{V}_A dP - \overline{S}_A dT + RT(d\ln\tilde{a}_A)_{PT} \\ &= RT\left(\frac{\partial \ln\tilde{a}_A}{\partial P}\right)_T dP + \left(\frac{\partial \mu_A}{\partial T}\right)_P dT \\ &\quad + RT(d\ln\tilde{a}_A)_{PT} \end{aligned}$$

In short, we have developed several alternative notations for the same physical quantities. From the dependence of chemical potential on pressure, and because $\tilde{\mu}_A^o$ is not a function of pressure, we have a very useful relationship:

$$\left(\frac{\partial \mu_A}{\partial P}\right)_T = RT\left(\frac{\partial \ln\tilde{a}_A}{\partial P}\right)_T = \overline{V}_A$$

From the definition of activity and the dependence of chemical potential on temperature, we have:

$$\left(\frac{\partial \mu_A}{\partial T}\right)_P = \left(\frac{\partial \tilde{\mu}_A^o}{\partial T}\right)_P + R\ln\tilde{a}_A + RT\left(\frac{\partial \ln\tilde{a}_A}{\partial T}\right)_P = -\overline{S}_A$$

From the dependence of chemical potential on the composition of the system, we have

$$\sum_{j=1}^{\omega}\left(\frac{\partial \mu_A}{\partial n_j}\right)_{PT} dn_j = (d\mu_A)_{PT} = RT(d\ln\tilde{a}_A)_{PT}$$

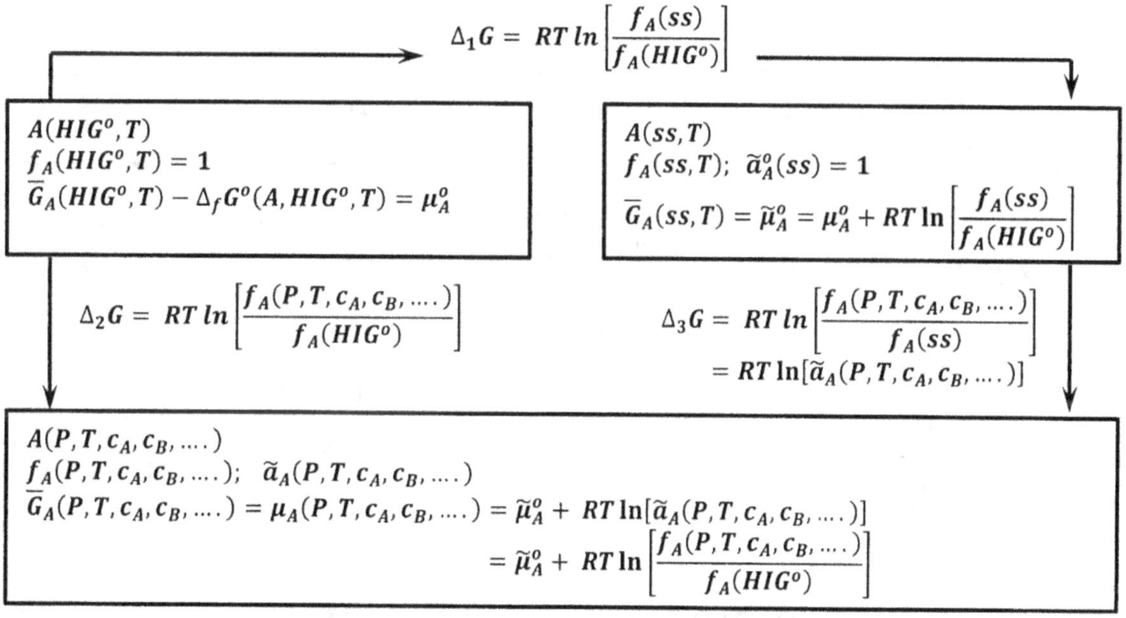

Figure 2. Relating the chemical activity of a substance to its activity.

This last equation shows explicitly that the activity of component A depends on all of the species present. The effects of interactions between A molecules and B molecules are represented in this sum by the term $(\partial\mu_A/\partial n_B)_{PT}$. When the effects of intermolecular interactions on the chemical potential are independent of the component concentrations, $(\partial\mu_A/\partial n_B)_{PT} = 0$, and the only surviving term is $(\partial\mu_A/\partial n_A)_{PT}$. If the interactions between A molecules and the rest of the system are constant over a range of concentrations of A, γ_A is constant over this range.

§14 Dependence of activity on temperature: relative partial molar enthalpies

Having found the activity of a component at one temperature, we want to be able to find it at a second temperature. The equation developed in §13 does not provide a practical way to find the temperature dependence of $\tilde{a}_A$ or $\ln \tilde{a}_A$. We can obtain a useful equation by rearranging the defining equation, taking the partial derivative of $\ln \tilde{a}_A$ with respect to temperature, and making use of the Gibbs-Helmholtz equation:

$$\left(\frac{\partial \ln \tilde{a}_A}{\partial T}\right)_P = \left[\frac{\partial}{\partial T}\left(\frac{\mu_A}{RT} - \frac{\tilde{\mu}_A^o}{RT}\right)\right]_P$$
$$= \frac{1}{R}\left[\left(\frac{\partial(\mu_A/T)}{\partial T}\right)_P - \left(\frac{\partial(\tilde{\mu}_A^o/T)}{\partial T}\right)_P\right]$$
$$= -\frac{\overline{H}_A}{RT^2} + \frac{\tilde{H}_A^o}{RT^2}$$

$\overline{H}_A$ is the partial molar enthalpy of component A as it is present in the system. $\tilde{H}_A^o$ is the partial molar enthalpy of component A in its activity standard state. Since this standard state need not correspond to any real system, $\tilde{H}_A^o$ can be the partial molar enthalpy of the substance in a hypothetical state.

In general, it is not possible to find $-\left(\overline{H}_A - \tilde{H}_A^o\right)$ by experimental measurement of the heat exchanged in a single-step process in which A passes from its state in the system of interest to its activity standard state. Instead, we devise a multi-step cycle in which we can determine the enthalpy change for each step. This cycle includes yet another state of substance A, which we call the reference state and whose molar enthalpy we designate as $\overline{H}_A^{ref}$. We devise this cycle to find two enthalpy changes. One is the enthalpy change that occurs when one mole of A passes from the system of interest to the reference state; this enthalpy change is represented by the difference $-\left(\overline{H}_A - \overline{H}_A^{ref}\right)$. The other is the enthalpy change that occurs when one mole of A passes from the activity standard state to the reference state. This enthalpy change is represented by the difference $-\left(\tilde{H}_A^o - \overline{H}_A^{ref}\right)$. The enthalpy change we seek is the difference between these two differences:

$$-\left(\overline{H}_A - \tilde{H}_A^o\right) = -\left(\overline{H}_A - \overline{H}_A^{ref}\right) + \left(\tilde{H}_A^o - \overline{H}_A^{ref}\right)$$

Because we explicitly choose the reference state so that these differences are experimentally measurable, it is useful to introduce still more terminology. We define the ***relative partial molar enthalpy*** of substance A, $\overline{L}_A$, as the difference between the partial molar enthalpy of A in the state of interest, $\overline{H}_A$, and the partial molar enthalpy of A in the reference state, $\overline{H}_A^{ref}$; that is, $\overline{L}_A = \overline{H}_A - \overline{H}_A^{ref}$ and $\tilde{L}_A^o = \tilde{H}_A^o - \overline{H}_A^{ref}$ where $\tilde{L}_A^o$ is the relative partial molar enthalpy of substance A in the activity standard state. Clearly, the values of $\overline{L}_A$ and $\tilde{L}_A^o$ depend on the choice of reference state. We choose the reference state so that $\overline{L}_A$ and $\tilde{L}_A^o$ can be measured directly. The quantity that we must evaluate experimentally in order to find

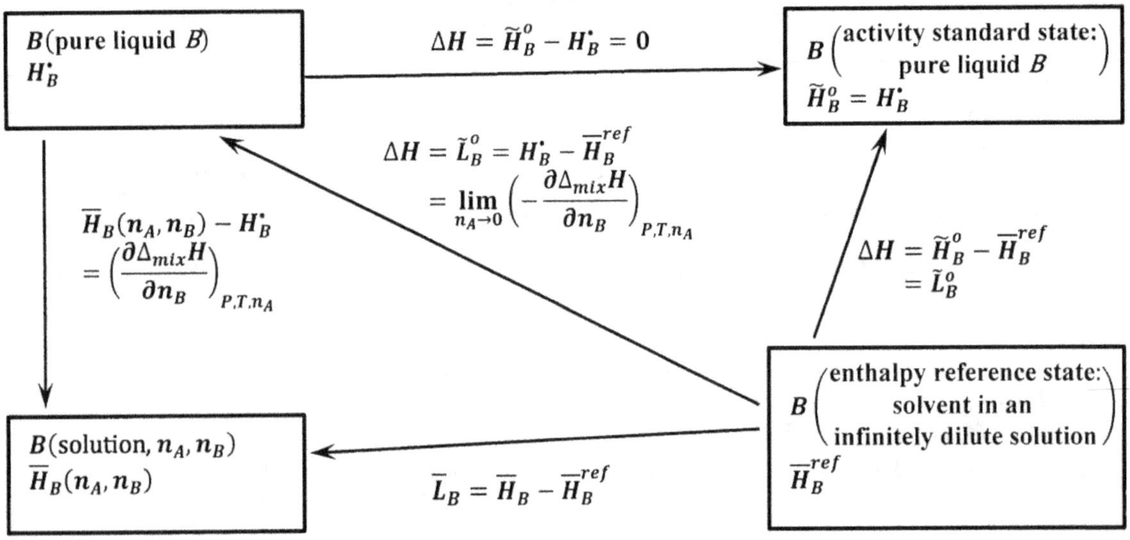

Figure 3. Cycles relating the molar enthalpy of a solvent in a solution to its molar enthalpies in the pure liquid, in an activity standard state, and in an enthalpy standard state.

$(\partial \ln \tilde{a}_A / \partial T)_P$ becomes

$$-\left(\overline{H}_A - \widetilde{H}_A^o\right) = -\left(\overline{L}_A - \widetilde{L}_A^o\right)$$

To see how these ideas and definitions can be given practical effect, let us consider a binary solution that comprises a relatively low concentration of solute, A, in a solvent, B, at a fixed pressure, normally one bar. We suppose that pure A is a solid and pure B is a liquid in the temperature range of interest. We need to choose an activity standard state and an enthalpy reference state for each substance. The most generally useful choices use the concept of an infinitely dilute solution. In an infinitely dilute solution, A molecules are dispersed so completely that they can interact only with B molecules. Consequently, the energy of the A molecules cannot change if additional pure solvent (initially at the same temperature and pressure) is added. Operationally then, we can recognize an infinitely dilute solution by mixing it with additional pure solvent; if no heat must be exchanged with the surroundings in order to keep the temperature constant, the original solution is infinitely dilute.

For the solvent, the concept of an infinitely dilute solution gives rise to the following choices, which are shown schematically in Figure 3:

For the **activity standard state of the solvent**, B, we choose pure liquid B. Then, the activity of pure liquid B is unity at any temperature. Letting $\widetilde{H}_B^o$ be the partial molar enthalpy of B in the activity standard state and $H_B^\bullet$ be the molar enthalpy of pure liquid B, we have $\widetilde{H}_B^o = H_B^\bullet$.

For the **enthalpy reference state of the solvent**, we choose B in an infinitely dilute solution. We represent the partial molar enthalpy of B in this infinitely dilute solution by $\overline{H}_B^{ref}$.

For the solute, the infinitely dilute solution involves the following choices, which are shown schematically in Figure 4:

For the **activity standard state of the solute**, we choose the hypothetical solution in which the concentration of A is one molal, and the activity of A is unity, but all of the effects of intermolecular interactions are the same as they are in an infinitely dilute solution. We represent the partial molar enthalpy of A in the activity standard state by $\widetilde{H}_A^o$.

For the **enthalpy reference state of the solute**, we choose the infinitely dilute solution and designate the partial molar enthalpy of A in this reference state by $\overline{H}_A^{ref}$. Since all of the intermolecular interactions are the same in the enthalpy reference state as they are in the activity standard state, there can be no energy change when one mole of A goes from one of these states to the other. It follows that $\widetilde{H}_A^o = \overline{H}_A^{ref}$.

It follows from these choices that the relative partial molar enthalpies of A and B in their activity standard states are $\widetilde{L}_A^o = \widetilde{H}_A^o - \overline{H}_A^{ref} = 0$ and

$$\widetilde{L}_B^o = \widetilde{H}_B^o - \overline{H}_B^{ref} = \lim_{n_A \to 0} \left(-\frac{\partial \Delta_{mix} H}{\partial n_B}\right)_{P,T,n_A}$$

The relevant process for which we can measure an enthalpy change is the isothermal mixing of n_A moles of pure solid A with n_B moles of pure liquid B to form a solution:

$$n_A A(\text{pure solid}) + n_B B(\text{pure liquid}) \xrightarrow{\Delta_{mix} H} A/B(\text{solution}, n_A, n_B)$$

Letting the enthalpy of mixing be $\Delta_{mix} H$ and the enthalpy of the resulting solution be H, we have

$$H = \Delta_{mix} H + n_A H_A^\bullet + n_B H_B^\bullet$$

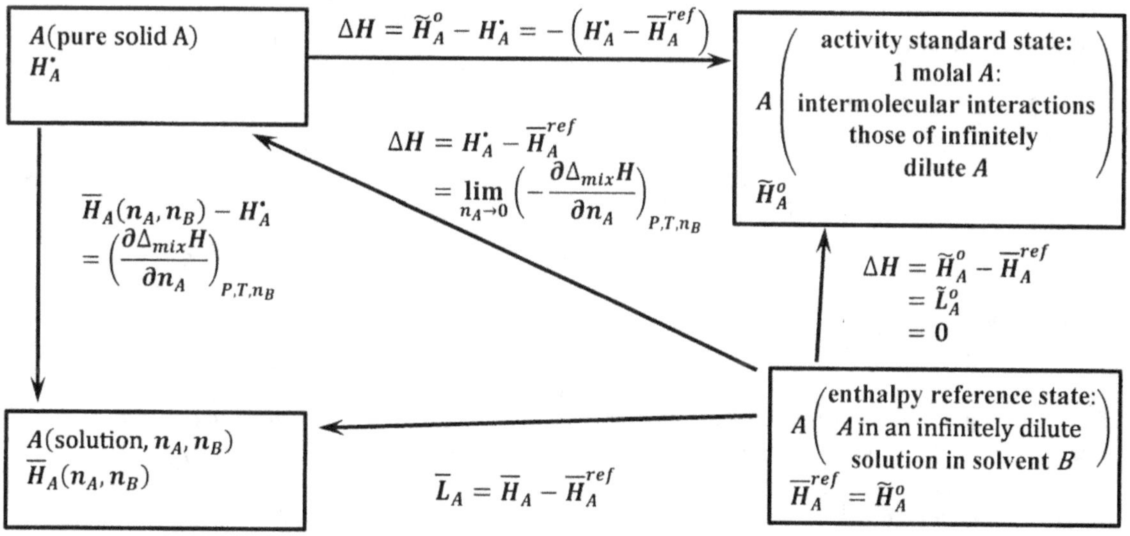

Figure 4. Cycles relating the molar enthalpy of a solute in a solution to its molar enthalpies in the pure solid, in an activity standard state, and in an enthalpy standard state.

Then

$$\overline{H}_A = \left(\frac{\partial H}{\partial n_A}\right)_{PT} = \left(\frac{\partial \Delta_{mix} H}{\partial n_A}\right)_{PT} + H_A^{\bullet}$$

and

$$\overline{H}_B = \left(\frac{\partial H}{\partial n_B}\right)_{PT} = \left(\frac{\partial \Delta_{mix} H}{\partial n_B}\right)_{PT} + H_B^{\bullet}$$

where, of course, $H, \Delta_{mix}H, \overline{H}_A$, and $\overline{H}_B$ are all functions of n_A and n_B. The partial molar enthalpies in the reference states are the limiting values of $\overline{H}_A$ and $\overline{H}_B$ as $n_A \to 0$. That is,

$$\overline{H}_A^{ref} = \lim_{n_A \to 0} \left(\frac{\partial \Delta_{mix} H}{\partial n_A}\right)_{P,T,n_B} + H_A^{\bullet}$$

and

$$\overline{H}_B^{ref} = \lim_{n_B \to 0} \left(\frac{\partial \Delta_{mix} H}{\partial n_B}\right)_{P,T,n_A} + H_B^{\bullet}$$

When we base the enthalpy reference states on the infinitely dilute solution, we find for the solute

$$\begin{aligned} -\left(\overline{H}_A - \widetilde{H}_A^o\right) &= -\left(\overline{H}_A - \overline{H}_A^{ref}\right) + \left(\widetilde{H}_A^o - \overline{H}_A^{ref}\right) \\ &= -\left(\overline{L}_A - \widetilde{L}_A^o\right) \\ &= -\overline{L}_A \\ &= -\left(\frac{\partial \Delta_{mix} H}{\partial n_A}\right)_{P,T,n_B} + \lim_{n_A \to 0} \left(\frac{\partial \Delta_{mix} H}{\partial n_A}\right)_{P,T,n_B} \end{aligned}$$

and for the solvent

$$\begin{aligned} -\left(\overline{H}_B - \widetilde{H}_B^o\right) &= -\left(\overline{H}_B - \overline{H}_B^{ref}\right) + \left(\widetilde{H}_B^o - \overline{H}_B^{ref}\right) \\ &= -\left(\overline{L}_B - \widetilde{L}_B^o\right) \\ &= -\left(\frac{\partial \Delta_{mix} H}{\partial n_B}\right)_{P,T,n_A} \end{aligned}$$

The temperature dependence of the activities becomes

$$\left(\frac{\partial \ln \tilde{a}_A}{\partial T}\right)_P = \frac{1}{RT^2}\left[-\left(\frac{\partial \Delta_{mix} H}{\partial n_A}\right)_{P,T,n_B} + \lim_{n_A \to 0} \left(\frac{\partial \Delta_{mix} H}{\partial n_A}\right)_{P,T,n_B}\right]$$

and

$$\left(\frac{\partial \ln \tilde{a}_B}{\partial T}\right)_P = -\frac{1}{RT^2}\left(\frac{\partial \Delta_{mix} H}{\partial n_B}\right)_{P,T,n_A}$$

To a good first approximation, we can measure $\Delta_{mix}H$ as a function of composition at a single temperature, determine $\overline{L}_A$ and $\overline{L}_B - \widetilde{L}_B^o$ at that temperature, and assume that these values are independent of temperature. For a more exact treatment, we can measure $\Delta_{mix}H$ as a function of composition at several temperatures and find $\overline{L}_A$ and $\overline{L}_B - \widetilde{L}_B^o$ as functions of temperature. It proves to be useful to define the ***relative partial molar heat capacity*** of A, to which we give the symbol, $\overline{J}_A$, as the temperature derivative of $\overline{L}_A$:

$$\overline{J}_A = \left(\frac{\partial \overline{L}_A}{\partial T}\right)_P$$

To illustrate the use of these ideas, let us suppose that we measure the enthalpy of mixing of solute A in 1 kg water (solvent B). We make this measurement for several quantities of A at each of several temperatures between 273.15 K and 293.15 K. For each experiment in this series, n_B is 55.51 mole and n_A is equal to the molality of A, $\underline{m}$, in the solution. We fit the experimental data to empirical equations. Let us suppose that the enthalpy of mixing data at any given temperature are described adequately by the equation $\Delta_{mix}H = \alpha_1 \underline{m} + \alpha_2 \underline{m}^2$ and that α_1 and α_2 depend linearly on T according to

$$\alpha_1 = \beta_{11} + \beta_{12}(T - 273.15)$$

and

$$\alpha_2 = \beta_{21} + \beta_{22}(T - 273.15)$$

Then

$$\begin{aligned} \left(\frac{\partial \Delta_{mix} H}{\partial n_A}\right)_{P,T,n_B} &= \left(\frac{\partial \Delta_{mix} H}{\partial \underline{m}}\right)_{P,T,n_B} \\ &= \alpha_1 + 2\alpha_2 \underline{m} \end{aligned}$$

and

$$\lim_{n_A \to 0} \left(\frac{\partial \Delta_{mix} H}{\partial n_A}\right)_{P,T,n_B} = \alpha_1$$

so that

$$\begin{aligned} \overline{L}_A &= \left(\frac{\partial \Delta_{mix} H}{\partial n_A}\right)_{P,T,n_B} - \lim_{n_A \to 0} \left(\frac{\partial \Delta_{mix} H}{\partial n_A}\right)_{P,T,n_B} \\ &= 2\alpha_2 \underline{m} \end{aligned}$$

and

$$\begin{aligned} \left(\frac{\partial \ln \tilde{a}_A}{\partial T}\right)_P &= -\frac{\overline{L}_A}{RT^2} \\ &= -\frac{2\alpha_2 \underline{m}}{RT^2} \end{aligned}$$

and

$$\begin{aligned} \overline{J}_A &= \left(\frac{\partial \overline{L}_A}{\partial T}\right)_P \\ &= 2\underline{m}\left(\frac{\partial \alpha_2}{\partial T}\right)_P \\ &= 2\underline{m}\beta_{22} \end{aligned}$$

In the experiments of this illustrative example, n_B is constant. This might make it seem that we would have to do an additional set of experiments, a set in which n_B

is varied, in order to find $\bar{L}_B$, $(\partial \ln \tilde{a}_B/\partial T)_P$, and $\bar{J}_B$. However, this is not the case. Since $\bar{L}_A$ and $\bar{L}_B$ are partial molar quantities, we have

$$n_A d\bar{L}_A + n_B d\bar{L}_B = 0$$

so that

$$d\bar{L}_B = -\left(\frac{n_A}{n_B}\right) d\bar{L}_A$$

and we can find $\bar{L}_B$ from

$$\bar{L}_B(\underline{m}) - \bar{L}_B(0) = \bar{L}_B(\underline{m}) = \int_0^{\underline{m}} -\left(\frac{2\underline{m}\alpha_2}{55.51}\right) d\underline{m} = -\frac{\underline{m}^2\alpha_2}{55.51}$$

and

$$\bar{J}_B = \left(\frac{\partial \bar{L}_B}{\partial T}\right)_P = -\frac{\underline{m}^2}{55.51}\left(\frac{\partial \alpha_2}{\partial T}\right)_{P,n_A} = -\frac{\underline{m}^2\beta_{22}}{55.51}$$

Problems

1. When we express the energy of a system as a function of entropy, volume, and composition, we have $E = E(S, V, n_1, n_2, \dots, n_\omega)$. Since S and V are extensive variables, we have $\lambda E = E(\lambda S, \lambda V, \lambda n_1, \lambda n_2, \dots, \lambda n_\omega)$. Find $(\partial(\lambda E)/\partial \lambda)_{SV}$. From this result, show that

$$G = \sum_{j=1}^{\omega} \mu_j n_j$$

2. When we express the energy of a system as a function of pressure, temperature, and composition, we have $E = E(P, T, n_1, n_2, \dots, n_\omega)$. Because P and T are independent of λ, $\lambda E = E(P, T, \lambda n_1, \lambda n_2, \dots, \lambda n_\omega)$. Show that

$$E = \sum_{j=1}^{\omega} \bar{E}_j n_j$$

3. From $E = E(P, T, n_1, n_2, \dots, n_\omega)$ and the result in problem 2, show that

$$\left[\left(\frac{\partial H}{\partial T}\right)_P - P\left(\frac{\partial V}{\partial T}\right)_P\right] dT - \left[P\left(\frac{\partial V}{\partial P}\right)_T + T\left(\frac{\partial V}{\partial T}\right)_P\right] dP = \sum_{j=1}^{\omega} n_j\, d\bar{E}_j$$

Note that at constant pressure and temperature,

$$\sum_{j=1}^{\omega} n_j\, d\bar{E}_j = 0$$

4. If pressure and temperature are constant, $E = E(n_1, n_2, \dots, n_\omega)$ and $\lambda E = E(\lambda n_1, \lambda n_2, \dots, \lambda n_\omega)$. Show that $\sum_{j=1}^{\omega} n_j\, d\bar{E}_j = 0$ follows from these relationships.

5. A solution contains n_1 moles of component 1, n_2 moles of component 2, n_3 moles of component 3, *etc.* Let $n = n_1 + n_2 + n_3 + \dots$ The mole fraction of component j is $x_j = n_j/n$. Show that

$$\left(\frac{\partial x_j}{\partial n_j}\right) = \frac{n - n_j}{n^2}$$

and, for $j \neq k$,

$$\left(\frac{\partial x_j}{\partial n_k}\right) = \frac{-n_j}{n^2}$$

What are

$$\left(\frac{\partial x_1}{\partial n_1}\right)$$

and

$$\left(\frac{\partial x_2}{\partial n_2}\right)$$

if the solution has only two components?

6. For any extensive state function, $Y(P, T, n_1, n_2, \dots, n_\omega)$, the arguments developed in this chapter lead, at constant P and T, to the equations

$$Y = n_1\bar{Y}_1 + n_2\bar{Y}_2 + \dots + n_\omega\bar{Y}_\omega$$

and

$$0 = n_1 d\bar{Y}_1 + n_2 d\bar{Y}_2 + \dots + n_\omega d\bar{Y}_\omega$$

Where $\bar{Y}_j$ is the partial molar quantity $(\partial Y/\partial n_j)_{P,T,n_{m\neq j}}$.

(a) Prove that

$$0 = x_1 d\bar{Y}_1 + x_2 d\bar{Y}_2 + \dots + x_\omega d\bar{Y}_\omega$$

(b) Prove that

$$0 = n_1\left(\frac{\partial \bar{Y}_1}{\partial n_1}\right) + n_2\left(\frac{\partial \bar{Y}_2}{\partial n_2}\right) + \dots + n_\omega\left(\frac{\partial \bar{Y}_\omega}{\partial n_\omega}\right)$$

(c) Prove that

$$0 = x_1\left(\frac{\partial \bar{Y}_1}{\partial x_1}\right) + x_2\left(\frac{\partial \bar{Y}_2}{\partial x_2}\right) + \dots + x_\omega\left(\frac{\partial \bar{Y}_\omega}{\partial x_\omega}\right)$$

7. The enthalpy of mixing is measured in a series of experiments in which solid solute, A, dissolves to form an aqueous solution. These enthalpy data are represented well by empirical equations $\Delta_{mix}H = \alpha_1 \underline{m} + \alpha_2 \underline{m}^2$, $\alpha_1 = \beta_{11} + \beta_{12}(T - 273.15)$ and $\alpha_2 = \beta_{21} + \beta_{22}(T - 273.15)$ with

$\beta_{11} = 10.0 \text{ kJ molal}^{-1}$

$\beta_{12} = -0.14 \text{ kJ molal}^{-2}\ K^{-1}$

$\beta_{21} = -3.00 \text{ kJ molal}^{-1}$

$\beta_{22} = -0.040 \text{ kJ molal}^{-2}\ K^{-1}$

Find $\overline{L}_A$, $\overline{L}_{H_2O}$, $\overline{J}_A$, and $\overline{J}_{H_2O}$ as functions of $\underline{m}_A$ and T. Find $\overline{L}_A$, $\overline{L}_{H_2O}$, $\overline{J}_A$, and $\overline{J}_{H_2O}$ for a one molal solution at 209 K. What is the value of

$$\ln \frac{\tilde{a}_A(1 \text{ molal}, 290 \text{ K})}{\tilde{a}_A(1 \text{ molal}, 273.15 \text{ K})}$$

Notes

[1] We can make other assumptions. It is possible to describe an inhomogeneous system as a collection of many macroscopic, approximately homogeneous regions.

15

Chemical Potential, Fugacity, Activity, and Equilibrium

In Chapters 11-14, we define chemical potential, fugacity, and activity. We find numerous relationships among these quantities. In §1 and §2, we summarize the principal relationships between chemical potential and fugacity and between chemical potential and activity. Thereafter, we introduce some basic ideas about the chemical potentials, fugacities, and activities of liquids, solids, solvents, and solutes. We use these ideas to relate standard Gibbs free energy changes to fugacities and activities in systems at equilibrium.

§1 The chemical potential and fugacity of a gas

The third law and the fugacity of a pure real gas.– In Chapter 11, we introduce the fugacity as a measure of the difference between the molar Gibbs free energy of a real gas, $\overline{G}(P,T)$ at pressure P, and that of the pure gas in its hypothetical ideal-gas standard state at the same temperature. We choose the standard Gibbs free energy of formation, $\Delta_f G^o(HIG^o,T)$, to be the Gibbs free energy of the real gas in its hypothetical ideal-gas standard state. Letting the gas be A, we find

$$\overline{G}(P,T) = \Delta_f G^o(HIG^o,T) + RT \ln\left[\frac{f_A^\bullet(P)}{f_A(HIG^o)}\right]$$

(real gas)

where the fugacity depends on pressure according to

$$\ln\left[\frac{f_A^\bullet(P)}{f_A(HIG^o)}\right] = \ln\left[\frac{P}{P^o}\right] + \int_0^P \left[\frac{\overline{V}_A^\bullet}{RT} - \frac{1}{P}\right] dP$$

(real gas)

and $\overline{V}_A^\bullet$ is the molar volume of the pure real gas. (In Chapter 14, we introduce a solid-bullet superscript to indicate that a particular property is that of a pure substance.) Given $\Delta_f G^o(HIG^o,T)$ and an equation of state for the real gas, we can calculate the fugacity and molar Gibbs free energy of the real gas at any pressure.

The fugacity of a pure ideal gas.– For a pure ideal gas, we have

$$\frac{\overline{V}_A^\bullet}{RT} - \frac{1}{P} = 0$$

(ideal gas)

The fugacity becomes equal to the ideal-gas pressure

$$f_A^\bullet(P) = P$$

(ideal gas)

and the Gibbs free energy relationship becomes

$$\overline{G}_A(P,T) = \Delta_f G^o(A,P^o,T) + RT \ln\left[\frac{P}{P^o}\right]$$

(ideal gas)

For pure gases, the system pressure that appears in these equations, P, is the same thing as the pressure of the gas.

The fugacity of an ideal gas in a mixture.– In Chapter 13, we find that the molar Gibbs free energy of a component of an ideal gas mixture is unaffected by the presence of the other gases. For an ideal gas, A, present at mole fraction x_A, in a system whose pressure is P, the partial pressure is $P_A = x_A P$. Since the partial pressure is the pressure that the system would exhibit if only ideal gas A were present, the molar Gibbs free energy of an ideal gas in a mixture is

$$\overline{G}_A(x_A,P,T) = \Delta_f G^o(A,P^o,T) + RT \ln\left[\frac{x_A P}{P^o}\right]$$

(ideal gas)

The chemical potential and fugacity of real gases.– In Chapter 14, we introduce the chemical potential as the partial molar Gibbs free energy. The defining relationship is

$$\mu_A = \overline{G}_A = \left(\frac{\partial G}{\partial n_A}\right)_{P,T,n_{i\neq A}}$$

(any substance in any system)

When the system is a pure substance, the chemical potential is identical to the Gibbs free energy per mole of the pure substance at the same temperature and pressure. For the chemical potential of A in a system comprised of pure A, we can write

$$\mu_A^\bullet = \overline{G}_A^\bullet = \frac{G^\bullet}{n_A} = \frac{dG^\bullet}{dn_A}$$

(any system comprised of pure A)

From Euler's theorem, we find that the Gibbs free energy of any system is the composition-weighted sum of the chemical potentials of the substances present:

$$G = \sum_{i=1}^{\omega} n_i \mu_i$$

For a pure real gas, the partial molar Gibbs free energy and the molar Gibbs free energy are the same thing; we also write

$$\mu_A^{\bullet}(P,T) = \Delta_f G^o(A, HIG^o, T) + RT \ln\left[\frac{f_A^{\bullet}(P)}{f_A(HIG^o)}\right]$$

(pure real gas A)

and introducing $\mu_A^o(T) = \Delta_f G^o(A, HIG^o, T)$, we write

$$\mu_A^{\bullet}(P,T) = \mu_A^o(T) + RT \ln\left[\frac{f_A^{\bullet}(P)}{f_A(HIG^o)}\right]$$

(pure real gas A)

Since μ_A^o, $\Delta_f G^o(A)$, and $f_A(HIG^o)$ are defined to be properties of one mole of pure A, it is not necessary to include either the solid-bullet superscript or the solid over-bar in these symbols.

In §14-11, we find that the partial molar Gibbs free energy of a component of a real-gas mixture is

$$\mu_A(P,T) = \mu_A^o(T) + RT \ln\left[\frac{f_A(P)}{f_A(HIG^o)}\right]$$

(real gas A in a mixture)

where the fugacity of A, present at mole fraction x_A in a system whose pressure is P, is given by

$$RT \ln\left[\frac{f_A(P)}{f_A(HIG^o)}\right] = \ln\left[\frac{x_A P}{P^o}\right] + \int_0^P \left[\frac{\overline{V}_A}{RT} - \frac{1}{P}\right] dP$$

(real gas A in a mixture)

where $f_A(HIG^o) = P^o = 1$ bar. The partial molar volume is a function of the system's pressure, temperature, and composition; that is,

$$\overline{V}_A(P) = \overline{V}_A(P, T, x_A, x_B, \dots, x_\omega)$$

and the fugacity depends on the same variables,

$$f_A(P) = f_A(P, T, x_A, x_B, \dots, x_\omega)$$

If the system is a mixture of ideal gases,

$$V = (n_A + n_B + \cdots + n_\omega)\, RT/P$$

and

$$\overline{V}_A = (\partial V/\partial n_A)_{PTn_{m \neq A}} = RT/P$$

The integrand becomes zero, and the fugacity relationship reduces to the ideal-gas fugacity equation introduced in Chapter 13 and repeated above.

The fugacity of a gas in any system is a measure of the difference between its chemical potential in that system and its chemical potential in its hypothetical ideal-gas standard state at the same temperature. The chemical potential of A in a particular system, μ_A, is the change in the Gibbs free energy when the amounts of the elements that form one mole of A pass from their standard states as elements into the (very large) system as one mole of substance A.

§2 The chemical potential and activity of a gas

To make predictions about processes involving substance A, we need information about the chemical potential of A. Introducing the fugacity does not introduce new information; the fugacity is merely a convenient way to relate the chemical potential to the composition of the system. The fugacity relationship is valid whether we can actually measure μ_A and $\mu_A^o(HIG^o)$ or not. To use the relationship for practical calculations, we must know both, of course.

We introduce the activity function to cope with situations in which we cannot measure the fugacity. For volatile liquids—or solids—we can obtain the Gibbs free energy of formation for both the condensed phase and for the hypothetical ideal-gas standard state. In §4, we consider the relationship between the two.

The chemical activity of substance A measures the change in the chemical potential when one mole of A in some arbitrarily chosen standard state passes into a very large system of specified composition. We introduce

$$\mu_A = \tilde{\mu}_A^o + RT \ln\left[\frac{\tilde{a}_A}{\tilde{a}_A(ss)}\right]$$

where, as always,

$$\mu_A = \left(\frac{\partial G}{\partial n_A}\right)_{P,T,n_{i \neq A}}$$

We let the activity of A in the arbitrarily chosen standard state, designated "ss", be unity, so that $\tilde{a}_A(ss) = 1$ and the chemical potential of A in this standard state is $\tilde{\mu}_A^o$. The activity, $\tilde{a}_A$, is a function of the pressure, temperature, and composition of the system.

While we are free to choose any standard state we please for the activity of a gas, the hypothetical ideal-gas standard state is the most practical. In this case, the activity of a gas is given by

$$\tilde{a}_A(P) = \frac{f_A(P)}{f_A(HIG^o)}$$

and

$$\tilde{\mu}_A^o = \mu_A^o(HIG^o)$$

Then the only difference between fugacity and activity is that fugacity has the units of pressure, whereas activity is dimensionless. For any gas in any state, we have

$$\frac{\mu_A - \mu_A^o(HIG^o)}{RT} = \ln\left[\frac{f_A(P)}{f_A(HIG^o)}\right] = \ln[\tilde{a}_A(P)]$$

For an ideal-gas mixture whose pressure is P and in

which the mole fraction of A is x_A, we have

$$\tilde{a}_A(x_A, P) = \frac{f_A(x_A, P)}{f_A(HIG^o)} = \frac{x_A P}{P^o}$$

(ideal gas mixture)

§3 The pressure-dependence of the fugacity and activity of a condensed phase

So far, we have investigated fugacity and activity only for gases. Let us now consider a system that consists entirely of substance A present as either a pure liquid or a pure solid. We assume that the temperature is fixed and that the pressure on this condensed phase can be varied. For our present discussion, it does not matter whether the condensed phase is a liquid or a solid. For specificity, let us assume that it is a liquid. We can imagine that the pressure changes are effected with the pure substance contained in a cylinder that is sealed by a frictionless piston. We ask how the fugacity and activity vary when the system pressure changes. Let the molar volume of the pure condensed phase be $\bar{V}_A^\bullet$. Since the coefficient of isothermal compression is small for condensed phases, it is often adequate to assume that $\bar{V}_A^\bullet$ is a constant. We do so here.

We have developed several designations for the molar Gibbs free energy of the condensed phase. For pure liquid A at pressure P, we can write

$$\bar{G}^\bullet(A, \ell, P) = \bar{G}_A^\bullet(\ell, P) = \mu_A^\bullet(\ell, P)$$

(pure liquid)

For the pure liquid at a pressure of one bar, it is convenient to let the molar Gibbs free energy of the liquid be equal to the molar Gibbs free energy of formation. We let

$$\bar{G}^\bullet(A, \ell, P^o) = \bar{G}_A^\bullet(\ell, P^o) = \mu_A^\bullet(\ell, P^o) = \Delta_f G^o(A, \ell, P^o)$$

(pure liquid, $P = 1$ bar)

We also have several designations for the pressure dependence of the Gibbs free energy of this liquid. Since the system consists entirely of the pure liquid, we have

$$\begin{aligned}(\partial \bar{G}^\bullet(A, \ell, P)/\partial P)_T &= (\partial \bar{G}_A^\bullet(\ell, P)/\partial P)_T \\ &= (\partial \mu_A^\bullet(\ell, P)/\partial P)_T \\ &= RT\ (\ \partial\{\ln[f_A^\bullet(\ell, P)]\}/\partial P\ )_T \\ &= RT\ (\ \partial\{\ln[\tilde{a}_A^\bullet(\ell, P)]\}/\partial P\ )_T \\ &= \bar{V}_A^\bullet(\ell, P)dP\end{aligned}$$

We are free to choose any state of any system that contains A as the standard state for the activity of substance A. Often, it is convenient to let this standard state be the pure liquid (or the pure solid) at a pressure of one bar. The activity of A is unity in the standard state. Taking the partial molar volume to be constant, $\bar{V}_A^\bullet(\ell, P) = \bar{V}_A^\bullet$, and integrating between one bar and an arbitrary pressure, P, we can express the pressure dependence of the Gibbs free energy of the pure liquid in several equivalent ways:

$$\begin{aligned}\bar{G}_A^\bullet(\ell, P) &= \bar{G}^\bullet(A, \ell, P) - \bar{G}^\bullet(A, \ell, P^o) \\ &= \bar{G}^\bullet(A, \ell, P) - \Delta_f G^o(A, \ell) \\ &= \mu_A^\bullet(\ell, P) - \mu_A^\bullet(\ell, P^o) \\ &= RT \ln\left[\frac{f_A^\bullet(\ell, P)}{f_A^\bullet(\ell, P^o)}\right] \\ &= RT \ln[\tilde{a}_A^\bullet(\ell, P)] \\ &= \bar{V}_A^\bullet(P - P^o)\end{aligned}$$

From the last equations, we see that the activity and fugacity of the pure liquid vary with the system pressure as

$$\tilde{a}_A^\bullet(\ell, P) = \frac{f_A^\bullet(\ell, P)}{f_A^\bullet(\ell, P^o)} = \exp\left[\frac{\bar{V}_A^\bullet(P - P^o)}{RT}\right]$$

At ordinary temperatures and pressures, $\bar{V}_A^\bullet(P - P^o) \ll RT$. In consequence, the system pressure must become much greater than one bar before the exponential term becomes significantly different from one. Thus, the activity of a condensed phase is approximately unity until the system reaches high pressures. At pressures near one bar, the fugacity of a condensed phase is only a weak function of pressure.

This argument provides rigorous justification for treating the activities (or concentrations) of pure solids and liquids as constants when we use equilibrium constant data to calculate the compositions of systems that are at equilibrium. (We address this issue previously in §5-17 and §13-8.)

§4 Standard states for the fugacity and activity of a pure liquid

If substance A is a liquid at one bar and the temperature of interest, pure liquid A is the standard state for the calculation of the enthalpy and Gibbs free energy of formation. From thermal measurements, we can find the standard Gibbs free energy of formation of this liquid, $\Delta_f G^o(A, \ell) = \mu_A^\bullet(\ell, P^o)$. If we can measure the vapor pressure of the substance and find an equation of state that describes the behavior of the real vapor, we can also find its fugacity and the standard Gibbs free energy of formation of its hypothetical ideal gas, $\Delta_f G^o(A, HIG^o) = \mu_A^o(HIG^o)$. From the principle that the chemical potential of substance A is the same in any two phases that are at equilibrium, it follows that the fugacity is the same in each phase.

If we choose the hypothetical ideal-gas standard state as the standard state for the activity of A, then the activity and fugacity are the same thing, and the standard state chemical potential is the same thing as the Gibbs free energy of formation of the hypothetical ideal gas.

$$\mu_A^o = \tilde{\mu}_A^o = \Delta_f G^o(A, HIG^o)$$

(activity standard state is the hypothetical ideal gas)

Alternatively, we can choose the pure liquid as the standard state for the activity of A. In this case, there are two further options: We can choose the pure liquid either at one bar pressure, P^o, or at its equilibrium vapor pressure, $P_{vp}^\bullet$. If we choose the pure liquid at P^o, we have

$$\tilde{\mu}_A^o = \mu_A^\bullet(\ell, P^o) = \Delta_f G^o(A, \ell, P^o)$$
(activity standard state is the pure liquid at P^o)

In Chapter 16, we see that the pure liquid at $P_{vp}^\bullet$ proves to be the most generally useful choice for A in a solution. For the pure liquid at $P_{vp}^\bullet$, we have

$$\tilde{\mu}_A^o = \mu_A^\bullet(\ell, P_{vp}^\bullet)$$
(activity standard state is the pure liquid at $P_{vp}^\bullet$)

Evidently, it is useful to be able to relate the quantities $\Delta_f G^o(A, HIG^o)$, $\mu_A^\bullet(\ell, P^o) = \Delta_f G^o(A, \ell, P^o)$, and $\mu_A^\bullet(\ell, P_{vp}^\bullet)$ to one another.

The difference between the Gibbs free energy of formation of the ideal gas in its hypothetical ideal-gas state and the Gibbs free energy of formation of the liquid in its standard state is a quantity that we can call the standard Gibbs free energy of vaporization, $\Delta_{vap} G^o(A, P^o)$, because both the initial and final states are at a pressure of one bar. At an arbitrary temperature, a liquid at one bar is not at equilibrium with its own ideal gas at one bar, and the standard Gibbs free energy of vaporization is not zero. We have

$$\begin{aligned}\Delta_{vap} G^o(A, P^o) &= \Delta_f G^o(A, HIG^o) - \Delta_f G^o(A, \ell, P^o)\\ &= \mu_A^o - \mu_A^\bullet(\ell, P^o)\\ &= RT \ln\left[\frac{f_A(HIG^o)}{f_A^\bullet(\ell, P^o)}\right]\end{aligned}$$

Figure 1 describes a reversible process that takes one mole of A from the pure liquid state at one bar to its hypothetical ideal-gas standard state. We first reversibly decrease the pressure applied to pure liquid A until we reach its equilibrium vapor pressure, $P_{vp}^\bullet$, at the temperature of interest. The molar Gibbs free energy change for this process is

$$\begin{aligned}\Delta_{press} G(A, \ell) &= \mu_A^\bullet(\ell, P_{vp}^\bullet) - \Delta_f G^o(A, \ell, P^o)\\ &= \mu_A^\bullet(\ell, P_{vp}^\bullet) - \mu_A^\bullet(\ell, P^o)\\ &= \int_{P^o}^{P_{vp}^\bullet} \bar{V}_A^\bullet(\ell)\, dP\end{aligned}$$

To reach the hypothetical ideal-gas standard state, we then reversibly and isothermally evaporate one mole of the liquid to its real gas. The Gibbs free energy change for this reversible process is zero, $\Delta_{vap} G(A, P_{vp}^\bullet) = 0$. Finally, we isothermally and reversibly expand the real gas to an arbitrarily low pressure, P^*, conceptually convert the real gas to an ideal gas, and compress this ideal gas from P^* to one bar. The Gibbs free energy change for these latter steps is

$$\begin{aligned}\Delta_{press} G(A, g) &= RT \ln\left[\frac{f_A(HIG^o)}{f_A^\bullet(RG, P_{vp}^\bullet)}\right]\\ &= -RT \ln\left[\frac{P_{vp}^\bullet}{P^o}\right] - RT \int_0^{P_{vp}^\bullet} \left[\frac{\bar{V}_A^\bullet(g)}{RT} - \frac{1}{P}\right] dP\end{aligned}$$

so that we can also express the standard Gibbs free energy of vaporization as

$$\Delta_{vap} G^o(A, P^o) = \Delta_{press} G(A, \ell) + \Delta_{press} G(A, g)$$

$$= \int_{P^o}^{P_{vp}^\bullet} \bar{V}_A^\bullet(\ell)\, dP - RT \ln\left[\frac{P_{vp}^\bullet}{P^o}\right] - RT \int_0^{P_{vp}^\bullet} \left[\frac{\bar{V}_A^\bullet(g)}{RT} - \frac{1}{P}\right] dP$$

Equating expressions for $\Delta_{vap} G^o(A, P^o)$, we find

$$\begin{aligned}&\Delta_f G^o(A, HIG^o) - \Delta_f G^o(A, \ell, P^o)\\ &= RT \ln\left[\frac{f_A(HIG^o)}{f_A^\bullet(\ell, P^o)}\right]\\ &= \int_{P^o}^{P_{vp}^\bullet} \bar{V}_A^\bullet(\ell)\, dP - RT \ln\left[\frac{P_{vp}^\bullet}{P^o}\right] - RT \int_0^{P_{vp}^\bullet} \left[\frac{\bar{V}_A^\bullet(g)}{RT} - \frac{1}{P}\right] dP\end{aligned}$$

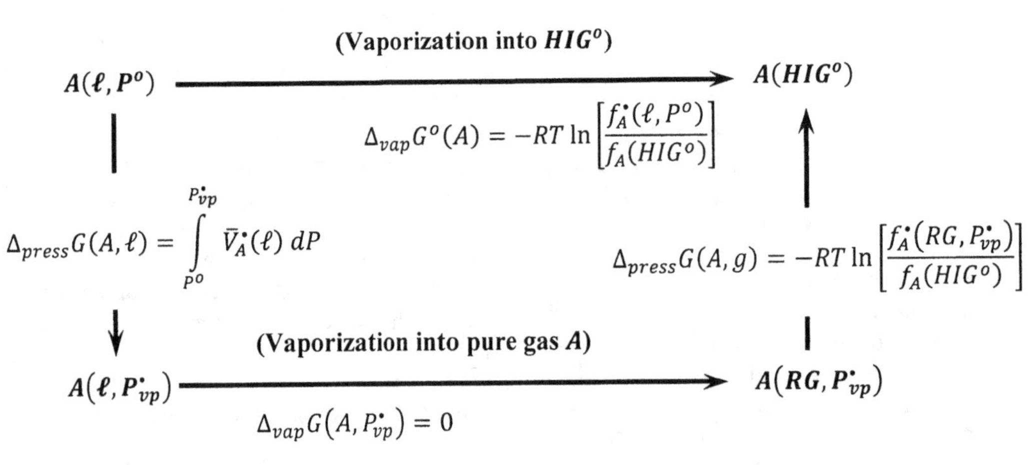

Figure 1. Cycle that takes a pure liquid to its hypothetical ideal gas state.

Thus, we can find the standard Gibbs free energy of formation of the hypothetical ideal gas from the standard Gibbs free energy of formation of the liquid, the equilibrium vapor pressure of the pure substance, and the equation of state of the pure gas:

$$\begin{aligned}&\Delta_f G^o(A, HIG^o) - \Delta_f G^o(A, \ell, P^o)\\&= \Delta_f G^o(A, HIG^o) - \mu_A^\bullet(\ell, P^o)\\&= \int_{P^o}^{P_{vp}^\bullet} \bar{V}_A^\bullet(\ell)\, dP - RT \ln\left[\frac{P_{vp}^\bullet}{P^o}\right] - RT \int_0^{P_{vp}^\bullet} \left[\frac{\bar{V}_A^\bullet(g)}{RT} - \frac{1}{P}\right] dP\end{aligned}$$

If the vapor of the substance behaves as an ideal gas, the last integral vanishes. If we also neglect the integral of the molar volume of the liquid, we have

$$\begin{aligned}\Delta_f G^o(A, HIG^o) &= \Delta_f G^o(A, \ell, P^o) - RT \ln\left[\frac{P_{vp}^\bullet}{P^o}\right]\\&= \mu_A^\bullet(\ell, P^o) - RT \ln\left[\frac{P_{vp}^\bullet}{P^o}\right]\end{aligned}$$

(ideal gas)

and $f_A^\bullet(\ell, P^o) = P_{vp}^\bullet$.

In this development, we suppose that we know $\Delta_f G^o(A, \ell, P^o)$ from thermal measurements. We can calculate the Gibbs free energy difference between the liquid and hypothetical ideal-gas standard states if we have an equation of state for the vapor. The chemical potential in the hypothetical ideal-gas standard state is

$$\mu_A^o(HIG^o) = \Delta_f G^o(A, HIG^o)$$

and the chemical potential of the pure liquid

$$\mu_A^\bullet(\ell, P^o) = \Delta_f G^o(A, \ell, P^o)$$

is expressed as a function of the fugacity of the pure liquid:

$$\mu_A^\bullet(\ell, P^o) = \mu_A^o(HIG^o) + RT \ln\left[\frac{f_A^\bullet(\ell, P^o)}{f_A(HIG^o)}\right]$$

The activity formalism provides an alternative way to express the same information. When we choose the pure liquid at one bar as the activity standard state; we set $\tilde{a}_A(\ell, P^o) = 1$. For this activity scale, the standard chemical potential becomes the Gibbs free energy of formation of the pure liquid, $\tilde{\mu}_A^o(\ell) = \Delta_f G^o(A, \ell, P^o)$. Since the chemical potential of the hypothetical ideal-gas standard state is $\mu_A^o(HIG^o) = \Delta_f G^o(A, HIG^o)$, the activity relationship becomes

$$\begin{aligned}\Delta_f G^o(A, HIG^o) - \Delta_f G^o(A, \ell, P^o) &=\\&= \mu_A^o(HIG^o) - \tilde{\mu}_A^o(\ell)\\&= -\left(\mu_A^\bullet(\ell, P^o) - \mu_A^o(HIG^o)\right)\\&= RT \ln[\tilde{a}_A(HIG^o)]\end{aligned}$$

Comparison with the previous equation shows that

$$\ln[\tilde{a}_A(HIG^o)] = \ln\left[\frac{f_A(HIG^o)}{f_A^\bullet(\ell, P^o)}\right]$$

If the pure liquid is at equilibrium with a gas mixture at P^o in which the mole fraction of A is x_A^{eq}, the fugacity of the pure liquid is equal to the fugacity of the gas in the mixture; that is,

$$f_A^\bullet(\ell, P^o) = f_A\left(RG, x_A^{eq}, P^o\right)$$

so that

$$\ln[\tilde{a}_A(HIG^o)] = \ln\left[\frac{f_A(HIG^o)}{f_A\left(RG, x_A^{eq}, P^o\right)}\right]$$

and

$$\tilde{a}_A(HIG^o) = \frac{f_A(HIG^o)}{f_A\left(RG, x_A^{eq}, P^o\right)}$$

Finally, let us take the activity standard state to be pure liquid A at 1 bar and find the activity of A in an arbitrary real-gas mixture whose pressure is P and in which the mole fraction of A is x_A. Let us represent this state as $A(RG, x_A, P)$. The activity of real gas A in this state is

$$\tilde{a}_A(RG, x_A, P) = \frac{f_A(RG, x_A, P)}{f_A^\bullet(\ell, P^o)} = \frac{f_A(RG, x_A, P)}{f_A\left(RG, x_A^{eq}, P^o\right)}$$

and the chemical potential is

$$\mu_A(RG, x_A, P) = \tilde{\mu}_A^o(\ell) + RT \ln[\tilde{a}_A(RG, x_A, P)]$$

If the real gas that is present at mole fraction x_A in a system whose pressure is P can be treated approximately as an ideal gas, these gas-fugacity terms can be approximated as

$$\begin{aligned}f_A(RG, x_A, P) &\approx f_A(IG, x_A, P)\\&= x_A P\end{aligned}$$

and

$$\begin{aligned}f_A\left(RG, x_A^{eq}, P^o\right) &\approx f_A\left(IG, x_A^{eq}, P^o\right)\\&= f_A\left(IG, P_{vp}^\bullet\right)\\&= P_{vp}^\bullet\end{aligned}$$

The activity becomes

$$\begin{aligned}\tilde{a}_A(RG, x_A, P) &\approx \tilde{a}_A(IG, x_A, P)\\&\approx x_A P / P_{vp}^\bullet\end{aligned}$$

and the chemical potential becomes

$$\begin{aligned}\mu_A(RG, x_A, P) &\approx \mu_A(IG, x_A, P)\\&= \tilde{\mu}_A^o(\ell) + RT \ln\left(\frac{x_A P}{P_{vp}^\bullet}\right)\end{aligned}$$

§5 The chemical potential, fugacity and activity of a pure solid

The relationship between the standard Gibbs free energy of formation of a substance whose standard state is a solid and the Gibbs free energy of the substance in its hypothetical ideal-gas standard state is essentially the same as described in the previous section for a liquid. In each case, to find the fugacity of the condensed phase in its standard state, it is necessary to find a reversible path that takes the condensed-phase substance to its hypothetical ideal-gas standard state. If the solid substance has a significant vapor (sublimation) pressure, the paths described for a liquid in the previous section are also available for the solid. Otherwise, it may be possible to determine the Gibbs free energy change along some more complicated path.

Of course, whether the Gibbs free energy of formation for the hypothetical ideal-gas standard state can be evaluated or not, the fugacity and activity relationships remain valid. For substance A in another state—in which A need not be pure and the pressure is generally not one bar—we have

$$\mu_A = \mu_A^o + RT \ln \left[\frac{f_A(P,T,x_A,x_B,\dots.)}{f_A(HIG^o)} \right]$$

where $\mu_A^o = \Delta_f G^o(A, HIG^o, T)$, and the fugacity of substance A, $f_A(P,T,x_A,x_B,\dots.)$, is simply an alternative expression of the difference between the chemical potential of the substance, as it occurs in the system, and its chemical potential in the hypothetical ideal-gas standard state. We write $f_A(P,T,x_A,x_B,\dots.)$ to indicate that the state of the system is specified by its pressure, temperature, and composition.

When the fugacity is difficult to measure, the activity function becomes essential. Choosing the standard state for the activity to be pure solid A in the same standard state that we use for the Gibbs free energy of formation, we have

$$\tilde{\mu}_A^o(s) = \Delta_f G^o(A, s)$$

and

$$\tilde{\mu}_A^o(s) = \mu_A^o + RT \ln \left[\frac{f_A^\bullet(s, P^o)}{f_A(HIG^o)} \right]$$

Then, for substance A in an arbitrary state at the temperature of interest:

$$\begin{aligned}
\mu_A &= \mu_A^o + RT \ln \left[\frac{f_A(P,T,x_A,x_B,\dots.)}{f_A(HIG^o)} \right] \\
&= \tilde{\mu}_A^o(s) - RT \ln \left[\frac{f_A^\bullet(s, P^o)}{f_A(HIG^o)} \right] \\
&\qquad + RT \ln \left[\frac{f_A(P,T,x_A,x_B,\dots.)}{f_A(HIG^o)} \right] \\
&= \tilde{\mu}_A^o(s) + RT \ln \left[\frac{f_A(P,T,x_A,x_B,\dots.)}{f_A^\bullet(s, P^o)} \right] \\
&= \tilde{\mu}_A^o(s) + RT \ln \tilde{a}_A
\end{aligned}$$

From one perspective, the activity function is simply a mathematical device that expresses the chemical potential relative to an arbitrarily chosen reference state. If we can measure this difference experimentally, we can find $\tilde{a}_A$ whether we can measure the fugacity of A or not. For clarity, we designate the chemical potential in the reference state as $\tilde{\mu}_A^o(s)$. When we let $\tilde{\mu}_A^o(s)$ be the Gibbs free energy of formation of the substance in the arbitrarily chosen reference state and let the activity in this reference state be unity, μ_A is the chemical potential difference between the substance in the state of interest and the chemical potential of its constituent elements in their standard states at the same temperature.

§6 Chemical potential, fugacity, and equilibrium

In Chapter 13, we develop the relationship between the standard Gibbs free energy change for a reaction and the equilibrium constant for that reaction, under the assumption that all of the substances involved in the reaction behave ideally. In the gas phase, they behave as ideal gases; when dissolved in a solution, their concentrations are proportional to their mole fractions in a gas phase at equilibrium with the solution.

We can now repeat this development using the fugacities instead of the pressures of the reacting species. Let the reaction be $aA + bB \rightarrow cC + dD$. We introduce the reactions that create the reactants and the products in their hypothetical ideal-gas standard states from their elements in their standard states. The Gibbs free energy change for creating the reactants in their hypothetical ideal-gas standard states from the elements in their standard states is

$$a\Delta_f G^o(A, HIG^o) + b\Delta_f G^o(B, HIG^o)$$

The Gibbs free energy change for creating the products in their standard states, from the same set of elements, is

$$c\Delta_f G^o(C, HIG^o) + d\Delta_f G^o(D, HIG^o)$$

Next, we introduce a set of processes, each of which adds a further quantity of a reactant or product to a very large system at equilibrium. That is, we transfer an additional a moles of pure A from its hypothetical ideal-gas standard state into a very large system in which its fugacity is $f_A(P,T,x_A,x_B,\dots.)$. The Gibbs free energy change for this process is

$$a\, RT \ln \left[\frac{f_A(P,T,x_A,x_B,\dots.)}{f_A(HIG^o)} \right]$$

Corresponding processes add b moles of pure B at fugacity $f_B(P,T,x_A,x_B,\dots.)$, etc. These processes are diagrammed in Figure 2. Since the fugacity of a substance in any state is a rigorous measure of the difference between its chemical potential in that state and its chemical potential in its hypothetical ideal-gas standard state, these Gibbs free energy changes are exact. Since the very large system is at equilibrium, there is no Gibbs free energy change when a moles of A and b moles of B react according to $aA + bB \rightarrow cC + dD$.

Let us compute the Gibbs free energy change in a clockwise direction around the reversible cycle in Figure

2. The elements in their standard states are converted first to isolated products, then to components of the large equilibrium system, then to separated reactants, and finally back to the elements in their standard states. We have

$$\begin{aligned} 0 = {} & c\,\Delta_f G^o(C, HIG^o) + d\,\Delta_f G^o(D, HIG^o) \\ & - a\,\Delta_f G^o(A, HIG^o) \\ & - b\,\Delta_f G^o(B, HIG^o) \\ & + c\,RT \ln\left[\frac{f_C(P, T, x_A, x_B, \dots.)}{f_C(HIG^o)}\right] \\ & + d\,RT \ln\left[\frac{f_D(P, T, x_A, x_B, \dots.)}{f_D(HIG^o)}\right] \\ & - a\,RT \ln\left[\frac{f_A(P, T, x_A, x_B, \dots.)}{f_A(HIG^o)}\right] \\ & - b\,RT \ln\left[\frac{f_B(P, T, x_A, x_B, \dots.)}{f_B(HIG^o)}\right] \end{aligned}$$

To express the fugacity ratios more compactly, we let

$$f_A = \frac{f_A(P, T, x_A, x_B, \dots.)}{f_A(HIG^o)}$$

etc. Letting

$$\begin{aligned} \Delta_r G^o = {} & c\,\Delta_f G^o(C, HIG^o) + d\,\Delta_f G^o(D, HIG^o) \\ & - a\,\Delta_f G^o(A, HIG^o) \\ & - b\,\Delta_f G^o(B, HIG^o) \end{aligned}$$

the Gibbs free energy around this reversible cycle simplifies to

$$\Delta_r G^o = -RT \ln \frac{f_C^c f_D^d}{f_A^a f_B^b}$$

We can express the criterion for equilibrium as

$$K_f = \frac{f_C^c f_D^d}{f_A^a f_B^b}$$

where

$$K_f = \exp\left(\frac{-\Delta_r G^o}{RT}\right)$$

We introduce the subscript, "f", to indicate that the equilibrium constant, K_f, is a function of the fugacities of the reacting substances. These relationships parallel those that we found for equilibrium among ideal gases, with real-gas fugacities replacing ideal-gas pressures.

As we did when we considered ideal-gas equilibria, let us suppose that the very large equilibrium system contains a liquid phase in which the reactants and products are soluble. The reaction can also occur in this liquid phase, and this liquid-phase reaction must be at equilibrium. Since the system is at equilibrium, each chemical species must have the same chemical potential in the solution as it does in the gas phase. Hence, the fugacities and the fugacity-based equilibrium constant are the same in both phases.

At this point, we have obtained—in principle—a complete solution to the problem of predicting the equilibrium position for any reaction. If we can find the Gibbs free energy of formation of each substance in its hypothetical ideal-gas standard state, and we can find its fugacity as a function of the composition and pressure of the system in which the reaction occurs, we can find the equilibrium constant and the equilibrium composition of the system.

In practice, a great many substances are non-

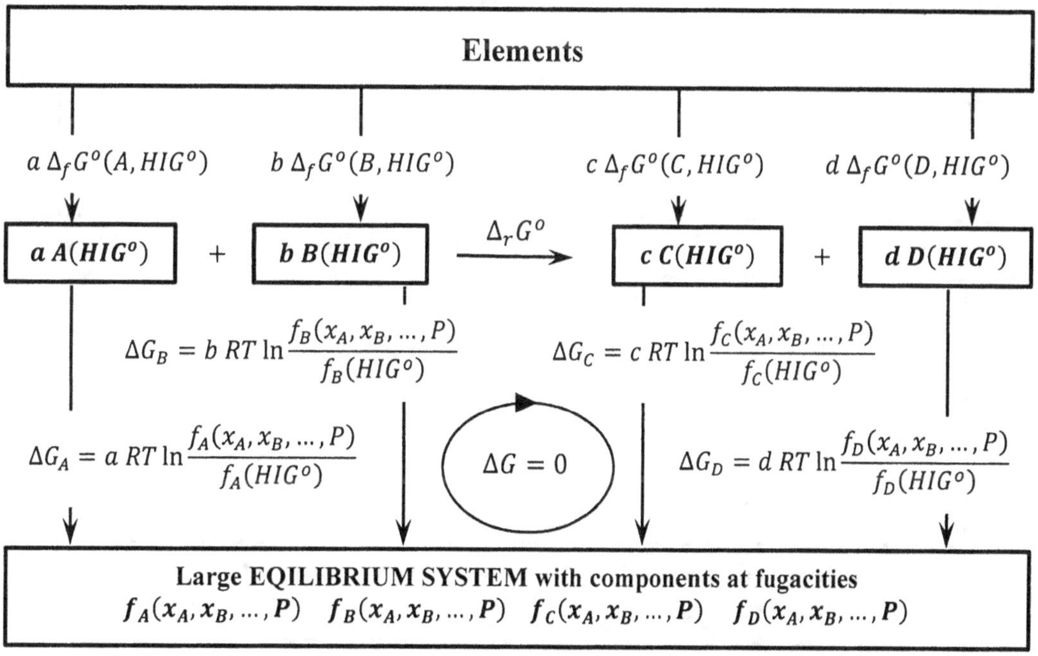

Figure 2. Relating fugacities in an equilibrium system to the Gibbs free energies of formation of the components in their hypothetical ideal gas states.

volatile. The Gibbs free energy of formation of their hypothetical ideal-gas standard states and their fugacities cannot be measured. For such substances, we have recourse to other standard states and use activities to express the equilibrium constant.

§7 Chemical potential, activity, and equilibrium

When we choose a standard state for the activity of a substance, we want the chemical potential that we calculate from the measured activity of a substance in a particular system to be the Gibbs free energy difference for the formation of the substance in that system from its constituent elements in their standard states.

Thus we must be able to measure the Gibbs free change for the process that produces the substance in its activity standard state from its constituent elements; this is the quantity that we designate as $\Delta_f \tilde{G}^o(A, ss) = \tilde{\mu}_A^o$. We must also be able to measure the difference between the chemical potential of the substance in this standard state and its chemical potential in the system of interest; this is the quantity that we designate as $RT \ln \tilde{a}_A$. The sum

$$\mu_A = \tilde{\mu}_A^o + RT \ln \tilde{a}_A$$

is then the desired Gibbs free energy change for formation of the substance in the system of interest.

When it is convenient to choose the activity standard state to be the standard state of the pure liquid or the pure solid, we have

$$\Delta_f \tilde{G}^o(A, ss) = \tilde{\mu}_A^o = \Delta_f G^o(A, \ell)$$

or

$$\Delta_f \tilde{G}^o(A, ss) = \tilde{\mu}_A^o = \Delta_f G^o(A, s)$$

For other choices, we may be unable to measure the difference between the chemical potential of the substance in the activity standard state and that of its constituent elements in their standard states; that is, the value of $\Delta_f \tilde{G}^o(A, ss) = \tilde{\mu}_A^o$ may be unknown. Nevertheless, we can describe the Gibbs free energy changes in a cycle that goes from the elements in their standard states to the chemical species in the equilibrium system and back.

For the reaction $aA + bB \rightarrow cC + dD$, this cycle is shown in Figure 3. Summing the Gibbs free energy changes in a clockwise direction around this cycle, we have

$$\begin{aligned} 0 = c\, \Delta_f \tilde{G}^o(C, ss) + d\, \Delta_f \tilde{G}^o(D, ss) - a\, \Delta_f \tilde{G}^o(A, ss) \\ - b\, \Delta_f \tilde{G}^o(B, ss) \\ + c\, RT \ln \tilde{a}_C^c + d\, RT \ln \tilde{a}_D^d - a\, RT \ln \tilde{a}_A^a - b\, RT \ln \tilde{a}_B^b \end{aligned}$$

Writing $\Delta_r \tilde{G}^o$ to emphasize that the standard Gibbs free energy change for the reaction is now a difference between the chemical potentials of reactants and products in their activity standard states, we have

$$\begin{aligned} \Delta_r \tilde{G}^o &= c\, \Delta_f \tilde{G}^o(C, ss) + d\, \Delta_f \tilde{G}^o(D, ss) \\ &\quad - a\, \Delta_f \tilde{G}^o(A, ss) - b\, \Delta_f \tilde{G}^o(B, ss) \\ &= \Delta_r \tilde{\mu}^o \end{aligned}$$

and the equation for the Gibbs free energy change around the cycle becomes

$$\begin{aligned} \Delta_r \tilde{G}^o &= \Delta_r \tilde{\mu}^o \\ &= -RT \ln \frac{\tilde{a}_C^c \tilde{a}_D^d}{\tilde{a}_A^a \tilde{a}_B^b} \end{aligned}$$

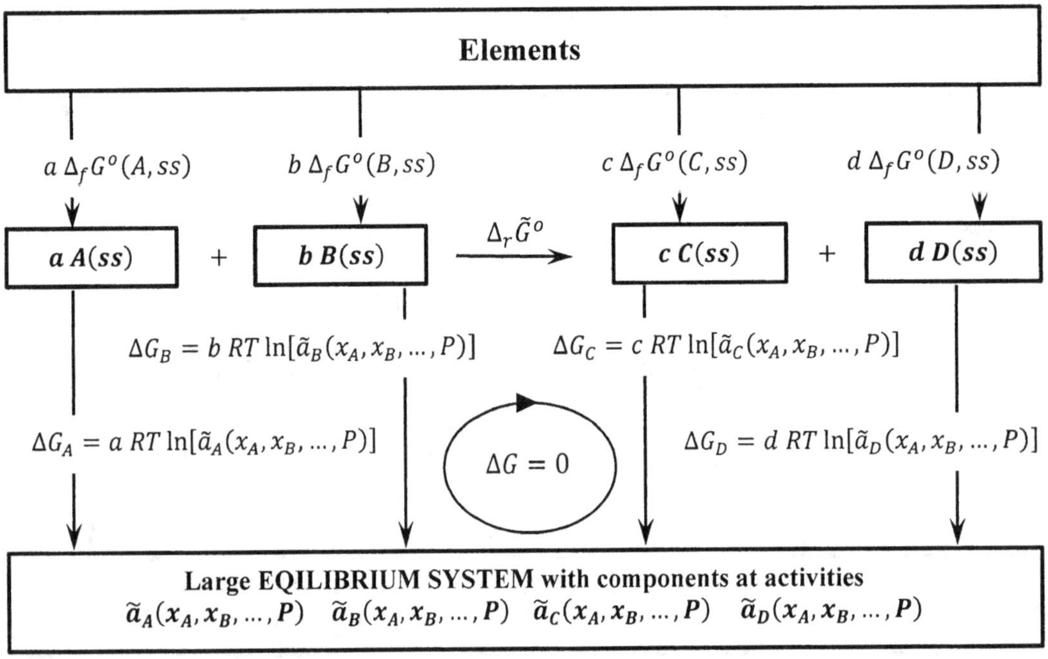

Figure 3. Relating activities in an equilibrium system to the Gibbs free energies of formation of the components in their activity standard states.

Since the chemical reaction is at equilibrium, the activities term is constant. We have

$$K_a = \frac{\tilde{a}_C^c \tilde{a}_D^d}{\tilde{a}_A^a \tilde{a}_B^b}$$

and

$$K_a = \exp\left(\frac{-\Delta_r \tilde{G}^o}{RT}\right) = \exp\left(\frac{-\Delta_r \tilde{\mu}^o}{RT}\right)$$

We append the subscript, "a", to indicate that the equilibrium constant, K_a, is expressed in terms of the activities of the reacting substances. This cycle demonstrates the underlying logic for our calculation of K_a from our definition of activity, $\mu_A = \tilde{\mu}_A^o + RT \ln \tilde{a}_A$, our definition of $\Delta_r \mu$, and the fact that

$$\Delta_r \mu = \sum_{j=1}^{\omega} \mu_j \nu_j = 0$$

is a criterion for equilibrium.

Introducing activity coefficients, we have

$$K_a = \frac{\tilde{a}_C^c \tilde{a}_D^d}{\tilde{a}_A^a \tilde{a}_B^b} = \left(\frac{c_C^c c_D^d}{c_A^a c_B^b}\right)\left(\frac{\gamma_C^c \gamma_D^d}{\gamma_A^a \gamma_B^b}\right)$$

When we know K_a and can estimate the activity coefficients, we can predict the thermodynamic properties of a system whose component concentrations are known. In general, to determine K_a and the activity coefficients is more difficult than to determine the equilibrium concentrations. Over narrow ranges of concentrations it is often adequate to assume that the activity-coefficient function,

$$K_\gamma = \frac{\gamma_C^c \gamma_D^d}{\gamma_A^a \gamma_B^b}$$

is approximately constant. When this is the case, the concentration function,

$$K_c = \frac{c_C^c c_D^d}{c_A^a c_B^b}$$

is approximately constant. Then, a direct measurement of K_c in one equilibrium system makes it possible to predict the position of equilibrium in other similar systems.

The relationship between the equilibrium constant and the standard Gibbs free energy change for a reaction is extremely useful. If we can calculate the standard Gibbs free energy change from tabulated values, we can find the equilibrium constant and predict the position of equilibrium for a particular system. Conversely, if we can measure the equilibrium constant, we can find the standard Gibbs free energy change, $\Delta_r \tilde{G}^o$.

This brings us back to a central challenge in our development. We now have rigorous relationships between the Gibbs free energy and the fugacity or activity of a substance. To use these relationships, we must be able to relate the fugacity or activity to the concentration of a substance in any particular system that we want to study. For many common systems, this is difficult. In Chapter 16, we discuss some basic approaches to accomplishing it.

§8 The rate of Gibbs free energy change with extent of reaction

In §13-5, we demonstrate that the Gibbs free energy change for a reaction among ideal gases is the same thing as the rate at which the Gibbs free energy of the system changes with the extent of reaction. That is, for an ideal-gas reaction at constant temperature and pressure, we find $\Delta_r G = (\partial G / \partial \xi)_{TP}$. We can now show that this conclusion is valid for any reaction.

With the introduction of the activity function, we have developed a very general expression for the Gibbs free energy of any substance in any system. For substance A at a fixed temperature, we have

$$\overline{G}_A = \mu_A = \tilde{\mu}_A^o + RT \ln \tilde{a}_A$$

For a reaction that we describe with generalized substances and stoichiometric coefficients as

$$|\nu_1| X_1 + |\nu_2| X_2 + \cdots + |\nu_i| X_i \rightarrow |\nu_j| X_j + |\nu_k| X_k + \cdots + |\nu_\omega| X_\omega$$

we can write the Gibbs free energy change in several equivalent ways:

$$\Delta_r G = \sum_{j=1}^{\omega} \nu_j \overline{G}_j = \Delta_r \mu = \sum_{j=1}^{\omega} \nu_j \mu_j \leq 0$$

The Gibbs free energy of the system is a function of temperature, pressure, and composition,

$$G = G(P, T, n_1, n_2, \dots, n_\omega)$$

To introduce the dependence of the Gibbs free energy of the system on the extent of reaction, we use the stoichiometric relationships $n_j = n_j^o + \nu_j \xi$. (n_j is the number of moles of the jth reacting species; n_j^o is the number of moles of the jth reacting species when ξ =0. If the kth substance does not participate in the reaction, $\nu_k = 0$.) Then,

$$G = G(P, T, n_1^o + \nu_1 \xi,\ n_2^o + \nu_2 \xi, \dots, n_\omega^o + \nu_\omega \xi)$$

At constant temperature, pressure, and composition, the dependence of the Gibbs free energy on the extent of reaction is

$$\left(\frac{\partial G}{\partial \xi}\right)_{PTn_m} = \sum_{j=1}^{\omega}\left(\frac{\partial G}{\partial(n_j^o + \nu_j\xi)}\right)_{PTn_{m\neq j}}\left(\frac{\partial(n_j^o + \nu_j\xi)}{\partial \xi}\right)_{PTn_m} = \sum_{j=1}^{\omega}\nu_j\mu_j$$

It follows that

$$\left(\frac{\partial G}{\partial \xi}\right)_{PTn_m} = \Delta_r G \leq 0$$

expresses the thermodynamic criteria for change when the process is a chemical reaction.

If a reacting system is not at equilibrium, the extent of reaction is time-dependent. We see that the Gibbs free energy of a reacting system depends on time according to

$$\frac{dG}{dt} = \left(\frac{\partial G}{\partial \xi}\right)_{PTn_m}\left(\frac{d\xi}{dt}\right) = \Delta_r\mu\left(\frac{d\xi}{dt}\right)$$

Problems

Use data from the table below to find the thermodynamic properties requested in problems 1 to 7.

Properties	CH_3OH	CH_3CH_2OH
Density, g cm^{-3} at 20 C	0.7914	0.7893
Mol mass, g mol^{-1}	32.04	46.07
bp, C	64.6	78.2
$\Delta_f G^o(300K, HIG^o)$ kJ mol^{-1}	–159.436	–162.934
Vapor pressure at 320 K, bar	0.5063	0.2764
Virial coefficient, B, m^3 mol^{-1}	-1.421×10^{-3}	-2.710×10^{-3}

1. Find the chemical potentials of the pure gases, assuming that they are ideal, taking the hypothetical ideal gas standard state as the standard state for activity ($f = \tilde{a} = P$).

2. Find the chemical potentials of the mixed gases, assuming that they are ideal, taking the hypothetical ideal gas standard state as the standard state for activity ($f = \tilde{a}_A = x_A P$).

3. Find the chemical potentials of the mixed gases, assuming them to obey the Virial equation,

$$PV/RT = 1 + (BP/RT),$$

assuming that the partial molar volumes in the mixture are equal to the partial molar volumes of the pure gases at the same pressure, and taking the hypothetical ideal gas standard state as the standard state for activity.

4. Find the standard chemical potentials of the pure liquids at 320 K, assuming that the gases behave ideally.

5. Find the standard chemical potentials of the pure liquids at 320 K, assuming that the gases obey the Virial equation.

6. Find the chemical potential of the pure liquids as a function of pressure, assuming that the partial molar volumes of the pure liquids are constant and that the gases obey the Virial equation.

7. Find the activity and chemical potential of the pure liquids at 101 bar, taking the pure liquids at 1 bar as the standard state for activity.

8. A system is created by mixing one mole of gas A with one mole of gas C. Reaction occurs according to the stoichiometry $A + C \rightleftharpoons D$. Assume that the behaviors of these gases in their equilibrium mixture are adequately approximated by the Virial equations $P\overline{V}_A/RT = 1 + (B_A P/RT)$, etc.

(a) Show that the fugacity of gas A is given by

$$\ln f_A = \ln\left(\frac{x_A P}{P^o}\right) + \frac{B_A P}{RT}$$

(b) Write an equation for $\mu_A = \mu_A(x_A, P)$ at constant temperature, T.

(c) Write an equation for $\Delta_r\mu$.

(d) Assume that $\Delta_r\mu^o = \mu_D^o - \mu_A^o - \mu_C^o = 1000$ J mol^{-1}. If all three gases behave ideally ($B_A = B_C = B_D = 0$), what is K_P for this reaction at 300 K? At equilibrium at 300 K and 1 bar, what are the mole fractions of A, C, and D?

(e) Under the assumptions in part (d), what are the equilibrium mole fractions of A, C, and D at 300 K and 10 bar?

(f) Suppose that, contrary to the assumptions in (d) and (e), the Virial coefficients are not zero and that $B_A = B_C = B_D = -10^{-3}$ m^3 mol^{-1}. At equilibrium at 300 K and 1 bar, what are the mole fractions of A, C, and D?

(g) Under the assumptions in part (f), what are the equilibrium mole fractions of A, C, and D at 300 K and 10 bar?

9. Suppose that the reaction $A + C \rightleftharpoons D$ occurs in an inert solvent and that it is convenient to express concentrations as molarities. A frequently convenient choice of activity standard state for a solute is a hypothetical one-molar solution in which the chemical potential of the solute is equal to the chemical potential of the solute in a very ("infinitely") dilute solution in the same solvent. Then $\tilde{a}_A = [A]\gamma_A$, and $\gamma_A \to 1$ as $[A] \to 0$. The thermodynamic equilibrium constant becomes

$$K_a = \frac{\tilde{a}_D}{\tilde{a}_A \tilde{a}_C} = \frac{[D]}{[A][C]} \frac{\gamma_D^d}{\gamma_A^a \gamma_C^c} = K_c K_\gamma$$

where we introduce

$$K_c = \frac{[D]}{[A][C]}$$

and

$$K_\gamma = \frac{\gamma_D^d}{\gamma_A^a \gamma_C^c}$$

In a very dilute solution, $K_\gamma \to 1$ and $K_c = K_a$. Therefore, we can estimate K_a by finding the limiting value of K_c as all of the concentrations become very small. From values of K_c at higher concentrations, we can develop an empirical equation for K_γ. The form of this equation can be anything that can adequately represent the experimental data. Note, however, that finding an empirical model for K_γ does not solve the problem of finding empirical models for γ_A, γ_C, and γ_D individually.

(a) Given that the hypothetical one-molar solution is chosen to be the activity standard state for all three species, what is the physical significance of $\Delta_r\tilde{\mu}^o$?

(b) A simple function that has the properties required of γ_A is $\gamma_A = \alpha^{[A]}$, where α is a constant. Represent , γ_C and , γ_D by similar functions and show that this leads to $\ln K_\gamma = \beta_A [A] + \beta_C[C] + \beta_D[D]$, where β_A, β_C, and β_D are constants.

(c) A series of solutions is prepared. The equilibrium concentrations of A, B, and D in these solutions are given below. Calculate K_c for each solution. Estimate K_a and the parameters β_A, β_C, and β_D in the equation of part (b).

$[A]$	[B]	[C]
1.96×10^{-3}	1.96×10^{-3}	3.84×10^{-5}
7.85×10^{-3}	1.85×10^{-3}	1.45×10^{-4}
3.94×10^{-2}	1.44×10^{-3}	5.57×10^{-4}
1.99×10^{-1}	7.15×10^{-4}	1.29×10^{-3}
9.98×10^{-1}	2.84×10^{-4}	1.72×10^{-3}
1.85×10^{-3}	7.85×10^{-3}	1.45×10^{-4}
1.44×10^{-3}	3.94×10^{-2}	5.60×10^{-4}
6.97×10^{-4}	1.99×10^{-1}	1.30×10^{-3}
2.38×10^{-4}	9.98×10^{-1}	1.76×10^{-3}
9.32×10^{-3}	7.32×10^{-3}	6.78×10^{-4}
3.24×10^{-2}	3.04×10^{-2}	9.57×10^{-3}
1.06×10^{-1}	1.04×10^{-1}	9.59×10^{-2}

(d) Using the values you find in part (c), estimate the equilibrium concentrations of A, B, and D when a solution is prepared by mixing one mole of A with one mole of C and sufficient solvent to make 1 L of solution at equilibrium.

10. An ester, RCO_2R', undergoes hydrolysis in an ether solvent:

$$RCO_2R' + H_2O \rightleftharpoons RCO_2H + HOR'$$

We can express the activity of any of these species as the product of a concentration (in any convenient units) and an activity coefficient. When all of the reactants and products are present at low concentrations, the activity coefficients are approximately unity. The standard state for each species becomes a hypothetical solution of unit concentration in which the chemical potential (per mole) of that species is the same as its chemical potential in an arbitrarily (infinitely) dilute solution. A solution is prepared by mixing 2×10^{-3} mole of the ester and 10^{-1} mole water in sufficient ether to make 1 L of solution. When equilibrium is reached, the acid and alcohol concentrations are 9.66×10^{-3} molar.

(a) What is the equilibrium constant for this reaction?

(b) A solution is prepared by mixing 2×10^{-2} mole of the acid, 3×10^{-2} mole of the alcohol, and 10^{-1} mole of water in sufficient ether to make 1 L of solution. When equilibrium is reached, what is the concentration of the ester?

(c) With this choice of the standard states, what physical process does the standard chemical-potential change, $\Delta\tilde{\mu}^o$, describe?

16

The Chemical Activities of the Components of a Solution

§1 Solutions whose components are in equilibrium with their own gases

We are frequently interested in equilibrium processes that occur in a solution at a constant temperature. If we are able to find the activities of the species making up the solution, we can describe the thermodynamics of such processes. Many experimental methods have been developed for the measurement of the activities of species in solution. In general, the accurate measurement of chemical activities is experimentally exacting. In this chapter, we consider some of the basic concepts involved. We focus primarily on molecular solvents and solutes; that is, neutral molecules that exist as such in solution. We introduce a simplified model, called the ideal solution model, which is often a useful approximation, particularly for dilute solutions. In §16-§18, we touch on the special issues that arise when we consider the activities of dissolved ions.

One way to find activities is to find the composition and pressure of the gas phase that is in equilibrium with the solution. If the gases are not ideal, we also need experimental data on the partial molar volumes of the components in the gas phase. Collecting such data is feasible for solutions of volatile molecular liquids. For solutions of electrolytes or other non-volatile components, other methods are required.

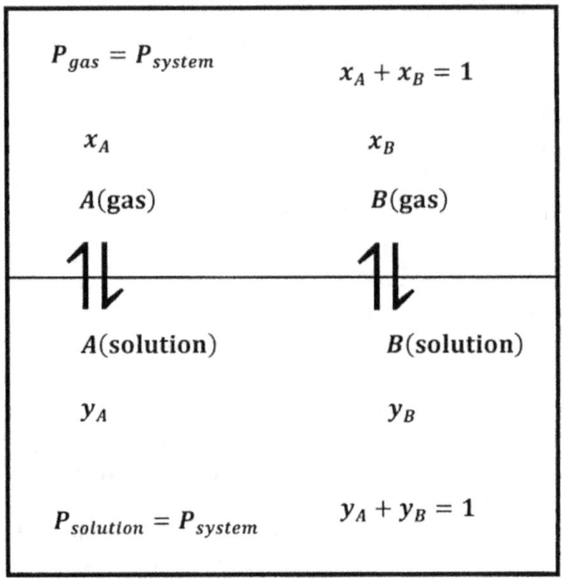

Figure 2. Liquid–gas equilibrium for a solution.

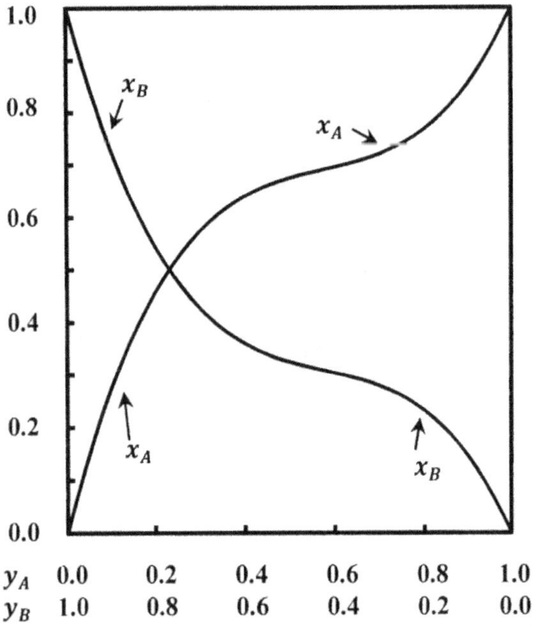

Figure 1. Equilibrium mole fractions in gas and liquid phases.

The curves sketched in Figure 1 describe a system containing components A and B. The mole fractions in the solution and the mole fractions in the gas are related in a non-linear way. Let the mole fractions in the gas be x_A and x_B; let those in the solution be y_A and y_B. We have $x_A + x_B = 1$ and $y_A + y_B = 1$. At equilibrium, both phases are at the same pressure, P.

We imagine obtaining the data we need about this system by preparing many mixtures of A and B. Beginning with an entirely liquid system at some applied pressure, we slowly decrease the applied pressure until the applied pressure becomes equal to the equilibrium pressure, P, and the liquid begins to vaporize. Figure 2 shows this system schematically. We determine the compositions of the gas and liquid phases by chemical analysis; for each system, we determine P, x_A, x_B, y_A, and y_B. From these data we can develop empirical equations that express P, x_A, and $x_B$as functions of y_A; that is, we have $P = P(y_A)$, $x_A = x_A(y_A)$, and $x_B = x_B(y_A)$. Finally, we can find the value of the products $x_A P$ and $x_B P$. Figure 3 illustrates a possible function $P = P(y_A)$ and the

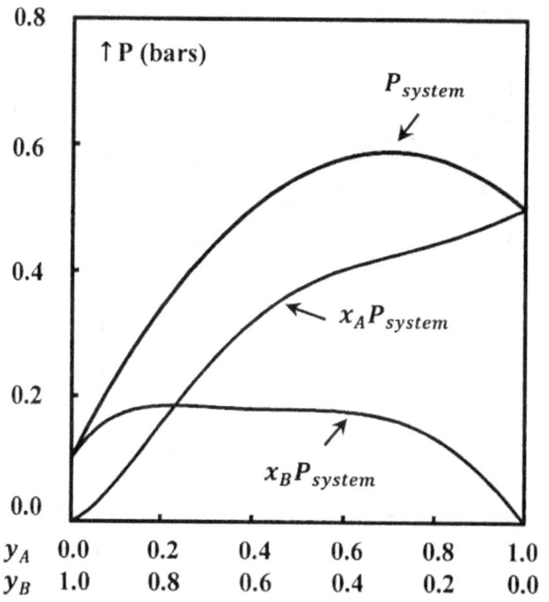

Figure 3. Equilibrium pressures over a solution.

products $x_A P$, and $x_B P$, when the gas-phase mole fractions depend on y_A as shown in Figure 1.

With the hypothetical ideal gas standard state as the standard state for A in the gas phase, we see in §14-11 that the chemical potential of A in the gas phase is

$$\mu_A(g,P,x_A,x_B) = \Delta_f G^o(A,HIG^o) + RT\ln\left[\frac{x_A P}{P^o}\right] + RT\int_0^P \left(\frac{\bar{V}_A(g)}{RT} - \frac{1}{P}\right) dP$$

(any gas; activity standard state is HIG°)

where $\bar{V}_A(g)$ is the partial molar volume and x_A is the mole fraction of A in the gaseous mixture. The fugacity and activity of A in the gas phase are given by

$$\ln[\tilde{a}_A, P, x_A, x_B] = \ln\left[\frac{f_A(g,P,x_A,x_B)}{f_A(HIG^o)}\right] = \ln\left[\frac{x_A P}{P^o}\right] + \int_0^P \left(\frac{\bar{V}_A(g)}{RT} - \frac{1}{P}\right) dP$$

and the standard state chemical potential is

$$\mu_A^o = \Delta_f G^o(A,HIG^o)$$

We want to express the chemical potential of A in the liquid solution using the properties of the solution. To do so, we introduce the chemical activity of component A. We write $\tilde{a}_A(P,y_A,y_B)$ to represent the activity of A in a solution at pressure P and in which the composition is specified by the mole fractions y_A and y_B. If it suits our purposes, we are free to choose a standard state for the activity of A in the liquid solution that is different from the standard state we choose for A in the gas phase.

For reasons that become apparent below, it is often useful to choose the standard state for the activity of A in the liquid solution to be pure liquid A at its equilibrium vapor pressure, $P_A^\bullet$. We represent the chemical potential of A in this standard state by $\tilde{\mu}_A^o(\ell, P_A^\bullet)$. Note that this state is not identical to the standard state for the pure liquid, for which the pressure is one bar and the chemical potential is $\Delta_f G^o(A,\ell)$. The chemical potential and the activity of A in the solution are related by

$$\mu_A(P,y_A,y_B) = \tilde{\mu}_A^o(\ell,P_A^\bullet) + RT\ln[\tilde{a}_A(P,y_A,y_B)]$$

(any solution; activity standard state for A in solution is pure liquid A at its equilibrium vapor pressure)

Since the system is at equilibrium, we have

$$\mu_A(g,P,x_A,x_B) = \mu_A(P,y_A,y_B)$$

Equating our equations for these quantities, we find

$$\ln[\tilde{a}_A(P,y_A,y_B)] = \frac{\Delta_f G^o(A,HIG^o) - \tilde{\mu}_A^o(\ell,P_A^\bullet)}{RT} + \ln\left[\frac{x_A P}{P^o}\right] + \int_0^P \left(\frac{\bar{V}_A(g)}{RT} - \frac{1}{P}\right) dP$$

This equation gives the activity of A in a liquid solution whose state is specified by the liquid-phase mole fraction y_A. In particular, it must give the activity of A in the "solution" for which $y_A = 1$ and $y_B = 0$; of course, this "solution" is pure liquid A. At equilibrium with pure liquid A, the gas phase contains pure gaseous A; therefore, we have $x_A = 1$ and $P = P_A^\bullet$. The gas-phase partial molar volume is that for the pure gas, $\bar{V}_A^\bullet(g)$. Moreover, this "solution" is the standard state for the activity of component A, for which the activity of A is unity; that is, $\tilde{a}_A(P,y_A,y_B) = \tilde{a}_A(P_A^\bullet,1,0) = 1$. Making these substitutions into our equation for $\ln[\tilde{a}_A(P,y_A,y_B)]$ and rearranging, we find

$$\frac{\Delta_f G^o(A,HIG^o) - \tilde{\mu}_A^o(\ell,P_A^\bullet)}{RT} = -\ln\left[\frac{P_A^\bullet}{P^o}\right] - \int_0^{P_A^\bullet} \left(\frac{\bar{V}_A^\bullet(g)}{RT} - \frac{1}{P}\right) dP$$

Substituting this result into our general equation for $\ln[\tilde{a}_A(P,y_A,y_B)]$ we find a completely general function for the activity of component A.

$$\ln[\tilde{a}_A(P,y_A,y_B)] = \ln\left[\frac{x_A P}{P_A^\bullet}\right] - \int_0^{P_A^\bullet} \left(\frac{\bar{V}_A^\bullet(g)}{RT} - \frac{1}{P}\right) dP + \int_0^P \left(\frac{\bar{V}_A(g)}{RT} - \frac{1}{P}\right) dP$$

(any solution; x_A is the mole fraction in the gas; the standard state for A in solution is the pure liquid at its equilibrium vapor pressure)

In such a system, the roles of solute and solvent are interchangeable. Interchanging the labels "A" and "B" gives an equation for the activity of component B.

As circumstances warrant, several approximations can be applied to this result. When the partial molar volume of A in the gas, $\bar{V}_A(g)$, is not available, some approximation is required. Perhaps the least drastic approximation is that introduced in §14-11. We equate the unknown partial molar volume to the partial molar volume of the pure real gas at the same system pressure. Setting $\bar{V}_A(g) = \bar{V}_A^\bullet(g)$, we have

$$\ln[\tilde{a}_A(P, y_A, y_B)] = \ln\left[\frac{x_A P}{P_A^\bullet}\right] + \int_{P_A^\bullet}^{P} \left(\frac{\bar{V}_A^\bullet(g)}{RT} - \frac{1}{P}\right) dP$$

Other approximations lead to greater simplifications. In the following sections, we discuss several. All of them assume that the components behave ideally in the gas phase. In this case, the integrals in our general equation for $\ln[\tilde{a}_A(P, y_A, y_B)]$ vanish. Then,

$$\ln[\tilde{a}_A(P, y_A, y_B)] = \ln\left[\frac{x_A P}{P_A^\bullet}\right]$$

(solution;
x_A is the mole fraction in the ideal gas
that is at equilibrium with the solution)

Our general result gives the activity of component A in solution using the mole fraction of its own vapor in the equilibrium system. When we have an empirical function, $x_A = x_A(y_A)$, that relates the mole fraction in the gas to that in the solution, we can make this substitution and express the activity of A in the solution using its concentration in the solution. In the next several sections, we develop some basic methods for finding $x_A = x_A(y_A)$.

§2 Raoult's law and ideal solutions

An ***ideal solution*** is a homogeneous liquid solution that is at equilibrium with an ideal-gas solution in which the vapor pressure of each component satisfies Raoult's law[1]. Since the gas is ideal, the partial pressure of A is $P_A = x_A P$. Raoult's law asserts a relationship among the gas- and solution-phase mole fractions of A, the vapor pressure of the pure liquid, and the pressure of the system:

$$P_A = x_A P = y_A P_A^\bullet$$

(Raoult's law)

For a binary mixture of A and B that satisfies Raoult's law, we have also that $P_B = x_B P = y_B P_B^\bullet$, and the total pressure becomes $P = P_A + P_B = y_A P_A^\bullet + y_B P_B^\bullet$. The lines sketched in Figure 4 show how P_A, P_B, and P vary with the solution-phase composition when the solution is ideal.

When the standard state for A in solution is taken to be pure liquid A at its equilibrium vapor pressure, substitution of Raoult's law into the results in §1 gives the activity of component A in an ideal solution as

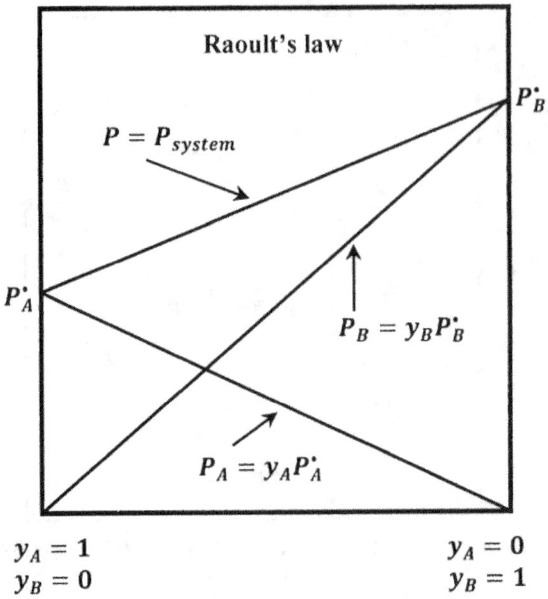

Figure 4. Equilibrium pressures over an ideal solution.

$$\ln[\tilde{a}_A(P, y_A, y_B)] = \ln\left[\frac{x_A P}{P_A^\bullet}\right] = \ln\left[\frac{y_A P_A^\bullet}{P_A^\bullet}\right] = \ln y_A$$

and

$$\tilde{a}_A(P, y_A, y_B) = y_A$$

(ideal solution, Raoult's law)

In general, the activity and chemical potential of a component depend on pressure. If the solution is ideal, we see that the system pressure is fixed by $P = y_A P_A^\bullet + y_B P_B^\bullet$, and the pure-component vapor pressures depend only on temperature. Since for the binary solution, $y_B = 1 - y_A$, we can write the chemical potential of component A as

$$\mu_A(P, y_A, y_B) = \mu_A(y_A) = \tilde{\mu}_A^o(\ell, P_A^\bullet) + RT \ln y_A$$

(ideal solution)

We can also use relationships we develop earlier to find another representation for $\tilde{\mu}_A^o(\ell, P_A^\bullet)$. The chemical potential of A in the liquid phase is the same as in the gas. Using the chemical potential for A in the gas phase that we find in §1, we have

$$\begin{aligned}\mu_A(P, y_A, y_B) &= \mu_A(g, P, x_A, x_B) \\ &= \Delta_f G^o(A, HIG^o) + RT \ln\left[\frac{x_A P}{P^o}\right] \\ &= \Delta_f G^o(A, HIG^o) + RT \ln\left[\frac{P_A^\bullet}{P^o}\right] \\ &\quad + RT \ln y_A\end{aligned}$$

and hence,

$$\tilde{\mu}_A^o(\ell, P_A^\bullet) = \Delta_f G^o(A, HIG^o) + RT \ln \left[\frac{P_A^\bullet}{P^o}\right]$$

In §15-4, we find, for an ideal gas,

$$\Delta_f G^o(A, HIG^o) + RT \ln \left[\frac{P_A^\bullet}{P^o}\right] = \Delta_f G^o(A, \ell) + \int_{P^o}^{P_A^\bullet} \bar{V}_A^\bullet(\ell)\, dP$$

so that the chemical potential of the pure liquid at its vapor pressure is also given by

$$\tilde{\mu}_A^o(\ell, P_A^\bullet) = \Delta_f G^o(A, \ell) + \int_{P^o}^{P_A^\bullet} \bar{V}_A^\bullet(\ell)\, dP$$

The integral is the difference between the Gibbs free energy of the pure liquid at its vapor pressure and that of the pure liquid at $P^o = 1$ bar. Note that we can obtain the same result much more simply by integrating $(dG_A^\bullet)_T = \bar{V}_A^\bullet dP$ between the same two states. In §15-3, we see that the value of the integral is usually negligible. To a good approximation, we have

$$\tilde{\mu}_A^o(\ell, P_A^\bullet) \approx \Delta_f G^o(A, \ell)$$

(ideal solution)

§3 Expressing the activity coefficient as a deviation from Raoult's law

If the components behave ideally in the gas phase and if pure liquid A at its equilibrium vapor pressure, $P_A^\bullet$, is the standard state for the activity of A in solution, we find in §2 that the activity of component A is $\tilde{a}_A(P, y_A, y_B) = x_A P / P_A^\bullet$. Experimentally, we determine a relationship between the mole fractions of A in the gas and liquid phases. By expressing this relationship as the function $x_A = x_A(y_A)$, we can express the activity as a function of y_A. If Raoult's law is obeyed, we have seen that this function is $x_A(y_A) = y_A P_A^\bullet / P$, and the activity is $\tilde{a}_A(P, y_A, y_B) = y_A$.

If Raoult's law is not obeyed, we must find an alternative function that adequately describes the experimentally observed relationship between x_A and y_A. As we note in §14-10, we want to construct this function so that it approaches y_A whenever the behavior of the solution approaches the behavior of an ideal solution. We can accomplish this by defining the activity coefficient for component A, $\gamma_A = \gamma_A(P, y_A, y_B)$, by the equation

$$\tilde{a}_A(P, y_A, y_B) = y_A \gamma_A(P, y_A, y_B)$$

(Raoult's law activity)

where the argument lists serve to emphasize that the activity and the activity coefficient are functions of the same thermodynamic variables.

Dropping the argument lists and equating the two activity relationships, we have

$$\frac{x_A P}{P_A^\bullet} = y_A \gamma_A$$

so that the activity coefficient is

$$\gamma_A = \frac{x_A P}{y_A P_A^\bullet}$$

(Raoult's Law activity coefficient)

Since we are using pure liquid A at its equilibrium vapor pressure as the standard state for component A, the chemical potential can be expressed as

$$\mu_A(P, y_A, y_B) = \tilde{\mu}_A^o(\ell, P_A^\bullet) + RT \ln y_A \gamma_A$$

Introduction of the activity coefficient adds nothing to our store of information about the system. It merely provides a convenient way to recast the available information, so that the solute mole fraction, y_A, becomes the independent variable in the chemical-potential equation. (For a one-phase two-component system, γ_A is completely determined by the temperature, system pressure, and y_A. Then y_A is the concentration variable in the chemical-potential equation. If there are more than two components, additional concentration variables are required to specify the composition of the system and the values of $\tilde{a}_A$ and γ_A.)

In summary, since the gas is ideal, the partial pressure of A above the solution is $P_A = x_A P$, whether the solution is ideal or not. If the solution is ideal, we have $y_A P_A^\bullet = x_A P$, so that $y_A P_A^\bullet / P^o = x_A P / P^o$. If the solution is not ideal, we introduce the activity coefficient, γ_A, as the "fudge factor" that makes $\gamma_A y_A P_A^\bullet = x_A P$ true. That is, the activity coefficient is just the actual value of the partial pressure of ideal gas A, $x_A P$, divided by the value it would have if the solution were ideal, $y_A P_A^\bullet$. The activity coefficient corrects for the departure of the real solution from the behavior that Raoult's law predicts for the ideal solution. When we define the activity coefficient so that $\tilde{a}_A = y_A \gamma_A$, we have $\tilde{a}_A P_A^\bullet = x_A P$, so that $\tilde{a}_A P_A^\bullet / P^o = x_A P / P^o$, thus preserving the form of the ideal solution result—with $\tilde{a}_A$ replacing y_A. For an ideal solution, $\gamma_A = 1$, and our result for the real solution activity reduces to the ideal solution activity.

§4 Henry's law and the fugacity and activity of a solution component

In describing the activities of solution components, we have taken the standard state of component A to be the pure liquid at its equilibrium vapor pressure, $P_A^\bullet$, at the temperature of the solution. We can also express the activity using ***Henry's law***. Henry's law states that the partial pressure of a component above its solution is directly proportional to the concentration of the component. We can choose any convenient unit to express the solute concentration. The value of the proportionality constant depends on this choice. Using mole fraction as the unit of concentration, Henry's law is

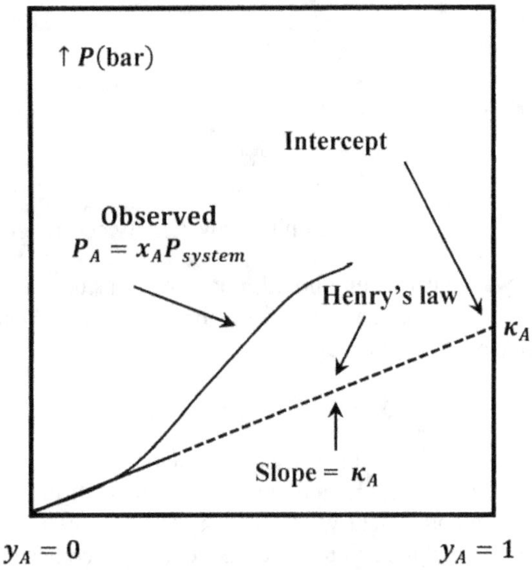

Figure 5. Henry's law for a solute.

$$P_A = x_A P = \kappa_A y_A \qquad \text{(Henry's law)}$$

where the proportionality constant, κ_A, is called the ***Henry's-law constant.*** (When we write $P_A = x_A P$, we implicitly assume that the gas-phase components of the equilibrium system behave as ideal gases.) Henry's law is more general than Raoult's law. Indeed, Raoult's law is a special case of Henry's law; if the solute obeys Raoult's law, the Henry's-law constant is $P_A^\bullet$.

The value of the Henry's-law constant depends on the components and the temperature. Experimentally, the value of the constant is determined by finding the slope of a plot of P_A versus y_A in the limit as $y_A \to 0$. The sketch in Figure 5 illustrates the relationship between the P_A curve and the Henry's-law tangent to it at $y_A = 0$. Note that the slope of the tangent line at $y_A = 0$ is equal to its intercept at $y_A = 1$.

In practice, a sufficiently low concentration of any non-electrolyte component, in any liquid solution, obeys Henry's law. For this reason, we refer to the component that obeys Henry's law as the solute. We designate the higher concentration component as the solvent. The universal validity of Henry's law as a low-concentration approximation has a simple physical interpretation. The solute vapor pressure depends upon the net effects of solute–solute, solute–solvent, and solvent–solvent intermolecular forces.

If all of these intermolecular forces are the same, the intermolecular interactions that determine the gas-phase composition are the same for solvent molecules as they are for solute molecules; the vapor pressures of the solvent and the solute are the same; the gas above the solution has the same composition as the solution; the partial pressure of the solute is proportional to the solute concentration, y_A; the proportionality constant is the pure-solvent vapor pressure, $P_A^\bullet$; and the solution obeys Raoult's law. However, if the intermolecular forces are not all the same, their net effect changes as the solute concentration changes. As the solute concentration increases, the effects of solute–solute interactions become increasingly important. If these are different from the effects of solute–solvent and solvent–solvent interactions, the solute partial pressure is not proportional to the solute concentration.

Conversely, at some sufficiently low concentration, solute molecules are so far apart that the effects of solute–solute interactions become negligible. Only solute–solvent and solvent–solvent interactions affect the solute vapor pressure. Because these remain constant as the solute concentration decreases further, the solute partial pressure is proportional to the solute concentration in this low-concentration regime. However, if the effects of solute–solvent interactions are different from those of solvent–solvent interactions, the pure-solvent vapor pressure, $P_A^\bullet$, is not the proportionality constant.

We can assume the existence of a hypothetical ideal gas state for any substance, even a substance that has no measureable vapor pressure at any attainable temperature. We can assume that the substance has a well-defined Gibbs free energy of formation in this state even though there may be no possibility of measuring its value. Likewise, we can assume that any solute exerts some partial pressure over its solution. We consider that this partial pressure has some finite value, even if it is much too small to measure. It follows that the fugacity of the solute has a finite value. Henry's law implies that the fugacity is proportional to the solute concentration, at least in the limit of arbitrarily low concentration. Expressing the concentration of solute A as its mole fraction, we have

$$\begin{aligned} f_A(P, y_A, y_B) &= P_A/P^o \\ &= x_A P/P^o \\ &= \kappa_A y_A/P^o \end{aligned}$$

and the chemical potential of the solute is

$$\mu_A(P, y_A, y_B) = \Delta_f G^o(A, HIG^o) + RT \ln f_A\,(P, y_A, y_B)$$

If the solute behaves ideally in the gas phase, Henry's law leads to simple expressions for its chemical potential and activity. The development of these expressions is very similar to the corresponding development from Raoult's law. The essential differences arise from the introduction of a different standard state for the activity of the solute. We begin with our basic equation for the chemical potential of A in the gas phase:

$$\begin{aligned} \mu_A(g, P, x_A, x_B) = \Delta_f G^o(A, HIG^o) + RT \ln \left[\frac{x_A P}{P^o}\right] \\ + \int_0^P \left(\frac{\bar{V}_A(g)}{RT} - \frac{1}{P}\right) dP \end{aligned}$$

The integral term vanishes because we assume ideal gas behavior. We write $\tilde{a}_A(\text{solute}, P, y_A, y_B)$ to represent the activity of A in a solution at pressure P and in which the composition is specified by the mole fractions y_A and y_B.

We choose a hypothetical liquid to be the standard

state for the activity of the solute. This hypothetical liquid is pure liquid A at the vapor pressure it would exhibit if it followed Henry's law over the entire range of possible system compositions. This pressure is equal to its Henry's law constant κ_A. (See Figure 5.) Let us denote the chemical potential of this standard state by $\tilde{\mu}_A^o(Hyp\ \ell, \kappa_A)$. The chemical potential of solute A and the activity of A in the solution are related by

$$\mu_A(P, y_A, y_B) = \tilde{\mu}_A^o(Hyp\ \ell, \kappa_A) + RT \ln \tilde{a}_A\,(P, y_A, y_B)$$

and since $\mu_A(g, P, x_A, x_B) = \mu_A(P, y_A, y_B)$, and the gas is ideal, we have

$$\ln[\tilde{a}_A(P, y_A, y_B)] = \frac{\Delta_f G^o(A, HIG^o) - \tilde{\mu}_A^o(Hyp\ \ell, \kappa_A)}{RT} + \ln\left[\frac{x_A P}{P^o}\right]$$

Except for the solute standard state, this is same as the equation that we develop in §1.

This equation must give the activity of solute A in its standard state, which is pure hypothetical liquid at a pressure equal to the Henry's law constant: $P = P_A = \kappa_A$. In this state, $x_A = y_A = 1$ and $x_B = y_B = 0$. In its standard state, the activity of solute A is unity; we have $\tilde{a}_A(P, y_A, y_B) = \tilde{a}_A(\kappa_A, 1{,}0) = 1$. Making these substitutions and rearranging, we find

$$\frac{\Delta_f G^o(A, HIG^o) - \tilde{\mu}_A^o(Hyp\ \ell, \kappa_A)}{RT} = -\ln\left[\frac{\kappa_A}{P^o}\right]$$

Substituting this result into our general equation for $\ln[\tilde{a}_A(\text{solute}, P, y_A, y_B)]$, we find that the activity of solute A is

$$\tilde{a}_A(P, y_A, y_B) = \frac{x_A P}{\kappa_A}$$
(any solution, ideal gas)

The chemical potential of the standard-state hypothetical pure liquid whose vapor pressure at T is κ_A is

$$\tilde{\mu}_A^o(Hyp\ \ell, \kappa_A) = \Delta_f G^o(A, HIG^o) + RT \ln\left[\frac{\kappa_A}{P^o}\right]$$

This is a general result for the activity of solute A when the standard state is the hypothetical pure liquid whose pressure is κ_A, and A behaves ideally in the gas phase. If Henry's law is obeyed, we have $x_A P = y_A \kappa_A$. Substituting, we find

$$\tilde{a}_A(P, y_A, y_B) = y_A$$
(Henry's law is obeyed)

The Henry's law development and the Raoult's law development give the same value for the chemical activity. However, the standard states are different.

If the solute obeys Raoult's law, the standard state we choose for the solute is the pure solute at its equilibrium vapor pressure; in this state, the pure solute is in equilibrium with its own gas. Real substances can satisfy this condition. If the solute obeys Henry's law, the standard state we choose for the solute is a hypothetical pure-liquid solute at a pressure equal to κ_A. In this state, we assume that the hypothetical liquid is in equilibrium with its own gas, also at pressure κ_A. The hypothetical standard-state liquid is a substance in which the interactions among A molecules have the same effects as the interactions, in a very dilute solution, between A molecules and the B molecules that comprise the solvent. If solutions of A and B are described poorly by Raoult's law, the vapor pressure of pure liquid A, $P_A^\bullet$, is likely to be very different from the vapor pressure, κ_A, of the hypothetical standard-state liquid that we define using Henry's law.

Our development produces a model for the chemical potential of the solute in which the activity is equal to the solute mole fraction. At the same mole fraction, every solute has the same activity. The chemical potentials of different solutes vary because the chemical potential in its standard state is different for every solute. We conclude that we can let $\tilde{a}_A = y_A$ for any sufficiently dilute solute, even when it is not feasible to measure the chemical potential of the solute in its standard state experimentally.

§5 Expressing the activity coefficient as a deviation from Henry's law

Even if Henry's law is valid only for solute concentrations very close to zero, we can use it to express the activity of the real system as a function of solute concentration. Let us suppose that we have data on the mole fraction of A, x_A, in a gas whose pressure is P and which is at equilibrium with a solution in which its mole fraction is y_A. In the preceding section, we find that we can choose the solute's standard state so that its activity in any state is $\tilde{a}_A = x_A P/\kappa_A$. Introducing the activity coefficient, defined by $\tilde{a}_A = y_A \gamma_A$, we have $x_A P/\kappa_A = y_A \gamma_A$. The activity coefficient is

$$\gamma_A = \frac{x_A P}{y_A \kappa_A}$$
(Henry's law activity coefficient)

and the chemical potential is

$$\mu_A = \tilde{\mu}_A^o(Hyp\ \ell, \kappa_A) + RT \ln y_A \gamma_A$$

Just as when we define the activity coefficient using the deviation from Raoult's law, this development provides a way to recast the available information in a way that makes the solute mole fraction, y_A, the independent variable in the chemical-potential equation. In §4, we note that Raoult's law is the special case of Henry's law in which $P_A^\bullet = \kappa_A$. If we make this substitution into the Henry's-law based activity coefficient, we recover the Raoult's-law based activity coefficient.

§6 Henry's law and the hypothetical one-molal standard state

When the solution is dilute, it is often convenient to use the molality of the solute rather than its mole fraction. We define the molality as the number of moles solute, A, per kilogram of solvent, B. We use $\underline{m}_A$ to represent the molality of A and $\overline{M}_B$ to represent the gram-molar mass of B. A solution that contains n_A and n_B moles of A and B, respectively, contains a mass of solvent given in kilograms by

$$m_B = \frac{n_B \overline{M}_B}{1000}$$

Then the molality of A is

$$\underline{m}_A = \frac{n_A}{m_B} = \frac{1000 n_A}{\overline{M}_B n_B}$$

so that

$$n_A = \frac{\underline{m}_A n_B \overline{M}_B}{1000}$$

The mole fraction of A is given by

$$y_A = \frac{n_A}{n_A + n_B} = \frac{\underline{m}_A \overline{M}_B}{\underline{m}_A \overline{M}_B + 1000}$$

In dilute solution, where $n_A \ll n_B$ and $\underline{m}_A \overline{M}_B \ll 1000$, the mole fraction and molality of the solute are related by

$$y_A \approx \left(\frac{\overline{M}_B}{1000}\right) \underline{m}_A$$

(dilute solution)

Using this approximation, assuming that solute A obeys Henry's law and that gas A behaves ideally, we have

$$\begin{aligned}
\mu_A(P, \underline{m}_A) &= \mu_A(P, y_A, y_B) \\
&= \mu_A(g, P, x_A, x_B) \\
&= \Delta_f G^o(A, HIG^o) + RT \ln \left[\frac{x_A P}{P^o}\right] \\
&= \Delta_f G^o(A, HIG^o) + RT \ln \left[\frac{\kappa_A y_A}{P^o}\right] \\
&= \Delta_f G^o(A, HIG^o) + RT \ln \left[\frac{\kappa_A \overline{M}_B}{1000 P^o}\right] \\
&\quad + RT \ln \underline{m}_A
\end{aligned}$$

(solute A obeys Henry's law)

When Henry's law is not obeyed over the composition range of interest, it is often convenient to choose the standard state of the solute to be a one-molal solution of a hypothetical substance that obeys Henry's law with the Henry's law constant κ_A. Then the activity of this solution is unity, and its chemical potential is the chemical potential of A in this hypothetical standard state. Letting this standard-state chemical potential be $\tilde{\mu}_A^o(Hyp\ 1\ \underline{m}\text{ solute}, P)$, we have

$$\begin{aligned}
&\tilde{\mu}_A^o(Hyp\ 1\ \underline{m}\text{ solute}, P) \\
&\qquad = \Delta_f G^o(A, HIG^o) + RT \ln \left[\frac{\kappa_A \overline{M}_B}{1000 P^o}\right]
\end{aligned}$$

The chemical potential of a substance that satisfies Henry's law is

$$\mu_A(HL, P, \underline{m}_A) = \tilde{\mu}_A^o(Hyp\ 1\ \underline{m}\text{ solute}, P) + RT \ln \underline{m}_A$$

If A behaves as an ideal gas and the solution is dilute ($\underline{m}_A \overline{M}_B \ll 1000$), but $\underline{m}_A$ is above the range in which Henry's law is obeyed, we introduce the Henry's law activity coefficient, $\gamma_A = \tilde{a}_A / \underline{m}_A$, to measure the departure of the real solution behavior from that predicted by Henry's law. Then the chemical potential of A in any solution is

$$\begin{aligned}
\mu_A(P, \underline{m}_A) &= \tilde{\mu}_A^o(Hyp\ 1\ \underline{m}\text{ solute}, P) + RT \ln \tilde{a}_A \\
&= \tilde{\mu}_A^o(Hyp\ 1\ \underline{m}\text{ solute}, P) + RT \ln \underline{m}_A \gamma_A \\
&= \tilde{\mu}_A^o(Hyp\ 1\ \underline{m}\text{ solute}, P) + RT \ln \gamma_A \\
&\quad + RT \ln \underline{m}_A
\end{aligned}$$

and the logarithm of the activity coefficient measures the difference between the chemical potential of the real solute and that of a solute that obeys Henry's law over an extended concentration range:

$$\ln \gamma_A = \frac{\mu_A(P, \underline{m}_A) - \mu_A(HL, P, \underline{m}_A)}{RT}$$

In §19, we consider the determination of Henry's law-based activity coefficients further.

§7 Finding the activity of a solute from the activity of the solvent

We have seen that the activity of any component of an equilibrium system contains information about the activities of all the other components. From

$$d\mu = -\overline{S} dT + \overline{V} P + RT d \ln \tilde{a}$$

and the Gibbs-Duhem equation, we can find a general relationship among the activities. Substituting $d\mu_A$ and $d\mu_B$ into the Gibbs-Duhem equation, we have, for a two-component solution,

$$\begin{aligned}
-SdT + VdP &= n_A d\mu_A + n_B d\mu_B \\
&= n_A(-\overline{S}_A dT + \overline{V}_A P + RT d \ln \tilde{a}_A) \\
&\quad + n_B(-\overline{S}_B dT + \overline{V}_B P + RT d \ln \tilde{a}_B) \\
&= -(n_A \overline{S}_A + n_B \overline{S}_B) dT \\
&\quad + (n_A \overline{V}_A + n_B \overline{V}_B) dP + n_A RT d \ln \tilde{a}_A \\
&\quad + n_B RT d \ln \tilde{a}_B
\end{aligned}$$

Since $S = n_A\overline{S}_A + n_B\overline{S}_B$ and $V = n_A\overline{V}_A + n_B\overline{V}_B$, this simplifies to

$$0 = n_A d \ln \tilde{a}_A + n_B d \ln \tilde{a}_B$$

or, dividing by $n_A + n_B$,

$$0 = y_A d \ln \tilde{a}_A + y_B d \ln \tilde{a}_B$$

For simplicity, let us consider a system in which a non-volatile solute, A, is dissolved in a volatile solvent, B. Measuring the pressure of the system and applying the equations that we developed in §1 for volatile component A to the volatile solvent, B, in the present system, we can determine the activity of the solvent, B. Let us use mole fractions to measure concentrations and take pure liquid B at its equilibrium vapor pressure as the activity standard state for both liquid- and gas-phase B. When B is in its standard state, we have $x_A = 0$, $x_B = 1$, and $\overline{V}_B(g) = \overline{V}_B^\bullet(g)$. Then, since the solute is non-volatile, we can determine the activity of the solvent, B, from the pressure of the system. We have

$$\ln[\tilde{a}_B(P, y_A, y_B)] = RT \ln\left[\frac{P}{P_B^\bullet}\right] - \int_{P_B^\bullet}^{P} \left(\frac{\overline{V}_B^\bullet(g)}{RT} - \frac{1}{P}\right) dP$$

Assuming that the integral makes a negligible contribution to the activity, we have

$$\begin{aligned} \tilde{a}_B(P, y_A, y_B) &= \tilde{a}_B \\ &= y_B \gamma_B(P, y_A, y_B) \\ &= \frac{P}{P_B^\bullet} \end{aligned} \quad \text{(solvent)}$$

so that

$$\begin{aligned} \gamma_B(P, y_A, y_B) &= \gamma_B \\ &= \frac{P}{y_B P_B^\bullet} \end{aligned} \quad \text{(solvent)}$$

Since the gas-phase concentration of A is immeasurably small, we must determine its activity indirectly. Let the standard state for solute activity be the hypothetical pure liquid, $y_A = 1$, whose equilibrium vapor pressure is equal to the Henry's law constant of solute A. (We can determine the solute's activity without measuring its Henry's law constant.) We have

$$\mu_A(P, y_A, y_B) = \tilde{\mu}_A^o(Hyp\ \ell, \kappa_A) + RT \ln[\tilde{a}_A(P, y_A, y_B)]$$

where

$$\begin{aligned} \tilde{a}_A(P, y_A, y_B) &= \tilde{a}_A \\ &= y_A \gamma_A \end{aligned} \quad \text{(solute)}$$

Since we are able to measure the activity of the solvent, we can determine the activity of the solute from the relationship $0 = y_A d \ln \tilde{a}_A + y_B d \ln \tilde{a}_B$. Rearranging, we have

$$\begin{aligned} d \ln \tilde{a}_A &= -\frac{y_B}{y_A} d \ln \tilde{a}_B \\ &= -\left(\frac{1 - y_A}{y_A}\right) d \ln \tilde{a}_B \end{aligned}$$

For two solutions in which the mole fractions of A are y_A and $y_A^\#$, and in which the activities of A and B are $\tilde{a}_A$, $\tilde{a}_B$, $\tilde{a}_A^\#$, and $\tilde{a}_B^\#$, we have

$$\ln \frac{\tilde{a}_A}{\tilde{a}_A^\#} = -\int_{y_A^\#}^{y_A} \left(\frac{1 - y_A}{y_A}\right) d \ln \tilde{a}_B$$

Graphically, the integral is the area under a plot of $-(1 - y_A)/y_A$ versus $\ln \tilde{a}_B$, from $y_A^\#$ to y_A.

Typically, we are interested in solutions for which $y_A \ll y_B$. In the limit as the solution becomes very dilute, the activity, mole fraction, and activity coefficient of the solvent, B, all approach unity: $\tilde{a}_B \to 1$, $y_B \to 1$, and $\gamma_B \to 1$. The activity of the solute, A, approaches the mole fraction of A. As a matter of experience, the approach is asymptotic: as the mole fraction approaches zero, $y_A \to 0$, the solute activity coefficient approaches unity, $\gamma_A \to 1$, and does so asymptotically, so that $\tilde{a}_A \to y_A$. For dilute solutions, $\ln \tilde{a}_A \to -\infty$ and $\ln \gamma_A \to 0$ asymptotically. In consequence,

$$\lim_{y_A \to 0} (d \ln \gamma_A) = 0$$

Because the activity coefficient approaches a finite limit while the activity does not, we can express the solute's activity most simply by finding the solute's activity coefficient. Since $\tilde{a}_A = y_A \gamma_A$ and $\tilde{a}_B = y_B \gamma_B = (1 - y_A)\gamma_B$, we have

$$\begin{aligned} 0 &= y_A d \ln \tilde{a}_A + y_B d \ln \tilde{a}_B \\ &= y_A d \ln y_A \gamma_A + y_B d \ln y_B \gamma_B \\ &= y_A d \ln y_A + y_A d \ln \gamma_A + y_B d \ln y_B + y_B d \ln \gamma_B \\ &= (dy_A + dy_B) + y_A d \ln \gamma_A + y_B d \ln \gamma_B \\ &= y_A d \ln \gamma_A + y_B d \ln \gamma_B \end{aligned}$$

(Since $y_A + y_B = 1$, we have $dy_A + dy_B = 0$.) We can rearrange this to

$$d \ln \gamma_A = -\left(\frac{1 - y_A}{y_A}\right) d \ln \gamma_B$$

As the solute concentration approaches zero, $(1 - y_A)/y_A$ becomes arbitrarily large. However, since $\lim_{y_A \to 0} (d \ln \gamma_A) = 0$, it follows that

$$\lim_{y_A \to 0} d \ln \gamma_B = 0$$

We see that the solvent activity coefficient also approaches unity asymptotically as the solute concentration goes to zero. The solute activity coefficient at any $y_A > 0$ is then given by

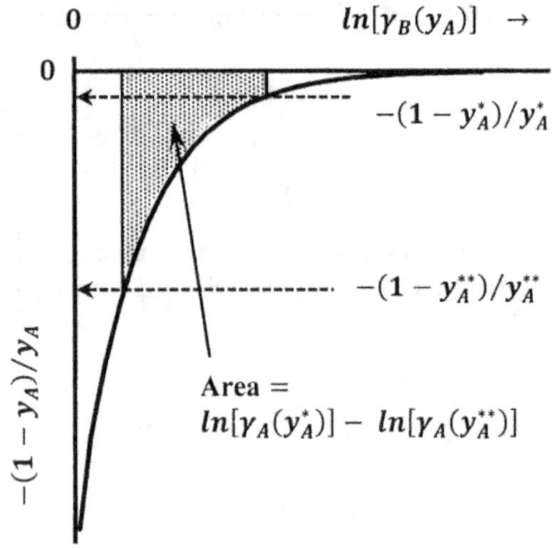

Figure 6. Graphical representation of $ln\,\gamma$ for the solute when γ of the solvent is known.

$$\int_0^{y_A} d\ln[\gamma_A(y_A)] = \ln[\gamma_A(y_A)]$$

$$= \int_0^{y_A} -\left(\frac{1-y_A}{y_A}\right) d\ln[\gamma_B(y_A)]$$

As sketched in Figure 6, the latter integral is the area under a graph of $-(1-y_A)/y_A$ *versus* $\ln[\gamma_B(y_A)]$, between $y_A = 0$ and y_A. Since $\gamma_A(y_A) \to 1$ as $y_A \to 0$, this integral must remain finite even though $-(1-y_A)/y_A \to -\infty$ as $y_A \to 0$. This can occur, because $\lim_{y_A \to 0} d\ln\gamma_B = 0$, as we observe above. Nevertheless, the fact that the integrand is unbounded can limit the accuracy of the necessary integration. For accurate measurement of the solute activity coefficient, it is important to obtain solvent-activity data at the lowest possible solute concentration.

The most desirable situation is to collect solvent-activity data down to solute concentrations at which the solvent activity coefficient, γ_B, becomes unity. If $\gamma_B(y_A^{\#}) = 1$ when the solute concentration is $y_A^{\#}$, $\ln[\gamma_A(y_A)]$ can be evaluated with $y_A^{\#}$, rather than zero, as the lower limit of integration. In some cases, $\ln[\gamma_A(y_A^{\Delta})]$ may be known from some other measurement at a particular concentration, y_A^{Δ}; if so, we can find $\ln[\gamma_A(y_A)]$ by carrying out the numerical integration between the limits y_A^{Δ} and y_A.

If the measurement of γ_B cannot be extended to values of y_A at which $\gamma_B(y_A) = 1$, we must find an empirical function, call it $f(y_A)$, that fits the experimental values of $\ln[\gamma_B(y_A)]$, for the smallest values of y_A. (That is, the empirical function is $f(y_A) = \ln[\gamma_B(y_A)]$.) The differential of $f(y_A)$ is then a mathematical model for $d\ln[\gamma_B(y_A)]$ over the region of low solute concentrations. Letting $y_A^{\#}$ be the smallest solute concentration for which the solvent activity can be determined, we can integrate, using the function for $d\ln[\gamma_B(y_A)]$ that we derive from this model, to estimate $\ln[\gamma_B(y_A^{\#})]$. Uncertainty about the accuracy of the mathematical model becomes a significant source of uncertainty in the calculated values of γ_A.

Of course, if we can find an analytical function that provides a good mathematical model for all of the solvent-activity data, the differential of this function can be used in the integral to evaluate $\ln[\gamma_A(y_A)]$ over the entire range of the experimental data. If necessary, the evaluation of this integral can be accomplished using numerical methods.

It is essential that any empirical function, $f(y_A) = \ln[\gamma_B(y_A)]$, have the correct mathematical properties over the concentration range to which it is applied. If it is to be used to extend the integration to $y_A = 0$, $f(y_A)$ must satisfy $f(0) = 0$ and $df(0) = 0$. This is a significant condition. For example, consider the approximation

$$f(y_A) = \ln[\gamma_B(y_A)]$$
$$= c_B y_A^{\alpha_B}$$

This model gives

$$\frac{d\ln[\gamma_A(y_A)]}{dy_A} = -(1-y_A)c_B\alpha_B y_A^{\alpha_B-2}$$

and

$$\lim_{y_A\to 0}\frac{d\ln[\gamma_A(y_A)]}{dy_A} = 0$$

requires $\alpha_B > 2$.

§8 When the solute obeys Henry's law, the solvent obeys Raoult's law

In §4, we conclude that any sufficiently dilute solute obeys Henry's law, and define a hypothetical, pure-liquid standard state that makes the solute activity equal to its mole fraction, $\tilde{a}_A(P, y_A, y_B) = y_A$. In §7, we find that the mole fractions and activities of the components of any binary solution are related by $y_A d\ln\tilde{a}_A + y_B d\ln\tilde{a}_B = 0$. For a solute that obeys Henry's law, we have

$$d\ln\tilde{a}_B = -\left(\frac{y_A}{y_B}\right)d\ln y_A$$
$$= -\left(\frac{y_A}{y_B}\right)(d\ln y_B)\left(\frac{d\ln y_A}{d\ln y_B}\right)$$
$$= -\left(\frac{y_A}{y_B}\right)(d\ln y_B)\left(\frac{dy_A/y_A}{dy_B/y_B}\right)$$
$$= -\left(\frac{y_A}{y_B}\right)(d\ln y_B)\left(\frac{-dy_B/y_A}{dy_B/y_B}\right)$$
$$= d\ln y_B$$

This result follows for any choice of standard state for the activity of solvent B. It is satisfied by $\tilde{a}_B = k y_B$, where k is a constant. It is valid even if A is completely nonvolatile. When gas-phase B behaves as an ideal gas, and we choose the ideal gas at P^o as the standard state for both gas- and solution-phase B, we have

$$\begin{aligned}\tilde{a}_B(\text{gas}) &= f_B \\ &= P_B/P^o \\ &= x_B P/P^o\end{aligned}$$

Since the standard states are the same, the fugacity and activity of B in solution are the same as they are in the gas phase above it. We have $\tilde{a}_B(\text{solution}) = ky_B = x_B P/P^o$. To find k, we consider the system comprised of pure B, for which $y_B = x_B = 1$ and $P = P_B^\bullet$. Substituting, we find $k = P_B^\bullet/P^o$. With this value for k,

$$\begin{aligned}\tilde{a}_A(\text{solution}) &= P_B^\bullet y_B/P^o \\ &= x_B P/P^o\end{aligned}$$

so that $y_B P_B^\bullet = x_B P$. This is Raoult's law. Thus when the solute obeys Henry's law and the solvent behaves as an ideal gas in the gas phase above its solution, the solvent obeys Raoult's law.

Evidently the converse is also true. If the solvent obeys Raoult's law, $y_B P_B^\bullet = x_B P$. With pure ideal gas B as the standard state for B in both the gas phase and the solution phase, we have

$$\begin{aligned}\tilde{a}_B(\text{solution}) &= \tilde{a}_B(\text{gas}) \\ &= f_B \\ &= P_B/P^o \\ &= x_B P/P^o \\ &= y_B(P_B^\bullet/P^o)\end{aligned}$$

so that

$$d\ln \tilde{a}_B(\text{solution}) = d\ln y_B$$

From $y_A d\ln \tilde{a}_A + y_B d\ln \tilde{a}_B = 0$ and $d\ln \tilde{a}_B = d\ln y_B$, we have

$$\begin{aligned}d\ln \tilde{a}_A &= -\left(\frac{y_B}{y_A}\right) d\ln y_B \\ &= -\left(\frac{y_B}{y_A}\right)\left(\frac{dy_B}{y_B}\right) \\ &= \frac{dy_A}{y_A} \\ &= d\ln y_A\end{aligned}$$

so that $\tilde{a}_A(\text{solution}) = ky_A$, where k is a constant. When we choose the standard state such that $\tilde{a}_A(\text{ss, solution}) = 1$ when $y_A = 1$, we find $k = 1$ and $\tilde{a}_A(\text{solution}) = y_A$. The activity of the solute is related to its fugacity and the fugacity of its standard state by

$$\tilde{a}_A(\text{solution}) = y_A = \frac{f_A(\text{solution})}{f_A(ss, \text{solution})}$$

When $y_A = 1$, the fugacity is that of the standard state, which is a system of the hypothetical pure liquid in equilibrium with its own ideal gas. Letting the pressure of this ideal gas be κ_A, we have $f_A(ss, \text{solution}) = \kappa_A$, so that $f_A(\text{solution}) = \kappa_A y_A$, which is equal to the fugacity of the gas with which it is at equilibrium. The fugacity of the ideal gas is $x_A P$, so that $x_A P = \kappa_A y_A$. This is Henry's law. Thus, if solvent B obeys Raoult's law, solute A obeys Henry's law.

§9 Properties of ideal solutions

We have found the chemical potential of any component in an ideal solution. Now let us find some other thermodynamic properties of an ideal solution. The value of an extensive thermodynamic property of the solution will be the sum of the values of that property for the separate pure components plus the change that occurs when these components are mixed. (The initial state of the system comprises the pure, separate components at a particular temperature and pressure. The mixed state is a homogeneous liquid solution at the same temperature and pressure.) If the solution contains n_A moles of component A and n_B moles of component B, the Gibbs free energy is

$$G_{\text{solution}}(n_A, n_B) = n_A \mu_A + n_B \mu_B$$

Dividing through by $n_A + n_B$ to find the Gibbs free energy of the mixture per mole of solution, we have

$$\begin{aligned}\overline{G}_{\text{solution}}(y_A, y_B) &= y_A \mu_A + y_B \mu_B \\ &= y_A \Delta_f G^o(A) + y_A RT \ln y_A + \\ &\qquad y_B \Delta_f G^o(B) + y_B RT \ln y_B\end{aligned}$$

To make this mixture, we need y_A moles of A and y_B moles of B. The Gibbs free energy of these amounts of unmixed pure A and B, each in its standard state, is

$$\overline{G}_{\text{initial}} = y_A \Delta_f G^o(A) + y_B \Delta_f G^o(B)$$

For the process of mixing pure A and pure B, each originally in its standard state, to form one mole of an ideal solution, the Gibbs free energy change is

$$\Delta_{\text{mix}}\overline{G} = \overline{G}_{\text{solution}} - \overline{G}_{\text{initial}} = y_A RT \ln y_A + y_B RT \ln y_B$$

In §13-3, we found this same relationship for mixing ideal gases:

$$\Delta_{\text{mix}}\overline{G}(\text{gas}) = x_A RT \ln x_A + x_B RT \ln x_B$$

From $(\partial \Delta_{\text{mix}} G/\partial T)_P = -\Delta_{\text{mix}} S$, we find

$$\Delta_{\text{mix}}\overline{S} = -y_A R \ln y_A - y_B R \ln y_B$$

and the entropy of the liquid solution is

$$\overline{S}_{\text{solution}} = y_A S_A^o + y_B S_B^o - y_A R \ln y_A - y_B R \ln y_B$$

From $(\partial \Delta_{\text{mix}} G/\partial P)_T = \Delta_{\text{mix}} V$, we find

$$\Delta_{\text{mix}}\overline{V} = 0$$

and from $\Delta_{\text{mix}} H = \Delta_{\text{mix}} G + T\Delta_{\text{mix}} S$, we find

$$\Delta_{\text{mix}}\overline{H} = 0$$

Thus, $\Delta_{\text{mix}}\overline{S}$, $\Delta_{\text{mix}}\overline{V}$, and $\Delta_{\text{mix}}\overline{H}$ for forming an ideal solution are identical also to the relationships we found for mixing ideal gases.

These results have an important physical interpretation. That $\Delta_{\text{mix}}\overline{V} = 0$ implies that the molecules of A and the molecules of B occupy the same volume in the mixture as they do in the pure state. From $\Delta_{\text{mix}}\overline{V} = 0$ and $\Delta_{\text{mix}}\overline{H} = 0$, it follows that $\Delta_{\text{mix}}\overline{E} = 0$ at constant pressure. In turn, this implies that the forces between an A molecule and a B molecule are the same as the forces between two A molecules or between two B molecules. If the force of attraction between an A molecule and a B molecule were stronger than that between two A molecules or between two B molecules, molecules in the mixture would be—on average—closer together in the mixture than in the separate components; we would find $\Delta_{\text{mix}}\overline{V} < 0$. Moreover, the potential energy of the mixed state would be lower than that of the separate components; the mixing process would evolve heat at constant temperature; we would find $\Delta_{\text{mix}}\overline{E} < 0$.

Conversely, if the repulsive force between an A molecule and a B molecule were stronger than the repulsive forces between two A molecules or between two B molecules, the average separation would be greater in the mixture; we would find $\Delta_{\text{mix}}\overline{V} > 0$. The potential energy of the mixed state would be greater than that of the separate components; the mixing process would consume heat at constant temperature; we would find $\Delta_{\text{mix}}\overline{E} > 0$.

In an ideal gas, molecules do not interact at all. In an ideal solution, the molecules must interact, because only their mutual attraction can keep them in the liquid state. The ideal solution behaves ideally not because the intermolecular interactions are zero but rather because the intermolecular interactions are the same for all of the kinds of molecules present in the mixture. This interpretation implies that the vapor pressures of the pure

components of an ideal solution should be equal. Even for solutions that follow Raoult's law quite closely, this expected equality is often imperfectly realized. Not surprisingly, ideal-solution behavior is best exhibited when the components are isotopically-substituted versions of the same compound.

In an ideal solution, the activities of the components are equal to their mole fractions. The activity of the solvent depends only on the solvent mole fraction. The properties of the solvent in an ideal solution are independent of the specific substance that comprises the solute; they depend only on the concentration of solute particles present. Systems in which this is a useful approximation are sufficiently common that their properties are given a special name: A ***colligative property*** of a solution is a property that depends only on the concentration of solute particles and not on the specific chemical properties of the solute. We expect this approximation to become better as the solute concentration approaches zero. When a solute obeys Raoult's law or Henry's law, its effects on the thermodynamic properties of the solvent depend only on the concentration of the solute. Consequently, Raoult's law and Henry's law prove to be useful when we seek to model colligative properties.

In §10 through §14, we evaluate five colligative properties: boiling-point elevation, freezing-point depression, osmotic pressure, solid-solute solubility and gas-solute solubility. We derive the first three of these properties from the perspective that they enable us to determine the molar mass of solutes. However, boiling-point elevation, freezing-point depression, and osmotic pressure are important methods for the measurement of activity coefficients in non-ideal solutions. To illustrate the measurement of activity coefficients, we develop a more detailed analysis of freezing-point depression in §15.

§10 Colligative properties: boiling-point elevation

The system we envision when we talk about boiling-point elevation is described schematically in Figure 7. We consider a solution of two components, A and B. The mole fractions of A and B, y_A and y_B, specify the composition of the solution. We suppose that one of the components is present at a low concentration. We call this component the solute, and designate it as compound A. Under these assumptions, we have $y_A \approx 0$ and $y_B = 1 - y_A \approx 1$. We assume further that A is nonvolatile, by which we mean that the vapor pressure of pure A, P_A^*, is very small. Then the second component, B, comprises most of the material of the system. We call component B the solvent. We suppose that the A–B solution is in equilibrium with a gas phase. In principle, molecules of both components are present in this gas. Since we assume that essentially no component A is present in the gas phase, we have $x_A = 0$ and $x_B = 1$. We assume also that gas-phase B behaves as an ideal gas and solute A obeys Henry's law.

When we measure the boiling point of a liquid system, we find the temperature at which the vapor pressure

Boiling-point elevation

$$P = P_B = y_B P_B^*$$

$$d\mu_{B,\text{gas}} = -\overline{S}_{B,\text{gas}} dT_{bp} \qquad B(\text{gas})$$

$$y_A \quad A(\text{solution}) \qquad y_B \quad B(\text{solution})$$

$$d\mu_{B,\text{solution}} = -\overline{S}_{B,\text{solution}} dT_{bp} + RT_{bp}(d\ln y_B)$$

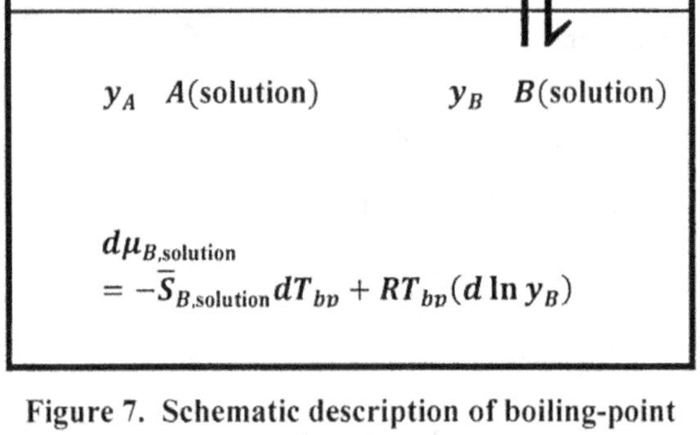

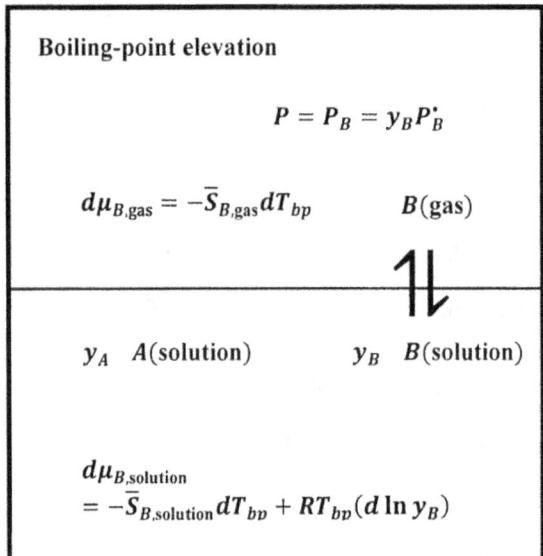

Figure 7. Schematic description of boiling-point elevation.

of the system becomes equal to a specified value. For the normal boiling point, this pressure is 1 atmosphere, or 1.01325 bars. At the boiling point, liquid-phase solvent is in equilibrium with gas-phase solvent, so that the chemical potential of liquid-phase solvent is equal to the chemical potential of gas-phase solvent. That is, we have $\mu_{B,\text{solution}} = \mu_{B,\text{gas}}$. We want to describe the change in the equilibrium position that occurs when there is an incremental change in the solute concentration, dy_A, while the pressure of the system remains constant. If the system is to remain at equilibrium, $\mu_{B,\text{solution}} = \mu_{B,\text{gas}}$ must remain true. It follows that the chemical potentials of the two phases must change in tandem. Continued equilibrium implies that $d\mu_{B,\text{solution}} = d\mu_{B,\text{gas}}$ when the solute concentration changes by dy_A.

We can analyze the boiling-point elevation phenomenon for any fixed pressure at which pure liquid B can be at equilibrium with pure gas B. Let us designate the fixed pressure as $P^{\#}$. We designate the boiling-point temperature of pure solvent B, at $P^{\#}$, as T_B; thus, $P^{\#} = P_B^{\bullet}(T_B)$, where $P_B^{\bullet}(T_B)$ designates the equilibrium vapor pressure of pure solvent B at temperature T_B. Our goal is to find the temperature at which a binary solution is in equilibrium with pure gas B at the fixed pressure $P^{\#}$. We let T_{bp} be the boiling temperature of the solution at $P^{\#}$. The composition of the solution is specified by the solute concentration, $y_A = 1 - y_B$.

Since we assume that the solute obeys Henry's law, we choose the standard state for solute A to be the pure hypothetical liquid A whose vapor pressure is κ_A at T. We suppose that κ_A is exceedingly small. From §4, we then have $\tilde{a}_{A,\text{solution}} = y_A$, so that

$$d\ln\tilde{a}_{A,\text{solution}} = d\ln y_A$$

at any temperature. From §8, we have

$$d\ln\tilde{a}_{B,\text{solution}} = d\ln y_B$$

Pure liquid-phase solvent is at equilibrium with gas-phase solvent at $P^{\#} = P_B^{\bullet}(T_B)$ and T_B. We imagine that we create a solution by adding a small amount of solute A, making the concentrations of solute and solvent y_A and $y_B = 1 - y_A$, respectively. We maintain the system pressure constant at $P^{\#}$, while changing the temperature to maintain equilibrium between gas-phase and solution-phase solvent B. The new temperature is T_{bp}.

The pressure of gas-phase B is constant at $P^{\#}$. The temperature goes from T_B to T_{bp}. We choose the activity standard state to be pure gas B at $P^{\#}$ and T. This means that the activity of the pure gas is unity at every temperature, so that $d\ln\tilde{a}_{B,\text{gas}} = 0$.

It is worthwhile to note that we can arrive at this conclusion from a different perspective: From §14-14, the incremental change in the activity is

$$d\ln\tilde{a}_{B,\text{gas}} = \left(-\frac{\overline{H}_B}{RT^2} + \frac{\widetilde{H}_B^o}{RT^2}\right)dT$$

where $\overline{H}_B$ is the partial molar enthalpy of gas-phase B at T, and $\widetilde{H}_B^o$ is the partial molar enthalpy of B in its activity standard state at T. Since we assume that the gas phase is essentially pure B, we have $\overline{H}_B = \widetilde{H}_B^o$ and, again, $d\ln\tilde{a}_{B,\text{gas}} = 0$.

From §14-3, we have the general result that

$$d\mu_B = \overline{V}_B dP - \overline{S}_B dT + RT(d\ln\tilde{a}_B)$$

The system pressure and temperature are $P = P^{\#}$ and $T = T_{bp}$. For both the gas phase and the solution phase, we have $dP = 0$ and $dT = dT_{bp}$. Since $d\ln\tilde{a}_{B,\text{gas}} = 0$, we have

$$d\mu_{B,\text{gas}} = -\overline{S}_{B,\text{gas}} dT_{bp}$$

Since $d\ln\tilde{a}_{B,\text{solution}} = d\ln y_B$, we have

$$d\mu_{B,\text{solution}} = -\overline{S}_{B,\text{solution}} dT_{bp} + RT_{bp}(d\ln y_B)$$

The chemical potential of the pure, constant-pressure, gas-phase solvent depends only on temperature. The chemical potential of the constant-pressure, solution-phase solvent depends on temperature and solute concentration. Equilibrium is maintained if

$$d\mu_{B,\text{solution}} = d\mu_{B,\text{gas}}$$

Substituting, we have

$$-\overline{S}_{B,\text{solution}} dT_{bp} + RT_{bp}(d\ln y_B) = -\overline{S}_{B,\text{gas}} dT_{bp}$$

Since $y_B = 1 - y_A \approx 1$,

$$d\ln y_B = d\ln(1 - y_A) = -dy_A/(1 - y_A) \approx -dy_A$$

The relationship $d\mu_{B,\text{solution}} = d\mu_{B,\text{gas}}$ becomes

$$-\overline{S}_{B,\text{solution}} dT_{bp} - RT_{bp} dy_A = -\overline{S}_{B,\text{gas}} dT_{bp}$$

or

$$dy_A = \left(\frac{\overline{S}_{B,\text{gas}} - \overline{S}_{B,\text{solution}}}{RT_{bp}}\right) dT_{bp}$$

We consider systems in which the boiling point of the solution, T_{bp}, is little different from the boiling point of the pure solvent, T_B. Then, $T_B \approx T_{bp}$, and $T_{bp}/T_B \approx 1$. We let $\Delta T = T_{bp} - T_B$, where $|\Delta T| \ll T_B$. Since the solution is almost pure B, the partial molar entropy of B in the solution is approximately that of pure B. Consequently, this partial molar entropy difference is, to a good approximation, just the entropy of vaporization of the solvent, at equilibrium, at the boiling point for the specified system pressure, $P^{\#} = P_B^{\bullet}(T_B)$. Then, since the vaporization of pure B at $P^{\#}$ and T_B is a reversible process,

$$\begin{aligned}\left(\overline{S}_{B,\text{gas}} - \overline{S}_{B,\text{solution}}\right)_{P^{\#},T_{bp}} &\approx \left(\bar{S}^{\bullet}_{B,\text{gas}} - \bar{S}^{\bullet}_{B,\text{liquid}}\right)_{P^{\#},T_B} \\ &= \Delta_{\text{vap}} S_B \\ &= \Delta_{\text{vap}} H_B / T_B\end{aligned}$$

so that

$$dy_A = \left(\frac{\Delta_{\text{vap}} H_B}{RT_{bp}T_B}\right) dT_{bp}$$

In the solution, the solute mole fraction is y_A; in the pure solvent, it is zero. At $P^{\#}$ and T_B, $\Delta_{\text{vap}} H_B$ is a constant. Integrating, between the limits $(0, T_B)$ and (y_A, T_{bp}), we have

$$\int_0^{y_A} dy_A = \frac{\Delta_{\text{vap}} H_B}{RT_B} \int_{T_B}^{T_{bp}} \frac{dT_{bp}}{T_{bp}}$$

and

$$y_A = \frac{\Delta_{\text{vap}} H_B}{RT_B} \ln \frac{T_{bp}}{T_B}$$

Introducing the approximation $\ln x \approx x - 1$, which is valid for $x \approx 1$, we have

$$\begin{aligned} y_A &= \frac{\Delta_{\text{vap}} H_B}{RT_B} \left(\frac{T_{bp}}{T_B} - 1\right) \\ &= \frac{\Delta_{\text{vap}} H_B}{RT_B^2} \Delta T \end{aligned}$$

Solving for ΔT,

$$\Delta T = \left(\frac{RT_B^2}{\Delta_{\text{vap}} H_B}\right) y_A$$

Since the enthalpy of vaporization and the mole fraction are both greater than zero, $\Delta T = T_{bp} - T_B > 0$; that is, the addition of a non-volatile solute increases the boiling point of a liquid system. By measuring ΔT, we can find y_A; if we know the molar mass of the solvent, we can calculate the number of moles of solute in the solution. If we know the mass of the solute used to prepare the solution, we can calculate the molar mass of the solute.

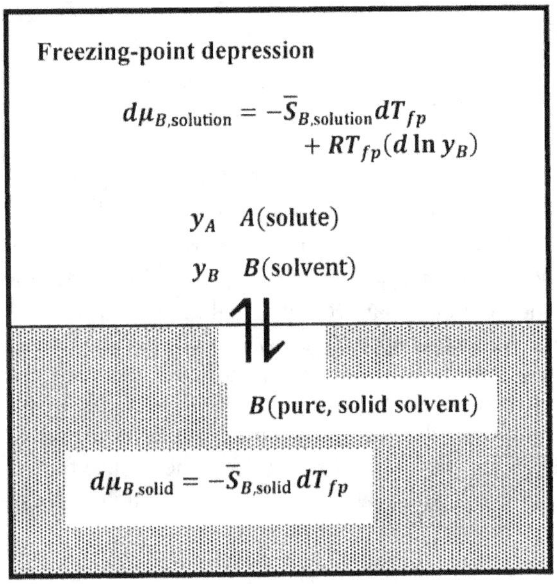

Figure 8. Schematic description of freezing-point depression.

Frequently it is useful to express the solute concentration as a molality rather than a mole fraction. Using the dilute-solution relationship between mole fraction and molality from §6, $y_A = \overline{M}_B \underline{m}_A / 1000$, the boiling-point elevation becomes:

$$\Delta T = \left(\frac{RT_B^2}{\Delta_{\text{vap}} H_B}\right)\left(\frac{\overline{M}_B}{1000}\right) \underline{m}_A$$

Our theory predicts that the boiling-point elevation observed for a given solvent is proportional to the solute concentration and independent of the molecular characteristics of the solute. Experiments validate this prediction; however, its accuracy decreases as the solute concentration increases. Letting

$$\kappa_B = \frac{RT_B^2}{\Delta_{\text{vap}} H_B}$$

and

$$\kappa_B^* = \frac{RT_B^2 \overline{M}_B}{1000\ \Delta_{\text{vap}} H_B}$$

we have $\Delta T = \kappa_B y_A$ and $\Delta T = \kappa_B^* \underline{m}_A$. We call κ_B or κ_B^* the ***boiling-point** (or boiling-temperature) **elevation constant*** for solvent B. For practical determination of molecular weights, we usually find κ_B or κ_B^* by measuring the increase in the boiling point of a solution of known composition.

§11 Colligative properties: freezing-point depression

The boiling point of a pure solvent, at a specified pressure, is the temperature at which the chemical potential of the pure solvent gas is equal to the chemical potential of the pure solvent liquid. The boiling point of a solution that contains a nonvolatile solute is the temperature at which the chemical potential of the pure solvent gas is equal to the chemical potential of the solvent in the solution. In the preceding section, we found that the boiling point of the solution is greater than the boiling point of the pure solvent. The temperature difference is the boiling-point elevation.

Similarly, the freezing point of a pure solvent, at a specified pressure, is the temperature at which the chemical potential of the pure-solid solvent is equal to the chemical potential of the pure-liquid solvent. The freezing point of a solution is the temperature at which the chemical potential of the pure-solid solvent is equal to the chemical potential of the solution-phase solvent. We find that the freezing point of the solution is less than the freezing point of the pure solvent. The temperature difference is the freezing-point depression.

In the boiling-point elevation case, we assume that the pure solvent gas contains no solute. In the freezing-point depression case, we assume that the pure solvent solid contains no solute. We find the relationship between composition and freezing-point depression by an argument very similar to that for boiling-point elevation. The equilibrium state in the freezing-point depression experiment is described schematically in Figure 8.

Since the equilibrium temperature decreases as the solute concentration increases, we can realize the equilibrium state experimentally by slowly cooling a solution of the specified composition. We determine the temperature at which the first, very small, crystal of solid solvent forms. Because the pure solvent freezes at a higher temperature than any solution, the first crystal formed is nearly pure solid solvent. Since this first crystal is very small, its formation does not change the composition of the solution significantly. Hence, the solution is in equilibrium with pure solid solvent at this temperature; we call this temperature the freezing point of the solution.

In practice, it is common to determine the melting point of a solid mixture rather than the freezing point of the liquid solution. The temperature of the solid mixture is increased slowly. As the mixture melts to form a homogeneous solution, the solute–solvent ratio in the melt approaches the ratio in which the mixture was prepared. When the last bit of solid melts, the composition of the solution is known from the manner of preparation. This last bit of solid melts at the highest temperature of any part of the mixture. It contains therefore the smallest proportion of solute. If this last bit of solid is in fact pure solvent, the temperature at which the last solid melts is the freezing point of the liquid solution. In the limit that the freezing-point and melting-point experiments are carried out reversibly, the state of the freezing-point system just after the first bit of solid freezes is the same as the state of the melting-point system just before the last bit of solid melts.

We again specify the composition of the solution by the mole fractions of A and B. We let the solute be compound A, and assume that its concentration is low. We let the solute concentration be y_A, where $y_A = 1 - y_B$, $y_A \approx 0$ and $y_B = 1 - y_A \approx 1$. The A–B solution is in equilibrium with pure solid B. We want to find the temperature at which these phases are in equilibrium. At this temperature, $\mu_{B,\text{solution}} = \mu_{B,\text{solid}}$, and hence $d\mu_{B,\text{solution}} = d\mu_{B,\text{solid}}$ for any change that takes the system to a new equilibrium state.

We can analyze the freezing-point depression phenomenon for any fixed pressure at which pure-liquid B can be in equilibrium with pure-solid B. Let us designate the fixed pressure as $P^\#$ and the freezing-point temperature of pure-liquid B, at $P^\#$, as T_F. Our goal is to find the temperature at which a binary solution is in equilibrium with pure-solid B at the fixed pressure $P^\#$. We let T_{fp} be the freezing-point temperature of the solution at $P^\#$. We base our analysis on the assumption that A obeys Henry's law.

We let the pure-solid solvent be the standard state for the solid solvent (see §15-5). Then, at every temperature, $\mu_{B,\text{solid}} = \tilde{\mu}^o_{B,\text{solid}}$, and

$$\begin{aligned}\mu_{B,\text{solid}} - \tilde{\mu}^o_{B,\text{solid}} &= RT \ln \tilde{a}_{B,\text{solid}} \\ &= 0\end{aligned}$$

At every temperature, $\tilde{a}_{B,\text{solid}} = 1$ so that $\left(\ln \tilde{a}_{B,\text{solid}}\right)_{PT} = 0$. The system pressure is constant at $P^\#$, so $dP = 0$. In the general expression

$$d\mu_{B,\text{solid}} = \overline{V}_{B,\text{solid}} dP - \overline{S}_{B,\text{solid}} dT + RT\left(d \ln \tilde{a}_{B,\text{solid}}\right)_{PT}$$

only the term in dT is non-zero. Recognizing that dT is the change in the freezing-point temperature at $P^\#$, for some change in the chemical potential of pure solid B, we have

$$d\mu_{B,\text{solid}} = -\bar{S}^\bullet_{B,\text{solid}}\, dT_{fp}$$

We let the pure-liquid solvent at its equilibrium vapor pressure be the standard state for the solution-phase solvent (see §2). We designate this equilibrium vapor pressure as $P^\bullet_B\left(T_{fp}\right)$. Now, since we ultimately find $T_{fp} < T_F$, the pure liquid freezes spontaneously at T_{fp}. The standard state for the liquid solvent is therefore a hypothetical state; it is a pure, super-cooled liquid. The properties of this hypothetical liquid can be estimated from our theory; however, except possibly in unusual circumstances, they cannot be measured directly. Since we assume that the solute obeys Henry's law, we have from §8 that $d \ln \tilde{a}_{B,\text{solution}} = d \ln y_B$. Thus, while the activity of the pure-solid solvent is constant, the activity of the solvent in the solution varies with the solute concentration. We have

$$d\mu_{B,\text{solution}} = -\overline{S}_{B,\text{solution}}\, dT_{fp} + RT_{fp}(d \ln y_B)_{PT}$$

Using $d \ln y_B \approx -dy_A$, the relationship $d\mu_{B,\text{solution}} = d\mu_{B,\text{solid}}$ becomes

$$-\overline{S}_{B,\text{solution}}\, dT_{fp} - RT_{fp} dy_A = -\bar{S}^\bullet_{B,\text{solid}}\, dT_{fp}$$

or

$$dy_A = -\left(\frac{\overline{S}_{B,\text{solution}} - \bar{S}^\bullet_{B,\text{solid}}}{RT_{fp}}\right) dT_{fp}$$

We consider systems in which the freezing point of the solution, T_{fp}, is little different from the freezing point of the pure solvent, T_F. Then, $T_F \approx T_{fp}$, and $T_{fp}/T_F \approx 1$. We let $|\Delta T| = T_F - T_{fp}$, where $|\Delta T| \ll T_F$. Since the solution is almost pure B, the partial molar entropy of B in the solution is approximately that of pure liquid B. Consequently, the partial molar entropy difference is, to a good approximation, just the entropy of fusion of the pure solvent, at equilibrium, at the freezing point for the specified system pressure. That is,

$$\begin{aligned}\left(\overline{S}_{B,\text{solution}} - \bar{S}^\bullet_{B,\text{solid}}\right)_{P^\#,T_{fp}} &\approx \left(\bar{S}^\bullet_{B,\text{liquid}} - \bar{S}^\bullet_{B,\text{solid}}\right)_{P^\#,T_F} \\ &= \Delta_{\text{fus}} S_B \\ &= \Delta_{\text{fus}} H_B / T_F\end{aligned}$$

so that

$$dy_A = -\left(\frac{\Delta_{\text{fus}}H_B}{RT_{fp}T_F}\right)dT_{fp}$$

At $P^{\#}$ and T_F, $\Delta_{\text{fus}}H_B$ is a constant. In the solution, the solute mole fraction is y_A; in the pure solid solvent, it is zero. Integrating between the limits $(0, T_F)$ and (y_A, T_{fp}), we have

$$\int_0^{y_A} dy_A = -\frac{\Delta_{\text{fus}}H_B}{RT_F}\int_{T_F}^{T_{fp}} \frac{dT_{fp}}{T_{fp}}$$

and

$$y_A = -\frac{\Delta_{\text{fus}}H_B}{RT_F}\ln\frac{T_{fp}}{T_F}$$

Introducing $\ln x \approx x - 1$, we have

$$y_A = -\frac{\Delta_{\text{fus}}H_B}{RT_F}\left(\frac{T_{fp}}{T_F} - 1\right) = \frac{\Delta_{\text{fus}}H_B}{RT_F^2}\Delta T$$

Solving for ΔT,

$$\Delta T = \left(\frac{RT_F^2}{\Delta_{\text{fus}}H_B}\right)y_A$$

The fusion process is endothermic, and $\Delta_{\text{fus}}H_B > 0$. Therefore, we find $\Delta T = T_F - T_{fp} > 0$; that is, the addition of a solute decreases the freezing point of a liquid. The depression of the freezing point is proportional to the solute concentration.

Since measurement of ΔT enables us to find y_A, freezing-point depression—like boiling point elevation—enables us to determine the molar mass of a solute. In our discussion of boiling-point elevation, we noted that it is often convenient to express the concentration of a dilute solute in units of molality rather than mole fraction. This applies also to freezing-point depression. Likewise, for practical applications, we usually find the freezing-point depression constant by measuring the depression of the freezing point of a solution of known composition.

§12 Colligative properties: osmotic pressure

The phenomena of boiling-point elevation and freezing-point depression involve relationships between composition and equilibrium temperature—at constant system pressure. We turn now to a phenomenon, ***osmotic pressure***, which involves a relationship between composition and equilibrium pressure—at constant system temperature.

To analyze boiling-point elevation, we equate the chemical potential of the solvent in two subsystems, a solution and the gas phase above it. To analyze freezing-point depression, we equate the chemical potential of the solvent in solution and solid subsystems. Similarly, to analyze osmotic pressure, we equate the chemical potential of the pure solvent—at one pressure—to the chemical potential of the solvent in a solution—at a second pressure. We find that equilibrium can be obtained only when the pressure in the solution subsystem exceeds the pressure in the solvent subsystem. The difference between these two pressures is the osmotic pressure.

Osmotic pressure

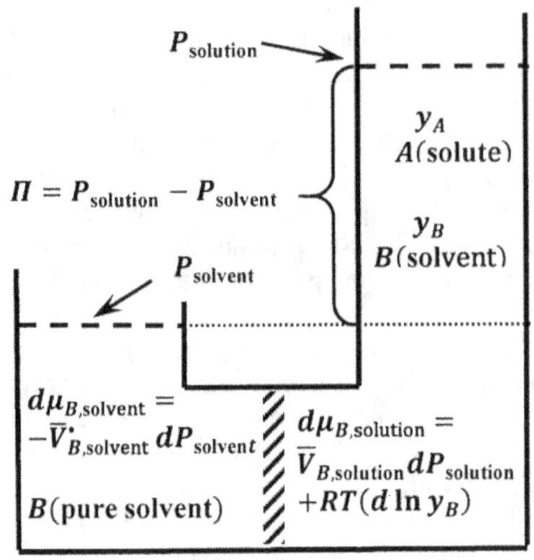

Figure 9. Schematic description of osmotic pressure.

In the boiling-point elevation and freezing-point depression phenomena, the subsystems are separated by a phase boundary. In the osmotic pressure phenomenon, a pure solvent phase is separated from a solution phase by a ***semi-permeable membrane***. A semi-permeable membrane allows free passage to solvent molecules; however, solute molecules cannot pass through it. In practice, the semi-permeable membrane is a material that is penetrated by pores, or channels, whose cross-sectional dimensions are nearly as small as typical solvent molecules. Solvent molecules can diffuse through these pores and pass from one side of the membrane to the other. With such a membrane, we can satisfy the osmotic pressure conditions by choosing a solute whose molecules are larger than the pore diameters, because large molecules will be unable to pass through the pores. In practice, the solute in osmotic pressure experiments is typically a polymer or a biologically derived molecule of high molecular weight. Osmotic pressure measurements have been an important source of data on the molar masses of such substances.

The osmotic pressure experiment is described schematically in Figure 9. The semi-permeable membrane must be sufficiently robust to support the pressure drop between the two subsystems. At constant pressure, mixing of the two subsystems is a spontaneous process. Were we to remove the membrane and the pressure drop that it supports, the subsystems would mix to form a single, more dilute solution. We see therefore that there is a tendency for net migration of solvent molecules from the solvent side of the membrane to the solution side. We can oppose this tendency by applying additional pressure on the solution side. Evidently, for any given solution composition, there will be an applied pressure at which the subsystems are in equilibrium with one another.

We let the pure-liquid solvent at its equilibrium vapor pressure be the standard state for both the pure-liquid and the solution-phase solvent (see §2). For the two subsystems to be in equilibrium, we must have $\mu_{B,\text{soluton}} = \mu_{B,\text{solvent}}$. For any change that takes one equilibrium state to another, we have $d\mu_{B,\text{soluton}} = d\mu_{B,\text{solvent}}$. Since the pure solvent subsystem contains only B, we have $\tilde{a}_{B,\text{solvent}} = \text{constant}$ so that $d \ln \tilde{a}_{B,\text{solvent}} = 0$. Since the temperature is constant, we have $dT = 0$. For the solvent subsystem, the general expression for $d\mu_{B,\text{solvent}}$ reduces to

$$d\mu_{B,\text{solvent}} = \bar{V}^{\bullet}_{B,\text{solvent}} dP_{\text{solvent}}$$

For the solution subsystem, $dT = 0$. Assuming the solvent in the solution obeys Raoult's law, we have $\tilde{a}_{B,\text{solution}} = y_B$. The general equation for $d\mu_{B,\text{soluton}}$ reduces to

$$d\mu_{B,\text{soluton}} = \overline{V}_{B,\text{solution}} dP_{\text{solution}} + RT(d \ln y_B)$$

Using $d \ln y_B \approx -dy_A$, the relationship $d\mu_{B,\text{solution}} = d\mu_{B,\text{solvent}}$ becomes

$$\overline{V}_{B,\text{solution}} dP_{\text{solution}} - RT dy_A = \bar{V}^{\bullet}_{B,\text{solvent}} dP_{\text{solvent}}$$

The molar volume of a liquid is nearly independent of the system pressure. Because the solution is nearly pure solvent, the molar volume of B in the solution is approximately equal to the molar volume of pure solvent B. Letting $\overline{V}_{B,\text{solution}} = \bar{V}^{\bullet}_{B,\text{solvent}} = \bar{V}^{\bullet}_B$, this becomes

$$\begin{aligned} dy_A &= \left(\frac{\bar{V}^{\bullet}_B}{RT}\right)(dP_{\text{solution}} - dP_{\text{solvent}}) \\ &= \left(\frac{\bar{V}^{\bullet}_B}{RT}\right) d(P_{\text{solution}} - P_{\text{solvent}}) \end{aligned}$$

This pressure difference is the osmotic pressure; it is often represented by the Greek alphabet capital pi: $\Pi = P_{\text{solution}} - P_{\text{solvent}}$. The osmotic pressure of the pure solvent must be zero; that is, $\Pi = 0$ when $y_A = 0$. Integrating between the limits $(0,0)$ and (y_A, Π), we have

$$\int_0^{y_A} dy_A = \frac{\bar{V}^{\bullet}_B}{RT} \int_0^{\Pi} d\Pi$$

and

$$y_A = \frac{\bar{V}^{\bullet}_B \Pi}{RT}$$

or

$$\Pi \bar{V}^{\bullet}_B = y_A RT$$

From this equation, we see that the osmotic pressure must be positive; that is, at equilibrium, the pressure on the solution must be greater than the pressure on the solvent: $\Pi = P_{\text{solution}} - P_{\text{solvent}} > 0$.

The osmotic pressure equation can be put into an easily remembered form. For $n_A \ll n_B$, $y_A = n_A/(n_A + n_B) \approx n_A/n_B$. With this substitution, $\Pi(n_B \bar{V}^{\bullet}_B) = n_A RT$, but since $\bar{V}^{\bullet}_B$ is the molar volume of pure B, $V = n_B \bar{V}^{\bullet}_B$ is just the volume of the solvent and essentially the same as the volume of the solution. The osmotic pressure equation has the same form as the ideal gas equation:

$$\begin{aligned} \Pi(n_B \bar{V}^{\bullet}_B) &= \Pi V \\ &= n_A RT \end{aligned}$$

§13 Colligative properties: solubility of a solid solute in an ideal solution

Although the result has few practical applications, we can also use these ideas to calculate the solubility of a solid solute in an ideal solution. The arguments are similar to those we used to estimate the freezing-point depression of a solution. The freezing point of a solution is the temperature at which the solution is in equilibrium with its pure-solid solvent. The solubility of a solute is the mole fraction of the solute in a solution that is at equilibrium with pure-solid solute. In this analysis, we assume that the solid phase is pure solute. Our analysis does not apply to a solid solution in equilibrium with a liquid solution. The properties of the solvent have no role in our description of the solid–ideal-solution equilibrium state. Consequently, our analysis produces a model in which the solubility of an ideal solute depends only on the properties of the solute; for a given solute, the ideal-solution solubility is the same in every solvent.

We specify the composition of the solution by the mole fractions of A and B, again letting the solute be compound A. When we consider freezing-point depression, an A–B solution of specified composition is in equilibrium with pure solid B, and we want to find the equilibrium temperature. When we consider solute solubility, the A–B solution is in equilibrium with pure solid A at a specified pressure, $P^{\#}$, and temperature, T; we want to find the equilibrium composition. Since pure solid A is present, the temperature must be less than the melting point of pure A. We let the melting point of the pure solute be T_{FA}, at the specified system pressure.

The activity of pure solid A and the system pressure are both constant; we have $d \ln \tilde{a}_{A,\text{solid}} = 0$, $dP = 0$, and

$$d\mu_{A,\text{solid}} = -\bar{S}^{\bullet}_{A,\text{solid}} dT$$

In the saturated ideal solution in equilibrium with this solid, we have $\tilde{a}_{A,\text{solution}} = y_A$, $dP = 0$, and

$$d\mu_{A,\text{solution}} = -\overline{S}_{A,\text{solution}} dT + RT(d \ln y_A)$$

The relationship $d\mu_{A,\text{solution}} = d\mu_{A,\text{solid}}$ becomes

$$-\overline{S}_{A,\text{solution}} dT + RT(d \ln y_A) = -\bar{S}^{\bullet}_{A,\text{solid}} dT$$

and

$$d \ln y_A = \left(\frac{\overline{S}_{A,\text{solution}} - \bar{S}^{\bullet}_{A,\text{solid}}}{RT} \right) dT$$

Now, $\overline{S}_{A,\text{solution}} - \bar{S}^{\bullet}_{A,\text{solid}}$ is the entropy change for the reversible (equilibrium) process in which one mole of pure solid A dissolves in a very large volume of a saturated solution; the mole fraction of A in this solution is constant at y_A. During this process, the pressure and temperature are constant at $P^{\#}$ and T. Letting the heat absorbed by the system during this process be $q^{rev}_{P^{\#}}$, we have

$$\left(\overline{S}_{A,\text{solution}} - \bar{S}^{\bullet}_{A,\text{solid}}\right)_{P^{\#},T} = q^{rev}_{P^{\#}}/T$$

The heat absorbed is also expressible as the difference between the partial molar enthalpy of A in the solution and that of the pure solid; that is,

$$q^{rev}_{P^{\#}} = \left(\overline{H}_{A,\text{solution}} - \overline{H}^{\bullet}_{A,\text{solid}}\right)_{P^{\#},T}$$

One of the properties of an ideal solution is that the enthalpy of mixing is zero. Thus, the partial molar enthalpy of A in an ideal solution is independent of y_A, so that the partial molar enthalpy of A in an ideal solution is the same as the partial molar enthalpy of pure liquid A; that is, $\overline{H}_{A,\text{solution}} = \overline{H}^{\bullet}_{A,\text{liquid}}$, and

$$\begin{aligned} q^{rev}_{P^{\#}} &= \left(\overline{H}_{A,\text{solution}} - \overline{H}^{\bullet}_{A,\text{solid}}\right)_{P^{\#},T} \\ &= \left(\overline{H}^{\bullet}_{A,\text{liquid}} - \overline{H}^{\bullet}_{A,\text{solid}}\right)_{P^{\#},T} \\ &= \left(\Delta_{\text{fus}} H_A\right)_{P^{\#},T} \\ &\approx \left(\Delta_{\text{fus}} H_A\right)_{P^{\#},T_{FA}} \end{aligned}$$

Then,

$$\left(\overline{S}_{A,\text{solution}} - \bar{S}^{\bullet}_{A,\text{solid}}\right)_{P^{\#},T} \approx \frac{\left(\Delta_{\text{fus}} H_A\right)_{P^{\#},T_{FA}}}{T}$$

Dropping the subscript information and replacing the approximate equality, we have

$$d \ln y_A = \left(\frac{\Delta_{\text{fus}} H_A}{RT^2} \right) dT$$

At $P^{\#}$ and T_{FA}, $\Delta_{\text{fus}} H_A$ is a property of pure A and is independent of the solution composition. When the pure solid solute melts at T_{FA}, the solute mole fraction is unity in the liquid phase with which it is in equilibrium: At T_{FA}, $y_A = 1$. At temperature T, y_A is the solute mole fraction in the liquid-phase solution that is at equilibrium with the pure-solid solute. Integrating between the limits $(1, T_{FA})$ and (y_A, T), we have

$$\int_1^{y_A} d \ln y_A = \frac{\Delta_{\text{fus}} H_A}{R} \int_{T_{FA}}^{T} \frac{dT}{T^2}$$

and

$$\ln y_A = \frac{-\Delta_{\text{fus}} H_A}{R} \left(\frac{1}{T} - \frac{1}{T_{FA}} \right)$$

For a given solute, $\Delta_{\text{fus}} H_A$ and T_{FA} are fixed and are independent of the characteristics of the solvent. The mole fraction of A in the saturated solution depends only on temperature. Since $\Delta_{\text{fus}} H_A > 0$ and $T < T_{FA}$, we find $\ln y_A < 0$. Therefore, we find that $y_A < 1$, as it must be. However, y_A increases, with T, implying that the solubility of a solid increases as the temperature increases, as we usually observe.

§14 Colligative properties: solubility of a gas

A similar analysis yields an equation for the solubility of a gas in a liquid solvent as a function of temperature at a fixed pressure, $P^{\#}$. We refer to the gas component as A and the liquid component as B. We assume that solvent B is nonvolatile, so that the gas phase with which it is in equilibrium is essentially pure gaseous solute A. We again find that the properties of the solvent have no role in our model, and the solubility of gas A is the same in every solvent.

We assume that low concentrations of the solute obey Henry's law and choose the solution-phase standard state for solute A to be the pure hypothetical liquid A whose vapor pressure is κ_A at T. From §4, we then have $\tilde{a}_{A,\text{solution}} = y_A$, so that $d \ln \tilde{a}_{A,\text{solution}} = d \ln y_A$ at any temperature. Substituting into the general equation

$$d\mu_A = \overline{V}_A dP - \overline{S}_A dT + RT(d \ln \tilde{a}_A)$$

we have

$$d\mu_{A,\text{solution}} = -\overline{S}_{A,\text{solution}}\, dT + RT(d \ln y_A)$$

The pressure of gas-phase A is constant at $P^{\#}$, and $dP = 0$. We choose the gas-phase activity standard state to be pure gas A at $P^{\#}$ and T. Since this makes the activity of the pure gas unity at any temperature, we have $d \ln \tilde{a}_{B,gas} = 0$. Substituting, we have

$$d\mu_{A,\text{gas}} = -\overline{S}_{A,\text{gas}}\, dT$$

Any constant-pressure process that maintains equilibrium between gas-phase A and solution-phase A must involve the same change in the chemical potential of A in each phase, so that $d\mu_{A,\text{gas}} = d\mu_{A,\text{solution}}$, and

$$-\overline{S}_{A,\text{gas}}\, dT = -\overline{S}_{A,\text{solution}}\, dT + RT(d \ln y_A)$$

so that

$$d \ln y_A = -\frac{\left(\overline{S}_{A,\text{gas}} - \overline{S}_{A,\text{solution}}\right)}{RT}\, dT$$

The difference $\overline{S}_{A,\text{gas}} - \overline{S}_{A,\text{solution}}$ is the entropy change for the equilibrium—and hence reversible—process in which one mole of substance A originally in solution vaporizes into a gas phase consisting of essentially pure gas A while the system is at the constant pressure $P^{\#}$. Let us designate the enthalpy change for this reversible process at $P^{\#}$ and T as $\Delta_{\text{vap}} \overline{H}_{A,\text{solution}}$. Then, we have

$$\overline{S}_{A,\text{gas}} - \overline{S}_{A,\text{solution}} = \frac{\Delta_{\text{vap}}\overline{H}_{A,\text{solution}}}{T}$$

so that

$$d\ln y_A = -\frac{\Delta_{\text{vap}}\overline{H}_{A,\text{solution}}}{RT^2}\,dT$$

Since enthalpy changes are generally relatively insensitive to temperature, we expect that, at least over small ranges of y_A and T, $\Delta_{\text{vap}}\overline{H}_{A,\text{solution}}$ is approximately constant. Since the vaporization process takes A from a state in which it has some of the characteristics of a liquid into a gaseous state, we can be confident that $\Delta_{\text{vap}}\overline{H}_{A,\text{solution}} > 0$. This conclusion implies that

$$\frac{d\ln y_A}{dT} < 0$$

Thus, our thermodynamic model leads us to the conclusion that the solubility of gas A decreases as the temperature increases. That the solubilities of gases generally decrease with increasing temperature is a well-known experimental observation. It stands in contrast to the observation that the solubilities of liquid or solid—at $P^{\#}$ and T—substances generally increase with increasing temperature. Our analysis of gas solubility provides a satisfying theoretical interpretation for an experimental observation which otherwise appears to be counterintuitive.

The meaning of $\Delta_{\text{vap}}\overline{H}_{A,\text{solution}}$ is unambiguous. Our analysis enables us to measure it by experimentally measuring y_A as a function of T. We can estimate $\Delta_{\text{vap}}\overline{H}_{A,\text{solution}}$ from another perspective: When we consider the "solution" in which $y_A = 1$, the vaporization process is the vaporization of liquid A into a gas phase of pure A at $P^{\#}$ and T . Since we assume that A is stable as a gas at T, the boiling point of pure liquid A must be less than T at $P^{\#}$ and the vaporization A must be a spontaneous process at $P^{\#}$ and T. The enthalpy of vaporization datum which is most accessible for liquid A is that for the reversible vaporization at one atmosphere and the normal boiling point, T_B, which we designate as $\Delta_{\text{vap}}H_A^o$. If we stipulate that $P^{\#}$ is one atmosphere; assume that our solubility equation remains valid as y_A increases from $y_A \approx 0$ to $y_A = 1$; and assume that the enthalpy of vaporization is approximately constant between the boiling point of A and T, we have $\Delta_{\text{vap}}\overline{H}_{A,\text{solution}} = \Delta_{\text{vap}}H_A^o$. Then,

$$d\ln y_A = -\frac{\Delta_{\text{vap}}H_A^o}{RT^2}\,dT$$

and

$$\int_1^{y_A} d\ln y_A = \int_{T_B}^{T} -\frac{\Delta_{\text{vap}}H_A^o}{RT^2}\,dT$$

so that

$$\ln y_A = \frac{\Delta_{\text{vap}}H_A^o}{R}\left(\frac{1}{T} - \frac{1}{T_B}\right)$$

Viewed critically, the accuracy of the approximation $\Delta_{\text{vap}}\overline{H}_{A,\text{solution}} \approx \Delta_{\text{vap}}H_A^o$ is dubious. The assumptions we make to reach it are essentially equivalent to assuming that the cohesive forces in solution are about the same between A molecules as they are between A molecules and B molecules. We expect this approximation to be more accurate the more closely the solution exhibits ideal behavior. However, if solvent B is to satisfy our assumption that the solvent is nonvolatile, the cohesive interactions between B molecules must be greater than those between A molecules, and this not consistent with ideal-solution behavior.

§15 Solvent activity coefficients from freezing-point depression measurements

The analysis of freezing-point depression that we present in §11 introduces a number of simplifying assumptions. We now undertake a more rigorous analysis of this phenomenon. This analysis is of practical importance. Measuring the freezing-point depression of a solution is one way that we can determine the activity and the activity coefficient of the solvent component. As we see in §7, if we have activity coefficients for the solvent over a range of solute concentrations, we can use the Gibbs-Duhem equation to find activity coefficients for the solute. Freezing-point depression measurements have been used extensively to determine the activity coefficients of aqueous solutes by measuring the activity of water in their solutions.

As in our earlier discussion of freezing-point depression, the equilibrium system is a solution of solute A in solvent B, which is in phase equilibrium with pure solid solvent B. Our present objective is to determine the activity of the solvent in its solutions at the melting point of the pure solvent. Having obtained this information, we can use the Gibbs-Duhem relationship to find the activity of the solute, as a function of solute concentration, at the melting point of the pure solvent. Once we have the solute activity at the melting point of the pure solvent, we can use the methods developed in §14-14 to find the solute activity in a solution at any higher temperature.

In §14-14, we find the temperature dependence of the natural logarithm of the chemical activity of a component of a solution. For a particular choice of activity standard states and enthalpy reference states, we develop a method to obtain the experimental data that we need to apply this equation. For brevity, let us refer to these choices as the infinite dilution standard states. In order to determine the activity of a solvent in its solutions at the melting point of the pure solvent, it is useful to define an additional standard state for the solvent. At temperatures below the normal melting point, which we again designate as T_F, we let the activity standard state of the solvent be pure solid B. Above the melting point, we use the infinite dilution standard state that we define in §14-14; that is, we let the activity standard state of the solvent be pure liquid solvent B.

At and below the melting point, T_F, the activity standard state for the solvent, B, is pure solid B. At and

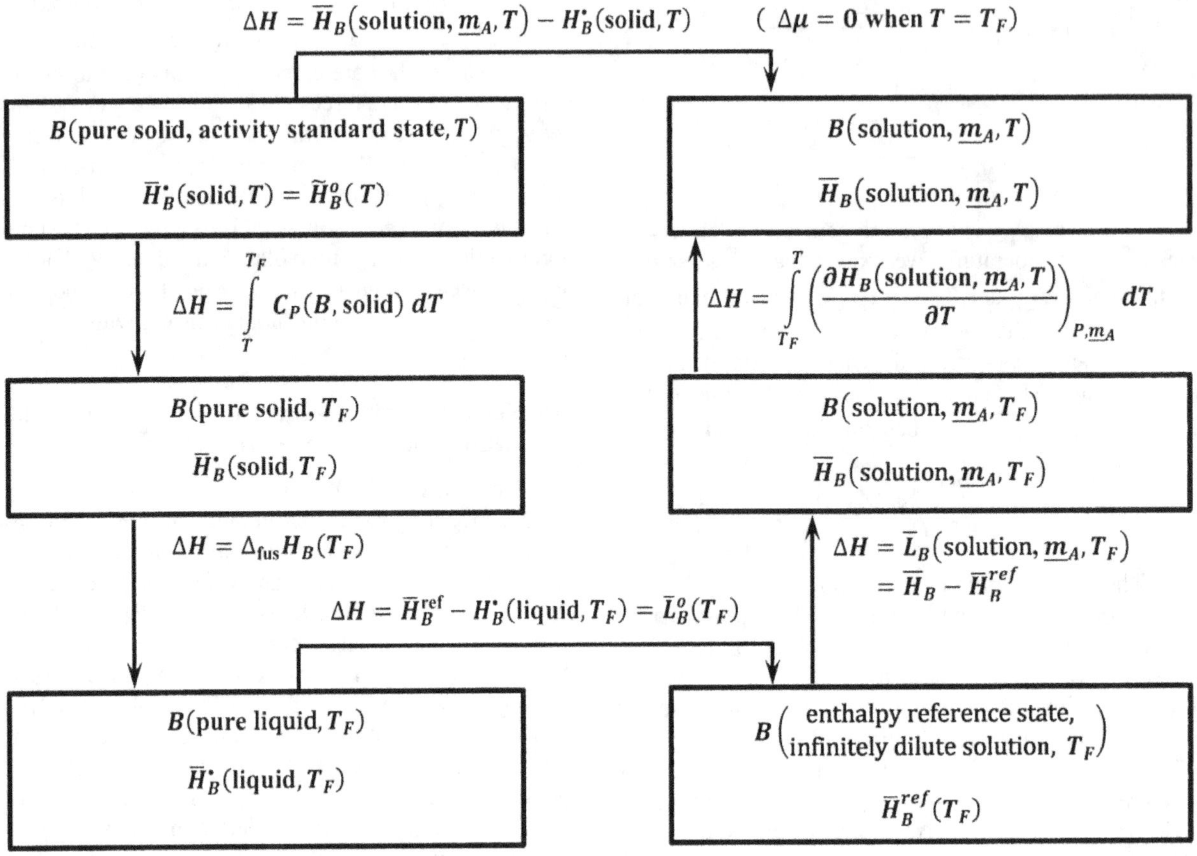

Figure 10. Enthalpy cycle for the temperature dependence of $\ln \tilde{a}_B$.

above the melting point, the activity standard state for the solvent is pure liquid B. At the melting point, pure solid solvent is in equilibrium with pure liquid solvent, which is also the solvent in an infinitely dilute solution. At T_F, the activity standard state chemical potentials of the pure solid solvent, the pure liquid solvent, and the solvent in an infinitely dilute solution are all the same. It follows that the value that we obtain for the activity of the solvent at T_F, for any particular solution, will be the same whether we determine it from measurements below T_F using the pure solid standard state or from measurements above T_F using the infinitely dilute solution standard state.

Now let us consider the chemical potential of liquid solvent B in a solution whose composition is specified by the molality of solute A, $\underline{m}_A$, when the activity standard state is pure solid B. We want to find this chemical potential at temperatures in the range $T_{fp} < T < T_F$, where T_{fp} is the freezing point of the solution whose composition is specified by $\underline{m}_A$. In this temperature range, we have

$$\mu_B(\text{solution}, \underline{m}_A, T) = \tilde{\mu}_B^{o}(\text{pure solid}, T) + RT \ln \tilde{a}_B (\text{solution}, \underline{m}_A, T)$$

Using the Gibbs-Helmholtz equation, we obtain

$$\left(\frac{\partial \ln \tilde{a}_B (\text{solution}, \underline{m}_A, T)}{\partial T}\right)_{P,\underline{m}_A} = \frac{-\overline{H}_B(\text{solution}, \underline{m}_A, T)}{RT^2} + \frac{\widetilde{H}_B^{o}(T)}{RT^2} = \frac{-\left[\overline{H}_B(\text{solution}, \underline{m}_A, T) - \overline{H}_B^{\bullet}(\text{solid}, T)\right]}{RT^2}$$

where we have $\widetilde{H}_B^{o}(T) = \overline{H}_B^{\bullet}(\text{solid}, T)$, because pure solid B is the activity standard state. Using the ideas developed in §14-14, we can use the thermochemical cycle shown in Figure 10 to evaluate

$$\overline{H}_B(\text{solution}, \underline{m}_A, T) - \overline{H}_B^{\bullet}(\text{solid}, T)$$

In this cycle, $\Delta_{\text{fus}}\overline{H}_B(T_F)$ is the molar enthalpy of fusion of pure B at the melting point, T_F. $\overline{L}_B(\text{solution}, \underline{m}_A, T)$ is the relative partial molar enthalpy of B at T_F in a solution whose composition is specified by $\underline{m}_A$. The only new quantity in this cycle is

$$\int_{T_f}^{T} \left(\frac{\partial \overline{H}_B(\text{solution}, \underline{m}_A, T)}{\partial T}\right)_{P,\underline{m}_A} dT$$

We can use the relative partial molar enthalpy of the solution to find it. By definition,

$$\overline{L}_B(\text{solution}, \underline{m}_A, T) = \left(\frac{\partial H(\text{solution}, \underline{m}_A, T)}{\partial n_B}\right)_{P,T,n_A} - \overline{H}_B^{\text{ref}}(T)$$

or, dropping the parenthetical information,

$$\overline{L}_B = \overline{H}_B - \overline{H}_B^{\text{ref}}(T)$$

so that

$$\left(\frac{\partial \overline{L}_B}{\partial T}\right)_P = \left(\frac{\partial \overline{H}_B}{\partial T}\right)_P - \left(\frac{\partial \overline{H}_B^{\text{ref}}}{\partial T}\right)_P$$

In §14-14, we introduce the relative partial molar heat capacity,

$$\overline{J}_B(T) = \left(\frac{\partial \overline{L}_B}{\partial T}\right)_P$$

Since the infinitely dilute solution is the enthalpy reference state for B in solution, we expect the molar enthalpy of pure liquid B to be a good approximation to the partial molar enthalpy of liquid B in the enthalpy reference state. Then, $\left(\partial \overline{H}_B^{\text{ref}}/\partial T\right)_P$ is just the molar heat capacity of pure liquid B, $C_P(B,\text{liquid}, T)$. (See problem 16-11.) We find

$$\left(\frac{\partial \overline{H}_B}{\partial T}\right)_P = \overline{J}_B(T) + C_P(B, \text{liquid}, T)$$

Using this result, the enthalpy changes around the cycle in Figure 10 yield

$$\begin{aligned}
&\overline{H}_B(\text{solution}, \underline{m}_A, T) - \overline{H}_B^{\bullet}(\text{solid}, T)\\
&= \int_T^{T_F} C_P(B,\text{solid}, T)\, dT + \Delta_{\text{fus}}\overline{H}_B(T_F) - L_B^{o}(T_F)\\
&\qquad + \overline{L}_B(\underline{m}_A, T_F)\\
&\qquad + \int_{T_F}^{T} \left[\overline{J}_B(T) + C_P(B, \text{liquid}, T)\right] dT\\
&= \Delta_{\text{fus}}\overline{H}_B(T_F) + \overline{L}_B(\underline{m}_A, T_F) - L_B^{o}(T_F)\\
&\qquad - \int_T^{T_F} \left[C_P(B, \text{liquid}, T)\right.\\
&\qquad \left. - C_P(B,\text{solid}, T) + \overline{J}_B\right] dT
\end{aligned}$$

Since we know how to determine $\overline{J}_B$ and the heat capacities as functions of temperature, we can evaluate this integral to obtain a function of temperature. For present purposes, let us assume that $\overline{J}_B$ and the heat capacities are essentially constant and introduce the abbreviation $\Delta C_P = C_P(B, \text{liquid}, T) - C_P(B,\text{solid}, T)$, so that

$$\int_T^{T_F} \left[C_P(B, \text{liquid}, T) - C_P(B,\text{solid}, T) + \overline{J}_B\right] dT = \left(\Delta C_P + \overline{J}_B\right)(T_F - T)$$

The temperature derivative of $\ln \tilde{a}_B\left(\text{solution}, \underline{m}_A, T\right)$ becomes

$$\left(\frac{\partial \ln \tilde{a}_B\left(\text{solution}, \underline{m}_A, T\right)}{\partial T}\right)_{P,\underline{m}_A} = \frac{-\Delta_{\text{fus}}\overline{H}_B(T_F) - \overline{L}_B(\underline{m}_A, T_F) + L_B^{o}(T_F)}{RT^2} + \frac{\left(\Delta C_P + \overline{J}_B\right)(T_F - T)}{RT^2}$$

T_{fp} is the freezing point of the solution whose composition is specified by $\underline{m}_A$. At T_{fp} the solvent in this solution is at equilibrium with pure solid solvent. Hence, the chemical potential of the solution solvent is equal to that of the pure-solid solvent. Then, because the pure solid is the activity standard state for both solution solvent and pure-solid solvent at T_{fp}, the activity of the solution solvent is equal to that of the pure-solid solvent. Because the pure solid is the activity standard state, the solvent activity is unity at T_{fp}. This means that we can integrate the temperature derivative from T_{fp} to T_F to obtain

$$\begin{aligned}
&\int_{T_{fp}}^{T_F} d\ln \tilde{a}_B\left(\text{solution}, \underline{m}_A, T\right)\\
&= \ln \tilde{a}_B\left(\text{solution}, \underline{m}_A, T_F\right)\\
&= \left(\frac{\Delta_{\text{fus}}\overline{H}_B(T_F) + \overline{L}_B(\underline{m}_A, T_F) + L_B^{o}(T_F)}{R}\right)\left(\frac{1}{T_F} - \frac{1}{T_{fp}}\right)\\
&\qquad - \left(\frac{\Delta C_P + \overline{J}_B}{R}\right)\left(1 - \frac{T_F}{T_{fp}} + \ln\left(\frac{T_F}{T_{fp}}\right)\right)
\end{aligned}$$

Thus, from the measured freezing point of a solution whose composition is specified by $\underline{m}_A$, we can calculate the activity of the solvent in that solution at T_F.

Several features of this result warrant mention. It is important to remember that we obtained it by assuming that $\Delta C_P + \overline{J}_B(T)$ is a constant. This is usually a good assumption. It is customary to express experimental results as values of the freezing-point depression, $\Delta T = T_F - T_{fp}$. The activity equation becomes

$$\begin{aligned}
&\ln \tilde{a}_B\left(\text{solution}, \underline{m}_A, T_F\right)\\
&= -\left(\frac{\Delta_{\text{fus}}\overline{H}_B(T_F) + \overline{L}_B(\underline{m}_A, T_F) + L_B^{o}(T_F)}{RT_F T_{fp}}\right)\Delta T\\
&\qquad + \left(\frac{\Delta C_P + \overline{J}_B}{R}\right)\left(\frac{\Delta T}{T_{fp}} - \ln\left(1 + \frac{\Delta T}{T_{fp}}\right)\right)
\end{aligned}$$

The terms involving $\overline{L}_B$, L_B^o, ΔC_P, and $\overline{J}_B$ are often negligible, particularly when the solute concentration is low. When $T_F/T_{fp} \approx 1$, that is, when the freezing-point depression is small, the coefficient of $\Delta C_P + \overline{J}_B$ is approximately zero. When these approximations apply, the activity equation is approximated by

$$\ln \tilde{a}_B \left(\text{solution}, \underline{m}_A, T_F\right) = -\left(\frac{\Delta_{\text{fus}}\overline{H}_B(T_F)}{RT_F^2}\right)\Delta T$$

§16 Electrolytic solutions

Thus far in our discussion of solute activities, we have assumed that the solute is a molecular species whose chemical structure is unchanged when the pure substance dissolves. This is not the case when salts dissolve in water and other polar solvents. A pure solid salt exists as a lattice of charged ions, rather than electrically neutral molecular moieties, and its solutions contain solvated ions. Since salt solutions conduct electricity, we often call them ***electrolytic solutions***. Solutions of salts in water are extremely important from both practical and theoretical standpoints. Accordingly, we focus our discussion on aqueous solutions; however, the ideas that we develop apply to salt solutions in any solvent that supports the formation of solvated ions.

We can apply the concepts that we develop in this chapter to measure the activities of aqueous salt solutions. When we do so, we find new features. These features arise from the formation of aquated ionic species and from electrical interactions among these species. In this chapter, we consider only the most basic issues that arise when we investigate the activities of dissolved salts. We consider only ***strong electrolytes***; that is, salts that are completely dissociated in solution. In this section, we briefly review the qualitative features of such solutions.

Departure from Henry's law behavior begins at markedly lower concentrations when the solute is a salt than when it is a neutral molecular species. This general observation is easily explained: Departures from Henry's law are caused by interactions among solution species. For neutral molecules separated by a distance r, the variation of the interaction energy with distance is approximately proportional to r^{-6}. This means that only the very closest molecules interact strongly with one another. For ions, Coulomb's law forces give rise to interaction energies that vary as r^{-1}. Compared to neutral molecules, ions interact with one another at much greater distances, so that departures from Henry's law occur at much lower concentrations.

Our qualitative picture of an aqueous salt solution is that the cations and anions that comprise the solid salt are separated from one another in the solution. Both the cations and the anions are surrounded by layers of loosely bound water molecules. The binding results from the electrical interaction between the ions and the water-molecule dipole. The negative (oxygen) end of the water dipole is preferentially oriented toward cations and the positive (hydrogen) end is preferentially oriented toward anions.

In aqueous solution, simple metallic cations are coordinated to a first layer of water molecules that occupy well-defined positions around the cation. In this layer, the bonding can have a covalent component. Such combinations of metal and coordinated water molecules are called ***aquo complexes***. For most purposes, we can consider that the aquo complex is the cationic species in solution. Beyond the layer of coordinated water molecules, a second layer of water molecules is less tightly bound. The positions occupied by these molecules are more variable. At still greater distances, water molecules interact progressively more weakly with the central cation. In general, when we consider the water molecules that surround a given anion, we find that even the closest solvent molecules do not occupy well-defined positions.

In any macroscopic quantity of solution, each ion has a specific average concentration. On a microscopic level, the Coulomb's law forces between dissolved ions operate to make the relative locations of cations and anions less random. It is useful to think about a spherical volume that surrounds a given ion. We suppose that the diameter of this sphere is several tens of nanometers. Within such a sphere centered on a particular cation, the concentration of anions will be greater than the average concentration of anions; the concentration of cations will be less than the average concentration of cations. Likewise, within a microscopic sphere centered on a given anion, the concentration of cations will be greater than the average concentration of cations; the concentration of anions will be below average.

As the concentration of a dissolved salt increases, distinguishable species can be formed in which a cation and an anion are nearest neighbors. We call such species ***ion pairs***. At sufficiently high salt concentrations, a significant fraction of the ions can be found in such ***ion-pair complexes***. Compared to other kinds of chemical bonds, ion-pair bonds are weak. The ion-pair bond is labile; the lifetime of a given ion pair is short. At still higher salt concentrations, the formation of significant concentrations of higher aggregates becomes possible. Characterizing the species present in an electrolytic solution becomes progressively more difficult as the salt concentration increases.

§17 Activities of electrolytes: the mean activity coefficient

We can find the activity of a salt in its aqueous salt solutions. For example, we can measure the freezing point depression for aqueous solutions of sodium chloride, find the activity of water in these solutions as a function of the sodium chloride concentration, and use the Gibbs-Duhem equation to find the activity of the dissolved sodium chloride as a function of its concentration. When we do so, we find some marked differences from our observations on molecular solutes.

For molecular solutes, the activity approaches the solute concentration as the concentration approaches zero; that is, the activity coefficient for a molecular solute approaches unity as the concentration approaches zero. For sodium chloride, and other 1:1 electrolytes, we find that the activity we measure in this way approaches the square of the solute concentration as the concentration approaches zero. For other salts, the measured activity approaches other powers of the solute concentration as the concentration approaches zero.

The dissociation of the solid salt into solvated ions explains these observations. Let us consider a solution made by dissolving n moles of a salt, A_pB_q, in n_{solvent} moles of solvent. (Let A be the cation and B the anion.) For present purposes, the cation and anion charges are not important. We use p and q to designate the composition of the salt. Typically, we are interested in dilute solutions, and it is convenient to use the hypothetical one-molal solution as the standard state for the activity of a solute species. We can represent the Gibbs free energy of this solution as

$$G = n_{\text{solvent}}\mu_{\text{solvent}} + n\mu_{A_pB_q}$$

where μ_{solvent} and $\mu_{A_pB_q}$ are the partial molar Gibbs free energies of the solvent and the solute in the solution. We can also write

$$\mu_{A_pB_q} = \tilde{\mu}^o_{A_pB_q} + RT\ln\tilde{a}_{A_pB_q}$$

where $\tilde{\mu}^o_{A_pB_q}$ is the partial molar Gibbs free energy when $\tilde{a}_{A_pB_q} = \underline{m}_{A_pB_q}\gamma_{A_pB_q} = 1$ in the activity standard state of the salt.

We assume that A_pB_q is a strong electrolyte; its solution contains np moles of the cation, A, and nq moles of the anion, B. In principle, we can also represent the Gibbs free energy of the solution as

$$G = n_{\text{solvent}}\mu_{\text{solvent}} + np\mu_A + nq\mu_B$$

and the individual-ion chemical potentials as $\mu_A = \tilde{\mu}^o_A + RT\ln\tilde{a}_A$ and $\mu_B = \tilde{\mu}^o_B + RT\ln\tilde{a}_B$, where $\tilde{\mu}^o_A$ and $\tilde{\mu}^o_B$ are the partial molar Gibbs free energies of the ions A and B in their hypothetical one-molal activity standard states. Equating the two equations for the Gibbs free energy of the solution, we have

$$\mu_{A_pB_q} = p\mu_A + q\mu_B$$

and

$$\begin{aligned}\tilde{\mu}^o_{A_pB_q} &+ RT\ln\tilde{a}_{A_pB_q} \\ &= p\tilde{\mu}^o_A + RT\ln\tilde{a}^p_A + q\tilde{\mu}^o_B + RT\ln\tilde{a}^q_B\end{aligned}$$

While it is often experimentally challenging to do so, we can measure $\tilde{\mu}^o_{A_pB_q}$ and $\tilde{a}_{A_pB_q}$. In principle, the meanings of the individual-ion activities, $\tilde{a}_A$ and $\tilde{a}_B$, and their standard-state chemical potentials, $\tilde{\mu}^o_A$ and $\tilde{\mu}^o_B$, are unambiguous; however, since we cannot prepare a solution that contains cation A and no anion, we cannot make measurements of $\tilde{a}_A$ or $\tilde{\mu}^o_A$ that are independent of the properties of B, or some other anion. Consequently, we must adopt some conventions to relate these properties of the ions, which we cannot measure, to those of the salt solution, which we can.

The universally adopted convention for the standard chemical potentials is to equate the sum of the standard chemical potentials of the constituent ions to that of the salt. We can think of this as assigning an equal share of the standard-state chemical potential of the salt to each of its ions; that is, we let

$$\tilde{\mu}^o_A = \tilde{\mu}^o_B = \frac{\tilde{\mu}^o_{A_pB_q}}{p+q}$$

Then, $\tilde{\mu}^o_{A_pB_q} = p\tilde{\mu}^o_A + q\tilde{\mu}^o_B$, and the activities of the individual ions are related to that of the salt by

$$\tilde{a}_{A_pB_q} = \tilde{a}^p_A\,\tilde{a}^q_B$$

We can develop the convention for the activities of the individual ions by representing each activity as the product of a concentration and an activity coefficient. That is, we represent the activity of each individual ion in the same way that we represent the activity of a molecular solute. In effect, this turns the problem of developing a convention for the activities of the individual ions into the problem of developing a convention for their activity coefficients. Using the hypothetical one-molal standard state for each ion, we write $\tilde{a}_A = \underline{m}_A\gamma_A$ and $\tilde{a}_B = \underline{m}_B\gamma_B$, where $\underline{m}_A, \gamma_A, \underline{m}_B$, and γ_B are the molalities and activity coefficients for ions A and B, respectively. Let the molality of the salt, A_pB_q, be $\underline{m}$. Then $\underline{m}_A = p\underline{m}$ and $\underline{m}_B = q\underline{m}$, and

$$\begin{aligned}\tilde{a}_{A_pB_q} &= \tilde{a}^p_A\,\tilde{a}^q_B \\ &= (p\underline{m}\gamma_A)^p(q\underline{m}\gamma_B)^q \\ &= (p^pq^q)\underline{m}^{p+q}\gamma^p_A\gamma^q_B\end{aligned}$$

Now we introduce the geometric mean of the activity coefficients γ_A and γ_B; that is, we define the geometric ***mean activity coefficient***, $\gamma_\pm$, by

$$\gamma_\pm = \left(\gamma^p_A\gamma^q_B\right)^{1/(p+q)}$$

The activity of the dissolved salt is then given by

$$\tilde{a}_{A_pB_q} = (p^pq^q)\underline{m}^{p+q}\gamma_\pm{}^{p+q}$$

The mean activity coefficient, $\gamma_\pm$, can be determined experimentally as a function of $\underline{m}_{A_pB_q}$, but the individual activity coefficients, γ_A and γ_B, cannot. It is common to present the results of activity measurements on electrolytic solutions as a table or a graph that shows the mean activity coefficient as a function of the salt molality.

While we cannot determine the activity or activity coefficient for an individual ion experimentally, no principle prohibits a theoretical model that estimates individual ion activities. Debye and Hückel developed such a theory. The Debye- Hückel theory gives reasonably accurate predictions for the activity coefficients of ions for solutions in which the total ion concentration is about

0.01 molal or less. We summarize the results of the Debye-Hückel theory in §18.

§18 Activities of electrolytes: the Debye-Hückel theory

In earlier sections, we introduce some basic methods for the experimental measurement of activities and activity coefficients. The Debye-Hückel theory leads to an equation for the activity coefficient of an ion in solution. The theory gives accurate values for the activity of an ion in very dilute solutions. As salt concentrations become greater, the accuracy of the Debye-Hückel model decreases. As a rough rule of thumb, the theory gives useful values for the activity coefficients of dissolved ions in solutions whose total salt concentrations are less than about 0.01 molal.[2] The theory is based on an electrostatic model. We describe this model and present the final result. We do not, however, present the argument by which the result is obtained.

We begin by reviewing some necessary ideas from electrostatics. When point charges q_1 and q_2 are embedded in a continuous medium, the Coulomb's law force exerted on q_1 by q_2 is

$$\vec{F}_{21} = \frac{q_1 q_2 \hat{r}_{21}}{4\pi\varepsilon_0 D r_{12}^2}$$

where ε_0 is a constant called the permittivity of free space, and D is a constant called the dielectric coefficient of the continuous medium. $\hat{r}_{21}$ is a unit vector in the direction from the location of q_2 to the location of q_1. When q_1 and q_2 have the same sign, the force is positive and acts to increase the separation between the charges. The force exerted on q_2 by q_1 is $\vec{F}_{12} = -\vec{F}_{21}$; the net force on the system of charges is $\vec{F}_{net} = \vec{F}_{12} + \vec{F}_{21} = 0$.

When the force is expressed in newtons, the point charges are expressed in coulombs, and distance is expressed in meters, $\varepsilon_0 = 8.854 \times 10^{-12}\ \mathrm{C^2N^{-1}m^{-2}}$. The dielectric coefficient is a dimensionless quantity whose value in a vacuum is unity. In liquid water at 25 C, $D = 78.4$ We are interested in the interactions between ions whose charges are multiples of the fundamental unit of charge, e. We designate the charge on a proton and an electron as e and $-e$, respectively, where $e = 1.602 \times 10^{-19}$ C. We express the charge on a cation, say A^{m+}, as $z_A e$, and that on an anion, say B^{n-}, as $z_B e$, where $z_A = +m > 0$ and $z_B = -n < 0$.

The Debye-Hückel theory models the environment around a particular central ion—the ion whose activity coefficient we calculate. We assume that the interactions between the central ion and all other ions result exclusively from Coulomb's law forces. We assume that the central ion is a hard sphere whose charge, q_C, is located at the center of the sphere. We let the radius of this sphere be a_C. Focusing on the central ion makes it possible to simplify the mathematics by fixing the origin of the coordinate system at the center of the central ion; as the central ion moves through the solution, the coordinate system moves with it. The theory develops a relationship between the activity coefficient of the central ion and the electrical work that is done when the central ion is brought into the solution from an infinite distance—where its potential energy is taken to be zero.

The theory models the interactions of the central ion with the other ions in the solution by supposing that, for every type of ion, k, in the solution, there is a spherically symmetric function, $\rho_k(r)$, which specifies the concentration of k-type ions at the location specified by r, for $r \geq a_C$. That is, we replace our model of mobile point-charge ions with a model in which charge is distributed continuously. The physical picture corresponding to this assumption is that the central ion remains discrete while all of the other ions are "ground up" into tiny charged bits that are spread smoothly—but not uniformly—throughout the solution that surrounds the central ion. The introduction of $\rho_k(r)$ changes our model from one involving point-charge neighbor ions—whose effects would have to be obtained by summing an impracticably large number of terms and whose locations are not well defined anyway—to one involving a mathematically continuous function. From this perspective, we adopt, for the sake of a quantitative mathematical treatment, a physical model that violates the atomic description of everything except the central ion.

It is useful to have a name for the collection of charged species around the central ion; we call it the ***ionic atmosphere***. The ionic atmosphere occupies a microscopic region around the central ion in which ionic concentrations depart from their macroscopic-solution values. The magnitudes of these departures depend on the sign and magnitude of the charge on the central ion.

The essence of the Debye-Hückel model is that the charge of the central ion gives rise to the ionic atmosphere. To appreciate why this is so, we can imagine introducing an uncharged moiety, otherwise identical to the central ion, into the solution. In such a process, no ionic atmosphere would form. As far as long-range Coulombic forces are concerned, no work would be done.

When we imagine introducing the charged central ion into the solution in this way, Coulombic forces lead to the creation of the ionic atmosphere. Since formation of the ionic atmosphere entails the separation of charge, albeit on a microscopic scale, this process involves electrical work. Alternatively, we can say that electrical work is done when a charged ion is introduced into a salt solution and that this work is expended on the creation of the ionic atmosphere.

In the Debye-Hückel model, this electrical work is the energy change associated with the process of solvating the ion. Since the reversible, non-pressure–volume work done in a constant-temperature, constant-pressure process is also the Gibbs free energy change for that process, the work of forming the ionic atmosphere is the same thing as the Gibbs free energy change for introducing the ion into the solution.

The Debye-Hückel theory makes these ideas quantitative by finding the work done in creating the ionic atmosphere. To do this, it proves to be useful to define a quantity that we call the ***ionic strength*** of the solution

By definition, the ionic strength is

$$I = \sum_{k=1}^{n} z_k^2 \underline{m}_k / 2$$

where the sum is over all of the ions present in the solution. The factor of $1/2$ is essentially arbitrary. We introduce it in order to make the ionic strength of a 1:1 electrolyte equal to its molality. (z_k is dimensionless.)

For the hypothetical one-molal standard state that we consider in §6, the activity coefficient for solute C, γ_C, is related to the chemical potential of the real substance, $\mu_C(P, \underline{m}_C)$, and that of a hypothetical ideal solute C at the same concentration, $\mu_C(\text{Hyp solute}, P, \underline{m}_C)$, by

$$\ln \gamma_C = \frac{\mu_C(P, \underline{m}_C) - \mu_C(\text{Hyp solute}, P, \underline{m}_C)}{RT}$$

The Debye-Hückel model equates this chemical-potential difference to the electrical work that accompanies the introduction of the central ion into a solution whose ionic strength is I. The final result is

$$\ln \gamma_C = \frac{-z_C^2 e^2 \kappa \overline{N}}{8\pi\varepsilon_0 D(1 + \kappa a_C)}$$

(While it is not obvious from our discussion, the parameter,

$$\kappa = \left(\frac{2e^2 \overline{N} d_w I}{\varepsilon_0 DkT} \right)^{1/2}$$

characterizes the ionic atmosphere around the central ion. The quantity d_w is the density of the pure solvent, which is usually water.)

For sufficiently dilute solutions, $1 + \kappa a_C \approx 1$. (See problem 14.) Introducing this approximation, substituting for κ, and dividing by 2.303 to convert to base-ten logarithms, we obtain the ***Debye-Hückel limiting law*** in the form in which it is usually presented:

$$\log_{10} \gamma_C = -A_\gamma z_C^2 I^{1/2}$$

Where

$$A_\gamma = \frac{(2d_w)^{1/2} \overline{N}^2}{2.303(8\pi)} \left(\frac{e^2}{\varepsilon_0 DRT} \right)^{3/2}$$

For aqueous solutions at 25 C, $A_\gamma = 0.510$.

The Debye-Hückel model finds the activity of an individual ion. In §18, we note that the activity of an individual ion cannot be determined experimentally. We introduce the mean activity coefficient, $\gamma_\pm$, for a strong electrolyte to as a way to express the departure of a salt solution from ideal-solution behavior. Adopting the hypothetical one-molal ideal-solution state as the standard state for the salt, $A_p B_q$, we develop conventions that express the Gibbs free energy of a real salt solution and find $\gamma_\pm = \left(\gamma_A^p \gamma_B^q\right)^{1/(p+q)}$. Using the Debye-Hückel limiting law values for the individual-ion activity coefficients, we find

$$\begin{aligned} \log_{10} \gamma_\pm &= \frac{p \log_{10} \gamma_A + q \log_{10} \gamma_B}{p+q} \\ &= \frac{-pA_\gamma z_A^2 I^{1/2} - qA_\gamma z_B^2 I^{1/2}}{p+q} \\ &= -\left(\frac{pz_A^2 + qz_B^2}{p+q} \right) A_\gamma I^{1/2} \\ &= -A_\gamma z_A z_B I^{1/2} \end{aligned}$$

where we use the identity

$$\frac{pz_A^2 + qz_B^2}{p+q} = -z_A z_B$$

(See problem 12.)

§19 Finding solute activity using the hypothetical one-molal standard state

In this chapter, we introduce several ways to measure the activities and chemical potentials of solutes. In §1–§6 we consider the determination of the activities and chemical potentials of solutes with measureable vapor pressures. To do so, we use the ideal behavior expressed by Raoult's Law and Henry's Law. In §15 we discuss the determination of solvent activity coefficients from measurements of the decrease in the freezing point of the solvent. In §7 we discuss the mathematical analysis by which we can obtain solute activity coefficients from measured solvent activity coefficients. Electrical potential measurements on electrochemical cells are an important source of thermodynamic data. In Chapter 17, we consider the use of electrochemical cells to measure the Gibbs free energy difference between two systems that contain the same substances but at different concentrations.

We define the activity of substance A in a particular system such that $\overline{G}_A = \mu_A = \tilde{\mu}_A^o + RT \ln \tilde{a}_A$. In the activity standard state the chemical potential is $\tilde{\mu}_A^o$ and the activity is unity, $\tilde{a}_A = 1$. It is often convenient to choose the standard state of the solute to be the hypothetical one-molal solution, particularly for relatively dilute solutions. In the hypothetical one-molal standard state, the solute molality is unity and the environment of a solute molecule is the same as its environment at infinite dilution. The solute activity is a function of its molality, $\tilde{a}_A(\underline{m}_A)$. We let the molality of the actual solution of unit activity be $\underline{m}_A^o$. That is, we let $\tilde{a}_A(\underline{m}_A^o) = 1$; consequently, we have $\mu_A(\underline{m}_A^o) = \tilde{\mu}_A^o$ even though the actual solution whose molality is $\underline{m}_A^o$ is not the standard state. To relate the solute activity and chemical potential in the actual solution to the solute molality, we must find the activity coefficient, γ_A, as a function of the solute molality,

$$\gamma_A = \gamma_A(\underline{m}_A)$$

Then

$$\tilde{a}_A(\underline{m}_A) = \underline{m}_A \gamma_A(\underline{m}_A)$$

and

$$\tilde{a}_A(\underline{m}_A^o) = \underline{m}_A^o \gamma_A(\underline{m}_A^o) = 1$$

To introduce some basic approaches to the determination of activity coefficients, let us assume for the moment that we can measure the actual chemical potential, μ_A, in a series of solutions where $\underline{m}_A$ varies. We have

$$\begin{aligned}\mu_A &= \tilde{\mu}_A^o + RT \ln \tilde{a}_A \\ &= \tilde{\mu}_A^o + RT \ln \underline{m}_A + RT \ln \gamma_A\end{aligned}$$

We know $\underline{m}_A$ from the preparation of the system—or by analysis. If we also $\tilde{\mu}_A^o$, we can calculate $\gamma_A(\underline{m}_A)$ from our experimental values of μ_A. If we don't know $\tilde{\mu}_A^o$, we need to find it before we can proceed. To find it, we recall that

$$\lim_{\underline{m}_A \to 0} RT \ln \gamma_A = 0$$

Then

$$\begin{aligned}\lim_{\underline{m}_A \to 0} (\mu_A - RT \ln \underline{m}_A) &= \lim_{\underline{m}_A \to 0} (\tilde{\mu}_A^o + RT \ln \gamma_A) \\ &= \tilde{\mu}_A^o\end{aligned}$$

and a plot of $(\mu_A - RT \ln \underline{m}_A)$ *versus* $\underline{m}_A$ will intersect the line $\underline{m}_A = 0$ at $\tilde{\mu}_A^o$.

Now, in fact, we can measure only Gibbs free energy differences. In the best of circumstances what we can measure is the difference between the chemical potential of A at two different concentrations. If we choose a reference molality, $\underline{m}_A^{\text{ref}}$, the chemical potential difference $\Delta\mu_A(\underline{m}_A) = \mu_A(\underline{m}_A) - \mu_A(\underline{m}_A^{\text{ref}})$ is a measurable quantity. A series of such results can be displayed as a plot of $\Delta\mu_A(\underline{m}_A)$ *versus* $\underline{m}_A$—or any other function of $\underline{m}_A$ that proves to suit our purposes. The reverence molality, $\underline{m}_A^{\text{ref}}$, can be chosen for experimental convenience.

If our theoretical structure is valid, the results are represented by the equations

$$\begin{aligned}\Delta\mu_A(\underline{m}_A) &= \mu_A(\underline{m}_A) - \mu_A(\underline{m}_A^{\text{ref}}) \\ &= RT \ln \frac{\tilde{a}_A(\underline{m}_A)}{\tilde{a}_A(\underline{m}_A^{\text{ref}})} \\ &= RT \ln \underline{m}_A + RT \ln \gamma_A(\underline{m}_A) \\ &\quad - RT \ln \tilde{a}_A(\underline{m}_A^{\text{ref}})\end{aligned}$$

When $\underline{m}_A = \underline{m}_A^o$, we have

$$\begin{aligned}\Delta\mu_A(\underline{m}_A^o) &= \mu_A(\underline{m}_A^o) - \mu_A(\underline{m}_A^{\text{ref}}) \\ &= \tilde{\mu}_A^o - \mu_A(\underline{m}_A^{\text{ref}}) \\ &= RT \ln \frac{\tilde{a}_A(\underline{m}_A^o)}{\tilde{a}_A(\underline{m}_A^{\text{ref}})} \\ &= -RT \ln \tilde{a}_A(\underline{m}_A^{\text{ref}})\end{aligned}$$

and

$$\Delta\mu_A(\underline{m}_A) - \Delta\mu_A(\underline{m}_A^o) = RT \ln \underline{m}_A + RT \ln \gamma_A(\underline{m}_A)$$

so that

$$RT \ln \gamma_A(\underline{m}_A) = -\Delta\mu_A(\underline{m}_A^o) + \Delta\mu_A(\underline{m}_A) - RT \ln \underline{m}_A$$

Since $\lim_{\underline{m}_A \to 0} \gamma_A(\underline{m}_A) = 1$, we have

$$\begin{aligned}0 &= \lim_{\underline{m}_A \to 0} RT \ln \gamma_A(\underline{m}_A) \\ &= -\Delta\mu_A(\underline{m}_A^o) + \lim_{\underline{m}_A \to 0} [\Delta\mu_A(\underline{m}_A) - RT \ln \underline{m}_A]\end{aligned}$$

Letting

$$\beta(\underline{m}_A) = \Delta\mu_A(\underline{m}_A) - RT \ln \underline{m}_A$$

and

$$\beta^o = \lim_{\underline{m}_A \to 0} [\Delta\mu_A(\underline{m}_A) - RT \ln \underline{m}_A]$$

we have

$$\Delta\mu_A(\underline{m}_A^o) = \beta^o$$

Then

$$\begin{aligned}RT \ln \gamma_A(\underline{m}_A) &= -\Delta\mu_A(\underline{m}_A^o) + \Delta\mu_A(\underline{m}_A) - RT \ln \underline{m}_A \\ &= -\beta^o + \Delta\mu_A(\underline{m}_A) - RT \ln \underline{m}_A\end{aligned}$$

so that we know both the activity coefficient, $\gamma_A = \gamma_A(\underline{m}_A)$, and the activity, $\tilde{a}_A(\underline{m}_A) = \underline{m}_A \gamma_A(\underline{m}_A)$, of A as a function of its molality. Consequently, we know the value of $\Delta\mu_A(\underline{m}_A) - \Delta\mu_A(\underline{m}_A^o)$ as a function of molality. Since this difference vanishes when $\underline{m}_A = \underline{m}_A^o$, we can find $\underline{m}_A^0$ from our experimental data. Finally, the activity equation becomes

$$RT \ln \tilde{a}_A(\underline{m}_A) = \Delta\mu_A(\underline{m}_A) - \beta^o$$

This analysis of the $\Delta\mu_A(\underline{m}_A)$ data assumes that we can find $\beta^o = \lim_{\underline{m}_A \to 0} [\Delta\mu_A(\underline{m}_A) - RT \ln \underline{m}_A]$. To find an accurate value for β^o, it is important to collect data for $\Delta\mu_A(\underline{m}_A)$ at the lowest possible values for $\underline{m}_A \mathrm{m_A}$. Inevitably, however, the experimental error in $\Delta\mu_A(\underline{m}_A)$ in creases as $\underline{m}_A$ decreases. Our theory requires that $\beta(\underline{m}_A) = \beta^o + f(\underline{m}_A)$, where $\lim_{\underline{m}_A \to 0} f(\underline{m}_A) = 0$, so that the graph of $\beta(\underline{m}_A)$ versus $f(\underline{m}_A)$ has an intercept at β^o. Accurate extrapolation of the data to the intercept at $\underline{m}_A = 0$ is greatly facilitated if we can choose $f(\underline{m}_A)$ so that the graph is linear. In practice, the increased experimental error in $\beta(\underline{m}_A)$ at the lowest values of $\underline{m}_A$ causes the uncertainty in the extrapolated value of β^o for a given choice of $f(\underline{m}_A)$ to be similar to the range of β^o values estimated using different functions. For some p in the range $0.5 < p < 2$, letting $f(\underline{m}_A) = \underline{m}_A^p$ often provides a fit that is as satisfactorily linear as the experimental uncertainty can justify.

This procedure yields the activity of A as a function of the solute molality. We obtain this function from measurements of $\Delta\mu_A(m_A) = \mu_A(m_A) - \mu_A(\underline{m}_A^{\text{ref}})$.

These measurements do not yield a value for $\mu_A(\underline{m}_A)$; what we obtain from our analysis is an alternative expression,

$$RT\ln\frac{\tilde{a}_A(\underline{m}_A)}{\tilde{a}_A(\underline{m}_A^{\text{ref}})}$$

for the chemical potential difference, $\mu_A(\underline{m}_A)-\mu_A(\underline{m}_A^{ref})$ between two states of the same substance. $\mu_A(\underline{m}_A)$ is the difference between the chemical potential of solute A at $\underline{m}_A$ and the chemical potential of its constituent elements in their standard states at the same temperature. To find this difference is a separate experimental undertaking. If, however, we can find $\mu_A(\underline{m}_A^*)$ for some $\underline{m}_A^*$, our activity equation yields $\tilde{\mu}_A^o$ as

$$\tilde{\mu}_A^0=\mu_A(\underline{m}_A^*)-RT\ln\tilde{a}_A(\underline{m}_A^*)$$

Finally, let us contrast this analysis to the analysis of chemical equilibrium that we discuss briefly in Chapter 15. In the present analysis, we use an extrapolation to infinite dilution to derive activity values from the difference between the chemical potentials of the same substance at different concentrations. In the chemical equilibrium analysis for $aA+bB\rightleftharpoons cD+dD$, we have

$$\begin{aligned}\Delta_r\mu&=\Delta_r\tilde{\mu}^o+RT\ln\frac{\tilde{a}_C^c\tilde{a}_D^d}{\tilde{a}_A^a\tilde{a}_B^b}\\&=\Delta_r\tilde{\mu}^o+RT\ln\frac{\underline{m}_C^c\underline{m}_D^d}{\underline{m}_A^a\underline{m}_B^b}+RT\ln\frac{\gamma_C^c\tilde{\gamma}_D^d}{\gamma_A^a\gamma_B^b}\end{aligned}$$

When the system is at equilibrium, we have $\Delta_r\mu=0$. Since $\lim_{\underline{m}_i\to 0}\gamma_i=1$, we have, in the limit that all of the concentrations go to zero in an equilibrium system,

$$\begin{aligned}0&=\lim_{\underline{m}_i\to 0}RT\ln\frac{\gamma_C^c\gamma_D^d}{\gamma_A^a\gamma_B^b}\\&=\Delta_r\tilde{\mu}^o+\lim_{\underline{m}_i\to 0}RT\ln\frac{\underline{m}_C^c\underline{m}_D^d}{\underline{m}_A^a\underline{m}_B^b}\end{aligned}$$

Letting

$$K_c=\frac{\underline{m}_C^c\underline{m}_D^d}{\underline{m}_A^a\underline{m}_B^b}$$

We have

$$\Delta_r\tilde{\mu}^o=-RT\lim_{\underline{m}_i\to 0}K_c$$

Since $\Delta_r\mu=0$ whenever the system is at equilibrium, measurement of K_c for any equilibrium state of the reaction yields the corresponding ratio of activity coefficients:

$$RT\ln\frac{\gamma_C^c\gamma_D^d}{\gamma_A^a\gamma_B^b}=-\Delta_r\tilde{\mu}^o-RT\ln K_c$$

Problems

1. At 100 C, the enthalpy of vaporization of water is 40.657 kJ mol^{-1}. Calculate the boiling-point elevation constant for water when the solute concentration is expressed in molality units.

2. At 0 C, the enthalpy of fusion of water is 6.009 kJ mol^{-1}. Calculate the freezing-point depression constant for water when the solute concentration is expressed in molality units.

3. A solution is prepared by dissolving 20.0 g of ethylene glycol (1,2-ethanediol) in 1 kg of water. Estimate the boiling point and the freezing point of this solution.

4. A biopolymer has a molecular weight of 250,000 dalton. At 300 K, estimate the osmotic pressure of a solution that contains 1 g of this substance in 10 mL of water.

5. Cyclohexanol melts at 25.46 C; the enthalpy of fusion is 1.76 kJ mol^{-1}. Estimate the freezing-point depression constant when the solute concentration is expressed as a mole fraction and when it is expressed in molality units. A solution is prepared by mixing 1 g of ethylene glycol with 50 g of liquid cyclohexanol. How much is the freezing point of this solution depressed relative to the freezing point of pure cyclohexanol?

6. Freezing-point depression data for numerous solutes in aqueous solution[3] are reproduced below. Calculate the freezing-point depression, $-\Delta T_{fp}/\underline{m}$, for each of these solutes. Compare these values to the freezing-point depression constant that you calculated in problem 2. Explain any differences.

Solute	**molality**	$-\Delta T_{fp}$**, K**
Acetone	0.087	0.16
ethanol	0.109	0.20
ethylene glycol	0.081	0.15
ammonia	0.295	0.55
glycerol	0.110	0.18
lithium chloride	0.119	0.42
nitric acid	0.080	0.28
potassium bromide	0.042	0.15
barium chloride	0.024	0.12

7. In a binary solution of solute A in solvent B, the mole fractions in the pure solvent are $y_A=0$ and $y_B=1$. We let the pure solvent be the solvent standard state; when $y_A=0$, $y_B=\tilde{a}_B=1$, and $\ln\tilde{a}_B=0$. What happens to the value of $\ln\tilde{a}_B$ as $y_A\to 1$? Sketch the graph of $(1-y_A)/y_A$ versus $\ln\tilde{a}_B$. For $0<y_A<y_A^*<1$, shade the area on this graph that represents the integral

$$\int_{y_A}^{y_A^*} \left(\frac{1-y_A}{y_A}\right) d\ln\gamma_A$$

Is this area greater or less than zero?

8. In a binary solution of solute A in solvent B, the activity coefficient of the solvent can be modeled by the equation $\ln\gamma_B(y_A) = c y_A^p$, where the constants c and p are found by using least squares to fit experimental data to the equation. Find an equation for $\ln\gamma_A(y_A)$. For $c = 8.4$ and $p = 2.12$, plot $\ln\gamma_B(y_A)$ and $\ln\gamma_A(y_A)$ versus y_A.

9. A series of solutions contains a non-volatile solute, A, dissolved in a solvent, B. At a fixed temperature, the vapor pressure of solvent B is measured for these solutions and for pure B ($y_A = 0$). At low solute concentrations, the vapor pressure varies with the solute mole fraction according to $P = P^\bullet(1-y_A)\exp(-\alpha y_A^\beta)$.

(a) If the pure solvent at one bar is taken as the standard state for liquid A, and gaseous B behaves as an ideal gas, now does the activity of solvent B vary with y_A?

(b) How does $\ln\gamma_B$ vary with y_A?

(c) Find $\ln\gamma_A(y_A)$.

10. In §14-14, we find for liquid solvent B,

$$\overline{L}_B^o = H_B^o - \overline{H}_B^{ref} = \lim_{T\to 0}\left(-\frac{\partial\Delta_{\text{mix}}\overline{H}}{\partial n_B}\right)_{P,T,n_A}$$

Since $\overline{H}_B^\bullet$ is the molar enthalpy of pure liquid B, we have

$$\left(\frac{\partial\overline{H}_B^\bullet}{\partial T}\right)_P = C_P(B,\text{liquid},T)$$

In §16-15, we set

$$\left(\frac{\partial\overline{H}_B^{ref}}{\partial T}\right)_P = C_P(B,\text{liquid},T)$$

Show that this is equivalent to the condition

$$\left(\frac{\partial\overline{L}_B^o}{\partial T}\right)_P \ll C_P(B,\text{liquid},T)$$

11. If $pz_A = -qz_B$, prove that

$$\frac{pz_A^2 + qz_B^2}{p+q} = -z_A z_B$$

12. At temperatures of 5 C, 25 C, and 45 C, evaluate Debye-Hückel parameter κ for aqueous sodium chloride solutions at concentrations of $10^{-3}\,\underline{m}$, $10^{-2}\,\underline{m}$, and $10^{-1}\,\underline{m}$.

13. Introducing the approximation $1 + \kappa a_c \approx 1$ produces the Debye-Hückel limiting law, which is strictly applicable only in the limiting case of an infinitely dilute solution. Introducing the approximation avoids the problem of choosing an appropriate value for a_c. If $a_c = 0.2$ nm, calculate $1 + \kappa a_c$ for aqueous solutions in which the ionic strength, I, is $10^{-3}\,\underline{m}$, $10^{-2}\,\underline{m}$, and $10^{-1}\,\underline{m}$. What does the result suggest about the ionic-strength range over which the limiting law is a good approximation?

14. The solubility product for barium sulfate, $K_{sp} = \tilde{a}_{Ba^{2+}}\tilde{a}_{SO_4^{2-}}$, is 1.08×10^{-10}. Estimate the solubility of barium sulfate in pure water and in $10^{-2}\,\underline{m}$ potassium perchlorate.

15. The enthalpy of vaporization[3] of n-butane at its normal boiling point, 272.65 K, is 22.44 kJ mol^{-1}. In the temperature range $273.15 < T < 349.15$ K, the solubility[3] of n-butane in water is given by

$$\ln y_A = A + \frac{100\,B}{T} + C\,\ln\left(\frac{T}{100}\right)$$

where $A = -102.029$, $B = 146.040\ \text{K}^{-1}$, and $C = 38.7599$. From the result we develop in §14, calculate $\Delta_{\text{vap}}\overline{H}_{A,\text{solution}}$ for n-butane at is normal boiling point. (Note that the normal boiling temperature is slightly below the temperature range to which the equation for $\ln y_A$ is valid.) Comment.

16. The enthalpy of vaporization[3] of molecular oxygen at its normal boiling point, 90.02 K, is 6.82 kJ mol^{-1}. In the temperature range $273.15 < T < 348.15$, the solubility[3] of oxygen in water is given by

$$\ln y_A = A + \frac{100\,B}{T} + C\,\ln\left(\frac{T}{100}\right)$$

where $A = -66.7354$, $B = 87.4755\ \text{K}^{-1}$, and $C = 24.4526$. From the result we develop in §14, calculate $\Delta_{\text{vap}}\overline{H}_{A,\text{solution}}$ for oxygen at 273.25 K and at its normal boiling point, 90.02 K. Comment.

Notes

[1] Raoult's law and ideal solutions can be defined using fugacities in place of partial pressures. The result is more general but—for those whose intuition has not yet embraced fugacity—less transparent.

[2] For a discussion of the concentration range in which the Debeye-Huckel model is valid and of various supplemental models that allow for the effects of forces that are specific to the chemical characteristics of the interacting ions, see Lewis and Randall, Pitzer and Brewer, ***Thermodynamics***, 2nd Edition, McGraw Hill Book Company, New York, 1961, Chapter 23.

[3] Data from ***CRC Handbook of Chemistry and Physics***, 79th Edition, David R. Lide, Ed., CRC Press, 1998-1999.

17

Electrochemistry

§1 Oxidation-reduction reactions

We find it useful to classify reactions according to the type of change that the reagents undergo. Many classification schemes exist, often overlapping one another. The three most commonly used classifications are acid–base reactions, substitution reactions, and oxidation–reduction reactions.

Acids and bases can be defined in several ways, the most common being the Brønsted-Lowry definition, in which acids are proton donors and bases are proton acceptors. The Brønsted-Lowry definition is particularly useful for reactions that occur in aqueous solutions. A prototypical example is the reaction of acetic acid with hydroxide ion to produce the acetate ion and water.

$$CH_3CO_2H + OH^- \rightarrow CH_3CO_2^- + H_2O$$

Here acetic acid is the proton donor, hydroxide ion is the proton acceptor. The products are also an acid and a base, since water is a proton donor and acetate ion is a proton acceptor.

When we talk about substitution reactions, we focus on a particular substituent in a chemical compound. The original compound is often called the substrate. In a substitution reaction, the original substituent is replaced by a different chemical moiety. A prototypical example is the displacement of one substituent on a tetrahedral carbon atom by a more nucleophilic group, as in the reaction of methoxide ion with methyl iodide to give dimethyl ether and iodide ion.

$$CH_3I + CH_3O^- \rightarrow CH_3OCH_3 + I^-$$

We could view a Brønsted-Lowry acid-base reaction as a substitution reaction in which one group (the acetate ion in the example above) originally bonded to a proton is replaced by another (hydroxide ion). Whether we use one classification scheme or another to describe a particular reaction depends on which is better suited to our immediate purpose.

In acid-base reactions and substitution reactions, we focus on the transfer of a chemical moiety from one chemical environment to another. In a large and important class of reactions we find it useful to focus on the transfer of one or more electrons from one chemical moiety to another. For example, copper metal readily reduces aqueous silver ion. If we place a piece of clean copper wire in an aqueous silver nitrate solution, reaction occurs according to the equation

$$2Ag^+ + Cu^0 \rightarrow 2Ag^0 + Cu^{2+}$$

We have no trouble viewing this reaction as the transfer of two electrons from the copper atom to the silver ions. In consequence, a cupric ion, formed at the copper surface, is released into the solution. Two atoms of metallic silver are deposited at the copper surface. Reactions in which electrons are transferred from one chemical moiety to another are called ***oxidation–reduction*** reactions, or ***redox*** reactions, for short.

We define oxidation as the loss of electrons by a chemical moiety. Reduction is the gain of electrons by a chemical moiety. Since a moiety can give up electrons only if they have some place to go, oxidation and reduction are companion processes. Whenever one moiety is oxidized, another is reduced. In the reduction of silver ion by copper metal, it is easy to see that silver ion is gaining electrons and copper metal is losing them. In other reactions, it is not always so easy to see which moieties are gaining and losing electrons, or even that electron transfer is actually involved. As an adjunct to our ideas about oxidation and reduction, we develop a scheme for formally assigning electrons to the atoms in a molecule or ion. This is called the ***oxidation state*** formalism and comprises a series of rules for assigning a number, which we call the oxidation state (or oxidation number), to every atom in the molecule. When we adopt this scheme, the redox character of a reaction is determined by which atoms increase their oxidation state and which decrease their oxidation state as a consequence of the reaction. Those whose oxidation state increases lose electrons and are ***oxidized***, while those whose oxidation state decreases gain electrons and are ***reduced***.

A process of electron loss is called an oxidation because reactions with elemental oxygen are viewed as prototypical examples of such processes. Since many observations are correlated by supposing that oxygen atoms in compounds are characteristically more electron-rich than the atoms in elemental oxygen, it is useful to regard a reaction of a substance with oxygen as a reaction in which the atoms of the substance surrender electrons to oxygen atoms. It is then a straightforward generalization to say that a substance is oxidized whenever it loses electrons, whether oxygen atoms or some other chemical moiety takes up those electrons. So, for example, the reaction of sodium metal with oxygen in a dry environment produces sodium oxide, Na_2O, in which the sodium is usefully viewed as carrying a positive charge. (The oxidation state of sodium is 1+; the oxidation state of oxygen is 2–.)

The conversion of a metal oxide to the corresponding metal is described as reducing the oxide. Since converting a metal oxide to the metal reverses the change that occurs when we oxidize it, generalization of this idea leads us to apply the term reduction to any process in which a chemical moiety gains electrons. It is a fortunate coincidence that a reduction process is one in which the oxidation number of an atom becomes smaller (more negative) and is therefore reduced, in the sense of being decreased.

Another feature of oxidation–reduction reactions, and one that relates to the utility of viewing these reactions in terms of electron gain and loss, emerges when we observe the reaction of aqueous silver ions with copper metal closely. As the reaction proceeds, the aqueous solution becomes blue as cupric ions accumulate. Long needle-like crystals of silver metal grow out from the copper surface. The simplest mechanism that we can imagine for the growth of well-formed silver crystals is that silver ions from the solution plate out on the surface of the growing silver crystal, accepting an electron from the metallic crystal as they do so. The silver metal acquires this electron from the copper metal, with which it is in contact, but at a large (on an atomic scale) distance from the site at which the new atom of silver is deposited. Evidently the processes of electron loss and gain that characterize an overall reaction can occur at different locations, if there is a suitable process for moving the electron from one location to the other.

§2 Electrochemical cells

We can extend the idea of carrying out the electron loss and electron gain steps in different physical locations. Suppose that the only aqueous species in contact with the silver metal are silver ions and nitrate ions; the silver metal is also in contact with a length of copper wire, whose other end dips into a separate reservoir containing an aqueous solution of sodium nitrate. This arrangement is sketched in Figure 1. When we create this arrangement, nothing happens. We do not see any visible change in the silver metal, and the water contacting the copper wire never turns blue. On the one hand, this result does not surprise us. We are accustomed to the idea that reactants must be able to contact one another in order for reaction to occur.

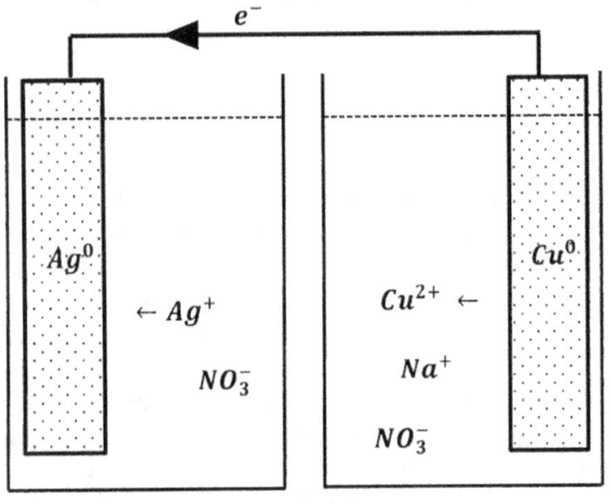

Figure 1. Electron transport between compartments is possible, but ion transport is not.

On the other hand, the original experiment really does show that silver ions can accept electrons in one location while copper atoms give them up in another, so long as we provide a metal bridge on which the electrons can move between the two locations. Why should this not continue to happen in the new experimental arrangement? In fact, it does. It is just that the reaction occurs to only a very small extent before stopping altogether. The reason is easy to appreciate. After a very small number of silver ions are reduced, the silver nitrate solution contains more nitrate ions than silver ions; the solution as a whole has a negative charge. In the other reservoir, a small number of cupric ions dissolve, but there is no increase in the number of counter ions, so this solution acquires a positive charge. These net charges polarize the metal that connects them; the metal has an excess of positive charge at the copper-solution end and an excess of negative charge at the silver-solution end. This polarization opposes the motion of a negatively charged electron from the copper-solution end toward the silver-solution end. When the polarization becomes sufficiently great, electron flow ceases and no further reaction can occur.

By this analysis, the anions that the cupric solution needs in order to achieve electroneutrality are present in the silver-ion solution. The reaction stops because the anions have no way to get from one solution to the other. Evidently, the way to make the reaction proceed is to modify the two-reservoir experiment so that nitrate ions can move from the silver-solution reservoir to the copper-solution reservoir. Alternatively, we could introduce a modification that allows copper ions to move in the opposite direction or one that allows both kinds of movement. We can achieve the latter by connecting the two solutions with a tube containing sodium nitrate solution, as diagrammed in Figure 2. Now, nitrate ions can move between the reservoirs and maintain electroneutrality in both of them. However, silver ions can also move between the reservoirs. When we do this experiment, we observe that electrons do flow through the wire, indicating that silver-ion reduction and copper-atom oxidation are occurring at the separated sites. However, after a short time, the solutions mix; silver ions migrate through the aqueous medium and react directly with the copper metal. Because the mixing is poorly controllable, the reproducibility of this experiment is poor.

Evidently, we need a way to permit the exchange of ions between the two reservoirs that does not permit the wholesale transfer of reactive species. One device that accomplishes this is called a ***salt bridge***. The requirement we face is that ions should be able to migrate from reservoir to reservoir so as to maintain electroneutrality. However, we do not want ions that participate in

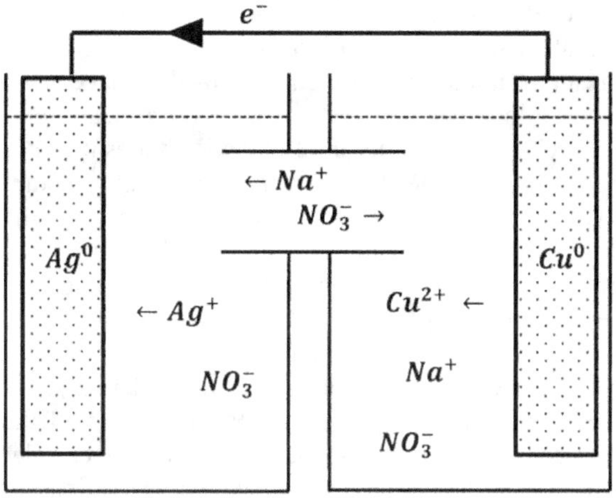

Figure 2. Electron transport and fluid flow between compartments are possible.

electrode reactions to migrate. A salt bridge is simply a salt solution that we use to connect the two reservoirs. To avoid introducing unwanted ions into the reservoir solutions, we prepare the salt-bridge solution using a salt whose ions are not readily oxidized or reduced. Alkali metal salts with nitrate, perchlorate, or halide anions are often used. To avoid mixing the reservoir solutions with the salt bridge solution, we plug each end of the salt bridge with a porous material that permits diffusion of ions but inhibits bulk movement of solution in or out of the bridge. The inhibition of bulk movement can be made much more effective by filling the bridge with a gel, so that the solution is unable to undergo bulk motion in any part of the bridge.

With a salt bridge in place, inert ions can move from one reservoir to the other to maintain electroneutrality. Under these conditions, we see an electrical current through the external circuit and a compensating diffusion of ions through the salt bridge. The salt bridge completes the circuit. Transport of electrons from one electrode to the other carries charge in one direction; motion of ionic species through the salt bridge carries negative charge

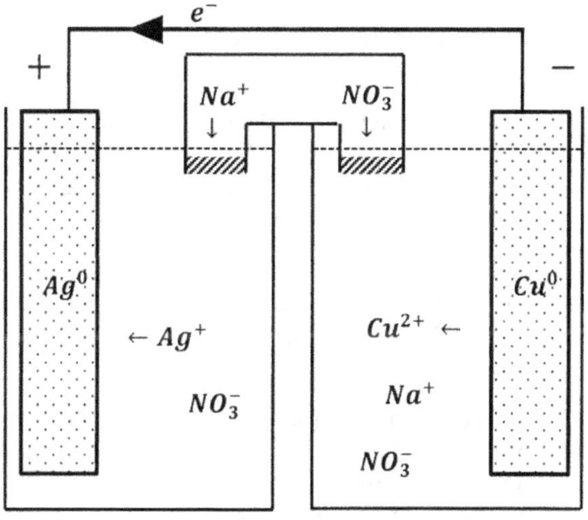

Figure 3. Electron transport and ion transport between compartments are possible.

through the solution in the opposite direction. This compensating ionic motion has anions moving opposite to the electron motion and cations moving in the same direction as the electrons.

We have just described one kind of electrochemical cell. As diagrammed in Figure 3, it has four principal features: two reservoirs within which reactions can occur, a wire through which electrons can pass from one reservoir to the other, and a salt bridge through which ionic species can pass. Many similar electrochemical cells can be constructed. The reservoirs can contain a wide variety of reagents. Because each reservoir must be able to exchange electrons with the connecting wire, each must contain an electrically conducting solid that serves as a ***terminal*** and a ***current collector***, and often participates in the chemical change as a reactant or as a catalyst. The combination of reagents and current collector is called a ***half-cell***. The current collector itself is called an ***electrode***, although this term is often applied to the whole half-cell as well. In this case, the wire is the ***external circuit***. In applications of chemical interest, the external circuit typically contains devices to measure the electrochemical cell's properties as a ***circuit element***.

If we view this electrochemical cell as a device for producing an electrical current, we see that it has a number of practical limitations. Two of the most important relate to the performance of the salt bridge. Whenever electrons move through the external circuit, the salt bridge must accept a charge-compensating number of ions from one reservoir and release the same quantity of ionic charge to the other reservoir. We construct the salt bridge so that ions can pass into and out of it only by diffusion. Consequently, the rate at which ions can diffuse through the salt bridge limits the rate at which electrons can flow through the external circuit. Since diffusion is a slow process over the macroscopic dimensions of the bridge, the cell can pass only a small current. From an electrical perspective, slow diffusion of ions through the salt bridge causes a surplus of positively charged species to develop at one end of the salt bridge and a surplus of negatively charged species to develop at the other. This charge imbalance means that there is a potential gradient across the salt bridge, whose effect is to oppose the flow of further current.

The second limitation of the cell attributable to the properties of the salt bridge is that the amount of current the cell can produce before its performance characteristics change dramatically is limited by the amount of inert salt in the bridge. After a relatively small charge passes through the cell, migration of reactive species from one reservoir to the other becomes significant. Effective electrochemical power sources must use other methods to separate reactants and products while allowing for the transport of ions between half-cells.

Despite these limitations, such electrochemical cells are very effective tools for the study of the thermodynamics of electrochemical reactions. The principle interaction between electrochemistry and thermodynamics revolves around the relationship between the free energy change for the reaction and the properties of the electrochemical cell viewed as a circuit element. In §14, we see that the Gibbs free energy change for the chemical reaction is proportional to the electrical potential that develops across the terminals of the corresponding electrochemical cell.

In experiments, we find that the electrical potential difference across a cell depends on the amount of current that is being drawn from the cell. Because the movement of ions and other substances within the cell is slow compared to the rate at which a wire can transfer electrons from one terminal to another, potential differences that develop within an operating cell decrease the electrical potential across the terminals. Only when the current being drawn from the cell is zero does the electrical potential correspond precisely to the Gibbs free energy change of the chemical reaction occurring in the cell. This should not surprise us. The experimental measurement of any entropy-dependent thermodynamic function must be made on a system that is undergoing reversible change. A reversible change in an electrochemical cell is a change in which the current flow is zero.

Measuring the electrical potential at zero current is experimentally straightforward, at least in principle. We connect the cell to some reference device that provides a known and variable electrical potential. The connection is made such that the electrical potential from the reference device opposes the potential from the electrochemical cell; that is, we connect the positive terminal of the reference device to the positive terminal of the cell, and the negative terminal of the device to the negative terminal of the cell. (See §7.) We then vary the potential of the reference device until current flow in the circuit stops. When this occurs the potential drop being supplied by the reference device must be precisely equal to the potential drop across the electrochemical cell, which is the datum we want.

In practice, the reference device is another "standard" electrochemical cell, whose potential drop is defined to have a particular value at specified conditions. Modern electronics make it possible to do the actual measurements with great sophistication. The necessary measurements can also be done with very basic equipment. The principles remain the same. In the basic experiment, a variable resistor is used to adjust the potential drop across the standard cell until it exactly matches that of the cell being studied. When this potential is reached, current flow ceases. Current flow is monitored using a sensitive galvanometer. It is not necessary to actually measure the current. Since we are interested in locating the potential drop at which the current flow is zero, it is sufficient to find the potential drop at which the galvanometer detects no current. The accuracy of the potential measurement depends on the stability of the standard cell potential, the accuracy of the variable resistor, and the sensitivity of the galvanometer.

§3 Defining oxidation states

We introduce oxidation states to organize our thinking about oxidation–reduction reactions and electrochemical cells. When we define oxidation states, we create a set of rules for allocating the electrons in a molecule or ion to the individual atoms that make it up. The definition of oxidation states is therefore an accounting exercise. The definition of oxidation states predates our ability to estimate electron densities through quantum mechanical calculations. As it turns out, however, the ideas that led to the oxidation state formalism are directionally correct; atoms that have high positive oxidation states according to the formalism also have relatively high positive charges by quantum mechanical calculation. In general, the absolute values of oxidation states are substantially larger than the absolute values of the partial charges found by quantum-mechanical calculation; however, there is no simple quantitative relationship between oxidation states and the actual distribution of electrons in real chemical moieties. It is a serious mistake to think that our accounting system provides a quantitative description of actual electron densities.

The rules for assigning oxidation states grow out of the primitive (and quantitatively incorrect) idea that oxygen atoms usually acquire two electrons and hydrogen atoms usually lose one electron in forming chemical compounds and ionic moieties. The rest of the rules derive from a need to recognize some exceptional cases and from applying the basic ideas to additional elements. The rules of the oxidation state formalism are these:

- For any element in any of its allotropic forms, the oxidation state of its atoms is zero.
- In any of its compounds, the oxidation state of an oxygen atom is 2–, ***except*** in compounds that contain an oxygen–oxygen bond, where the oxidation state of oxygen is 1–. The excepted compounds are named peroxides. Examples include sodium peroxide, Na_2O_2, and hydrogen peroxide, H_2O_2.
- In any of its compounds, the oxidation state of a hydrogen atom is 1+, ***except*** in compounds that contain a metal–hydrogen bond, where the oxidation state of hydrogen is 1–. The excepted compounds are named hydrides. Examples include sodium hydride, NaH, and calcium hydride, CaH_2.
- In any of their compounds, the oxidation states of alkali metal atoms (lithium, sodium, potassium, rubidium, cesium, and francium) are 1+. (There are exceptional cases, but we do not consider them.)
- In any of their compounds, the oxidation states of halogen atoms (fluorine, chlorine, bromine, iodine, and astatine) are 1–, ***except*** in compounds that contain a halogen–oxygen bond.
- The oxidation states of any other atoms in a compound are chosen so as to make the sum of the oxidation states in the chemical moiety equal to its charge. So, for a neutral molecule, the oxidation states sum to zero. For a monovalent anion, they sum to 1–, *etc.*

§4 Balancing oxidation–reduction reactions

Having defined oxidation states, we can now redefine an ***oxidation–reduction reaction*** as one in which at least one element undergoes a change of oxidation state. For example, in the reaction between permanganate ion and oxalate ion, the oxidation states of manganese and carbon atoms change. In the reactants, the oxidation state of manganese is 7+; in the products, it is 2+. In the reactants, the oxidation state of carbon is 3+; in the products, it is 4+.

$$\overset{7+}{MnO_4^-} + \overset{3+}{C_2O_4^{2-}} \rightarrow \overset{2+}{Mn^{2+}} + \overset{4+}{CO_2}$$

These oxidation state changes determine the stoichiometry of the reaction. In terms of the oxidation state formalism, each manganese atom gains five electrons and each carbon atom loses one electron. Thus the reaction must involve five times as many carbon atoms as manganese atoms. Allowing for the presence of two carbon atoms in the oxalate ion, conservation of electrons requires that the stoichiometric coefficients be

$$2\,MnO_4^- + 5\,C_2O_4^- \rightarrow 2\,Mn^{2+} + 10\,CO_2$$

Written this way, two MnO_4^- moieties gain ten electrons, and five $C_2O_4^{2-}$ moieties lose ten electrons. When we fix the coefficients of the redox reactants, we also fix the coefficients of the redox products. However, inspection shows that both charge and the number of oxygen atoms are out of balance in this equation.

The reaction occurs in acidic aqueous solution. This means that enough water molecules must participate in the reaction to achieve oxygen-atom balance. Adding eight water molecules to the product brings oxygen into balance. Now, however, charge and hydrogen atoms

$$2\,MnO_4^- + 5\,C_2O_4^- \rightarrow 2\,Mn^{2+} + 10\,CO_2 + 8\,H_2O$$

do not balance. Since the solution is acidic, we can bring hydrogen into balance by adding sixteen protons to the reactants. When we do so, we find that charge balances also.

$$2\,MnO_4^- + 5\,C_2O_4^- + 16\,H^+ \rightarrow 2\,Mn^{2+} + 10\,CO_2 + 8\,H_2O$$

Evidently, this procedure achieves charge balance because the oxidation state formalism enables us to find the correct stoichiometric ratio between oxidant and reductant.

We can formalize this thought process in a series of rules for balancing oxidation–reduction reactions. In doing this, we can derive some advantage from splitting the overall chemical change into two parts, which we call half-reactions. It is certainly not necessary to introduce half-reactions just to balance equations; the real advantage is that a half-reaction describes the chemical change in an individual half-cell. The rules for balancing oxidation–reduction reactions using half-cell reactions are these:

- Find the oxidation state of every atom in every reactant and every product.
- Write skeletal equations showing:
  - the oxidizing agent → its reduced product
  - the reducing agent → its oxidized product
- Balance the skeletal equations with respect to all elements other than oxygen and hydrogen.
- Add electrons to each equation to balance those gained or lost by the atoms undergoing oxidation-state changes.
- For a reaction occurring in acidic aqueous solution:
  - balance oxygen atoms by adding water to each equation.
  - balance hydrogen atoms by adding protons to each equation
- For a reaction occurring in basic aqueous solution, balance oxygen and hydrogen atoms by adding water to one side of each equation and hydroxide ion to the other.
  - The net effect of adding one water and one hydroxide is to increase by one the number of hydrogen atoms on the side to which the water is added.
  - Adding two hydroxide ions to one side and a water molecule to the other increases by one the number of oxygen atoms on the side to which hydroxide is added.
- Multiply each half-reaction by a factor chosen to make each of the resulting half-reactions contain the same number of electrons.
- Add the half-reactions to get a balanced equation for the overall chemical change. The electrons cancel. Often, some of the water molecules, hydrogen atoms, or hydroxide ions cancel also.

When we apply this method to the permanganate–oxalate reaction, we have

$$2\,MnO_4^- + 16\,H^+ + 10\,e^- \rightarrow 2\,Mn^{2+} + 8\,H_2O$$
reduction half-reaction

$$5\,C_2O_4^- \rightarrow 10\,CO_2 + 10\,e^-$$
oxidation half-reaction

$$2\,MnO_4^- + 5\,C_2O_4^- + 16\,H^+ \rightarrow 2\,Mn^{2+} + 10\,CO_2 + 8\,H_2O$$
balanced reaction

The half-reactions sum to the previously obtained result; the electrons cancel. For an example of a reaction in basic solution, consider the disproportionation of chloride dioxide to chlorite and chlorate ions:

$$\overset{4+}{ClO_2} \rightarrow \overset{3+}{ClO_2^-} + \overset{5+}{ClO_3^-}$$
skeletal reaction

$$ClO_2 + e^- \rightarrow ClO_2^-$$
reduction half-reaction

$$ClO_2 + 2OH^- \rightarrow ClO_3^- + H_2O + e^-$$
oxidation half-reaction

$$2\,ClO_2 + 2OH^- \rightarrow ClO_2^- + ClO_3^- + H_2O$$

balanced equation

§5 Electrical potential

Electrical potential is measured in volts. If a system comprising one coulomb of charge passes through a potential difference of one volt, one joule of work is done on the system. The work done on the system is equal to the change in the energy of the system. For Q coulombs passing through a potential difference of $\mathcal{E}$ volts, we have $\Delta E = w_{elec} = Q\mathcal{E}$. Whether this represents an increase or a decrease in the energy of the system depends on the sign of the charge and on the sign of the potential difference.

Electrical potential and gravitational potential are analogous. The energy change associated with moving a mass from one elevation to another in the earth's gravitational field is $\Delta E = w_{grav} = mgh_{final} - mgh_{initial} = m\Phi_{grav}$, where $\Phi_{grav} = g(h_{final} - h_{initial})$. Φ_{grav} is the gravitational potential difference. The role played by charge in the electrical case is played by mass in the gravitational case. The energies of these systems change because charge or mass moves in response to the application of a force. In the electrical case, the force is the electrical force that arises from the interaction between charges. In the gravitational case, the force is the gravitational force that arises from the interaction between masses. A notable difference is that mass is always a positive quantity, whereas charge can be positive or negative.

The distinguishing feature of an electrochemical cell is that there is an electrical potential difference between the two terminals. For any given cell, the magnitude of the potential difference depends on the magnitude of the current that is flowing. (Making the general problem even more challenging, we find that it depends also on the detailed history of the conditions under which electrical current has been drawn from the cell.) Fortunately, if we keep the cell's temperature constant and measure the potential at zero current, the electrical potential is constant. Under these conditions, the cell's characteristics are fixed, and potential measurements give reproducible results. We want to understand the origin and magnitude of this potential difference. Experimentally, we find:

- If we measure the zero-current electrical potential of the same cell at different temperatures, we find that this potential depends on temperature.
- If we prepare two cells with different chemical species, they exhibit different electrical potentials—except possibly for an occasional coincidence.
- If we prepare two cells with the same chemical species at different concentrations and measure their zero-current electrical potentials at the same temperature, we find that they exhibit different potentials.
- If we draw current from a given cell over a period of time, we find that there is a change in the relative amounts of the reagents present in the cell. Overall, a chemical reaction occurs; some reagents are consumed, while others are produced.

We can summarize these experimental observations by saying that the central issue in electrochemistry is the interrelation of three characteristics of an electrochemical cell: the electrical-potential difference between the terminals of the cell, the flow of electrons in the external circuit, and the chemical changes inside the cell that accompany this electron flow.

§6 Electrochemical cells as circuit elements

Suppose we use a wire to connect the terminals of the cell built from the silver–silver ion half-cell and the copper–cupric ion half-cell. This wire then constitutes the external circuit, the path that the electrons follow as chemical change occurs within the cell. When the external circuit is simply a low-resistance wire, the cell is short-circuited. The external circuit can be more complex. For example, when we want to know the direction of electron flow, we incorporate a galvanometer.

If the reaction between silver ions and copper metal is to occur, electrons must pass through the external circuit from the copper terminal to the silver terminal. An electron that is free to move in the presence of an electrical potential must move away from a region of more negative electrical potential and toward a region of more positive electrical potential. Since the electron-flow is away from the copper terminal and toward the silver terminal, the copper terminal must be electrically negative and the silver terminal must be electrically positive. Evidently, if we know the chemical reaction that occurs in an electrical cell, we can immediately deduce the direction of electron flow in the external circuit. Knowing the direction of electron flow in the external circuit immediately tells us which is the negative and which the positive terminal of the cell.

The converse is also true. If we know which cell terminal is positive, we know that electrons in the external circuit flow toward this terminal. Even if we know nothing about the composition of the cell, the fact that electrons are flowing toward a particular terminal tells us that the reaction occurring in that half-cell is one in which a solution species, or the electrode material, takes up electrons. That is to say, some chemical entity is reduced in a half-cell whose potential is positive. It can happen that we know the half-reaction that occurs in a given half-cell, but that we do not know which direction the reaction goes. For example, if we replace the silver–silver ion half cell with a similar cell containing an aqueous zinc nitrate solution and a zinc electrode, we are confident that the half-cell reaction is either

$$Zn^0 \rightarrow Zn^{2+} + 2\,e^-$$

or

$$Zn^{2+} + 2\,e^- \rightarrow Zn^0$$

When we determine experimentally that the copper

electrode is electrically positive with respect to the zinc electrode, we know that electrons are leaving the zinc electrode and flowing to the copper electrode. Therefore, the cell reaction must be

$$Zn^0 + Cu^{2+} \rightarrow Zn^{2+} + Cu^0$$

It is convenient to have names for the terminals of an electrochemical cell. One naming convention is to call one terminal the ***anode*** and the other terminal the ***cathode***. The definition is that the cathode is the electrode at which a reacting species is reduced. In the silver–silver ion containing cell, the silver electrode is the cathode. In the zinc–zinc ion containing cell, the copper electrode is the cathode. In these cells, the cathode is the electrically positive electrode. An important feature of these experiments is that the direction of the electrical potential in the external circuit is established by the reactions that occur spontaneously in the cells. The cells are sources of electrical current. Cells that operate to produce current are called ***galvanic*** cells.

§7 The direction of electron flow and its implications

We can incorporate another potential source into the external circuit of an electrochemical cell. If we do so in such a way that the two electrical potentials augment one another, as diagrammed in Figure 4, the potential drop around the new external circuit is the sum of the potential drops of the two sources taken independently. The direction of electron flow is unchanged. An electron anywhere in the external circuit is propelled in the same direction by either potential source. The effective potential difference in the composite circuit is the sum of the potentials that the sources exhibit when each acts alone.

Alternatively, we can connect the two potential sources so that they oppose one another, as diagrammed in Figure 5. Now an electron in the external circuit is pushed in one direction by one of the potential sources and in the opposite direction by the other potential source. The effective potential difference in the composite circuit is the difference between the potentials that the sources exhibit when each acts alone. In the composite circuit, the direction of electron flow is determined by the potential source whose potential difference is greater.

This has a dramatic effect on the direction of the reaction occurring in the weaker cell. In the composite cell, the direction of electron flow through the weaker cell is opposite to the direction of electron flow when the weaker cell is operating as a galvanic cell. Since the direction of electron flow in the external circuit determines the directions in which the half-reactions occur, the chemical reaction that occurs in the cell must occur in the opposite direction also. When the direction of current flow through a cell is determined by connection to a greater potential difference in this fashion, the cell is called an ***electrolytic*** cell. Reduction occurs at the negative terminal of an electrolytic cell. In an electrolytic cell,

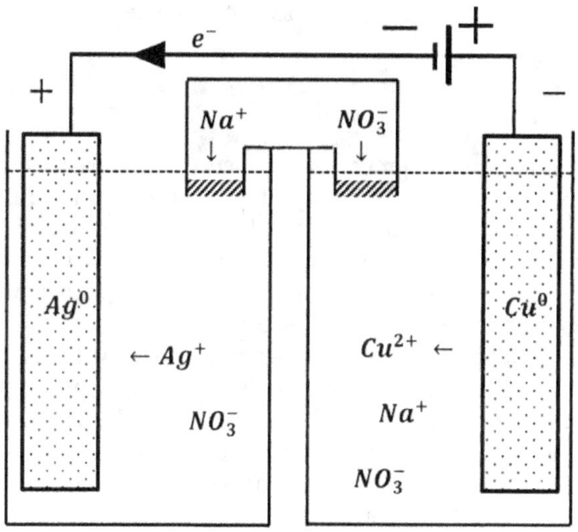

Figure 4. Applied potential augments the cell potential.

the cathode is the electrically negative electrode. The direction of current flow in any cell can be reversed by the application of a sufficiently large counter-potential.

When a cell operates as a source of current (that is, as a galvanic cell), the cell reaction is a spontaneous process. Since, as the cell reaction proceeds, electrons move through a potential difference in the external circuit, the reaction releases energy in the cell's surroundings. If the external circuit is simply a resistor, as when the terminals are short-circuited, the energy is released as heat. Let q be the heat released and let Q be the amount of charge passed through the external circuit in a time interval Δt. The heat-release rate is given by

$$\frac{q}{\Delta t} = \frac{\Delta E}{\Delta t} = \frac{Q\mathcal{E}}{\Delta t}$$

The electrical current is $I = Q/\Delta t$. If the resistor follows Ohm's law, $\mathcal{E} = IR$, where R is the magnitude of

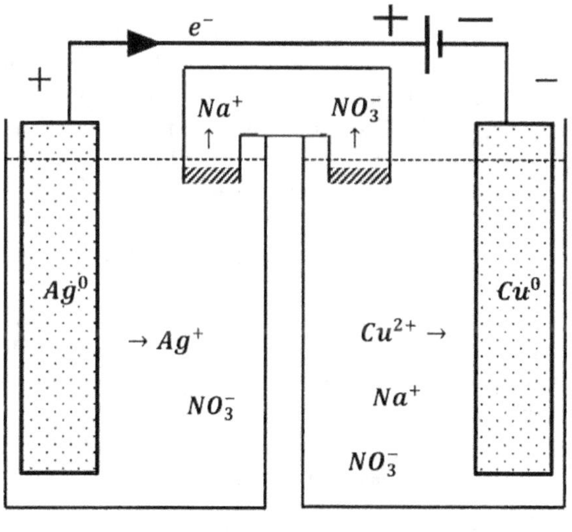

Figure 5. Applied potential opposes the cell potential.

the resistance, the heat release rate becomes

$$\frac{q}{\Delta t} = I^2 R$$

As the reaction proceeds and energy is dissipated in the external circuit, the ability of the cell to supply further energy is continuously diminished. The energy delivered to the surroundings through the external circuit comes at the expense of the cell's internal energy and corresponds to the depletion of the cell reactants.

When the chemical reaction occurring within a cell is driven by the application of an externally supplied potential difference, the opposite occurs. In the driven (electrolytic) cell, the direction of the cell reaction is opposite the direction of the spontaneous reaction that occurs when the cell operates galvanically. The electrolytic process produces the chemical reagents that are consumed in the spontaneous cell reaction. The external circuit delivers energy to the electrolytic cell, increasing its content of spontaneous-direction reactants and thereby increasing its ability to do work.

In summary, the essential difference between electrolytic and galvanic cells lies in the factor that determines the direction of current flow and, correspondingly, the direction in which the cell reaction occurs. In a galvanic cell, a spontaneous chemical reaction occurs and this reaction determines the direction of current flow and the signs of the electrode potentials. In an electrolytic cell, the sign of the electrode potentials is determined by an applied potential source, which determines the direction of current flow; the cell reaction proceeds in the non-spontaneous direction.

§8 Electrolysis and the faraday

Electrolytic cells are very important in the manufacture of products essential to our technology-intensive civilization. Only electrolytic processes can produce many materials, notably metals that are strong reducing agents. Aluminum and the alkali metals are conspicuous examples. Many manufacturing processes that are not themselves electrolytic utilize materials that are produced in electrolytic cells. These processes would not be possible if the electrolytic products were not available. For example, elemental silicon, the essential precursor of most contemporary computer chips, is produced from silicon tetrachloride by reduction with sodium.

$$SiCl_4 + 4\,Na^0 \rightarrow Si^0 + 4\,NaCl$$

(The silicon so produced is intensively refined, formed into large single crystals, and sliced into wafers before the chip-manufacturing process begins.) Elemental sodium is produced by the electrolysis of molten sodium chloride.

Successful electrolytic processes involve artful selection of the current-collector material and the reaction conditions. The design of the cell is often crucial. Since sodium metal reacts violently with water, we recognize immediately that electrolysis of aqueous sodium chloride solutions cannot produce sodium metal. What products are obtained depends on numerous factors, notably the composition of the electrodes, the concentration of the salt solution, and the potential that is applied to the cell.

Electrolysis of concentrated, aqueous, sodium chloride solutions is used on a vast scale in the ***chlor-alkali*** process for the co-production of chlorine and sodium hydroxide, both of which are essential for the manufacture of many common products.

$$2\,NaCl(aq) + 2\,H_2O(\ell) \rightarrow 2\,NaOH(aq) + Cl_2(g) + H_2(g)$$

Hydrogen is a by-product. The overall process does not involve sodium ion; rather, the overall reaction is an oxidation of chloride ion and a reduction of water.

$$2\,Cl^-(aq) \rightarrow Cl_2(g) + 2\,e^- \quad \text{oxidation half-reaction}$$

$$2\,H_2O(\ell) + 2\,e^- \rightarrow 2\,OH^-(aq) + H_2(g) \quad \text{reduction half-reaction}$$

The engineering difficulties associated with the chlor-alkali process are substantial. They occur because hydroxide ion reacts with chlorine gas; a practical cell must be designed to keep these two products separate. Commercially, two different designs have been successful. The diaphragm-cell process uses a porous barrier to separate the anodic and cathodic cell compartments. The mercury-cell process uses elemental mercury as the cathodic current collector; in this case, sodium ion is reduced, but the product is sodium amalgam (sodium–mercury alloy) not elemental sodium. Like metallic sodium, sodium amalgam reduces water, but the amalgam reaction is much slower. The amalgam is removed from the cell and reacted with water to produce sodium hydroxide and regenerate mercury for recycle to the electrolytic cell.

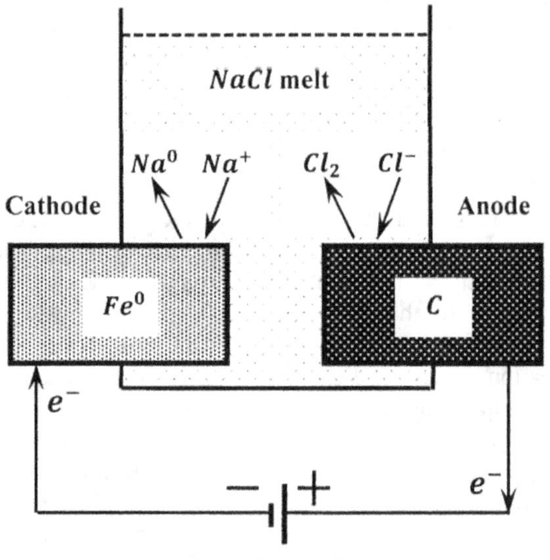

Figure 6. An electrolytic cell for the production of sodium metal.

Elemental sodium is manufactured by the electrolysis of molten sodium chloride. This is effected commercially using an iron cathode and a carbon anode. The reaction is

$$NaCl(\ell) \rightarrow Na^0 + Cl_2(g)$$

Such a cell is diagrammed in Figure 6. A mechanical barrier suffices to keep the products separate and prevent their spontaneous reaction back to the salt. A more significant problem in the design of the cell was to find an anode material that did not react with the chlorine produced. From the cell reaction, we see that one electron passes through the external circuit for every sodium atom that is produced. The charge that passes through the external circuit during the production of one mole of sodium metal is, therefore, the charge on one mole of electrons.

In honor of Michael Faraday, the magnitude of the charge carried by a mole of electrons is called the ***faraday***. The faraday constant is denoted by the symbol "$\mathcal{F}$." That is,

$$1\mathcal{F} = \frac{6.02214 \times 10^{23} \text{ electrons}}{\text{mol}} \times \frac{1.602187 \times 10^{-19}\, C}{\text{electron}}$$
$$= 96{,}485\ C\ \text{mol}^{-1}$$

The faraday is a useful unit in electrochemical calculations. The unit of electrical current, the ampere, is defined as the passage of one coulomb per second. Knowing the current in a circuit and the time for which it is passed, we can calculate the number of coulombs that are passed. Remembering the value of one faraday enables us to do stoichiometric calculations without bringing in Avogadro's number and the electron charge every time.

Tabulated information about the thermodynamic characteristics of half-reactions enable us to make useful predictions about what can and cannot occur in various cells that we might think of building. This information can be used to predict the potential difference that will be observed in a galvanic cell made by connecting two arbitrarily selected half-cells. In any electrolytic cell, more than one electron-transfer reaction can usually occur. In the chlor-alkali process, for example, water rather than chloride ion might be oxidized at the anode. In such cases, tabulated half-cell data enable us to predict which species can react at a particular applied potential.

§9 Electrochemistry and conductivity

From the considerations we have discussed, it is evident that any electrolytic cell involves a flow of electrons in an external circuit and a flow of ions within the materials comprising the cell. The function of the current collectors is to transfer electrons back and forth between the external circuit and the cell reagents.

The measurement of solution conductivity is a useful technique for determining the concentrations and mobilities of ions in solution. Since conductivity measurements involve the passage of electrical current through a liquid medium, the process must involve electrode reactions as well as motion of ions through the liquid. Normally, the electrode reactions are of little concern in conductivity measurements. The applied potential is made large enough to ensure that some electrode reaction occurs. When the liquid medium is water, the electrode reactions are usually the reduction of water at the cathode and its oxidation at the anode. The conductivity attributable to a given ionic species is approximately proportional to its concentration. In the absence of dissolved ions, little current is passed. For aqueous solutions, this just restates the familiar observation that pure water is a poor electrical conductor. When few ions are present, it is not possible to move charge through the cell quickly enough to support a significant current in the external circuit.

§10 The Standard Hydrogen Electrode (S.H.E.)

In §4, we introduce the idea of a half-reaction and a half-cell in the context of balancing equations for oxidation–reduction reactions. The real utility of these ideas is that they correspond to distinguishable parts of actual electrochemical cells. Information about the direction of a spontaneous reaction enables us to predict the relative electrical potentials of the half-cells that make up the corresponding electrochemical cell. Conversely, given information about the characteristic electrical potentials of half-cells, we can predict what chemical reactions can occur spontaneously. In short, there is a relationship between the electrical potential of an electrochemical cell at a particular temperature and pressure and the Gibbs free energy change for the corresponding oxidation–reduction reaction.

Since cell potentials vary with the concentrations of the reactive components, we can simplify our record-keeping requirements by defining standard reference conditions that apply to a standard electrode of any type. We adopt the convention that a standard electrochemical cell contains all reactive components at unit activity. The vast majority of electrochemical cells that have been studied contain aqueous solutions. In data tables, the activity standard state for solute species is nearly always the hypothetical one-molal solution. For many purposes, it is an adequate approximation to say that all solutes are present at a concentration of one mole per liter, and all reactive gases at a pressure of one bar. (In §15, we see that the dependence of cell potential on reagent concentration is logarithmic.) In §2 and §7, we discuss the silver–silver ion electrode; in this approximation, a standard silver–silver ion electrode is one in which the silver ion is present in the solution at a concentration of one mole per liter. Likewise, a standard copper–cupric ion electrode is one in which cupric ion is present in the solution at one mole per liter.

We also need to choose an arbitrary reference half-cell. The choice that has been adopted is the ***Standard***

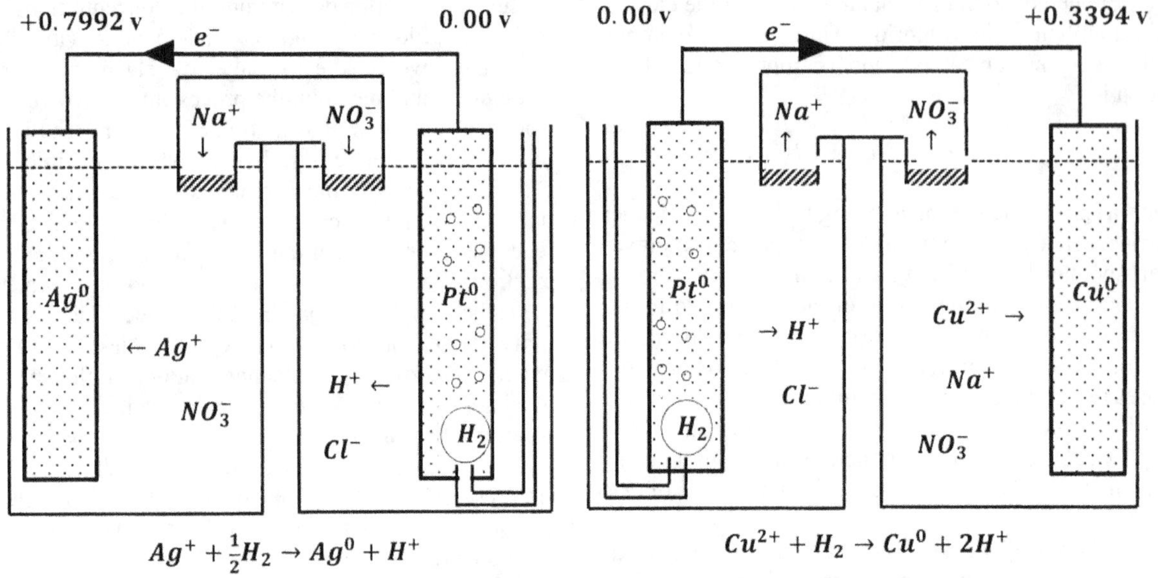

Figure 7. Two cells, each with a standard hydrogen electrode.

Hydrogen Electrode, often abbreviated the S.H.E. The S.H.E. is defined as a piece of platinum metal, immersed in a unit-activity aqueous solution of a protonic acid, and over whose surface hydrogen gas, at unit fugacity, is passed continuously. These concentration choices make the electrode a standard electrode. Frequently, it is adequate to approximate the S.H.E. composition by assuming that the hydrogen ion concentration is one molar and the hydrogen gas pressure is one bar. The half-reaction associated with the S.H.E. is

$$H^+ + e^- \to \frac{1}{2}H_2$$

We define the electrical potential of this half-cell to be zero volts.

§11 Half-reactions and half-cells

Let us consider some standard electrochemical cells we could construct using the S.H.E. Two possibilities are electrochemical cells in which the second electrode is the standard silver–silver ion electrode or the standard copper–cupric ion electrode. The diagrams in Figure 7 summarize the half-reactions and the electrical potentials that we find when we construct these cells.

We can also connect these cells so that the two S.H.E. are joined by one wire, while a second wire joints the silver and copper electrodes. This configuration is sketched in Figure 8. Whatever happens at one S.H.E. happens in the exact reverse at the other S.H.E. The net effect is essentially the same as connecting the silver–

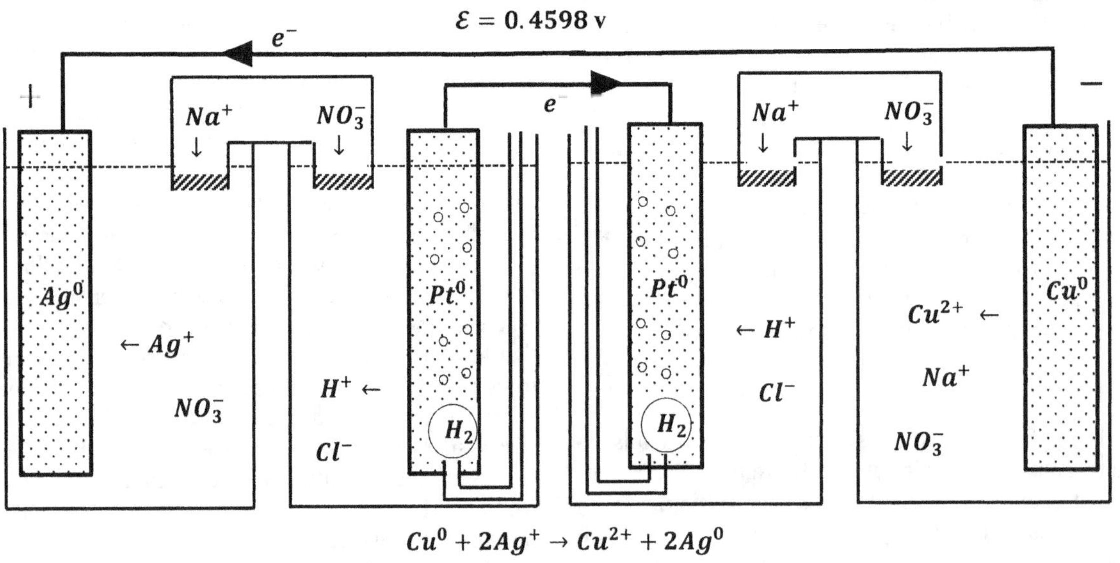

Figure 8. After connecting the two standard hydrogen electrodes, the combination becomes equivalent to two half cells connected to a salt bridge.

silver ion half-cell to the copper–cupric ion half-cell by a single salt bridge. If we did not already know what reaction occurs, we could figure it out from the information we have about how each of these two cells performs when it operates against the S.H.E.

§12 Standard electrode potentials

We adopt a very useful convention to tabulate the potential drops across standard electrochemical cells, in which one half-cell is the S.H.E. Since the potential of the S.H.E. is zero, we define the ***standard electrode potential***, $\mathcal{E}^o$, of any other standard half-cell (and its associated half-reaction) to be the potential difference when the half-cell operates spontaneously versus the S.H.E. The electrical potential of the standard half-cell determines both the magnitude and sign of the standard half-cell potential.

If the process that occurs in the half-cell reduces a solution species or the electrode material, electrons traverse the external circuit toward the half-cell. Hence, the electrical sign of the half-cell terminal is positive. By the convention, the algebraic sign of the cell potential is positive ($\mathcal{E}^o > 0$). If the process that occurs in the half-cell oxidizes a solution species or the electrode, electrons traverse the external circuit away from the half-cell and toward the S.H.E. The electrical sign of the half-cell is negative, and the algebraic sign of the cell potential is negative ($\mathcal{E}^o < 0$).

If we know the standard half-cell potential, we know the essential electrical properties of the standard half-cell operating spontaneously versus the S.H.E. at zero current. In particular, the algebraic sign of the standard half-cell potential tells us the direction of current flow and hence the direction of the reaction that occurs spontaneously.

An older convention associates the sign of the standard electrode potential with the direction in which an associated half-reaction is written. This convention is compatible with the definition we have chosen; however, it creates two ways of expressing the same information. The difference is whether we write the direction of the half-reaction with the electrons appearing on the right or on the left side of the equation.

When the half-reaction is written as a reduction process, with the electrons appearing on the left, the associated half-cell potential is called the ***reduction potential*** of the half-cell. Thus we would convey the information we have developed about the silver–silver ion and the copper–copper ion half cells by presenting the reactions and their associated potentials as

$$Ag^+ + e^- \rightarrow Ag^0 \qquad \mathcal{E}^0 = +0.7992 \text{ volts}$$

$$Cu^{2+} + 2\,e^- \rightarrow Cu^0 \qquad \mathcal{E}^0 = +0.3394 \text{ volts}$$

When the half-reaction is written as a reduction process, the sign of the electrode potential is the same as the sign of the electrical potential of the half-cell when the half-cell operates spontaneously versus the S.H.E. Thus, the reduction potential has the same algebraic sign as the electrode potential of our definition.

We can convey the same information by writing the half-reaction in the reverse direction; that is, as an oxidation process in the left-to-right direction so that the electrons appear on the right. The agreed-upon convention is that we reverse the sign of the half-cell potential when we reverse the direction in which we write the equation. When the half-reaction is written as an oxidation process, the associated half-cell potential is called the ***oxidation potential*** of the half-cell. Older tabulations of electrochemical data often present half-reactions written as oxidation processes, with the electrons on the right, and present the potential information using the oxidation potential convention.

$$Ag^+ + e^- \rightarrow Ag^0 \qquad \mathcal{E}^0 = +0.7992 \text{ volts}$$
reduction potential

$$Ag^0 \rightarrow Ag^+ + e^- \qquad \mathcal{E}^0 = -0.7992 \text{ volts}$$
oxidation potential

Note that, in the convention that we have adopted, the term half-cell potential always denotes the potential of the half-cell when it operates spontaneously versus the S.H.E. In this convention, we do not need to write the half-reaction in order to specify the standard potential. It is sufficient to specify the chemical constituents of the half-cell. This is achieved using another representational convention.

This cell-describing convention lists the active components of a half-cell, using a vertical line to indicate the presence of a phase boundary like that separating silver metal from an aqueous solution containing silver ion. The silver–silver ion cell is denoted $Ag^+ \mid Ag^0$. (Using the superscript zero on the symbol for elemental silver is redundant; however, it does promote clarity.) The copper–cupric ion cell is denoted $Cu^{2+} \mid Cu^0$. The S.H.E. is denoted $H^+ \mid H_2 \mid Pt^0$, reflecting the presence of three distinct phases in the operating electrode. A complete electrochemical cell can be described using this convention. When the complete cell contains a salt bridge, this is indicated with a pair of vertical lines, ||. A cell composed of a silver–silver ion half-cell and a S.H.E. is denoted $Pt^0 \mid H_2 \mid H^+ \| Ag^+ \mid Ag^0$. A further convention stipulates that the half-cell with the more positive electrode potential is written on the right. Under this convention, spontaneous operation of the standard full cell transfers electrons through the external circuit from the terminal shown on the left to the terminal shown on the right.

We can now present our information about the behavior of the silver–silver ion half-cell versus the S.H.E. by writing that the standard potential of the $Ag^+ \mid Ag^0$ half-cell is +0.7792 volts. The standard potential of the $Cu^{2+} \mid Cu^0$ half-cell is +0.3394 volts. The standard potential of the $H^+ \mid H_2 \mid Pt^0$ (the S.H.E.) half-cell is 0.0000 volts. Again, our definition of the standard electrode potential makes the sign of the standard electrode potential independent of the direction in which the equation of the corresponding half-reaction is written.

§13 Predicting the direction of spontaneous change

While our convention does not use the equation that we write for the half-reaction to establish the algebraic sign of the standard electrode potential, it is useful to associate the standard electrode potential with the half-reaction written as a reduction, that is, with the electrons written on the left side of the equation. We also establish the convention that reversing the direction of the half-reaction reverses the algebraic sign of its potential. When these conventions are followed, the overall reaction and the full-cell potential can be obtained by adding the corresponding half-cell information. If the resulting full-cell potential is greater than zero, the spontaneous overall reaction proceeds in the direction it is written, from left to right. If the full-cell potential is negative, the direction of spontaneous reaction is opposite to that written; that is, a negative full cell potential corresponds to the spontaneous reaction occurring from right to left. For example,

$$2\,Ag^{+} + 2\,e^{-} \rightarrow 2\,Ag^{0} \qquad \mathcal{E}^{0} = +0.7992 \text{ volts}$$

$$Cu^{0} \rightarrow Cu^{2+} + 2\,e^{-} \qquad \mathcal{E}^{0} = -0.3394 \text{ volts}$$

$$2\,Ag^{+} + Cu^{0} \rightarrow 2\,Ag^{0} + Cu^{2+} \qquad \mathcal{E}^{0} = +0.4598 \text{ volts}$$

yields the equation corresponding to the spontaneous reaction and a positive full-cell potential. Writing

$$2\,Ag^{0} \rightarrow 2\,Ag^{+} + 2\,e^{-} \qquad \mathcal{E}^{0} = -0.7992 \text{ volts}$$

$$Cu^{2+} + 2\,e^{-} \rightarrow Cu^{0} \qquad \mathcal{E}^{0} = +0.3394 \text{ volts}$$

$$2\,Ag^{0} + Cu^{2+} \rightarrow 2\,Ag^{+} + Cu^{0} \qquad \mathcal{E}^{0} = -0.4598 \text{ volts}$$

yields the equation for the non-spontaneous reaction and, correspondingly, the full-cell potential is less than zero.

Note that when we multiply a chemical equation by some factor, we do not apply the same factor to the corresponding potential. The electrical potential of the corresponding electrochemical cell is independent of the number of moles of reactants and products that we choose to write. The cell potential is an intensive property. It has the same value for a small cell as for a large one, so long as the other intensive properties (temperature, pressure, and concentrations) are the same.

§14 Cell potentials and the Gibbs free energy

In §11, we see that the electrical potential drop across the standard cell $Pt^{0} \mid H_2 \mid H^{+} \| Ag^{+} \mid Ag^{0}$ is 0.7992 volts. We measure this potential under conditions in which no current is flowing. That is, we find the counter-potential at which no current flows through the cell in either direction. An arbitrarily small change in the counter-potential away from this value, in either direction, is sufficient to initiate current flow. This means that the standard potential is measured when the cell is operating reversibly. By the definition of a standard cell, all of the reactants are at the standard condition of unit activity. If any finite current is drawn from a cell of finite size, the concentrations of the reagents will no longer be exactly the correct values for a standard cell. Nevertheless, we can calculate the energy that would be dissipated in the surroundings if the cell were to pass one mole of electrons (corresponding to consuming one mole of silver ions and one-half mole of hydrogen gas) through the external circuit while the cell conditions remain exactly those of the standard cell. This energy is

$$96{,}485 \text{ C mol}^{-1} \times 0.7992 \text{ V} = 77{,}110 \text{ J mol}^{-1}$$

The form in which this energy appears in the surroundings depends on the details of the external circuit. However, we know that this energy represents the reversible work done on electrons in the external circuit as they traverse the path from the anode to the cathode. We call this the electrical work. Above we describe this as the energy change for a hypothetical reversible process in which the composition of the cell does not change. We can also view it as the energy change per electron for one electron-worth of real process, multiplied by the number of electrons in a mole. Finally, we can also describe it as the reversible work done on electrons during the reaction of one mole of silver ions in an infinitely large standard cell.

The Gibbs free energy change for an incremental reversible process is $dG = VdP - SdT + dw_{NPV}$, where dw_{NPV} is the increment of non-pressure–volume work. In the case of an electrochemical cell, the electrical work is non-pressure–volume work. In the particular case of an electrochemical cell operated at constant temperature and pressure, $dP = dT = 0$, and $dG = dw_{NPV} = dw_{\text{elect}}$.

The electrical work is just the charge times the potential drop. Letting n be the number of moles of electrons that pass through the external circuit for one unit of reaction, the total charge is $Q = -n\mathcal{F}$, where $\mathcal{F}$ is one faraday. For a standard cell, the potential drop is $\mathcal{E}^{0}$, so the work done on the electrons is $Q\mathcal{E}^{0} = -n\mathcal{F}\mathcal{E}^{0}$. Since the standard conditions for Gibbs free energies are the same as those for electrical cell potentials, we have

$$w_{elect}^{rev} = \Delta_r G^{o} = -n\mathcal{F}\mathcal{E}^{o}$$

If the reaction occurs spontaneously when all of the reagents are in their standard states, we have $\mathcal{E}^{o} > 0$. For a spontaneous process, the work done on the system is less than zero, $w_{elect}^{rev} < 0$; the work done on the surroundings is $\hat{w}_{elect}^{rev} = -w_{elect}^{rev} > 0$; and the energy of the surroundings increases as the cell reaction proceeds. The standard potential is an intensive property; it is independent of the size of the cell and of the way we write the equation for the chemical reaction. However, the work and the Gibbs free energy change depend on the number of electrons that pass through the external circuit. We usually specify the number of electrons by specifying the chemical equation to which the Gibbs free energy change applies. That is, if the associated reaction is written as

$$Ag^+ + \tfrac{1}{2}H_2 \rightarrow Ag^0 + H^+$$

we understand that one mole of silver ions are reduced and one mole of electrons are transferred; $n = 1$ and $\Delta G^o = -\mathcal{F}\mathcal{E}^o$. If the reaction is written

$$2\,Ag^+ + H_2 \rightarrow 2\,Ag^0 + 2\,H^+$$

we understand that two moles of silver ions are reduced and two moles of electrons are transferred, so that $n = 2$ and $\Delta G^o = -2\mathcal{F}\mathcal{E}^o$.

The same considerations apply to measurement of the potential of electrochemical cells whose component are not at the standard condition of unit activity. If the cell is not a standard cell, we can still measure its potential. We use the same symbol to denote the potential, but we omit the superscript zero that denotes standard conditions. These are, of course, just the conventions we have been using to distinguish the changes in other thermodynamic functions that occur at standard conditions from those that do not. We have therefore, for the Gibbs free energy change for the reaction occurring in an electrochemical cell that is not at standard conditions,

$$w^{rev}_{elect} = \Delta_r G = -n\mathcal{F}\mathcal{E}$$

§15 The Nernst equation

In Chapter 14, we find that the Gibbs free energy change is a function of the activities of the reactants and products. For the general reaction $aA + bB \rightarrow cC + dD$ we have

$$\Delta_r G = \Delta_r G^o + RT \ln \frac{\tilde{a}_C^c \tilde{a}_D^d}{\tilde{a}_A^a \tilde{a}_B^b}$$

Using the relationship between cell potentials and the Gibbs free energy, we find

$$-n\mathcal{F}\mathcal{E} = -n\mathcal{F}\mathcal{E}^o + RT \ln \frac{\tilde{a}_C^c \tilde{a}_D^d}{\tilde{a}_A^a \tilde{a}_B^b}$$

or

$$\mathcal{E} = \mathcal{E}^o - \frac{RT}{n\mathcal{F}} \ln \frac{\tilde{a}_C^c \tilde{a}_D^d}{\tilde{a}_A^a \tilde{a}_B^b}$$

This is the **Nernst equation**. We derive it from our previous results for the activity dependence of the Gibbs free energy, which makes no explicit reference to electrochemical measurements at all. When we make the appropriate experimental measurements, we find that the Nernst equation accurately represents the temperature and concentration dependence of electrochemical-cell potentials.

Reagent activities are often approximated adequately by molalities or molarities, for solute species, and by partial pressures—expressed in bars—for gases. The activities of pure solid and liquid phases can be taken as unity. For example, if we consider the reaction

$$Ag^+ + \tfrac{1}{2}H_2 \rightarrow Ag^0 + H^+$$

it is often sufficiently accurate to approximate the Nernst equation as

$$\mathcal{E} = \mathcal{E}^o - \frac{RT}{n\mathcal{F}} \ln \frac{[H^+]}{[Ag^+]\,P_{H_2}^{1/2}}$$

§16 The Nernst equation for half-cells

If the S.H.E. is one of the half-cells, the corresponding Nernst equation can be viewed as a description of the other half-cell. Using the cell in which the silver–silver ion electrode opposes the S.H.E., as in the preceding example, the cell potential is the algebraic sum of the potential of the silver terminal and the potential of the platinum terminal. We can represent the potential of the silver–silver ion electrode as $\mathcal{E}_{Ag|Ag^+}$. Since the S.H.E. is always at standard conditions, its potential, which we can represent as $\mathcal{E}^o_{Pt|H_2|H^+}$, is zero by definition. The cell potential is

$$\mathcal{E} = \mathcal{E}_{Ag|Ag^+} + \mathcal{E}^o_{Pt|H_2|H^+}$$

The potential of the cell with both half-cells at standard conditions is

$$\mathcal{E}^o = \mathcal{E}^o_{Ag|Ag^+} + \mathcal{E}^o_{Pt|H_2|H^+}$$

and, again since the S.H.E. is at standard conditions, $\tilde{a}_{H^+} = 1$ and $P_{H_2} = 1$. Substituting into the Nernst equation for the full cell, we have

$$\mathcal{E}_{Ag|Ag^+} + \mathcal{E}^o_{Pt|H_2|H^+}$$
$$= \mathcal{E}^o_{Ag|Ag^+} + \mathcal{E}^o_{Pt|H_2|H^+} - \frac{RT}{\mathcal{F}} \ln \frac{1}{\tilde{a}_{Ag^+}}$$

or

$$\mathcal{E}_{Ag|Ag^+} = \mathcal{E}^o_{Ag|Ag^+} - \frac{RT}{\mathcal{F}} \ln \frac{1}{\tilde{a}_{Ag^+}}$$

where the algebraic signs of $\mathcal{E}_{Ag|Ag^+}$ and $\mathcal{E}^o_{Ag|Ag^+}$ correspond to writing the half-reaction in the direction $Ag^+ + e^- \rightarrow Ag^0$. Note that this is precisely the equation that we would obtain by writing out the Nernst equation corresponding to the chemical equation $Ag^+ + e^- \rightarrow Ag^0$.

To see how these various conventions work together, let us consider the oxidation of hydroquinone (H_2Q) to quinone (Q) by ferric ion in acidic aqueous solutions:

$$2\,Fe^{3+} + H_2Q \rightleftharpoons 2\,Fe^{2+} + Q + 2H^+$$

The quinone–hydroquinone couple is

$$O\text{=}\langle \rangle\text{=}O \;\; + 2H^+ + 2e^- \rightleftharpoons \;\; HO\text{-}\langle \rangle\text{-}OH$$

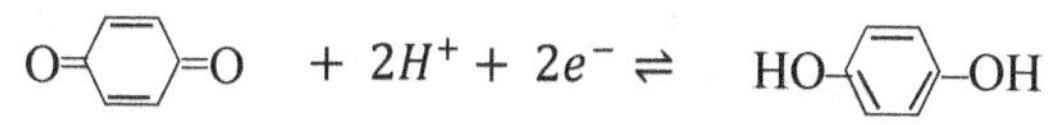

quinone (Q) hydroquinone (H_2Q)

and the ferric ion–ferrous ion couple is

$$Fe^{3+} + e^- \rightleftharpoons Fe^{2+}$$

The standard electrode potentials are $\mathcal{E}_{Pt|Q,H_2Q,H^+} = +0.699$ v and $\mathcal{E}_{Pt|Fe^{3+},Fe^{2+}} = +0.783$ v. In each case, the numerical value is the potential of a full cell in which the other electrode is the S.H.E. The algebraic sign of the half-cell potential is equal to the sign of the half-cell's electrical potential when it operates *versus* the S.H.E.

To carry out this reaction in an electrochemical cell, we can use a salt bridge to join a $Pt \mid Fe^{3+}, Fe^{2+}$ cell to a $Pt \mid Q, H_2Q, H^+$ cell. To construct a standard $Pt \mid Fe^{3+}, Fe^{2+}$ cell, we need only insert a platinum wire into a solution containing ferric and ferrous ions, both at unit activity. To construct a standard $Pt \mid Q, H_2Q, H^+$ cell, we insert a platinum wire into a solution containing quinone, hydroquinone, and hydronium ion, all at unit activity. For standard half-cells, the cathode and anode reactions are

$$Fe^{3+} + e^- \rightleftharpoons Fe^{2+}$$

and

$$H_2Q \rightleftharpoons Q + 2H^+ + 2e^-$$

We can immediately write the Nernst equation for each of these half-reactions as

$$\mathcal{E}_{Pt|Fe^{3+},Fe^{2+}} = \mathcal{E}^o_{Pt|Fe^{3+},Fe^{2+}} - \frac{RT}{\mathcal{F}} \ln \frac{\tilde{a}_{Fe^{2+}}}{\tilde{a}_{Fe^{3+}}}$$

and

$$\left(-\mathcal{E}_{Pt|Q,H_2Q,H^+}\right) = \left(-\mathcal{E}^o_{Pt|Q,H_2Q,H^+}\right) - \frac{RT}{2\mathcal{F}} \ln \frac{\tilde{a}_Q \tilde{a}^2_{H^+}}{\tilde{a}_{H_2Q}}$$

If we add the equations for these half-reactions, the result does not correspond to the original full-cell reaction, because the number of electrons does not cancel. This can be overcome by multiplying the ferric ion–ferrous ion half-reaction by two. What do we then do about the corresponding half-cell Nernst equation? Clearly, the values of $\mathcal{E}_{Pt|Fe^{3+},Fe^{2+}}$ and $\mathcal{E}^o_{Pt|Fe^{3+},Fe^{2+}}$ do not depend on the stoichiometric coefficients in the half-reaction equation. However, the activity terms in the logarithm's argument do, as does the number of electrons taking part in the half-reaction. We have

$$2Fe^{3+} + 2e^- \rightleftharpoons 2Fe^{2+}$$

with

$$\begin{aligned}\mathcal{E}_{Pt|Fe^{3+},Fe^{2+}} &= \mathcal{E}^o_{Pt|Fe^{3+},Fe^{2+}} - \frac{RT}{2\mathcal{F}} \ln \frac{\tilde{a}^2_{Fe^{2+}}}{\tilde{a}^2_{Fe^{3+}}} \\ &= \mathcal{E}^o_{Pt|Fe^{3+},Fe^{2+}} - \frac{RT}{\mathcal{F}} \ln \frac{\tilde{a}_{Fe^{2+}}}{\tilde{a}_{Fe^{3+}}}\end{aligned}$$

We see that we can apply any factor we please to the half-reaction. The Nernst equation gives the same dependence of the half-cell potential on reagent concentrations no matter what factor we choose. This is true also of the Nernst equation for any full-cell reaction. In the present example, adding the appropriate half-cell equations and their corresponding Nernst equations gives

$$2\,Fe^{3+} + H_2Q \rightleftharpoons 2\,Fe^{2+} + Q + 2H^+$$

and

$$\begin{aligned}\mathcal{E} &= \mathcal{E}_{Pt|Fe^{3+},Fe^{2+}} - \mathcal{E}_{Pt|Q,H_2Q,H^+} \\ &= \mathcal{E}^o_{Pt|Fe^{3+},Fe^{2+}} - \mathcal{E}^o_{Pt|Q,H_2Q,H^+} - \frac{RT}{2\mathcal{F}} \ln \frac{\tilde{a}^2_{Fe^{2+}}}{\tilde{a}^2_{Fe^{3+}}} \\ &\qquad - \frac{RT}{2\mathcal{F}} \ln \frac{\tilde{a}_Q \tilde{a}^2_{H^+}}{\tilde{a}_{H_2Q}} \\ &= \mathcal{E}^0 - \frac{RT}{2\mathcal{F}} \ln \frac{\tilde{a}_Q \tilde{a}^2_{H^+} \tilde{a}^2_{Fe^{2+}}}{\tilde{a}_{H_2Q} \tilde{a}^2_{Fe^{3+}}}\end{aligned}$$

§17 Combining two half-cell equations to obtain a new half-cell equation

The same chemical species can be a reactant or product in many different half-cells. Frequently, data on two different half-cells can be combined to give information about a third half-cell. Let us consider two half-cells that involve the ferrous ion, Fe^{2+}. Ferrous ion and elemental iron form a redox couple. The half-cell consists of a piece of pure ion in contact with aqueous ferrous ion at unit activity. Our notation for this half-cell and its potential are $Fe \mid Fe^{2+}$ and $\mathcal{E}_{Fe|Fe^{2+}}$. The corresponding half-reaction and its potential are

$$Fe^{2+} + 2e^- \rightleftharpoons Fe^0$$

and

$$\mathcal{E}_{Fe|Fe^{2+}} = \mathcal{E}^o_{Fe|Fe^{2+}} - \frac{RT}{2\mathcal{F}} \ln \frac{1}{\tilde{a}_{Fe^{2+}}}$$

Ferrous ion can also give up an electron at an inert electrode, forming ferric ion, Fe^{3+}. This process is reversible. Depending on the potential of the half-cell with which it is paired, the inert electrode can either accept an electron from the external circuit and deliver it to a ferric ion, or take an electron from a ferrous ion and deliver it to the external circuit. Thus, ferrous and ferric ions form a redox couple. Platinum metal functions as an inert electrode in this reaction. The half-cell consists of a piece of pure platinum in contact with aqueous ferrous and ferric ions, both present at unit activity. Our notation for this half-cell and potential are $Pt \mid Fe^{2+}, Fe^{3+}$ and $\mathcal{E}_{Pt|Fe^{2+},Fe^{3+}}$. The corresponding half-reaction and its potential are

$$Fe^{3+} + e^- \rightleftharpoons Fe^{2+}$$

and

$$\mathcal{E}_{Pt|Fe^{2+},Fe^{3+}} = \mathcal{E}^o_{Pt|Fe^{2+},Fe^{3+}} - \frac{RT}{\mathcal{F}} \ln \frac{\tilde{a}_{Fe^{2+}}}{\tilde{a}_{Fe^{3+}}}$$

We can add these two half-reactions, to obtain

$$Fe^{3+} + 3e^- \rightleftharpoons Fe^0$$

The Nernst equation for this half-reaction is

$$\mathcal{E}_{Fe|Fe^{3+}} = \mathcal{E}^o_{Fe|Fe^{3+}} - \frac{RT}{3\mathcal{F}} \ln \frac{1}{\tilde{a}_{Fe^{3+}}}$$

From our past considerations, both of these equations are clearly correct. However, in this case, the Nernst equation of the sum is not the sum of the Nernst equations. Nor should we expect it to be. The half-cell Nernst equations are really shorthand notation for the behavior of the half-cell when it is operated against a S.H.E. Adding half-cell Nernst equations corresponds to creating a new system by connecting the two S.H.E. electrodes of two separate full cells, as we illustrate in Figure 8. In the present instance, we are manipulating two half-reactions to obtain a new half-reaction; this manipulation does not correspond to any possible way of interconnecting the corresponding half-cells.

Nevertheless, if we know the standard potentials for the first two reactions ($\mathcal{E}^o_{Fe|Fe^{2+}}$ and $\mathcal{E}^o_{Pt|Fe^{2+},Fe^{3+}}$), we can obtain the standard potential for their sum ($\mathcal{E}^o_{Fe|Fe^{3+}}$). To do so, we exploit the relationship we found between electrical potential and Gibbs free energy. The first two reactions represent sequential steps that jointly achieve the same net change as the third reaction. Therefore, the sum of the Gibbs free energy changes for the first two reactions must be the same as the Gibbs free energy change for the third reaction. The standard potentials are not additive, but the Gibbs free energy changes are. We have

$Fe^{3+} + e^- \rightleftharpoons Fe^{2+}$	$\Delta G^o_{Fe^{3+}\to Fe^{2+}} = \mathcal{F}\mathcal{E}^o_{Pt\|Fe^{2+},Fe^{3+}}$
$Fe^{2+} + 2e^- \rightleftharpoons Fe^0$	$\Delta G^o_{Fe^{2+}\to Fe^0} = 2\,\mathcal{F}\mathcal{E}^o_{Fe\|Fe^{2+}}$
$Fe^{3+} + 3e^- \rightleftharpoons Fe^0$	$\Delta G^o_{Fe^{3+}\to Fe^0} = 3\,\mathcal{F}\mathcal{E}^o_{Fe\|Fe^{3+}}$

Since also

$$\Delta G^o_{Fe^{3+}\to Fe^{2+}} + \Delta G^o_{Fe^{2+}\to Fe^0} = \Delta G^o_{Fe^{3+}\to Fe^0}$$

we have

$$\mathcal{F}\mathcal{E}^o_{Pt|Fe^{2+},Fe^{3+}} + 2\mathcal{F}\mathcal{E}^o_{Fe|Fe^{2+}} = 3\mathcal{F}\mathcal{E}^o_{Fe|Fe^{3+}}$$

and

$$\mathcal{E}^o_{Fe|Fe^{3+}} = \frac{\mathcal{E}^o_{Pt|Fe^{2+},Fe^{3+}} + 2\mathcal{E}^o_{Fe|Fe^{2+}}}{3}$$

§18 The Nernst equation and the criterion for equilibrium

In §15 we find for the general reaction $aA + bB \to cD + dD$ that the Nernst equation is

$$\mathcal{E} = \mathcal{E}^o - \frac{RT}{n\mathcal{F}}\ln\frac{\tilde{a}_C^c\tilde{a}_D^d}{\tilde{a}_A^a\tilde{a}_B^b}$$

We now want to consider the relationship between the potential of an electrochemical cell and the equilibrium position of the cell reaction. If the potential of the cell is not zero, short-circuiting the terminals of the cell will cause electrons to flow in the external circuit and reaction to proceed spontaneously in the cell. Since a spontaneous reaction occurs, the cell is not at equilibrium with respect to the cell reaction.

As we draw current from any electrochemical cell, cell reactants are consumed and cell products are produced. Experimentally, we see that the cell voltage decreases continuously, and inspection of the Nernst equation shows that it predicts a potential decrease. Eventually, the voltage of a short-circuited cell decreases to zero. No further current is passed. The cell reaction stops; it has reached chemical equilibrium. If the cell potential is zero, the cell reaction must be at equilibrium, and *vice versa*.

We also know that, at equilibrium, the activity ratio that appears as the argument of the logarithmic term is a constant—the equilibrium constant. So when $\mathcal{E} = 0$, we have also that

$$K_a = \frac{\tilde{a}_C^c\tilde{a}_D^d}{\tilde{a}_A^a\tilde{a}_B^b}$$

Substituting these conditions into the Nernst equation, we obtain

$$0 = \mathcal{E}^o - \frac{RT}{n\mathcal{F}}\ln K_a$$

or

$$K_a = \exp\frac{(n\mathcal{F}\mathcal{E}^o)}{RT}$$

We can obtain this same result if we recall that $\Delta G^o = -RT\ln K_a$ and that $\Delta G^o = -n\mathcal{F}\mathcal{E}^o$. We can determine equilibrium constants by measuring the potentials of standard cells. Alternatively, we can measure an equilibrium constant and determine the potential of the corresponding cell without actually constructing it. Standard potentials and equilibrium constants are both measures of the Gibbs free energy change when the reaction occurs under standard conditions.

Problems

1. Balance the following chemical equations assuming that they occur in aqueous solution.

(a) $Cu^0 + Ag^+ \to Cu^{2+} + Ag^0$
(b) $Fe^{2+} + Cr_2O_7^{2-} \to Fe^{3+} + Cr^{3+}$
(c) $Cr^{2+} + Cr_2O_7^{2-} \to Cr^{3+}$
(d) $Cl_2 + Br^- \to Cl^- + Br_2$
(e) $ClO_3 \to Cl^- + ClO_4^-$
(f) $I^- + IO_3^- \to I_2$
(g) $I^- + O_2 \to I_2 + OH^-$ (basic solution)
(h) $H_2C_2O_4^{2-} + MnO_4^- \to CO_2 + Mn^{2+}$
(i) $Fe^{2+} + MnO_4^- \to Fe^{3+} + Mn^{2+}$
(j) $H_2O_2 + MnO_4^- \to Mn^{2+} + O_2$
(k) $PbO_2 + Pb^0 + H_2SO_4 \to PbSO_4$
(l) $Fe^{2+} \to Fe^0 + Fe^{3+}$

(m) $Cu^{2+} + Fe^0 \rightarrow Cu^0 + Fe^{3+}$
(n) $Al^0 + OH^- \rightarrow H_2 + Al(OH)_4^-$ (basic solution)
(o) $Au^0 + CN^- + O_2 \rightarrow Au(CN)_4^-$ (basic solution)
(p) $Cu^0 + HNO_3 \rightarrow Cu^{2+} + NO_2$
(q) $Al^0 + Fe_2O_3 \rightarrow Al_2O_3 + Fe^0$
(r) $I^- + H_2O_2 \rightarrow I_2 + OH^-$ (basic solution)
(s) $HFeO_4^- + Mn^{2+} \rightarrow Fe^{3+} + MnO_2$
(t) $Fe^{2+} + S_2O_8^{2-} \rightarrow Fe^{3+} + SO_4^{2-}$
(u) $Cu^+ + O_2 \rightarrow Cu^{2+} + OH^-$ (basic solution)

2. Calculate the equilibrium constant, K_a, for the reaction $Cu^0 + 2Ag^+ \rightarrow Cu^{2+} + 2Ag^0$. An excess of clean copper wire is placed in a 10^{-1} M silver nitrate solution. Assuming that molarities adequately approximate the activities of the ions, find the equilibrium concentrations of Ag^+ and Cu^{2+}.

3. The standard potentials for reduction of Fe^{2+} and Fe^{3+} to Fe^0 are

$$Fe^{2+} + 2e^- \rightarrow Fe^0 \qquad \mathcal{E}^o = -0.447 \text{ v}$$
$$Fe^{3+} + 3e^- \rightarrow Fe^0 \qquad \mathcal{E}^o = -0.037 \text{ v}$$

(a) Find the standard potential for the disproportionation of Fe^{2+} to Fe^{3+} and Fe^0: $Fe^{2+} \rightarrow Fe^0 + Fe^{3+}$.
(b) Find the standard half-cell potential for the reduction of Fe^{3+} to Fe^{2+}: $Fe^{3+} + e^- \rightarrow Fe^{2+}$.

4. The standard potential for reduction of tris-ethylenediamineruthenium (III) to tris-ethylenediamineruthenium (II) is

$$[Ru(en)_3]^{3+} + e^- \rightarrow [Ru(en)_3]^{2+} \quad \mathcal{E}^o = +0.210 \text{ v}$$

Half-cell potential data are given below for several oxidants. Which of them can oxidize $[Ru(en)_3]^{2+}$ to $[Ru(en)_3]^{3+}$ in acidic ($[H^+] \approx \tilde{a}_{H^+} = 10^{-1}$) aqueous solution?

(a) $UO_2^{2+} + e^- \rightarrow UO_2^+$ $\quad \mathcal{E}^o = +0.062$ v
(b) $[Ru(NH_3)_6]^{3+} + e^- \rightarrow [Ru(NH_3)_6]^{2+}$ $\quad \mathcal{E}^o = +0.10$ v
(c) $Cu^{2+} + e^- \rightarrow Cu^+$ $\quad \mathcal{E}^o = +0.153$ v
(d) $AgCl + e^- \rightarrow Ag^0 + Cl^-$ $\quad \mathcal{E}^o = +0.222$ v
(e) $Hg_2Cl_2 + 2e^- \rightarrow 2Hg^0 + 2Cl^-$ $\quad \mathcal{E}^o = +0.268$ v
(f) $AgCN + e^- \rightarrow Ag^0 + CN^-$ $\quad \mathcal{E}^o = -0.017$ v
(g) $SnO_2 + 4H^+ + 4e^- \rightarrow Sn^0 + 2H_2O$ $\quad \mathcal{E}^o = -0.117$ v

5. An electrochemical cell is constructed in which one half cell is a standard hydrogen electrode and the other is a hydrogen electrode immersed in a solution of $pH = 7$ ($[H^+] \approx \tilde{a}_{H^+} = 10^{-7}$). What is the potential difference between the terminals of the cell? What chemical change occurs in this cell?

6. The standard half-cell potential for the reduction of oxygen gas at an inert electrode (like platinum metal) is

$$O_2 + 4H^+ + 4e^- \rightarrow 2H_2O \qquad \mathcal{E}^o = +1.229 \text{ v}$$

An electrochemical cell is constructed in which one half cell is a standard hydrogen electrode and the other cell is a piece of platinum metal, immersed in a 1 $\underline{M}$ solution of $HClO_4$, which is continuously in contact with bubbles of oxygen gas at a pressure of 1 bar.
(a) What is the potential difference between the terminals of the cell? What chemical change occurs in this cell?
(b) The 1 $\underline{M}$ $HClO_4$ solution in part (a) is replaced with pure water ($[H^+] \approx \tilde{a}_{H^+} = 10^{-7}$). What is the potential difference between the terminals of this cell?
(c) The 1 $\underline{M}$ $HClO_4$ solution in part (a) is replaced with 1 $\underline{M}$ $NaOH$ ($[H^+] \approx \tilde{a}_{H^+} = 10^{-14}$). What is the potential difference between the terminals of this cell?

7. A variable electrical potential source is introduced into the external circuit of the cell in part (a) of problem 6. The negative terminal of the potential source is connected to the oxygen electrode and the positive terminal of the potential source is connected to the standard hydrogen electrode. If the applied electrical potential is 1.3 v, what chemical change occurs? What is the minimum electrical potential that must be applied to electrolyze water if the oxygen electrode contains a 1 $\underline{M}$ $HClO_4$ solution? A neutral ($pH = 7$) solution? A 1 $\underline{M}$ $NaOH$ solution?

8. Two platinum electrodes are immersed in 1 $\underline{M}$ $HClO_4$. What potential difference must be applied between these electrodes in order to electrolyze water? (Assume that $P_{O_2} = 1$ bar and $P_{H_2} = 1$ bar at their respective electrodes, as will be the case as soon as a few bubbles of gas have accumulated at each electrode.) What potential difference is required if the electrodes are immersed in pure water? In 1 $\underline{M}$ $NaOH$?

18

Quantum Mechanics and Molecular Energy Levels

§1 Energy distributions and energy levels

Beginning in Chapter 20, we turn our attention to the distribution of energy among the molecules in a closed system that is immersed in a constant-temperature bath, that is at equilibrium, and that contains a large number of molecules. We want to find the probability that the energy of a molecule in such a system is in a particular interval of energy values. This probability is also the fraction of the molecules whose energies are in the specified interval, since we assume that these statements mean the same thing for a system at equilibrium.

The probability that the energy of a particular molecule is in a particular interval is intimately related to the energies that it is possible for a molecule to have. Before we can make further progress in describing molecular energy distributions, we must discuss atomic and molecular energies. For our development of the Boltzmann equation, we need to introduce the idea of quantized energy states. This requires a short digression on the basic ideas of quantum mechanics and the quantized energy levels of atoms and molecules.

We have derived two expressions that relate the energy of a molecule to the probability that the molecule will have that energy. One follows from the barometric formula

$$\eta(h) = \eta(0)\exp\left(\frac{-mgh}{kT}\right) = \eta(0)\exp\left(\frac{-\epsilon_{potential}}{kT}\right)$$

in which the number density of molecules depends exponentially on their gravitational potential energy, mgh, and the reciprocal of the temperature. From the barometric formula, we can find the probability density function

$$\frac{df}{dh} = \frac{mg}{kT}\exp\left(\frac{-mgh}{kT}\right)$$

(See problem 22, Chapter 3.) The other is the Maxwell-Boltzmann distribution function

$$\frac{df}{dv} = 4\pi\left(\frac{m}{2\pi kT}\right)^{3/2} v^2\exp\left(\frac{-mv^2}{2kT}\right) = 4\pi\left(\frac{m}{2\pi kT}\right)^{3/2} v^2\exp\left(\frac{-\epsilon_{kinetic}}{2kT}\right)$$

in which the probability density of molecular velocities depends exponentially on their kinetic energies, $mv^2/2$, and the reciprocal of the temperature. We will see that this dependence is very general. Any time the molecules in a system can have a range of energies, the probability that a molecule has energy ϵ is proportional to $\exp(-\epsilon/kT)$. The exponential term, $\exp(-\epsilon/kT)$, is often called the Boltzmann factor.

We might try to develop a more general version of the Maxwell-Boltzmann distribution function by an argument that somehow parallels our derivation of the Maxwell-Boltzmann equation. It turns out that any such attempt is doomed to failure, because it is based on a fundamentally incorrect view of nature. In developing the barometric formula and the Maxwell-Boltzmann distribution, we assume that the possible energies are continuous; a molecule can be at any height above the surface of the earth, and its translational velocity can have any value. When we turn to the distribution of other ways in which molecules can have energy, we find that this assumption produces erroneous predictions about the behavior of macroscopic collections of molecules.

The failure of such attempts led Max Planck to the first formulation of the idea that energy is quantized. The spectrum of light emitted from glowing-hot objects (so-called "black bodies") depends on the temperature of the emitting object. Much of the experimentally observable behavior of light can be explained by the hypothesis that light behaves like a wave. Mechanical (matter-displacement) waves carry energy; the greater the amplitude of the wave, the more energy it carries. Now, light is a form of energy, and a spectrum is an energy distribution. It was a challenge to late nineteenth century physics to use the wave model for the behavior of light to predict experimentally observed emission spectra. This challenge went unmet until Planck introduced the postulate that such "black-body radiators" absorb or emit electromagnetic radiation only in discrete quantities, called quanta. Planck proposed that the energy of one such quantum is related to the frequency, ν, of the radiation by the equation $E = h\nu$, where the proportionality constant, h, is now called ***Planck's constant***. In Planck's model, the energy of an electromagnetic wave depends on its frequency, not its amplitude.

In the years following Planck's hypothesis, it became clear that many properties of atoms and molecules are incompatible with the idea that an atom or molecule can have any arbitrary energy. We obtain agreement between experimental observations and theoretical models only if we assume that atoms and molecules can have only very particular energies. This is observed most conspicuously in the interactions of atoms and molecules with electromagnetic radiation. One such interaction gives rise to a series of experimental observations known as the photoelectric effect. In order to explain the photoelectric effect, Albert Einstein showed that it is necessary to extend Planck's concept to assume that light itself is a stream of discrete energy quanta, called photons. In our present understanding, it is necessary to describe some of the properties of light as wave-like and some as particle-like.

In many absorption and emission spectra, we find that a given atom or molecule can emit or absorb electromagnetic radiation only at very particular frequencies. For example, the light emitted by atoms excited by an electrical discharge contains a series of discrete emission lines. When it is exposed to a continuous spectrum of frequencies, an atom is observed to absorb light at precisely the discrete frequencies that are observed in emission. Niels Bohr explained these observations by postulating that the electrons in atoms can have only particular energies. The absorption of visible light by atoms and molecules occurs when an electron takes up electromagnetic energy and moves from one discrete energy level to a second, higher, one. (Absorption of a continuous range of frequencies begins to occur only when the light absorbed provides sufficient energy to separate an electron from the original chemical species, producing a free electron and a positively charged ion. At the onset frequency, neither of the product species has any kinetic energy. Above the onset frequency, spectra are no longer discrete, and the species produced have increasingly greater kinetic energies.) Similar discrete absorption lines are observed for the absorption of infrared light and microwave radiation by diatomic or polyatomic gas molecules. Infrared absorptions are associated with vibrational motions, and microwave absorptions are associated with rotational motions of the molecule about its center of mass. These phenomena are explained by the quantum theory.

The quantized energy levels of atoms and molecules can be found by solving the Schroedinger equation for the system at hand. To see the basic ideas that are involved, we discuss the Schroedinger equation and some of the most basic approximations that are made in applying it to the description of atomic and molecular systems. But first, we should consider one more preliminary question: If the quantum hypothesis is so important to obtaining valid equations for the distribution of energies, why are the derivations of the Maxwell-Boltzmann equation and the barometric formula successful? Maxwell's derivation is successful because the quantum mechanical description of a molecule's translational kinetic energy is very well approximated by the assumption that the molecule's kinetic energy can have any value. In the language of quantum mechanics, the number of translational energy levels available to a molecule is very large, and successive energy levels are very close together—so close together that it is a good approximation to say that they are continuous. Similarly, the gravitational potential energies available to a molecule in the earth's atmosphere are well approximated by the assumption that they belong to a continuous distribution.

§2 Quantized energy: de Broglie's hypothesis and the Schroedinger equation

Subsequent to Planck's proposal that energy is quantized, the introduction of two further concepts led to the theory of quantum mechanics. The first was Einstein's relativity theory, and his deduction from it of the equivalence of matter and energy. The relativistic energy of a particle is given by $E^2 = p^2c^2 + m_0^2c^4$, where p is the momentum and m_0 is the mass of the particle when it is at rest. The second was de Broglie's hypothesis that any particle of mass m moving at velocity v, behaves like a wave. De Broglie's hypothesis is an independent postulate about the structure of nature. In this respect, its status is the same as that of Newton's laws or the laws of thermodynamics. Nonetheless, we can construct a line of thought that is probably similar to de Broglie's, recognizing that these are heuristic arguments and not logical deductions.

We can suppose that de Broglie's thinking went something as follows: Planck and Einstein have proposed that electromagnetic radiation—a wave-like phenomenon—has the particle-like property that it comes in discrete lumps (photons). This means that things we think of as waves can behave like particles. Conversely, the lump-like photons behave like waves. Is it possible that other lump-like things can behave like waves? In particular is it possible that material particles might have wave-like properties? If a material particle behaves like a wave, what wave-like properties should it exhibit?

Well, if we are going to call something a wave, it must have a wavelength, λ, a frequency, ν, and a propagation velocity, v, and these must be related by the equation $v = \lambda\nu$. The velocity of propagation of light is conventionally given the symbol c, so $c = \lambda\nu$. The Planck-Einstein hypothesis says that the energy of a particle (photon) is $E = h\nu = hc/\lambda$. Einstein proposes that the energy of a particle is given by $E^2 = p^2c^2 + m_0^2c^4$. A photon travels at the speed of light. This is compatible with other relativistic equations only if the rest mass of a photon is zero. Therefore, for a photon, we must have $E = pc$. Equating these energy equations, we find that the momentum of a photon is

$$p = h/\lambda$$

Now in a further exercise of imagination, we can suppose that this equation applies also to any mass moving with any velocity. Then we can replace p with mv, and write

$$mv = h/\lambda$$

We interpret this to mean that any mass, m, moving with velocity, v, has a wavelength, λ, given by

$$\lambda = h/mv$$

This is ***de Broglie's hypothesis***. We have imagined that de Broglie found it by a series of imaginative—and not entirely logical—guesses and suppositions. The illogical parts are the reason we call the result a hypothesis rather than a derivation, and the originality of the guesses and suppositions is the reason de Broglie's hypothesis was new. It is important physics, because it turns out to be experimentally valid. Very small particles do exhibit wave-like properties, and de Broglie's hypothesis correctly predicts their wavelengths.

In a similar vein, we can imagine that Schroedinger followed a line of thought something like this: de Broglie proposes that any moving particle behaves like a wave whose wavelength depends on its mass and velocity. If a particle behaves as a wave, it should have another wave property; it should have an amplitude. In general, the amplitude of a wave depends on location and time, but we are thinking about a rather particular kind of wave, a wave that—so to speak—stays where we put it. That is, our wave is supposed to describe a particle, and particles do not dissipate themselves in all directions like the waves we get when we throw a rock in a pond. We call a wave that stays put a standing wave; it is distinguished by the fact that its amplitude depends on location but not on time.

Mathematically, the amplitude of any wave can be described as a sum of (possibly many) sine and cosine terms. A single sine term describes a simple wave. If it is a standing wave, its amplitude depends only on distance, and its amplitude is the same for any two points separated by one wavelength. Letting the amplitude be ψ, this standing wave is described by $\psi(x) = A\sin(ax)$, where x is the location, expressed as a distance from the origin at $x = 0$. In this wave equation, A and a are parameters that fix the maximum amplitude and the wavelength, respectively. Requiring the wavelength to be λ means that $a\lambda = 2\pi$. (Since ψ is a sine function, it repeats every time its argument increases by 2π radians. We require that ψ repeat every time its argument increases by $a\lambda$ radians, which requires that $a\lambda = 2\pi$.) Therefore, we have

$$a = 2\pi/\lambda$$

and the wave equation must be

$$\psi(x) = A\sin(2\pi x/\lambda)$$

Equations, ψ, that describe standing waves satisfy the differential equation

$$\frac{d^2\psi}{dx^2} = -C\psi$$

where C is a constant. In the present instance, we see that

$$\begin{aligned}\frac{d^2\psi}{dx^2} &= -\left(\frac{2\pi}{\lambda}\right)^2 A\sin\left(\frac{2\pi x}{\lambda}\right)\\ &= -\left(\frac{2\pi}{\lambda}\right)^2 \psi\end{aligned}$$

From de Broglie's hypothesis, we have $\lambda = h/mv$, so that the constant C can be written as

$$\begin{aligned}C &= \left(\frac{2\pi}{\lambda}\right)^2\\ &= \left(\frac{2\pi mv}{h}\right)^2\\ &= \left(\frac{2\pi}{h}\right)^2 (2m)\left(\frac{mv^2}{2}\right)\\ &= \left(\frac{8\pi^2 m}{h^2}\right)\left(\frac{mv^2}{2}\right)\end{aligned}$$

Let T be the kinetic energy, $mv^2/2$, and let V be the potential energy of our wave-like particle. Then its energy is $E = T + V$, and we have $mv^2/2 = T = E - V$. The constant C becomes

$$\begin{aligned}C &= \left(\frac{8\pi^2 m}{h^2}\right)T\\ &= \left(\frac{8\pi^2 m}{h^2}\right)(E - V)\end{aligned}$$

Making this substitution for C, we find a differential equation that describes a standing wave, whose wavelength satisfies the de Broglie equation. This is the time-independent ***Schroedinger equation*** in one dimension:

$$\frac{d^2\psi}{dx^2} = -\left(\frac{8\pi^2 m}{h^2}\right)(E - V)\psi$$

or

$$-\left(\frac{h^2}{8\pi^2 m}\right)\frac{d^2\psi}{dx^2} + V\psi = E\psi$$

Often the latter equation is written as

$$\left[-\left(\frac{h^2}{8\pi^2 m}\right)\frac{d^2}{dx^2} + V\right]\psi = E\psi$$

where the expression in square brackets is called the ***Hamiltonian operator*** and abbreviated to H, so that the Schroedinger equation becomes simply, if cryptically,

$$H\psi = E\psi$$

If we know how the potential energy of a particle, V, depends on its location, we can write down the Hamiltonian operator and the Schroedinger equation that describe the wave properties of the particle. Then we need to find the wave equations that satisfy this differential equation. This can be difficult even when the Schroedinger equation involves only one particle. When we write the Schroedinger equation for a system containing multiple particles that interact with one another, as for example an atom containing two or more electrons, analytical solutions become unattainable; only approximate solutions are possible. Fortunately, a great deal can be done with approximate solutions.

The Schroedinger equation identifies the value of the wave function, $\psi(x)$, with the amplitude of the particle wave at the location x. Unfortunately, there is no physical interpretation for $\psi(x)$; that is, no measurable quantity corresponds to the value of $\psi(x)$. There is, however, a physical interpretation for the product $\psi(x)\psi(x)$ or $\psi^2(x)$. [More accurately, the product $\psi(x)\psi^*(x)$, where $\psi^*(x)$ is the complex conjugate of $\psi(x)$. In general, x is a complex variable.] $\psi^2(x)$ is the probability density function for the particle whose wave function is $\psi(x)$. That is, the product $\psi^2(x)dx$ is the probability of finding the particle within a small distance, dx, of the location x. Since the particle must be somewhere, we also have

$$1 = \int_{-\infty}^{+\infty} \psi^2(x)\, dx$$

§3 The Schroedinger equation for a particle in a box

A problem usually called the ***particle in a box*** provides a convenient illustration of the principles involved in setting up and solving the Schroedinger equation. Besides being a good illustration, the problem also proves to be a useful approximation to many physical systems. The statement of the problem is simple. We have a particle of mass m that is constrained to move only in one dimension. For locations in the interval $0 \le x \le \ell$, the particle has zero potential energy. For locations outside this range, the particle has infinite potential energy. Since the particle cannot have infinite energy, this means that it can never find its way into locations outside of the interval $0 \le x \le \ell$. We can think of this particle as a bead moving on a wire, with stops located on the wire at $x = 0$ and at $x = \ell$. We can also think of it as being confined to a one-dimensional box of length ℓ, which is the viewpoint represented by the name. The particle in a box model is diagrammed in Figure 1.

The potential energy constraints mean that the amplitude of the particle's wave function must be zero, $\psi(x) = 0$, when the value of x lies in the interval

$$-\infty < x < 0$$

or

$$\ell < x < +\infty$$

We assume that the probability of finding the particle cannot change abruptly when its location changes by an arbitrarily small amount. This means that the wave function must be continuous, and it follows that $\psi(0) = 0$ and $\psi(\ell) = 0$. Inside the box, the particle's Schroedinger equation is

$$-\left(\frac{h^2}{8\pi^2 m}\right)\frac{d^2\psi}{dx^2} = E\psi$$

and we seek those functions $\psi(x)$ that satisfy both this differential equation and the constraint equations

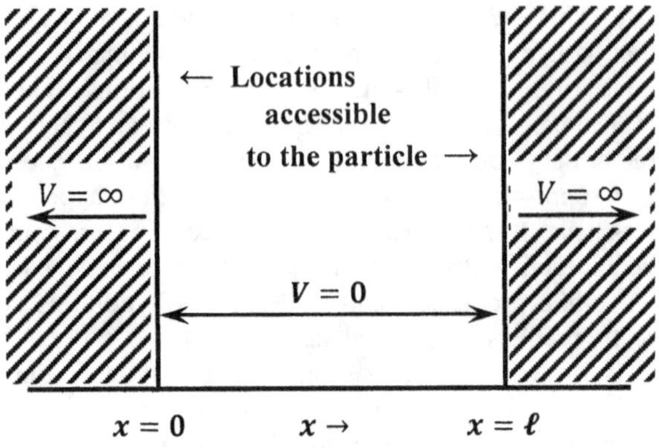

Figure 1. Potential energy versus distance for a particle in a box.

$\psi(0) = 0$ and $\psi(\ell) = 0$. It turns out that there are infinitely many such solutions, ψ_n, each of which corresponds to a unique energy level, E_n.

To find these solutions, we first guess—guided by our considerations in §2—that solutions will be of the form

$$\psi(x) = A\sin(ax) + B\cos(bx)$$

A solution must satisfy

$$\psi(0) = A\sin(0) + B\cos(0) = B\cos(0) = 0$$

so that $B = 0$. At the other end of the box, we must have

$$\psi(\ell) = A\sin(a\ell) = 0$$

which means that $a\ell = n\pi$, where n is any integer: $n = 1, 2, \ldots$. Hence, we have

$$a = n\pi/\ell$$

and the only equations of the proposed form that satisfy the conditions at the ends of the box are

$$\psi_n(x) = A\sin(n\pi x/\ell)$$

To test whether these equations satisfy the Schroedinger equation, we check

$$-\left(\frac{h^2}{8\pi^2 m}\right)\frac{d^2}{dx^2}\left[A\sin\left(\frac{n\pi x}{\ell}\right)\right] = E_n\psi_n$$

and find

$$\left(\frac{h^2}{8\pi^2 m}\right)\left(\frac{n\pi}{\ell}\right)^2\left[A\sin\left(\frac{n\pi x}{\ell}\right)\right] = \left(\frac{n^2h^2}{8m\ell^2}\right)\psi_n = E_n\psi_n$$

so that the wave functions $\psi_n(x) = A\sin(n\pi x/\ell)$ are indeed solutions and the energy, E_n, associated with the wave function $\psi_n(x)$ is

$$E_n = \frac{n^2h^2}{8m\ell^2}$$

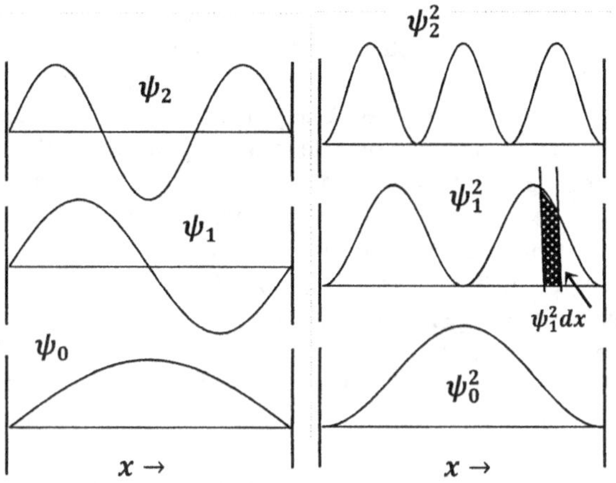

Figure 2. Wave functions for a particle in a box.

We see that the energy values are quantized; although there are infinitely many energy levels, E_n, only very particular real numbers—those given by the equation above—correspond to energies that the particle can have. If we sketch the first few wave functions, $\psi_n(x)$, we see that there are always $n-1$ locations inside the box at which $\psi_n(x)$ is zero. These locations are called nodes. Once we know n, we know the number of nodes, and we can sketch the general shape of the corresponding wave function. The first three wave functions and their squares are sketched in Figure 2.

At this point, we have found a complete set of infinitely many solutions, except for the parameter A. To determine A, we interpret $\psi^2(x)$ as a probability density function, and we require that the probability of finding the particle in the box be equal to unity. This means that

$$1 = \int_0^\ell A^2 \sin^2\left(\frac{n\pi x}{\ell}\right) dx$$

$$= A^2 \int_0^\ell \left[\frac{1}{2} - \frac{1}{2}\cos\left(\frac{2n\pi x}{\ell}\right)\right] dx$$

$$= A^2\left(\frac{\ell}{2}\right)$$

so that $A = \sqrt{2/\ell}$, and the final wave functions are

$$\psi_n(x) = \sqrt{\frac{2}{\ell}} \sin\left(\frac{n\pi x}{\ell}\right)$$

§4 The Schroedinger equation for a molecule

Molecules are composed of atoms, and atoms are composed of nuclei and electrons. When we consider the internal motions of molecules, we have to consider the motions of a large number of charged particles with respect to one another. In principle, we can write down the potential function (the V in the Schroedinger equation) that describes the Coulomb's law based potential energy of the system of charged particles. In principle, we can then solve the Schroedinger equation and obtain a series of wave functions, $\psi_n(x)$, and their corresponding energies, E_n, that completely characterize the motions of the molecule's constituent particles. Each of the E_n is an energy value that the molecule can have. Often we say that it is an energy level that the molecule can occupy.

Since every distance between two charged particles is a variable in the Schroedinger equation, the number of variables increases dramatically as the size of the molecule increases. The two-particle hydrogen-atom problem has been solved analytically. For any chemical species larger than the hydrogen atom, only approximate solutions are possible. Nevertheless, approximate results can be obtained to very high accuracy. Greater accuracy comes at the expense of more extensive calculations.

Let us look briefly at the more fundamental approximations that are made. One is called the ***Born-Oppenheimer approximation***; it states that the motions of the nuclei in a molecule are too slow to affect the motions of the electrons. This occurs because nuclei are much more massive than electrons. The Born-Oppenheimer approximation assumes that the electronic motions can be calculated as if the nuclei are fixed at their equilibrium positions without introducing significant error into the result. That is, there is an approximate wave function describing the motions of the electrons that is independent of a second wave function that describes the motions of the nuclei.

The mathematical description of the nuclear motions can be further simplified using additional approximations; we can separate the nuclear motions into translational, rotational, and vibrational modes. Translational motion is the three-dimensional displacement of an entire molecule. It can be described by specifying the motion of the molecule's center of mass. The motions of the constituent nuclei with respect to one another can be further subdivided: rotational motions change the orientation of the whole molecule in space; vibrational motions change distances between constituent nuclei.

The result is that the wave function for the molecule as a whole can be approximated as a product of a wave function ($\psi_{electronic}$ or ψ_e) for the electronic motions, a wave function ($\psi_{vibration}$ or ψ_v) for the vibrational motions, a wave function ($\psi_{rotation}$ or ψ_r) for the rotational motions, and a wave function ($\psi_{translation}$ or ψ_t) for the translational motion of the center of mass. We can write

$$\psi_{molecule} = \psi_e \psi_v \psi_r \psi_t$$

(None of this is supposed to be obvious. We are merely describing the essential results of a considerably more extensive development.)

When we write the Hamiltonian for a molecule under the approximation that the electronic, vibrational, rotational, and translational motions are independent of each other, we find that the Hamiltonian is a sum of terms. In some of these terms, the only independent variables are those that specify the locations of the electrons. We call these variables electronic coordinates. Some of the remaining terms involve only vibrational coordinates, some involve only rotational coordinates, and some involve only translational coordinates. That is, we find that the Hamiltonian for the molecule can be expressed as a sum of terms, each of which is the Hamiltonian for one of the kinds of motion:

$$H_{molecule} = H_e + H_v + H_r + H_t$$

where we have again abbreviated the subscripts denoting the various categories of motion.

Consequently, when we write the Schoedinger equation for the molecule in this approximation, we have

$$\begin{aligned}
&H_{molecule}\psi_{molecule} \\
&= (H_e + H_v + H_r + H_t)\psi_e\psi_v\psi_r\psi_t \\
&= \psi_v\psi_r\psi_t H_e\psi_e + \psi_e\psi_r\psi_t H_v\psi_v + \psi_e\psi_v\psi_t H_r\psi_r \\
&\qquad + \psi_e\psi_v\psi_r H_t\psi_t \\
&= \psi_v\psi_r\psi_t E_e\psi_e + \psi_e\psi_r\psi_t E_v\psi_v + \psi_e\psi_v\psi_t E_r\psi_r \\
&\qquad + \psi_e\psi_v\psi_r E_t\psi_t \\
&= (E_e + E_v + E_r + E_t)\psi_e\psi_v\psi_r\psi_t
\end{aligned}$$

We find that the energy of the molecule as a whole is simply the sum of the energies associated with the several kinds of motion

$$E_{molecule} = E_e + E_v + E_r + E_t$$

ψ_t, ψ_v, ψ_r, and ψ_e can be further approximated as products of wave functions involving still smaller numbers of coordinates. We can have a component wave function for every distinguishable coordinate that describes a possible motion of a portion of the molecule. The three translational modes are independent of one another. It is a good approximation to assume that they are also independent of the rotational and vibrational modes. Frequently, it is a good approximation to assume that the vibrational and rotational modes are independent of one another. We can deduce the number of one-dimensional wave functions that are required to give an approximate wave function that describes all of the molecular motions, because this will be the same as the number of coordinates required to describe the nuclear motions. If we have a collection of N atoms that are not bonded to one another, each atom is free to move in three dimensions. The number of coordinates required to describe their motion is $3N$. When the same atoms are bonded to one another in a molecule, the total number of motions remains the same, but it becomes convenient to reorganize the way we describe them.

First, we recognize that the atomic nuclei in a molecule occupy positions that are approximately fixed relative to one another. Therefore, to a good approximation, the motion of the center of mass is independent of the way that the atoms move relative to one another or relative to the center of mass. It takes three coordinates to describe the motion of the center of mass, so there are $3N - 3$ coordinates left over after this is done.

The number of rotational motions available to a molecule depends upon the number of independent axes about which it can rotate. We can imagine a rotation of a molecule about any axis we choose. In general, in three dimensions, we can choose any three non-parallel axes and imagine that the molecule rotates about each of them independently of its rotation about the others. If we consider a set of more than three non-parallel axes, we find that any of the axes can be expressed as a combination of any three of the others. This means that the maximum number of independent rotational motions for the molecule as a whole is three.

If the molecule is linear, we can take the molecular axis as one of the axes of rotation. Most conveniently, we can then choose the other two axes to be perpendicular to the molecular axis and perpendicular to each other. However, rotation about the molecular axis does not change anything about the molecule's orientation in space. If the molecule is linear, rotation about the molecular axis is not a rotation at all! So, if the molecule is linear, only two coordinates are required to describe all of the rotational motions, and there are $3N - 5$ coordinates left over after we allocate those needed to describe the translational and rotational motions.

The coordinates left over after we describe the translational and rotational motions must be used to describe the motion of the atoms with respect to one another. These motions are called vibrations, and hence the number of coordinates needed to describe the vibrations of a non-linear molecule is $3N - 6$. For a linear molecule, $3N - 5$ coordinates are needed to describe the vibrations.

§5 Solutions to Schroedinger equations for harmonic oscillators and rigid rotors

We can approximate the wave function for a molecule by partitioning it into wave functions for individual translational, rotational, vibrational, and electronic modes. The wave functions for each of these modes can be approximated by solutions to a Schroedinger equation that approximates that mode. Our objective in this chapter is to introduce the quantized energy levels that are found.

Translational modes are approximated by the particle in a box model that we discuss above.

Vibrational modes are approximated by the solutions of the Schroedinger equation for coupled harmonic oscillators. The vibrational motion of a diatomic molecule is approximated by the solutions of the Schroedinger equation for the vibration of two masses linked by a spring. Let the distance between the masses be r and the equilibrium distance be r_0. Let the reduced mass of the molecule be μ, and let the force constant for the spring be λ. From classical mechanics, the potential energy of the system is

$$V(r) = \frac{\lambda(r - r_0)^2}{2}$$

and the vibrational frequency of the classical oscillator is

$$\nu = \frac{1}{2\pi}\sqrt{\frac{\lambda}{\mu}}$$

The Schroedinger equation is

$$-\left(\frac{h^2}{8\pi^2\mu}\right)\frac{d^2\psi}{dr^2} + \frac{\lambda(r-r_0)^2}{2}\psi = E\psi$$

The solutions to this equation are wave functions and energy levels that constitute the quantum mechanical description of the classical harmonic oscillator. The energy levels are given by

$$E_n = h\nu\left(n + \frac{1}{2}\right)$$

where the quantum numbers, n, can have any of the values $n = 0, 1, 2, 3, \ldots$. The lowest energy level, that for which $n = 0$, has a non-zero energy; that is,

$$E_0 = h\nu/2$$

The quantum mechanical oscillator can have infinitely many energies, each of which is a half-integral multiple of the classical frequency, ν. Each quantum mechanical energy corresponds to a quantum mechanical frequency:

$$\nu_n = \nu\left(n + \frac{1}{2}\right)$$

A classical rigid rotor consists of two masses that are connected by a weightless rigid rod. The rigid rotor is a dumbbell. The masses rotate about their center of mass. Each ***two-dimensional rotational motion of a diatomic molecule*** is approximated by the solutions of the Schroedinger equation for the motion of a rigid rotor in a plane. The simplest model assumes that the potential term is zero for all angles of rotation. Letting I be the molecule's moment of inertia and φ be the rotation angle, the Schroedinger equation is

$$-\left(\frac{h^2}{8\pi^2 I}\right)\frac{d^2\psi}{d\varphi^2} = E\psi$$

The energy levels are given by

$$E_m = \frac{m^2h^2}{8\pi^2 I}$$

where the quantum numbers, m, can have any of the values $m = 1, 2, 3, \ldots$,(but not zero). Each of these energy levels is two-fold degenerate. That is, two quantum mechanical states of the molecule have the energy E_m.

The ***three-dimensional rotational motion of a diatomic molecule*** is approximated by the solutions of the Schroedinger equation for the motion of a rigid rotor in three dimensions. Again, the simplest model assumes that the potential term is zero for all angles of rotation. Letting θ and φ be the two rotation angles required to describe the orientation in three dimensions, the Schroedinger equation is

$$-\frac{h^2}{8\pi^2 I}\left(\frac{1}{\sin\theta}\frac{\partial}{\partial\theta}\left(\sin\theta\frac{\partial\psi}{\partial\theta}\right) + \frac{1}{\sin^2\theta}\frac{d^2\psi}{d\varphi^2}\right) = E\psi$$

The energy levels are given by

$$E_J = \frac{h^2}{8\pi^2 I}J(J+1)$$

where the quantum numbers, J, can have any of the values $J = 0, 1, 2, 3, \ldots$. E_J is $(2J+1)$-fold degenerate. That is, there are $2J+1$ quantum mechanical states of the molecule all of which have the same energy, E_J.

Equations for the rotational energy levels of larger molecules are more complex.

§6 Wave functions, quantum states, energy levels, and degeneracies

We approximate the wave function for a molecule by using a product of approximate wave functions, each of which models some subset of the motions that the molecule undergoes. In general, the wave functions that satisfy the molecule's Schroedinger equation are degenerate; that is, two or more of these wave functions have the same energy. (The one-dimensional particle in a box and the one-dimensional harmonic oscillator have non-degenerate solutions. The rigid-rotor in a plane has doubly degenerate solutions; two wave functions have the same energy. The J-th energy level of the three-dimensional rigid rotor is $(2J+1)$-fold degenerate; there are $(2J+1)$ wave functions whose energy is E_J.)

We use doubly subscripted symbols to represent the wave functions that satisfy the molecule's Schroedinger equation. We write $\psi_{i,j}$ to represent all of the molecular wave functions whose energy is ϵ_i. We let g_i be the number of wave functions whose energy is ϵ_i. We say that the energy level ϵ_i is g_i-fold degenerate. The wave functions

$$\psi_{i,1}, \psi_{i,2}, \ldots, \psi_{i,j}, \ldots, \psi_{i,g_i}$$

are all solution to the molecule's Schroedinger equation; we have

$$H_{molecule}\psi_{i,j} = \epsilon_i\psi_{i,j}$$

for $j = 1, 2, \ldots, g_i$. Every energy level ϵ_i is associated with g_i ***quantum states***. For simplicity, we can think of each of the g_i wave functions, $\psi_{i,j}$, as a quantum state; however, the molecule's Schroedinger equation is also satisfied by any set of g_i independent linear combinations of the $\psi_{i,j}$. For present purposes, all that matters is that there are g_i quantum-mechanical descriptions—quantum states—all of which have energy ϵ_i.

§7 Particle spins and statistics: Bose-Einstein and Fermi-Dirac statistics

Our goal is to develop the theory of statistical thermodynamics from Boltzmann statistics. In this chapter, we explore the rudiments of quantum mechanics in order to become familiar with the idea that we can describe a series of discrete energy levels for any given molecule. For our purposes, that is all we need. We should note, however, that we are not developing the full story about the relationship between quantum mechanics and statistical thermodynamics. The spin of a particle is an important quantum mechanical property. It turns out that quantum mechanical solutions depend on the spin of the particle being described. Particles with integral spins behave differently from particles with half-integral spins. When we treat the statistical distribution of these particles, we need to treat particles with integral spins differently from particles with half-integral spins. Particles with integral spins are said to obey ***Bose-Einstein statistics***; particles with half-integral spins obey ***Fermi-Dirac statistics***.

Fortunately, both of these treatments converge to the Boltzmann distribution if the number of quantum states available to the particles is much larger than the number of particles. For macroscopic systems at ordinary temperatures, this is the case. In Chapters 19 and 20, we introduce the ideas underlying the theory of statistical mechanics. In Chapter 21, we derive the Boltzmann distribution from a set of assumptions that does not correspond to either the Bose-Einstein or the Fermi-Dirac requirement. In Chapter 25, we derive the Bose-Einstein and Fermi-Dirac distributions and show how they become equivalent to the Boltzmann distribution for most systems of interest in chemistry.

19

The Distribution of Outcomes for Multiple Trials

§1 Distribution of results for multiple trials with two possible outcomes

Suppose that we have two coins, one minted in 2001 and one minted in 2002. Let the probabilities of getting a head and a tail in a toss of the 2001 coin be $P_{H,1}$ and $P_{T,1}$, respectively. We assume that these outcomes exhaust the possibilities. From the laws of probability, we have: $1 = (P_{H,1} + P_{T,1})$. For the 2002 coin, we have $1 = (P_{H,2} + P_{T,2})$. The product of these two probabilities must also be unity. Expanding this product gives

$$\begin{aligned} 1 &= (P_{H,1} + P_{T,1})(P_{H,2} + P_{T,2}) \\ &= P_{H,1}P_{H,2} + P_{H,1}P_{T,2} + P_{T,1}P_{H,2} + P_{T,1}P_{T,2} \end{aligned}$$

This equation represents the probability of a trial in which we toss the 2001 coin first and the 2002 coin second. The individual terms are the probabilities of the possible outcomes of such a trial. It is convenient to give a name to this latter representation of the product; we will call it the expanded representation of ***the total probability sum***.

Our procedure for multiplying two binomials generates a sum of four terms. Each term contains two factors. The first factor comes from the first binomial; the second term comes from the second binomial. Each of the four terms corresponds to a combination of an outcome from tossing the 2001 coin and an outcome from tossing the 2002 coin. Conversely, every possible combination of outcomes from tossing the two coins is represented in the sum. $P_{H,1}P_{H,2}$ represents the probability of getting a head from tossing the 2001 coin and a head from tossing the 2002 coin. $P_{H,1}P_{T,2}$ represents the probability of getting a head from tossing the 2001 coin and a tail from tossing the 2002 coin, *etc.* In short, there is a one-to-one correspondence between the terms in this sum and the possible combinations of the outcomes of tossing these two coins.

This analysis depends on our ability to tell the two coins apart. For this, the mint date is sufficient. If we toss the two coins simultaneously, the four possible outcomes remain the same. Moreover, if we distinguish the result of a first toss from the result of a second toss, etc., we can generate the same outcomes by using a single coin. If we use a single coin, we can represent the possible outcomes from two tosses by the ordered sequences HH, HT, TH, and TT, where the first symbol in each sequence is the result of the first toss and the second symbol is the result of the second toss. The ordered sequences HT and TH differ only in the order in which the symbols appear. We call such ordered sequences ***permutations***.

Now let us consider a new problem. Suppose that we have two coin-like slugs that we can tell apart because we have scratched a "1" onto the surface of one and a "2" onto the surface of the other. Suppose that we also have two cups, one marked "H" and the other marked "T." We want to figure out how many different ways we can put the two slugs into the two cups. We can also describe this as the problem of finding the number of ways we can assign two distinguishable slugs (objects) to two different cups (categories). There are four such ways: Cup H contains slugs 1 and 2; Cup H contains slug 1 and Cup T contains slug 2; Cup H contains slug 2 and Cup T contains slug 1; Cup T contains slugs 1 and 2.

We note that, given all of the ordered sequences for tossing two coins, we can immediately generate all of the ways that two distinguishable objects (numbered slugs) can be assigned to two categories (Cups H and T). For each ordered sequence, we assign the first object to the category corresponding to the first symbol in the sequence, and we assign the second object to the category corresponding to the second symbol in the sequence.

In short, there are one-to-one correspondences between the sequences of probability factors in the total probability sum, the possible outcomes from tossing two distinguishable coins, the possible sequences of outcomes from two tosses of a single coin, and the number of ways we can assign two distinguishable objects to two categories. (See Table 1.)

If the probability of tossing a head is constant, we have $P_{H,1} = P_{H,2} = P_H$ and $P_{T,1} = P_{T,2} = P_T$. Note that we are not assuming $P_H = P_T$. If we do not care about the order in which the heads and tails appear, we can simplify our equation for the product of probabilities to

$$1 = P_H^2 + 2P_HP_T + P_T^2$$

P_H^2 is the probability of tossing two heads, P_HP_T is the probability of tossing one head and one tail, and P_T^2 is the probability of tossing two tails. We must multiply the P_HP_T-term by two, because there are two two-coin outcomes and correspondingly two combinations, $P_{H,1}P_{T,2}$ and $P_{T,1}P_{H,2}$, that have the same probability, P_HP_T. Completely equivalently, we can say that the reason for multiplying the P_HP_T-term by two is that there are two permutations, HT and TH, which correspond to one head

Table 1.				
Problems	**Correspondences**			
Sequences of probability factors in the total probability sum	$P_{H,1}P_{H,2}$	$P_{H,1}P_{T,2}$	$P_{T,1}P_{H,2}$	$P_{T,1}P_{T,2}$
Probability factors for coins distinguished by identification numbers	$P_H P_H$	$P_H P_T$	$P_T P_H$	$P_T P_T$
Sequences from toss of a single coin	HH	HT	TH	TT
Assignments of two distinguishable objects to two categories	Cup H holds slugs 1 & 2	Cup H holds slug 1 Cup T holds slug 2	Cup H holds slug 2 Cup T holds slug 1	Cup T holds slugs 1 & 2

and one tail in successive tosses of a single coin.

We have lavished considerable attention on four related but very simple problems. Now, we want to extend this analysis—first to tosses of multiple coins and then to situations in which multiple outcomes are possible for each of many independent events. Eventually we will find that understanding these problems enables us to build a model for the behavior of molecules that explains the observations of classical thermodynamics.

If we extend our analysis to tossing n coins, which we label coins 1, 2, *etc.*, we find:

$$\begin{aligned} 1 &= (P_{H,1} + P_{T,1})(P_{H,2} + P_{T,2}) \dots (P_{H,n} + P_{T,n}) \\ &= (P_{H,1}P_{H,2} \dots P_{H,n}) + (P_{H,1}P_{H,2} \dots P_{H,i} \dots P_{T,n}) + \cdots \\ &\quad + (P_{T,1}P_{T,2} \dots P_{T,i} \dots P_{T,n}) \end{aligned}$$

We write each of the product terms in this expanded representation of the total-probability sum with the second index, r, increasing from 1 to n as we read through the factors, $P_{X,r}$, from left to right. Just as for tossing only two coins:

- Each product term is a sequence of probability factors that appears in the total probability sum.
- Each product term corresponds to a possible outcome from tossing n coins that are distinguished from one another by identification numbers.
- Each product term is equivalent to a possible outcome from repeated tosses of a single coin: the rth factor is P_H or P_T according as the rth toss produces a head or a tail.
- Each product term is equivalent to a possible assignment of n distinguishable objects to the two categories H and T.

In §3-9, we introduce the term ***population set*** to denote a set of numbers that represents a possible combination of outcomes. Here the possible combinations of outcomes are the numbers of heads and tails. If in five tosses we obtain 3 heads and 2 tails, we say that this group of outcomes belongs to the population set {3,2}. If in n tosses, we obtain n_H heads and n_T tails, this group of outcomes belongs to the population set $\{n_H, n_T\}$. For five tosses, the possible population sets are {5,0}, {4,1}, {3,2}, {2,3}, {1,4}, and {5,0}. Beginning in the next chapter, we focus on the energy levels that are available to a set of particles and on the number of particles that has each of the available energies. Then the number of particles, N_i, that have energy ϵ_i is the population of the ϵ_i-energy level. The set of all such numbers is the energy-level population set for the set of particles.

If we cannot distinguish one coin from another, the sequence $P_{H,1}P_{T,2}P_{H,3}P_{H,4}$ becomes $P_H P_T P_H P_H$. We say that $P_H P_T P_H P_H$ is distinguishable from $P_H P_H P_T P_H$ because the tails-outcome appears in the second position in $P_H P_T P_H P_H$ and in the third position in $P_H P_H P_T P_H$. We say that $P_{H,1}P_{T,2}P_{H,3}P_{H,4}$ and $P_{H,3}P_{T,2}P_{H,1}P_{H,4}$are indistinguishable, because both become $P_H P_T P_H P_H$. In general, many terms in the expanded form of the total probability sum belong to the population set corresponding to n_H heads and n_T tails. Each such term corresponds to a distinguishable permutation of n_H heads and n_T tails and the corresponding distinguishable permutation of P_H and P_T terms.

We use the notation $C(n_H, n_T)$ to denote the number of terms in the expanded form of the total probability sum in which there are n_H heads and n_T tails. $C(n_H, n_T)$ is also the number of distinguishable permutations of n_H heads and n_T tails or of n_H P_H-terms and n_T P_T-terms. The principal goal of our analysis is to find a general formula for $C(n_H, n_T)$. To do so, we make use of the fact that $C(n_H, n_T)$ is also the number of ways that we can assign n objects (coins) to two categories (heads or tails) in such a way that n_H objects are in one category (heads) and n_T objects are in the other category (tails). We also call $C(n_H, n_T)$ the number of ***combinations*** possible for distinguishable coins in the population set $\{n_H, n_T\}$.

The importance of $C(n_H, n_T)$ is evident when we recognize that, if we do not care about the sequence (permutation) in which a particular number of heads and tails occurs, we can represent the total-probability sum in a much compressed form:

$$\begin{aligned} 1 = P_H^n + nP_H^{n-1}P_T + \cdots + C(n_H, n_T)P_H^{n_H}P_T^{n_T} \\ + nP_H P_T^{n-1} + P_T^n \end{aligned}$$

In this representation, there are n terms in the total-probability sum that have $n_H = n-1$ and $n_T = 1$. These are the terms

$$P_{H,1}P_{H,2}P_{H,3} \dots P_{H,i} \dots P_{H,n-1}\boldsymbol{P_{T,n}}$$
$$P_{H,1}P_{H,2}P_{H,3} \dots P_{H,i} \dots \boldsymbol{P_{T,n-1}}P_{H,n}$$
$$P_{H,1}P_{H,2}P_{H,3} \dots \boldsymbol{P_{T,i}} \dots P_{H,n-1}P_{H,n}$$
$$\dots$$
$$P_{H,1}P_{H,2}\boldsymbol{P_{T,3}} \dots P_{H,i} \dots P_{H,n-1}P_{H,n}$$
$$P_{H,1}\boldsymbol{P_{T,2}}P_{H,3} \dots P_{H,i} \dots P_{H,n-1}P_{H,n}$$
$$\boldsymbol{P_{T,1}}P_{H,2}P_{H,3} \dots P_{H,i} \dots P_{H,n-1}P_{H,n}$$

Each of these terms represents the probability that $n-1$ heads and one tail will occur in the order shown. Each of these terms has the same value. Each of these terms is a distinguishable permutation of $n-1$ P_H terms and one P_T term. Each of these terms corresponds to a combination in which one of n numbered slugs is assigned to Cup T, while the remaining $n-1$ numbered slugs are assigned to Cup H. It is easy to see that there are n such terms, because each term is the product of n probabilities, and the tail can occur at any of the n positions in the product. If we do not care about the order in which heads and tails occur and are interested only in the value of the sum of these n terms, we can replace these n terms by the one term $nP_H^{n-1}P_T$. We see that $nP_H^{n-1}P_T$ is the probability of tossing $n-1$ heads and one tail, irrespective of which toss produces the tail.

There is another way to show that there must be n terms in the total-probability sum in which there are $n-1$ heads and one tail. This method relies on the fact that the number of such terms is the same as the number of combinations in which n distinguishable things are assigned to two categories, with $n-1$ of the things in one category and the remaining thing in the other category, $C(n-1,1)$. This method is a little more complicated, but it offers the great advantage that it can be generalized.

The new method requires that we think about all of the permutations we can create by reordering the results from any particular series of n tosses. To see what we have in mind when we say all of the permutations, let $P_{X,k}$ represent the probability of toss number k, where for the moment we do not care whether the outcome was a head or a tail. When we say all of the permutations, we mean the number of different ways we can order (permute) n different values $P_{X,k}$. It is important to recognize that one and only one of these permutations is a term in the total-probability sum, specifically:

$$P_{X,1}P_{X,2}P_{X,3} \dots P_{X,k} \dots P_{X,n}$$

in which the values of the second subscript are in numerical order. When we set out to construct all of these permutations, we see that there are n ways to choose the toss to put first and $n-1$ ways to choose the toss to put second, so there are $n(n-1)$ ways to choose the first two tosses. There are $n-2$ ways to choose the third toss, so there are $n(n-1)(n-2)$ ways to choose the first three tosses. Continuing in this way through all n tosses, we see that the total number of ways to order the results of n tosses is $n(n-1)(n-2)(n-3)\dots(3)(2)(1) = n!$

Next, we need to think about the number of ways we can permute n values $P_{X,k}$ if $n-1$ of them are $P_{H,1}$, $P_{H,2}, \dots, P_{H,r-1}, \dots, P_{H,r+1}, \dots, P_{H,n}$ and one of them is $P_{T,r}$, and we always keep the one factor $P_{T,r}$ in the same position. By the argument above, there are $(n-1)!$ ways to permute the values $P_{H,s}$ in a set containing $n-1$ members. So for every term (product of factors $P_{X,k}$) that occurs in the total-probability sum, there are $(n-1)!$ other products (other permutations of the same factors) that differ only in the order in which the $P_{H,s}$ appear. The single tail outcome occupies the same position in each of these permutations. If the rth factor in the term in the total probability sum is $P_{T,r}$, then $P_{T,r}$ is the rth factor in each of the $(n-1)!$ permutations of this term. This is an important point, let us repeat it in slightly different words: For every term that occurs in the total-probability sum, there are $(n-1)!$ permutations of the same factors that leave the heads positions occupied by heads and the tails position occupied by tails.

Equivalently, for every assignment of $n-1$ distinguishable objects to one of two categories, there are $(n-1)!$ permutations of these objects. There are $C(n-1,1)$ such assignments. Accordingly, there are a total of $(n-1)!\,C(n-1,1)$ permutations of the n distinguishable objects. Since we also know that the total number of permutations of n distinguishable objects is $n!$, we have

$$n! = (n-1)!\,C(n-1,1)$$

so that

$$C(n-1,1) = \frac{n!}{(n-1)!}$$

which is the same result that we obtained by our first and more obvious method.

The distinguishable objects within a category in a particular assignment can be permuted. We give these ***within-category permutations*** another name; we call them ***indistinguishable permutations***. (This terminology reflects our intended application, which is to find the number of ways n identical molecules can be assigned to a set of energy levels. We can tell two isolated molecules of the same substance apart only if they have different energies. We can distinguish molecules in different energy levels from one another. We cannot distinguish two molecules in the same energy level from one another. Two different permutations of the molecules within any one energy level are indistinguishable from one another.) For every term in the expanded representation of the total probability sum, indistinguishable permutations can be obtained by exchanging P_H factors with one another, or by exchanging P_T factors with one another, but not by exchanging P_H factors with P_T factors. That is, heads are exchanged with heads; tails are exchanged with tails; but heads are not exchanged with tails.

Now we can consider the general case. We let

$C(n_H, n_T)$ be the number of terms in the total-probability sum in which there are n_H heads and n_T tails. We want to find the value of $C(n_H, n_T)$. Let's suppose that one of the terms with n_H heads and n_T tails is

$$(P_{H,a}P_{H,b} \dots P_{H,m})(P_{T,r}P_{T,s} \dots P_{T,z})$$

where there are n_H indices in the set $\{a, b, \dots, m\}$ and n_T indices in the set $\{r, s, \dots, z\}$. There are $n_H!$ ways to order the heads outcomes and $n_T!$ ways to order the tails outcomes. So, there are $n_H!\,n_T!$ possible ways to order n_H heads and n_T tails outcomes. This is true for any sequence in which there are n_H heads and n_T tails; there will always be $n_H!\,n_T!$ permutations of n_H heads and n_T tails, whatever the order in which the heads and tails appear. This is also true for every term in the total-probability sum that contains n_H heads factors and n_T tails factors. The number of such terms is $C(n_H, n_T)$. For every such term, there are $n_H!\,n_T!$ permutations of the same factors that leave the heads positions occupied by heads and the tails positions occupied by tails.

Accordingly, there are a total of $n_H!\,n_T!\,C(n_H, n_T)$ permutations of the n distinguishable objects. The total number of permutations of n distinguishable objects is $n!$, so that

$$n! = n_H!\,n_T!\,C(n_H, n_T)$$

and

$$C(n_H, n_T) = \frac{n!}{n_H!\,n_T!}$$

Equivalently, we can construct a sum of terms, R, in which the terms are all of the $n!$ permutations of $P_{H,i}$ factors for n_H heads and $P_{T,j}$ factors for n_T tails. The value of each term in R is $P_H^{n_H} P_T^{n_T}$. So we have

$$R = n!\,P_H^{n_H} P_T^{n_T}$$

R contains all $C(n_H, n_T)$ of the $P_H^{n_H} P_T^{n_T}$-valued terms that appear in the total-probability sum. For each of these $P_H^{n_H} P_T^{n_T}$-valued terms there are $n_H!\,n_T!$ indistinguishable permutations that leave heads positions occupied by heads and tails positions occupied by tails. R will also contain all of the $n_H!\,n_T!$ permutations of each of these $P_H^{n_H} P_T^{n_T}$-valued terms. That is, every term in R is either a term in the expanded representation of the total probability sum or an indistinguishable permutation of such a term. It follows that R is also given by

$$R = n_H!\,n_T!\,C(n_H, n_T) P_H^{n_H} P_T^{n_T}$$

Equating these equations for R, we have

$$n!\,P_H^{n_H} P_T^{n_T} = n_H!\,n_T!\,C(n_H, n_T) P_H^{n_H} P_T^{n_T}$$

and, again,

$$C(n_H, n_T) = \frac{n!}{n_H!\,n_T!}$$

In summary: The total number of permutations is $n!$ The number of combinations of n distinguishable things in which n_H of them are assigned to category H and $n_T = n - n_H$ are assigned to category T is $C(n_H, n_T)$. (Every combination is a distinguishable permutation.) The number of indistinguishable permutations of the objects in each such combination is $n_H!\,n_T!$. The relationship among these quantities is

total number of permutations =

(number of distinguishable combinations) ×

(number of indistinguishable permutations for each distinguishable combination)

We noted earlier that $C(n_H, n_T)$ is the formula for the binomial coefficients. If we do not care about the order in which the heads and tails arise, the probability of tossing n_T tails and $n_H = n - n_T$ heads is

$$C(n_H, n_T) P_H^{n_H} P_T^{n_T} = \left(\frac{n!}{n_H!\,n_T!}\right) P_H^{n_H} P_T^{n_T}$$

and the sum of such terms for all $n + 1$ possible values of n_T in the interval $0 \le n_T \le n$ is the total probability for all possible outcomes from n tosses of a coin. This total probability must be unity. That is, we have

$$\begin{aligned} 1 &= (P_H + P_T)^n \\ &= \sum_{n_T=0}^{n} C(n_H, n_T) P_H^{n_H} P_T^{n_T} \\ &= \sum_{n_T=0}^{n} \left(\frac{n!}{n_H!\,n_T!}\right) P_H^{n_H} P_T^{n_T} \end{aligned}$$

For an unbiased coin, $P_H = P_T = 1/2$, and $P_H^{n_H} P_T^{n_T} = (1/2)^n$, for all n_T. This means that the probability of tossing n_H heads and n_T tails is proportional to $C(n_H, n_T)$ where the proportionality constant is $(1/2)^n$. The probability of $n^{\blacksquare}$ heads and $n - n^{\blacksquare}$ tails is the same as the probability of $n - n^{\blacksquare}$ heads and $n^{\blacksquare}$ tails.

Nothing in our development of the equation for the total probability requires that we set $P_H = P_T$, and in fact, the binomial probability relationship applies to any situation in which there are repeated trials, where each trial has two possible outcomes, and where the probability of each outcome is constant. If $P_H \neq P_T$, the symmetry observed for tossing coins does not apply, because

$$P_H^{n-n^{\blacksquare}} P_T^{n^{\blacksquare}} \neq P_H^{n^{\blacksquare}} P_T^{n-n^{\blacksquare}}$$

This condition corresponds to a biased coin.

Another example is provided by a spinner mounted at the center of a circle painted on a horizontal surface. Suppose that a pie-shaped section accounting for 25% of the circle's area is painted white and the rest is painted black. If the spinner's stopping point is unbiased, it will stop in the white zone with probability $P_W = 0.25$ and in the black zone with probability $P_B = 0.75$. After n spins, the probability of n_W white outcomes and n_B black outcomes is

$$\left(\frac{n!}{n_W!\,n_B!}\right)(0.25)^{n_W}(0.75)^{n_B}$$

After n spins, the sum of the probabilities for all possible combinations of white and black outcomes is

$$\begin{aligned}1 &= (P_W + P_B)^n\\ &= \sum_{n_B=0}^{n} C(n_W, n_B) P_W^{n_W} P_B^{n_B}\\ &= \sum_{n_B=0}^{n} \left(\frac{n!}{n_W!\,n_B!}\right) P_W^{n_W} P_B^{n_B}\\ &= \sum_{n_B=0}^{n} \left(\frac{n!}{n_W!\,n_B!}\right)(0.25)^{n_W}(0.75)^{n_B}\end{aligned}$$

§2 Distribution of results for multiple trials with three possible outcomes

Let us extend the ideas we have developed for binomial probabilities to the case where there are three possible outcomes for any given trial. To be specific, suppose we have a coin-sized object in the shape of a truncated right-circular cone, whose circular faces are parallel to each other. The circular faces have different diameters. When we toss such an object, allowing it to land on a smooth hard surface, it can wind up resting on the big circular face (***H***eads), the small circular face (***T***ails), or on the conical surface (***C***one-side). Let the probabilities of these outcomes in a single toss be P_H, P_T, and P_C, respectively. In general, we expect these probabilities to be different from one another; although, of course, we require $1 = (P_H + P_T + P_C)$.

Following our development for the binomial case, we want to write an equation for the total probability sum after n tosses. Let n_H, n_T, and n_C be the number of H, T, and C outcomes exhibited in $n_H + n_T + n_C = n$ trials. We let the probability coefficients be $C(n_H, n_T, n_C)$. The probability of n_H, n_T, n_C outcomes in n trials is

$$C(n_H, n_T, n_C) P_H^{n_H} P_T^{n_T} P_C^{n_C}$$

and the total probability is

$$\begin{aligned}1 &= (P_H + P_T + P_C)^n\\ &= \sum_{n_H, n_T, n_C} C(n_H, n_T, n_C) P_H^{n_H} P_T^{n_T} P_C^{n_C}\end{aligned}$$

where the summation is to be carried out over all combinations of integer values for n_H, n_T, and n_C, consistent with $n_H + n_T + n_C = n$.

To find $C(n_H, n_T, n_C)$, we proceed as before. We suppose that one of the terms with n_H heads, n_T tails, and n_C cone-sides is

$$(P_{H,a}P_{H,b}\dots P_{H,f})(P_{T,g}P_{T,h}\dots P_{T,m})(P_{C,p}P_{C,q}\dots P_{C,z})$$

where there are n_H indices in the set $\{a, b, \dots, f\}$, n_T indices in the set $\{g, h, \dots, m\}$, and n_C indices in the set {p, q,…, z}. There are $n_H!$ ways to order the heads outcomes, $n_T!$ ways to order the tails outcomes, and $n_C!$ ways to order the cone-sides outcomes. So, there are $n_H!\,n_T!\,n_C!$ possible ways to order n_H heads, n_T tails, and n_C cone-sides. There will also be $n_H!\,n_T!\,n_C!$ indistinguishable permutations of any combination (particular assignment) of n_H heads, n_T tails, and n_C cone-sides. There are $n!$ possible permutations of n probability factors and $C(n_H, n_T, n_C)$ distinguishable combinations with n_H heads, n_T tails, and n_C cone-sides. As before, we have

total number of permutations =

(number of distinguishable combinations) ×

(number of indistinguishable permutations for each distinguishable combination)

so that

$$n! = n_H!\,n_T!\,n_C!\,C(n_H, n_T, n_C)$$

and hence,

$$C(n_H, n_T, n_C) = \frac{n!}{n_H!\,n_T!\,n_C!}$$

Equivalently, we can construct a sum of terms, S, in which the terms are all of the $n!$ permutations of $P_{H,r}$ factors for n_H heads, $P_{T,s}$ factors for n_T tails, and $P_{C,t}$ factors for n_C cone-sides. The value of each term in S will be $P_H^{n_H} P_T^{n_T} P_C^{n_C}$. Thus, we have

$$S = n!\,P_H^{n_H} P_T^{n_T} P_C^{n_C}$$

S will contain all $C(n_H, n_T, n_C)$ of the distinguishable combinations n_H heads, n_T tails, and n_C cone-sides outcomes that give rise to $P_H^{n_H} P_T^{n_T} P_C^{n_C}$-valued terms. Moreover, S will also include all of the $n_H!\,n_T!\,n_C!$ indistinguishable permutations of each of these $P_H^{n_H} P_T^{n_T} P_C^{n_C}$-valued terms, and we also have

$$S = n_H!\,n_T!\,n_C!\,C(n_H, n_T, n_C) P_H^{n_H} P_T^{n_T} P_C^{n_C}$$

Equating these two expressions for S gives us the number of $P_H^{n_H} P_T^{n_T} P_C^{n_C}$-valued terms in the total-probability product, $C(n_H, n_T, n_C)$. That is,

$$\begin{aligned}S &= n!\,P_H^{n_H} P_T^{n_T} P_C^{n_C}\\ &= n_H!\,n_T!\,n_C!\,C(n_H, n_T, n_C) P_H^{n_H} P_T^{n_T} P_C^{n_C}\end{aligned}$$

and, again,

$$C(n_H, n_T, n_C) = \frac{n!}{n_H!\, n_T!\, n_C!}$$

In the special case that $P_H = P_T = P_C = 1/3$, all of the products $P_H^{n_H} P_T^{n_T} P_C^{n_C}$ will have the value $(1/3)^n$. Then the probability of any set of outcomes, $\{n_H, n_T, n_C,\}$, is proportional to $C(n_H, n_T, n_C)$ with the proportionality constant $(1/3)^n$.

§3 Distribution of results for multiple trials with many possible outcomes

It is now easy to extend our results to multiple trials with any number of outcomes. Let the outcomes be A, B, C, …., Z, for which the probabilities in a single trial are P_A, P_B, P_C,…P_Z. We again want to write an equation for the total probability after n trials. We let n_A, n_B, n_C,…n_Z be the number of A, B, C,…, Z outcomes exhibited in $n_A + n_B + n_C + \ldots + n_Z = n$ trials. If we do not care about the order in which the outcomes are obtained, the probability of n_A, n_B, n_C,…, n_Z outcomes in n trials is

$$C(n_A, n_B, n_C, \ldots, n_Z) P_A^{n_A} P_B^{n_B} P_C^{n_C} \ldots P_Z^{n_Z}$$

and the total probability sum is

$$1 = (P_A + P_B + P_C + \cdots + P_Z)^n$$

$$= \sum_{n_I} C(n_A, n_B, n_C, \ldots, n_Z) P_A^{n_A} P_B^{n_B} P_C^{n_C} \ldots P_Z^{n_Z}$$

where the summation is to be carried out over all combinations of integer values for n_A, n_B, n_C,…, n_Z consistent with $n_A + n_B + n_C + \ldots + n_Z = n$.

Let one of the terms for n_A A-outcomes, n_B B-outcomes, n_C C-outcomes, …, n_Z Z-outcomes, be

$$\left(P_{A,a} P_{A,b} \ldots P_{A,f}\right)\left(P_{B,g} P_{B,h} \ldots P_{B,m}\right) \times \left(P_{C,p} P_{C,q} \ldots P_{C,t}\right) \ldots \left(P_{Z,u} P_{Z,v} \ldots P_{Z,z}\right)$$

where there are n_A indices in the set $\{a, b, \ldots, f\}$, n_B indices in the set $\{g, h, \ldots, m\}$, n_C indices in the set $\{p, q, \ldots, t\}$, …, and n_Z indices in the set $\{u, v, \ldots, z\}$. There are $n_A!$ ways to order the A-outcomes, $n_B!$ ways to order the B-outcomes, $n_C!$ ways to order the C-outcomes, …, and $n_Z!$ ways to order the Z-outcomes. So, there are $n_A!\, n_B!\, n_C! \ldots n_Z!$ ways to order n_A A-outcomes, n_B B-outcomes, n_C C-outcomes, …, and n_Z Z-outcomes. The same is true for any other distinguishable combination; for every distinguishable combination belonging to the population set $\{n_A,\ n_B,\ n_C, \ldots,\ n_Z\}$ there are $n_A!\, n_B!\, n_C! \ldots n_Z!$ indistinguishable permutations. Again, we can express this result as the general relationship:

total number of permutations =

(number of distinguishable combinations) ×

(number of indistinguishable permutations for each distinguishable combination)

so that

$$n! = n_A!\, n_B!\, n_C! \ldots n_Z!\, C(n_A, n_B, n_C, \ldots, n_Z)$$

and

$$C(n_A, n_B, n_C, \ldots, n_Z) = \frac{n!}{n_A!\, n_B!\, n_C! \ldots n_Z!}$$

Equivalently, we can construct a sum, T, in which we add up all of the $n!$ permutations of $P_{A,a}$ factors for n_A A-outcomes, $P_{B,b}$ factors for n_B B-outcomes, $P_{C,c}$ factors for n_C C-outcomes, …, and $P_{Z,z}$ factors for n_Z Z-outcomes. The value of each term in T will be $P_A^{n_A} P_B^{n_B} P_C^{n_C} \ldots P_Z^{n_Z}$. So we have

$$T = n!\, P_A^{n_A} P_B^{n_B} P_C^{n_C} \ldots P_Z^{n_Z}$$

T will contain all $C(n_A, n_B, n_C, \ldots, n_Z)$ of the $P_A^{n_A} P_B^{n_B} P_C^{n_C} \ldots P_Z^{n_Z}$-valued products (distinguishable combinations) that are a part of the total-probability sum. Moreover, T will also include all of the $n_A!\, n_B!\, n_C! \ldots n_Z!$ indistinguishable permutations of each of these $P_A^{n_A} P_B^{n_B} P_C^{n_C} \ldots P_Z^{n_Z}$-valued products. Then we also have

$$T = n_A!\, n_B!\, n_C! \ldots n_Z!\, C(n_A, n_B, n_C, \ldots, n_Z) \times P_A^{n_A} P_B^{n_B} P_C^{n_C} \ldots P_Z^{n_Z}$$

Equating these two expressions for T gives us the number of $P_A^{n_A} P_B^{n_B} P_C^{n_C} \ldots P_Z^{n_Z}$-valued products

$$n!\, P_A^{n_A} P_B^{n_B} P_C^{n_C} \ldots P_Z^{n_Z} = n_A!\, n_B!\, n_C! \ldots n_Z! \times C(n_A, n_B, n_C, \ldots, n_Z) P_A^{n_A} P_B^{n_B} P_C^{n_C} \ldots P_Z^{n_Z}$$

and hence,

$$C(n_A, n_B, n_C, \ldots, n_Z) = \frac{n!}{n_A!\, n_B!\, n_C! \ldots n_Z!}$$

In the special case that $P_A = P_B = P_C = \cdots = P_Z$, all of the products $P_A^{n_A} P_B^{n_B} P_C^{n_C} \ldots P_Z^{n_Z}$ have the same value. Then, the probability of any set of outcomes, $\{n_A, n_B, n_C, \ldots, n_Z\}$, is proportional to $C(n_A, n_B, n_C, \ldots, n_Z)$.

§4 Stirling's approximation

The polynomial coefficient, C, is a function of the factorials of large numbers. Since $N!$ quickly becomes very large as N increases, it is often impractical to evaluate $N!$ from the definition,

$$N! = (N)(N-1)(N-2) \ldots (3)(2)(1)$$

Fortunately, an approximation, known as ***Stirling's formula*** or ***Stirling's approximation*** is available. Stirling's approximation is a product of factors. Depending on the application and the required accuracy, one or two of these factors can often be taken as unity. Stirling's approximation is

$$N! \approx N^N(2\pi N)^{1/2}\exp(-N)\exp\left(\frac{1}{12N}\right)$$
$$\approx N^N(2\pi N)^{1/2}\exp(-N)$$
$$\approx N^N\exp(-N)$$

In many statistical thermodynamic arguments, the important quantity is the natural logarithm of $N!$ or its derivative, $d\ln N!/dN$. In such cases, the last version of Stirling's approximation is usually adequate, even though it affords a rather poor approximation for $N!$ itself.

Problems

1. Leland got a train set for Christmas. It came with seven rail cars. (We say that all seven cars are "distinguishable.") Four of the rail cars are box cars and three are tank cars. If we distinguish between permutations in which the box cars are coupled (lined up) differently but not between permutations in which tank cars are coupled differently, how many ways can the seven cars be coupled so that all of the tank cars are together? What are they? What formula can we use to compute this number? (Hint: We can represent one of the possibilities as $b_1b_2b_3b_4T$. This is one of the possibilities in which the first four cars behind the engine are all box cars. There are 4! such possibilities; that is, there are 4! possible permutations for placing the four box cars.)

2. If we don't care about the order in which the box cars are coupled, and we don't care about the order in which the tank cars are coupled, how many ways can the rail cars in problem 1 be coupled so that all of the tank cars are together? What are they? What formula can we use to compute this number?

3. If we distinguish between permutations in which either the box cars or the tank cars in problem 1 are ordered differently, how many ways can the rail cars be coupled so that all of the tank cars are together? What formula can we use to compute this number?

4. How many ways can all seven rail cars in problem 1 be coupled if the tank cars need not be together?

5. If, as in the previous problem, we distinguish between permutations in which any of the rail cars are ordered differently, how many ways can the rail cars be coupled so that not all of the tank cars are together?

6. If we distinguish between box cars and tank cars, but we do not distinguish one box car from another box car, and we do not distinguish one tank car from another tank car, how many ways can the rail cars in problem 1 be coupled?

7. If Leland gets five flat cars for his birthday, he will have four box cars, three tank cars and five flat cars. How many ways will Leland be able to couple (permute) these twelve rail cars?

8. If we distinguish between box cars and tank cars, between box cars and flat cars, and between tank cars and flat cars, but we do not distinguish one box car from another box car, and we do not distinguish one tank car from another tank car, and we do not distinguish one flat car from another flat car, how many ways can the rail cars in problem seven be coupled? What formula can we use to compute this number?

9. We are given four distinguishable marbles, labeled $A-D$, and two cups, labeled 1 and 2. We want to explore the number of ways we can put two marbles in cup 1 and two marbles in cup 2. This is the number of combinations, $C(2,2)$, for the population set $N_1 = 2$, $N_2 = 2$.

(a) One combination is $[AB]_1[CD]_2$. Find the remaining combinations. What is $C(2,2)$?

(b) There are four permutations for the combination given in (a): $[AB]_1[CD]_2$; $[BA]_1[CD]_2$; $[AB]_1[DC]_2$; $[BA]_1[DC]_2$. Find all of the permutations for each of the remaining combinations.

(c) How many permutations are there for each combination?

(d) Write down all of the possible permutations of marbles $A-D$. Show that there is a one-to-one correspondence with the permutations in (b).

(e) Show that the total number of permutations is equal to the number of combinations times the number of permutations possible for each combination.

10. We are given seven distinguishable marbles, labeled $A-G$, and two cups, labeled 1 and 2. We want to find the number of ways we can put three marbles in cup 1 and four marbles in cup 2. That is, we seek $C(3,4)$, the number of combinations in which $N_1 = 3$ and $N_2 = 4$. $[ABC]_1[DEFG]_2$ is one such combination.

(a) How many different ways can these marbles be placed in different orders without exchanging any marbles between cup 1 and cup 2? (This is the number of permutations associated with this combination.)

(b) Find a different combination with $N_1 = 3$ and $N_2 = 4$.

(c) How many permutations are possible for the marbles in (b)? How many permutations are possible for any combination with $N_1 = 3$ and $N_2 = 4$?

(d) If $C(3,4)$ is the number of combinations in which $N_1 = 3$ and $N_2 = 4$, and if P is the number of permutations for each such combination, what is the total number of permutations possible for 7 marbles?

(e) How else can one express the number of permutations possible for 7 marbles?

(f) Equate your conclusions in (d) and (e). Find $C(3,4)$.

11.
(a) Calculate the probabilities of 0, 1, 2, 3, and 4 heads in a series of four tosses of an unbiased coin. The event of 2 heads is 20% of these five events. Note particularly the probability of the event: 2 heads in 4 tosses.
(b) Calculate the probabilities of 0, 1, 2, 3,…, 8, and 9 heads in a series of nine tosses of an unbiased coin. The events of 4 heads and 5 heads comprise 20% of these ten cases. Calculate the probability of 4 heads or 5 heads; i.e., the probability of being in the middle 20% of the possible events.
(c) Calculate the probabilities of 0, 1, 2, 3,…, 13, and 14 heads in a series of fourteen tosses of an unbiased coin. The events of 6 heads, 7 heads, and 8 heads comprise 20% of these fifteen cases. Calculate the probability of 6, 7, or 8 heads; i.e., the probability of being in the middle 20% of the possible events.
(d) What happens to the probabilities for the middle 20% of possible events as the number of tosses becomes very large? How does this relate to the fraction heads in a series of tosses when the total number of tosses becomes very large?

12. Let the value of the outcome heads be one and the value of the outcome tails be zero. Let the "score" from a particular simultaneous toss of n coins be

$$\text{score} = 1 \times \left(\frac{number\ of\ heads}{number\ of\ coins}\right) + 0 \times \left(\frac{number\ of\ tails}{number\ of\ coins}\right)$$

Let us refer to the distribution of scores from tosses of n coins as the "S_n distribution."
(a) The S_1 distribution comprises two outcomes: {1 head, 0 tail} and {0 head, 1 tail}.
What is the mean of the S_1 distribution?
(b) What is the variance of the S_1 distribution?
(c) What is the mean of the S_n distribution?
(d) What is the variance of the S_n distribution?

13. Fifty unbiased coins are tossed simultaneously.
(a) Calculate the probability of 25 heads and 25 tails.
(b) Calculate the probability of 23 heads and 27 tails.
(c) Calculate the probability of 3 heads and 47 tails.
(d) Calculate the ratio of your results for parts (a) and (b).
(e) Calculate the ratio of your results for parts (a) and (c).

14. For $N = 3, 6$ and 10, calculate
(a) The exact value of $N!$
(b) The value of $N!$ according to the approximation
$$N! \approx N^N (2\pi N)^{1/2} \exp(-N) \exp\left(\frac{1}{12N}\right)$$
(c) The value of N! according to the approximation
$$N! \approx N^N (2\pi N)^{1/2} \exp(-N)$$
(d) The value of N! according to the approximation
$$N! \approx N^N \exp(-N)$$
(e) The ratio of the value in (b) to the corresponding value in (a).
(f) The ratio of the value in (c) to the corresponding value in (a).
(g) The ratio of the value in (d) to the corresponding value in (a).
(h) Comment.

15. Find, $d \ln N!/dN$ using each of the approximations
$$N! \approx N^N (2\pi N)^{1/2} \exp(-N) \exp\left(\frac{1}{12N}\right)$$
$$\approx N^N (2\pi N)^{1/2} \exp(-N)$$
$$\approx N^N \exp(-N)$$
How do the resulting approximations for $d \ln N!/dN$ compare to one another as N becomes very large?

16. There are three energy levels available to any one molecule in a crystal of the substance. Consider a crystal containing 1000 molecules. These molecules are distinguishable because each occupies a unique site in the crystalline lattice. How many combinations (microstates) are associated with the population set $N_1 = 800$, $N_2 = 150$, $N_3 = 50$?

20

Boltzmann Statistics

§1 The independent-molecule approximation

In Chapter 18, our survey of quantum mechanics introduces the idea that a molecule can have any of an infinite number of discrete energies, which we can put in order starting with the smallest. We now turn our attention to the properties of a system composed of a large number of molecules. This multi-molecule system must obey the laws of quantum mechanics. Therefore, there exists a Schroedinger equation, whose variables include all of the inter-nucleus, inter-electron, and electron-nucleus distance and potential terms in the entire multi-molecule system. The relevant boundary conditions apply at the physical boundaries of the macroscopic system. The solutions of this equation include a set of infinitely many wave functions, $\Psi_{i,j}$, each describing a quantum mechanical state of the entire multi-molecule system. In general, the collection of elementary particles that can be assembled into a particular multi-molecule system can also be assembled into many other multi-molecule systems. For example, an equimolar mixture of CO and H_2O can be reassembled into a system comprised of equimolar CO_2 and H_2, or into many other systems containing mixtures of CO, H_2O, CO_2, and H_2. Infinitely many quantum-mechanical states are available to each of these multi-molecule systems.

For every such multi-molecule wave function, $\psi_{i,j}$, there is a corresponding system energy, E_i. In general, the system energy, E_i, is Ω_i-fold degenerate; there are Ω_i wave functions, $\Psi_{i,1}$, $\Psi_{i,2}$, ..., Ψ_{i,Ω_i}, whose energy is E_i. The wave functions include all of the interactions among the molecules of the system, and the energy levels of the system reflect all of these interactions. While generating and solving this multi-molecule Schroedinger equation is straightforward in principle, it is completely impossible in practice.

Fortunately, we can model multi-molecule systems in another way. The primary focus of chemistry is the study of the properties and reactions of molecules. Indeed, the science of chemistry exists, as we know it, only because the atoms comprising a molecule stick together more tenaciously than molecules stick to one another. (Where this is not true, we get macromolecular materials like metals, crystalline salts, *etc.*) This occurs because the energies that characterize the interactions of atoms within a molecule are much greater than the energies that characterize the interaction of one molecule with another. Consequently, the energy of the system can be viewed as the sum of two terms. One term is a sum of the energies that the component molecules would have if they were all infinitely far apart. The other term is a sum of the energies of all of the intermolecular interactions, which is the energy change that would occur if the molecules were brought from a state of infinite separation to the state of interest.

In principle, we can describe a multi-molecule system in this way with complete accuracy. This description has the advantage that it breaks a very large and complex problem into two smaller problems, one of which we have already solved: In Chapter 18, we see that we can approximate the quantum-mechanical description of a molecule and its energy levels by factoring molecular motions into translational, rotational, vibrational, and electronic components. It remains only to describe the intermolecular interactions. When intramolecular energies are much greater than intermolecular-interaction energies, it may be a good approximation to ignore the intermolecular interactions altogether. This occurs when we describe ideal gas molecules; in the limit that a gas behaves ideally, the force between any two of its molecules is nil.

In Chapter 23, we return to the idea of multi-molecule wave functions and energy levels. Meanwhile we assume that intermolecular interactions can be ignored. This is a poor approximation for many systems. However, it is a good approximation for many others, and it enables us to keep our description of the system simple while we use molecular properties in our development of the essential ideas of statistical thermodynamics.

We focus on developing a theory that gives the macroscopic thermodynamic properties of a pure substance in terms of the energy levels available to its individual molecules. To begin, we suppose that we solve the Schroedinger equation for an isolated molecule. In this Schroedinger equation, the variables include the inter-nucleus, inter-electron, and electron-nucleus distance and potential terms that are necessary to describe the molecule. The solutions are a set of infinitely many wave functions, $\psi_{i,j}$, each describing a different quantum-mechanical state of an isolated molecule. We refer to each of the possible wave functions as ***quantum state*** of the molecule. For every such wave function, there is a corresponding molecular energy, ϵ_i. Every unique molecular energy, ϵ_i, is called an ***energy level***. Several quantum states can have the same energy. When two or more quantum states have the same energy, we say that they belong to the same energy level, and the energy level is said to be ***degenerate***. In general, there are g_i quantum states that we can represent by the g_i wave functions, $\psi_{i,1}$, , $\psi_{i,2}$, ..., , ψ_{i,g_i}, each of whose energy is ϵ_i. The number of quantum states that have the same energy is called the ***degeneracy*** of the energy level. Figure 1 illustrates the terms we use to describe the quantum states and energy levels available to a molecule.

In our development of classical thermodynamics, we find it convenient to express the value of a thermodynamic property of a pure substance as the change that occurs during a formal process that forms one mole of the substance, in its standard state, from its unmixed constituent elements, in their standard states. In developing statistical thermodynamics, we find it convenient to express the value of a molecular energy, ϵ_i, as the change that occurs during a formal process that forms a molecule of the substance, in one of its quantum states, $\psi_{i,j}$, from its infinitely separated, stationary, constituent atoms. That is, we let the isolated constituent atoms be the reference state for the thermodynamic properties of a pure substance.

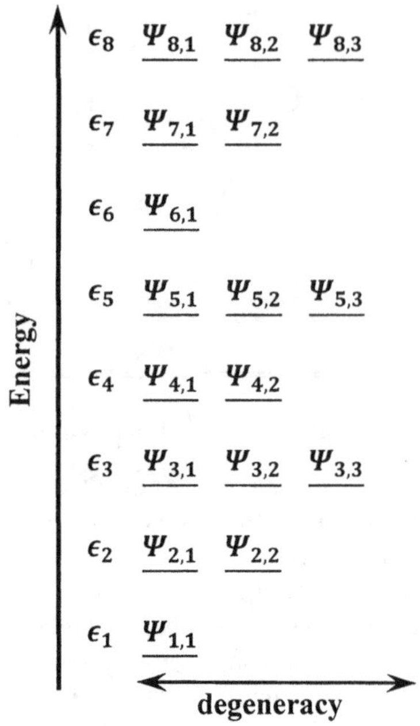

Figure 1. Quantum states and degenerate energy levels.

§2 The probability of an energy level at constant N, V, and T

If only pressure–volume work is possible, the state of a closed, reversible system can be specified by specifying its volume and temperature. Since the system is closed, the number, N, of molecules is constant. Let us consider a closed, equilibrated, constant-volume, constant-temperature system in which the total number of molecules is very large. Let us imagine that we can monitor the quantum state of one particular molecule over a very long time. Eventually, we are able to calculate the fraction of the elapsed time that the molecule spends in each of the quantum states. We label the available quantum states with the wave function symbols, $\psi_{i,j}$.

We assume that the fraction of the time that a molecule spends in the quantum state $\psi_{i,j}$ is the same thing as the probability of finding the molecule in quantum state $\psi_{i,j}$. We denote this probability as $\rho(\psi_{i,j})$. To develop the theory of statistical thermodynamics, we assume that this probability depends on the energy, and only on the energy, of the quantum state $\psi_{i,j}$. Consequently, any two quantum states whose energies are the same have the same probability, and the g_i-fold degenerate quantum states, $\psi_{i,j}$, whose energies are ϵ_i, all have the same probability. In our imaginary monitoring of the state of a particular molecule, we observe that the probabilities of two quantum states are the same if and only if their energies are the same; that is, we observe $\rho(\psi_{i,j}) = \rho(\psi_{k,m})$ if and only if $i = k$.

The justification for this assumption is that the resulting theory successfully models experimental observations. We can ask, however, why we might be led to make this assumption in the first place. We can reason as follows: The fact that we observe a definite value for the energy of the macroscopic system implies that quantum states whose energies are much greater than the average molecular energy must be less probable than quantum states whose energies are smaller. Otherwise, the sum of the energies of high-energy molecules would exceed the energy of the system. Therefore, we can reasonably infer that the probability of a quantum state depends on its energy. On the other hand, we can think of no plausible reason for a given molecule to prefer one quantum state to another quantum state that has the same energy.

This assumption means that a single function suffices to specify the probability of finding a given molecule in any quantum state, $\psi_{i,j}$, and the only independent variable is the quantum-state energy, ϵ_i. We denote the probability of a single quantum state, $\psi_{i,j}$, whose energy is ϵ_i, as $\rho(\epsilon_i)$. Since this is the probability of each of the g_i-fold degenerate quantum states, $\psi_{i,j}$, that have energy ϵ_i, the probability of finding a given molecule in any energy level, ϵ_i, is $P(\epsilon_i) = g_i\rho(\epsilon_i)$. We find it convenient to introduce "P_i" to abbreviate this probability; that is, we let

$$P_i = \sum_{j=1}^{g_i} \rho(\psi_{i,j}) = P(\epsilon_i) = g_i\rho(\epsilon_i)$$

(the probability of energy level ϵ_i)

There is a P_i for every energy level ϵ_i. P_i must be the same for any molecule, since every molecule has the same properties. If the population set $\{N_1^\bullet, N_2^\bullet, \ldots, N_i^\bullet, \ldots\}$ characterizes the equilibrium system, the fraction of the molecules that have energy ϵ_i is $N_i^\bullet/N$. (Elsewhere, an energy-level population set is often called a "distribution." Since we define a distribution somewhat differently, we avoid this usage.) Since the fraction of the molecules in an energy level at any instant of time is the same as the fraction of the time that one molecule spends in that energy level, we have

$$P_i = P(\epsilon_i) = g_i\rho(\epsilon_i) = \frac{N_i^\bullet}{N}$$

As long as the system is at equilibrium, this fraction is constant. In Chapter 21, we find an explicit equation for the probability function, $\rho(\epsilon_i)$.

The energy levels, ϵ_i, depend on the properties of the molecules. In developing Boltzmann statistics for non-interacting molecules, we assume that the probability of finding a molecule in a particular energy level is independent of the number of molecules present in the system. While P_i and $\rho(\epsilon_i)$ depend on the energy level, ϵ_i, neither depends on the number of molecules, N. If we imagine inserting a barrier that converts an equilibrated collection of molecules into two half-size collections, each of the new collections is still at equilibrium. Each contains half as many molecules and has half the total energy of the original. In our model, the fraction of the molecules in any given energy level remains constant. Consequently, the probabilities associated with each energy level remain constant. (In Chapter 25, we introduce Fermi-Dirac and Bose-Einstein statistics. When we must use either of these models to describe the system, P_i is affected by rules for the number of molecules that can occupy an energy level.)

The number of molecules and the total energy are extensive properties and vary in direct proportion to the size of the system. The probability, P_i, is an intensive variable that is a characteristic property of the macroscopic system. P_i is a state function. P_i depends on ϵ_i. So long as the thermodynamic variables that determine the state of the system remain constant, the ϵ_i are constant. For a given macroscopic system in which only pressure–volume work is possible, the quantum mechanical energy levels, ϵ_i, are constant so long as the system volume and temperature are constant. However, the ϵ_i are quantum-mechanical quantities that depend on our specification of the molecule and on the boundary values in our specification of the system. If we change any molecular properties or the dimensions of the system, the probabilities, P_i, change.

§3 The population sets of a system at equilibrium at constant N, V, and T

In developing Boltzmann statistics, we assume that we can tell different molecules of the same substance apart. We say that the molecules are ***distinguishable***. This assumption is valid for molecules that occupy lattice sites in a crystal. In a crystal, we can specify a particular molecule by specifying its position in the lattice. In other systems, we may be unable to distinguish between different molecules of the same substance. Most notably, we cannot distinguish between two molecules of the same substance in the gas phase. The fact that gas molecules are indistinguishable, while we assume otherwise in developing Boltzmann statistics, turns out to be a problem that is readily overcome. We discuss this in §24-2.

We want to model properties of a system that contains N, identical, distinguishable, non-interacting molecules. The solutions of the Schroedinger equation presume fixed boundary conditions. This means that the volume of this N-molecule system is constant. We assume also that the temperature of the N-molecule system is constant. Thus, our goal is a theory that predicts the properties of a system when N, V, and T are specified. When there are no intermolecular interactions, the energy of the system is just the sum of the energies of the individual molecules. If we know how the molecules are allocated among the energy levels, we can find the energy of the system. Letting N_i be the population of the energy level ϵ_i, any such allocation is a population set $\{N_1, N_2, \dots, N_i, \dots\}$. We have

$$N = \sum_{i=1}^{\infty} N_i$$

and the system energy is

$$E = \sum_{i=1}^{\infty} N_i\,\epsilon_i$$

Let us imagine that we can assemble a system with the molecules allocated among the energy levels in any way we please. Let $\{N_1^o, N_2^o, \dots, N_i^o, \dots\}$ represent an initial population set that describes a system that we assemble in this way. This population set corresponds to a well-defined system energy. We imagine immersing the container in a constant-temperature bath. Since the system can exchange energy with the bath, the molecules of the system gain or lose energy until the system attains the temperature of the bath in which it is immersed. As this occurs, the populations of the energy levels change. A series of different population sets characterizes the state of the system as it evolves toward thermal equilibrium. When the system reaches equilibrium, the population sets that characterize it are different from the initial one, $\{N_1^o, N_2^o, \dots, N_i^o, \dots\}$.

Evidently, the macroscopic properties of such a system also change with time. The changes in the macroscopic properties of the system parallel the changing energy-level populations. At thermal equilibrium, macroscopic properties of the system cease to undergo any further change. In §3-9, we introduce the idea that the most probable population set, which we denote as

$$\{N_1^\bullet, N_2^\bullet, \dots, N_i^\bullet, \dots\}$$

or its proxy,

$$\{NP(\epsilon_1), NP(\epsilon_2), \dots, NP(\epsilon_i), \dots\}$$

(where $N = N_1^\bullet + N_2^\bullet + \dots + N_i^\bullet + \dots$), is the best prediction we can make about the outcomes in a future set of experiments in which we find the energy of each of N different molecules at a particular instant. We hypothesize that the most probable population set specifies all of the properties of the macroscopic system in its equilibrium state. When we develop the logical consequences of this hypothesis, we find a theory that expresses macroscopic thermodynamic properties in terms of the energy levels available to individual molecules. In the end, the justification of this hypothesis is that it enables us to calculate thermodynamic properties that agree with experimental measurements made on macroscopic systems.

Our hypothesis asserts that the properties of the equilibrium state are the same as the properties of the system when it is described by the most probable population set. Evidently, we can predict the system's equilibrium state if we can find the equilibrium $N_i^{\bullet}$ values, and *vice versa*. To within an arbitrary factor representing its size, an equilibrated system can be completely described by its intensive properties. In the present instance, the fractions $N_1^{\bullet}/N, N_2^{\bullet}/N, \dots, N_i^{\bullet}/N, \dots$ describe the equilibrated system to within the factor, N, that specifies its size. Since we infer that $P_i = P(\epsilon_i) = N_i^{\bullet}/N$, the equilibrated system is also described by the probabilities $(P_1, P_2, \dots, P_i, \dots)$.

Our hypothesis does not assert that the most-probable population set is the only population set possible at equilibrium. A very large number of other population sets may describe an equilibrium system at different instants of time. However, when its state is specified by any such population set, the macroscopic properties of the system are indistinguishable from the macroscopic properties of the system when its state is specified by the most-probable population set. The most-probable population set characterizes the equilibrium state of the system in the sense that we can calculate the properties of the equilibrium state of the macroscopic system by using the single-molecule energy levels and the most probable population set—or its proxy. The relationship between a molecular energy level, ϵ_i, and its equilibrium population, $N_i^{\bullet}$, is called the ***Boltzmann equation***. From $P_i = N_i^{\bullet}/N$, we see that the Boltzmann equation specifies the probability of finding a given molecule in energy level ϵ_i.

Although we calculate thermodynamic properties from the most probable population set, the population set that describes the system can vary from instant to instant while the system remains at equilibrium. The central limit theorem enables us to characterize the amount of variation that can occur. When N is comparable to the number of molecules in a macroscopic system, the probability that variation among population sets can result in a macroscopically observable effect is vanishingly small. The hypothesis is successful because the most probable population set is an excellent proxy for any other population set that the equilibrium system is remotely likely to attain.

We develop the theory of statistical thermodynamics for N-molecule systems by considering the energy levels, ϵ_i, available to a single molecule that does not interact with other molecules. Thereafter, we develop a parallel set of statistical thermodynamic results by considering the energy levels, $\hat{E}_i$, available to a system of N molecules. These N-molecule-system energies can reflect the effects of any amount of intermolecular interaction. We can apply the same arguments to find that the Boltzmann equation also describes the equilibrium properties of systems in which intermolecular interactions are important. That is, the probability, $P_i(\hat{E}_i)$, that an N-molecule system has energy $\hat{E}_i$ is the same function of $\hat{E}_i$ as the molecular-energy probability, $P_i = P(\epsilon_i)$, is of ϵ_i.

When we finish our development based on single-molecule energy levels, we understand nearly all of the ideas that we need in order to complete the development for the energies of an N-molecule system. This development is an elegant augmentation of the basic argument called the ***ensemble treatment*** or the ***ensemble method***. The ensemble treatment is due to J. Willard Gibbs; we discuss it in Chapter 23. For now, we simply note that our approach involves no wasted effort. When we discuss the ensemble method, we use all of the ideas that we develop in this chapter and the next. The extension of these arguments that is required for the ensemble treatment is so straightforward as to be (almost) painless.

§4 How can infinitely many probabilities sum to unity?

There are an infinite number of successively greater energies for a quantum mechanical system. We infer that the probability that a given energy level is occupied is a property of the energy level. Each of the probabilities must satisfy $0 < P_i < 1$. When we sum the fixed probabilities associated with the energy levels, the sum contains an infinite number of terms. By the nature of probability, the sum of this infinite number of terms must be one:

$$\begin{aligned} 1 &= P_1 + P_2 + \cdots + P_i + \cdots \\ &= P(\epsilon_1) + P(\epsilon_2) + \cdots + P(\epsilon_i) + \cdots \\ &= \sum_{i=1}^{\infty} P(\epsilon_i) \end{aligned}$$

That is, the sum of the probabilities is an infinite series, which must converge: The sum of all of the occupancy probabilities must be unity. This can happen only if all later members of the series are very small. In the remainder of this chapter, we explore some of the thermodynamic ramifications of these facts. In the next chapter, we use this relationship to find the functional dependence of the P_i on the energy levels, ϵ_i. To obtain these results, we need to think further about the probabilities associated with the various population sets that can occur. Also, we need to introduce a new fundamental postulate.

To focus on the implications of this sum of probabilities, let us review geometric series. A geometric series is a sum of terms, in which each successive term is a multiple of its predecessor. A geometric series is an infinite sum that can converge:

$$T = a + ar + ar^2 + \cdots + ar^i \ldots$$
$$= a(1 + r + r^2 + \cdots + r^i + \cdots)$$
$$= a + a\sum_{i=1}^{\infty} r^i$$

Successive terms approach zero if $|r| < 1$. If $|r| \geq 1$, successive terms do not become smaller, and the sum does not have a finite limit. If $|r| \geq 1$, we say that the infinite series diverges.

We can multiply an infinite geometric series by its constant factor to obtain

$$rT = ar + ar^2 + ar^3 + \cdots + ar^i + \cdots$$
$$= a(r + r^2 + r^3 + \cdots + r^i + \cdots)$$
$$= a\sum_{i=1}^{\infty} r^i$$

If $|r| < 1$, we can subtract and find the value of the infinite sum:

$$T - rT = a$$

so that

$$T = a/(1-r)$$

In a geometric series, the ratio of two successive terms is $r^{n+1}/r^n = r$ The condition of convergence for a geometric series can also be written as

$$\left|\frac{r^{n+1}}{r^n}\right| < 1$$

We might anticipate that any other series also converges if its successive terms become smaller at least as fast as those of a geometric series. In fact, this is true and is the basis for the ***ratio test*** for convergence of an infinite series. If we represent successive terms in an infinite series as t_i, their sum is

$$T = \sum_{i=0}^{\infty} t_i$$

The ratio test is a theorem which states that the series converges, and T has a finite value, if

$$\lim_{n\to\infty}\left|\frac{t_{n+1}}{t_n}\right| < 1$$

One of our goals is to discover the relationship between the energy, ϵ_i, of a quantum state and the probability that a molecule will occupy one of the quantum states that have this energy, $P_i = g_i\rho(\epsilon_i)$. When we do so, we find that the probabilities for all of the quantum mechanical systems that we discuss in Chapter 18 satisfy the ratio test.

§5 The total probability sum at constant N, V, and T

In a collection of distinguishable independent molecules at constant N, V, and T, the probability that a randomly selected molecule has energy ϵ_i is P_i; we have $1 = P_1 + P_2 + \cdots + P_i + \cdots$. At any instant, every molecule in the N-molecule system has a specific energy, and the state of the system is described by a population set, $\{N_1, N_2, \ldots, N_i, \ldots.\}$, wherein N_i can have any value in the range $0 \leq N_i \leq N$, subject to the condition that

$$N = \sum_{i=1}^{\infty} N_i$$

The probabilities that we assume for this system of molecules have the properties we assume in Chapter 19 where we find the total probability sum by raising the sum of the energy-level probabilities to the Nth power.

$$1 = (P_1 + P_2 + \cdots + P_i + \cdots)^N$$
$$= \sum_{\{N_i\}} \frac{N!}{N_1!\, N_2! \ldots N_i! \ldots} P_1^{N_1} P_2^{N_2} \ldots P_i^{N_i} \ldots$$

The total-probability sum is over all possible population sets, $\{N_1, N_2, \ldots, N_i, \ldots.\}$,, which we abbreviate to $\{N_i\}$, in indicating the range of the summation. Each term in this sum represents the probability of the corresponding population set $\{N_1, N_2, \ldots, N_i, \ldots.\}$,. At any given instant, one of the possible population sets describes the way that the molecules of the physical system are apportioned among the energy levels. The corresponding term in the total probability sum represents the probability of this apportionment. It is not necessary that all of the energy levels be occupied. We can have $N_k = 0$, in which case $P_k^{N_k} = P_k^0 = 1$ and $N_k! = 1$. Energy levels that are not occupied have no effect on the probability of a population set. The unique population set

$$\{N_1^\bullet, N_2^\bullet, \ldots, N_i^\bullet, \ldots.\}$$

that we conjecture to characterize the equilibrium state is represented by one of the terms in this total probability sum. We want to focus on the relationship between a term in the total probability sum and the corresponding state of the physical system.

Each term in the total probability sum includes a probability factor, $P_1^{N_1} P_2^{N_2} \ldots P_i^{N_i} \ldots$ This factor is the probability that N_i molecules occupy each of the energy levels ϵ_i. This term is not affected by our assumption that the molecules are distinguishable. The probability factor is multiplied by the polynomial coefficient

$$\frac{N!}{N_1!\, N_2! \ldots N_i! \ldots}$$

This factor is the number of combinations of distinguishable molecules that arise from the population set $\{N_1, N_2, \ldots, N_i, \ldots\}$. It is the number of ways that the N distinguishable molecules can be assigned to the available energy levels so that N_1 of them are in energy level, ϵ_1, *etc.*

The combinations for the population set {3,2} are shown in Figure 2. The expression for the number of combinations takes the form it does only because the molecules can be distinguished from one another. To emphasize this point, let us find the number of combinations using the method we develop in Chapter 19. Briefly recapitulated, the argument is this:

Combination	ϵ_1	ϵ_2
1	ABC	DE
2	ABD	CE
3	ACD	BE
4	BCD	AE
5	ABE	CD
6	ACE	BD
7	BCE	AD
8	ADE	BC
9	BDE	AC
10	CDE	AB

There are 3!2!=12 permutations of the molecules for each of the 10 combinations. Hence, there are 120 permutations altogether. We also find 120 total permutations by the calculation 5!=120.

Figure 2. Combinations for the population set {3,2}.

- We can permute the N molecules in $N!$ ways. If we were to distinguish (as different combinations) any two permutations of all of the molecules, this would also be the number of combinations.

- In fact, however, we do not distinguish between different permutations of those molecules that are assigned to the same energy level. If the N_1 molecules assigned to the first energy level are $B, C, Q, \dots, X$, we do not distinguish the permutation $BCQ \dots X$ from the permutation $CBQ \dots X$ or from any other permutation of these N_1 molecules. Then the complete set of $N!$ permutations contains a subset of $N_1!$ permutations, all of which are equivalent because they have the same molecules in the first energy level. So the total number of permutations, $N!$, over-counts the number of combinations by a factor of $N_1!$ We can correct for this over-count by dividing by $N_1!$ That is, after correcting for the over-counting for the N_1 molecules in the first energy level, the number of combinations is $N!/N_1!$ (If all N of the molecules were in the first energy level, there would be only one combination. We would have $N = N_1$, and the number of combinations calculated from this formula would be $N!/N! = 1$, as required.)

- The complete set of $N!$ permutations also includes $N_2!$ permutations of the N_2 molecules in the second energy level. In finding the number of combinations, we want to include only one of these permutations, so correcting for the over-counting due to both the N_1 molecules in the first energy level and the N_2 molecules in the second energy level gives

$$\frac{N!}{N_1!\,N_2!}$$

- Continuing this argument through all of the occupied energy levels, we see that the total number of combinations is

$$C(N_1, N_2, \dots, N_i, \dots) = \frac{N!}{N_1!\,N_2!\dots N_i!\dots}$$

Because there are infinitely many energy levels and probabilities, P_i, there are infinitely many terms in the total-probability sum. Every energy available to the macroscopic system is represented by one or more terms in this total-probability sum. Since there is no restriction on the energy levels that can be occupied, there are an infinite number of such system energies. There is an enormously large number of terms each of which corresponds to an enormously large system energy. Nevertheless, the sum of all of these terms must be one. The P_i form a convergent series, and the total probability sum must sum to unity.

Just as the P_i series can converge only if the probabilities of high molecular energies become very small, so the total probability sum can converge only if the probabilities of high system energies become very small. If a population set has N_i molecules in the ith energy level, the probability of that population set is proportional to $P_i^{N_i}$. We see therefore, that the probability of a population set in which there are many molecules in high energy levels must be very small. Terms in the total probability sum that correspond to population sets with many molecules in high energy levels must be negligible. Equivalently, at a particular temperature, macroscopic states in which the system energy is anomalously great must be exceedingly improbable.

What terms in the total probability sum do we need to consider? Evidently from among the infinitely many terms that occur, we can select a finite subset whose sum is very nearly one. If there are many terms that are small and nearly equal to one another, the number of terms in this finite subset could be large. Nevertheless, we can see that terms in this subset must involve the largest possible P_i values raised to the smallest possible powers, N_i, consistent with the requirement that the N_i sum to N.

If an equilibrium macroscopic system could have only one population set, the probability of that population set would be unity. Could an equilibrium system be characterized by two or more population sets for appreciable fractions of an observation period? Would this require that the macroscopic system change its properties with time as it jumps from one population set to another? Evidently, it would not, since our observations of macroscopic systems show that the equilibrium properties are unique. A system that wanders between two (or more) macroscopically distinguishable states cannot be at equilibrium. We are forced to the conclusion that, if a macroscopic equilibrium system has multiple population sets with non-negligible probabilities, the macroscopic properties associated with each of these population sets must be indistinguishably similar. (The alternative is to abandon the theory, which is useful only if its microscopic description of a system makes useful predictions about the system's macroscopic behavior.)

To be a bit more precise about this, we recognize that our theory also rests on another premise: Any intensive macroscopic property of many independent molecules depends on the energy levels available to an individual molecule and the fraction of the molecules that populate each energy level. The average energy is a prime example. For the population set $\{N_1, N_2, \dots, N_i, \dots.\}$, the average molecular energy is

$$\overline{\epsilon} = \sum_{i=1}^{\infty} \left(\frac{N_i}{N}\right) \epsilon_i$$

We recognize that many population sets may contribute to the total probability sum at equilibrium. If we calculate essentially the same $\overline{\epsilon}$ from each of these contributing population sets, then all of the contributing population sets correspond to indistinguishably different macroscopic energies. We see in the next section that the central limit theorem guarantees that this happens whenever N is as large as the number of molecules in a macroscopic system.

§6 The most probable population set at constant N, V, and T

We are imagining that we can examine a collection of N distinguishable molecules and determine the energy of each molecule in the collection at any particular instant. If we do so, we find the population set, $\{N_1, N_2, \dots, N_i, \dots.\}$, that characterizes the system at that instant. In §3-9, we introduce the idea that the most probable population set, $\{N_1^\bullet, N_2^\bullet, \dots N_i^\bullet, .,,,\}$, or its proxy, $\{NP(\epsilon_1), NP(\epsilon_2), \dots, NP(\epsilon_i), \dots.\}$, is the best prediction we can make about the outcome of a future replication of this measurement. In §2, we hypothesize that the properties of the system when it is characterized by the most probable population set are indistinguishable from the properties of the system at equilibrium.

Now let us show that this hypothesis is implied by the central limit theorem. We suppose that the population set that characterizes the system varies from instant to instant and that we can find this population set at any given instant. The population set that we find at a particular instant comprises a random sample of N molecular energies. For this sample, we can find the average energy from

$$\overline{\epsilon} = \sum_{i=1}^{\infty} \left(\frac{N_i}{N}\right) \epsilon_i$$

The expected value of the molecular energy is

$$\langle \epsilon \rangle = \sum_{i=1}^{\infty} P_i \epsilon_i$$

It is important that we remember that $\overline{\epsilon}$ and $\langle \epsilon \rangle$ are not the same thing. There is a distribution of $\overline{\epsilon}$ values, one $\overline{\epsilon}$ value for each of the possible population sets $\{N_1, N_2, \dots, N_i, \dots.\}$. In contrast, when N, V, and T are fixed, the expected value, $\langle \epsilon \rangle$, is a constant; the value of $\langle \epsilon \rangle$ is completely determined by the values of the variables that determine the state of the system and fix the probabilities P_i. If our theory is to be useful, the value of $\langle \epsilon \rangle$ must be the per-molecule energy that we observe for the macroscopic system we are modeling.

According to the central limit theorem, the average energy of a randomly selected sample, $\overline{\epsilon}$, approaches the expected value for the distribution, $\langle \epsilon \rangle$, as the number of molecules in the sample becomes arbitrarily large. In the present instance, we hypothesize that the most probable population set, or its proxy, characterizes the equilibrium system. When N is sufficiently large, this hypothesis implies that the probability of the ith energy level is given by $P_i = N_i^\bullet/N$. Then the expected value of a molecular energy is

$$\langle \epsilon \rangle = \sum_{i=1}^{\infty} P_i \epsilon_i = \sum_{i=1}^{\infty} \left(\frac{N_i^\bullet}{N}\right) \epsilon_i$$

Since the central limit theorem asserts that $\overline{\epsilon}$ approaches $\langle \epsilon \rangle$ as N becomes arbitrarily large:

$$\begin{aligned} 0 &= \lim_{N\to\infty} \left(\overline{\epsilon} - \langle \epsilon \rangle\right) \\ &= \lim_{N\to\infty} \sum_{i=1}^{\infty} \left(\frac{N_i}{N} - P_i\right) \varepsilon_i \\ &= \lim_{N\to\infty} \sum_{i=1}^{\infty} \left(\frac{N_i}{N} - \frac{N_i^\bullet}{N}\right) \epsilon_i \end{aligned}$$

One way for the limit of this sum to be zero is for the limit of every individual term to be zero. If the ϵ_i were arbitrary, this would be the only way that the sum could always be zero. However, the ϵ_i and the P_i are related, so we might think that the sum is zero because of these relationships.

To see that the limit of every individual term must in fact be zero, we devise a new distribution. We assign a completely arbitrary number, X_i, to each energy level. Now the ith energy level is associated with an X_i as well as an ϵ_i. We have an X distribution as well as an energy distribution. We can immediately calculate the expected value of X. It is

$$\langle X \rangle = \sum_{i=1}^{\infty} P_i X_i$$

When we find the population set $\{N_1, N_2, \dots, N_i, \dots.\}$, we can calculate the corresponding average value of X. It is

$$\overline{X} = \sum_{i=1}^{\infty} \left(\frac{N_i}{N}\right) X_i$$

The central limit theorem applies to any distribution. So, it certainly applies to the X distribution; the average value of X approaches the expected value of X as N becomes arbitrarily large:

$$\begin{aligned} 0 &= \lim_{N\to\infty} \left(\overline{X} - \langle X \rangle\right) \\ &= \lim_{N\to\infty} \sum_{i=1}^{\infty} \left(\frac{N_i}{N} - P_i\right) X_i \\ &= \lim_{N\to\infty} \sum_{i=1}^{\infty} \left(\frac{N_i}{N} - \frac{N_i^{\bullet}}{N}\right) X_i \end{aligned}$$

Now, because the X_i can be chosen completely arbitrarily, the only way that the limit of this sum can always be zero is that every individual term becomes zero.

In the limit as $N \to \infty$, we find that

$$N_i/N \to N_i^{\bullet}/N$$

As the number of molecules in the equilibrium system becomes arbitrarily large, the fraction of the molecules in each energy level at an arbitrarily selected instant approaches the fraction in that energy level in the equilibrium-characterizing most-probable population set, $\{N_1^{\bullet}, N_2^{\bullet}, \dots N_i^{\bullet} \dots\}$. In other words, the only population sets that we have any significant chance of observing in a large equilibrium system are population sets whose occupation fractions, N_i/N, are all very close to those, $N_i^{\bullet}/N$, in the equilibrium-characterizing population set. Estimating P_i as the ratio N_i/N gives essentially the same result whichever of these population sets we use. Below, we see that the ϵ_i and the P_i determine the thermodynamic properties of the system. Consequently, when we calculate any observable property of the macroscopic system, each of these population sets gives the same result.

Since the only population sets that we have a significant chance of observing are those for which

$$N_i/N \approx N_i^{\bullet}/N$$

we frequently say that we can ignore all but the most probable population set. What we have in mind is that the most probable population set is the only one we need in order to calculate the macroscopic properties of the equilibrium system. We are incorrect, however, if we allow ourselves to think that the most probable population set is necessarily much more probable than any of the others. Nor does the fact that the N_i/N are all very close to the $N_i^{\bullet}/N$ mean that the N_i are all very close to the $N_i^{\bullet}$. Suppose that the difference between the two ratios is 10^{-10}. If $N = 10^{20}$, the difference between N_i and $N_i^{\bullet}$ is 10^{10}, which probably falls outside the range of values that we usually understand by the words "very close."

We develop a theory that includes a mathematical model for the probability that a molecule has any one of its quantum-mechanically possible energies. It turns out that we are frequently interested in macroscopic systems in which the number of energy levels greatly exceeds the number of molecules. For such systems, we find $NP_i \ll 1$, and it is no longer possible to say that a single most-probable population set, $\{N_1^{\bullet}, N_2^{\bullet}, \dots N_i^{\bullet}, \dots\}$, describes the equilibrium state of the system. When it is very unlikely that any energy level is occupied by more than one molecule, the probability of any population set in which any N_i is greater than one becomes negligibly small. We can approximate the total probability sum as

$$\begin{aligned} 1 &= (P_1 + P_2 + \cdots + P_i + \cdots)^N \\ &\approx \sum_{\{N_i\}} N!\, P_1^{N_1} P_2^{N_2} \dots P_i^{N_i} \dots \end{aligned}$$

However, the idea that the proxy, $\{NP(\epsilon_1), NP(\epsilon_2), \dots, NP(\epsilon_i), \dots.\}$, describes the equilibrium state of the system remains valid. In these circumstances, a great many population sets can have essentially identical properties; the properties calculated from any of these are indistinguishable from each other and indistinguishable from the properties calculated from the proxy. Since the equilibrium properties are fixed, the value of these extended products is fixed. For any of the population sets available to such a system at equilibrium, we have

$$P_1^{N_1} P_2^{N_2} \dots P_i^{N_i} \dots = P_1^{NP_1} P_2^{NP_2} \dots P_i^{NP_i} \dots = \text{constant}$$

It follows that, for some constant, c, we have

$$c = \sum_{i=1}^{\infty} NP_i \ln P_i = N \sum_{i=1}^{\infty} P_i \ln P_i$$

As it evolves, we see that the probability of finding a molecule in an energy level is the central feature of our theory.

§7 The microstates of a given population set

Thus far, we have considered only the probabilities associated with the assignments of distinguishable molecules to the allowed energy levels. In §2, we introduce the hypothesis that all of the g_i degenerate quantum states with energy ϵ_i are equally probable, so that the probability that a molecule has energy ϵ_i is $P_i = P(\epsilon_i) = g_i \rho(\epsilon_i)$. Making this substitution, the total probability sum becomes

$$\begin{aligned} 1 &= (P_1 + P_2 + \cdots + P_i + \cdots)^N \\ &= \sum_{\{N_i\}} \frac{N!}{N_1!\, N_2! \dots N_i! \dots} P_1^{N_1} P_2^{N_2} \dots P_i^{N_i} \dots \\ &= \sum_{\{N_i\}} \frac{N!\, g_1^{N_1} g_2^{N_2} \dots g_i^{N_i} \dots}{N_1!\, N_2! \dots N_i! \dots} \rho(\epsilon_1)^{N_1} \rho(\epsilon_2)^{N_2} \dots \rho(\epsilon_i)^{N_i} \dots. \end{aligned}$$

$$= \sum_{\{N_i\}} N! \prod_{i=1}^{\infty} \left(\frac{g_i^{N_i}}{N_i!} \right) \rho(\epsilon_i)^{N_i}$$

$$= \sum_{\{N_i\}} W \prod_{i=1}^{\infty} \rho(\epsilon_i)^{N_i}$$

where we use the notation

$$a_1 \times a_2 \times \dots a_i \times \dots a_\omega \times = \prod_{i=1}^{\omega} a_i$$

for extended products and introduce the function

$$\begin{aligned} W &= W(N_i, g_i) \\ &= W(N_1, g_1, N_2, g_2, \dots, N_i, g_i, \dots .) \\ &= N! \prod_{i=1}^{\infty} \left(\frac{g_i^{N_i}}{N_i!} \right) \\ &= C(N_1, N_2, \dots, N_i, \dots) \prod_{i=1}^{\infty} g_i^{N_i} \end{aligned}$$

For reasons that become clear later, W is traditionally called the ***thermodynamic probability***. This name is somewhat unfortunate, because W is distinctly different from an ordinary probability.

In §5, we note that $P_1^{N_1} P_2^{N_2} \dots P_i^{N_i}$ is the probability that N_i molecules occupy each of the energy levels ϵ_i and that $N!/(N_1!\,N_2! \dots N_i! \dots)$ is the number of combinations of distinguishable molecules that arise from the population set $\{N_1, N_2, \dots, N_i, \dots\}$. Now we observe that the extended product

$$\rho(\epsilon_1)^{N_1} \rho(\epsilon_2)^{N_2} \dots \rho(\epsilon_i)^{N_i} \dots .$$

is the probability of any one assignment of the distinguishable molecules to quantum states such that N_i molecules are in quantum states whose energies are ϵ_i. Since a given molecule of energy ϵ_i can be in any of the g_i degenerate quantum states, the probability that it is in the energy level ϵ_i is g_i-fold greater that the probability that it is in any one of these quantum states.

We call a particular assignment of distinguishable molecules to the available quantum states a ***microstate***. For any population set, there are many combinations. When energy levels are degenerate, each combination gives rise to many microstates. The factor $\rho(\epsilon_1)^{N_1} \rho(\epsilon_2)^{N_2} \dots \rho(\epsilon_i)^{N_i} \dots .$ is the probability of any one microstate of the population set $\{N_1, N_2, \dots, N_i, \dots\}$. Evidently, the thermodynamic probability

$$W = N! \prod_{i=1}^{\infty} \left(\frac{g_i^{N_i}}{N_i!} \right)$$

is the total number of microstates of that population set.

To see directly that the number of microstates is $N! \prod_{i=1}^{\infty} (g_i^{N_i}/N_i!)$, let us consider the number of ways we can assign N distinguishable molecules to the quantum states when the population set is $\{N_1, N_2, \dots, N_i, \dots\}$ and energy level ϵ_i is g_i-fold degenerate. We begin by assigning the N_1 molecules in energy level ϵ_1. We can choose the first molecule from among any of the N distinguishable molecules and can choose to place it in any of the g_1 quantum states whose energy is ϵ_1. The number of ways we can make these choices is Ng_1. We can choose the second molecule from among the $N-1$ remaining distinguishable molecules. In Boltzmann statistics, we can place any number of molecules in any quantum state, so there are again g_1 quantum states in which we can place the second molecule. The total number of ways we can place the second molecule is $(N-1)g_1$.

The number of ways the first and second molecules can be chosen and placed is therefore $N(N-1)g_1^2$. We find the number of ways that successive molecules can be placed in the quantum states of energy ϵ_1 by the same argument. The last molecule whose energy is ϵ_1 can be chosen from among the $(N - N_1 + 1)$ remaining molecules and placed in any of the g_1 quantum states. The total number of ways of placing the N_1 molecules in energy level ϵ_1 is $N(N-1)(N-2)\dots(N-N_1+1)g_1^{N_1}$.

This total includes all possible orders for placing every set of N_1 distinguishable molecules into every possible set of quantum states. However, the order doesn't matter; the only thing that affects the state of the system is which molecules go into which quantum state. (When we consider all of the ways our procedure puts all of the molecules into any of the quantum states, we find that any assignment of molecules A, B, and C to any particular set of quantum states occurs six times. Selections in the orders A,B,C; A,C,B; B,A,C; B,C,A; C,A,B; and C,B,A all put the same molecules in the same quantum states.) There are $N_1!$ orders in which our procedure chooses the N_1 molecules; to correct for this, we must divide by $N_1!$, so that the total number of assignments we want to include in our count is

$$N(N-1)(N-2)\dots(N-N_1+1)g_1^{N_1}/N_1!$$

The first molecule that we assign to the second energy level can be chosen from among the $N - N_1$ remaining molecules and placed into any of the g_2 quantum states whose energy is ϵ_2. The last one can be chosen from among the remaining $(N - N_1 - N_2 + 1)$ molecules. The number of assignments of the N_2 molecules to g_2-fold degenerate quantum states whose energy is ϵ_2 is

$$(N-N_1)(N-N_1-1)\dots(N-N_1-N_2+1)g_2^{N_2}/N_2!$$

When we consider the number of assignments of molecules to quantum states with energies ϵ_1 and ϵ_2 we have

$$\begin{aligned} N(N-1)\dots(N-N_1+1)(N-N_1)(N-N_1-1)\dots \\ \times (N-N_1-N_2+1)\left(\frac{g_1^{N_1}}{N_1!} \right)\left(\frac{g_2^{N_2}}{N_2!} \right) \end{aligned}$$

Microstates	$\psi_{1,1}$	$\psi_{1,2}$	$\psi_{2,1}$
1	AB		C
2	A	B	C
3	B	A	C
4		AB	C
5	AC		B
6	A	C	B
7	C	A	B
8		AC	B
9	BC		A
10	B	C	A
11	C	B	A
12		BC	A

Figure 3. Microstates for {2,1} with quantum states $\psi_{1,1}$, $\psi_{1,2}$, and $\psi_{2,1}$.

Let the last energy level to contain any molecules be ϵ_ω. The number of ways that the N_ω molecules can be assigned to the quantum states with energy ϵ_ω is $N_\omega(N_\omega - 1)\dots(1)g_\omega^{N_\omega}/N_\omega!$ The total number of microstates for the population set $\{N_1, N_2, \dots, N_i, \dots\}$ becomes

$$N(N-1)\dots(N-N_1)(N-N_1-1)\dots$$
$$\times (N_\omega)(N_\omega - 1)\dots(1)\prod_{i=1}^{\infty}\left(\frac{g_i^{N_i}}{N_i!}\right)$$
$$= N!\prod_{i=1}^{\infty}\left(\frac{g_i^{N_i}}{N_i!}\right)$$

When we consider Fermi-Dirac and Bose-Einstein statistics, it is no longer true that the molecules are distinguishable. For Fermi-Dirac statistics, no more than one molecule can be assigned to a particular quantum state. For a given population set, Boltzmann, Fermi-Dirac, and Bose-Einstein statistics produce different numbers of microstates.

It is helpful to have notation that enables us to specify different combinations and different microstates. If ϵ_i is the energy associated with the wave equation that describes a particular molecule, it is convenient to say that the molecule is in energy level ϵ_i; that is, its quantum state is one of those that has energy ϵ_i. Using capital letters to represent molecules, we indicate that molecule A is in energy level ϵ_i by writing $\epsilon_i(A)$. To indicate that A, B, and C are in ϵ_i, we write $\epsilon_i(A, B, C)$. Similarly, to indicate that molecules D and E are in ϵ_k, we write $\epsilon_k(D, E)$. For this system of five molecules, the assignment $\epsilon_i(A, B, C)\epsilon_k(D, E)$ represents one of the possible combinations. The order in which we present the molecules that have a given energy is immaterial: $\epsilon_i(A, B, C)\epsilon_k(D, E)$ and $\epsilon_i(C, B, A)\epsilon_k(E, D)$ represent the same combination. When any one molecule is distinguishable from others of the same substance, assignments in which a given molecule has different energies are physically different and represent different combinations. The assignments $\epsilon_i(A, B, C)\epsilon_k(D, E)$ and $\epsilon_i(D, B, C)\epsilon_k(A, E)$ represent different combinations. In Figure 2, we represent these assignments more schematically.

Any two assignments in which a particular molecule occupies different quantum states give rise to different microstates. If the ith energy level is three-fold degenerate, a molecule in any of the quantum states $\psi_{i,1}$, $\psi_{i,2}$, or $\psi_{i,3}$ has energy ϵ_i. Let us write

$$\psi_{i,1}(A, B)\psi_{i,2}(C)\psi_{k,1}(DE)$$

to indicate the microstate arising from the combination $\epsilon_i(A, B, C)\epsilon_k(D, E)$ in which molecules A and B occupy $\psi_{i,1}$, molecule C occupies $\psi_{i,2}$, and molecules D and E occupy $\psi_{k,1}$. Then,

$$\psi_{i,1}(A, B)\psi_{i,2}(C)\psi_{k,1}(DE)$$
$$\psi_{i,1}(B, C)\psi_{i,2}(A)\psi_{k,1}(DE)$$
$$\psi_{i,1}(A)\psi_{i,2}(B, C)\psi_{k,1}(DE)$$

are three of the many microstates arising from the combination $\epsilon_i(A, B, C)\epsilon_k(D, E)$. Figure 3 shows all of the microstates possible for the population set $\{2,1\}$ when the quantum states of a molecule are $\psi_{1,1}$, $\psi_{1,2}$, and $\psi_{2,1}$.

§8 The probabilities of microstates that have the same energy

In §2, we introduce the assumption that, for a molecule in a constant-N-V-T system, for which the g_i and ϵ_i are fixed, the probability of a quantum state, $\rho(\epsilon_i)$, depends only on its energy. It follows that two or more quantum states that have the same energy must have equal probabilities. We accept the idea that the probability depends only on energy primarily because we cannot see any reason for a molecule to prefer one state to another if both states have the same energy.

We extend this thinking to multi-molecule systems. If two microstates have the same energy, we cannot see any reason for the system to prefer one rather than the other. In a constant-N-V-T system, in which the total energy is not otherwise restricted, each microstate of $\{N_1, N_2, \dots, N_i, \dots.\}$ occurs with probability $\rho(\epsilon_1)^{N_1}\rho(\epsilon_2)^{N_2}\dots\rho(\epsilon_i)^{N_i}\dots.$, and each microstate of $\{N_1^\#, N_2^\#, \dots, N_I^\#, \dots\}$ occurs with probability $\rho(\epsilon_1)^{N_1^\#}\rho(\epsilon_2)^{N_2^\#}\dots\rho(\epsilon_i)^{N_i^\#}\dots.$ When the energies of these population sets are equal, we infer that these probabilities are equal, and their value is a constant of the system. That is,

$$\rho(\epsilon_1)^{N_1}\rho(\epsilon_2)^{N_2}\dots\rho(\epsilon_i)^{N_i}\dots.$$
$$= \rho(\epsilon_1)^{N_1^\#}\rho(\epsilon_2)^{N_2^\#}\dots\rho(\epsilon_i)^{N_i^\#}\dots.$$
$$= \rho_{MS,N,E}$$
$$= \text{constant}$$

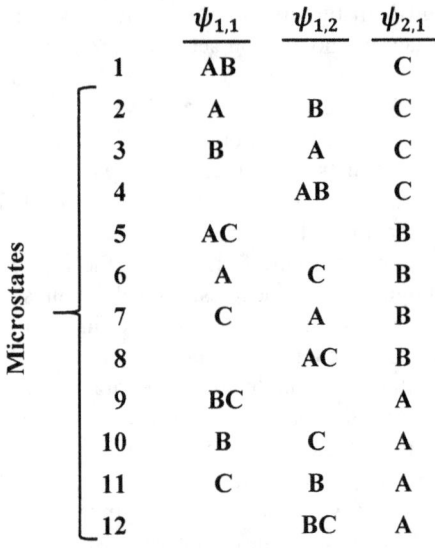

where we introduce $\rho_{MS,N,E}$ to represent the probability of a microstate of a system of N molecules that has total energy E. If $E = E^{\#}$, then $\rho_{MS,N,E} = \rho_{MS,N,E^{\#}}$.

When we think about it critically, the logical basis for this equal-probability idea is not very impressive. While the idea is plausible, it is not securely rooted in any particular empirical observation or prior postulate. The equal-probability idea is useful only if it leads us to theoretical models that successfully mirror the behavior of real macroscopic systems. This it does. Accordingly, we recognize that the equal-probability idea is really a fundamental postulate about the behavior of quantum-mechanical systems. It is often called the ***principle of equal a priori probabilities***:

> **For a particular system, all microstates that have the same energy have the same probability.**

Our development of statistical thermodynamics relies on the principle of equal *a priori* probabilities. For now, let us summarize the important relationships that the principle of equal *a priori* probabilities imposes on our microscopic model for the probabilities of two population sets of a constant-N-V-T system that have the same energy:

- A given population set $\{N_1, N_2, \dots, N_i, \dots .\}$ gives rise to $W(N_i, g_i)$ microstates, and each of these microstates has energy

$$E = \sum_{i=1}^{\infty} N_i \epsilon_i$$

- A second population set, $\{N_1^{\#}, N_2^{\#}, \dots, N_I^{\#}, \dots\}$, that has the same energy need not—and usually will not—give rise to the same number of microstates. In general, for two such population sets,

$$W(N_i, g_i) \neq W(N_i^{\#}, g_i)$$

- However, because each microstate of either population set has the same energy, we have

$$E = \sum_{i=1}^{\infty} N_i \epsilon_i = \sum_{i=1}^{\infty} N_i^{\#} \epsilon_i$$

- The probability of a microstate of a given population set $\{N_1, N_2, \dots, N_i, \dots .\}$ depends only on its energy:

$$\rho(\epsilon_1)^{N_1} \rho(\epsilon_2)^{N_2} \dots \rho(\epsilon_i)^{N_i} \dots . = \rho_{MS,N,E} = \text{constant}$$

§9 The probabilities of the population sets of an isolated system

In principle, the energy of an equilibrium system that is in contact with a constant-temperature heat reservoir can vary slightly with time. In contrast, the energy of an isolated system is constant. A more traditional and less general statement of the equal *a priori* probability principle focuses on isolated systems, for which all possible microstates necessarily have the same energy:

> **All microstates of an isolated (constant energy) system occur with equal probability.**

If we look at the fraction of the molecules of an isolated system that are in each microstate, we expect to find that these fractions are approximately equal. In consequence, for an isolated system, the probability of a population set, $\{N_1, N_2, \dots, N_i, \dots .\}$, is proportional to the number of microstates, $W(N_i, g_i)$, to which that population set gives rise.

In principle, the population sets of a constant-N-V-T system can be significantly different from those of a constant-N-V-E system. That is, if we move an isolated system, whose temperature is T, into thermal contact with a heat reservoir at constant-temperature T, the population sets that characterize the system can change. In practice, however, for a system containing a large number of molecules, the population sets that contribute to the macroscopic properties of the system must be essentially the same.

The fact that the same population sets are important in both systems enables us to make two further assumptions that become important in our development. We assume that the proportionality between the probability of a population set and $W(N_i, g_i)$, which is strictly true only for a constant-N-V-E system, is also true for the corresponding constant-N-V-T system. We also assume that the probabilities of a quantum state, $\rho(\epsilon_i)$, and a microstate, $\rho_{MS,N,E}$, which we defined for the constant-N-V-T system, are the same for the corresponding constant-N-V-E system.

Let us see why we expect the same population sets to dominate the macroscopic properties of otherwise identical constant-energy and constant-temperature systems. Suppose that we isolate a constant-N-V-T system in such a way that the total energy, $E = \sum_{i=1}^{\infty} N_i \epsilon_i$, of the isolated system is exactly equal to the expected value, $\langle E \rangle = N \sum_{i=1}^{\infty} P_i \epsilon_i$, of the energy of the system when its temperature is constant. What we have in mind is a *gedanken* experiment, in which we monitor the energy of the thermostatted system as a function of time, waiting for an instant in which the system energy, $E = \sum_{i=1}^{\infty} N_i \epsilon_i$, is equal to the expected value of the system energy, $\langle E \rangle$. When this occurs, we instantaneously isolate the system.

We suppose that the isolation process is accomplished before any molecule can experience an energy change, so that the population set that characterizes the system immediately afterwards is the same as the one that characterizes it before. After isolation, of course, the molecules can exchange energy with one another, and many population sets may be available to the system.

Clearly, the value of every macroscopic property of the isolated system must be the same as its observable value in the original constant-temperature system. Our microscopic description of it is different. Every population set that is available to the isolated system has energy $E = \langle E \rangle$, and gives rise to

$$W(N_i, g_i) = N! \prod_{i=1}^{\infty} \left(\frac{g_i^{N_i}}{N_i!} \right)$$

microstates. At the same temperature, each of these microstates occurs with the same probability. Since the isolated-system energy is $\langle E \rangle$, this probability is $\rho_{MS,N,\langle E \rangle}$. The probability of an available population set is $W(N_i, g_i)\rho_{MS,N,\langle E \rangle}$.

Since the temperature can span a range of values centered on $\langle T \rangle$, where $\langle T \rangle$ is equal to the temperature of the original constant-N-V-T system, there is a range of $\rho_{MS,N,\langle E \rangle}$ values spanning the (small) range of temperatures available to the constant-energy system. Summing over all of the population sets that are available to the isolated system, we find

$$1 = \sum_{\{N_i\}, E=\langle E \rangle, T=\langle T \rangle} W(N_i, g_i)\rho_{MS,N,\langle E \rangle} + \sum_{\{N_i\}, E=\langle E \rangle, T\neq\langle T \rangle} W(N_i, g_i)\rho_{MS,N,\langle E \rangle}$$

The addition of "$E = \langle E \rangle$" beneath the summation sign emphasizes that the summation is to be carried out over the population sets that are consistent with both the molecule-number and total-energy constraints and no others. The total probability sum breaks into two terms, one spanning population sets whose temperature is exactly $\langle T \rangle$ and another spanning all of the other population sets. (Remember that the $\rho(\epsilon_i)$ are temperature dependent.)

The population sets available to the isolated system are slightly different from those available to the constant-temperature system. In our microscopic model, only population sets that have exactly the right total energy can occur in the isolated system. Only population sets that have exactly the right temperature can occur in the constant-temperature system.

Summing over all of the population sets that are available to the constant-temperature system, we partition the total probability sum into two terms:

$$1 = \sum_{\{N_i\}, E=\langle E \rangle, T=\langle T \rangle} W(N_i, g_i)\rho_{MS,N,\langle E \rangle} + \sum_{\{N_i\}, E\neq\langle E \rangle, T=\langle T \rangle} W(N_i, g_i)\rho_{MS,N,\langle E \rangle}$$

From the central limit theorem, we expect the constant-energy system to have (relatively) few population that fail to meet the condition $E = \langle E \rangle$. Likewise, we expect the constant temperature system to have (relatively) few population sets that fail to meet the condition $T = \langle T \rangle$. The population sets that satisfy both of these criteria must dominate both sums. For the number of molecules in macroscopic systems, we expect the approximation to the total probability sum

$$1 = \sum_{\{N_i\}, E} W(N_i, g_i)\rho_{MS,N,\langle E \rangle}$$

$$\approx \sum_{\{N_i\}, E=\langle E \rangle, T=\langle T \rangle} W(N_i, g_i)\rho_{MS,N,\langle E \rangle}$$

to be very good. The same population sets dominate both the constant-temperature and constant-energy systems. Each system must have a most probable population set, $\{N_1^\bullet, N_2^\bullet, \ldots N_i^\bullet, \ldots\}$. If these are not identically the same set, they must be so close that the same macroscopic properties are calculated using either one.

Thus, the central limit theorem implies that the total probability sum, which we develop for the constant-temperature system, also describes the constant-energy system, so long as the number of molecules in the system is sufficiently large.

Now, two aspects of this development warrant elaboration. The first is that the probability of population sets that have energies and temperature that satisfy $E = \langle E \rangle$ and $T = \langle T \rangle$ *exactly* may actually be much less than one. The second is that constant-energy and constant-temperature systems are creatures of theory. No real system can actually have an *absolutely* constant energy or temperature.

Recognizing these facts, we see that when we stipulate $E = \langle E \rangle$ or $T = \langle T \rangle$, what we really mean is that $E = \langle E \rangle \pm \delta E$ and $T = \langle T \rangle \pm \delta T$, where the intervals $\pm\delta E$ and $\pm\delta T$ are vastly smaller than any differences we could actually measure experimentally. When we write $E \neq \langle E \rangle$ and $T \neq \langle T \rangle$, we really intend to specify energies and temperatures that fall outside the intervals $E = \langle E \rangle \pm \delta E$ and $T = \langle T \rangle \pm \delta T$. If the system contains sufficiently many molecules, the population sets whose energies and temperatures fall within the intervals $E = \langle E \rangle \pm \delta E$ and $T = \langle T \rangle \pm \delta T$ account for nearly all of the probability—no matter how small we choose δE and δT. All of the population sets whose energies and temperatures fall within the intervals $E = \langle E \rangle \pm \delta E$ and $T = \langle T \rangle \pm \delta T$ correspond to the same macroscopically observable properties.

§10 Entropy and equilibrium in an isolated system

In an isolated system, the probability of population set $\{N_1, N_2, \ldots, N_i, \ldots\}$ is $W(N_i, g_i)\rho_{MS,N,\langle E \rangle}$, where $\rho_{MS,N,\langle E \rangle}$ is a constant. It follows that $W = W(N_i, g_i)$ is proportional to the probability that the system is in one of the microstates associated with the population set $\{N_1, N_2, \ldots, N_i, \ldots\}$. Likewise, $W^\# = W(N_i^\#, g_i)$ is proportional to the probability that the system is in one of the microstates associated with the population set $\{N_1^\#, N_2^\#, \ldots N_i^\#, \ldots\}$. Suppose that we observe the isolated system for a long time. Let F be the fraction of the time that the system is in microstates of population set $\{N_1, N_2, ..., N_i, ...\}$ and $F^{\#}$ be the fraction of the time that the system is in microstates of the population set $\{N_1^{\#}, N_2^{\#}, ... N_i^{\#}, ...\}$. The principle of equal *a priori* probabilities implies that we would find

$$\frac{F^{\#}}{F} = \frac{W^{\#}}{W}$$

Suppose that $W^{\#}$ is much larger than W. This means there are many more microstates for $\{N_1^{\#}, N_2^{\#}, ... N_i^{\#}, ...\}$ than there are for $\{N_1, N_2, ..., N_i, ...\}$. The fraction of the time that the population set $\{N_1^{\#}, N_2^{\#}, ... N_i^{\#}, ...\}$ characterizes the system is much greater than the fraction of the time $\{N_1, N_2, ..., N_i, ...\}$ characterizes it. Alternatively, if we examine the system at an arbitrary instant, we are much more likely to find the population set $\{N_1^{\#}, N_2^{\#}, ... N_i^{\#}, ...\}$ than the population set $\{N_1, N_2, ..., N_i, ...\}$. The larger $W(N_1, g_1, N_2, g_2, ..., N_i, g_i, ...)$, the more likely it is that the system will be in one of the microstates associated with the population set $\{N_1, N_2, ..., N_i, ...\}$. In short, W predicts the state of the system; it is a measure of the probability that the macroscopic properties of the system are those of the population set $\{N_1, N_2, ..., N_i, ...\}$.

If an isolated system can undergo change, and we re-examine it at after a few molecules have moved to different energy levels, we expect to find it in one of the microstates of a more-probable population set; that is, in one of the microstates of a population set for which W is larger. At still later times, we expect to see a more-or-less smooth progression: the system is in microstates of population sets for which the values of W are increasingly larger. This can continue only until the system occupies one of the microstates of the population set for which W is a maximum or a microstate of one of the population sets whose macroscopic properties are essentially the same as those of the constant-N-V-E population set for which W is a maximum.

Once this occurs, later inspection may find the system in other microstates, but it is overwhelmingly probable that the new microstate will still be one of those belonging to the largest-W population set or one of those that are macroscopically indistinguishable from it. Any of these microstates will belong to a population set for which W is very well approximated by $W(N_1^{\bullet}, g_1, N_2^{\bullet}, g_2, ..., N_i^{\bullet}, g_i, ...)$. Evidently, the largest-W population set characterizes the equilibrium state of the either the constant-N-V-T system or the constant--N-V-E system. Either system can undergo change until W reaches a maximum. Thereafter, it is at equilibrium and can undergo no further macroscopically observable change.

Boltzmann recognized this relationship between W, the thermodynamic probability, and equilibrium. He noted that the unidirectional behavior of W in an isolated system undergoing spontaneous change is like the behavior we found for the entropy function. Boltzmann proposed that, for an isolated (constant energy) system, S and W are related by the equation $S = k \ln W$, where k is Boltzmann's constant. This relationship associates an entropy value with every population set. For an isolated macroscopic system, equilibrium corresponds to a state of maximum entropy. In our microscopic model, equilibrium corresponds to the population set for which W is a maximum. By the argument we make in §6, this population set must be well approximated by the most probable population set, $\{N_1^{\bullet}, N_2^{\bullet}, ... N_i^{\bullet}, ,,,,\}$. That is, the entropy of the equilibrium state of the macroscopic system is

$$\begin{aligned} S &= k \ln W_{max} \\ &= k \ln \frac{N!}{N_i^{\bullet}!\, N_i^{\bullet}! \ldots N_i^{\bullet}! \ldots} + k \sum_{i=1}^{\infty} N_i^{\bullet} \ln g_i \end{aligned}$$

This equation can be taken as the definition of entropy. Clearly, this definition is different from the thermochemical definition, $S = q^{rev}/T$. We can characterize—imperfectly—the situation by saying that the two definitions provide alternative scales for measuring the same physical property. As we see below, our statistical theory enables us to define entropy in still more ways, all of which prove to be functionally equivalent. Gibbs characterized these alternatives as "entropy analogues;" that is, functions whose properties parallel those of the thermochemically defined entropy.

We infer that the most probable population set characterizes the equilibrium state of either the constant-temperature or the constant-energy system. Since our procedure for isolating the constant-temperature system affects only the thermal interaction between the system and its surroundings, the entropy of the constant-temperature system must be the same as that of the constant-energy system. Using $N_i^{\bullet} = NP_i = Ng_i\rho(\epsilon_i)$ and assuming that the approximation $\ln N_i^{\bullet}! = N_i^{\bullet} \ln N_i^{\bullet} - N_i^{\bullet}$ is adequate for all of the energy levels that make a significant contribution to S, substitution shows that the entropy of either system depends only on probabilities:

$$\begin{aligned} S &= kN \ln N - kN - k \sum_{i=1}^{\infty} [NP_i \ln(NP_i) - NP_i] \\ &\qquad + k \sum_{i=1}^{\infty} NP_i \ln g_i \\ &= kN \ln N - \mathrm{kN} \\ &\qquad - kN \sum_{i=1}^{\infty} [P_i \ln(N) + P_i \ln P_i - P_i - P_i \ln g_i] \\ &= k(N \ln N - N) - k(N \ln N - N) \sum_{i=1}^{\infty} P_i \\ &\qquad - kN \sum_{i=1}^{\infty} P_i \,[\ln P_i - \ln g_i] \\ &= -kN \sum_{i=1}^{\infty} P_i \ln \rho(\epsilon_i) \end{aligned}$$

The entropy per molecule, S/N, is proportional to the expected value of $\ln \rho(\epsilon_i)$; Boltzmann's constant is the proportionality constant. At constant temperature, $\rho(\epsilon_i)$ depends only on ϵ_i. The entropy per molecule depends only on the quantum state properties, g_i and ϵ_i.

§11 Thermodynamic probability and equilibrium in an isomerization reaction

To relate these ideas to a change in a more specific macroscopic system, let us consider isomeric substances A and B. (We consider this example further in Chapter 21.) In principle, we can solve the Schroedinger equation for a molecule of isomer A and for a molecule of isomer B. We obtain all possible energy levels for a molecule of each isomer.[1] If we list these energy levels in order, beginning with the lowest, some of these levels belong to isomer A and the others belong to isomer B.

Now let us consider a mixture of N_A molecules of A and N_B molecules of B. We suppose that individual molecules are distinguishable and that intermolecular interactions can be ignored. Since a group of atoms that can form an A molecule can also form a B molecule, every energy level is accessible to this group of atoms; that is, we can view both sets of energy levels as being available to the atoms that make up the molecules. For a given system energy, there will be many population sets in which only the energy levels belonging to isomer A are occupied. For each of these population sets, there is a corresponding thermodynamic probability, W. Let W_A^{max} be the largest of these thermodynamic probabilities. Similarly, there will be many population sets in which only the energy levels corresponding to isomer B are occupied. Let W_B^{max} be the largest of the thermodynamic probabilities associated with these population sets. Finally, there will be many population sets in which the occupied energy levels belong to both isomer A and isomer B. Let $W_{A,B}^{max}$ be the largest of the thermodynamic probabilities associated with this group of population sets.

Now, W_A^{max} is a good approximation to the number of ways that the atoms of the system can come together to form isomer A. W_B^{max} is a good approximation to the the number of ways that the atoms of the system can come together to form isomer B. At equilibrium, therefore, we expect

$$K = \frac{N_B}{N_A} = \frac{W_B^{max}}{W_A^{max}}$$

If we consider the illustrative—if somewhat unrealistic—case of isomeric molecules whose energy levels all have the same degeneracy ($g_i = g$ for all i), we can readily see that the equilibrium system must contain some amount of each isomer. For a system containing N molecules, $N!\,g^N$ is the numerator in each of the thermodynamic probabilities W_A^{max}, W_B^{max}, and $W_{A,B}^{max}$. The denominators are different. The denominator of $W_{A,B}^{max}$ must contain terms, $N_i!$, for essentially all of the levels represented in the denominator of W_A^{max}. Likewise, it must contain terms, $N_j!$, for essentially all of the energy levels represented in the denominator of W_B^{max}. Then the denominator of $W_{A,B}^{max}$ is a product of $N_k!$ terms that are generally smaller than the corresponding factorial terms in the denominators of W_A^{max} and W_B^{max}. As a result, the denominators of W_A^{max} and W_B^{max} are larger than the denominator of $W_{A,B}^{max}$. In consequence, $W_{A,B}^{max} > W_A^{max}$ and $W_{A,B}^{max} > W_B^{max}$. (See problems 5 and 6.)

If we create the system as a collection of A molecules, or as a collection of B molecules, redistribution of the sets of atoms among all of the available energy levels must eventually produce a mixture of A molecules and B molecules. Viewed as a consequence of the principle of equal *a priori* probabilities, this occurs because there are necessarily more microstates of the same energy available to some mixture of A and B molecules than there are microstates available to either A molecules alone or B molecules alone. Viewed as a consequence of the tendency of the isolated system to attain the state of maximum entropy, this occurs because $k \ln W_{A,B}^{max} > k \ln W_A^{max}$ and $k \ln W_{A,B}^{max} > k \ln W_B^{max}$.

§12 The degeneracy of the energy of an isolated system

In §9, we find that the sum of the probabilities of the population sets of an isolated system is $1 = \sum_{\{N_i\},E} W(N_i, g_i)\rho_{MS,N,E}$. By the principle of equal *a priori* probabilities, $\rho_{MS,N,E}$ is a constant, and it can be factored out of the sum. We have

$$1 = \rho_{MS,N,E} \sum_{\{N_i\},E} W(N_i, g_i)$$

Moreover, the sum of the thermodynamic probabilities over all allowed population sets is just the number of microstates that have energy E. This sum is just the ***degeneracy of the system energy***, E. The symbol Ω_E is often given to this system-energy degeneracy. That is,

$$\Omega_E = \sum_{\{N_i\},E} W(N_i, g_i)$$

The sum of the probabilities of the population sets of an isolated system becomes

$$1 = \rho_{MS,N,E}\Omega_E$$

In §9, we infer that

$$\rho_{MS,N,E} = \prod_{i=1}^{\infty} \rho(\epsilon_i)^{N_i}$$

so we have

$$1 = \Omega_E \prod_{i=1}^{\infty} \rho(\epsilon_i)^{N_i}$$

§13 The degeneracy of an isolated system and its entropy

In §10, we observe that the entropy of an isolated equilibrium system can be defined as $S = k \ln W_{max}$. In §12, we see that the system-energy degeneracy is a sum of terms, one of which is $W_{max} = W(N_i^{\bullet}, g_i)$. That is, we have

$$\Omega_E = W_{max} + \sum_{\{N_i\} \neq \{N_i^{\bullet}\}, E_{total}} W(N_i, g_i)$$

where the last sum is taken over all energy-qualifying population sets other than the most-probable population set.

Let us now consider the relative magnitude of Ω_E and W_{max}. Clearly, $\Omega_E \geq W_{max}$. If only one population set is consistent with the total-molecule and total-energy constraints of the isolated system, then $\Omega_E = W_{max}$. In general, however, we must expect that there will be many, possibly an enormous number, of other population sets that meet the constraints. Ultimately, the relative magnitude of Ω_E and W_{max} depends on the energy levels available to the molecules and the number of molecules in the system and so could be almost anything. However, rather simple considerations lead us to expect that, for most macroscopic collections of molecules, the ratio $\alpha = \Omega_E / W_{max}$ will be much less than W_{max}. That is, although the value of α may be very large, for macroscopic systems we expect to find $\alpha \ll W_{max}$. If $\Omega_E = W_{max}$, then $\alpha = 1$, and $\ln \alpha = 0$.

Because W for any population set that contributes to Ω_E must be less than or equal to W_{max}, the maximum value of α must be less than the number of population sets which satisfy the system constraints. For macroscopic systems whose molecules have even a modest number of accessible energy levels, calculations show that W_{max} is a very large number indeed. Calculation of α for even a small collection of molecules is intractable unless the number of accessible molecular energy levels is small. Numerical experimentation on small systems, with small numbers of energy levels, shows that the number of qualifying population sets increases much less rapidly than W_{max} as the total number of molecules increases. Moreover, the contribution that most qualifying population sets make to Ω_E is much less than W_{max}.

For macroscopic systems, we can be confident that W_{max} is enormously greater than α. Hence Ω_E is enormously greater than α. When we substitute for W_{max} in the isolated-system entropy equation, we find

$$\begin{aligned} S &= k \ln W_{max} \\ &= k \ln(\Omega_E / \alpha) \\ &= k \ln \Omega_E - k \ln \alpha \\ &\approx k \ln \Omega_E \end{aligned}$$

where the last approximation is usually very good.

In many developments, the entropy of an isolated system is defined by the equation $S = k \ln \Omega_E$ rather than the equation we introduced first, $S = k \ln W_{max}$. From the considerations above, we expect the practical consequences to be the same. In §14, we see that the approximate equality of $\ln W_{max}$ and $\ln \Omega_E$ is a mathematical consequence of our other assumptions and approximations.

§14 Effective equivalence of the isothermal and constant-energy conditions

In principle, an isolated system is different from a system with identical macroscopic properties that is in equilibrium with its surroundings. We emphasize this point, because this distinction is important in the logic of our development. However, our development also depends on the assumption that, when N is a number that approximates the number molecules in a macroscopic system, the constant-temperature and constant-energy systems are functionally equivalent.

In §9, we find that any calculation of macroscopic properties must produce the same result whether we consider the constant-temperature or the constant-energy system. The most probable population set, $\{N_1^{\bullet}, N_2^{\bullet}, \dots N_i^{\bullet}, \dots\}$, provides an adequate description of the macroscopic state of the constant-temperature system precisely because it is representative of all the population sets that contribute significantly to the total probability of the constant-temperature system. The effective equivalence of the constant-temperature and constant-energy systems ensures that the most probable population set is also representative of all the population sets that contribute significantly to the total probability of the constant-energy system.

In §12, we see that the essential equivalence of the isothermal and constant-energy systems means that we have

$$1 = \Omega_E \prod_{i=1}^{\infty} \rho(\epsilon_i)^{N_i^{\bullet}}$$

Taking logarithms of both sides, we find

$$\ln \Omega_E = -\sum_{i=1}^{\infty} N_i^{\bullet} \ln \rho(\epsilon_i)$$

From $S = k \ln \Omega_E$, it follows that

$$S = -k \sum_{i=1}^{\infty} N_i^{\bullet} \ln \rho(\epsilon_i)$$

For the constant-temperature system, we have $N_i^{\bullet} = NP_i$. When we assume that the equilibrium constant-temperature and constant-energy systems are essentially equivalent, the entropy of the N-molecule system becomes

$$\begin{aligned} S &= -k \sum_{i=1}^{\infty} N_i^{\bullet} \ln \rho(\epsilon_i) \\ &= -kN \sum_{i=1}^{\infty} P_i \ln \rho(\epsilon_i) \end{aligned}$$

so that we obtain the same result from assuming that $S = k \ln \Omega_E$ as we do in §10 from assuming that $S = k \ln W_{max}$. Under the approximations we introduce, $\ln \Omega_E$ and $\ln W_{max}$ evaluate to the same thing.

Problems

1. Three non-degenerate energy levels are available to a set of five distinguishable molecules, $\{A, B, C, D, E\}$. The energies of these levels are 1, 2, and 3, in arbitrary units. Find all of the population sets that are possible in this system. For each population set, find the system energy, E, and the number of microstates, W. For each system energy, E, list the associated population sets and the total number of microstates. How many population sets are there? What is W_{max}? If this system is isolated with $E = 10$, how many population sets are possible? What is Ω_E for $E = 10$?

2. For the particle in a box, the allowed energies are proportional to the squares of the successive integers. What population sets are possible for the distinguishable molecules, $\{A, B, C, D, E\}$, if they can occupy three quantum states whose energies are 1, 4, and 9? For each population set, find the system energy, E, and the number of microstates. For each system energy, E, list the associated population sets and the total number of microstates. How many population sets are there? What is W_{max}? If this system is isolated with $E = 24$, how many population sets are possible? What is Ω_E for $E = 24$?

3. Consider the results you obtained in problem 2. In general, when the allowed energies are proportional to the squares of successive integers, how many population sets do you think will be associated with each system energy?

4.
(a) Compare W for the population set $\{3,3,3\}$ to W for the population set $\{2,5,2\}$. The energy levels are non-degenerate.
(b) Consider an N-molecule system that has a finite number, M, of quantum states. Show that W is (at least locally) a maximum when $N_1 = N_2 = \cdots = N_M = N/M$. (Hint: Let $U = N/M$, and assume that N can be chosen so that U is an integer. Let

$$W_U = N!\Big/\left[U!\,U!\prod_{i=1}^{i=M-2} U!\right]$$

and let

$$W_O = N!\Big/\left[(U+1)!\,(U-1)!\prod_{i=1}^{i=M-2} U!\right]$$

Show that $W_O/W_U < 1$.)

5. The energy levels available to isomer A are $\epsilon_0 = 1$, $\epsilon_2 = 2$, and $\epsilon_4 = 3$, in arbitrary units. The energy levels available to isomer B are $\epsilon_1 = 2$, $\epsilon_3 = 3$, and $\epsilon_5 = 4$. The energy levels are non-degenerate.
(a) A system contains five molecules. The energy of the system is 10. List the population sets that are consistent with $N = 5$ and $E = 10$. Find W for each of these population sets. What are $W_{A,B}^{max}$, W_A^{max}, and W_B^{max}? What is the total number of microstates, $\Omega_{A,B}$, available to the system in all of the cases in which A and B molecules are present? What is the ratio $\Omega_{A,B}/W_{A,B}^{max}$?
(b) Repeat this analysis for a system that contains six molecules and whose energy is 12.
(c) Would the ratio $\Omega_{A,B}/W_{A,B}^{max}$ be larger or smaller for a system with $N = 50$ and $E = 100$?
(d) What would happen to this ratio if the number of molecules became very large, while the average energy per molecule remained the same?

6. In §11, we assume that all of the energy levels available to an isomeric pair of molecules have the same degeneracy. We then argue that the thermodynamic probabilities of a mixture of the isomers must be greater than the thermodynamic probability of either pure isomer: $W_{A,B}^{max} > W_A^{max}$ and $W_{A,B}^{max} > W_B^{max}$. Implicitly, we assume that many energy levels are multiply occupied: $N_i > 1$ for many energy levels ϵ_i. Now consider the case that $g_i > 1$ for most ϵ_i, but that nearly all energy levels are either unoccupied or contain only one molecule: $N_i = 0$ or $N_i = 1$. Show that under this assumption also, we must have $W_{A,B}^{max} > W_A^{max}$ and $W_{A,B}^{max} > W_B^{max}$.

Notes

[1]The statistical-mechanical procedures that have been developed for finding the energy levels available to a molecule express molecular energies as the difference between the molecule energy and the energy that its constituent atoms have when they are motionless. This is usually effected in two steps. The molecular energy levels are first expressed relative to the energy of the molecule's own lowest energy state. The energy released when the molecules is formed in its lowest energy state from the isolated constituent atoms is then added. The energy of each level is then equal to the work done on the component atoms when they are brought together from infinite separation to form the molecule in that energy level. (Since energy is released in the formation of a stable molecule, the work done on the atoms and the energy of the resulting molecule are less than zero.) In our present discussion, we suppose that we can solve the Schroedinger equation to find the energies of the allowed quantum states. This corresponds to choosing the isolated constituent electrons and nuclei as the zero of energy for both isomers.

21

The Boltzmann Distribution Function

§1 Finding the Boltzmann equation

The probabilities of the energy levels of a constant-temperature system at equilibrium must depend only on the intensive variables that serve to characterize the equilibrium state. In §20-8, we introduce the principle of equal *a priori* probabilities, which asserts that any two microstates of an isolated system have the same probability. From the central limit theorem, we infer that an isolated system is functionally equivalent to a constant-temperature system when the system contains a sufficiently large number of molecules. From these ideas, we can now find the relationship between the energy values, ϵ_i, and the corresponding probabilities, $P_i = P(\epsilon_i) = g_i \rho(\epsilon_i)$.

Let us consider the microstates of an isolated system whose energy is $E^\#$. For any population set, $\{N_1, N_2, \dots, N_i, \dots\}$, that has energy $E^\#$, the following relationships apply.

- The sum of the energy-level populations is the total number of molecules:

$$N = N_1 + N_2 + \cdots + N_i = \sum_{j=1}^{\infty} N_j$$

- The energy of the system is the sum of the energies of its constituent molecules:

$$E^\# = N_1\epsilon_1 + N_2\epsilon_2 + \cdots + N_i\epsilon_i = \sum_{j=1}^{\infty} N_j\,\epsilon_j$$

- The product of powers of quantum-state probabilities is a constant:

$$\rho(\epsilon_1)^{N_1}\rho(\epsilon_1)^{N_1}\dots\rho(\epsilon_1)^{N_1}\dots = \kappa$$

or, equivalently,

$$\begin{aligned} N_1 \ln\rho(\epsilon_1) + N_2 \ln\rho(\epsilon_2) + \cdots + N_i \ln\rho(\epsilon_i) + \dots \\ = \sum_{i=1}^{\infty} N_i \ln\rho(\epsilon_i) \\ = \ln\kappa \end{aligned}$$

- For the system at constant temperature, the sum of the energy-level probabilities is one. When we infer that the constant-temperature system and the isolated system are functionally equivalent, we assume that this is true also for the isolated system:

$$\begin{aligned} 1 &= P(\epsilon_1) + P(\epsilon_2) + \cdots + P(\epsilon_i) + \cdots \\ &= \sum_{j=1}^{\infty} P(\epsilon_j) \end{aligned}$$

We want to find a function, $\rho(\epsilon)$, that satisfies all four of these conditions. One way is to keep trying functions that look like they might work until we find one that does. A slightly more sophisticated version of this approach is to try the most general possible version of each such function and see if any set of restrictions will make it work. We could even try an infinite series. Suppose that we are clever (or lucky) enough to try the series solution

$$\begin{aligned} \ln\rho(\epsilon) &= c_0 + c_1\epsilon + \cdots + c_i\epsilon^i + \cdots \\ &= \sum_{k=0}^{\infty} c_k\,\epsilon^k \end{aligned}$$

Then the third condition becomes

$$\begin{aligned} \ln\kappa &= \sum_{i=1}^{\infty} N_i \ln\rho\,(\epsilon_i) \\ &= \sum_{i=1}^{\infty} N_i \sum_{k=0}^{\infty}\left[c_k\epsilon_i^k\right] \\ &= \sum_{k=0}^{\infty}\sum_{i=1}^{\infty} c_k N_i \epsilon_i^k \\ &= c_0 \sum_{i=1}^{\infty} N_i\,\epsilon_i^0 + c_1 \sum_{i=1}^{\infty} N_i\,\epsilon_i^1 + \cdots + c_k \sum_{k=2}^{\infty}\sum_{i=1}^{\infty} N_i\epsilon_i^k \\ &\quad + \cdots \\ &= c_0 N + c_1 E^\# + \cdots + c_k \sum_{k=2}^{\infty}\sum_{i=1}^{\infty} N_i\epsilon_i^k + \cdots \end{aligned}$$

We see that the coefficient of c_0 is N and the coefficient of c_1 is the total energy, $E^\#$. Therefore, the sum of the first two terms is a constant. We can make the trial function satisfy the third condition if we set $c_k = 0$ for all $k > 1$. We find

$$\ln \kappa = \sum_{i=1}^{\infty} N_i \ln \rho\,(\epsilon_i)$$
$$= \sum_{i=1}^{\infty} N_i\,(c_0 + c_1\epsilon_i)$$

The last equality is satisfied if, for each quantum state, we have

$$\ln \rho\,(\epsilon_i) = c_0 + c_1\epsilon_i$$

or

$$\rho(\epsilon_i) = \alpha \exp(c_1\epsilon_i)$$

where $\alpha = \exp(c_0)$. Since the ϵ_i are positive and the probabilities $\rho(\epsilon_i)$ lie in the interval $0 < \rho(\epsilon_i) < 1$, we must have $c_1 < 0$. Following custom, we let $c_1 = -\beta$, where β is a constant, and $\beta > 0$. Then,

$$\rho(\epsilon_i) = \alpha \exp(-\beta\epsilon_i)$$

and

$$P_i = g_i\rho(\epsilon_i)$$
$$= \alpha g_i \exp(-\beta\epsilon_i)$$

The fourth condition is that the energy-level probabilities sum to one. Using this, we have

$$1 = \sum_{i=1}^{\infty} P(\epsilon_i)$$
$$= \alpha \sum_{i=1}^{\infty} g_i \exp(-\beta\epsilon_i)$$

The sum of exponential terms is so important that it is given a name. It is called the ***molecular partition function***. It is often represented by the letter "z." Letting

$$z = \sum_{i=1}^{\infty} g_i \exp(-\beta\epsilon_i)$$

we have

$$\alpha = \frac{1}{\sum_{i=1}^{\infty} g_i \exp(-\beta\epsilon_i)} = z^{-1}$$

Thus, we have the Boltzmann probability:

$$P(\epsilon_i) = g_i\rho(\epsilon_i)$$
$$= \frac{g_i \exp(-\beta\epsilon_i)}{\sum_{i=1}^{\infty} g_i \exp(-\beta\epsilon_i)}$$
$$= \frac{g_i}{z} \exp(-\beta\epsilon_i)$$

The probability of an energy level depends only on its degeneracy, g_i, its energy, ϵ_i, and the constant β. Since the equilibrium-characterizing population set is determined by the probabilities, we have $P_i = N_i^\bullet/N$, and

$$\frac{N_i^\bullet}{N} = \frac{g_i}{z} \exp(-\beta\epsilon_i)$$

In §2, we develop Lagrange's method of undetermined multipliers. In §3, we develop the same result by applying Lagrange's method to our model for the probabilities of the microstates of an isolated system. That is, we find the Boltzmann probability equation by applying Lagrange's method to the entropy relationship,

$$S = -Nk \sum_{i=1}^{\infty} P_i \ln \rho(\epsilon_i)$$

that we first develop in §20-11. In §4, we find the Boltzmann probability equation by using Lagrange's method to find the values of $N_i^\bullet$ that produce the largest possible value for W_{max} in an isolated system. This argument requires us to assume that there is a very large number of molecules in each of the occupied energy levels of the most probable population set. Since our other arguments do not assume anything about the magnitude of the various $N_i^\bullet$, it is evident that some of the assumptions we make when we apply Lagrange's method to find the $N_i^\bullet$ are not inherent characteristics of our microscopic model.

§2 Lagrange's method of undetermined multipliers

Lagrange's method of undetermined multipliers is a method for finding the minimum or maximum value of a function subject to one or more constraints. A simple example serves to clarify the general problem. Consider the function

$$z = z_0 \exp\,(x^2 + y^2)$$

where z_0 is a constant. This function is a surface of revolution, which is tangent to the plane $z = z_0$ at $(0,0,z_0)$. The point of tangency is the minimum value of z. At any other point in the xy-plane, $z(x,y)$ is greater than z_0. If either x or y becomes arbitrarily large, z does also. If we project a contour of constant z onto the xy-plane, the projection is a circle of radius $r = (x^2 + y^2)^{1/2}$.

Suppose that we introduce an additional condition; we require $y = 1 - x$. Then we ask for the smallest value of z consistent with this constraint. In the xy-plane the constraint is a line of slope -1 and intercept 1. A plane that includes this line and is parallel to the z-axis intersects the function z. As sketched in Figure 1, this intersection is a curve. Far away from the origin, the value of z at which the intersection occurs is large. Nearer the origin, the value of z is smaller, and there is some (x,y) at which it is a minimum. Our objective is to find this minimum.

There is a straightforward solution of this problem; we can substitute the constraint equation for y into the equation for z, making z a function of only one variable, x. We have

$$z = z_0 \exp\,(x^2 + (1-x)^2)$$
$$= z_0 \exp\,(2x^2 - 2x + 1)$$

To find the minimum, we equate the derivative to zero,

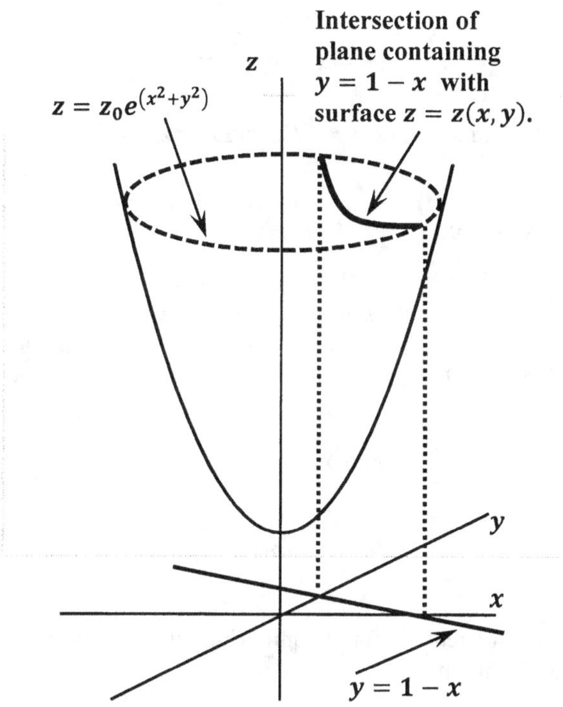

Figure 1. A suface and a constraint equation.

giving

$$\begin{aligned} 0 &= \frac{dz}{dx} \\ &= (4x - 2)z_0 \exp\,(2x^2 - 2x + 1) \end{aligned}$$

so that the minimum occurs at $x = 1/2$, y$= 1/2$, and

$$z = z_0 \exp\,(1/2)$$

Solving such problems by elimination of variables can become difficult. Lagrange's method of undetermined multipliers is a general method, which is usually easy to apply and which is readily extended to cases in which there are multiple constraints. We can see how Lagrange's method arises by thinking further about our particular example. We can imagine that we "walk" along the constraint line in the xy-plane and measure the z that is directly overhead as we progress. The problem is to find the minimum value of z that we encounter as we proceed along the line. This perspective highlights the central feature of the problem: While it is formally a problem in three dimensions (x, y, and z), the introduction of the constraint makes it a two-dimensional problem. We can think of one dimension as a displacement along the line $y = 1 - x$, from some arbitrary starting point on the line. The other dimension is the perpendicular distance from the xy-plane to the intersection with the surface z.

The relevant part of the xy-plane is just the one-dimensional constraint line. We can recognize this by parameterizing the line. Let t measure location on the line relative to some initial point at which $t = 0$. Then we have $x = x(t)$ and $y = y(t)$ and $z(x, y) = z\big(x(t), y(t)\big) = z(t)$. The point we seek is the one at which $dz/dt = 0$.

Now let us examine a somewhat more general problem. We want a general way to find the values (x, y) that minimize (or maximize) a function $h = h(x, y)$ subject to a constraint of the form $c = g(x, y)$, where c is a constant. As in our example, this constraint requires a solution in which (x, y) are on a particular line. If we parameterize this problem, we have

$$h = h(x, y) = h\big(x(t), y(t)\big) = h(t)$$

and

$$c = g(x, y) = g\big(x(t), y(t)\big) = g(t)$$

Because c is a constant, $dc/dt = dg/dt = 0$. The solution we seek is the point at which h is an extremum. At this point, $dh/dt = 0$. Therefore, at the point we seek, we have

$$\frac{dh}{dt} = \left(\frac{\partial h}{\partial x}\right)_y \frac{dx}{dt} + \left(\frac{\partial h}{\partial y}\right)_x \frac{dy}{dt} = 0$$

and

$$\frac{dg}{dt} = \left(\frac{\partial g}{\partial x}\right)_y \frac{dx}{dt} + \left(\frac{\partial g}{\partial y}\right)_x \frac{dy}{dt} = 0$$

We can multiply either of these equations by any factor, and the product will be zero. We multiply dg/dt by λ (where $\lambda \neq 0$) and subtract the result from dh/dt. Then, at the point we seek,

$$\begin{aligned} 0 &= \frac{dh}{dt} - \lambda\frac{dg}{dt} \\ &= \left(\frac{\partial h}{\partial x} - \lambda\frac{\partial g}{\partial x}\right)_y \frac{dx}{dt} + \left(\frac{\partial h}{\partial y} - \lambda\frac{\partial g}{\partial y}\right)_x \frac{dy}{dt} \end{aligned}$$

Since we can choose $x(t)$ and $y(t)$ any way we please, we can insure that $dx/dt \neq 0$ and $dy/dt \neq 0$ at the solution point. If we do so, the terms in parentheses must be zero at the solution point.

Conversely, setting

$$\left(\frac{\partial h}{\partial x} - \lambda\frac{\partial g}{\partial x}\right)_y = 0$$

and

$$\left(\frac{\partial h}{\partial y} - \lambda\frac{\partial g}{\partial y}\right)_x = 0$$

is sufficient to insure that

$$\frac{dh}{dt} = \lambda\frac{dg}{dt}$$

Since $dg/dt = 0$, these conditions insure that $dh/dt = 0$. This means that, if we can find a set $\{x, y, \lambda\}$ satisfying

$$\left(\frac{\partial h}{\partial x} - \lambda\frac{\partial g}{\partial x}\right)_y = 0$$

and

$$\left(\frac{\partial h}{\partial y} - \lambda\frac{\partial g}{\partial y}\right)_x = 0$$

and

$$c - g(x, y) = 0$$

then the values of x and y must be those make $h(x, y)$ an extremum, subject to the constraint that $c = g(x, y)$. We have not shown that the set $\{x, y, \lambda\}$ exists, but we have shown that if it exists, it is the desired solution.

A useful mnemonic simplifies the task of generating the family of equations that we need to use Lagrange's method. The mnemonic calls upon us to form a new function, which is a sum of the function whose extremum we seek and a series of additional terms. There is one additional term for each constraint equation. We generate this term by putting the constraint equation in the form $c - g(x, y) = 0$ and multiplying by an undetermined parameter. For the case we just considered, the mnemonic function is

$$F_{mn} = h(x, y) + \lambda\big(c - g(x, y)\big)$$

We can generate the set of equations that describe the solution set, $\{x, y, \lambda\}$, by equating the partial derivatives of F_{mn} with respect to x, y, and λ to zero. That is, the solution set satisfies the simultaneous equations

$$\frac{\partial F_{mn}}{\partial x} = 0$$

$$\frac{\partial F_{mn}}{\partial y} = 0$$

and

$$\frac{\partial F_{mn}}{\partial \lambda} = 0$$

If there are multiple constraint equations, $c_\lambda - g_\lambda(x, y) = 0$, $c_\alpha - g_\alpha(x, y) = 0$, and $c_\beta - g_\beta(x, y) = 0$, then the mnemonic function is

$$F_{mn} - h(x, y) + \lambda\big(c_\lambda - g_\lambda(x, y)\big) + \alpha\big(c_\alpha - g_\alpha(x, y)\big) + \beta\left(c_\beta - g_\beta(x, y)\right)$$

and the simultaneous equations that represent the constrained extremum are $\partial F_{mn}/\partial x = 0$, $\partial F_{mn}/\partial y = 0$, $\partial F_{mn}/\partial \lambda = 0$, $\partial F_{mn}/\partial \alpha = 0$, and $\partial F_{mn}/\partial \beta = 0$.

To illustrate the use of the mnemonic, let us return to the example with which we began. The mnemonic equation is

$$F_{mn} = z_0 \exp\left(x^2 + y^2\right) + \lambda(1 - x - y)$$

so that

$$\frac{\partial F_{mn}}{\partial x} = 2xz_0 \exp\left(x^2 + y^2\right) - \lambda = 0$$
$$\frac{\partial F_{mn}}{\partial y} = 2yz_0 \exp\left(x^2 + y^2\right) - \lambda = 0$$

and

$$\frac{\partial F_{mn}}{\partial \lambda} = 1 - x - y = 0$$

which yield $x = 1/2$, y= $1/2$, and $\lambda = z_0 \exp(1/2)$.

§3 Deriving the Boltzmann equation from $S = -Nk\sum_{i=1}^{\infty} P_i \ln \rho(\epsilon_i)$

In §20-10 and §20-14, we develop the relationship between the system entropy and the probabilities of a microstate, $\rho(\epsilon_i)$, and an energy level, $P_i = g_i\rho(\epsilon_i)$, in our microscopic model. We find

$$S = -Nk\sum_{i=1}^{\infty} P_i \ln \rho(\epsilon_i)$$
$$= -Nk\sum_{i=1}^{\infty} g_i\rho(\epsilon_i) \ln \rho(\epsilon_i)$$

For an isolated system at equilibrium, the entropy must be a maximum, and hence

$$-\sum_{i=1}^{\infty} g_i\rho(\epsilon_i) \ln \rho(\epsilon_i)$$

must be a maximum. We can use Lagrange's method to find the dependence of the quantum-state probability on its energy. The $\rho(\epsilon_i)$ must be such as to maximize $-\sum_{i=1}^{\infty} g_i\rho(\epsilon_i) \ln \rho(\epsilon_i)$ subject to the constraints

$$1 = \sum_{i=1}^{\infty} P_i$$
$$= \sum_{i=1}^{\infty} g_i\rho(\epsilon_i)$$

and

$$\langle\epsilon\rangle = \sum_{i=1}^{\infty} P_i\epsilon_i$$
$$= \sum_{i=1}^{\infty} g_i\varepsilon_i\rho(\epsilon_i)$$

where $\langle\epsilon\rangle$ is the expected value of the energy of one molecule. The mnemonic function becomes

$$F_{mn} = -\sum_{i=1}^{\infty} g_i\rho(\epsilon_i) \ln \rho(\epsilon_i) + \alpha^*\left(1 - \sum_{i=1}^{\infty} g_i\rho(\epsilon_i)\right) + \beta\left(\langle\epsilon\rangle - \sum_{i=1}^{\infty} g_i\varepsilon_i\rho(\epsilon_i)\right)$$

Equating the partial derivative with respect to $\rho(\epsilon_i)$ to zero,

$$\frac{\partial F_{mn}}{\partial \rho(\epsilon_i)} = -g_i \ln \rho(\epsilon_i) - g_i - \alpha^* g_i - \beta g_i \epsilon_i$$
$$= 0$$

so that

$$\rho(\epsilon_i) = \exp(-\alpha^* - 1)\exp(-\beta\epsilon_i)$$

From

$$1 = \sum_{i=1}^{\infty} P_i = \sum_{i=1}^{\infty} g_i \rho(\epsilon_i)$$

the argument we use in §1 again leads to the partition function, z, and the Boltzmann equation

$$P_i = g_i \rho(\epsilon_i) = z^{-1} g_i \exp(-\beta\epsilon_i)$$

§4 Deriving the Boltzmann equation from W$_{max}$

In §20-9, we find that the probability of the population set $\{N_1, N_2, \dots, N_i, \dots\}$ in an isolated system is

$$\rho_{MS,N,E}\, N! \prod_{i=1}^{\infty} \frac{g_i^{N_i}}{N_i!}$$

The thermodynamic probability

$$W(N_i, g_i) = N! \prod_{i=1}^{\infty} \frac{g_i^{N_i}}{N_i!}$$

is the number of microstates of the population set. $\rho_{MS,N,E}$ is the constant probability of any one microstate. In consequence, as we see in §20-10, the probability of a population set is proportional to its thermodynamic probability, $W(N_i, g_i)$. It follows that the most probable population set is that for which $W(N_i, g_i)$ is a maximum. Our microscopic model asserts that the most probable population set, $\{N_1^\bullet, N_2^\bullet, \dots, N_i^\bullet, \dots\}$, characterizes the equilibrium state, because the equilibrium system always occupies the either the most probable population set or another population set whose macroscopic properties are indistinguishable from those of the most probable one.

Evidently, the equilibrium-characterizing population set is the one for which $W(N_i, g_i)$, or $\ln W(N_i, g_i)$, is a maximum. Let us assume that the N_i are very large so that we can treat them as continuous variables, and we can use Stirling's approximation for $N_i!$. Then we can use Lagrange's method of undetermined multipliers to find the most probable population set by finding the set, $\{N_1, N_2, \dots, N_i, \dots\}$, for which $\ln W(N_i, g_i)$ is a maximum, subject to the constraints $N = \sum_{i=1}^{\infty} N_i$ and $E = \sum_{i=1}^{\infty} N_i \epsilon_i$. From our definition of the system, both N and E are constant. The mnemonic function is

$$F_{mn} = \ln\left(\frac{N!\, g_1^{N_1} g_2^{N_2} \dots g_i^{N_i} \dots}{N_1!\, N_2! \dots N_i! \dots}\right) + \alpha\left(N - \sum_{i=1}^{\infty} N_i\right) + \beta\left(E - \sum_{i=1}^{\infty} N_i \epsilon_i\right)$$
$$\approx N \ln N - N - \sum_{i=1}^{\infty} N_i \ln N_i + \sum_{i=1}^{\infty} N_i + \sum_{i=1}^{\infty} N_i \ln g_i + \alpha\left(N - \sum_{i=1}^{\infty} N_i\right) + \beta\left(E - \sum_{i=1}^{\infty} N_i \epsilon_i\right)$$

Taking the partial derivative with respect to N_i gives

$$\frac{\partial F_{mn}}{\partial N_i} = -N_i\left(\frac{1}{N_1}\right) - \ln N_i + 1 + \ln g_i - \alpha - \beta\epsilon_i$$
$$= -\ln N_i + \ln g_i - \alpha - \beta\epsilon_i$$

from which we have, for the population set with the largest possible thermodynamic probability,

$$-\ln N_i^\bullet + \ln g_i - \alpha - \beta\epsilon_i = 0$$

or

$$N_i^\bullet = g_i \exp(-\alpha)\exp(-\beta\epsilon_i)$$

We can again make use of the constraint on the total number of molecules to find $\exp(-\alpha)$:

$$N = \sum_{i=1}^{\infty} N_i^\bullet = \exp(-\alpha) \sum_{i=1}^{\infty} g_i \exp(-\beta\epsilon_i)$$

so that $\exp(-\alpha) = N z^{-1}$, where z is the partition function, $z = \sum_{i=1}^{\infty} g_i \exp(-\beta\epsilon_i)$. Therefore, in the most probable population set, the number of molecules having energy ϵ_i is

$$N_i^\bullet = N z^{-1} g_i \exp(-\beta\epsilon_i)$$

The fraction with this energy is

$$N_i^\bullet / N = z^{-1} g_i \exp(-\beta\epsilon_i)$$

This fraction is also the probability of finding an arbitrary molecule in one of the quantum states whose energy is ϵ_i. When the isolated system and the corresponding constant-temperature system are functionally equivalent, this probability is P_i. As in the two previous analyses, we have $P_i = g_i \rho(\epsilon_i) = z^{-1} g_i \exp(-\beta\epsilon_i)$.

This derivation of Boltzmann's equation from W_{max} is the most common introductory treatment. It relies on the assumption that all of the N_i are large enough to justify treating them as continuous variables. This assumption proves to be invalid for many important systems. (For ideal gases, we find that $N_i = 0$ or $N_i = 1$ for nearly all of the very large number of energy levels that are available to a given molecule.) Nevertheless, the result obtained is clearly correct; not only is it the same as the result of our two previous arguments, but also it leads to satisfactory agreement between microscopic models and the macroscopic properties of a wide variety of systems.

§5 Partition functions and equilibrium: Isomeric molecules

In §20-11, we discuss chemical equilibrium between isomers from the perspective afforded by Boltzmann's definition of entropy. Now, let us consider equilibrium in this system from the perspective afforded by the energy-level probabilities. Let us assign even-integer labels to energy levels of isomer A and odd-integer labels to energy levels of isomer B. A group of atoms that can arrange itself into either a molecule of A or a molecule of B can occupy any of these energy levels. The partition function for this group of molecules to which all energy levels are available is

$$z_{A+B} = \sum_{i=1}^{\infty} g_i \exp(-\beta\epsilon_i)$$

The fraction of molecules in the first (odd) energy level associated with molecules of isomer B is

$$N_1^{\bullet}/N_{A+B} = g_1(z_{A+B})^{-1} \exp(-\beta\epsilon_1)$$

and the fraction in the next is

$$N_3^{\bullet}/N_{A+B} = g_3(z_{A+B})^{-1} \exp(-\beta\epsilon_3)$$

The total number of B molecules is

$$N_B^{\bullet} = \sum_{i\ odd} N_i$$

so that the fraction of all of the molecules that are B molecules is

$$N_B^{\bullet}/N_{A+B} = (z_{A+B})^{-1} \sum_{i\ odd} g_i\, exp(-\beta\epsilon_i) = z_B/z_{A+B}$$

Likewise, the fraction that is A molecules is

$$N_A^{\bullet}/N_{A+B} = (z_{A+B})^{-1} \sum_{i\ even} g_i\, exp(-\beta\epsilon_i) = z_A/z_{A+B}$$

The equilibrium constant for the equilibrium between A and B is

$$K_{eq} = \frac{N_B^{\bullet}}{N_A^{\bullet}} = \frac{z_B}{z_A}$$

We see that the equilibrium constant for the isomerization reaction is simply equal to the ratio of the partition functions of the isomers.

It is always true that the equilibrium constant is a product of partition functions for reaction-product molecules divided by a product of partition functions for reactant molecules. However, the partition functions for the various molecules must be expressed with a common zero of energy. Choosing the infinitely separated component atoms as the zero-energy state for every molecule assures that this is the case. However, it is often convenient to express the partition function for a molecule by measuring each molecular energy level, ϵ_i, relative to the lowest energy state of that isolated molecule. When we do this, the zero of energy is different for each molecule.

To adjust the energies in a molecule's partition function so that they are expressed relative to the energy of the molecule's infinitely separated atoms, we must add to each molecular energy the energy required to take the molecule from its lowest energy state to its isolated component atoms. If z is the partition function when the ϵ_i are measured relative to the lowest energy state of the isolated molecule, $\Delta\epsilon$ is the energy released when the isolated molecule is formed from its component atoms, and z^* is the partition function when the ϵ_i are measured relative to the molecule's separated atoms, we have $z^* = z\exp(-\Delta\epsilon/kT)$.

§6 Finding β and the thermodynamic functions for distinguishable molecules

All of a substance's thermodynamic functions can be derived from the molecular partition function. We begin with the entropy. We consider closed (constant N) systems of independent, distinguishable molecules in which only pressure–volume work is possible. In §20-10 and §20-14, we find that two different approaches give the entropy of this system, $S = -Nk\sum_{i=1}^{\infty} P_i \ln \rho(\epsilon_i)$. In §1, §3, and §4, we find that three different approaches give the Boltzmann equation, $P_i = g_i\rho(\epsilon_i) = z^{-1}g_i \exp(-\beta\epsilon_i)$. We have

$$\ln \rho(\epsilon_i) = -\ln z - \beta\epsilon_i$$

Substituting, and recognizing that the energy of the N-molecule system is $E = N\langle\epsilon\rangle$, we find that the entropy of the system is

$$\begin{aligned} S &= kN \sum_{i=1}^{\infty} P_i\, [\ln z + \beta\epsilon_i] \\ &= kN \ln z \sum_{i=1}^{\infty} P_i + k\beta N \sum_{i=1}^{\infty} P_i\, \epsilon_i \\ &= kN \ln z + k\beta E \end{aligned}$$

In §10-1, we find that the fundamental equation implies that

$$\left(\frac{\partial E}{\partial S}\right)_V = T$$

Since the ϵ_i are fixed when the volume and temperature of the system are fixed, ln z is constant when the volume and temperature of the system are constant. Differentiating $S = kN \ln z + k\beta E$ with respect to S at constant V, we find

$$1 = k\beta \left(\frac{\partial E}{\partial S}\right)_V$$
$$= k\beta T$$

so that

$$\beta = \frac{1}{kT}$$

This is an important result: Because we have now identified all of the parameters in our microscopic model, we can write the results we have found in forms that are more useful:

$$z = \sum_{i=1}^{\infty} g_i \exp\left(\frac{-\epsilon_i}{kT}\right)$$

(molecular partition function)

$$P_i = g_i \rho(\epsilon_i)$$
$$= z^{-1} g_i \exp\left(\frac{-\epsilon_i}{kT}\right)$$

(Boltzmann's equation)

$$S = kN \ln z + \frac{E}{T}$$

(entropy of an N-molecule system)

To express the system energy in terms of the molecular partition function, we first observe that

$$E = N\langle\epsilon\rangle$$
$$= N \sum_{i=1}^{\infty} P_i \epsilon_i$$
$$= Nz^{-1} \sum_{i=1}^{\infty} g_i \epsilon_i \exp\left(\frac{-\epsilon_i}{kT}\right)$$

Then we observe that

$$\left(\frac{\partial \ln z}{\partial T}\right)_V = z^{-1} \sum_{i=1}^{\infty} g_i \left(\frac{\epsilon_i}{kT^2}\right) \exp\left(\frac{-\epsilon_i}{kT}\right)$$
$$= \left(\frac{1}{NkT^2}\right) Nz^{-1} \sum_{i=1}^{\infty} g_i \epsilon_i \exp\left(\frac{-\epsilon_i}{kT}\right)$$
$$= \frac{E}{NkT^2}$$

The system energy becomes

$$E = NkT^2 \left(\frac{\partial \ln z}{\partial T}\right)_V$$

(energy of an N-molecule system)

By definition, $A = E - TS$. Rearranging our entropy result, $S = kN \ln z + E/T$, we have $E - TS = -NkT \ln z$. Thus,

$$A = -NkT \ln z$$

(Helmholtz free energy of an N-molecule system)

From $dA = -SdT - PdV$, we have

$$\left(\frac{\partial A}{\partial V}\right)_T = -P$$

(Here, of course, P is the pressure of the system, not a probability.) Differentiating $A = -NkT \ln z$ with respect to V at constant T, we find

$$P = NkT \left(\frac{\partial \ln z}{\partial V}\right)_T$$

(pressure of an N-molecule system)

The pressure–volume product becomes

$$PV = NkTV \left(\frac{\partial \ln z}{\partial V}\right)_T$$

Substituting into $H = E + PV$, the enthalpy becomes

$$H = NkT \left[T \left(\frac{\partial \ln z}{\partial T}\right)_V + V \left(\frac{\partial \ln z}{\partial V}\right)_T\right]$$

(enthalpy of an N-molecule system)

The Gibbs free energy is given by $G = A + PV$. Substituting, we find

$$G = -NkT \ln z + NkTV \left(\frac{\partial \ln z}{\partial V}\right)_T$$

(Gibbs free energy of an N-molecule system)

The chemical potential can be found from

$$\mu = \left(\frac{\partial A}{\partial n}\right)_{V,T}$$

At constant volume and temperature, $kT \ln z$ is constant. Substituting $N = n\overline{N}$ into $A = -NkT \ln z$ and taking the partial derivative, we find

$$\mu = -\overline{N}kT \ln z$$
$$= -RT \ln z$$

(chemical potential of distinguishable molecules)

In statistical thermodynamics we frequently express the chemical potential per molecule, rather than per mole; then,

$$\mu = \left(\frac{\partial A}{\partial N}\right)_{V,T}$$

and

$$\mu = -kT \ln z$$

(chemical potential per molecule)

§7 The microscopic model for reversible change

Now let us return to the closed (constant-N) system to develop another perspective on the dependence of its macroscopic thermodynamic properties on the molecular energy levels and their probabilities. We undertake to describe the system using volume and temperature as the independent variables. In thinking about the energy-level probabilities, we stipulate that any parameters that affect the state of the system remain constant. Specifically, we mean that any parameters that appear in the Schroedinger equation remain constant. For example, the energy levels of a particle in a box depend on the mass of the particle and the length of the box. Any such parameter is called an ***exogenous*** variable. If we change an exogenous variable (say the length of the box) by a small amount, all of the energy levels change by a small amount, and all of the probabilities change by a small amount. The energy levels and their probabilities are smooth functions of the exogenous variable. If ξ is the exogenous variable, we have

$$P_i = P(\epsilon_i) = g_i\rho\big(\epsilon_i(\xi)\big)$$

A change in the exogenous variable corresponds to a reversible macroscopic process.

For a particle in a box, the successive ψ_i are functions that depend on the quantum number, i, and the length of the box, ℓ. When we change the length of the box, the wave function and its associated energy both change. Both are continuous functions of the length of the box. The energy is

$$\epsilon_i = \frac{i^2h^2}{8m\ell^2}$$

Changing the length of the box is analogous to changing the volume of a system. A reversible volume change entails work. We see that changing the length of the box does work on the particle-in-a-box, just as changing the volume of a three-dimensional system does work on the system.

Temperature plays a central role in the description of equilibrium from the macroscopic perspective. We can see that temperature enters the description of equilibrium from the microscopic perspective through its effect on the probability factors. When we increase the temperature of a system, its energy increases. The average energy of its molecules increases. The probability of an energy level must depend on temperature. Evidently, the probabilities of energy levels that are higher than the original average energy increase when the temperature increases. The probabilities of energy levels that are lower than the original average energy decrease when the temperature increases. The effects of heat and work on the energy levels and their equilibrium populations are diagrammed in Figure 2.

If our theory is to be useful, the energy we measure for a macroscopic system must be indistinguishably close to the expected value of the system energy as calculated from our microscopic model:

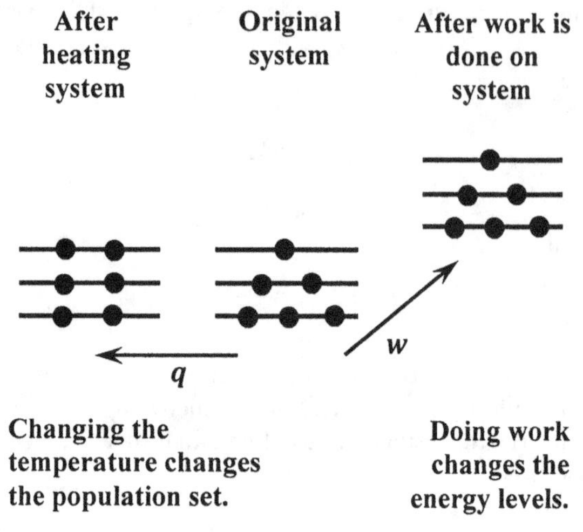

Figure 2. The effects of heat and work on energy levels and their populations.

$$\begin{aligned} E_{\text{experiment}} &\approx \langle E\rangle \\ &= N\langle\epsilon\rangle \\ &= N\sum_{i=1}^{\infty} P_i\epsilon_i \end{aligned}$$

We can use this equation to relate the probabilities, P_i, to other thermodynamic functions. Dropping the distinction between the experimental and expected energies, and assuming that the ϵ_i and the P_i are continuous variables, we find the total differential

$$dE = N\sum_{i=1}^{\infty}\epsilon_i dP_i + N\sum_{i=1}^{\infty}P_i d\epsilon_i$$

This equation is important because it describes a reversible macroscopic process in terms of the microscopic variables ϵ_i and P_i.

Let us consider the first term. Since N is a constant, we have from $N_i^\bullet = P_iN$ that $dN_i^\bullet = NdP_i$. Substituting, we have

$$\begin{aligned} (dE)_{\epsilon_i} &= N\sum_{i=1}^{\infty}\epsilon_i dP_i \\ &= \sum_{i=1}^{\infty}\epsilon_i\, dN_i^\bullet \end{aligned}$$

This asserts that the energy of the system changes if we redistribute the molecules among the various energy levels. If the redistribution takes molecules out of lower energy levels and puts them into higher energy levels, the energy of the system increases. This is our statistical-mechanical picture of the shift in the equilibrium position that occurs when we heat a system of independent molecules; the allocation of molecules among the available energy levels shifts to put more molecules in higher energy levels and fewer in lower ones. This corresponds to an increase in the temperature of the macroscopic system.

In terms of the macroscopic system, the first term represents an increment of heat added to the system in a reversible process; that is,

$$dq^{rev} = N\sum_{i=1}^{\infty} \epsilon_i dP_i$$

The second term, $N\sum_{i=1}^{\infty} P_i d\epsilon_i$, is a contribution to the change in the energy of the system from reversible changes in the energy of the various quantum states, while the number of molecules in each quantum state remains constant. This term corresponds to a process in which the quantum states (and their energies) evolve in a continuous way as the state of the system changes. The second term represents an increment of work done on the system in a reversible process; that is

$$dw^{rev} = N\sum_{i=1}^{\infty} P_i d\epsilon_i$$

Evidently, the total differential expression for dE is the fundamental equation of thermodynamics expressed in terms of the variables we use to characterize the molecular system. It enables us to relate the variables that characterize our microscopic model of the molecular system to the variables that characterize the macroscopic system.

For a system in which the reversible work is pressure–volume work, the energy levels depend on the volume. At constant temperature we have

$$\begin{aligned} dw^{rev} &= -PdV \\ &= N\sum_{i=1}^{\infty} P_i d\epsilon_i \\ &= N\sum_{i=1}^{\infty} P_i \left(\frac{\partial \epsilon_i}{\partial V}\right)_T dV \end{aligned}$$

so that the system pressure, P, is related to the energy-level probabilities, P_i, as

$$P = -N\sum_{i=1}^{\infty} P_i \left(\frac{\partial \epsilon_i}{\partial V}\right)_T$$

To evaluate the pressure, we must know how the energy levels depend on the volume of the system.

The first term relates the entropy to the energy-level probabilities. Since $dq^{rev} = TdS = N\sum_{i=1}^{\infty} \epsilon_i dP_i$, we have

$$dS = \frac{N}{T}\sum_{i=1}^{\infty} \epsilon_i dP_i$$

From the Boltzmann distribution function we have $P_i = z^{-1} g_i \exp(-\epsilon_i/kT)$, or

$$\epsilon_i = -kT\ln P_i + kT\ln g_i - kT\ln z$$

Substituting into our expression for dS, we find

$$\begin{aligned} dS = -Nk\sum_{i=1}^{\infty}(\ln P_i)\, dP_i + Nk\sum_{i=1}^{\infty}(\ln g_i)\, dP_i \\ - Nk(\ln z)\sum_{i=1}^{\infty} dP_i \end{aligned}$$

Since $\sum_{i=1}^{\infty} P_i = 1$, we have $\sum_{i=1}^{\infty} dP_i = 0$, and the last term vanishes. Also,

$$\begin{aligned} \sum_{i=1}^{\infty} d(P_i \ln P_i) &= \sum_{i=1}^{\infty}(\ln P_i) dP_i + \sum_{i=1}^{\infty} dP_i \\ &= \sum_{i=1}^{\infty}(\ln P_i) dP_i \end{aligned}$$

so that

$$dS = -Nk\sum_{i=1}^{\infty} d(P_i \ln P_i) + Nk\sum_{i=1}^{\infty}(\ln g_i)\, dP_i$$

At any temperature, the probability ratio for any two successive energy levels is

$$\frac{P_{i+1}(T)}{P_i(T)} = \frac{P_{i+1}}{P_i} = \frac{g_{i+1}}{g_i}\exp\left(\frac{-(\epsilon_{i+1} - \epsilon_i)}{kT}\right)$$

In the limit as the temperature goes to zero,

$$\frac{P_{i+1}}{P_i} \to 0$$

It follows that $P_1(0) = 1$ and $P_i(0) = 0$ for $i > 1$. Integrating from $T = 0$ to T, the entropy of the system goes from $S(0) = S_0$ to $S(T)$, and the energy-level probabilities go from $P_i(0)$ to $P_i(T)$. We have

$$\begin{aligned} \int_{S_0}^{S(T)} dS = -Nk\sum_{i=1}^{\infty}\int_{P_i(0)}^{P_i(T)} d(P_i \ln P_i) \\ + Nk\sum_{i=1}^{\infty}\int_{P_i(0)}^{P_i(T)} (\ln g_i) dP_i \end{aligned}$$

so that

$$\begin{aligned} S(T) - S_0 = -Nk\sum_{i=1}^{\infty} P_i(T)\ln P_i(T) \\ + NkP_1(0)\ln P_1(0) \\ + Nk\sum_{i=1}^{\infty}(\ln g_i) P_i(T) \\ - Nk(\ln g_1) P_1(0) \end{aligned}$$

Since $P_1(0) = 1$, $\ln P_1(0)$ vanishes. The entropy change becomes

$$S(T) - S_0 = -Nk \sum_{i=1}^{\infty} P_i\,[\ln P_i - \ln g_i] - Nk \ln g_1$$
$$= -Nk \sum_{i=1}^{\infty} P_i \ln \rho(\epsilon_i) - Nk \ln g_1$$

We have $S_0 = Nk \ln g_1$. If $g_1 = 1$, the lowest energy level is non-degenerate, and $S_0 = 0$; then we have

$$S = -Nk \sum_{i=1}^{\infty} P_i \ln \rho(\epsilon_i)$$

This is the entropy of an N-molecule, constant-volume, constant-temperature system that is in thermal contact with its surroundings at the same temperature. We obtain this same result in §20-10 and §20-14 by arguments in which we assume that the system is isolated. In all of these arguments, we assume that the constant-temperature system and its isolated counterpart are functionally equivalent; that is, a group of population sets that accounts for nearly all of the probability in one system also accounts for nearly all of the probability in the other.

Because we obtain this result by assuming that the system is composed of N, independent, non-interacting, distinguishable molecules, the entropy of this is system is N times the entropy contribution of an individual molecule. We can write

$$S_{\text{molecule}} = -k \sum_{i=1}^{\infty} P_i \ln \rho(\epsilon_i)$$

§8 The third law of thermodynamics

In §7, we obtain the entropy by a definite integration. We take the lower limits of integration, at $T = 0$, as $P_1(0) = 1$ and $P_i(0) = 0$, for $i > 1$. In doing so, we apply the third law of thermodynamics, which states that the entropy of a perfect crystal can be chosen to be zero when the temperature is at absolute zero. The idea behind the third law is that, at absolute zero, the molecules of a crystalline substance all are in the lowest energy level that is available to them. The probability that a molecule is in the lowest energy state is, therefore, $P_1 = 1$, and the probability that it is any higher energy level, $i > 1$, is $P_i = 0$.

While the fact is not relevant to the present development, we note in passing that the energy of a perfect crystal is not zero at absolute zero. While all of the constituent particles will be in their lowest vibrational energy levels at absolute zero, the energies of these lowest vibrational levels are not zero. In the harmonic oscillator approximation, the lowest energy possible for each oscillator is $h\nu/2$. (See §18-5).

By a perfect crystalline substance we mean one in which the lowest energy level is non-degenerate; that is, for which $g_1 = 1$. We see that our entropy equation conforms to the third law when we let

$$S_0 = Nk \ln g_1$$

so that $S_0 = 0$ when $g_1 = 1$.

Let us consider a crystalline substance in which the lowest energy level is degenerate; that is, one for which $g_1 > 1$. This substance is not a perfect crystal. In this case, the temperature-zero entropy is

$$S_0 = Nk \ln g_1 > 0$$

The question arises: How can we determine whether a crystalline substance is a perfect crystal? In Chapter 11, we discuss the use of the third law to determine the absolute entropy of substances at ordinary temperatures. If we assume that the substance is a perfect crystal at zero degrees when it is not, our theory predicts a value for the absolute entropy at higher temperatures that is too small, because it does not include the term $S_0 = Nk \ln g_1$. When we use this too-small absolute entropy value to calculate entropy changes for processes involving the substance, the results do not agree with experiment.

Absolute entropies based on the third law have been experimentally determined for many substances. As a rule, the resulting entropies are consistent with other experimentally observed entropy changes. In some cases, however, the assumption that the entropy is zero at absolute zero leads to absolute entropy values that are not consistent with other experiments. In these cases, the absolute entropies can be brought into agreement with other entropy measurements by assuming that, indeed, $g_1 > 1$ for such substances. In any particular case, the value of g_1 that must be used is readily reconciled with other information about the substance.

For example, the third law entropy for carbon monoxide must be calculated taking $g_1 = 2$ in order to obtain a value that is consistent with other entropy measurements. This observation is readily rationalized. In perfectly crystalline carbon monoxide, all of the carbon monoxide molecules point in the same direction, as sketched in Figure 11-2. However, the two ends of the carbon monoxide molecule are very similar, with the consequence that the carbon monoxide molecules in the crystal point randomly in either of two directions. Thus there are two (approximately) equally energetic states for a carbon monoxide molecule in a carbon monoxide crystal at absolute zero, and we can take $g_1 = 2$. (We are over-simplifying here. We explore this issue further in §22-7.)

§9 The partition function for a system of N molecules

At a given temperature, the Boltzmann equation gives the probability of finding a molecule in any of the energy levels that the molecule can occupy. Throughout our development, we assume that there are no energies of interaction among the molecules of the system. The molecular partition function contains information about the energy levels of only one molecule. We obtain equations for the thermodynamic functions of an N-molecule system in terms of this molecular partition function. However, since these results are based on assigning the same isolated-molecule energy levels to each of the molecules, they do not address the real-system situation in which intermolecular interactions make important contributions to the total energy of the system.

As we mention in §20-1 and §20-3, the ensemble theory of statistical thermodynamics extends our arguments to express the thermodynamic properties of a macroscopic system in terms of all of the total energies that are available to the macroscopic system. The molecular origins of the energies of the system enter the ensemble treatment only indirectly. The theory deals with the relationships between the possible values of the energy of the system and its thermodynamic state. How molecular energy levels and intermolecular interactions give rise to these values of the system energy becomes a separate issue. Fortunately, ensemble theory just reuses—from a different perspective—all of the ideas we have just studied. The result is just the Boltzmann equation, again, but now the energies that appear in the partition function are the possible energies for the collection of N molecules, not the energies available to a single molecule.

Problems

1. Consider a system with three non-degenerate quantum states having energies $\epsilon_1 = 0.9\ kT$, $\epsilon_2 = 1.0\ kT$, and $\epsilon_3 = 1.1\ kT$. The system contains $N = 3 \times 10^{10}$ molecules. Calculate the partition function and the number of molecules in each quantum state when the system is at equilibrium. This is the equilibrium population set $\{N_1^\bullet, N_2^\bullet, N_3^\bullet\}$. Let W_{mp} be the number of microstates associated with the equilibrium population set. Consider the population set when 10^{-5} of the molecules in ϵ_2 are moved to each of ϵ_1 and ϵ_3. This is the population set $\{N_1^\bullet + 10^{-5}N_2^\bullet,\ N_2^\bullet - 2 \times 10^{-5},\ N_3^\bullet + 10^{-5}N_2^\bullet\}$. Let W be the number of microstates associated with this non-equilibrium population set.

(a) What percentage of the molecules are moved in converting the first population set into the second?

(b) How do the energies of these two populations sets differ from one another?

(c) Find W_{mp}/W. Use Stirling's approximation and carry as many significant figures as your calculator will allow. You need at least six.

(d) What does this calculation demonstrate?

2. Find the approximate number of energy levels for which $\epsilon < kT$ for a molecule of molecular weight 40 in a box of volume $10^{-6}\ \text{m}^3$ at 300 K.

3. The partition function plays a central role in relating the probability of finding a molecule in a particular quantum state to the energy of that state. The energy levels available to a particle in a one-dimensional box are

$$\epsilon_n = \frac{n^2 h^2}{8m\ell^2}$$

where m is the mass of the particle and ℓ is the length of the box. For molecular masses and boxes of macroscopic lengths, the factor $h^2/8m\ell^2$ is a very small number. Consequently, the energy levels available to a molecule in such a box can be considered to be effectively continuous in the quantum number, n. That is, the partition function sum can be closely approximated by an integral in which the variable of integration, n, runs from 0 to ∞.

(a) Obtain a formula for the partition function of a particle in a one-dimensional box. Integral tables give

$$\int_0^\infty \exp(-an^2)\,dn = \sqrt{\pi/4a}$$

(b) The expected value of the energy of a molecule is given by

$$\langle\epsilon\rangle = kT^2\left(\frac{\partial \ln z}{\partial T}\right)_V$$

What is $\langle\epsilon\rangle$ for a particle in a box?

(c) The relationship between the partition function and the per-molecule Helmholtz free energy is $A = -kT\ln z$. For a molecule in a one-dimensional box, we have $dA = -SdT - \rho\ell$, where ρ is the per-molecule "pressure" on the ends of the box and ℓ is the length of the box. (The increment of work associated with changing the length of the box is $dw = -\rho\, d\ell$. In this relationship, $d\ell$ is the incremental change in the length of the box and ρ is the one-dimensional "pressure" contribution from each molecule. ρ is, of course, just the force required to push the end of the box outward by a distance $d\ell$. $\rho d\ell$ is the one-dimensional analog of PdV.) For the one-dimensional system, it follows that

$$\rho = -\left(\frac{\partial A}{\partial \ell}\right)_T$$

Use this information to find ρ for a molecule in a one-dimensional box.

(d) We can find ρ for a molecule in a one-dimensional box in another way. The per-molecule contribution to the pressure of a three-dimensional system is related to the energy-level probabilities, P_i, by

$$P_{\text{molecule}}^{\text{system}} = -\sum_{n=1}^{\infty} P_n \left(\frac{\partial \epsilon_n}{\partial V}\right)_T$$

By the same argument we use for the three-dimensional case, we find that the per-molecule contribution to the "pressure" inside a one-dimensional box is

$$\rho = -\sum_{n=1}^{\infty} P_n \left(\frac{\partial \epsilon_n}{\partial \ell}\right)_T$$

From the equation for the energy levels of a particle in a one dimensional box, find an equation for

$$\left(\frac{\partial \epsilon_n}{\partial \ell}\right)_T$$

(Hint: We can express this derivative as a simple multiple of ϵ_n.)

(e) Using your result from part (d), show that the per molecule contribution, ρ, to the "one-dimensional pressure" of N molecules in a one-dimensional box is

$$\rho = 2\langle\epsilon\rangle/\ell$$

(f) Use your results from parts (b) and (e) to express ρ as a function of k, T, and ℓ.

(g) Let Π be the pressure of a system of N molecules in a one-dimensional box. From your result in part (c) or part (f), give an equation for Π. Show how this equation is analogous to the ideal gas equation.

22

Some Basic Applications of Statistical Thermodynamics

§1 Interpreting the partition function

When it is a good approximation to say that the energy of a molecule is the sum of translational, rotational, vibrational, and electronic components, we have

$$\epsilon_{i,j,k,m} = \epsilon_{t,i} + \epsilon_{r,j} + \epsilon_{v,k} + \epsilon_{e,m}$$

where the indices i, j, k, and m run over all possible translational, rotational, vibrational, and electronic quantum states, respectively. Then the partition function for the molecule can be expressed as a product of the individual partition functions z_t, z_r, z_v, and z_e; that is,

$$z_{\text{molecule}} =$$

$$= \sum_t \sum_r \sum_v \sum_e g_{t,i}\, g_{r,j} g_{v,k} g_{e,m} \exp\left(\frac{-\epsilon_{i,j,k,m}}{kT}\right)$$

$$= \sum_t g_{t,i}\, exp\left(\frac{-\epsilon_{t,i}}{kT}\right) \sum_r g_{r,j}\, exp\left(\frac{-\epsilon_{r,j}}{kT}\right)$$

$$\times \sum_v g_{v,k}\, exp\left(\frac{-\epsilon_{v,k}}{kT}\right) \sum_e g_{e,m}\, exp\left(\frac{-\epsilon_{e,m}}{kT}\right)$$

$$= z_t z_r z_v z_e$$

The magnitude of an individual partition function depends on the magnitudes of the energy levels associated with that kind of motion. Table 1 gives the contributions made to their partition functions by levels that have various energy values.

Table 1.

ϵ_i	$\frac{-\epsilon_i}{kT}$	$\exp\left(\frac{-\epsilon_i}{kT}\right)$	Type of Motion
$10^{-2}\,kT$	-10^{-2}	0.990	
$10^{-1}\,kT$	-10^{-1}	0.905	
kT	-1	0.365	translational
$5\,kT$	-5	0.0067	rotational
$10\,kT$	-10	4.5×10^{-5}	vibration
$100\,kT$	-100	3.7×10^{-44}	electronic

We see that only quantum states whose energy is less than kT can make substantial contributions to the magnitude of a partition function. Very approximately, we can say that the partition function is equal to the number of quantum states for which the energy is less than kT. Each such quantum state will contribute approximately one to the sum that comprises the partition function; the contribution of the corresponding energy level will be approximately equal to its degeneracy. If the energy of a quantum state is large compared to kT, the fraction of molecules occupying that quantum state will be small. This idea is often expressed by saying that such states are "unavailable" to the molecule. It is then said that the value of the partition function is approximately equal to the number of available quantum states. When most energy levels are non-degenerate, we can also say that the value of the partition function is approximately equal to the number of available energy levels.

§2 Conditions under which integrals approximate partition functions

The Boltzmann equation gives the equilibrium fraction of particles in the ith energy level, ϵ_i, as

$$\frac{N_i^\bullet}{N} = \frac{g_i}{z} \exp\left(\frac{-\epsilon_i}{kT}\right)$$

so the fraction of particles in energy levels less than ϵ_n is

$$f(\epsilon_n) = z^{-1} \sum_{i=1}^{n-1} g_i \exp\left(\frac{-\epsilon_i}{kT}\right)$$

where $z = \sum_{i=1}^{\infty} g_i \exp(\epsilon_i/kT)$. We can represent either of these sums as the area under a bar graph, where the height and width of each bar are $g_i \exp(\epsilon_i/kT)$ and unity, respectively. If g_i and ϵ_i can be approximated as continuous functions, this area can be approximated as the area under the continuous function $y(i) = g_i \exp(\epsilon_i/kT)$. That is,

$$\sum_{i=1}^{n-1} g_i \exp\left(\frac{-\epsilon_i}{kT}\right) \approx \int_{i=0}^{n} g_i \exp\left(\frac{-\epsilon_i}{kT}\right) di$$

To evaluate this integral, we must know how both g_i and ϵ_i depend on the quantum number, i.

Let us consider the case in which $g_i = 1$ and look at the constraints that the ϵ_i must satisfy in order to make the integral a good approximation to the sum. The graphical description of this case is sketched in Figure 1. Since $\epsilon_i > \epsilon_{i-1} > 0$, we have

$$e^{-\epsilon_{i-1}/kT} - e^{-\epsilon_i/kT} > 0$$

For the integral to be a good approximation, we must have $e^{-\epsilon_{i-1}/kT} \gg e^{-\epsilon_{i-1}/kT} - e^{-\epsilon_i/kT} > 0$, which means that

$$1 \gg 1 - e^{-\Delta\epsilon/kT} > 0$$

where $\Delta\epsilon = \epsilon_i - \epsilon_{i-1}$. Now,

$$e^x \approx 1 + x + \frac{x^2}{2!} + \frac{x^3}{3!} + \cdots$$

so that the approximation will be good if

$$1 \gg 1 - \left(1 - \frac{\Delta\epsilon}{kT} + \cdots\right)$$

or

$$1 \gg \frac{\Delta\epsilon}{kT}$$

or

$$kT \gg \Delta\epsilon$$

We can be confident that the integral is a good approximation to the exact sum whenever there are many pairs of energy levels, ϵ_i and ϵ_{i-1}, that satisfy the condition $\Delta\epsilon = \epsilon_i - \epsilon_{i-1} \ll kT$. If there are many energy levels that satisfy $\epsilon_i \ll kT$, there are necessarily many intervals, $\Delta\epsilon$, that satisfy $\Delta\epsilon \ll kT$. In short, if a large number of the energy levels of a system satisfy the criterion $\epsilon \ll kT$, we can use integration to approximate the sums that appear in the Boltzmann equation. In §24-3, we use this approach and the energy levels for a particle in a box to find the partition function for an ideal gas.

§3. Probability density functions from the energies of classical-mechanical models

Guided by our development of the Maxwell-Boltzmann probability density function for molecular velocities, we could postulate that similar probability density functions apply to other energies derived from classical-mechanical models for molecular motion. We will see that this can indeed be done. The results correspond to the results that we get from the Boltzmann equation, where we assume for both derivations that many energy levels satisfy $\epsilon \ll kT$. The essential point is that, at a sufficiently high temperature, the behavior predicted by the quantum mechanical model and that predicted from classical mechanics converge. This high-temperature approximation is a good one for translational motions but a very poor one for vibrational motions. These results further illuminate the differences between the classical-mechanical and the quantum-mechanical models for the behavior of molecules.

Let us look at how we can generate probability density functions based on the energies of classical-mechanical models for molecular motions. In the classical mechanical model, a particle moving in one dimension with velocity v has kinetic energy $mv^2/2$. From the discussion above, if many velocities satisfy $kT \gg mv^2/2$, we can postulate a probability density function of the form

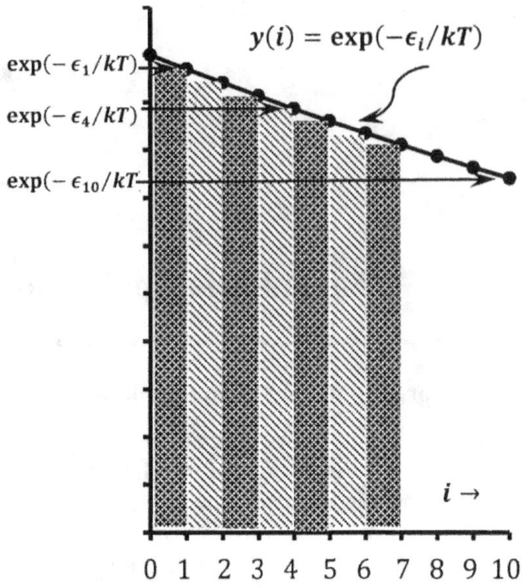

Figure 1. Approximating the partition function as an integral.

$$\frac{df}{dv} = B_{\text{trans}} \exp\left(\frac{-mv^2}{2kT}\right)$$

where B_{trans} is fixed by the condition

$$\int_{-\infty}^{\infty} \left(\frac{df}{dv}\right) dv = B_{\text{trans}} \int_{-\infty}^{\infty} \exp\left(\frac{-mv^2}{2kT}\right) dv = 1$$

Evidently, this postulate assumes that each velocity constitutes a quantum state and that the degeneracy is the same for all velocities. This assumption is successful for one-dimensional translation, but not for translational motion in two or three dimensions. The definite integral is given in Appendix D. We find

$$B_{\text{trans}} = (m/2\pi k\,T)^{1/2}$$

and

$$\frac{df}{dv} = \left(\frac{m}{2\pi kT}\right)^{1/2} \exp\left(\frac{-mv^2}{2kT}\right)$$

With $m/kT = \lambda$, this is the same as the result that we obtain in §4-4. With B_{trans} in hand, we can calculate the average energy associated with the motion of a gas molecule in one dimension

$$\langle\epsilon\rangle = \int_{-\infty}^{\infty} \left(\frac{mv^2}{2}\right)\left(\frac{df}{dv}\right) dv$$
$$= \left(\frac{m^3}{8\pi kT}\right)^{1/2} \int_{-\infty}^{\infty} v^2 \exp\left(\frac{-mv^2}{2kT}\right) dv$$

This definite integral is also given in Appendix D. We find

$$\langle \epsilon_{\text{trans}} \rangle = \frac{kT}{2}$$

We see that we can obtain the average kinetic energy for one degree of translational motion by a simple argument that uses classical-mechanical energies in the Boltzmann equation. We can make the same argument for each of the other two degrees of translational motion. We conclude that each degree of translational freedom contributes $kT/2$ to the average energy of a gas molecule. For three degrees of translational freedom, the total contribution is $3kT/2$, which is the result that we first obtained in §2-10.

Now let us consider a classical-mechanical model for a rigid molecule rotating in a plane. The classical kinetic energy is $\epsilon_{\text{rot}} = I\omega^2/2$, where I is the molecule's moment of inertia about the axis of rotation, and ω is the angular rotation rate. This has the same form as the translational kinetic energy, so if we assume $kT \gg I\omega^2/2$ and a probability density function of the form

$$\frac{df}{d\omega} = B_{\text{rot}} \exp\left(\frac{-I\omega^2}{2kT}\right)$$

finding B_{rot} and $\langle \epsilon_{\text{rot}} \rangle$ follows exactly as before, and the average rotational kinetic energy is

$$\langle \epsilon_{\text{rot}} \rangle = kT/2$$

for a molecule with one degree of rotational freedom.

For a classical harmonic oscillator, the vibrational energy has both kinetic and potential energy components. They are $mv^2/2$ and $kx^2/2$ where v is the oscillator's instantaneous velocity, x is its instantaneous location, and k is the force constant. Both of these have the same form as the translational kinetic energy equation. If we can assume that $kT \gg mv^2/2$, that $kT \gg kx^2/2$, and that the probability density functions are

$$\frac{df}{dv} = B_{\text{vib}}^{\text{kinetic}} \exp\left(\frac{-mv^2}{2kT}\right)$$

and

$$\frac{df}{dx} = B_{\text{vib}}^{\text{potential}} \exp\left(\frac{-kx^2}{2kT}\right)$$

the same arguments show that the average kinetic energy and the average potential energy are both $kT/2$:

$$\langle \epsilon_{\text{vib}}^{\text{kinetic}} \rangle = kT/2$$

and

$$\langle \epsilon_{\text{vib}}^{\text{potential}} \rangle = kT/2$$

so that the average total vibrational energy is

$$\langle \epsilon_{\text{vib}}^{\text{total}} \rangle = kT$$

In summary, because the energy for translational motion in one dimension, the energy for rotational motion about one axis, the energy for vibrational kinetic energy in one dimension, and the energy for vibrational potential energy in one dimension all have the same form ($\epsilon = Xu^2$) each of these modes can contribute $kT/2$ to the average energy of a molecule. For translation and rotation, the total is $kT/2$ for each degree of translational or rotational freedom. For vibration, because there is both a kinetic and a potential energy contribution, the total is kT per degree of vibrational freedom.

Let us illustrate this for the particular case of a nonlinear, triatomic molecule. From our discussion in §18-4, we see that there are three degrees of translational freedom, three degrees of rotational freedom, and three degrees of vibrational freedom. The contributions to the average molecular energy are

$3(kT/2)$	from translation
$+3(kT/2)$	from rotation
$+3kT$	from vibration
$= 6kT$	in total

Since the heat capacity is

$$C_V = \left(\frac{\partial \epsilon}{\partial T}\right)_v$$

each translational degree of freedom can contribute $k/2$ to the heat capacity. Each rotational degree of freedom can also contribute $k/2$ to the heat capacity. Each vibrational degree of freedom can contribute k to the heat capacity. It is important to remember that these results represent upper limits for real molecules. These limits are realized at high temperatures, or more precisely, at temperatures where many energy levels, ϵ_i, satisfy $\epsilon_i \ll kT$

§4 Partition functions and average energies at high temperatures

It is enlightening to find the integral approximations to the partition functions and average energies for our simple quantum-mechanical models of translational, rotational, and vibrational motions. In doing so, however, it is important to remember that the use of integrals to approximate Boltzmann-equation sums assumes that there are a large number of energy levels, ϵ_i, for which $\epsilon_i \ll kT$. If we select a high enough temperature, the energy levels for any motion will always satisfy this condition. The energy levels for translational motion satisfy this condition even at sub-ambient temperatures. This is the reason that Maxwell's derivation of the probability density function for translational motion is successful.

Rotational motion is an intermediate case. At subambient temperatures, the classical-mechanical derivation can be inadequate; at ordinary temperatures, it is a good approximation. This can be seen by comparing the classical-theory prediction to experimental values for diatomic molecules. For diatomic molecules, the classical model predicts a constant-volume heat capacity of $5k/2$ from 3 degrees of translational and 2 degrees of rotational freedom. Since this does not include the contributions from vibrational motions, constant-volume heat capacities for diatomic molecules must be greater than $5k/2$ if both the translational and rotational contributions are accounted for by the classical model. For diatomic molecules at 298 K, the experimental values are indeed somewhat larger than $5k/2$. (Hydrogen is an exception; its value is $2.47\ k$.)

Vibrational energies are usually so big that only a minor fraction of the molecules can be in higher vibrational levels at reasonable temperatures. If we try to increase the temperature enough to make the high-temperature approximation describe vibrational motions, most molecules decompose. Likewise, electronic partition functions must be evaluated from the defining equation.

The high-temperature limiting average energies can also be calculated from the Boltzmann equation and the appropriate quantum-mechanical energies. Recall that we find the following quantum-mechanical energies for simple models of translational, rotational, and vibrational motions:

Translation

$$\epsilon_{\text{trans}}^{(n)} = \frac{n^2h^2}{8m\ell^2}$$

(n = 1, 2, 3, ... Derived for a particle in a box)

Rotation

$$\epsilon_{\text{rot}}^{(m)} = \frac{m^2h^2}{8\pi^2 I}$$

(m = 1, 2, 3, ... Derived for rotation about one axis—each energy level is doubly degenerate)

Vibration

$$\epsilon_{\text{vibration}}^{(n)} = h\nu\left(n+\frac{1}{2}\right)$$

(n = 0, 1, 2, 3, ... Derived for simple harmonic motion in one dimension)

When we assume that the temperature is so high that many ϵ_i are small compared to kT, we find the following high-temperature limiting partition functions for these motions:

$$\begin{aligned} z_{\text{translation}} &= \sum_{n=1}^{\infty} \exp\left(\frac{-n^2h^2}{8m\ell^2kT}\right) \\ &\approx \int_0^{\infty} \exp\left(\frac{-n^2h^2}{8m\ell^2kT}\right) dn \\ &= \left(\frac{2\pi mkT\ell^2}{h^2}\right)^{1/2} \end{aligned}$$

$$\begin{aligned} z_{\text{rotation}} &= \sum_{m=1}^{\infty} 2\exp\left(\frac{-m^2h^2}{8\pi^2 IkT}\right) \\ &\approx 2\int_0^{\infty} \exp\left(\frac{-m^2h^2}{8\pi^2 IkT}\right) dn \\ &= \left(\frac{8\pi^3 IkT}{h^2}\right)^{1/2} \end{aligned}$$

$$\begin{aligned} z_{\text{vibration}} &= \sum_{n=0}^{\infty} \exp\left(\frac{-h\nu}{kT}\left(n+\frac{1}{2}\right)\right) \\ &\approx \int_0^{\infty} \exp\left(\frac{-h\nu}{kT}\left(n+\frac{1}{2}\right)\right) dn \\ &= \frac{kT}{h\nu}\exp\left(\frac{-h\nu}{2kT}\right) \end{aligned}$$

We can then calculate the average energy for each mode as

$$\langle\epsilon\rangle = z^{-1}\int_0^{\infty} \epsilon_n \exp\left(\frac{-\epsilon_n}{kT}\right) dn$$

and find

$$\begin{aligned} \langle\epsilon_{\text{translation}}\rangle &= z_{\text{translation}}^{-1}\int_0^{\infty}\left(\frac{n^2h^2}{8m\ell^2}\right)\exp\left(\frac{-n^2h^2}{8m\ell^2kT}\right) dn \\ &= \frac{kT}{2} \end{aligned}$$

$$\begin{aligned} \langle\epsilon_{\text{rotation}}\rangle &= z_{\text{rotation}}^{-1}\int_0^{\infty} 2\left(\frac{m^2h^2}{8\pi^2 I}\right)\exp\left(\frac{-m^2h^2}{8\pi^2 IkT}\right) dm \\ &= \frac{kT}{2} \end{aligned}$$

$$\begin{aligned} \langle\epsilon_{\text{vibration}}\rangle &= z_{\text{vibration}}^{-1} \\ &\quad \times \int_0^{\infty} h\nu\left(n+\frac{1}{2}\right)\exp\left(\frac{-h\nu}{kT}\left(n+\frac{1}{2}\right)\right) dn \\ &= kT + \frac{h\nu}{2} \\ &\approx kT \end{aligned}$$

where the last approximation assumes that $h\nu/2 \ll kT$. In the limit as $T \to 0$, the average energy of the vibrational mode becomes just $h\nu/2$. This is just the energy of the lowest vibrational state, implying that all of the molecules are in the lowest vibrational energy level at absolute zero.

§5 Energy levels for a three-dimensional harmonic oscillator

One of the earliest applications of quantum mechanics was Einstein's demonstration that the union of statistical mechanics and quantum mechanics explains the temperature variation of the heat capacities of solid materials. In §7-14, we note that the heat capacities of solid materials approach zero as the temperature approaches absolute zero. We also review the law of Dulong and Petit, which describes the limiting heat capacity of many solid elements at high (ambient) temperatures. The Einstein model accounts for both of these observations.

The physical model underlying Einstein's development is that a monatomic solid consists of atoms vibrating about fixed points in a lattice. The particles of this solid are distinguishable from one another, because the location of each lattice point is uniquely specified. We suppose that the vibration of any one atom is independent of the vibrations of the other atoms in the lattice. We assume that the vibration results from a Hooke's Law restoring force

$$\vec{F} = -\lambda\vec{r} = -\lambda(x\vec{\imath} + y\vec{\jmath} + z\vec{k})$$

that is zero when the atom is at its lattice point, for which $\vec{r} = (0,0,0)$. The potential energy change when the atom, of mass m, is driven from its lattice point to the point $\vec{r} = (x,y,x)$ is

$$\begin{aligned} V &= \int_{\vec{r}=\vec{0}}^{\vec{r}} -\vec{F}\cdot d\vec{r} \\ &= \lambda\int_{x=0}^{x} xdx + \lambda\int_{y=0}^{y} ydy + \lambda\int_{z=0}^{z} zdz \\ &= \lambda\frac{x^2}{2} + \lambda\frac{y^2}{2} + \lambda\frac{z^2}{2} \end{aligned}$$

The Schroedinger equation for this motion is

$$\begin{aligned} -\frac{h^2}{8\pi^2 m}\left[\frac{\partial^2\psi}{\partial x^2} + \frac{\partial^2\psi}{\partial y^2} + \frac{\partial^2\psi}{\partial z^2}\right] + \lambda\left[\frac{x^2}{2} + \frac{y^2}{2} + \frac{z^2}{2}\right]\psi \\ = \epsilon\psi \end{aligned}$$

where ψ is a function of the three displacement coordinates; that is $\psi = \psi(x,y,z)$. We assume that motions in the x-, y-, and z-directions are completely independent of one another. When we do so, it turns out that we can express the three-dimensional Schroedinger equation as the sum of three one-dimensional Schroedinger equations

$$\begin{aligned} &\left[-\frac{h^2}{8\pi^2 m}\frac{\partial^2\psi_x}{\partial x^2} + \lambda\frac{x^2\psi_x}{2}\right] \\ &+\left[-\frac{h^2}{8\pi^2 m}\frac{\partial^2\psi_y}{\partial y^2} + \lambda\frac{y^2\psi_y}{2}\right] \\ &+\left[-\frac{h^2}{8\pi^2 m}\frac{\partial^2\psi_z}{\partial z^2} + \lambda\frac{z^2\psi_z}{2}\right] \\ &= \epsilon\,\psi_x + \epsilon\,\psi_y + \epsilon\,\psi_z \end{aligned}$$

where any wave function $\psi_x^{(n)}$ is the same function as $\psi_y^{(n)}$ and $\psi_z^{(n)}$, and the corresponding energies $\epsilon_x^{(n)}$, $\epsilon_y^{(n)}$, and $\epsilon_z^{(n)}$ have the same values. The energy of the three-dimensional atomic motion is simply the sum of the energies for the three one-dimensional motions. That is, $\epsilon_{n,m,p} = \epsilon_x^{(n)} + \epsilon_y^{(m)} + \epsilon_z^{(p)}$, which, for simplicity, we also write as $\epsilon_{n,m,p} = \epsilon_n + \epsilon_m + \epsilon_p$.

§6 Energy and heat capacity of the "Einstein crystal"

In §4, we find an approximate partition function for the harmonic oscillator at high temperatures. Because it is a geometric series, the partition function for the harmonic oscillator can also be obtained exactly at any temperature. By definition, the partition function for the harmonic oscillator is

$$\begin{aligned} z &= \sum_{n=0}^{\infty} \exp\left(\frac{-h\nu}{kT}\left(n+\frac{1}{2}\right)\right) \\ &= \exp\left(\frac{-h\nu}{2kT}\right)\sum_{n=0}^{\infty}\exp\left(\frac{-nh\nu}{kT}\right) \\ &= \exp\left(\frac{-h\nu}{2kT}\right)\sum_{n=0}^{\infty}\left[\exp\left(\frac{-h\nu}{kT}\right)\right]^n \end{aligned}$$

This is just the infinite sum

$$z = a\sum_{n=0}^{\infty} r^n = \frac{a}{1-r}$$

with

$$a = \exp\left(\frac{-h\nu}{2kT}\right)$$

and

$$r = \exp\left(\frac{-h\nu}{kT}\right)$$

Hence, the exact partition function for the one-dimensional harmonic oscillator is

$$z = \frac{\exp(-h\nu/2kT)}{1-\exp(-h\nu/kT)}$$

The partition function for vibration in each of the other two dimensions is the same. To get the partition function for oscillation in all three dimensions, we must sum over all possible combinations of the three energies. Distinguishing the energies associated with motion in the x-, y-, and z-directions by the subscripts n, m, and p, respectively, we have for the three-dimensional harmonic oscillator:

$$\begin{aligned} z_{3D} &= \sum_{p=0}^{\infty}\sum_{m=0}^{\infty}\sum_{n=0}^{\infty}\exp\left[\frac{-(\epsilon_n+\epsilon_m+\epsilon_p)}{kT}\right] \\ &= \sum_{p=0}^{\infty}\exp\frac{-\epsilon_p}{kT}\sum_{m=0}^{\infty}\exp\frac{-\epsilon_m}{kT}\sum_{n=0}^{\infty}\exp\frac{-\epsilon_n}{kT} \\ &= z^3 \end{aligned}$$

Hence,

$$z_{3D} = \left[\frac{\exp(-h\nu/2kT)}{1-\exp(-h\nu/kT)}\right]^3$$

and the energy of a crystal of N, independent, distinguishable atoms is

$$\begin{aligned} E &= N\langle\epsilon\rangle \\ &= NkT^2\left(\frac{\partial \ln z_{3D}}{\partial T}\right)_V \\ &= \frac{3Nh\nu}{2}+\frac{3Nh\nu\exp(-h\nu/kT)}{1-\exp(-h\nu/kT)} \end{aligned}$$

Taking the partial derivative with respect to temperature gives the heat capacity of this crystal. The molar heat capacity can be expressed in two ways that are useful for our purposes:

$$\begin{aligned} C_V &= \left(\frac{\partial \overline{E}}{\partial T}\right)_V \\ &= 3\overline{N}k\left(\frac{h\nu}{kT}\right)^2\left[\frac{\exp(-h\nu/kT)}{(1-\exp(-h\nu/kT))^2}\right] \\ &= 3\overline{N}k\left(\frac{h\nu}{kT}\right)^2\left[\frac{\exp(h\nu/kT)}{(\exp(h\nu/kT)-1)^2}\right] \end{aligned}$$

Consider the heat capacity at high temperatures. As the temperature becomes large, $h\nu/kT$ approaches zero. Then

$$\exp\left(\frac{h\nu}{kT}\right)\approx 1+\frac{h\nu}{kT}$$

Using this approximation in the second representation of C_V gives for the high temperature limit

$$\begin{aligned} C_V &\approx 3\overline{N}k\left(\frac{h\nu}{kT}\right)^2\left[\frac{1+h\nu/kT}{(1+h\nu/kT-1)^2}\right] \\ &\approx 3\overline{N}k\left(1+\frac{h\nu}{kT}\right) \\ &\approx 3\overline{N}k \\ &= 3R \end{aligned}$$

Since C_V and C_P are about the same for solids at ordinary temperatures, this result is essentially equivalent to the law stated by Dulong and Petit. Indeed, it suggests that the law would be more accurate if stated as a condition on C_V rather than C_P, and this proves to be the case.

At low temperatures, $h\nu/kT$ becomes arbitrarily large and $\exp(-h\nu/kT)$ approaches zero. From the first representation of C_V, we see that

$$\lim_{T\to 0}\left(\frac{\partial \overline{E}}{\partial T}\right)_V = C_V = 0$$

In §10-9, we see that $C_P - C_V \to 0$ as $T \to 0$. Hence, the theory also predicts that $C_P \to 0$ as $T \to 0$, in agreement with experimental results.

The Einstein model assumes that energy variations in a solid near absolute zero are entirely due to variations in the vibrational energy. From the assumption that all of these vibrational motions are characterized by a single frequency, it predicts the limiting values for the heat capacity of a solid at high and low temperatures. At intermediate temperatures, the quantitative predictions of the Einstein model leave room for improvement. An important refinement developed by Peter Debye assumes a spectrum of vibrational frequencies and results in excellent quantitative agreement with experimental values at all temperatures.

We can give a simple qualitative interpretation for the result that heat capacities decrease to zero as the temperature goes to absolute zero. The basic idea is that, at a sufficiently low temperature, essentially all of the molecules in the system are in the lowest available energy level. Once essentially all of the molecules are in the lowest energy level, the energy of the system can no longer decrease in response to a further temperature decrease. Therefore, in this temperature range, the heat capacity is essentially zero. Alternatively, we can say that as the temperature approaches zero, the fraction of the molecules that are in the lowest energy level approaches one, and the energy of the system of N molecules approaches the smallest value it can have.

The weakness in this qualitative view is that there is always a non-zero probability of finding molecules in a higher energy level, and this probability changes as the temperature changes. To firm up the simple picture, we need a way to show that the energy decreases more rapidly than the temperature near absolute zero. More precisely, we need a way to show that

$$\lim_{T\to 0}\left(\frac{\partial \overline{E}}{\partial T}\right)_V = C_V = 0$$

Since the Einstein model produces this result, it constitutes a quantitative validation of our qualitative model.

§7 Applications of other entropy relationships

In most cases, calculation of the entropy from information about the energy levels of a system is best accomplished using the partition function. Occasionally other entropy relationships are useful. We illustrate this by using the entropy relationship

$$S = -Nk\sum_{i=1}^{\infty} g_i\rho(\epsilon_i)\ln\rho(\epsilon_i) + Nk\ln g_1$$

to find the entropy of an N-molecule disordered crystal at absolute zero. To be specific, let us consider a crystal of carbon monoxide.

We can calculate the entropy of carbon monoxide at absolute zero from either of two perspectives. Let us first assume that the energy of a molecule is almost completely independent of the orientations of its neighbors in the crystal. Then the energy of any molecule in the crystal is essentially the same in either of the two orientations available to it. In this model for the system, we consider that there are two, non-degenerate, low-energy quantum states available to the molecule. We suppose that all other quantum states lie at energy levels whose probabilities are very small when the temperature is near absolute zero. We have $g_1 = g_2 = 1$, $\epsilon_2 \approx \epsilon_1$. Near absolute zero, we have $\rho(\epsilon_2) \approx \rho(\epsilon_1) \approx 1/2$; for $i > 2$, $\rho(\epsilon_i) \approx 0$. The entropy becomes

$$\begin{aligned} S &= -Nk \sum_{i=1}^{\infty} g_i \rho(\epsilon_i) \ln \rho(\epsilon_i) + Nk \ln g_1 \\ &= -Nk \left(\frac{1}{2}\right) \ln \left(\frac{1}{2}\right) - Nk \left(\frac{1}{2}\right) \ln \left(\frac{1}{2}\right) \\ &= -Nk \ln \left(\frac{1}{2}\right) \\ &= Nk \ln 2 \end{aligned}$$

Alternatively, we can consider that there is just one low-energy quantum state available to the molecule but that this quantum state is doubly degenerate. In this model, the energy of the molecule is the same in either of the two orientations available to it. We have $g_1 = 2$. Near absolute zero, we have $\rho(\epsilon_1) \approx 1$; for $i > 1$, $\rho(\epsilon_i) \approx 0$. The summation term vanishes, and the entropy becomes

$$S = Nk \ln g_1 = Nk \ln 2$$

Either perspective implies the same value for the zero-temperature entropy of the N-molecule crystal.

Either of these treatments involves a subtle over-simplification. In our first model, we recognize that the carbon monoxide molecule must have a different energy in each of its two possible orientations in an otherwise perfect crystal. The energy of the orientation that makes the crystal perfect is slightly less than the energy of the other orientation. We introduce an approximation when we say that $\rho(\epsilon_2) \approx \rho(\epsilon_1) \approx 1/2$. However, if ϵ_2 is not exactly equal to ϵ_1, this approximation cannot be valid at an arbitrarily low temperature. To see this, we let the energy difference between these orientations be $\epsilon_2 - \epsilon_1 = \Delta\epsilon > 0$. At relatively high temperatures, at which $\Delta\epsilon \ll kT$, we have

$$\frac{\rho(\epsilon_2)}{\rho(\epsilon_1)} = \exp\left(\frac{-\Delta\epsilon}{kT}\right) \approx 1$$

and $\rho(\epsilon_2) \approx \rho(\epsilon_1) \approx 1/2$. At such temperatures, the system behaves as if the lowest energy level were doubly degenerate, with $\epsilon_2 = \epsilon_1$. However, since T can be arbitrarily close to zero, this condition cannot always apply. No matter how small $\Delta\epsilon$ may be, there are always temperatures at which $\Delta\epsilon \gg kT$ and at which we have

$$\frac{\rho(\epsilon_2)}{\rho(\epsilon_1)} \approx 0$$

This implies that the molecule should always adopt the orientation that makes the crystal perfectly ordered when the temperature becomes sufficiently close to zero. This conclusion disagrees with the experimental observations.

Our second model assumes that the energy of a carbon monoxide molecule is the same in either of its two possible orientations. However, its interactions with the surrounding molecules cannot be exactly the same in each orientation; consequently, its energy cannot be exactly the same. From first principles, therefore, our second model cannot be strictly correct.

To resolve these apparent contradictions, we assume that the rate at which a carbon monoxide molecule can change its orientation within the lattice depends on temperature. For some temperature at which $\Delta\epsilon \ll kT$, the reorientation process occurs rapidly, and the two orientations are equally probable. As the temperature decreases, the rate of reorientation becomes very slow. If the reorientation process effectively ceases to occur while the condition $\Delta\epsilon \ll kT$ applies, the orientations of the component molecules remain those that occur at higher temperatures no matter how much the temperature decreases thereafter. This is often described by saying that molecular orientations become "frozen." The zero-temperature entropy of the system is determined by the energy-level probabilities that describe the system at the temperature at which reorientation effectively ceases to occur.

Problems

1. The gravitational potential energies available to a molecule near the surface of the earth are $\epsilon(h) = mgh$. Each height, h, corresponds to a unique energy, so we can infer that the degeneracy of $\epsilon(h)$ is unity. Derive the probability density function for the distribution of molecules in the earth's atmosphere. (See problem 19 in chapter 3.)

2. The value of the molecular partition function approximates the number of quantum states that are available to the molecule and whose energy is less than kT. How many such quantum states are available to a molecule of molecular weight 40 that is confined in a volume of 10^{-6} m^3 at 300 K?

23

The Ensemble Treatment

§1 Ensembles of N-molecule systems

When we begin our discussion of Boltzmann statistics in Chapter 20, we note that there exists, in principle, a Schroedinger equation for an N-molecule system. For any particular set of boundary conditions, the solutions of this equation are a set of infinitely many wave functions, $\Psi_{i,j}$, for the N-molecule system. For every such wave function, there is a corresponding system energy, E_i. The wave functions reflect all of the attractive and repulsive interactions among the molecules of the system. Likewise, the energy levels of the system reflect all of these interactions.

In §20-12, we introduce the symbol Ω_E to denote the degeneracy of the energy, E, of an N-molecule system. Because the constituent molecules are assumed to be distinguishable and non-interacting, we have

$$\Omega_E = \sum_{\{N_i\},E} W(N_i, g_i)$$

In the solution of the Schroedinger equation for a system of N interacting molecules, each system-energy level, E_i, can be degenerate. We again let Ω denote the degeneracy of an energy level of the system. We use Ω_i (rather than Ω_{E_i}) to represent the degeneracy of E_i. It is important to recognize that the symbol "Ω_i" now denotes an intrinsic quantum-mechanical property of the N-particle system.

In Chapters 21 and 22, we denote the parallel properties of an individual molecule by $\psi_{i,j}$ for the molecular wave functions, ϵ_i for the corresponding energy levels, and g_i for the degeneracy of the ith energy level. We imagine creating an N-molecule system by collecting N non-interacting molecules in a fixed volume and at a fixed temperature.

In exactly the same way, we now imagine collecting $\widehat{N}$ of these N-molecule, constant-volume, constant-temperature systems. An aggregate of many multi-molecule systems is called an ***ensemble***. Just as we assume that no forces act among the non-interacting molecules we consider earlier, we assume that no forces act among the systems of the ensemble. However, as we emphasize above, our model for the systems of an ensemble recognizes that intermolecular forces among the molecules of an individual system can be important. We can imagine specifying the properties of the individual systems in a variety of ways. A collection is called a ***canonical ensemble*** if each of the systems in the ensemble has the same values of N, V, and T. (The sense of this name is that by specifying constant N, V, and T, we create the ensemble that can be described most simply.)

The canonical ensemble is a collection of $\widehat{N}$ identical systems, just as the N-molecule system is a collection of N identical molecules. We imagine piling the systems that comprise the ensemble into a gigantic three-dimensional stack. We then immerse the entire stack—the ensemble—in a constant temperature bath. The ensemble and its constituent systems are at the constant temperature T. The volume of the ensemble is $\widehat{N}V$. Because we can specify the location of any system in the ensemble by specifying its x-, y-, and z-coordinates in the stack, the individual systems that comprise the ensemble are distinguishable from one another. Thus the ensemble is analogous to a crystalline N-molecule system, in which the individual molecules are distinguishable from one another because each occupies a particular location in the crystal lattice, the entire crystal is at the constant temperature, T, and the crystal volume is NV_{molecule}.

Since the ensemble is a conceptual construct, we can make the number of systems in the ensemble, $\widehat{N}$, as large as we please. Each system in the ensemble will have one of the quantum-mechanically allowed energies, E_i. We let the number of systems that have energy E_1 be $\widehat{N}_1$. Similarly, we let the number with energy E_2 be $\widehat{N}_2$, and the number with energy E_i be $\widehat{N}_i$. Thus at any given instant, the ensemble is characterized by a population set, $\{\widehat{N}_1, \widehat{N}_2, \dots, \widehat{N}_i, \dots\}$, in exactly the same way that an N-molecule system is characterized by a population set, $\{N_1, N_2, \dots, N_i, \dots\}$. We have

$$\widehat{N} = \sum_{i=1}^{\infty} \widehat{N}_i$$

While all of the systems in the ensemble are immersed in the same constant-temperature bath, the energy of any one system in the ensemble is completely independent of the energy of any other system. This means that the total energy of the ensemble, $\widehat{E}$, is given by

$$\widehat{E} = \sum_{i=1}^{\infty} \widehat{N}_i E_i$$

Correspondences		
Property	**System**	**Ensemble**
Quantum entity	***Molecule*** at fixed volume and temperature	***System*** comprising a collection of N molecules at fixed volume and temperature
Aggregate of quantum entities	***System*** comprising a collection of N molecules at fixed volume and temperature	***Ensemble*** comprising $\hat{N}$ systems each of which contains N molecules
Number of quantum entities in aggregate	N	$\hat{N}$
Wave functions/quantum states	ψ_i	Ψ_i
Energy levels	ϵ_i	E_i
Energy level degeneracies	g_i	Ω_i
Probability that an energy level is occupied	P_i	$\hat{P}_i$
Number of quantum entities in the ith energy level	N_i	$\hat{N}_i$
Probability that a quantum state is occupied	$\rho(\epsilon_i)$	$\hat{\rho}(E_i)$
Energy of the aggregate's k^{th} population set	$E_k = \sum N_{k,i}\epsilon_i$	$\hat{E}_k = \sum \hat{N}_{k,i}\epsilon_i$
Expected value of the energy of the aggregate	$\langle E \rangle = N \sum P_i \epsilon_i$	$\langle \hat{E} \rangle = \hat{N} \sum \hat{P}_i E_i$

The population set, $\{\hat{N}_1, \hat{N}_2, \dots, \hat{N}_i, \dots\}$, that characterizes the ensemble is not constant in time. However, by the same arguments that we apply to the N-molecule system, there is a population set

$$\{\hat{N}_1^{\bullet}, \hat{N}_2^{\bullet}, \dots, \hat{N}_i^{\bullet}, \dots\}$$

which characterizes the ensemble when it is at equilibrium in the constant-temperature bath.

We define the probability, $\hat{P}_i$, that a system of the ensemble has energy E_i to be the fraction of the systems in the ensemble with this energy, when the ensemble is at equilibrium at the specified temperature. Thus, by definition, $\hat{P}_i = \hat{N}_i^{\bullet}/\hat{N}$. We define the probability that a system is in one of the states, $\Psi_{i,j}$, with energy E_i, as

$$\hat{\rho}(E_i) = \frac{\hat{P}_i}{\Omega_i}$$

The method we have used to construct the canonical ensemble insures that the entire ensemble is always at the specified temperature. If the component systems are at equilibrium, the ensemble is at equilibrium. The expected value of the ensemble energy is

$$\langle \hat{E} \rangle = \hat{N} \sum_{i=1}^{\infty} \hat{P}_i E_i = \sum_{i=1}^{\infty} \hat{N}_i^{\bullet} E_i$$

Because the number of systems in the ensemble, $\hat{N}$, is very large, we know from the central limit theorem that any observed value for the ensemble energy will be indistinguishable from the expected value. To an excellent approximation, we have at any time, $\hat{E} = \langle \hat{E} \rangle$ and $\hat{N}_i^{\bullet} = \hat{N}_i$

The table above summarizes the terminology that we have developed to characterize molecules, N-molecule systems, and $\hat{N}$-system ensembles of N-molecule systems.

We can now apply to an ensemble of $\hat{N}$, distinguishable, non-interacting systems the same logic that we applied to a system of N, distinguishable, non-interacting molecules. The probability that a system is in one of the energy levels is

$$1 = \hat{P}_1 + \hat{P}_2 + \dots + \hat{P}_i + \dots$$

The total probability sum for the constant-temperature ensemble is

$$\begin{aligned} 1 &= \left(\hat{P}_1 + \hat{P}_2 + \dots + \hat{P}_i + \dots\right)^{\hat{N}} \\ &= \sum_{\{\hat{N}_i\}} \hat{W}\left(\hat{N}_i, \Omega_i\right) \hat{\rho}(E_1)^{\hat{N}_1} \hat{\rho}(E_2)^{\hat{N}_2} \dots \hat{\rho}(E_i)^{\hat{N}_i} \dots \end{aligned}$$

where

$$\hat{W}\left(\hat{N}_i, \Omega_i\right) = \hat{N}! \prod_{i=1}^{\infty} \frac{\Omega_i^{\hat{N}_i}}{\hat{N}_i!}$$

Moreover, we can imagine instantaneously isolating the ensemble from the temperature bath in which it is immersed. This is a wholly conceptual change, which we effect by replacing the fluid of the constant-temperature bath with a solid blanket of insulation. The ensemble is then an isolated system whose energy, $\hat{E}$, is constant. Every system of the isolated ensemble is immersed in a constant-temperature bath, where the constant-temperature bath consists of the $\hat{N}-1$ systems that make up the rest of the ensemble. This is an important feature of the ensemble treatment. It means that any conclusion we reach about the systems of the constant-energy ensemble is also a conclusion about each of the $\hat{N}$ identical, constant-temperature systems that comprise the isolated, constant-energy ensemble.

Only certain population sets, $\{\hat{N}_1, \hat{N}_2, \dots, \hat{N}_i, \dots\}$, are consistent with the fixed value, $\hat{E}$, of the isolated ensemble. For each of these population sets, there are $\hat{W}(\hat{N}_i, \Omega_i)$ system states. The probability of each of these system states is proportional to $\hat{\rho}(E_1)^{\hat{N}_1}\hat{\rho}(E_2)^{\hat{N}_2}\dots\hat{\rho}(E_i)^{\hat{N}_i}\dots$. By the principle of equal *a priori* probability, every system state of the fixed-energy ensemble occurs with equal probability. We again conclude that the population set that characterizes the equilibrium state of the constant-energy ensemble, $\{\hat{N}_1^\bullet, \hat{N}_2^\bullet, \dots, \hat{N}_i^\bullet, \dots\}$, is the one for which $\hat{W}$ or $\ln \hat{W}$ is a maximum, subject to the constraints

$$\hat{N} = \sum_{i=1}^{\infty} \hat{N}_i$$

and

$$\hat{E} = \sum_{i=1}^{\infty} \hat{N}_i E_i$$

The fact that we can make $\hat{N}$ arbitrarily large ensures that any term, $\hat{N}_i^\bullet$, in the equilibrium-characterizing population set can be very large, so that $\hat{N}_i^\bullet$ can be found using Stirling's approximation and Lagrange's method of undetermined multipliers. We have the mnemonic function

$$F_{mn} = \hat{N}\ln\hat{N} - \hat{N} + \sum_{i=1}^{\infty}\left(\hat{N}_i \ln \Omega_i - \hat{N}_i \ln \hat{N}_i + \hat{N}_i\right) + \alpha\left(\hat{N} - \sum_{i=1}^{\infty}\hat{N}_i\right) + \beta\left(\hat{E} - \sum_{i=1}^{\infty}\hat{N}_i E_i\right)$$

so that

$$\left(\frac{\partial F_{mn}}{\partial \hat{N}_i^\bullet}\right)_{j\neq i} = \ln \Omega_i - \frac{\hat{N}_i^\bullet}{\hat{N}_i^\bullet} - \ln \hat{N}_i^\bullet + 1 - \alpha - \beta E_i = 0$$

and

$$\ln \hat{N}_i^\bullet = \ln \Omega_i - \alpha - \beta E_i$$

or

$$\hat{N}_i^\bullet = \Omega_i \exp(-\alpha) \exp -\beta E_i$$

When we make use of the constraint on the total number of systems in the ensemble, we have

$$\hat{N} = \sum_{i=1}^{\infty} \hat{N}_i^\bullet = \exp(-\alpha)\sum_{i=1}^{\infty}\Omega_i \exp(-\beta E_i)$$

so that

$$\exp(-\alpha) = \hat{N}Z^{-1}$$

where the partition function for a system of N possibly-interacting molecules is

$$Z = \sum_{i=1}^{\infty}\Omega_i \exp(-\beta E_i)$$

The probability that a system has energy E_i is equal to the equilibrium fraction of systems in the ensemble that have energy E_i, so that

$$\hat{P}_i = \frac{\hat{N}_i^\bullet}{\hat{N}} = \frac{\Omega_i \exp(-\beta E_i)}{Z}$$

§2 The ensemble entropy and the value of β

At equilibrium, the entropy of the $\hat{N}$-system ensemble, S_{ensemble}, must be a maximum. By arguments that parallel those in Chapter 20, $\hat{W}$ is a maximum for the ensemble population set that characterizes this equilibrium state. Applying the Boltzmann definition to the ensemble, the ensemble entropy is $S_{ensemble} = k \ln \hat{W}_{\text{max}}$. Since all $\hat{N}$ systems in the ensemble have effectively the same entropy, S, we have $S_{ensemble} = \hat{N}S$. When we assume that $\hat{W}_{\text{max}}$ occurs for the equilibrium population set, $\{\hat{N}_1^\bullet, \hat{N}_2^\bullet, \dots, \hat{N}_i^\bullet, \dots\}$, we have

$$\hat{W}_{\text{max}} = \hat{N}! \prod_{i=1}^{\infty}\frac{\Omega_i^{\hat{N}_i^\bullet}}{\hat{N}_i^\bullet !}$$

so that

$$S_{ensemble} = \hat{N}S = k\ln\hat{N}! + k\sum_{i=1}^{\infty}\hat{N}_i^\bullet \ln \Omega_i - k\sum_{i=1}^{\infty}\ln\left(\hat{N}_i^\bullet !\right)$$

From the Boltzmann distribution function, $\hat{N}_i^\bullet/\hat{N} = Z^{-1}\Omega_i \exp(-\beta E_i)$, we have

$$\ln \Omega_i = \ln Z + \ln \hat{N}_i^\bullet + \beta E_i - \ln \hat{N}$$

Substituting, and introducing Stirling's approximation, we find

$$\hat{N}S = k\hat{N}\ln\hat{N} - k\hat{N} + k\sum_{i=1}^{\infty}\hat{N}_i^\bullet\left(\ln Z + \ln\hat{N}_i^\bullet + \beta E_i - \ln\hat{N}\right) - k\sum_{i=1}^{\infty}\left(\hat{N}_i^\bullet \ln \hat{N}_i^\bullet - \hat{N}_i^\bullet\right)$$

$$= \hat{N}k\ln Z + k\beta\sum_{i=1}^{\infty}\hat{N}_i^{\bullet}E_i$$

Since $\sum_{i=1}^{\infty}\hat{N}_i^{\bullet}E_i$ is the energy of the $\hat{N}$-system ensemble and the energy of each system is the same, we have

$$\sum_{i=1}^{\infty}\hat{N}_i^{\bullet}E_i = E_{ensemble} = \hat{N}E$$

Substituting, we find

$$S = k\beta E + k\ln Z$$

where S, E, and Z are the entropy, energy, and partition function for the N-molecule system. From the fundamental equation, we have

$$\left(\frac{\partial E}{\partial S}\right)_V = T$$

Differentiating $S = k\beta E + k\ln Z$ with respect to entropy at constant volume, we find

$$1 = k\beta\left(\frac{\partial E}{\partial S}\right)_V$$

and it follows that

$$\beta = \frac{1}{kT}$$

We have, for the N-molecule system

$$Z = \sum_{i=1}^{\infty}\Omega_i\exp\left(\frac{-E_i}{kT}\right)$$

(System partition function)

$$\hat{P}_i = Z^{-1}\Omega_i\exp\left(\frac{-E_i}{kT}\right)$$

(Boltzmann's equation)

$$S = \frac{E}{T} + k\ln Z$$

(Entropy of the N-molecule system)

§3 The thermodynamic functions of the N-molecule system

With the results of §2 in hand, we can find the other thermodynamic functions for the N-molecule system from the equations for Z and $\hat{P}_i$ by the arguments we use in Chapters 20 and 21. Let us summarize these arguments. From

$$E = \sum_{i=1}^{\infty}\hat{P}_i E_i$$

we have

$$dE = \sum_{i=1}^{\infty}E_i d\hat{P}_i + \sum_{i=1}^{\infty}\hat{P}_i\, dE_i$$

We associate the first term with dq^{rev} and the second term with $dw = -PdV$; that is,

$$\begin{aligned} dq^{rev} &= TdS \\ &= \sum_{i=1}^{\infty}E_i d\hat{P}_i \\ &= -kT\sum_{i=1}^{\infty}\ln\left(\frac{\hat{P}_i}{\Omega_i}\right)d\hat{P}_i - kT\ln Z\sum_{i=1}^{\infty}d\hat{P}_i \end{aligned}$$

Where we substitute

$$E_i = -kT\ln\left(\frac{\hat{P}_i}{\Omega_i}\right) - kT\ln Z$$

which we obtain by taking the natural logarithm of the partition function. Since $\sum_{i=1}^{\infty}d\hat{P}_i = 0$, we have for each system,

$$\begin{aligned} dS &= -k\sum_{i=1}^{\infty}\ln\left(\frac{\hat{P}_i}{\Omega_i}\right)d\hat{P}_i \\ &= -k\sum_{i=1}^{\infty}\left\{\Omega_i d\left(\frac{\hat{P}_i}{\Omega_i}\ln\frac{\hat{P}_i}{\Omega_i}\right) - d\hat{P}_i\right\} \\ &= -k\sum_{i=1}^{\infty}d\left(\hat{P}_i\ln\frac{\hat{P}_i}{\Omega_i}\right) \end{aligned}$$

The system entropy, S, and the system-energy-level probabilities, $\hat{P}_i$, are functions of temperature. Integrating from $T = 0$ to T and choosing the lower limits for the integrations on the right to be $\hat{P}_1(0) = 1$ and $\hat{P}_i(0) = 0$ for $i > 1$, we have

$$\int_{S_0}^{S}dS = -k\sum_{i=1}^{\infty}\int_{\hat{P}_i(0)}^{\hat{P}_i(T)}d\left(\hat{P}_i\ln\frac{\hat{P}_i}{\Omega_i}\right)$$

Letting $\hat{P}_i(T) = \hat{P}_i$, the result is

$$\begin{aligned} S - S_0 &= -k\hat{P}_1\ln\frac{\hat{P}_1}{\Omega_1} + k\ln\frac{1}{\Omega_1} - k\sum_{i=2}^{\infty}\hat{P}_i\ln\frac{\hat{P}_i}{\Omega_i} \\ &= -k\sum_{i=1}^{\infty}\hat{P}_i\ln\frac{\hat{P}_i}{\Omega_i} - k\ln\Omega_1 \end{aligned}$$

From the partition function, we have

$$\ln\left(\frac{\hat{P}_i}{\Omega_i}\right) = -\frac{E_i}{kT} + \ln Z$$

so that

$$\begin{aligned} S - S_0 &= -k\sum_{i=1}^{\infty}\hat{P}_i\left(-\frac{E_i}{kT} + \ln Z\right) - k\ln\Omega_1 \\ &= \frac{1}{T}\sum_{i=1}^{\infty}\hat{P}_i E_i + k\ln Z\sum_{i=1}^{\infty}\hat{P}_i - k\ln\Omega_1 \\ &= \frac{E}{T} + k\ln Z - k\ln\Omega_1 \end{aligned}$$

We take the system entropy at absolute zero, S_0, to be

$$S_0 = k \ln \Omega_1$$

If the lowest energy state is non-degenerate, $\Omega_1 = 1$, and $S_0 = 0$, so that

$$S(T) = \frac{E}{T} + k \ln Z$$

As in §21-6, we observe that

$$E = \sum_{i=1}^{\infty} \hat{P}_i E_i = Z^{-1} \sum_{i=1}^{\infty} \Omega_i E_i \exp\left(\frac{-E_i}{kT}\right)$$

and that

$$\left(\frac{\partial \ln Z}{\partial T}\right)_V = Z^{-1} \sum_{i=1}^{\infty} \Omega_i \left(\frac{E_i}{kT^2}\right) \exp\left(\frac{-E_i}{kT}\right) = \frac{E}{kT^2}$$

so that

$$E = kT^2 \left(\frac{\partial \ln Z}{\partial T}\right)_V$$

From $A = E - TS$ and the entropy equation, $S = E/T + k \ln Z$, the Helmholtz free energy of the system is

$$A = -kT \ln Z$$

For the system pressure, we find from

$$P = -\left(\frac{\partial A}{\partial V}\right)_T$$

that

$$P = kT \left(\frac{\partial \ln Z}{\partial V}\right)_T$$

From $H = E + PV$, we find

$$H - kT^2 \left(\frac{\partial \ln Z}{\partial T}\right)_V + VkT \left(\frac{\partial \ln Z}{\partial V}\right)_T$$

and from $G = E + PV - TS$, we find

$$G = VkT \left(\frac{\partial \ln Z}{\partial V}\right)_T - kT \ln Z$$

For the chemical potential per molecule in the N-molecule system, we obtain

$$\mu = \left(\frac{\partial A}{\partial N}\right)_{VT} = -kT \left(\frac{\partial \ln Z}{\partial N}\right)_{VT}$$

Thus, we have found the principle thermodynamic functions for the N-molecule system expressed in terms of $\ln Z$ and its derivatives. The system partition function, Z, depends on the energy levels available to the N-molecule system. The thermodynamic functions we have obtained are valid for any system, including systems in which intermolecular forces make large contributions to the system energy. Of course, the system partition function, Z, must accurately reflect the effects of these forces.

In Chapter 24 we find that the partition function, Z, for a system of N, distinguishable, non-interacting molecules is related in a simple way to the molecular partition function, z. We find $Z = z^N$. When we substitute this result for Z into the system partition functions developed above, we recover the same results that we developed in Chapters 20 and 21 for the thermodynamic properties of a system of N, distinguishable, non-interacting molecules.

24

Indistinguishable Molecules: Statistical Thermodynamics of Ideal Gases

The ensemble analysis shows that the thermodynamic functions for an N-molecule system can be developed from the principles of statistical mechanics whether the molecules of the system interact or not. The theory is valid irrespective of the strengths of inter-molecular attractions and repulsions. However, to carry out numerical calculations, it is necessary to know the energy levels for the N-molecule system. For systems in which the molecules interact, obtaining useful approximations to these levels is a difficult problem. As a result, many applications assume that the molecules do not interact with one another. In this chapter we apply the results from the ensemble theory to the particular case of ideal gases.

§1 The partition function for N distinguishable, non-interacting molecules

In Chapter 21, our analysis of a system of N distinguishable and non-interacting molecules finds that the system entropy is given by

$$S = \frac{E}{T} + Nk \ln z = \frac{E}{T} + k \ln z^N$$

where E is the system energy and z is the molecular partition function. From ensemble theory, we found

$$S = \frac{E}{T} + k \ln Z$$

where Z is the partition function for the N-molecule system. Comparison implies that, for a system of N, distinguishable, non-interacting molecules, we have

$$Z = z^N$$

We can obtain this same result by writing out the energy levels for the system in terms of the energy levels of the distinguishable molecules that make up the system. First we develop the obvious notation for the energy levels of the individual molecules. We let the energy levels of the first molecule be the set $\{\epsilon_{1,i}\}$, the energy levels of the second molecule be the set $\{\epsilon_{2,i}\}$, and so forth to the last molecule for which the energy levels are the set $\{\epsilon_{N,i}\}$. Thus, the ith energy level of the rth molecule is $\epsilon_{r,i}$. We let the corresponding energy-level degeneracy be $g_{r,i}$ and the partition function for the rth molecule be z_r. Since all of the molecules are identical, each has the same set of energy levels; that is, we have $\epsilon_{p,i} = \epsilon_{r,i}$ and $g_{p,i} = g_{r,i}$ for any two molecules, p and r, and any energy level, i. It follows that the partition function is the same for every molecule

$$\begin{aligned} z_1 = z_2 = \cdots = z_j = \cdots = z_N \\ = z \\ = \sum_{i=1}^{\infty} g_{r,i} \exp\left(\frac{-\epsilon_{r,i}}{kT}\right) \end{aligned}$$

so that

$$z_1 z_2 \ldots z_r \ldots z_N = z^N$$

We can write down the energy levels available to the system of N distinguishable, non-interacting molecules. The energy of the system is just the sum of the energies of the constituent molecules, so the possible system energies consist of all of the possible sums of the distinguishable-molecule energies. Since there are infinitely many molecular energies, there are infinitely many system energies.

$$\begin{aligned} E_1 &= \epsilon_{1,1} + \epsilon_{2,1} + \cdots + \epsilon_{r,1} + \cdots + \epsilon_{N,1} \\ E_2 &= \epsilon_{1,2} + \epsilon_{2,1} + \cdots + \epsilon_{r,1} + \cdots + \epsilon_{N,1} \\ E_3 &= \epsilon_{1,3} + \epsilon_{2,1} + \cdots + \epsilon_{r,1} + \cdots + \epsilon_{N,1} \\ &\cdots \\ E_m &= \epsilon_{1,i} + \epsilon_{2,j} + \cdots + \epsilon_{r,k} + \cdots + \epsilon_{N,p} \\ &\cdots \end{aligned}$$

The product of the N molecular partition functions is

$$\begin{aligned} z_1 z_2 \ldots z_r \ldots z_N = \sum_{i=1}^{\infty} g_{1,i} \exp\left(\frac{-\epsilon_{1,i}}{kT}\right) \\ \times \sum_{j=1}^{\infty} g_{2,j} \exp\left(\frac{-\epsilon_{2,j}}{kT}\right) \times \ldots \times \sum_{k=1}^{\infty} g_{r,k} \exp\left(\frac{-\epsilon_{r,k}}{kT}\right) \times \\ \ldots \times \sum_{p=1}^{\infty} g_{N,p} \exp\left(\frac{-\epsilon_{N,p}}{kT}\right) \end{aligned}$$

$$\begin{aligned} = \sum_{i=1}^{\infty} \sum_{j=1}^{\infty} \ldots \sum_{k=1}^{\infty} \ldots \sum_{p=1}^{\infty} g_{1,i} g_{2,j} \ldots g_{r,k} \ldots g_{N,p} \\ \times \exp\left[\frac{-\left(\epsilon_{1,i} + \epsilon_{2,j} + \cdots + \epsilon_{r,k} + \cdots + \epsilon_{N,p}\right)}{kT}\right] \end{aligned}$$

The sum in each exponential term is just the sum of N single-molecule energies. Moreover, every possible combination of N single-molecule energies occurs in one of the exponential terms. Each of these possible combinations is a separate energy level available to the system of N distinguishable molecules.

The system partition function is

$$Z = \sum_{i=1}^{\infty} \Omega_i \exp\left(\frac{-E_i}{kT}\right)$$

The ith energy level of the system is the sum

$$E_i = \epsilon_{1,v} + \epsilon_{2,w} + \cdots + \epsilon_{r,k} + \cdots + \epsilon_{N,y}$$

The degeneracy of the ith energy level of the system is the product of the degeneracies of the molecular energy levels that belong to it. We have

$$\Omega_i = g_{1,v} g_{2,w} \cdots g_{r,k} \cdots g_{N,y}$$

Thus, by a second, independent argument, we see that

$$\begin{aligned} z_1 z_2 \dots z_r \dots z_N &= z^N \\ &= Z \end{aligned}$$

(N distinguishable, non-interacting molecules)

§2 The partition function for N <u>indistinguishable,</u> non-interacting molecules

In all of our considerations to this point, we focus on systems in which the molecules are distinguishable. This effectively confines the practical applications to crystalline solids. Since there is no way to distinguish one molecule of a given substance from another in the gas phase, it is evident that the assumptions we have used so far do not apply to gaseous systems. The number and importance of practical applications increases dramatically if we can extend the theory to describe the behavior of ideal gases.

We might suppose that distinguishability is immaterial—that there is no difference between the behavior of a system of distinguishable particles and an otherwise-identical system of indistinguishable particles. Indeed, this is an idea well worth testing. We know the partition function for a particle in box, and we have every reason to believe that this should be a good model for the partition function describing the translational motion of a gas particle. If an ideal gas behaves as a collection of N distinguishable particles-in-a-box, the translational partition of the gas is just z^N. Thermodynamic properties calculated on this basis for, say, argon should agree with those observed experimentally. Indeed, when the comparison is made, this theory gives some properties correctly. The energy is correct; however, the entropy is not.

Thus, experiment demonstrates that the partition function for a system of indistinguishable molecules is different from that of an otherwise-identical system of distinguishable molecules. The reason for this becomes evident when we compare the microstates available to a system of distinguishable molecules to those available to a system of otherwise-identical indistinguishable molecules. Consider the distinguishable-molecule microstate whose energy is

$$E_i = \epsilon_{1,v} + \epsilon_{2,w} + \cdots + \epsilon_{r,k} + \cdots + \epsilon_{N,y}$$

As a starting point, we assume that every molecule is in a different energy level. That is, all of the N energy levels, $\epsilon_{i,j}$, that appear in this sum are different. For the case in which the molecules are distinguishable, we can write down additional microstates that have this same energy just by permuting the energy values among the N molecules. (A second microstate with this energy is $E_i = \epsilon_{1,w} + \epsilon_{2,v} + \cdots + \epsilon_{r,k} + \cdots + \epsilon_{N,y}$.) Since there are $N!$ such permutations, there are a total of $N!$ quantum states that have this same energy, and each of them appears as an exponential term in the product $z_1 z_2 \dots z_r \dots z_N = z^N$.

If, however, the N molecules are indistinguishable, there is no way to tell one of these $N!$ assignments from another. They all become the same thing. All we know is that some one of the N molecules has the energy ϵ_w, another has the energy ϵ_v, *etc.* This means that there is only one way that the indistinguishable molecules can have the energy E_i. It means also that the difference between the distinguishable-molecules case and the indistinguishable-molecules case is that, while they contain the same system energy levels, each level appears $N!$ more times in the distinguishable-molecules partition function than it does in the indistinguishable-molecules partition function. We have

$$\begin{aligned} Z_{indistinguishable} &= \frac{1}{N!} Z_{distinguishable} \\ &= \frac{1}{N!} z^N \end{aligned}$$

In the next section, we see that nearly all of the molecules in a sample of gas must have different energies, so that this relationship correctly relates the partition function for a single gas molecule to the partition function for a system of N indistinguishable gas molecules.

Before seeing that nearly all of the molecules in a macroscopic sample of gas actually do have different energies, however, let us see what happens if they do not. Suppose that just two of the indistinguishable molecules have the same energy. Then there are not $N!$ permutations of the energies among the distinguishable molecules; rather there are only $N!/2!$ such permutations. In this case, the relationship between the system and the molecular partition functions is

$$\begin{aligned} Z_{indistinguishable} &= \frac{2!}{N!} Z_{distinguishable} \\ &= \frac{2!}{N!} z^N \end{aligned}$$

For the population set $\{N_1, N_2, \dots, N_r, \dots, N_\omega\}$ the relationship is

$$Z_{indistinguishable} = \frac{N_1!\, N_2! \dots N_r! \dots N_\omega!}{N!} z^N$$

which is much more complex than the case in which all molecules have different energies. Of course, if we extend the latter case, so that the population set consists of N energy levels, each occupied by at most one molecule, the relationship reverts to the one with which we began.

$$\begin{aligned} Z_{indistinguishable} &= \frac{1}{N!}\left(\prod_{i=1}^{\infty} N_i!\right) z^N \\ &= \frac{1}{N!}\left(\prod_{i=1}^{\infty} 1\right) z^N \\ &= \frac{1}{N!} z^N \end{aligned}$$

§3 Occupancy probabilities for translational energy levels

The particle in a box is a quantum mechanical model for the motion of a point mass in one dimension. In §18-3, we find that the energy levels are

$$\epsilon_n = \frac{n^2h^2}{8m\ell^2}$$

so that the partition function for a particle in a one-dimensional box is

$$z = \sum_{n=1}^{\infty} \exp\left(\frac{-n^2h^2}{8mkT\ell^2}\right)$$

When the mass approximates that of a molecule, the length of the box is macroscopic, and the temperature is not extremely low, there are a very large number of energy levels for which $\epsilon_n < kT$. When this is the case, we find in §22-4 that this sum can be approximated by an integral to obtain an expression for z in closed form:

$$z \approx \int_0^{\infty} \exp\left(\frac{-n^2h^2}{8mkT\ell^2}\right) dn = \left(\frac{2\pi mkT}{h^2}\right)^{1/2} \ell$$

A particle in a three-dimensional rectangular box is a quantum mechanical model for an ideal gas molecule. The molecule moves in three dimensions, but the component of its motion parallel to any one coordinate axis is independent of its motion parallel to the others. This being the case, the kinetic energy of a particle in a three-dimensional box can be modeled as the sum of the energies for motion along each of the three independent coordinate axes that describe the translational motion of the particle. Taking the coordinate axes parallel to the faces of the box and labeling the lengths of the sides ℓ_x, ℓ_y, and ℓ_z, the energy of the particle in the three-dimensional box becomes $\epsilon = \epsilon_x + \epsilon_y + \epsilon_z$, and the three-dimensional partition function becomes

$$\begin{aligned} z_t &= \sum_{n_x=1}^{\infty}\sum_{n_y=1}^{\infty}\sum_{n_z=1}^{\infty} \exp\left[\left(\frac{-h^2}{8mkT}\right)\left(\frac{n_x^2}{\ell_x^2}+\frac{n_y^2}{\ell_y^2}+\frac{n_z^2}{\ell_z^2}\right)\right] \\ &= \sum_{n_x=1}^{\infty} \exp\left(\frac{-n_x^2h^2}{8mkT\ell_x^2}\right) \sum_{n_y=1}^{\infty} \exp\left(\frac{-n_y^2h^2}{8mkT\ell_y^2}\right) \\ &\quad \times \sum_{n_z=1}^{\infty} \exp\left(\frac{-n_z^2h^2}{8mkT\ell_z^2}\right) \end{aligned}$$

or, recognizing this as the product of three one-dimensional partition functions, $z_t = z_x z_y z_z$. Approximating each of these as an integral gives

$$\begin{aligned} z_t &= \left(\frac{2\pi mkT}{h^2}\right)^{3/2} \ell_x\ell_y\ell_z \\ &= \left(\frac{2\pi mkT}{h^2}\right)^{3/2} V \end{aligned}$$

where the volume of the container is $V = \ell_x\ell_y\ell_z$.

Let us estimate a lower limit for the molecular partition function for the translational motion of a typical gas at ambient temperature. The partition function increases with volume, V, so we want to select a volume that is near the smallest volume a gas can have. We can estimate this as the volume of the corresponding liquid at the same temperature. Let us calculate the molecular translational partition function for a gas whose molar mass is 0.040 kg in a volume of 0.020 L at 300 K. We find $z_t = 5 \times 10^{27}$.

Given z_t, we can estimate the probability that any one of the energy levels available to this molecule is occupied. For any energy level, the upper limit to the term $\exp(-\epsilon_i/kT)$ is one. If the quantum numbers n_x, n_y, and n_z are different from one another, the corresponding molecular energy is non-degenerate. To a good approximation, we have $g_i = 1$. We find

$$\frac{N_i}{N} = \frac{g_i \exp(-\epsilon_i/kT)}{z_t} < \frac{1}{z_t} = 2 \times 10^{-28}$$

We calculate $N_i \approx 1 \times 10^{-4}$. When a mole of this gas occupies 0.020 L, the system density approximates that of a liquid. Therefore, even in circumstances selected to minimize the number of energy levels, there is less than one gas molecule per ten thousand energy levels.

For translational energy levels of gas molecules, it is an excellent approximation to say that each molecule occupies a different translational energy level. This is a welcome result, because it assures us that the translational partition function for a system containing a gas of N indistinguishable non-interacting molecules is just

$$Z_t = \frac{1}{N!}\left(\frac{2\pi mkT}{h^2}\right)^{3N/2} V^N$$

So that Z_t is the translational partition function for a system of N ideal gas molecules.

We derive Z_t from the assumption that every equilibrium population number, $N_i^{\bullet}$, for the molecular energy levels satisfies $N_i^{\bullet} \leq 1$. We use Z_t and the ensemble-treatment results that we develop in Chapter 23 to find thermodynamic functions for the N-molecule ideal-gas system. The ensemble development assumes that the number of systems, $\widehat{N}_i^{\bullet}$, in the ensemble that have energy E_i is very large. Since the ensemble is a creature of our imaginations, we can imagine that $\widehat{N}$ is as big as it needs to be in order that $\widehat{N}_i^{\bullet}$ be big enough. The population sets $\{N_i^{\bullet}\}$ and $\{\widehat{N}_i^{\bullet}\}$ are independent; they characterize different distributions. The fact that $N_i^{\bullet} \leq 1$ is irrelevant when we apply Lagrange's method to find the distribution function for $\widehat{N}_i^{\bullet}$, the partition function Z_t, and the thermodynamic functions for the system. Consequently, the ensemble treatment enables us to find the partition function for an ideal gas, Z_{IG}, by arguments that avoid the questions that arise when we apply Lagrange's method to the distribution of molecular translational energies.

§4 The separable-modes molecular model

At this point in our development, we have a theory that gives the thermodynamic properties of a polyatomic ideal gas molecule. To proceed, however, we must know the energy of every quantum state that is available to the molecule. There is more than one way to obtain this information. We will examine one important method—one that involves a further idealization of molecular behavior.

We have made great progress by using the ideal gas model, and as we have noted repeatedly, the essential feature of the ideal gas model is that there are no attractive or repulsive forces between its molecules. Now we assume that the molecule's translational, rotational, vibrational, and electronic motions are independent of one another. We could say that this idealization defines super-ideal gas molecules; not only does one molecule not interact with another molecule, an internal motion of one of these molecules does not interact with the other internal motions of the same molecule!

The approximation that a molecule's translational motion is independent of its rotational, vibrational, and electronic motions is usually excellent. The approximation that its intramolecular rotational, vibrational and electronic motions are also independent proves to be surprisingly good. Moreover, the very simple quantum mechanical systems that we describe in Chapter 18 prove to be surprisingly good models for the individual kinds of intramolecular motion. The remainder of this chapter illustrates these points.

In Chapter 18, we note that a molecule's wave function can be approximated as a product of a wave function for rotations, a wave function for vibrations, and a wave function for electronic motions. (As always, we are simply quoting quantum mechanical results that we make no effort to derive; we begin with the knowledge that the quantum mechanical problems have been solved and that the appropriate energy levels are available for our use.) Our goal is to see how we can apply the statistical mechanical results we have obtained to calculate the thermodynamic properties of ideal gases. To illustrate the essential features, we consider diatomic molecules. The same considerations apply to polyatomic molecules; there are additional complications, but none that introduce new principles.

For diatomic molecules, we need to consider the energy levels for translational motion in three dimensions, the energy levels for rotation in three dimensions, the energy levels for vibration along the inter-nuclear axis, and the electronic energy states.

§5 The partition function for a gas of indistinguishable, non-interacting, separable-modes molecules

We represent the successive molecular energy levels as ϵ_i and the successive translational, rotational, vibrational, and electronic energy levels as $\epsilon_{t,a}$, $\epsilon_{r,b}$, $\epsilon_{v,c}$, and $\epsilon_{e,d}$. Now the first subscript specifies the energy mode; the second specifies the energy level. We approximate the successive energy levels of a diatomic molecule as

$$\begin{aligned} \epsilon_1 &= \epsilon_{t,1} + \epsilon_{r,1} + \epsilon_{v,1} + \epsilon_{e,1} \\ \epsilon_2 &= \epsilon_{t,2} + \epsilon_{r,1} + \epsilon_{v,1} + \epsilon_{e,1} \\ &\cdots \\ \epsilon_i &= \epsilon_{t,a} + \epsilon_{r,b} + \epsilon_{v,c} + \epsilon_{e,d} \\ &\cdots \end{aligned}$$

In §22-1, we find that the partition function for the molecule becomes

$$\begin{aligned} z &= \sum_{a=1}^{\infty}\sum_{b=1}^{\infty}\sum_{c=1}^{\infty}\sum_{d=1}^{\infty} g_{t,a}\, g_{r,b} g_{v,c} g_{e,d} \\ &\qquad \times \exp\left[\frac{-\left(\epsilon_{t,a} + \epsilon_{r,b} + \epsilon_{v,c} + \epsilon_{e,d}\right)}{kT}\right] \\ &= z_t z_r z_v z_e \end{aligned}$$

where z_t, z_r, z_v, and z_e are the partition functions for the individual kinds of motion that the molecule undergoes; they are sums over the corresponding energy levels for the molecule. This is essentially the same argument that we use in §22-1 to show that the partition function for an N-molecule system is a product of N molecular partition functions: $Z = z^N$.

We are now able to write the partition function for a gas containing N molecules of the same substance. Since the molecules of a gas are indistinguishable, we use the relationship

$$Z_{indistinguishable} = \frac{1}{N!} z^N = \frac{1}{N!}\left(z_t z_r z_v z_e\right)^N$$

To make the notation more compact and to emphasize that we have specialized the discussion to the case of an ideal gas, let us replace "$Z_{indistinguishable}$" with "Z_{IG}". Also, recognizing that $N!$ enters the relationship because of molecular indistinguishability, and molecular indistinguishability arises because of translational motion, we regroup the terms, writing

$$Z_{\mathrm{IG}} = \left[\frac{(z_t)^N}{N!}\right](z_r)^N(z_v)^N(z_e)^N$$

Our goal is to calculate the thermodynamic properties of the ideal gas. These properties depend on the natural logarithm of the ideal-gas partition function. This is a sum of terms:

$$\ln Z_{IG} = \ln\left[\frac{(z_t)^N}{N!}\right] + N\ln z_r + N\ln z_v + N\ln z_e$$

In our development of classical thermodynamics, we find it convenient to express the properties of substance on a per-mole basis. For the same reasons, we focus on evaluating $\ln Z_{IG}$ for one mole of gas; that is, for the case that N is Avogadro's number, $\overline{N}$. We now examine the relationships that enable us to evaluate each of these contributions to $\ln Z_{IG}$.

§6 The translational partition function of an ideal gas

We can make use of Stirling's approximation to write the translational contribution to $\ln Z_{IG}$ per mole of ideal gas. This is

$$\ln\left[\frac{(z_t)^{\overline{N}}}{\overline{N}!}\right] = \overline{N}\ln z_t - \overline{N}\ln\overline{N} + \overline{N}$$
$$= \overline{N} + \overline{N}\ln\frac{z_t}{\overline{N}}$$

(We omit the other factors in Stirling's approximation. Their contribution to the thermodynamic values we calculate is less than the uncertainty introduced by the measurement errors in the molecular parameters we use.) In §3 we find the molecular partition function for translation:

$$z_t = \left(\frac{2\pi mkT}{h^2}\right)^{3/2} V$$

For one mole of an ideal gas, $\overline{V} = \overline{N}kT/P$. The translational contribution to the partition function for one mole of an ideal gas becomes

$$\ln\left[\frac{(z_t)^{\overline{N}}}{\overline{N}!}\right] = \overline{N} + \overline{N}\ln\left[\left(\frac{2\pi mkT}{h^2}\right)^{3/2}\frac{\overline{V}}{\overline{N}}\right]$$
$$= \overline{N} + \overline{N}\ln\left[\left(\frac{2\pi mkT}{h^2}\right)^{3/2}\frac{kT}{P}\right]$$

§7 The electronic partition function of an ideal gas

Our quantum-mechanical model for a diatomic molecule takes the zero of energy to be the infinitely separated atoms at rest—that is, with no kinetic energy. The

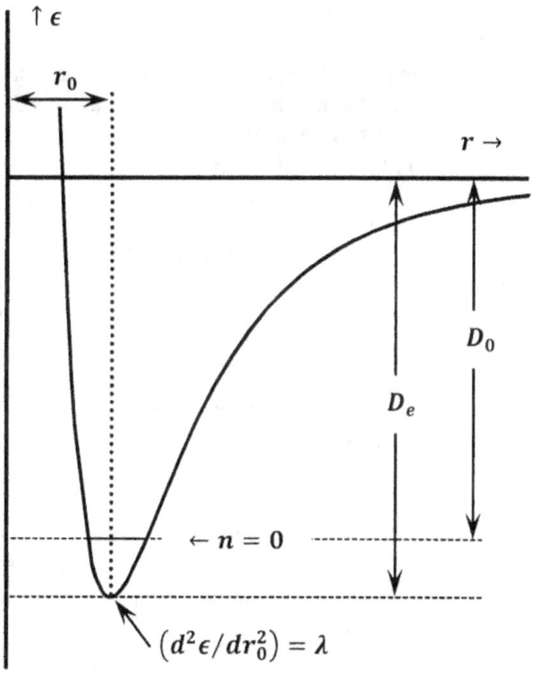

Figure 1. Equilibrium dissociation energy, D_e, and spectroscopic dissociation energy, D_0.

electrical interactions among the nuclei and electrons are such that, as the atoms approach one another, a bond forms and the energy of the two-atom system decreases. At some inter-nuclear distance, the energy reaches a minimum; at shorter inter-nuclear distances, the repulsive interactions between nuclei begin to dominate, and the energy increases. We can use quantum mechanics to find the wave function and energy of the molecule when the nuclei are separated to any fixed distance. By repeating the calculation at a series of inter-nuclear distances, we can find the distance at which the molecular energy is a minimum. We take this minimum energy as the electronic energy of the molecule, and the corresponding inter-nuclear distance as the bond length. This is the energy of the lowest electronic state of the molecule. The lowest electronic state is called the ground state.

Excited electronic states exist, and their energies can be estimated from spectroscopic measurements or by quantum mechanical calculation. For most molecules, these excited electronic states are at much higher energy than the ground state. When we compare the terms in the electronic partition function, we see that

$$\exp(-\epsilon_{e,1}/kT) \gg \exp(-\epsilon_{e,2}/kT)$$

The term for any higher energy level is insignificant compared to the term for the ground state. The electronic partition function becomes just

$$z_e = g_1\exp(-\epsilon_{e,1}/kT)$$

The ground-state degeneracy, g_1, is one for most molecules. For unusual molecules the ground-state degeneracy can be greater; for molecules with one unpaired electron, it is two.

The energy of the electronic ground state that we obtain by direct quantum mechanical calculation includes the energy effects of the motions of the electrons and the energy effects from the electrical interactions among the electrons and the stationary nuclei. Because we calculate it for stationary nuclei, the electronic energy does not include the energy of nuclear motions. The ground state electronic energy is the energy released when the atoms come together from infinite separation to a state in which they are at rest at the equilibrium inter-nuclear separation. This is just minus one times the work required to separate the atoms to an infinite distance, starting from the inter-nuclear separation with the smallest energy. On a graph of electronic (or potential) energy versus inter-nuclear distance, the ground state energy is just the depth of the energy well measured from the top down $(\epsilon_{e,1} < 0)$. The work required to separate one mole of these molecules into their constituent atoms is called the equilibrium dissociation energy, and conventionally given the symbol D_e. These definitions mean that $D_e > 0$ and $D_e = -\overline{N}\epsilon_{e,1}$.

In practice, the energy of the electronic ground state is often estimated from spectroscopic measurements. By careful study of its spectra, it is possible to find out how much energy must be added, as a photon, to cause a molecule to dissociate into atoms. Expressed per mole, this energy is called the spectroscopic dissociation energy, and it is conventionally given the symbol D_0. These spectroscopic measurements involve the absorption of photons by real molecules. Before they absorb the photon, these molecules already have energy in the form of vibrational and rotational motions. So the real molecules that are involved in any spectroscopic measurement have energies that are greater than the energies of the hypothetical motionless-atom molecules at the bottom of the potential energy well. This means that less energy is required to separate the real molecule than is required to separate the hypothetical molecule at the bottom of the well. For any molecule, $D_e > D_0$.

To have the lowest possible energy, a real molecule must be in its lowest rotational and lowest vibrational energy levels. As turns out, a molecule can have zero rotational energy, but its vibrational energy can never be zero. In §8 we review the harmonic oscillator approximation. In its lowest vibrational energy level ($n = 0$), a diatomic molecule's minimum vibrational energy is $h\nu/2$. D_0 and ν can be estimated from spectroscopic experiments. We estimate

$$\epsilon_{e,1} = -\frac{D_e}{\overline{N}} = -\left(\frac{D_0}{\overline{N}} + \frac{h\nu}{2}\right)$$

and the molecular electronic partition function becomes

$$z_e = g_1 \exp\left(\frac{D_0}{\overline{N}kT} + \frac{h\nu}{2kT}\right)$$

or

$$z_e = g_1 \exp\left(\frac{D_0}{RT} + \frac{h\nu}{2kT}\right)$$

§8 The vibrational partition function of a diatomic ideal gas

We base the electronic potential energy for a diatomic molecule on a model in which the nuclei are stationary at the bottom of the electronic potential energy well. We now want to expand this model to include vibrational motion of the atoms along the line connecting their nuclei. It is simple, logical, and effective to model this motion using the quantum mechanical treatment of the classical (Hooke's law) harmonic oscillator.

A Hooke's law oscillator has a location, r_0, at which the restoring force, $F(r_0)$, and the potential energy, $\epsilon(r_0)$, are zero. As it is displaced from r_0, the oscillator experiences a restoring force that is proportional to the magnitude of the displacement, $dF = -\lambda\, dr$. Then, we have

$$\int_{r_0}^{r} dF = -\lambda \int_{r_0}^{r} dr$$

so that $F(r) - F(r_o) = -\lambda(r - r_0)$. Since $F(r_o) = 0$, we have $F(r) = -\lambda(r - r_0)$. The change in the oscillator's potential energy is proportional to the square of the displacement,

$$\epsilon(r) - \epsilon(r_o) = \int_{r_0}^{r} -F\, dr = \lambda \int_{r_0}^{r} (r - r_0) dr = \frac{\lambda}{2}(r - r_0)^2$$

Since we take $\epsilon(r_o) = 0$, we have $\epsilon(r) = \lambda(r - r_0)^2/2$. Taking the second derivative, we find

$$\frac{d^2\epsilon}{dr^2} = \lambda$$

Therefore, if we determine the electronic potential energy function accurately near r_0, we can find λ from its curvature at r_0.

In Chapter 18, we note that the Schroedinger equation for such an oscillator can be solved and that the resulting energy levels are given by $\epsilon_n = h\nu(n + 1/2)$ where ν is the vibrational frequency. The relationship between frequency and force constant is

$$\nu = \frac{1}{2\pi}\sqrt{\frac{\lambda}{m}}$$

where the oscillator consists of a single moving mass, m.

In the case where masses m_1 and m_2 oscillate along the line joining their centers, it turns out that the same equations describe the relative motion, if the mass, m, is replaced by the reduced mass

$$\mu = \frac{m_1 m_2}{m_1 + m_2}$$

Therefore, in principle, we can find the characteristic frequency, ν, of a diatomic molecule by accurately calculating the dependence of the electronic potential energy on r in the vicinity of r_0. When we know ν, we know the vibrational energy levels available to the molecule. Alternatively, as discussed in §7, we can obtain information about the molecule's vibrational energy levels from its infrared absorption spectrum and use these data to find ν. Either way, once we know ν, we can evaluate the vibrational partition function. We have

$$z_v = \sum_{n=0}^{\infty} \exp\left[-\frac{h\nu}{kT}\left(n + \frac{1}{2}\right)\right] = \frac{\exp(-h\nu/2kT)}{1 - \exp(-h\nu/kT)}$$

where we take advantage of the fact that the vibrational partition function is the sum of a geometric series, as we show in §22-6.

§9 The rotational partition function of a diatomic ideal gas

For a diatomic molecule that is free to rotate in three dimensions, we can distinguish two rotational motions; however, their wave equations are intertwined, and the quantum mechanical result is that there is one set of degenerate rotational energy levels. The energy levels are

$$\epsilon_{r,J} = \frac{J(J+1)h^2}{8\pi^2 I}$$

with degeneracies $g_J = 2J + 1$, where $J = 0, 1, 2, 3, \ldots$. (Recall that I is the moment of inertia, defined as $I = \sum m_i r_i^2$, where r_i is the distance of the ith nucleus from the molecule's center of mass. For a diatomic molecule, XY, whose internuclear distance is r_{XY}, the values of r_X and r_Y must satisfy the conditions $r_X + r_Y = r_{XY}$ and $m_X r_X = m_Y r_Y$. From these relationships, it follows that the moment of inertia is $I = \mu r_{XY}^2$, where μ is the reduced mass.) For heteronuclear diatomic molecules, the rotational partition function is

$$z_r = \sum_{J=0}^{\infty} (2J+1) \exp\left[\frac{J(J+1)h^2}{8\pi^2 IkT}\right]$$

For homonuclear diatomic molecules, there is a complication. This complication occurs in the quantum mechanical description of the rotation of any molecule for which there is more than one indistinguishable orientation in space. When we specify the locations of the atoms in a homonuclear diatomic molecule, like H_2, we must specify the coordinates of each atom. If we rotate this molecule by 360° in a plane, the molecule and the coordinates are unaffected. If we rotate it by only 180° in a plane, the coordinates of the nuclei change, but the rotated molecule is indistinguishable from the original molecule. Our mathematical model distinguishes the 180°-rotated molecule from the original, unrotated molecule, but nature does not.

This means that there are twice as many energy levels in the mathematical model as actually occur in nature. The rotational partition function for a homonuclear diatomic molecule is exactly one-half of the rotational partition function for an "otherwise identical" heteronuclear diatomic molecule. To cope with this complication in general, it proves to be useful to define a quantity that we call the symmetry number for any molecule. The symmetry number is usually given the symbol σ; it is just the number of ways that the molecule can be rotated into indistinguishable orientations. For a homonuclear diatomic molecule, $\sigma = 2$; for a heteronuclear diatomic molecule, $\sigma = 1$.

Making use of the symmetry number, the rotational partition function for any diatomic molecule becomes

$$z_r = \left(\frac{1}{\sigma}\right) \sum_{J=0}^{\infty} (2J+1) \exp\left[\frac{J(J+1)h^2}{8\pi^2 IkT}\right]$$

For most molecules at ordinary temperatures, the lowest rotational energy level is much less than kT, and this infinite sum can be approximated to good accuracy as the corresponding integral. That is

$$z_r = \left(\frac{1}{\sigma}\right) \int_{J=0}^{\infty} (2J+1) \exp\left[\frac{J(J+1)h^2}{8\pi^2 IkT}\right] dJ$$

Initial impressions notwithstanding, this integral is easily evaluated. The substitutions $a = h^2/8\pi^2 IkT$ and $u = J(J+1)$ yield

$$z_r = \left(\frac{1}{\sigma}\right) \int_{u=0}^{\infty} \exp(-au)\, du = \left(\frac{1}{\sigma}\right)\left(\frac{1}{a}\right) = \frac{8\pi^2 IkT}{\sigma h^2}$$

To see that this is a good approximation for most molecules at ordinary temperatures, we calculate the successive terms in the partition function of the hydrogen molecule at 25 C. The results are shown in Table 1. We choose hydrogen because the energy difference between successive rotational energy levels becomes greater the smaller the values of I and T. Since hydrogen has the smallest angular momentum of any molecule, the integral approximation will be less accurate for hydrogen than for any other molecule at the same temperature. For hydrogen, summing the first seven terms in the exact calculation gives $z_{\text{rotation}} = 1.87989$, whereas the approximate calculation gives 1.70284. This difference corresponds to a difference of 245 J in the rotational contribution to the standard Gibbs free energy of molecular hydrogen.

Table 1.

Rotational Partition Function Contributions for Molecular Hydrogen at 298 K

J	$z_J = \frac{(2J+1)}{\sigma} exp\left(-\frac{J(J+1)h^2}{8\pi^2 IkT}\right)$	$z_r \approx \sum z_J$
0	0.50000	0.50000
1	0.83378	1.33378
2	0.42935	1.76313
3	0.10323	1.86637
4	0.01267	1.87904
5	0.00082	1.87986
6	0.00003	1.87989

Table 2.

Data[1] for the calculation of partition functions for $H_2(g)$, $I_2(g)$, and $HI(g)$

Compound	Molar mass, g	D_0, kJ mol^{-1}	ν, hertz	r_{XY}, m
H_2	2.016	432.073	1.31948×10^{14}	7.4144×10^{-11}
I_2	253.82	148.81	6.43071×10^{12}	2.666×10^{-10}
HI	127.918	294.67	6.69227×10^{13}	1.60916×10^{-10}

§10 The Gibbs free energy for one mole of an ideal gas

In our discussion of ensembles, we find that the thermodynamic functions for a system can be expressed as functions of the system's partition function. Now that we have found the molecular partition function for a diatomic ideal gas molecule, we can find the partition function, Z_{IG}, for a gas of N such molecules. From this system partition function, we can find all of the thermodynamic functions for this N-molecule ideal-gas system. The system entropy, energy, and partition function are related to each other by the equation

$$S = \frac{E}{T} + k \ln Z_{IG}$$

Rearranging, and adding $(PV)_{\text{system}}$ to both sides, we find the Gibbs free energy

$$\begin{aligned} G &= E - TS + (PV)_{\text{system}} \\ &= (PV)_{\text{system}} - kT \ln Z_{IG} \end{aligned}$$

For a system of one mole of an ideal gas, we have $(PV)_{\text{system}} = \overline{N}kT$. If the ideal gas is diatomic, we can substitute the molecular partition functions developed above to find

$$\begin{aligned} G_{IG} &= \overline{N}kT - kT \ln Z_{IG} \\ &= \overline{N}kT - k\text{T}\ln\left[\frac{(z_t)^{\overline{N}}}{\overline{N}!}\right] - \overline{N}kT \ln z_r - \overline{N}kT \ln z_v \\ &\quad - \overline{N}kT \ln z_e \\ &= \overline{N}kT - \overline{N}kT - \overline{N}kT \ln\left[\left(\frac{2\pi mkT}{h^2}\right)^{3/2} \frac{kT}{P}\right] \\ &\quad - \overline{N}kT \ln\left(\frac{8\pi^2 IkT}{\sigma h^2}\right) \\ &\quad - \overline{N}kT \ln\left(\frac{\exp(-h\nu/2kT)}{1-\exp(-h\nu/kT)}\right) \\ &\quad - \overline{N}kT \ln\left(\frac{D_0}{RT} + \frac{h\nu}{2kT}\right) \end{aligned}$$

For the standard Gibbs free energy of an ideal gas, we define the pressure to be one bar. Introduction of this condition ($P = P^o = 1$ bar $= 10^5$ Pa) and further simplification gives

$$\begin{aligned} G_{IG}^o &= -RT \ln\left[\left(\frac{2\pi mkT}{h^2}\right)^{3/2} \frac{kT}{P^o}\right] - RT \ln\left(\frac{8\pi^2 IkT}{\sigma h^2}\right) \\ &\quad - RT \ln\left(\frac{\exp(-h\nu/2kT)}{1-\exp(-h\nu/kT)}\right) \\ &\quad - RT\left(\frac{D_0}{RT} + \frac{h\nu}{2kT}\right) \end{aligned}$$

In this form, the successive terms represent, respectively, the translational, rotational, vibrational, and electronic contributions to the Gibbs free energy. Further simplification results because vibrational and electronic contributions from terms involving $h\nu/2kT$ cancel. This is a computational convenience. Factoring out RT,

$$\begin{aligned} G_{IG}^o = -RT \Big\{ &\ln\left[\left(\frac{2\pi mkT}{h^2}\right)^{3/2} \frac{kT}{P^o}\right] + \ln\left(\frac{8\pi^2 IkT}{\sigma h^2}\right) \\ &- \ln(1-\exp(-h\nu/kT)) + \frac{D_0}{RT} \Big\} \end{aligned}$$

§11 The standard Gibbs free energy for $H_2(g)$, $I_2(g)$, and $HI(g)$

For many diatomic molecules, the data needed to calculate G_{IG}^o are readily available in various compilations. For illustration, we consider the molecules H_2, I_2, and HI. The necessary experimental data are summarized in Table 2. The terms in the simplified equation for the standard Gibbs free energy at 298.15 K are given in Table 3. Finally, the standard molar Gibbs Free Energies at 298.15 K are summarized in Table 4.

Table 3. Gibbs free energy components				
Compound	$\ln\left[\left(\frac{2\pi mkT}{h^2}\right)^{3/2}\frac{kT}{P^o}\right]$	$\ln\left(\frac{8\pi^2 IkT}{\sigma h^2}\right)$	$-\ln\left(1-e^{-h\nu/kT}\right)$	$\frac{D_0}{RT}$
H_2	126.23929	0.6312■	0.0000	174.295
I_2	133.49256	7.9273	0.4388	60.0289
HI	132.46470	3.4604	0.0000_2	118.868

■Calculated as a sum of terms (see Table 1) rather than as the integral approximation.

These results can be used to calculate the standard Gibbs free energy change, at 298.15 K, for the reaction $H_2(g) + I_2(g) \rightarrow 2HI(g)$. We find

$$\Delta_r G^o_{298} = 2G^o(HI, g, 298.15\text{ K}) - G^o(H_2, g, 298.15\text{ K}) - G^o(I_2, g, 298.15\text{ K}) = -16.20\text{ kJ}$$

§12 The Gibbs free energy change for forming HI(g) from H_2(g) and I_2(g)

The standard Gibbs free energies of formation[1] for $HI(g)$ and $I_2(g)$ are 1.7 kJ mol^{-1} and 19.3 kJ mol^{-1}, respectively. Calculation of the Gibbs free energy of this reaction from thermochemical data gives $\Delta_r G^o(298.15\text{ K}) = -15.9$ kJ. The difference between this value and the value calculated above is 0.3 kJ. The magnitude of this difference is consistent with the number of significant figures given for the tabulated thermochemical data. However, some error results because we have used the simplest possible quantum mechanical models for rotational and vibrational motions. The accuracy of the statistical –mechanical calculation can be increased by using models in which the vibrational oscillator does not follow Hooke's law exactly and in which the rotating molecule is not strictly rigid.

§13 The reference state for molecular partition functions

In §11 and §12, we see that the standard Gibbs free energy, G^o, that we calculate from our statistical thermodynamic model is not the same quantity as the Gibbs free energy of formation, $\Delta_f G^o$. Nevertheless, these calculations show that we can use the statistical-thermodynamic Gibbs free energies of the reacting species to calculate the Gibbs free energy change for a reaction in exactly the same way that we use the corresponding Gibbs free energies of formation.

The use of Gibbs free energies of formation for these calculations is successful because we measure all Gibbs free energies of formation relative to the Gibbs free energies of the constituent elements in their standard states. By convention, we set the standard-state Gibbs free energies of the elements equal to zero, but this is incidental; our method is successful because the Gibbs free energies of the constituent elements cancel out when we calculate the Gibbs free energy change for a reaction from the Gibbs free energies of formation of the reacting species.

Table 4. Calculated Gibbs free energies.	
Compound	**$G^o_{298\text{ K}}$, kJ mol^{-1}**
H_2	−746.577
I_2	−500.471
HI	−631.622

Our statistical-mechanical Gibbs free energies represent the Gibbs free energy change for a different process. They correspond to the formation of the molecule from its isolated constituent atoms. The isolated constituent atoms are the reference state for our statistical-mechanical calculation of standard molar Gibbs free energies. We choose the Gibbs free energies of the isolated atoms to be zero. (Whatever Gibbs free energies we might assign to the isolated atoms, they cancel out when we calculate the Gibbs free energy change for a reaction from the statistical-thermodynamic Gibbs free energies of the reacting species.)

When we sum the component energies of our model for a diatomic molecule, we have $\epsilon_{\text{molecule}} = \epsilon_t + \epsilon_r + \epsilon_v + \epsilon_e$. The smallest of these quantum mechanically allowed values for $\epsilon_{\text{molecule}}$ is particularly significant in our present considerations. Once we have created this molecule in its lowest energy state, we can consider that we can get it into any other state just by adding energy to it. When the isolated constituent atoms are the reference state, the value of the lowest-energy state of the molecule is the energy exchanged with the surroundings when the molecule is formed in this state from the constituent atoms.

Reviewing our models for the motions that a diatomic molecule can undergo, we see that the translational and rotational energies can be zero. The smallest vibrational energy is $h\nu/2$, and the smallest electronic energy is

$$-\left(\frac{D_0}{\overline{N}}+\frac{h\nu}{2}\right)$$

The minimum molecular energy is $\epsilon_{molecule}^{minimum} = -D_0/\overline{N} < 0$. Since $D_0/\overline{N}$ is the energy required to just separate the diatomic molecule into its constituent elements, the end product of this process is two stationary atoms, situated at an infinite distance from one another. Conversely, $\epsilon_{molecule}^{minimum}$ is the energy released when the stationary constituent atoms approach one another from infinite separation to form the diatomic molecule in its lowest energy state. The reference state for the statistical-mechanical calculation of molecular thermodynamic properties is a set of isolated constituent atoms that have no kinetic energy. The stipulation that the reference-state atoms have no kinetic energy is often expressed by saying that the reference state is the constituent atoms at the absolute zero of temperature.

Problems

1. The partition function, Z, for a system of N, distinguishable, non-interacting molecules is $Z = z^N$, where z is the molecular partition function, $z = \sum g_i \exp(-\epsilon_i/kT)$, and the ϵ_i and g_i are the energy levels available to the molecule and their degeneracies. Show that the thermodynamic functions for the N-molecule system depend on the molecular partition function as follows:

(a) $E = NkT^2\left(\frac{\partial \ln z}{\partial T}\right)_V$

(b) $S = NkT\left(\frac{\partial \ln z}{\partial T}\right)_V + Nk\ln z$

(c) $A = -NkT\ln z$

(d) $P_{\text{system}} = NkT\left(\frac{\partial \ln z}{\partial V}\right)_T$

(e) $H = NkT^2\left(\frac{\partial \ln z}{\partial T}\right)_V + NkTV\left(\frac{\partial \ln z}{\partial V}\right)_T$

(f) $G = -NkT\ln z + NkTV\left(\frac{\partial \ln z}{\partial V}\right)_T$

2. When the number of available quantum states is much larger than the number of molecules, the partition function, Z, for a system of N, indistinguishable, non-interacting molecules is $Z = z^N/N!$, where z is the molecular partition function, $z = \sum g_i \exp(-\epsilon_i/kT)$, and the ϵ_i and g_i are the energy levels available to the molecule and their degeneracies. Show that the thermodynamic functions for the N-molecule system depend on the molecular partition function as follows:

(a) $E = NkT^2\left(\frac{\partial \ln z}{\partial T}\right)_V$

(b) $S = Nk\left[T\left(\frac{\partial \ln z}{\partial T}\right)_V + \ln\frac{z}{N} + 1\right]$

(c) $A = -NkT\left[1 + \ln\frac{z}{N}\right]$

(d) $P_{\text{system}} = NkT\left(\frac{\partial \ln z}{\partial V}\right)_T$

(e) $H = NkT^2\left(\frac{\partial \ln z}{\partial T}\right)_V + NkTV\left(\frac{\partial \ln z}{\partial V}\right)_T$

(f) $G = -NkT\left[1 + \ln\frac{z}{N} + V\left(\frac{\partial \ln z}{\partial V}\right)_T\right]$

3. The molecular partition function for the translational motion of an ideal gas is

$$z_t = \left(\frac{2\pi mkT}{h^2}\right)^{3/2} V$$

The partition function for a gas of N, monatomic, ideal-gas molecules is $Z = z_t^N/N!$. Show that the thermodynamic functions are as follows:

(a) $E = \frac{3}{2}NkT$

(b) $S = Nk\left[\frac{5}{2} + \ln\frac{z}{N}\right]$

(c) $A = -NkT\left[1 + \ln\frac{z}{N}\right]$

(d) $P_{\text{system}} = \frac{NkT}{V}$

(e) $H = \frac{5}{2}NkT$

(f) $G = -NkT\ln\frac{z}{N}$

4. Find E, S, A, H, and G for one mole of Xenon at 300 K and 1 bar.

Notes

[1] Data from the ***Handbook of Chemistry and Physics***, 79th Ed., David R. Linde, Ed., CRC Press, New York, 1998.

25

Bose-Einstein and Fermi-Dirac Statistics

§1 Quantum statistics

In developing the theory of statistical thermodynamics and the Boltzmann distribution function, we assume that molecules are distinguishable and that any number of molecules in a system can have the same quantum mechanical description. These assumptions are not valid for many chemical systems. Fortunately, it turns out that more rigorous treatment[1,2] of the conditions imposed by quantum mechanics usually leads to the same conclusions as the Boltzmann treatment. The Boltzmann treatment can become inadequate when the system consists of low-mass particles (like electrons) or when the system temperature is near absolute zero.

In this chapter, we introduce some modifications that make our statistical model more rigorous. We consider systems that contain large numbers of particles. We address the effects that the principles of quantum mechanics have on the equilibrium states that are possible, but we continue to assume that the particles do not otherwise exert forces on one another. We derive distribution functions for statistical models that satisfy quantum-mechanical restrictions on the number of particles that can occupy a particular quantum state. Our primary objective is to demonstrate that the more rigorous models reduce to the Boltzmann distribution function for most chemical systems at common laboratory conditions.

We have been using the quantum mechanical result that the discrete energy levels of a molecule or other particle can be labeled $\epsilon_1, \epsilon_2, \ldots, \epsilon_i, \ldots.$ We have assumed that we can put any number of identifiable particles into any of these energy levels. We have assumed also that we can distinguish one particle from another, so that we can know the energy of any particular particle. In fact, we may not be able to tell the particles apart. In this case, we can know how many particles have a given energy, but we cannot distinguish the particles that have this energy from one another. Moreover, there is a quantum-mechanical theorem about the number of particles that can occupy a quantum state. If the particles have integral $(0,\ 1,\ 2, \ldots.)$ spin, any number of them can occupy the same quantum state. Such particles are said to follow ***Bose-Einstein statistics***. If on the other hand, the particles have half-integral $(1/2,\ 3/2,\ 5/2, \ldots.)$ spin, then only one of them can occupy a given quantum state. Such particles are said to follow ***Fermi-Dirac statistics***.

Protons, neutrons, and electrons all have spin $1/2$. The spin of an atom or molecule is just the sum of the spins of its constituent elementary particles. If the number of protons, neutrons, and electrons is odd, the atom or molecule obeys Fermi-Dirac statistics. If it is even, the atom or molecule obeys Bose-Einstein statistics. For most molecules at temperatures that are not too close to absolute zero, the predicted difference in behavior is negligible. However, the isotopes of helium provide an important test of the theory. Near absolute zero, the behavior of He^3 differs markedly from that of He^4. The difference is consistent with the expected difference between the behavior of a spin-$1/2$ particle (He^3) and that of a spin-0 particle (He^4).

§2 Fermi-Dirac statistics and the Fermi-Dirac distribution function

Let us consider the total probability sum for a system of particles that follows Fermi-Dirac statistics. As before, we let $\epsilon_1, \epsilon_2, \ldots, \epsilon_i, \ldots.$ be the energies of the successive energy levels. We let $g_1, g_2, \ldots, g_i, \ldots.$ be the degeneracies of these levels. We let $N_1, N_2, \ldots, N_i, \ldots.$ be the number of particles in all of the degenerate quantum states of a given energy level. The probability of finding a particle in a quantum state depends on the number of particles in the system; we have $\rho(N_i, \epsilon_i)$ rather than $\rho(\epsilon_i)$. Consequently, we cannot generate the total probability sum by expanding an equation like $1 = (P_1 + P_2 + \cdots + P_i + \cdots)^N$. However, we continue to assume:

- A finite subset of the population sets available to the system accounts for nearly all of the probability when the system is held in a constant-temperature environment.
- Essentially the same finite subset of population sets accounts for nearly all of the probability when the system is isolated.
- All of the microstates that have a given energy have the same probability. We let this probability be $\rho^{FD}_{MS,N,E}$.

As before, the total probability sum will be of the form

$$1 = \sum_{\{N_i\}} W^{FD}(N_i, \epsilon_i)\, \rho^{FD}_{MS,N,E}$$

Each such term reflects the fact that there are

$W^{FD}(N_i, \epsilon_i)$ ways to put N_1 particles in the g_1 quantum states of energy level ϵ_1, and N_2 particles in the g_2 quantum states of energy level ϵ_2, and, in general, N_i particles in the g_i quantum states of energy level ϵ_i. Unlike Boltzmann statistics, however, the probabilities are different for successive particles, so the coefficient W^{FD} is different from the polynomial coefficient, or thermodynamic probability, W. Instead, we must discover the number of ways to put N_i indistinguishable particles into the g_i-fold degenerate quantum states of energy ϵ_i when a given quantum state can contain at most one particle.

These conditions can be satisfied only if $g_i \geq N_i$. If we put N_i of the particles into quantum states of energy ϵ_i, there are

- g_i ways to place the first particle, but only
- $g_i - 1$ ways to place the second, and
- $g_i - 2$ ways to place the third, and
- ...
- $g_i - (N_i - 1)$ ways to place the last one of the N_i particles.

This means that there are

$$(g_i)(g_i - 1)(g_i - 2) \dots (g_i - (N_i + 1)) =$$

$$= \frac{(g_i)(g_i - 1)(g_i - 2) \dots (g_i - (N_i + 1))(g_i - N_i) \dots (1)}{(g_i - N_i)!}$$

$$= \frac{g_i!}{(g_i - N_i)!}$$

ways to place the N_i particles. Because the particles cannot be distinguished from one another, we must exclude assignments which differ only by the way that the N_i particles are permuted. To do so, we must divide by $N_i!$. The number of ways to put N_i indistinguishable particles into g_i quantum states with no more than one particle in a quantum state is

$$\frac{g_i!}{(g_i - N_i)!\, N_i!}$$

The number of ways to put indistinguishable Fermi-Dirac particles of the population set $\{N_1, N_2, \dots, N_i, \dots\}$ into the available energy states is

$$W^{FD}(N_i, g_i) = \left[\frac{g_1!}{(g_1 - N_1)!\, N_1!}\right] \times \left[\frac{g_2!}{(g_2 - N_2)!\, N_2!}\right] \times \dots \times \left[\frac{g_i!}{(g_i - N_i)!\, N_i!}\right] \times \dots$$

$$= \prod_{i=1}^{\infty} \left[\frac{g_i!}{(g_i - N_i)!\, N_i!}\right]$$

so that the total probability sum for a Fermi-Dirac system becomes

$$1 = \sum_{\{N_j\}} \prod_{i=1}^{\infty} \left[\frac{g_i!}{(g_i - N_i)!\, N_i!}\right] [\rho^{FD}(\epsilon_i)]^{N_i}$$

To find the Fermi-Dirac distribution function, we seek the population set $\{N_1, N_2, \dots, N_i, \dots\}$ for which W^{FD} is a maximum, subject to the constraints

$$N = \sum_{i=1}^{\infty} N_i$$

and

$$E = \sum_{i=1}^{\infty} N_i \epsilon_i$$

The mnemonic function becomes

$$F_{mn}^{FD} = \sum_{i=1}^{\infty} \ln g_i! - \sum_{i=1}^{\infty} [(g_i - N_i) \ln(g_i - N_i) - (g_i - N_i)] - \sum_{i=1}^{\infty} [N_i \ln N_i - N_i] + \alpha \left[N - \sum_{i=1}^{\infty} N_i\right] + \beta \left[E - \sum_{i=1}^{\infty} N_i \epsilon_i\right]$$

We seek the $N_i^{\bullet}$ for which F_{mn}^{FD} is an extremum; that is, the $N_i^{\bullet}$ satisfying

$$0 = \frac{\partial F_{mn}^{FD}}{\partial N_i} = \frac{g_i - N_i^{\bullet}}{g_i - N_i^{\bullet}} + \ln(g_i - N_i^{\bullet}) - 1 - \frac{N_i^{\bullet}}{N_i^{\bullet}} - \ln N_i^{\bullet} + 1 - \alpha - \beta\epsilon_i$$
$$= \ln(g_i - N_i^{\bullet}) - \ln N_i^{\bullet} - \alpha - \beta\epsilon_i$$

Solving for $N_i^{\bullet}$, we find

$$N_i^{\bullet} = \frac{g_i e^{-\alpha} e^{-\beta\epsilon_i}}{1 + e^{-\alpha} e^{-\beta\epsilon_i}}$$

or, equivalently,

$$\frac{N_i^{\bullet}}{g_i} = \frac{1}{1 + e^{\alpha} e^{\beta\epsilon_i}}$$

If $1 \gg e^{-\alpha} e^{-\beta\epsilon_i}$ (or $1 \ll e^{\alpha} e^{\beta\epsilon_i}$), the Fermi-Dirac distribution function reduces to the Boltzmann distribution function. It is easy to see that this is the case. From

$$N_i^{\bullet} = \frac{g_i e^{-\alpha} e^{-\beta\epsilon_i}}{1 + e^{-\alpha} e^{-\beta\epsilon_i}} \approx g_i e^{-\alpha} e^{-\beta\epsilon_i}$$

and $N = \sum_{i=1}^{\infty} N_i^{\bullet}$, we have

$$N = e^{-\alpha} \sum_{i=1}^{\infty} g_i e^{-\beta\epsilon_i} = e^{-\alpha} z$$

It follows that $e^{\alpha} = z/N$. With $\beta = 1/kT$, we recognize that $N_i^{\bullet}/N$ is the Boltzmann distribution. For occupied energy levels, $e^{-\beta\epsilon_i} = e^{-\epsilon_i/kT} \approx 1$; otherwise, , $e^{-\beta\epsilon_i} = e^{-\epsilon_i/kT} < 1$. This means that the Fermi-Dirac distribution simplifies to the Boltzmann distribution whenever $1 \gg e^{-\alpha}$. We can illustrate that this is typically the case by considering the partition function for an ideal gas.

Using the translational partition function for one mole of a monatomic ideal gas from §24-3, we have

$$\begin{aligned} e^{\alpha} &= \frac{z_t}{N} \\ &= \left[\frac{2\pi mkT}{h^2}\right]^{3/2} \frac{\overline{V}}{\overline{N}} \\ &= \left[\frac{2\pi mkT}{h^2}\right]^{3/2} \frac{kT}{P^0} \end{aligned}$$

For an ideal gas of molecular weight 40 at 300 K and 1 bar, we find $e^{\alpha} = 1.02 \times 10^7$ and $e^{-\alpha} = 9.77 \times 10^{-8}$. Clearly, the condition we assume in demonstrating that the Fermi-Dirac distribution simplifies to the Boltzmann distribution is satisfied by molecular gases at ordinary temperatures. The value of e^{α} decreases as the temperature and the molecular weight decrease. To find $e^{\alpha} \approx 1$ for a molecular gas, it is necessary to consider very low temperatures.

Nevertheless, the Fermi-Dirac distribution has important applications. The behavior of electrons in a conductor can be modeled on the assumption that the electrons behave as a Fermi-Dirac gas whose energy levels are described by a particle-in-a-box model.

§3 Bose-Einstein statistics and the Bose-Einstein distribution function

For particles that follow Bose-Einstein statistics, we let the probability of a microstate of energy E in an N-particle system be $\rho^{BE}_{MS,N,E}$. For an isolated system of Bose-Einstein particles, the total probability sum is

$$1 = \sum_{\{N_i\}} W^{BE}(N_i, g_i)\rho^{BE}_{MS,N,E}$$

We need to find $W^{BE}(N_i, g_i)$, the number of ways to assign indistinguishable particles to the quantum states, if any number of particles can occupy the same quantum state.

We begin by considering the number of ways that N_i particles can be assigned to the g_i quantum states associated with the energy level ϵ_i. We see that the fewest number of quantum states that can be used is one; we can put all of the particles into one quantum state. At the other extreme, we cannot use more than the N_i quantum states that we use when we give each particle its own quantum state. We can view this problem as finding the number of way we can draw as many as g_i boxes around N_i points. Let us create a scheme for drawing such boxes. Suppose we have a linear frame on which there is a row of locations. Each location can hold one particle. The frame is closed at both ends. Between each successive pair of particle-holding locations, there is a slot, into

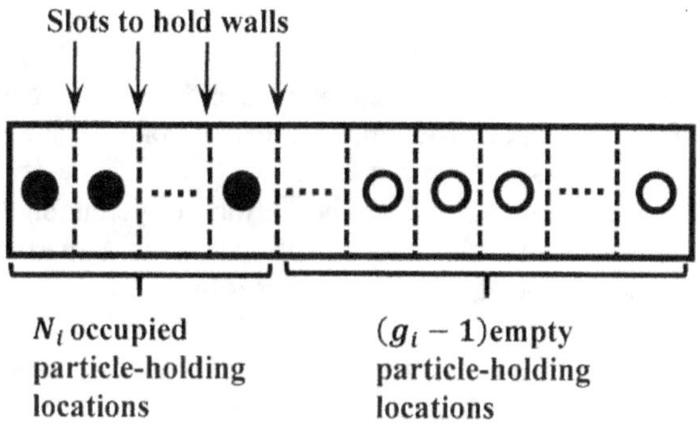

Figure 1. Scheme to assign Bose-Einstein particles to degenerate energy levels.

which a wall can be inserted. This frame is sketched in Figure 1. When we insert $(g_i - 1)$ walls into these slots, the frame contains g_i boxes. We want to be able to insert the walls so that the N_i particles are distributed among the g_i boxes in such a way that we can have any desired number of particles in any desired number of boxes. (Of course, placement of the walls is subject to the constraints that we use at most g_i boxes and exactly N_i particles.) We can achieve this by constructing the frame to have $(N_i + g_i - 1)$ particle-holding locations. To see this, we think about the case that requires the largest number of particle-holding locations. This is the case in which all N_i particles are in one box. (See Figure 2.) For this case, we need N_i occupied locations and $(g_i - 1)$ unoccupied locations.

Now we consider the number of ways that we can insert $(g_i - 1)$ walls into the $(N_i + g_i - 1)$ slots. The first wall can go into any of $(N_i + g_i - 1)$ slots. The second can go into any of $\left(N_i + g_i - 1 - (-1)\right)$ or $(N_i + g_i - 2)$ slots. The last wall can go into any of $\left(N_i + g_i - 1 - (g_i - 2)\right)$ or $(N_i + 1)$ slots. The total number of ways of inserting the $(g_i - 1)$ walls is therefore

$$\begin{aligned} &(N_i + g_i - 1)(N_i + g_i - 2) \dots (N_i + 1) \\ &= \frac{(N_i + g_i - 1)(N_i + g_i - 2) \dots (N_i + 1)(N_i) \dots (2)(1)}{N_i!} \\ &= \frac{(N_i + g_i - 1)!}{N_i!} \end{aligned}$$

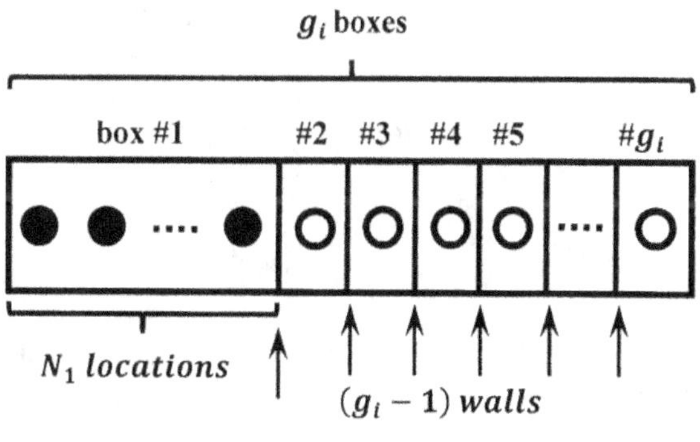

Figure 2. Maximum size frame for N_i particles in g_i locations.

This total is greater than the answer we seek, because it includes all permutations of the walls. It does not matter whether the first, the second, or the last wall occupies a given slot. Therefore, the expression we have obtained over-counts the quantity we seek by the factor $(g_i - 1)!$, which is the number of ways of permuting the $(g_i - 1)$ walls. We have therefore that the N_i particles can be assigned to g_i quantum states in

$$\frac{(N_i + g_i - 1)!}{N_i!\,(g_i - 1)!}$$

ways, and hence

$$\begin{aligned} W^{BE}(N_i, g_i) &= \left[\frac{(N_1 + g_1 - 1)!}{(g_1 - 1)!\,N_1!}\right] \times \left[\frac{(N_2 + g_2 - 1)!}{(g_2 - 1)!\,N_2!}\right] \\ &\quad \times \ldots \times \left[\frac{(N_i + g_i - 1)!}{(g_i - 1)!\,N_i!}\right] \times \ldots \\ &= \prod_{i=1}^{\infty} \left[\frac{(N_i + g_i - 1)!}{(g_i - 1)!\,N_i!}\right] \end{aligned}$$

so that the total probability sum for a Bose-Einstein system becomes

$$1 = \sum_{\{N_j\}} \prod_{i=1}^{\infty} \left[\frac{(N_i + g_i - 1)!}{(g_i - 1)!\,N_i!}\right] [\rho^{BE}(\epsilon_i)]^{N_i}$$

To find the Bose-Einstein distribution function, we seek the population set $\{N_1, N_2, \ldots, N_i, \ldots\}$ for which W^{BE} is a maximum, subject to the constraints

$$N = \sum_{i=1}^{\infty} N_i$$

and

$$E = \sum_{i=1}^{\infty} N_i\,\epsilon_i$$

The mnemonic function is

$$\begin{aligned} F_{mn}^{BE} = \sum_{i=1}^{\infty} [&(N_i + g_i - 1)\ln(N_i + g_i - 1) \\ &- (N_i + g_i - 1) - N_i \ln N_1 + N_i \\ &- (g_i - 1)\ln(g_i - 1) + (g_i - 1)] \\ &+ \alpha\left(N - \sum_{i=1}^{\infty} N_i\right) \\ &+ \beta\left(E - \sum_{i=1}^{\infty} N_i \epsilon_i\right) \end{aligned}$$

We seek the $N_i^{\bullet}$ for which F_{mn}^{BE} is an extremum; that is, the $N_i^{\bullet}$ satisfying

$$\begin{aligned} 0 &= \frac{\partial F_{mn}^{BE}}{\partial N_i^{\bullet}} \\ &= \frac{N_i^{\bullet} + g_i - 1}{N_i^{\bullet} + g_i - 1} + \ln(N_i^{\bullet} + g_i - 1) - 1 - \frac{N_i^{\bullet}}{N_i^{\bullet}} - \ln N_i^{\bullet} \\ &\quad + 1 - \alpha - \beta\epsilon_i \\ &= -\ln \frac{N_i^{\bullet}}{N_i^{\bullet} + g_i - 1} - \alpha - \beta\epsilon_i \end{aligned}$$

Solving for $N_i^{\bullet}$, we find

$$N_i^{\bullet} = \frac{(g_i - 1)e^{-\alpha}e^{-\beta\epsilon_i}}{1 - e^{-\alpha}e^{-\beta\epsilon_i}} \approx \frac{g_i e^{-\alpha}e^{-\beta\epsilon_i}}{1 - e^{-\alpha}e^{-\beta\epsilon_i}}$$

where the last expression takes advantage of the fact that g_i is usually a very large number, so the error introduced by replacing $(g_i - 1)$ by g_i is usually negligible. If $1 \gg e^{-\alpha}e^{-\beta\epsilon_i}$, the Bose-Einstein distribution function reduces to the Boltzmann distribution function. As we find in §2, this is always the case for a molecular gas at ambient temperatures.

Notes

[1]Richard C. Tolman, ***The Principles of Statistical Mechanics***, Dover Publications, Inc., New York, 1979, pp 367-378. (This is a republication of the book originally published in 1938 by Oxford University Press.)

[2]Malcom Dole, ***Introduction to Statistical Thermodynamics***, Prentice Hall, Inc., New York, 1954, pp 206-215.

Appendix A

Standard atomic weights 1999†

[Scaled to $A_r(^{12}C) = 12$, where ^{12}C is a neutral atom in its nuclear and electronic ground state]

At No	Name	Symbol	At Weight	At No	Name	Symbol	At Weight
1	Hydrogen	H	1.007 94	53	Iodine	I	126.904 47
2	Helium	He	4.002 602	54	Xenon	Xe	131.293
3	Lithium	Li	[6.941(2)]	55	Cesium	Cs	132.905 45
4	Beryllium	Be	9.012 182	56	Barium	Ba	137.327
5	Boron	B	10.811	57	Lanthanum	La	138.9055
6	Carbon	C	12.0107	58	Cerium	Ce	140.116
7	Nitrogen	N	14.0067	59	Praseodymium	Pr	140.907 65
8	Oxygen	O	15.9994	60	Neodymium	Nd	144.24
9	Fluorine	F	18.998 4032	61	Promethium	Pm^{145}	144.9127
10	Neon	Ne	20.1797	62	Samarium	Sm	150.36
11	Sodium	Na	22.989 770	63	Europium	Eu	151.964
12	Magnesium	Mg	24.3050	64	Gadolinium	Gd	157.25
13	Aluminum	Al	26.981 538	65	Terbium	Tb	158.925 34
14	Silicon	Si	28.0855	66	Dysprosium	Dy	162.50
15	Phosphorus	P	30.973 761	67	Holmium	Ho	164.930 32
16	Sulfur	S	32.065	68	Erbium	Er	167.259
17	Chlorine	Cl	35.453	69	Thulium	Tm	168.934 21
18	Argon	Ar	39.948	70	Ytterbium	Yb	173.04
19	Potassium	K	39.0983	71	Lutetium	Lu	174.967
20	Calcium	Ca	40.078	72	Hafnium	Hf	178.49
21	Scandium	Sc	44.955 910	73	Tantalum	Ta	180.9479
22	Titanium	Ti	47.867	74	Tungsten	W	183.84
23	Vanadium	V	50.9415	75	Rhenium	Re	186.207
24	Chromium	Cr	51.9961	76	Osmium	Os	190.23
25	Manganese	Mn	54.938 049	77	Iridium	Ir	192.217
26	Iron	Fe	55.845	78	Platinum	Pt	195.078
27	Cobalt	Co	58.933 200	79	Gold	Au	196.966 55
28	Nickel	Ni	58.6934	80	Mercury	Hg	200.59
29	Copper	Cu	63.546	81	Thallium	Tl	204.3833
30	Zinc	Zn	65.39	82	Lead	Pb	207.2
31	Gallium	Ga	69.723	83	Bismuth	Bi	208.980 38
32	Germanium	Ge	72.64	84	Polonium*	Po^{210}	209.9829
33	Arsenic	As	74.921 60	85	Astatine*	At^{210}	209.9871
34	Selenium	Se	78.96	86	Radon*	Rn^{222}	222.0176
35	Bromine	Br	79.904	87	Francium*	Fr^{223}	223.0197
36	Krypton	Kr	83.80	88	Radium*	Ra^{226}	226.0254
37	Rubidium	Rb	85.4678	89	Actinium*	Ac^{227}	227.0277
38	Strontium	Sr	87.62	90	Thorium*	Th	232.0381
39	Yttrium	Y	88.905 85	91	Protactinium*	Pa	231.035 88
40	Zirconium	Zr	91.224	92	Uranium*	U	238.028 91
41	Niobium	Nb	92.906 38	93	Neptunium*	Np^{237}	237.0482
42	Molybdenum	Mo	95.94	94	Plutonium*	Pu^{244}	244.0642
43	Technetium*	Tc^{98}	97.9072	95	Americium*	Am^{243}	243.0614
44	Ruthenium	Ru	101.07	96	Curium*	Cm^{247}	247.0704
45	Rhodium	Rh	102.905 50	97	Berkelium*	Bk^{247}	247.0703
46	Palladium	Pd	106.42	98	Californium*	Cf^{251}	251.0796
47	Silver	Ag	107.8682	99	Einsteinium*	Es^{252}	252.0830
48	Cadmium	Cd	112.411	100	Fermium*	Fm^{257}	257.0951
49	Indium	In	114.818	101	Mendelevium*	Md^{258}	258.0984
50	Tin	Sn	118.710	102	Nobelium*	No^{259}	259.1010
51	Antimony	Sb	121.760	103	Lawrencium*	Lr^{262}	262.1097
52	Tellurium	Te	127.60	104	Rutherfordium*	Rf^{261}	261.1088

†This table is slightly modified from that given in "ATOMIC WEIGHTS OF THE ELEMENTS, 1999" published by the International Union of Pure and Applied Chemistry, Inorganic Chemistry Division, Commission on Atomic Weights and Isotopic Abundances. Elements 105 – 118 are omitted. The Commission's report was prepared for publication by T. B. Coplen, U.S. Geological Survey, 431 National Center, Reston, Virginia 20192, USA. See http://www.physics.curtin.edu.au/iupac/docs/Atwt1999.doc

*Element has no stable nuclides. Three such elements (Th, Pa, and U) have a characteristic terrestrial isotopic composition, and for these an atomic weight is tabulated. When the element symbol is listed with an atomic number, that isotope has the longest half-life and its atomic weight is tabulated.

Appendix B

Fundamental Constants†

Quantity	Symbol	Value	Unit
speed of light in a vacuum	c	299 792 458	$m\ s^{-1}$
Planck constant	h	$6.626\ 068\ 76 \times 10^{-34}$	J s
elementary charge	e	$1.602\ 176\ 462 \times 10^{-19}$	C
electron mass	m_e	$9.109\ 381\ 88 \times 10^{-31}$	kg
proton mass	m_p	$1.672\ 621\ 58 \times 10^{-27}$	kg
Avogadro constant	$\overline{N}$	$6.022\ 141\ 99 \times 10^{23}$	mol^{-1}
Faraday constant	F	96 485.341 5	$C\ mol^{-1}$
molar gas constant	R	8.314 472	$J\ mol^{-1}\ K^{-1}$
	R	$8.314\ 472 \times 10^{-2}$	$bar\ L\ mol^{-1}\ K^{-1}$
	R	$8.205\ 745 \times 10^{-2}$	$atm\ L\ mol^{-1}\ K^{-1}$
	R	1.987 206	$cal\ mol^{-1}\ K^{-1}$
Boltzmann constant	k	$1.380\ 650\ 3 \times 10^{-23}$	$J\ K^{-1}$
standard acceleration of free fall at the earth's surface†	g	9.806 650	$m\ s^{-2}$

†Source: 1998 CODATA recommended values. Peter J. Mohr and Barry N. Taylor, ***Reviews of Modern Physics***, Vol. 72, No. 2, pp. 351-495, 2000. See www.physics.nist.gov/constants.

† While it is included here for convenience, g is not a Fundamental Constant. Value from David R. Lide, ***CRC Handbook of Chemistry and Physics***, 79th Ed., CRC Press, 1999-2000, pp. 1-27, reproduced from ***NIST Special Publication 811, Guide for the Use of the International System of Units*** (Superintendent of Documents, U. S. Government Printing Office, 1991).

Appendix C

Units and Conversion Factors

1 m	=	10^{-10} Å
	=	3.280 840 ft
	=	39.370 08 in
1 kg	=	2.204 622 lb
1 lb	=	453.592 5 g
$1 m^3$	=	10^3 L
	=	35.314 67 ft^3
	=	264.172 0 US liquid gallon
1 N	=	1 kg m s^{-2}
	=	1 Pa m^2
	=	10^{-5} bar m^2
	=	10^5 dyne (g cm s^{-2})
	=	0.138 255 0 poundal
1 bar	=	10^5 Pa
	=	10^5 N m^{-2}
	=	0.986 92 atm
	=	750.064 torr
1 atm	=	101 325 Pa
	=	1.013 25 bar
1 J	=	1 N m
	=	1 Pa m^3
	=	1 W s
	=	2 C V
	=	10^7 erg (g cm^2 s^{-2})
	=	10^{-5} bar m^3
	=	10^{-2} bar L
	=	$9.869\,23 \times 10^{-3}$ L atm
	=	0.239 0 cal
1 cal	=	4.184 J
1 eV	=	$1.602\,176\,462 \times 10^{-19}$ J (Energy released when one electron experiences a potential change of one volt.)

Conversion factors in this table are given by, or calculated from, values given by David R. Lide, ***CRC Handbook of Chemistry and Physics***, 79th Ed., CRC Press, 1999-2000, pp. 1-24 to 1-31, reproduced from ***NIST Special Publication 811, Guide for the Use of the International System of Units*** (Superintendent of Documents, U. S. Government Printing Office, 1991).

Appendix D

Some Important Definite Integrals

We frequently need the values of the definite integrals below. These values are available in standard tables. Note that integrands involving even powers of the argument are even functions; integrands involving odd powers are odd functions. (A function, $f(x)$, is even if $f(x) = f(-x)$; it is odd if $f(x) = -f(-x)$.) The integrals are given over the interval $0 < x < \infty$. For integrands that are even functions, the integrals over the interval $-\infty < x < \infty$ are twice the integrals over the interval $0 < x < \infty$. For integrands that are odd functions, the integrals over the interval $-\infty < x < \infty$ are zero.

$$\int_0^\infty \exp(-ax^2)\,dx = \frac{1}{2}\sqrt{\frac{\pi}{a}}$$

$$\int_0^\infty x\exp(-ax^2)\,dx = \frac{1}{2a}$$

$$\int_0^\infty x^2\exp(-ax^2)\,dx = \frac{1}{4}\sqrt{\frac{\pi}{a^3}}$$

$$\int_0^\infty x^3\exp(-ax^2)\,dx = \frac{1}{2a^2}$$

$$\int_0^\infty x^4\exp(-ax^2)\,dx = \frac{3}{8}\sqrt{\frac{\pi}{a^5}}$$

$$\int_0^\infty x^6\exp(-ax^2)\,dx = \frac{15}{16}\sqrt{\frac{\pi}{a^7}}$$

Index

www.ingramcontent.com/pod-product-compliance
Lightning Source LLC
Chambersburg PA
CBHW081309170526
45166CB00011B/3453